WORLD *of*

WORLD *of*

K. Lee Lerner and Brenda Wilmoth Lerner, *Editors*

Volume 2

M-Z

General Index

Detroit • New York • San Diego • San Francisco • Cleveland • New Haven, Conn. • Waterville, Maine • London • Munich

THOMSON
GALE

World of Earth Science

K. Lee Lerner and Brenda Wilmoth Lerner

Project Editor
Ryan L. Thomason

Editorial
Deirdre S. Blanchfield, Madeline Harris, Kate Kretschmann, Michael D. Lesniak, Kimberley A. McGrath, Brigham Narins, Mark Springer

Permissions
Shalice Shah-Caldwell

Imaging and Multimedia
Robert Duncan, Leitha Etheridge-Sims, Lezlie Light, Kelly A. Quin, Barbara J. Yarrow

Product Design
Michael Logusz, Tracey Rowens

Manufacturing
Wendy Blurton, Evi Seoud

For more information contact
The Gale Group, Inc.
27500 Drake Rd.
Farmington Hills, MI 48331-3535
Or you can visit our Internet site at
http://www.gale.com

ISBN 0-7876-7739-6 (set)
0-7876-7740-X (vol. 1)
0-7876-7741-8 (vol. 2)

LIBRARY OF CONGRESS CATALOGING-IN-PUBLICATION DATA

World of earth science / K. Lee Lerner and Brenda Wilmoth Lerner, editors.
p. cm.
Includes bibliographical references and index.
ISBN 0-7876-7739-6 (set)—ISBN 0-7876-7740-X (v. 1)—ISBN 0-7876-7741-8 (v. 2)
1. Earth sciences—Encyclopedias. I. Lerner, K. Lee. II. Lerner, Brenda Wilmoth.
QE5 .W59 2003
550'.3—dc21 2002012069

Printed in the United States of America
10 9 8 7 6 5 4 3 2 1

Contents

M

N

O

T

U

V

INTRODUCTION

As of June 2002, astronomers had discovered more than 100 other planets orbiting distant suns. With advances in technology, that number will surely increase during the opening decades of the twenty-first century. Although our explorations of the Cosmos hold great promise of future discoveries, among all of the known worlds, Earth remains unique. Thus far it is the only known planet with blue skies, warm seas, and life. Earth is our most tangible and insightful laboratory, and the study of Earth science offers us precious opportunities to discover many of the most fundamental laws of the Universe.

Although Earth is billions of years old, geology—literally meaning the study of Earth—is a relatively new science, having grown from seeds of natural science and natural history planted during the Enlightenment era of the eighteenth and nineteenth centuries. In 1807, the founding of the Geological Society of London, the first learned society devoted to geology, marked an important turning point for the science (some say its nascence). In the beginning, geologic studies were mainly confined to the study of minerals (mineralogy), strata (stratigraphy), and fossils (paleontology), and hotly debated issues of the day included how well new geologic findings fit into religious models of creation. In less than two centuries, geology has matured to embrace the most fundamental theories of physics and chemistry—and broadened in scope to include the diverse array of subdisciplines that comprise modern Earth science.

Modern geology includes studies in seismology (earthquake studies), volcanology, energy resources exploration and development, tectonics (structural and mountain building studies), hydrology and hydrogeology (water-resources studies), geologic mapping, economic geology (e.g., mining), paleontology (ancient life studies), soil science, historical geology and stratigraphy, geological archaeology, glaciology, modern and ancient climate and ocean studies, atmospheric sciences, planetary geology, engineering geology, and many other subfields. Although some scholars have traditionally attempted to compartmentalize geological sciences into subdisciplines, the modern trend is to incorporate a holistic view of broader Earth science issues. The incorporation of once-diverse fields adds strength and additional relevance to geoscience studies.

World of Earth Science is a collection of 650 entries on topics covering a diversity of geoscience related interests—from biographies of the pioneers of Earth science to explanations of the latest developments and advances in research. Despite the complexities of terminology and advanced knowledge of mathematics needed to fully explore some of the topics (e.g., seismology data interpretation), every effort has been made to set forth entries in everyday language and to provide accurate and generous explanations of the most important terms. The editors intend *World of Earth Science* for a wide range of readers. Accordingly, *World of Earth Science* articles are designed to instruct, challenge, and excite less experienced students, while providing a solid foundation and reference for more advanced students.

World of Earth Science has attempted to incorporate references and basic explanations of the latest findings and applications. Although certainly not a substitute for in-depth study of important topics, we hope to provide students and readers with the basic information and insights that will enable a greater understanding of the news and stimulate critical thinking regarding current events (e.g., the ongoing controversy over the storage of radioactive waste) that are relevant to the geosciences.

The broader and intellectually diverse concept of Earth science allows scientists to utilize concepts, techniques, and modes of thought developed for one area of the science, in the quest to solve problems in other areas. Further, many geological problems are interrelated and a full exploration of a particular phenomenon or problem demands overlap between subdisciplines. For this reason, many curricula in geological sciences at universities stress a broad geologic education to prepare graduates for the working world, where they may be called upon to solve many different sorts of problems.

World of Earth Science is devoted to capturing that sense of intellectual diversity. True to the modern concept of Earth science, we have deliberately attempted to include some of the most essential concepts to understanding Earth as a dynamic body traveling through space and time.

Although no encyclopedic guide to concepts, theories, discoveries, pioneers, issues and ethics related to Earth science could hope to do justice to any one of those disciplines in two volumes, we have attempted to put together a coherent collection of topics that will serve not only to ground students in the essential concepts, but also to spur interest in the many diverse areas of this increasing critical set of studies.

In addition to topics related to traditional geology and meteorology, we have attempted to include essential concepts in physics, chemistry, and astronomy. We have also attempted to include topical articles on the latest global positioning (GPS), measurement technologies, ethical, legal, and social issues and topics of interest to a wide audience. Lastly, we have attempted to integrate and relate topics to the intercomplexities of economics and geopolitical issues.

Such a multifaceted and "real world" approach to the geosciences is increasingly in demand. In the recent past, geologic employment was dominated by the petroleum industry and related geologic service companies. In the modern world, this is no longer so. Mining and other economic geology occupations (e.g., prospecting and exploration), in former days plentiful, have also fallen away as major employers. Environmental geology, engineering geology, and ground water related jobs are more common employment opportunities today. As these fields are modern growth areas with vast potential, this trend will likely hold true well into the future. Many modern laws and regulations require that licensed, professional geologists supervise all or part of key tasks in certain areas of engineering geologic work and environmental work. It is common for professional geologists and professional engineers to work together on such projects, including construction site preparation, waste disposal, ground-water development, engineering planning, and highway construction. Many federal, state, and local agencies employ geologists, and there are geologists as researchers and teachers in most academic institutions of higher education.

Appropriate to the diversity of Earth science, we attempted to give special attention to the contributions by women and scientists of diverse ethnic and cultural backgrounds. In addition, we have included special articles written by respected experts that are specifically intended to make *World of Earth Science* more relevant to those with a general interest in the historical and/or geopolitical topics aspects of Earth science.

The demands of a dynamic science and the urgency of many questions related to topics such as pollution, global warming, and ozone depletion place heightened demands on both general and professional students of geosciences to increasingly broaden the scope and application of their knowledge.

For example, geological investigations of ancient and modern disasters and potential disasters are important—and often contentious—topics of research and debate among geologists today. Among the focus areas for these studies are earthquake seismicity studies. While much work continues in well-known problem areas like southern California, Mexico City, and Japan, less well-known, but potentially equally dangerous earthquake zones like the one centered near New Madrid, Missouri (not far from Memphis, Tennessee and St. Louis, Missouri) now receive significnt research attention. Geologists cannot prevent earthquakes, but studies can help predict earthquake events and help in planning the design of earthquake-survivable structures. Another focus of study is upon Earth's volcanoes and how people may learn to live and work around them. Some volcanoes are so dangerous that no one should live near them, but others are more predictable. Earthquake prediction and planning for eruptions is going on today by looking at the geologic record of past eruptions and by modern volcano monitoring using thermal imaging and tilt or motion-measuring devices. Other foci of disaster prevention research include river-flood studies, studies of slope stability (prevention of mass movement landslides), seismic sea-wave (tsunami) studies, and studies of possible asteroid or comet impacts.

Aside from geologic studies of disaster, there is a side of geology centered upon providing for human day-to-day needs. Hydrology is an interdisciplinary field within geology that studies the relationship of water, the earth, and living things. A related area, hydrogeology, the study of ground water, has undergone a revolution recently in the use of computer modeling to help understand flow paths and characteristics. These studies of water flow on the surface and in the subsurface connect with other subdisciplines of geology, such as geomorphology (the study of landforms, many of which are formed by water flow), river hydrology, limnology (study of lakes), cave and sinkhole (karst) geology, geothermal energy, etc. Geologic studies related to human and animal health (i.e., medical geology) are becoming very common today. For example, much work is currently devoted to tracing sources of toxic elements like arsenic, radon, and mercury in rock, soil, air, water, and groundwater in many countries, including the United States. There has been a major effort on the part of medical geologists to track down dangerous mineral species of asbestos (not all asbestos is harmful) and determine how best to isolate or remove the material. Atmospheric scientists have been at work for some years on the issue of air-born pathogens, which ride across oceans and continents born on fine soil particles lifted by winds.

Geologists are also focused on study of the past. Today, paleobotany and palynology (study of fossil spores and pollen) complement traditional areas of invertebrate and vertebrate paleontogy. Recent discoveries such as small, feathered dinosaurs and snakes with short legs are helping fill in the ever-shrinking gaps within the fossil record of evolution of life on Earth. Paleontologic studies of extinction, combined with evidence of extraterrestrial bombardment, suggest that mass death and extinction of species on Earth at times in the past has come to us from the sky. In a slightly related area, geoarchaeology, the geologic context of archaeological remains and the geologic nature of archaeological artifacts remains key to interpreting details about the pre-historic human past. Careful study of drilling records of polar ice sheets, deep-sea sediments, and deep lake sediments has recently revealed that many factors, including subtle variations in some of Earth's orbital parameters (tilt, wobble, and shape of orbit around the Sun), has had a profound, cyclical effect upon Earth's climate in the past (and is continuing today). Paleontologic studies, combined with geologic investigations on temperature sensitive ratios of certain isotopes (e.g. O^{16}/O^{18}), have helped unlock mysteries of climate change on Earth (i.e., the greenhouse to icehouse vacillation through time).

Earth science studies are, for the first time, strongly focused on extraterrestrial objects as well. Voyages of modern exploratory spacecraft missions to the inner and outer planets have sent back a wealth of images and data from the eight major planets and many of their satellites. This has allowed a new field, planetary geology, to take root. The planetary geologist is engaged in photo-geologic interpretation of the origin of surface features and their chronology. Planetary geologic studies have revealed some important comparisons and contrasts with Earth. We know, for example, some events that affected our entire solar system, while the effects of other events were unique to certain planets and satellites. In addition, planetary geologists have found that impact-crater density is important for determining relative age on many planets and satellites. As a result, Earth is no longer the only planet with a knowable geologic time scale.

The geosciences have undergone recent revolutions in thought that have profoundly influenced and advanced human understanding of Earth. Akin to the fundamental and seminal concepts of cosmology and nucleosynthesis, beginning during the 1960s and continuing today, the concept of plate tectonics has revolutionized geologic thought and interpretation. Plate tectonics, the concept that the rigid outer part of Earth's crust is subdivided into plates, which move about on the surface (and have moved about on the surface for much of geologic time) has some profound implications for all of geology. This concept helps explain former mysteries about the distribution of volcanoes, earthquakes, and mountain chains. Plate tectonics also helps us understand the distribution of rocks and sediments on the sea floor, and the disparity in ages between continents and ocean floors. Plate motion, which has been documented through geologic time, helps paleontologists explain the distribution of many fossil species and characteristics of their ancient climates. Plate tectonic discoveries have caused a rewriting of historical geology textbooks in recent years.

Although other volumes are chartered to specifically explore ecology related issues, the topics included in *World of Earth Science* were selected to provide a solid geophysical foundation for ecological or biodiversity studies. We have specifically included a few revolutionary and controversial concepts, first written about in a comprehensive way during the 1970s, such as the Gaia hypothesis. Simply put, Gaia is the notion that all Earth systems are interrelated and interconnected so that a change in one system changes others. It also holds the view that Earth functions like a living thing. Gaia, which is really a common-sense philosophic approach to holistic Earth science, is at the heart of the modern environmental movement, of which geology plays a key part.

Because Earth is our only home, geoscience studies relating meteor impacts and mass extinction offer a profound insight into delicate balance and the tenuousness of life. As Carl Sagan wrote in *Pale Blue Dot: A Vision of the Human Future in Space*, "The Earth is a very small stage in a vast cosmic arena." For humans to play wisely upon that stage, to secure a future for the children who shall inherit Earth, we owe it to ourselves to become players of many parts, so that our repertoire of scientific knowledge enables us to use reason and intellect in our civic debates, and to understand the complex harmonies of Earth.

K. Lee Lerner & Brenda Wilmoth Lerner, editors
London
May, 2002

How to Use the Book

The articles in the book are meant to be understandable by anyone with a curiosity and willingness to explore topics in Earth science. Cross-references to related articles, definitions, and biographies in this collection are indicated by **bold-faced type**, and these cross-references will help explain, expand, and enrich the individual entries.

This first edition of *World of Earth Science* has been designed with ready reference in mind:

- **Entries are arranged alphabetically**, rather than by chronology or scientific field.
- **Bold-faced terms** direct reader to related entries.
- **"See also" references** at the end of entries alert the reader to related entries not specifically mentioned in the body of the text.
- A **Sources Consulted** section lists the most worthwhile print material and web sites we encountered in the compilation of this volume. It is there for the inspired reader who wants more information on the people and discoveries covered in this volume.
- The **Historical Chronology** includes many of the significant events in the advancement of the diverse disciplines of Earth science. The most current entries date from just days before *World of Earth Science* went to press.
- A **comprehensive General Index** guides the reader to topics and persons mentioned in the book. Bolded page references refer the reader to the term's full entry.

A detailed understanding of physics and chemistry is neither assumed nor required for *World of Earth Science*. In preparing this text, the editors have attempted to minimize the incorporation of mathematical formulas and to relate physics concepts in non-mathematical language. Accordingly, students and other readers should not be intimidated or deterred by chemical nomenclature. Where necessary, sufficient information regarding atomic or chemical structure is provided. If desired, more information can easily be obtained from any basic physics or chemistry textbook.

For those readers interested in more information regarding physics related topics, the editors recommend Gale's *World of Physics* as an accompanying reference. For those readers interested in a more comprehensive treatment of chemistry, the editors recommend Gale's *World of Chemistry*.

In an attempt to be responsive to advisor's requests and to conform to standard usage within the geoscience community, the editors elected to make an exception to previously used style guidelines regarding geologic time. We specifically adopted the convention to capitalize applicable eons, eras, periods and epochs. For example, Cenozoic Era, Tertiary Period, and Paleocene Epoch are intentionally capitalized.

Advisory Board

In compiling this edition, we have been fortunate in being able to rely upon the expertise and contributions of the following scholars who served as academic and contributing advisors for *World of Earth Science*, and to them we would like to express our sincere appreciation for their efforts to ensure that *World of Earth Science* contains the most accurate and timely information possible:

Cynthia V. Burek, Ph.D.
Environment Research Group, Biology Department
Chester College, England, U.K.

Nicholas Dittert, Ph.D.
Institut Universitaire Européen de la Mer
University of Western Brittany, France

William J. Engle. P.E.
Exxon-Mobil Oil Corporation (Rt.)
New Orleans, Louisiana

G. Thomas Farmer, Ph.D., R.G.
Earth & Environmental Sciences Division,
Los Alamos National Laboratory
Los Alamos, New Mexico

Lyal Harris, Ph.D.
Tectonics Special Research Center, Dept. of Geology & Geophysics
University of Western Australia
Perth, Australia

Alexander I. Ioffe, Ph.D.
Senior Scientist, Geological Institute of the Russian Academy of Sciences
Moscow, Russia

David T. King, Jr., Ph.D.
Professor, Dept. of Geology
Auburn University
Auburn, Alabama

Cherry Lewis, Ph.D.
Research Publicity Officer
University of Bristol
Bristol, England, U.K.

Eric v.d. Luft, Ph.D., M.L.S.
Curator of Historical Collections
S.U.N.Y. Upstate Medical University
Syracuse, New York

Jascha Polet, Ph.D.
Research Seismologist, Caltech Seismological Laboratory,
California Institute of Technology
Pasadena, California

Yavor Shopov, Ph.D.
Professor of Geology & Geophysics
University of Sofia
Sofia, Bulgaria

Acknowledgments

In addition to our advisors and contributing advisors, it has been our privilege and honor to work with the following contributing writers and scientists who represent scholarship in the geosciences that spans five continents:

Molly Bell, Ph.D.; Alicia Cafferty; John Cubit, Ph.D.; Laurie Duncan, Ph.D.; John Engle; Agnes Galambosi; Larry Gilman, Ph.D.; David Goings, Ph.D.; Brooke Hall, Ph.D.; William Haneberg, Ph.D.; Michael Lambert; Adrienne Wilmoth Lerner, (Graduate Student, Department of History, Vanderbilt University), Lee Wilmoth Lerner, Jill Liske, M.Ed.; Robert Mahin, Ph.D.; Kelli Miller; William Phillips, Ph.D.; William Rizer, Ph.D.; Jerry Salvadore, Ph.D.; and David Tulloch, Ph.D.

Many of the academic advisors for *World of Earth Science*, along with others, authored specially commissioned articles within their field of expertise. The editors would like to specifically acknowledge the following special contributions:

Cynthia V. Burek, Ph.D.
History of geoscience: Women in the history of geoscience

Nicholas Dittert, Ph.D.
Scientific data management in Earth sciences

William J. Engle. P.E.
Petroleum, economic uses of
Petroleum extraction

G. Thomas Farmer, Ph.D., R.G.
Hydrogeology

Lyal Harris, Ph.D.
Supercontinents

Alexander I. Ioffe, Ph.D.
Bathymetric mapping

David T. King, Jr., Ph.D.
Geologic time
Stratigraphy
Uniformitarianism

Cherry Lewis, Ph.D.
The biography of Author Holmes

Yavor Shopov, Ph.D.
Paleoclimate

The editors wish to gratefully acknowledge Dr. Eric v.d. Luft for his diligent and extensive research related to his compilation of selected biographies for *World of Earth Science*.

The editors also gratefully acknowledge Dr. Yavor Shopov's generous and significant contribution of photographs for *World of Earth Science* and Dr. David King's guidance, comments, and contributions to the introduction.

The editors thank Ms. Kelly Quin and others representing the Gale Imaging Team for their guidance through the complexities and difficulties related to graphics. Last, but certainly not least, the editors thank Mr. Ryan Thomason, whose dedication, energy, and enthusiasm made a substantial difference in the quality of *World of Earth Science*.

Because Earth is theirs to inherit, the editors lovingly dedicate this book to their children, Adrienne, Lee, Amanda, and Adeline. *Per ardua ad astra.*

Cover

The image on the cover depicts an example of several geologic cross sections of strata, illustrating the fundamental laws of geology.

M

MACH SCALE • *see* SOUND TRANSMISSION

MAFIC

Igneous rocks are classified by geologists using various schemes. One of the several schemes based on chemical composition divides igneous rocks into four categories according to silica (**silicon** dioxide, SiO_2) content: (1) Rocks containing more than 66% silica are **silicic**. (2) Rocks containing 52–66% silica are classified as intermediate. (3) Rocks containing 45–52% silica are mafic. (4) Rocks containing less than 45% silica are ultramafic. The term acidic is sometimes used as a synonym for silicic and the terms basic and ferromagnesian as synonyms for mafic. Mafic is an invented adjective based on the chemical symbols for magnesium (Ma) and **iron** (Fe): Ma-Fe-ic, mafic. Mafic is sometimes used as a synonym for "dark-colored" when discussing the appearance of **minerals**.

Some mafic and ultramafic rocks are found on Earth's surface. However, because magnesium and iron are denser than silica, mafic rocks are denser than silicic rocks and tend to sink below them. This density difference explains the dependence of Earth's composition on depth. Earth's core consists mostly of fairly pure metal (iron and nickel); surrounding the core is the mantle, a layer consisting mostly of ultramafic **rock** (**metals** mixed with silica). The outermost layer of the earth, the **crust**, consists of two basic types of crust, one primarily mafic (oceanic crust) and the other primarily silicic (continental crust). Oceanic crust, which is only about 4 miles (6 km) thick, consists mostly of **basalt**, a mafic rock. As oceanic crust inches away from its point of origin at a mid-ocean ridge, its underside cools the ultramafic mantle rocks over which it slides. These cooled mantle rocks stick to the underside of the oceanic crust, thickening it over time. The oceanic crust is thus weighed down by an increasingly thick undercoating of cooled ultramafic mantle rock as it ages. This cool undercoating is denser than the chemically identical but hotter mantle rocks below. Eventually it becomes heavy enough to drag the oceanic crust right down into the mantle, as occurs at a spontaneous **subduction zone**. The continents, in contrast, are silicic, and float permanently on the mantle. Mafic oceanic crust is spontaneously subducted into the mantle after at most 200 million years, while the continents have never been subducted in the three or four billion years since they were formed.

By weight, Earth consists mostly of mafic and ultramafic rocks, but silicic rocks are far more abundant on Earth's surface. Mafic rocks commonly found on the surface include basalt, pyroxene, and biotite; common ultramafic rocks are dunite and **peridotite**

See also Earth, interior structure; Felsic; Sea-floor spreading

MAGELLAN, FERDINAND (1480-1521)

Portuguese mariner, explorer

Ferdinand Magellan was the first explorer to lead an expedition that circumnavigated the globe. Like many of his contemporaries, Magellan underestimated the size of the **oceans**, and thought he could find a faster route to the Spice Islands by sailing west. He began his voyage in September of 1519 with five ships. After an arduous voyage, only one ship returned. Magellan, the expedition leader, was not onboard.

Magellan was born in Portugal in 1480. His parents were low-ranking nobles, active in the Portuguese royal court. Through his education at court, Magellan learned navigation. He attained the rank of squire while in royal service as a merchant marine clerk. He joined Francisco de Almeida's voyages to explore the eastern coast of **Africa** in 1505 and 1506. By 1509, Magellan had traveled to Africa, Turkey, and India. In 1511, Magellan ventured to the Far East on a Portuguese expedition to Malaysia. Magellan returned to **Europe** but, soon after arriving home, then departed to fight for Portuguese interests in Morocco. He was wounded, and left the royal service soon after. He then turned his attention to gaining a charter for a fleet of his own, in hopes of returning to the Far East. In 1517, he began lobbying the Portuguese crown to fund a large expedition. He was denied a ship from the Portuguese crown, and then turned to the rival king of Spain.

Interested in Magellan's proposal to find a faster shipping route to the Far East, the Spanish king granted Magellan abundant funds. With the money, Magellan purchased five ships: the *Conception*, the *Santiago*, the *San Antonio*, the *Trinidad*, and the *Victoria*. The fleet left harbor in September of 1519 with 275 men and adequate provisions for only a few months.

From the start of the voyage, Magellan's fleet was plagued by problems. Magellan himself was Portuguese, but he was sailing under the Spanish flag. The rival nations were competing for trade routes and land in the New World, as well as for control of the **seas** in general. Thus, Magellan needed to avoid armed Portuguese ships, as well as Portuguese controlled ports in the New World. This limited the places where Magellan and his crew could stop to restock provisions, and made them wary of crossing Portuguese trade routes.

Magellan's Spanish captains, who sailed the other four ships, threatened his command of the fleet. On November 20, 1519, when a plot to mutiny against Magellan, organized by the captain of the *San Antonio*, Juan de Cartegena, was discovered, Cartegena was relieved of his command and imprisoned aboard the *Victoria*.

When Magellan set forth to discover an expedient trade route to the Spice Islands, he knew he would have to either find a passage through the New World, or sail around it. However, Magellan made two fatal miscalculations. He thought that both the New World (the landmass of the Americas) and the Pacific Ocean were much smaller than they actually are. The crew did not have adequate supplies, and had to make frequent stops to restock provisions on the ships. They spent several months on the open seas, and many sailors fell victim to scurvy, typhus, and various fevers. The extended duration of the voyage, coupled with the appalling conditions onboard, further disposed the crew against Magellan.

The voyage itself was arduous. Magellan did not reach the coast of Brazil until the December of 1519. He anchored off of the Portuguese port of Rio de Janeiro, but because of hostile relations between Spain and Portugal, kept most of the men onboard the ships. The fleet then sailed along the coast of **South America** looking for an inland passage, but as the **weather** grew colder and seas rougher, the fleet anchored and wintered in Patagonia (present-day southern Argentina). While in Patagonia, another mutiny was attempted. As an attempt to quell dissent in the fleet, Magellan executed some rebels and marooned the leaders of the insurrection when the fleet departed. Magellan sent the *Santiago* ahead to scout for a passage through the continent, but the ship sank in rough seas. Soon after, the remainders of the fleet departed to look for a passage to the Pacific. They arrived at the southern tip of South America in October. Magellan named the connecting waters the Strait of All Saints, but the strait now bears his name. Frightened of a longer and more grueling voyage ahead, the captain of the *San Antonio* turned his boat and sailed back towards Spain.

The remaining three ships reached the Pacific, but there were no navigational charts of the entire ocean. Magellan assumed the ocean was rather small, and predicted that the journey to the Spice Islands would take little more than a week. After three months, the crew reached the island of Guam. Without the food stores that were aboard the *San Antonio*, the remaining sailors lived off of rats, hard tack, sawdust, and any fish they could catch. Magellan anchored in Guam for several weeks to let his beleaguered crew recover. The crew then continued on to the Philippines. There, Magellan established good relations with the local king, but he and his men became involved in a tribal dispute. Several men were wounded and killed in the fighting, including Magellan. He died on April 27, 1521.

Though Magellan never fully circumnavigated the globe himself, the expedition he began did accomplish that monumental task. Stripped of her crew, the *Conception* was intentionally burned. The surviving 120 men of Magellan's crew, in two ships, departed the Philippines in May. Sebastian del Cano assumed control over the expedition. The two vessels reached the Spice Islands. Cano decided that the chances of one ship making it back to Spain were greater if the boats took different routes. Carrying a full hull of valuable cargo, the *Trinidad* sailed east, and the *Victoria* continued westward. The *Trinidad* was captured by the Portuguese, but the *Victoria* returned to Spain, with only 18 crewmembers left. Magellan's flagship was the first to circumnavigate the earth.

See also History of exploration II (Age of exploration); Oceans and seas

MAGMA

In **geology**, magma refers to molten **rock** deep within Earth that consists of liquids, gases, and particles of rocks and **crystals**. Magma has been observed in the form of hot **lava** and the various rocks made from the solidification of magma. Geologists have created magmas (artificial melts) in the laboratory to learn more about the physical conditions in which magma originated and its composition. Magma is the source of **igneous rocks**; it can intrude or force itself into surrounding rock where it cools and eventually hardens. These rocks are called intrusive igneous rocks. If magma rises all the way to Earth's surface it will extrude (push out), flowing or erupting out at the surface as lava, forming extrusive igneous rock (also called volcanic rock). Magma and the rocks it creates have similar chemical compositions.

Magma is generated within Earth's mantle, the thick layer between Earth's **crust** and outer core. Rock found deep within the crust is extremely hot, soft, and pliable, but rock does not become liquid until much deeper in the upper mantle. Pockets, or chambers of magma, can originate at various depths within the earth. The composition of the magma varies and indicates the source materials and depth from which they originated. **Silicon** dioxide (SiO_2) is the predominant ingredi-

The magma chamber inside Mt. St. Helens has gradually produced this lava dome inside the caldera of the volcano. In time, the volcano will erupt again in a violent explosion as the pressure within the magma chamber becomes too great to be contained. ***AP/Wide World. Reproduced by permission.***

ent in magma. Other ingredients include **aluminum** oxide, **iron**, magnesium, calcium, sodium, potassium, titanium, manganese, phosphorus, and **water**.

There are three basic types of magma, each having a characteristic origin and composition: basaltic (the most common, originating in the lower crust/upper mantle), rhyolitic (originates in the oceanic crust), and andesitic (most originate in the continental crust). New magma is formed by rocks **melting** when they sink deep into the mantle at subduction zones. The chemical composition, **temperature**, and the amount of dissolved liquids and gases determine the viscosity of magma. The more fluid a lava mixture is, the lower the viscosity. As magma or a lava flow cools, the mixture becomes more viscous, making it move slowly. Magmas having a higher silica (SiO_2) content are very viscous and move very slowly.

Magma has the tendency to rise because it weighs less than surrounding hard rock (liquids are less dense than solids) and because of the pressure caused by extreme temperature. The pressure is reduced as magma rises toward the surface. Dissolved gases come out of solution and form bubbles. The bubbles expand, making the magma even less dense, causing the magma to rise faster. The magma exerts a great deal of pressure on weak spots and fills up any cracks produced by the continual shifting of the earth's crust. On its way up toward the surface, magma can melt adjacent rock, which provides a suitable environment for the development of metamorphic rocks. When magma erupts as lava, its gases are released at the surface into the atmosphere or can be trapped in the molten rock and cause "air bubbles" in rock. The gases can also create violent explosions, throwing debris for miles around.

See also Volcanic eruptions; Volcanic vent

MAGMA CHAMBER

A **magma** chamber is a reservoir of molten **rock** that is the source of **lava** in a volcanic eruption. Magma chambers are typically located a few kilometers below the surface. A magma chamber is created by a mantle plume, or upwelling of heat from the earth's mantle. This delivers heat and molten rock upwards to the base of the **crust**. Because magma is less dense than solid rock, it will rise through fractures. Dissolved gases under pressure also help to force the magma upwards. If the migrating magma can no longer find a path upwards, it may gather into a reservoir and form a magma chamber. Continual migration of more magma into the chamber will cause the pressure within the chamber to increase. At some point, the magma chamber may not be able to contain the pressure and the magma, and it may breech. The magma is then driven upward through a conduit such as a volcanic neck or **fissure**, resulting in the eruption of the **volcano**.

Sometimes, a magma chamber will experience a cessation in the influx of magma before it is able to release into a volcanic eruption. It this case, the magma may begin to cool and crystallize in place. As it does so, mineral **crystals** form and settle to the bottom while the remaining magma, depleted of the components that formed the initial **minerals**, pools at the top. Certain minerals will form first according to **Bowen's reaction series**, with crystallization becoming successively more silica rich. This process is known as fractional crystallization and results in a stratified igneous body, known as a layered intrusion. The Skaergaard intrusion in Greenland is a well known layered intrusion.

See also Volcanic eruptions

Magnetic field

Earth acts as though it were a huge dipole magnet with the positive and negative poles near the North and South Poles. This does not mean that Earth is literally a dipole magnet—there are too many variations in the field—but that the best fit for a model of the field is two poles of a magnet, rather than a quadrupole, or other shape. The magnetic field of the earth allows magnetic compasses to work, making navigation much easier. It also molds the configuration of Van Allen belts, bands of high-energy charged particles around the earth's atmosphere.

Most of Earth's magnetic field (90%) occurs below the surface and possibly exists because Earth's core doesn't move at the same rate as the earth's mantle (the layer between the earth's core and its **crust**). The external 10% of the field is generated by movement of ions in the upper atmosphere.

The earth's magnetic field may help some animals navigate as they migrate. People have been using magnetic compasses for navigation since the fifteenth century. Because it has been so important for navigation, the magnetic field has been mapped all over the surface of the earth.

The magnetic field can also be used in other ways. For example, an instrument called a geomagnetic electrokinetograph determines the direction and speed of ocean currents while a ship is moving by measuring the voltage induced in the moving conductive sea **water** by the magnetic field of the earth.

The earth's magnetic field can change quickly and temporarily or slowly and permanently, depending on the cause of the change. The magnetic field can change very quickly, within an hour, in **magnetic storms**. These occur when the magnetic field is disturbed by sunspots, which send **clouds** of charged particles into Earth's atmosphere. (These same protons and electrons excite **oxygen**, nitrogen, and hydrogen atoms in the upper atmosphere, causing the **aurora borealis** and aurora australis.) These disturbances can be measured all over the globe and can cause static on radio stations.

The orientation of the magnetic field also changes slowly over centuries. In the planet's lifetime, the magnetic field has changed and even reversed (north pole becomes south and vice-versa) several times. Evidence for this is seen in reversed paleomagnetism of some sedimentary and igneous **rock**. In the 1960s, scientists showed that rocks formed at a particular interval in **geologic time** all indicate a magnetic field with the same orientation; older or younger rocks may show a reversed orientation. The cause of these paleomagnetic reversals is not yet known.

Today, the magnetic poles are not at the same place as the poles of the earth's rotational axis. Therefore, "magnetic north" is not quite the direction of "true north." The difference is known as the magnetic declination. Accordingly, scientists have established a series of geomagnetic coordinates, including **latitude and longitude**. These are centered on the magnetic dipole of the earth and designed (like geographic **latitude** and **longitude**) as though the earth were a perfect sphere.

See also Earth, interior structure; Earth (planet); Polar axis and tilt

Magnetic storm • *see* Coronal ejections and magnetic storms

Magnetism and magnetic properties

Magnetism is a property of matter and it occurs in different forms and degrees in various Earth materials that act as conductors and insulators. For example, at low temperatures, metallic systems exhibit either superconducting or magnetic order. The degree of magnetism of a substance is due to the intrinsic magnetic dipole moment of its electrons. The degree of magnetism is also called magnetization and it is defined as the net magnetic dipole moment of the substance per unit volume.

In the nineteenth century, **Michael Faraday** was the first to start classifying substances according to their magnetic properties. Faraday classified them as either diamagnetic or paramagnetic and he based his classification on the force exerted on the materials when placed in an inhomogeneous **magnetic field**.

Diamagnetic substances have a negative magnetic susceptibility, (i.e., they are materials in which the magnetization and magnetic field are opposite). The electrons in the atoms of diamagnetic materials are all paired and there is no intrinsic magnetic moment. When a material is placed into a magnetic field its atoms acquire an induced magnetic moment pointing in a direction opposite to that of the external field and the material becomes magnetic.

The diamagnetic field produced opposes the external field, although this diamagnetic field is very weak (except in superconductors). If the atoms of a material have no magnetic moment of their own, then diamagnetism is the only magnetic property of the material and the material is called diamagnetic. Copper exhibits such diamagnanetism.

Paramagnetic substances have a weak positive magnetic susceptibility and their atoms usually have unpaired electrons of the same spin. Some **metals**, rare earth, and actinides are paramagnetic. All the magnetic moments of the electrons in their atoms do not completely cancel out, and each **atom** has a magnetic moment. Such materials thus have a permanent magnetic moment and they can interact with a magnetic field. An external magnetic field tends to align the magnetic moments in the direction of the applied field, but thermal motion tends to randomize the directions. If only relatively small fractions of the atoms are aligned with the field, then the magnetization obeys Curie's law. Curie's law states that if the applied magnetic field is increased, the magnetization of the material also increases. This is because a stronger magnetic field will align a greater quantity of dipoles. Curie's law also states that the magnetization decreases with increasing **temperature**. The magnetic field produced by the aligned magnetic moments of paramagnetic materials strengthens the external field, but at standard temperatures it averages no more than 10 times stronger than a diamagnetic field and is, therefore, still very weak.

Ferromagnetic materials have the highest magnetic susceptibilities. In these materials, the spins of neighboring atoms do align even in the absence of an externally applied field through a quantum effect known as exchange coupling. Besides **iron**, examples of ferromagnetic materials are nickel, cobalt, and alnico, an aluminum-nickel-cobalt **alloy**. In these materials, all metals, the electrons give rise to permanent dipole moments that can align with those of their neighbors, creating magnetic domains that produce a magnetic field. Above a certain temperature, called the Curie temperature, a ferromagnetic material ceases to be ferromagnetic because the addition of thermal energy increases the motion of the atoms, thus destroying the alignment of the dipole moments. The material then becomes paramagnetic with weak magnetic susceptibility. The magnetic domains of ferromagnetic materials allow them to be turned into permanent magnets. If a ferromagnetic material is placed in a strong magnetic field, its magnetic domains converge into large domains aligned with the externally applied field. Upon removal of the external field, the electrons maintain the alignment and the magnetism remains.

An example of a device incorporating a diamagnet is the metal detector. In this instrument, the magnetic field is generated by an electromagnet, which then forms eddy currents. The magnetic fields from the induced currents are in turn picked up by the metal detector in the form of small currents.

See also Atomic theory; Atoms; Coronal ejections and magnetic storms; Dating methods; Electricity and magnetism; Ferromagnetic; Geographic and magnetic poles; Magnetic field; Paleomagnetics

MANTLE • *see* EARTH, INTERIOR STRUCTURE

MANTLE PLUMES

Convection in Earth's mantle created by the dissipation of internal heat produces up-welling hot columns called mantle plumes and cold, sinking sheets. Numerical modeling suggests the presence of three types of mantle plumes. Regular mantle plumes originate from the core-mantle boundary (a depth of approximately 1,802 mi [2,900 km]) and may be stable for several hundred million years. Such plumes act as fixed reference frames for plate motion. A second type of plume, also originating from the core-mantle boundary, can be bent and move relative to the global circulation in the mantle. Several mantle plumes may also collide to form superplumes. Superplumes rising from the core-mantle boundary may produce additional, secondary plumes that develop above a 416 mi (670 km) boundary layer in Earth's mantle.

Mantle plumes impinge on the base of Earth's **lithosphere** in all plate tectonic settings and result in surface uplift of up to 875 yd (800 m), lithospheric thinning, extensional stress fields, and a thermal anomaly centered on the plume. Heating the base of the lithosphere by mantle plumes may lead to **partial melting** and the formation of **mafic** (i.e., **iron** and magnesium-rich) **magma**. Magma may intrude into fractures formed from extension of Earth's upper, brittle **crust** above mantle plumes to form mafic **dike** swarms. For example, diabase dikes of the Mackenzie dike swarm in north-western Canada that extend for over 1,243 mi (2,000 km) are thought to result from a single mantle plume source. Dikes typically radiate from a point centered above a mantle plume. Radiating arms of dike swarms from different continents have been used to help reconstruct past continent configurations.

Magma may also be extruded as **lava** flows on Earth's surface to form flood basalts over areas 621 mi (1,000 km) or more across. For example, the Paraná and Etendeka volcanics represent pre-breakup volcanism on the South American and African margins of the Tristan Plume and volcanics of the Deccan Traps in western India result from **melting** due to the Reunion Plume. Mantle plumes in the early stages of Earth's history are likely to have been stronger and hotter. Mafic and ultramafic volcanics called komatiites, within Archaean and Paleoproterozoic greenstone belts in **Australia**, Canada, the Baltic Shield and China, have been attributed to mantle plume sources. Some granitoid plutons in Archaean greenstone belts may also be indirectly related to crustal melting by mantle plumes.

Mantle plumes constitute a driving force in the fragmentation and **rifting** apart of continents. For example, the separation of **South America** from **Africa** and Greater India from Australia and **Antarctica** during the break-up of the supercontinent Gondwanaland is interpreted as resulting from rifts linking areas above several mantle plumes. Mantle plume-related rifts are typified by triple junctions where rifts, normal faults and dikes define arms at approximately 120° to each other that intersect above the mantle plume. Frequently, continental breakup and formation of oceanic crust occurs along two of the rift arms, whereas the third arm may be less developed, and constitute a failed rift or aulacogen. For example, a plume-related triple junction occurs over the Afar Plume, above which the Red Sea and Golf of Aden Rifts (along which there is active seafloor spreading) and the eastern, Ethiopian branch of the East African Rift (an intra-continental rift system) intersect. Not all plume-related rifts, however, define triple junctions and four or more rift arms may sometimes be present.

Mantle plumes may play an important role in the formation of mineral deposits e.g., nickel, chromium, platinum, palladium, diamonds, rare earth elements, tin, tantalum, niobium, copper, lead, and zinc. Such deposits may be related to alkaline magmatic fluids associated with mantle plumes, as well as being controlled by extensional structures due to plume-related stresses.

Mantle plumes are not unique to Earth. On Venus, where there is no evidence for **plate tectonics** as is known on Earth, deep rifts called chasmata and prominent radiating fracture systems from central volcanic peaks called novae develop above mantle plumes. Circular to elliptical volcano-tectonic features 37–1,616 mi (60–2,600 km) in diameter called coronae may also be sited above mantle plumes, or result from rifts linking several mantle plumes. Prominent concentric faults rimming smaller coronae are thought to form as a result of collapse due to withdrawal of magma produced by melting above a mantle plume.

See also Convergent plate boundary; Divergent plate boundary; Earth, interior structure; Geothermal deep ocean vents; Hawaiian Island formation; Hotspots; Volcanic eruptions

MAPPING TECHNIQUES

Geological maps portray the distribution of different **rock** types, the location of faults, **shear zones** and **folds**, and the orientation of primary and structural features. Mines, quarries, mineral occurrences, fossil localities, geochronological sampling sites, oil and **water** wells may also be shown. Geological maps illustrate rock relationships that enable the depositional, intrusive, and structural history of an **area** to be established, and the three-dimensional geometry to be visualized. They provide fundamental information for mineral and **petroleum** exploration and for hydrological and environmental investigations.

In small areas such as exploration tenements where detailed maps are required, it is common practice to undertake grid mapping. After a grid is surveyed and pegged, the geologist carries out detailed traverses along grid lines. Rock types, lithological contacts, and alteration are noted and structural measurements made using a compass-clinometer. Information may be recorded by hand on traverse maps or collected digitally. A complete map is compiled by interpolating between gridlines, collecting additional data where necessary. Aerial photographs, **satellite** or other remotely sensed images (as discussed below) serve as the base for recording regional map data. Digital data recorders integrated with **GPS** (global positioning system) location measurements enable lithological and structural data to also be digitally recorded in the field. Data can be directly input into a **GIS** (geographic information system) or custom computer package that enables different attributes to be displayed on a map and spatially analyzed.

A three dimensional, exaggerated view of the landscape is created when pairs of overlapping aerial photographs are viewed through a stereoscope. Stereographic views of aerial photographs assist in the identification and classification of **landforms**, the interpretation of rock types based on characteristic outcrop or **weathering** patterns, and the recognition of tectonic and intrusive structures. Areas where rock formations crop out can also be identified. Where aerial photographs record data in the visible (and sometimes infrared) parts of the **electromagnetic spectrum**, earth-sensing satellites collect data for several different wavelengths or bands. Some bands or ratios of bands highlight vegetation, whereas others respond to differences in water content of soils or different rock types. Individual bands, or ratio of bands are assigned to red, green or blue channels of image processing systems to produce false color images. There has been a marked increase in the resolution of commercially available satellite imagery from 262 ft (80 m) in early Landsat imagery to 2.3 ft (0.70 m) (panchromatic) and 9.2 ft (2.8 m) (multispectral) with the Quickbird-2 satellite. Several other satellites have resolutions of approximately 3.3 ft (1 m) (panchromatic) and between 6.6 and 16.4 ft (2–5 m) (multispectral). Whereas standard satellite imagery uses approximately seven bands, in hyper spectral **remote sensing**, data is collected simultaneously in over 200 narrow, contiguous spectral bands from sensors in high-flying aircraft or satellites. For example, NASA's AVIRIS (Airborne Visible/InfraRed Imaging Spectrometer) maps a strip 6.8 mi (11 km) wide with a ground resolution of approximately 21.8 yd (20 m). Hyperspectral data can be calibrated to distinguish different rock types. Bands that correspond to specific wavelengths at which **minerals** reflect or absorb energy are used to map the distribution of individual minerals, including clays and alteration minerals developed around mineral deposits. Side-looking airborne radar (SLAR) on aircraft or satellites transmits microwave energy and records the energy obliquely reflected from the ground. Radar imagery is useful in mapping areas covered by cloud that obscures normal satellite sensors and in structural mapping, especially in highly vegetated areas. Radar imagery currently has a resolution of 26.2 ft (8 m) or more, although 9.8 ft (3 m) resolution imagery will soon be available.

In contrast to the above techniques that record portions of the electromagnetic spectrum, magnetic and radiometric methods record differences in rock composition. The magnetic signature of a rock is due to the amount of magnetic minerals such as magnetite it contains. Some rock types (e.g., **mafic** volcanics and **banded iron formations**) have a high magnetite content and create magnetic highs. Other rocks low in magnetite (such as quartzite and shale) produce magnetic lows. Faults may be imaged due to magnetite destruction as a result of weathering or alteration during fluid flow along them. Local magnetic highs may indicate addition of magnetic minerals during mineralization and so provide exploration targets. Linear, crosscutting magnetic highs commonly represent mafic dikes. Aeromagnetic data is obtained by flying low altitude, closely spaced parallel paths with an aeroplane or helicopter mounted with (or trailing) a magnetic sensor. Ground magnetic data is recorded using sensors on a tall pole, either handheld or vehicle mounted. Data values are interpolated between recorded measurements. Raw magnetic data is generally first processed to appear as if the inducing **magnetic field** had a 90° inclination (a process called reduction to the pole). This simplifies magnetic anomalies and centers anomalies over the causative rock body. The distribution of different rock types, position of contacts, and form of folds and other structures can be interpreted from contoured or digitally processed magnetic images. Digital enhancements (such as artificial **sun** angles and vertical derivatives) are used to highlight faults, shear zones, and/or lithological contacts. Aeromagnetic data allows geological maps to be made in areas of no outcrop, even below **lakes** or superficial cover.

Airborne gamma ray spectrometry is a technique that measures variations in the potassium, thorium, and uranium content of rocks using sodium iodide crystal detectors mounted in aircraft (radiometric surveys are often carried out at the same time as aeromagnetic surveys). Radiometric data is presented as either individual images portraying the relative amount of each element or images of various element ratios. Radiometric images are useful in mapping compositional variations in granitic and high-grade metamorphic rocks (especially in areas with little or no transported sedimentary cover), rock types such as carbonatite that have unusual amounts or proportions of potassium, thorium and uranium, and alteration zones around mineralized areas. They can also be used to map the distribution of sediments derived by the weathering of granitic and other radiogenic source rocks.

While computer-enhanced images of remotely sensed data are increasingly used in geological interpretation, detailed fieldwork by geologists still provides the backbone for the creation of geological maps.

See also Bathymetric mapping; Cartography; Geologic map; Petroleum, detection; Physical geography; RADAR and SONAR; Remote sensing

Marble

Marble is metamorphosed **limestone**, that is, limestone that has been melted and allowed to resolidify. If the original limestone is a calcite limestone, then the marble is a calcite marble (i.e., mostly $CaCO_3$); if the original limestone is a dolomitic limestone, then the marble is a dolomitic or magnesian marble (i.e., mostly $CaMg(CO_3)_2$). In nongeological contexts the term marble is often used to refer to any hard, calcite **rock** that can be cut or polished, including some unmetamorphosed limestones. In **geology**, however, it is reserved strictly for metamorphosed limestones.

Certain marbles have been valued since antiquity for sculpture and for architectural uses. The marbles prized for statuary are usually quite pure (i.e., white in color and free from inclusions or marks) and reflect light softly or semi-translucently due to their property of allowing some incident light to penetrate to a depth of about an inch (1–2.5 cm) before reflecting it.

Some marbles that show colorful patterning are used for decorative architecture. Patterning in marble arises from various trace **minerals**, most often silicates (e.g., **quartz**, **olivine**, garnet), **graphite**, pyrite, and organic substances. The **magma** responsible for metamorphosing the original limestone may also contribute impurities.

Wrinkled thin layers that show in cross-section as sinuous lines are common in marbles. These layers are termed stylolites. Stylolites consist of silicates or other accessory minerals and are usually darker than the surrounding marble. They do not form as sedimentary layers in the original limestone, but result from the selective removal of limestone by **water**. Calcite is a highly soluble mineral; when part of the original limestone is dissolved by infiltrating water, the fine particles that are left are compacted into an irregular layer or stylolite. Comparison of accessory mineral concentrations in adjacent marble and in stylolites shows that 40% or more of a limestone bed may be dissolved in the process of forming stylolites.

Calcite marble, like any other calcite rock, effervesces vigorously (yielding **carbon dioxide** [CO_2]) when tested with hydrochloric acid. Dolomitic marble effervesces more weakly. Otherwise, they are difficult to distinguish.

See also Field methods in geology; Industrial minerals

Marinas trench • *see* Ocean trenches

Marine transgression and marine regression

Marine transgression occurs when an influx of the sea covers areas of previously exposed land. The reverse process, called marine regression, takes place when areas of submerged seafloor are exposed above sea level by basinward migration of a shoreline. Landward displacement of coastal and marine sedimentary environments accompanies transgression, and a shift from shallow **water** and terrestrial sediments, to deeper-water sedimentary facies, called onlap, indicates a transgression in a vertical succession of sedimentary strata. A shift from deeper marine sediments to terrestrial and fluvial sediments, or offlap, likewise suggests a basinward migration of the shoreline, or a marine regression.

The pattern of onlap and offlap preserved along a continental margin tells its history of alternating transgression and regression. An array of interacting processes determines the position of a shoreline at a specific location, and the geometry of continental margin strata records the combined effects of these interactions. Fluctuation of absolute, global sea level resulting from cyclical growth and decay of Earth's **polar ice** caps, called eustasy, is only one of the many factors that determine sea level relative to a specific coastal segment. Rates of sediment supply and transport, three-dimensional patterns of deposition and **erosion**, and crustal subsidence and uplift all influence the geometry of onlap and offlap at a particular location. Attempts to define the history of global eustatic sea level change by interpreting the stratigraphic geometry of individual continental margins have been largely unsuccessful, and the difficulty of this scientific problem has underscored the complexity of the systems that regulate shoreline migration.

The study of strata deposited along continental margins under the influence of cyclical Earth processes such as eustatic sea level change is a branch of **stratigraphy** called sequence stratigraphy. Development of this geologic subdiscipline in the early 1970s is largely attributed to **petroleum** industry researchers who first used seismic reflection profiles to map the distribution of oil and gas bearing strata in sedimentary basins. Because relative sea level change determines the location and geometry of these oil-bearing strata along continental margins, sequence stratigraphy is a powerful predictive tool for oil and gas exploration.

See also Beach and shoreline dynamics; Ice ages; Petroleum, detection; Petroleum, history of exploration

MARS • *see* SOLAR SYSTEM

MASS MOVEMENT

Mass movement refers to the downslope movement of **soil**, **regolith**, or **rock** under the influence of **gravity** and without the aid of a transporting medium such as **water**, **ice**, or air. The term is synonymous with **mass wasting** and stands in contrast to mass transport, in which the same kinds of material are transported by water, ice, or air.

Mass movement can occur by a variety of processes including landsliding in all of its forms, **creep**, and solifluction. Rates of mass movement can range from a few millimeters per year in the case of creep or solifluction to tens of meters per second in the case of **catastrophic mass movements** such as debris avalanches. Debris and mud (or earth) flows are generally considered to be forms of mass movement because they are comprised primarily of solid material with only a small proportion of water.

Both mass movement and mass transport are naturally occurring processes that contribute to the cycle of tectonic uplift, **erosion**, transportation, and deposition of sediments. They are responsible for the **topography** of mountain ranges and river canyons that has developed over **geologic time**. Since the Industrial Revolution, however, humans have become increasingly significant agents of mass movement and transport. Catastrophic mass movements at Elm, Frank, and Vaiont were triggered by human activity on or near potentially unstable slopes; the failure of hydraulic structures such as Teton and St. Francis dams have produced major **floods** with great erosional power; and open pit mining involves the movement of cubic kilometers of material over decades of operation. Agriculture is also a large, but subtle contributor to mass movement, because exposed and tilled soil is much more easily eroded than that in its natural state. Recent estimates suggest that humans are currently responsible for the movement of about 37 billion tons of soil and rock per year, and that the cumulative amount of soil and rock moved by humans is the equivalent of a mountain range that is 2.5 miles (4 km) high by 62 miles (100 km) long by 24.8 miles (40 km) wide.

See also Debris flow; Landslide; Mud flow; Rockfall; Slump

MASS WASTING

Mass wasting, or **mass movement**, is the process that moves Earth materials down a slope, under the influence of **gravity**. Mass wasting processes range from violent landslides to imperceptibly slow **creep**. Mass wasting decreases the steepness of slopes, leaving them more stable. While **ice** formation or **water** infiltration in sediments or rocks may aid mass wasting, the driving force is gravity. All mass wasting is a product of one or more of the following mass wasting processes: flow, fall, slide, or **slump**.

The four processes of mass wasting are distinguished based on the nature of the movement that they produce. Flow involves the rapid downslope movement of a chaotic mass of material. Varying amounts of water may be involved. A **mud flow**, for example, contains a large amount of water and involves the movement of very fine-grained Earth materials. Fall involves very rapid downslope movement of Earth materials as they descend (free fall) from a cliff. Ignoring **wind** resistance, falling materials accelerate at 32 ft/sec^2 (9.8 m/sec^2)—the average gravitational force of the earth. Slides result when a mass of material moves downslope, as a fairly coherent mass, along a planar surface. Slumps are similar to slides, but occur along a curved (concave-upward) surface and move somewhat more slowly.

Consider a chunk of **rock** currently attached to a jagged outcrop high on a mountain. It will move to the sea as a result of three processes: **weathering**, mass wasting, and **erosion**.

On warm days, water from **melting** snow trickles into a crack which has begun to form between this chunk and the rest of the mountain. Frigid nights make this water freeze again, and its expansion will widen and extend the crack. This and other mechanical, biological, and chemical processes (such as the growth of roots, and the dissolution of the more soluble components of rock) break apart **bedrock** into transportable fragments. This is called weathering.

Once the crack extends through it and the chunk has been completely separated from the rest of the mountain, it will fall and join the pile of rocks, called talus, beneath it that broke off the mountain previously. This pile of rocks is called a **talus pile**. This movement is an example of mass wasting, known as a **rockfall**. As the rocks in the talus pile slip and slide, adjusting to the weight of the overlying rocks, the base of the talus pile extends outward and eventually all the rocks making up the pile will move down slope a little bit to replace those below that also moved downslope. This type of mass movement is known as rock creep, and a talus pile that is experiencing rock creep is called a rock glacier.

In the valley at the bottom of this mountain, there may be a river or a glacier removing material from the base of the talus slope and transporting it away. Removal and transport by a flowing medium (**rivers**, **glaciers**, wind) is termed erosion.

These processes occur in many other situations. A river erodes by cutting a valley through layers of rock, transporting that material using flowing water. This erosion would result in deep canyons with vertical walls if the erosion by the river were the only factor. Very high, vertical walls, however, leave huge masses of rock unsupported except by the cohesive strength of the material of which they are made. At some point, the stresses produced by gravity will exceed the strength of the rock and an **avalanche** (another type of mass movement) will result. This will move some of the material down the slope into the river where erosion will carry it away.

Erosion and mass wasting work together by transporting material away. Erosion produces and steepens slopes, which are then reduced by mass wasting. The steepness of a natural slope depends on the size and shape of the material making up the slope and environmental factors, principally water content. Most people learn about this early in life, playing in a sandbox or on the beach. If dry **sand** is dumped from a bucket, it forms a conical hill. The more sand dumped, the larger the hill becomes, but the slope of the hill stays the same. Digging into the bottom of the hill causes sand to avalanche down into the hole you are trying to make. Loose, dry sand flows easily, and will quickly re-establish its preferred slope whenever anything is done to steepen it. The flow of sand is a simple example of mass wasting.

If sand is moist, the slope of a sand pile can be higher. A sandcastle can have vertical walls of moist sand when it is built in the morning, but, as the afternoon wears on and the sand dries out, it eventually crumbles and collapses (mass wastes) until a stable slope forms. This is because the water makes the sand more cohesive. With the proper moisture content, there will be both water and air between most of the grains of sand. The boundary between the water and the air has surface tension—the same surface tension that supports water striders or pulls liquids up a capillary tube. In moist sand, surface tension holds the grains together like a weak cement.

However, if sand becomes saturated with water (that is, its pores become completely water-filled as they are in quick sand), then the sand will flow in a process known as lateral spreading. Water-saturated sand flows because the weight of the sand is supported (at least temporarily) by the water, and so the grains are not continuously in contact. The slope of a pile of sand is dependent on water content, and either too little or too much water lowers the stable slope. This illustrates how slope stability is a function of water content.

The steepest slope that a material can have is called the angle of repose. Any loose pile of sediment grains has an angle of repose. As grain size increases, the angle of repose also increases. Talus slopes high on mountain sides may consist of large, angular boulders and can have slopes of up to 45°, whereas fine sand has an angle of repose of 34°. This is the slope that you can see inside a sand-filled hourglass. In nature, however, slopes less than the angle of repose are common because of wind activity and similar environmental processes.

A typical sand dune has a gentle slope on the windward side where erosion by the wind is responsible for the slope. On the leeward side, where sand falls freely, it usually maintains a slope close to the angle of repose. As with loose deposits of particles on land, similar conditions exist if they are under water, although stable slopes are much gentler. When sudden mass wasting events occur under water, large quantities of material may end up being suspended in the water producing turbidity currents that complicate the picture. Such currents occur because a mass of water with sediment suspended in it is denser than the clear water surrounding it, so it sinks, moving down the slope, eroding as it goes. Still, the initial adjustment of the slope was not the result of these currents, so the mechanism that produces turbidity currents is an example of mass wasting.

Most slopes in nature are on materials that are not loose collections of grains. They occur on bedrock or on soils that are bound together by organic or other material. Yet, many of the principles used to explain mass wasting in aggregates still apply. Instead of mass wasting taking place as an avalanche, however, it results from a portion of the slope breaking off and sliding down the hill. These events are usually called landslides, or avalanches—if they are large and damaging—or slumps if they are smaller.

If the gravitational forces acting on a mass of material are greater than its strength, a fracture will develop, separating the mass from the rest of the slope. Usually this fracture will be nearly vertical near the top of the break, curving to a much lower angle near the bottom of the break. Such events can be triggered by an increase in the driving forces (for example, the weight of the slope), a decrease in the strength of the material, or both.

Even solid rocks contain pores, and many of these pores are interconnected. It is through such pores that water and oil move toward wells. Below the **water table**, all the pores are filled with water with no surface tension to eliminate. It might seem that rocks down there would not be affected by rainfall at the surface. As the rains come, however, the water table rises, and the additional water increases the pressure in the fluids in the pores below. This increase in pore pressure pushes adjacent rock surfaces apart, reducing the friction between them, which lowers the strength of the rock and makes it easier for fractures to develop. Elevated pore pressures are implicated in many dramatic mass wasting events.

When mass wasting by flow occurs so slowly that it cannot be observed, it is called creep. Most vegetated slopes in humid climates are subject to **soil** creep, and there are many indicators that it occurs. Poles and fence posts often tip away from a slope a few years after they are placed. Trees growing on a slope usually have trunks with sharp curves at their bases. Older trees are bent more than younger ones. All this occurs because the upper layers of soil and weathered rock move gradually down the slope while deeper layers remain relatively fixed. This tips inanimate objects such as power poles.

It would tip trees, too, except that they grow toward the **Sun**, keeping the trunk growing vertically, and so a bend develops.

This gradual downslope movement requires years to result in significant transport, but because it occurs over a great portion of the surface of the earth it is responsible for most mass wasting.

See also Catastrophic mass movements

MATTHEWS, DRUMMOND (1931-1997)

English marine geophysicist

Drummond "Drum" Matthews had a long, outstanding career in **geology** and geophysics, contributing to the fundamental understanding of the structure and **evolution** of the earth's **crust**.

Matthews grew up near the sea, at Porlock in Somerset, England, and developed a lifelong love of the ocean. He attended Bryanston School in Dorset, before a term in the Royal Navy. He then studied at King's College, Cambridge, specializing in geology and petrology. After graduation, Matthews spent two years (1955–57) in the Falkland Islands, as part of the Dependencies Survey (later the British Antarctic Survey), before returning to Cambridge to complete a Ph.D. in marine geophysics.

In 1962, as part of the International Indian Ocean Expedition, Matthews made a small but detailed survey of a ridge in the north-west Indian Ocean that showed large areas of the seafloor magnetized in opposite polarities. This was to prove a key piece in the puzzle of seafloor creation and the theory of **plate tectonics**. As early as 1915, **Alfred Wegener** (1880–1930) had proposed that there had once been a supercontinent, which he named Pangea, that had slowly moved apart. However, Wegener could not explain how such continental drift occurred, and so his theory was not well received. In the early sixties Harry Hess (1906–1969) hypothesized that seafloor spreading was responsible for the motion of the continents. In 1963, Matthews, with his first graduate student, **Fred J. Vine** (1939-), published a paper, "Magnetic Anomalies Over Ocean Ridges," in *Nature*. In this work, the scientists proposed an idea that, if confirmed, would provide strong support for the seafloor spreading hypothesis. It had long been suspected, but not proven, that the earth's **magnetic field** has undergone a number of reversals in polarity in its long history. Vine and Matthews suggested that if ocean ridges were the sites of seafloor creation, and the earth's magnetic field does reverse, then new **lava** emerging would produce **rock** magnetized in the current magnetic field of the earth. Older rock would have an opposing polarity, depending on when it had been created. By 1966, further studies confirmed the theory for all **mid-ocean ridges**. This provided compelling evidence for sea floor spreading, and an explanation of the mechanism of continental drift.

From 1960 to 1966, Matthews was a Senior Assistant in Research in the Department of Geodesy and Geophysics, and a Research Fellow of King's College, Cambridge. He became an Assistant Director of Research at Cambridge in 1966, and was appointed Head of the Marine Geophysics Group. During his time as Head, the Group contributed to over 70 scientific expeditions and published nearly 200 academic papers, working in areas as diverse as the North Sea, the Eastern Mediterranean, and the Gulf of Oman. In 1971, he was appointed Reader in Marine Geology at Cambridge.

From 1979, Matthews began to study deep crustal seismics that allowed research into the structure and evolution of continental crust. He helped found the British Institutions Reflection Profiling Syndicate (BIRPS), and became its first Director in 1982. BIRPS revealed previously unknown structures in the lower crust and upper mantle. He left BIRPS in 1990, taking early retirement as the result of ill health.

Matthews received many honors and awards recognizing his contributions to geology and geophysics, including the Chapman Medal of the Royal Astronomical Society (1973, with Fred Vine), the Bigsby Medal of the Geological Society of London (1975), the Arthur L. Day prize and lectureship of the National Academy of Sciences (1975, with Vine), the Hughes Medal of the Royal Society (1982, with Vine), the International Balzan Prize (1981, with Vine and Dan McKenzie), and the G. P. Woollard Award of the Geological Society of America (1986). He was a Fellow of the Royal Society from 1974, and was made a Fellow of the American Geophysical Union in 1982. Matthews died at the age of 66 after a long battle with diabetes and a resulting heart condition.

See also Continental drift theory; Earth, interior structure

MAUPERTUIS, PIERRE-LOUIS MOREAU DE (1698-1759)

French astronomer, mathematician, and biologist

A mathematician, biologist, and astronomer, Pierre-Louis Moreau de Maupertuis was a strong proponent of Sir Issac Newton's theory of gravitation, helped confirm Newton's theory on the exact shape of the earth, and formulated the principle of least action in **physics**. Born in Saint Malo, France, Maupertuis had a wide range of scientific interests. As a biologist, he wrote *Systéme de la Nature*, in which he provided the first accurate scientific record of a dominant hereditary trait transmitted among humans. He also introduced the theory of the survival of the fittest in his *Essai de Cosmologie*, a theory that Charles Darwin later expounded to wide acceptance.

Maupertuis may be best known for his formulation in 1744 of the principle of least action, also known as the minimum principle or Maupertuis' principle. Essentially, the principle states that any change that occurs in the universe and nature, such as a moving body or light rays, changes in the most economical path possible. For example, bubbles form in a shape that presents the smallest surface for a given volume of air. In *Essai de Cosmologie*, Maupertuis presented his theory as something that might help prove the existence of God by unifying the laws of the universe. In 1736, Maupertuis led a famous expedition to Lapland near the North Pole that proved Newton's theory that the earth is an oblate sphere (flattened at the poles). The proof was accomplished by measuring the length of degree along a meridian and comparing the findings with the findings of another expedition near the equator in Peru performing similar measurements.

Despite his many accomplishments, Maupertuis was considered arrogant by many of his fellow countrymen. Eventually, Maupertuis became a target of German mathematician Samuel Koenig, who accused him of plagiarism, and of the French author Voltaire (1694–1778), whose satirical writings about Maupertuis were so savage that Maupertuis eventually left France. Maupertuis died in virtual exile in Basel, Switzerland, in the home of Swiss mathematician Johann Bernoulli (1667–1748).

See also Earth (planet); Electromagnetic spectrum

MAXWELL, JAMES CLERK (1831-1879)

Scottish physicist

James Clerk Maxwell was a physicist who introduced a new paradigm with his electromagnetic theory, influencing generations of researchers. Maxwell was without a doubt a child prodigy. At an early age, he solved geometric problems and wrote explanations that intrigued academics. Just as he considered how charged particles interact with their surrounding **area**, one might consider the interaction of the conditions of his inherent nature and the environment of his early childhood. Maxwell's life could make a good case study for the strength of the influences of heredity compared to environment as he had strong influences from both sources.

James Clerk Maxwell was a descendent of the Clerk and the Maxwell families, both with distinguished heritages. His father inherited a house in Edinburgh and land in the countryside. Maxwell was born in 1831 in Edinburgh while his parents were waiting for their country house to be built. They moved shortly after he was born. His father was a lawyer but was not very aggressive in pursuing new business. John Clerk Maxwell enjoyed studying science and building mechanical devices. As young as three years old, James was following his father insisting to know how everything worked. He was very close to his father all of his life. Maxwell's mother died suddenly when he was eight years old. For two years after his mother's death, he was educated by a series of tutors, but none were found suitable for Maxwell and his unique way of learning. His father and his aunt arranged for him to begin studies at the Edinburgh Academy. At the academy, Maxwell started to show his true capabilities and his classmates were less cruel.

In 1847, at age 16, Maxwell began his college studies at the University of Edinburgh. He spent three years there and during this time, he contributed two papers to the Edinburgh Royal Society. When he finished his studies at Edinburgh, his father sent him to Peterhouse, but shortly after beginning there, he transferred to Trinity where he believed he had a better chance for a fellowship. Maxwell studied at Trinity from early 1851 until he graduated in 1854. After graduation, he was awarded the fellowship. Maxwell then applied for a position at Marischal College to be close to his ailing father. However, his father did not live much longer. After his father's death in April 1855, he accepted the position at Marischal.

In 1858, he married the well-educated Katherine Dewar. Two years later, he had to leave Marischal, the victim of an institutional merger. He was immediately invited to teach at King's College, London. It was in London that he did his most prominent work. He remained there until he resigned his post (probably due to exhaustion) in the spring of 1865. He spent most of the next five years at his country home writing a book on his theory. He considered himself retired.

To stay involved in academia, Maxwell did consulting work for Cambridge. His encouraging of Cambridge to offer courses on heat and electromagnetism directly influenced the foundation of the Cavendish Laboratory. It was only natural that the first Cavendish professorship should be offered to him and he accepted. During his eight years as Cavendish professor, he worked to prepare for publication the experiment papers Henry Cavendish had written. It is well accepted that this self-imposed responsibility was influential in bringing due respect to Cavendish's work. In May 1879, as the school year wound down, it was obvious to many that Maxwell's health was beginning to fail. He tried to return to Cambridge in the autumn, but he could scarcely walk. Maxwell died the same year of abdominal cancer at the age of 48.

Maxwell's work leading to his kinetic theory of gases and his theory of electromagnetic fields was a logical advance from James Prescott Joule's work. Both researchers measured the velocity of gas molecules and both recognized that heat was not the fluid that it once was thought to be. The importance of Maxwell's work was the direction that it gave to new understanding. Joule showed only the scientific community what was possible to measure and what might be proven. Maxwell went forward with detailed mathematical models that left no holes unfilled, with one important exception. Maxwell used statistics to show the high probability that proposed laws would direct the behavior of matter. Discussing the probability of natural law took science away from determinism. This opened the door for the modern study of **physics**. Albert Einstein's theory of relativity and the recently nurtured **chaos theory** could not have been developed except for this new philosophical direction.

Maxwell began measuring the average velocity of a gas molecule with the objective to investigate whether the perceived random order of its movement could be predicted with some degree of accuracy. What he found was that the greater the velocity of the molecules, the greater the heat generated. There was a direct relationship between the amount of movement among the molecules and the amount of heat in a gas. In this experimental demonstration, heat was shown undeniably to be a property of particle movement and not a fluid flowing from one object to another. Furthermore, Maxwell's findings showed that the movement of particles could be controlled through increasing or reducing heat.

Maxwell understood Michael Faraday's theory of electric and magnetic fields. He worked to demonstrate what Faraday could not explain himself through complex calculations. Assuming that the **space** surrounding a charged particle contained a field of force, Maxwell created a mathematical model demonstrating all the possible phenomena of electric and magnetic fields. Through this model, Maxwell demonstrated that the electric and magnetic fields worked together. He coined the term "electromagnetic" to name this new breakthrough.

This discovery is important for **chemistry** because it ultimately led to the discovery of the electron. Joseph John Thomson discovered the electron when he was investigating the effects of the electromagnetic field on gases, applying the principles that Maxwell had established. Research on the effects of light on elements was furthered by Maxwell's work. His subsequent work on the velocity of the oscillation of electromagnetic fields demonstrated that light should be considered a form of electromagnetic radiation.

See also Atomic structure; Electromagnetic spectrum; Quantum electrodynamics (QED); Relativity theory

MEANDERS • *see* CHANNEL PATTERNS

MEDIAL MORAINE • *see* MORAINES

MEDITERRANEAN SEA • *see* OCEANS AND SEAS

MELTING • *see* FREEZING AND MELTING

MENDELEEV, DMITRY (1834-1907)

Russian chemist

One of the most unlikely success stories in the history of **chemistry** is that of Dmitry Ivanovich Mendeleev (also Mendeléev, Mendeleef, and Mendeleeff). Mendeleev was born in Tobolsk in western Siberia on February 8, 1834. He was the youngest child in a family of either 14 or 17 children (records do not agree). His father, a teacher at the Tobolsk gymnasium (high school) lost his job after he became blind when Dmitry was still quite young. His mother tried to take over support of the family by building a glassworks in the nearby town of Axemziansk.

Mendeleev was an average student. He learned science from a brother-in-law who had been exiled to Siberia because of revolutionary activities in Moscow. Dmitry completed high school at the age of 16, but only after the family had experienced further misfortune—the death of his father and destruction of his mother's glassworks by fire. In 1850, his mother decided to see that her two youngest children received a college education. She and the children traveled by horse first to Moscow, then on to St. Petersburg. Through the efforts of a family friend, she was able to enroll Dmitry at the Central Pedagogical Institute in St. Petersburg. A few months later, Mendeleev's mother died.

Mendeleev graduated from the Pedagogical Institute in 1855 and then traveled to France and Germany for graduate study. While at Heidelberg with Robert Bunsen, he discovered the phenomenon of critical **temperature**, the highest temperature at which a liquid and its vapor can exist in equilibrium. Credit for this discovery is usually given to Thomas Andrews (1813–1885) who made the same discovery independently two years later.

In 1861, Mendeleev returned to St. Petersburg, where he became professor of chemistry at the Technological Institute. Six years later, he was also appointed professor of general chemistry at the University of St. Petersburg, a post he held until 1890. In that year, he resigned his university appointment in a dispute with the Minister of Education. Three years later he was appointed Director of the Bureau of Weights and Measures, a post he held until his death on February 2, 1907. Mendeleev is remembered as a brilliant scholar, interesting teacher, and prolific writer. Besides his career in chemistry, he was interested in art, education, and economics. He was a man of strong opinions who was not afraid to express them, even when they might offend others. He was apparently bypassed for a few academic appointments and honors because of his irascible nature.

The achievement with which Mendeleev's name will forever be associated was his development of the periodic law. In 1868, he set out to write a textbook in chemistry, *Principles in Chemistry*, that was later to become a classic in the field. Mendeleev wanted to find some organizing principle on which he could base his discussion of Earth's 63 **chemical elements** then known. After attending the Karlsruhe Congress in 1860, he thought that the atomic weights of the elements might provide that organizing principle. He began by making cards for each of the known elements. On each card, he recorded an element's atomic weight, valence, and other chemical and physical properties. Then he tried arranging the cards in various ways to see if any pattern emerged. Mendeleev was apparently unaware of similar efforts to arrange the elements according to their weights made by J. A. R. Newlands (1838–1898) only a few years earlier.

Eventually he was successful. He saw that, when the elements were arranged in ascending order according to their weights, their properties repeated in a predictable, orderly manner. That is, when the cards were laid out in sequence, from left to right, the properties of the tenth element (sodium) were similar to those of the second element (lithium), the properties of the eleventh element (magnesium) were similar to those of the third element (beryllium), and so on.

When Mendeleev arranged all 63 elements according to their weights, he found a few places in which the law appeared to break down. For example, tellurium and iodine were in the wrong positions when arranged according to their weights. Mendeleev solved this problem by inverting the two elements, that is, by placing them where they ought to be according to their properties, even if they were no longer in the correct sequence according to their weights.

Mendeleev hypothesized that the atomic weights for these two elements had been incorrectly determined. He happened to be incorrect in this assumption, and it was not until Henry Moseley discovered atomic numbers in 1914 that the real explanation for inversion was found.

Mendeleev made one other critical hypothesis. He found three places in the **periodic table** where elements appeared to be missing. The blank spaces occurred when Mendeleev insisted on keeping elements with like properties underneath each other in the table, regardless of their weights. He predicted not only that the three missing elements would be found, but also what the properties of those elements would be.

Mendeleev's law was soon vindicated when the three missing elements were found in 1875 (gallium), 1879 (scandium), and 1885 (germanium).

See also Alkaline Earth metals; Atomic mass and weight; Atomic theory; Atoms; Bohr model of the atom; Chemical bonds and physical properties; Chemical elements; Geochemistry

MERCATOR, GERHARD (1512-1594)

Flemish cartographer, geographer, and mathematician

Mercator, the world's most influential mapmaker, modernized **cartography** according to mathematical principles, facilitated navigation by charts, invented the **projection** that bears his name, and coined the term "atlas" to refer to a book of maps.

Born as Gerhardus Cremer, or Kremer, the son of a shoemaker, on March 5, 1512 in Rupelmonde, Flanders, he began using the Latin form of his surname upon entering the University of Louvain in 1530. Both "mercator" in Latin and "cremer" in Flemish, which is cognate with "Kramer" in German, mean "merchant." Raised by his uncle, Gisbert Mercator, who intended him for the Roman Catholic priesthood through the Brethren of the Common Life, he attended secondary school at 'sHertogenbosch and was apparently quite pious. His religious doubts began when his philosophical and theological studies at Louvain prompted him to consider whether biblical and ancient Greek cosmologies could be reconciled. These questions gradually expanded into concern for accurate geography in support of cosmological beliefs. After receiving his M.A. from Louvain in 1532, he studied mathematics, **astronomy**, and geography privately under Reiner Gemma Frisius (1508–1555). Also in the early 1530s he acquired skill as an engraver from Gaspar Van der Heyden (Gaspar à Myrica), a goldsmith in the town of Louvain.

In 1536, Mercator created his first important cartographic work: a globe. His reputation grew internationally over the next few years, especially through his maps of Palestine in 1537, the world in 1538, and Flanders in 1540, as well as his celestial globe of 1537 and his terrestrial globe of 1541. His multifaceted expertise as artist, surveyor, instrument maker, geographer, and mathematician all contributed to his fame. About this time he began experimenting with new projections for maps.

Mercator's frequent travels in search of geographical data aroused suspicion, especially when he, a Catholic, ventured into Protestant lands. In 1544, he was arrested for heresy and jailed for seven months. Although he was released through the intercession of the University of Louvain, the whole experience soured him on the Low Countries and Catholicism. He soon converted to Protestantism and, in 1552, moved to Duisburg, a Protestant enclave in northern Germany, where Duke William of Cleves (1516–1592), brother of the fourth wife of English King Henry VIII, planned to establish a university. The university lasted from 1555 to 1818, but was refounded in 1972. Since 1994, its official name in German has been Gerhard-Mercator-Universität-Gesamthochschule Duisburg.

Mercator served in Duisburg as the duke's cosmographer. He ran his own shop, hired his own artisans, published his own books, and did some teaching. Through the patronage of the court of Cleves, the last four decades of Mercator's life were secure, happy, and productive. Among his best works of this period was his 1554 map of **Europe**.

In 1569, he first published maps based on "Mercator projection." In the systematic vocabulary of cartography, "projection" is a technique or strategy for representing the curved surface of the world or any part of it on the flat surface of a map. There are three general types of projections: cylindrical, conic, and azimuthal, which respectively project the surface of a sphere onto a cylinder, cone, or plane. Some aspect is gained and some aspect is lost with each type. Mercator projection is the best known and most useful of the cylindrical projections. It shows all meridians of longitude as if they were parallel to each other, and thus not converging at the poles; and all parallels of latitude as straight-line segments of equal length, increasing in distance from each other as their distance from the equator increases. The main advantage of Mercator projection is in marine navigation, because all direct sailing courses can be imposed as straight lines. Its main disadvantage is that representations of land areas become more distorted the closer they are to the poles, because of unnatural east-west enlargement.

In 1585, Mercator used the word "atlas" for his book of maps, taking it from the name of the ancient Greek mythological Titan who carried the sky on his shoulders. This gigantic atlas, begun in the 1570s but still unfinished when he died in Duisburg on December 2, 1594, contained corrected versions of the ancient maps of **Ptolemy** (ca. 130) and detailed, up-to-date maps of many parts of Europe.

See also Latitude and longitude; Mapping techniques; Surveying instruments

MERCURY • *see* SOLAR SYSTEM

MERIDIANS • *see* LATITUDE AND LONGITUDE

MESA • *see* LANDFORMS

MESOSPHERE

Based on the vertical **temperature** distribution in Earth's atmosphere, four semi-horizontal layers or "spheres" can be distinguished: the **troposphere**, **stratosphere**, mesosphere, and thermosphere,. These layers are separated by "pauses," where no change in the temperature occurs with altitude change: the tropopause (between the troposphere and the stratosphere), the stratopause (between the stratosphere and the mesosphere), and the mesopause (between the mesosphere and the **thermosphere**). The stratosphere and mesosphere together are called the middle atmosphere, and their region also overlaps with the **ionosphere**, which is a region defined on the basis of the electric charges of the particles there.

The mesosphere, which means middle sphere, is the third layer of Earth's atmosphere, between the stratosphere, and the thermosphere. It is located from about 55 kilometers (35 miles) to 85 kilometers (54 miles) above the surface of Earth. Temperature here decreases with height, so within the mesosphere it is warmest at its lowest level (–5°C, or 23°F), and becomes coldest at its highest level (–80°C, or –112°F). Depending on **latitude** and season, temperatures in the upper mesosphere can be as low as –140°C (–220°F). The temperature in the mesosphere is lower than the temperature of the troposphere or stratosphere, which makes the mesosphere the coldest among the atmospheric layers. It is colder then Antarctica's lowest recorded temperature, and it is cold enough to freeze **water** vapor into **ice clouds**, which can be seen mostly after sunset.

Although the air in the mesosphere is relatively mixed, it is very thin, resulting in low **atmospheric pressure**. At this height, not only concentrations of **ozone** and water vapor are negligible, air in the mesosphere contains much less **oxygen** than in the troposphere. The mesosphere is also the layer in which many meteors burn up when they enter the earth's atmosphere, as a result of the collision with some of the gas particles present in this layer.

See also Atmospheric composition and structure; Stratosphere and stratopause; Thermosphere

Mesozoic Era

In **geologic time**, the Mesozoic Era, the second era in the **Phanerozoic Eon**, spans the time between roughly 250 million years ago (mya) and 65 mya.

The Mesozoic Era contains three geologic time periods including the **Triassic Period** (250 mya to approximately 206 mya), **Jurassic Period** (206 mya to approximately 144 mya), and the **Cretaceous Period** (144 mya to 65 mya).

The Mesozoic Era begins at the end of the **Permian Period** of the **Paleozoic Era**. The Mesozoic Era's Cretaceous Period ends with the K-T boundary or **K-T event**.

During the Mesozoic Era the Pangaean supercontinent spanned Earth's equatorial regions and separated the Panthalassic Ocean and the Tethys Ocean basins. At the start of the Mesozoic Era there was little differentiation or separation between the continental **crust** that would eventually form the North American, European, South American, and African Continents.

Driven by **plate tectonics**, by the middle of the Mesozoic Era (approximately 170 mya) the North American and European continents diverged and the earliest form of the Atlantic Ocean emerged between the continents. At mid-Mesozoic Era, although still united along a broad region, what would become the South American and African Plates became distinguishable in a form similar to the modern continents. **North America** and **South America** remained united by a dry strip of land similar to the isthmus connection that exists today.

Late in the Mesozoic Era, an increase in sea level allowed the confluence of the now distinguishable Pacific Ocean and Atlantic Ocean to provide a wide **water** barrier between North and South America. Much of what are now the eastern and middle portions of the United States was flooded. By the end of the Mesozoic Era, water separated South America from **Africa**. The Australian and Antarctic continents were clearly articulated and the Antarctic continent began a southward migration to the south polar region.

The Mesozoic Era began with a mass extinction and ended with mass extinction. At the end of the Paleozoic Era, almost 80% of marine species became extinct. It would not be until well into the Mesozoic Era that marine life recovered and new reef-building corals evolved. Reptiles dominated the land. Accordingly, the Mesozoic Era is often termed "The Age of Reptiles."

Mesozoic essentially means "middle animals" and marked a fundamental high point in the number and types of species on Earth. Dinosaurs evolved to rule the Mesozoic Era but non-avian species of dinosaurs became extinct as part of the mass extinction that marked the end of the Mesozoic and start of the **Cenozoic Era**.

The landscape of the Mesozoic Era was also marked by substantial changes in vegetative patterns that altered erosional patterns involved in **landscape evolution**. During the Mesozoic Era, both gymnosperm (conifers, etc.) and subsequently angiosperm plants evolved in forms comparable to their modern form. Plant growth also allowed the subsequent development of extensive **coal** beds.

Like the Paleozoic, the Mesozoic Era closed with an episode of extinction. More than 70% of all existing life forms became extinct by the Mesozoic Era–Cenozoic Era boundary (also known as the Cretaceous-Tertiary boundary), including virtually all of the dinosaurs.

During the Mesozoic Era large meteor impacts were frequent. The impact at the start of the Mesozoic Era has been estimated to be of such force as to be able to creating a 350 km **impact crater**, The K-T event crater (i.e., the Chicxulub crater) measures 170 km in diameter.

See also Archean; Cambrian Period; Cenozoic Era; Dating methods; Devonian Period; Eocene Epoch; Evolution, evidence of; Fossil record; Fossils and fossilization; Geologic time; Historical geology; Holocene Epoch; Miocene Epoch; Mississippian Period; Oligocene Epoch; Ordovician Period; Paleocene Epoch; Pennsylvanian Period; Phanerozoic Eon; Pleistocene Epoch; Pliocene Epoch; Precambrian; Proterozoic Eon; Quaternary Period; Silurian Period; Supercontinents; Tertiary Period

METALS

Metals are rarely encountered in their elemental state in nature. They must first be extracted from the ground as an ore, which is then treated to release the metal. Some metals may be extracted from their natural state by electrolysis (for example sodium), while others may need more drastic treatment (such as **iron** or zinc). **Precious metals** (e.g., gold, silver, platinum) are relatively rare and as a result have a high value in the marketplace.

There are approximately 90 elements that can be described as metals (the number can fluctuate slightly depending upon the precise definition of a metal used to categorize the elements). Regardless, they all have various characteristics in common ranging from bonding to chemical nature.

In general, metals are elements that conduct **electricity**, are malleable, and are ductile. Another group of elements—the metalloid or semi-metal elements—share some properties with the metals and some with the nonmetals. There are eight of these elements and they are semiconductors.

Metals are usually solids at standard **temperature** and pressure (STP). One exception to this is mercury, which is a liquid at STP. As is to be expected from the fact that most metals are solids at STP, the majority of metals also have **melting** and boiling points that are high.

Electrical current is not the only thing that metals conduct. They are also efficient conductors of heat.

Metals have a shape that can be easily changed by hammering (i.e., they are malleable). Metals are also ductile (i.e., they can be drawn out into a long wire). With the exception of gold and copper, metals are silvery gray in color and all metals take a polish well.

Chemically, the atoms of a metallic element are bonded to their neighbors by metallic bonds, producing a giant metallic lattice structure. Metallic elements have relatively few electrons in their outermost shells. When metallic bonds are formed, these outermost electrons are lost into a pool of free or mobile electrons. Thus, the metallic lattice structure is actually comprised of positive ions packed closely together and a pool of freely moving electrons surrounding them. These free electrons are referred to as delocalized because they are not restricted to orbiting one particular ion or **atom**. This pool of delocalized electrons allows a metal to conduct an electrical charge because the electrons are free to move. Alloys also have this type of bonding, allowing them to conduct electricity as well. For this same reason metals are also good conductors of heat.

The close packing of the metal ions (the ions are packed as close as they can possibly be) explains the high density of the majority of the metals. A metal with a high molecular mass will have a greater density than one with a lower molecular mass even though their atomic radii may be similar. **Lead** and **aluminum** have similar atomic sizes, but lead has a much larger molecular mass and consequently it has a much higher density.

The malleability of metals is due to the regular arrangement of ions within the metallic lattice. The bonds holding the lattice in place are strong, but they are somewhat flexible. Layers of ions can slip over each other without the structure of the molecule being destroyed. This also explains why metals are ductile. Both of these characteristics are more noticeable when the metal is hot. The metallic lattice is also responsible for the appearance of some metals. When a metal is examined under a microscope, it is seen to have a crystalline structure that is made of regions called grains. The smaller the grain size the more closely packed are the ions of the metal and the stronger and harder it is. If hot metal is allowed to cool slowly, the resulting grains are large, making a metal that is easy to shape. This process is known as annealing. When a hot metal is cooled quickly, the **crystals** produced are small. When this cooling is carried out in **water**, it is called quenching. Quenching will produce a metal that is strong, hard, and brittle.

Many materials are referred to as metals when in fact they are not. The true metals are actually elements, whereas the false metals are alloys—composites of elements. For example, the element iron is a metal. Steel, however, is not a true metal, since it is an **alloy** containing a mixture of iron and carbon—the relative ratios of the two materials control the physical characteristics of the product. Alloys have different properties than the materials from which they are produced, so by careful blending the exact properties required can be manufactured.

Metals have a wide range of uses. For example, copper is used to conduct electricity in cables (an excellent conductor and very ductile). Tin is used to coat cans for food storage (non-poisonous and **corrosion** resistant). Aluminum is used as kitchen foil (high malleability). Iron is used as a fencing material (easily workable and relatively resistant to corrosion). Alloys made from metals have different uses. Steel—an alloy of iron and carbon—is used widely in the construction industry because of it great strength and ease with which it can be initially formed to create specific shaped structural components (e.g., beams, girders, etc.) Solder—tin and lead—is used for joining metals together because of its low melting point. Brass—copper and zinc—is fashioned into ornaments, buttons, and screws because of its high strength, low weight, and corrosion resistance.

See also Atomic theory; Atoms; Electricity and magnetism; Minerals

METAMORPHIC ROCK

Metamorphic **rock** is rock that has changed from one type of rock into another. The word metamorphic (from Greek) means "of changing form." Metamorphic rock is produced from either igneous rock (rock formed from the cooling and hardening of **magma**) or sedimentary rock (rock formed from compressed and solidified layers of organic or inorganic matter). Most of Earth's **crust** is made up of metamorphic rock. Igneous and **sedimentary rocks** become metamorphic rock as a result of intense heat from magma and pressure from tectonic shifting. Although the rock becomes extremely hot and under a great deal of pressure it does not melt. If the rock melted, the process would result in igneous, not metamorphic rock. Metamorphic alteration of rock causes the texture and/or mineral composition to change. New textures are formed from a process called recrystallization. New **minerals** (which are simply various combinations of elements) are created when elements recombine.

There are two basic types of metamorphic rock: regional and thermal. Regional metamorphic rock, found mainly in mountainous regions, is formed mainly by pressure, as opposed to heat. Different amounts of pressure produce different types of rock. The greater the pressure, the more drastic the change. Also, the deeper the rock the higher the **temperature**, which adds to the potential for diverse changes. For example, a pile of mud can turn into shale (a fine-grained sedimentary rock) with relatively low pressure, about 3 mi (5 km) down into the earth. With more pressure and some heat, shale can transform into **slate** and mica. Metamorphic rock found closer to Earth's surface, or produced by low pressure, characteristically splits or flakes into layers of varying thickness. This is called foliation. Slate is often used as roofing tiles and paving stones. With lots of pressure and increasing heat, rock called **schist** forms. Schist, which is a medium-grained regional metamorphic rock also has a tendency to split in layers, is subjected to high temperatures, and often contains **crystals**, such as garnets. **Gneiss** (pronounced "nice") is formed by a higher pressure and temperature than schist. These rocks are coarse grained and, although layered as schist is, do not split easily. Essentially, metamorphic rocks are made of the same minerals as the original rock or parent rock but the various minerals have been rearranged to make a new rock.

Thermal metamorphic rock, also called contact metamorphic rock, is formed not only by considerable pressure but, more importantly, by intense heat. Imagine molten rock pushing up into Earth's crust. The incredible pressure fills any empty space, every nook and cranny, with molten rock. This intense heat causes the surrounding rock to completely recrystallize. During recrystallization, the chemical composition regroups to form a new rock. An example of this type of thermal metamorphic rock is **marble**, which is actually **limestone** whose calcite has recrystallized. **Sandstone** made mostly of **quartz** fragments recrystallizes into quartzite. Thermal metamorphic rocks are not as common or plentiful as regional metamorphic rocks. Sometimes a metamorphic rock can become metamorphosed. This is known as polymetamorphism.

See also Metamorphism

Metamorphism

Metamorphism refers to the physical and chemical changes that rocks undergo when exposed to conditions of high **temperature**, high pressure, or some combination thereof. Rocks that have undergone metamorphism exhibit chemical and structural changes that result from the partial or complete recrystallization of **minerals** within them. These transformations occur while the **rock** is in the solid state, i.e., no **melting** occurs during metamorphism. The conditions of high temperature and pressure under which metamorphism occurs are typically the result of processes such as mountain building, plate convergence, volcanism, and **sedimentation**.

Any type of rock may be metamorphosed and several agents can be involved in altering a parent rock into its metamorphic product. The composition of the parent rock limits the mineral composition of the product, although subsurface gases and fluids may contribute new elements. Thermal energy at depth, either from the **geothermal gradient** or from plutonic activity, may provide the energy for recrystallization of the rock. As the temperature increases, volatile components such as **water** and **carbon dioxide** can be released causing chemical changes to the minerals within the parent rock. In addition, the temperature increase may cause the rock to behave plastically in response to stresses acting on it, frequently resulting in a contorted appearance. Pressure on the parent rock may be a result of the overlying rock, known as lithostatic or confining pressure, or may be due to forces acting in a particular direction due to tectonic activity, known as directed pressure. Pressures within the rock may cause the instability of certain minerals in favor of those that are more stable under the new conditions. The pressure may also be localized on irregularities on the boundaries of individual grains. Recrystallization of a rock undergoing directed pressure typically results in the development of a foliated rock fabric, in which the axes of the minerals are aligned with the differential pressures based on the stability of the crystal lattice to those pressures. The development of such **crystals** during metamorphism may be heavily influenced by amount of time that the rock is exposed to the conditions. The mobilization of ions that supports crystal growth within the rock can require extensive periods of time to produce larger mineral grains.

Metamorphism may occur in a number of forms, each having different results and areal extent. Contact metamorphism is the baking of **country rock** immediately adjacent to an intruded **magma** body. This type of metamorphism, also known as thermal metamorphism, is caused by the high temperatures associated with an igneous intrusion. The rock is altered only in a zone, called an aureole, which can range from a few centimeters to several hundred meters in width. These zones may occur very near the surface and pressure plays an insignificant role in the process. In the case of cataclastic or dynamic metamorphism, rocks in a localized zone undergo mechanical disruption without significant mineralogical change. This is a near-surface phenomenon that is often associated with faulting and occurs at low temperature. Regional metamorphism, as the name suggests, encompasses large areas and is associated with large mountain building and plutonic events. Relatively high temperature and intense, directed pressures are common in this process. The differential stress associated with regional, or dynamothermal, metamorphism frequently yields foliated rock.

See also Metamorphic rock; Shock metamorphism

Meteoroids and meteorites

The word meteor is derived from the Greek *meteron*, meaning something high up. Today, however, the term is used to describe the light phenomena produced by the entry of a meteoroid into Earth's atmosphere. A meteoroid is defined to be any solid object moving in interplanetary **space** that is much larger than an **atom** or a molecule, but smaller than a few meters in diameter. A visual meteor, or shooting star, is produced whenever a meteoroid is vaporized in Earth's upper atmosphere. If a meteoroid survives its passage through the atmosphere without being fully vaporized and falls to the ground, it is a called a meteorite.

Upon entering Earth's upper atmosphere, a meteoroid begins to collide with an ever-increasing number of air molecules. These collisions will both slow the meteoroid down and heat its surface layers. At the same time the meteoroid is being decelerated, that energy is transferred from the meteoroid to the surrounding air. Some of the meteoroid's lost energy is transformed into light; it is this light that we observe as a meteor. As the meteoroid continues its journey through the atmosphere, its surface layers become so hot that vaporization begins. Continued heating causes more and more surface mass to be lost in a process known as ablation, and ultimately the meteoroid is completely vaporized.

The amount of surface heating that a meteoroid experiences is proportional to its surface **area**, and consequently very small meteoroids are not fully vaporized in the atmosphere. The size limit below which vaporization is no longer important is about 0.0004 in (0.01 mm). The smallest of meteoroids can safely pass through Earth's atmosphere without much physical alteration, and they may be collected as micrometeorites at Earth's surface. It is estimated that 22,000 tons (20,000 metric tons) of micrometeoritic material fall to Earth every year.

Visual meteors (shooting stars) are produced through the vaporization of millimeter-sized meteoroids. The speed with which meteoroids enter Earth's atmosphere varies from a minimum of 7 mi/sec (11 km/sec) to a maximum of 45 mi/sec (72 km/sec). The meteoroid ablation process typically begins at heights ranging 62–71 mi (100–115 km) above the earth's surface, and the whole meteoroid is usually vaporized by the time it has descended to a height of 43.5 mi (70 km).

Astronomers have found that the visually observed meteors are derived from two meteoroid populations: a continuously active, but sporadic, background and a number of specific sources called meteoroid streams.

On any clear night of the year, an observer can expect to see 10–12 sporadic meteors per hour. Sporadic meteors can appear from any part of the sky, and about 500,000 sporadic meteoroids enter Earth's atmosphere every day.

Meteor activity is often described in terms of the number of meteors observed per hour. The observed hourly rate of meteors will be dependent upon the prevalent visual conditions. Factors such as the presence of a full **moon**, local light pollution, and **clouds** will reduce the meteor count and, hence, lower the observed hourly rate. Astronomers often quote a corrected hourly rate which describes the number of meteors that an observer would see, each hour, if the observing conditions were perfect.

Observations have shown that the corrected hourly rate of sporadic meteors varies in a periodic fashion during the course of a day. On a typical clear night, the hourly rate of sporadic meteors is at a minimum of about six meteors per hour at 6 P.M.. The hourly rate climbs steadily during the night until it reaches a maximum of about 16 meteors per hour around 4 A.M.

This daily variation in the hourly rate of sporadic meteors is due to Earth's **rotation** in its orbit about the **Sun**. In the evening, a sporadic meteoroid has to catch up with Earth if it is to enter the atmosphere and be seen. This is because at about 6 P.M. local time an observer will be on that part of Earth's surface which is trailing in the direction of Earth's motion. In the early morning, however, the observer will be on the leading portion of Earth's surface, and consequently Earth will tend to "sweep up" all the meteoroids in its path. An observer will typically see two to three times more sporadic meteors per hour in the early morning than in the early evening; and will see them at higher speeds relative to Earth.

Meteor showers occur when Earth passes through the tube-like structure of meteoroids left in the wake of a comet. Such meteoroid tubes—or as they are more commonly called, meteoroid streams—are formed after a comet has made many repeated passages by the Sun. Meteoroid streams are composed of silicate (i.e., rocky) grains that were once embedded in the surface ices of a parent comet. Grains are released from a comet's nucleus whenever solar heating causes the surface ices to sublimate. New grains are injected into the meteoroid stream each time the comet passes close by the Sun.

The individual dust grains (technically meteoroids once they have left the comet) move along orbits that are similar to that of the parent comet. Gradually, over the course of several hundreds of years, the meteoroids form a diffuse shell of material around the whole orbit of the parent comet. Provided that the stream meteoroids are distributed in a reasonably uniform manner, a meteor shower will be seen each year when the earth passes through the stream. The shower occurs at the same time each year because the position at which the meteoroid stream intersects Earth's orbit does not vary much from one year to the next. There are long-term variations, however, and the days during which a shower is active will eventually change.

When Earth passes through a meteoroid stream, the meteoroids are moving through space along nearly parallel paths. Upon entering Earth's atmosphere, however, a perspective effect causes the shower meteors to apparently originate from a small region of the sky; this region is called the radiant.

The radiant is typically just a few degrees across when projected onto the night sky. A meteor shower is usually, but not always, named after the constellation in which the radiant falls on the night of the shower maximum. The Orionid meteor shower, for example, is so named because on the night of the shower maximum (October 21st) the stream radiant is located in the constellation of Orion. Some meteor showers are named after bright stars. The Eta Aquarid meteor shower, for example, is so named because on the night of the shower maximum (May 3rd) the radiant is close to the seventh brightest star in the constellation of Aquarius (by convention the brightest stars in a constellation are labeled after the Greek alphabet, and accordingly, the seventh letter in the Greek alphabet is eta).

Probably the best-known meteor shower is the one known as the Perseid shower. This shower reaches its peak on the night of August 12th each year, but meteors can be observed from the stream for several weeks on either side of the maximum. The shower's radiant first appears in the constellation of Andromedia in mid-July, and by late August it has moved into the constellation of Camelopardalis. The radiant is in the constellation of Perseus on the night of the shower maximum.

The steady eastward drift of the radiant across the night sky is due to the motion of Earth through the Perseid meteoroid stream. The nearly constant year-to-year activity associated with the Perseid meteor shower indicates that the stream is very old. Essentially, the earth encounters about the same number of Perseid meteoroids each year even though it is sampling different segments of the stream. Since 1988, however, higher than normal meteor rates have been observed about 12 hours before the time of the traditional shower maximum (August 12th). This short-lived period (approximately half an hour) of high activity is caused by new meteoroids that were ejected from the stream's parent comet, Comet Swift-Tuttle, in 1862. Comet Swift-Tuttle last rounded the Sun in late 1992, and it is expected that higher than normal meteor rates will be visible half-a-day before the time of the "traditional" Perseid maximum until around 2015.

Another meteor shower known as the Leonid occurs every year in November, caused by the tail of comet Tempel-Tuttle, which passes through the inner **solar system** every 32–33 years. Such a year was 1998; on November 17 and 18, 1998, observers on Earth saw as many as 200 meteors an hour. The shower was so intense that it generated widespread concern about the disruption of global telecommunications and the possible damage or destruction of space telescopes. Partly as a result of careful preparation by **satellite** and **telescope** engineers, however, concerns appeared to be minimal.

If a meteoroid is to survive its passage through Earth's atmosphere to become a meteorite, it must be both large and dense. If these physical conditions are not met, it is more than likely that the meteoroid, as it ploughs through Earth's atmosphere, will either crumble into many small fragments, or it will be completely vaporized before it hits Earth's surface. Most of the meteoroids that produce meteorites are believed to be asteroidal in origin. In essence, they are the small fragmentary chips thrown off when two minor planets (**asteroids**) collide. Meteorites are very valuable then, for bringing samples of asteroidal material to Earth. A few very rare meteorite samples are believed to have come from the planet Mars and the Moon. It is believed that these rare meteorite specimens characterize material that was ejected from the surfaces of Mars and the Moon during the formation of large impact craters.

Accurate orbits are presently known for just four recovered meteorites (the Pibram meteorite, which fell in the Czech Republic in 1959; the Lost City meteorite, which fell in Oklahoma in 1970; the Innisfree meteorite, which fell in Alberta, Canada, in 1977; and the Peekskill meteorite, which fell in New York State in 1992). All four of these meteorites have orbits that extend to the main asteroid belt between the planets Mars and Jupiter.

Meteorites are superficially described as being either falls or finds. A meteorite fall is scientifically more useful than a find because the exact time that it hit Earth's surface is known. Finds, on the other hand, are simply that—meteorites that have been found by chance. The largest meteorite find to date is that of the 66 ton (60 metric ton) Hoba meteorite in South **Africa**. Meteorites are either named after the specific geographic location in which they fall, or after the postal station nearest to the site of the fall.

An analysis of meteorite fall statistics suggests that about 30,000 meteorites of mass greater than 3.5 oz (100 g) fall to Earth each year. Of these meteorites, the majority weigh just a few hundred grams, only a few (about 5,000) weigh more than 2.2 lb (1 kg), and fewer still (about 700) weigh more than 22 lb (10 kg). In general, the number of meteoroids hitting the earth's atmosphere increases with decreasing meteoroid mass: milligram meteoroids, for example, are about a million times more common than meteoroids weighing a kilogram.

Meteorites are classified according to the amount of silicate and metallic nickel-iron that they contain. Three main meteorite types are recognized; these are the irons, the stones, and the stony-irons. The **iron** meteorites consist almost entirely of nickel-iron, while the stone meteorites are mostly silicates. The stony-iron meteorites contain both nickel-iron and silicates. The stony meteorites are further divided into chondrites and achondrites. The term chondrite is applied if the meteorite is composed of many small, rounded fragments (called chondrules) bound together in a silicate matrix. If no chondrules are present then the meteorite is an achondrite. Most (about 85%) of the stony meteorites are chondrites. Meteorite fall statistics indicates that about 96% of meteorites are stony, 3% are irons and 1% are stony-irons.

Even though many thousands of meteorites fall to Earth each year it is rare for one to hit a human being. The chances of a human fatality resulting from the fall of a meteorite have been calculated as one death, somewhere in the world, every 52 years. Thankfully, no human deaths from falling meteorites have been reported this century. A woman in Sylacauga, Alabama, was injured, however, by a 8.6 lb (3.9 kg) meteorite that crashed through the roof of her house in 1954. Another close call occurred in August of 1991, when a small meteorite plunged to the ground just a few meters away from two boys in Noblesville, Indiana.

In contrast to the situation with human beings, meteorite damage to buildings is much more common—the larger an object is the more likely it will be hit by a meteorite. A farm building, for example, was struck by a meteorite fragment in St. Robert, Quebec in June of 1994. Likewise, in August 1992, a small village in Uganda was showered by at least 50 meteorite fragments. Two of the meteorites smashed through the roof of the local railway station, one meteorite pierced the roof of a cotton factory, and another fragment hit an oil storage facility. One of the more spectacular incidents of meteorite-sustained damage in recent times is that of the Peekskill meteorite which fell in October of 1992 and hit a parked car.

See also Astronomy; Barringer meteor crater; Impact crater; Murchison meteorite; Space and planetary geology

METEOROLOGY

Meteorology is a science that studies the processes and phenomena of the atmosphere. Accordingly, a person who studies the atmosphere is called a meteorologist. Meteorology consists of many areas: physical meteorology, dealing with physical aspects of the atmosphere such as rain or cloud formation, or rainbows and mirages; synoptic meteorology, the analysis and forecast of large-scale **weather** systems; dynamic meteorology, which is based on the laws of theoretical **physics**; climatology, the study of the climate of an **area**; aviation meteorology, researching weather information for aviation; **atmospheric chemistry**, examining the chemical composition and processes in the atmosphere; atmospheric optics, analyzing the optical phenomena of the atmosphere such as halos or rainbows; or agricultural meteorology, studying the relationship between weather and vegetation. While meteorology usually refers to the study of the earth's atmosphere, atmospheric science includes the study of the atmospheres of all the planets in the **solar system**.

Greek philosopher and scientist Aristotle (384–322 B.C.) is considered the father of meteorology, because he was the first one to use the word meteorology in his book *Meteorologica* around 340 B.C., summarizing the knowledge of that time about atmospheric phenomena. He speculatively wrote about clouds, rain, snow, **wind**, and climatic changes, and although many of his findings later proved to be incorrect, many of them were insightful. The title of the book refers to all the things being in the sky or falling from there, which at that time was called a meteor.

Although systematic weather data recording began about the fourteenth century, the lack of weather measuring instruments made only some visual observations possible at that time. The real scientific study of atmospheric phenomena started later with the invention of devices to measure weather data: the thermometer in about 1600 for measuring **temperature**, the barometer for measuring **atmospheric pressure** in 1643, the anemometer for measuring wind speed in 1667, and the hair hygrometer for measuring **humidity** in 1780. In 1802, the first cloud classification system was formulated, and in 1805, a wind scale was first introduced. These measuring instruments and new ideas made possible gathering of actual data from the atmosphere giving the basis for scientific theories for properties of the atmosphere (pressure, temperature, humidity, etc.) and its governing physical laws.

In the early 1840s, the first **weather forecasting** services started with the invention of the telegraph transporting meteorological information. At that time, meteorology was still in the descriptive phase, still on an empirical basis with little scientific theories and calculations involved, although weather maps could be drawn, and storm systems and surface wind patterns were being recognized.

Meteorology became more scientific only around World War One, when Norwegian physicist Vilhelm Bjerknes (1862–1951) introduced a modern meteorological theory stating that weather patterns in the temperate middle latitudes are the results of the interaction between warm and cold air masses. His description of atmospheric phenomena and forecasting techniques were based on the laws of physics, exploring the science of dynamic meteorology, assuming that knowing about the atmospheric conditions now, and knowing the governing physical laws for its movements, predictions for the future are possible.

By the 1940s, upper-level measurements of pressure, temperature, wind, and humidity clarified more about the vertical properties of the atmosphere. In 1946, the process of **cloud seeding** was invented which made possible some weather modification experiments. In the 1950s, radar became important for detecting **precipitation** of a remote area. Also in the 1950s, with the invention of the computer, weather forecasting became not only quicker but also more reliable, because the computers could solve the mathematical equations of the atmospheric models much faster than manually before. In 1960, the first meteorological **satellite** was launched to provide 24-hour monitoring of weather events worldwide.

These satellites now give three-dimensional data to high-speed computers for faster and more precise weather predictions. These days the computers are capable of plotting the observation data, and solving huge models not only for short-time weather forecasting, but also climatic models on time scales of centuries, for climate change studies. Meteorology has come a long way since Aristotle. Even so, the computers still have their capacity limits, the models are still with many uncertainties, and the effects of the atmosphere on our complex society and environment can be serious. Many complicated issues remain at the forefront of meteorology—including air pollution, **global warming**, El Niño events, climate change, the **ozone** hole, acid rain—making meteorology today a scientific area still riddled with many challenges and unanswered questions.

See also Air masses and fronts; Atmospheric circulation; Atmospheric composition and structure; Atmospheric inversion layers; Atmospheric lapse rate; Atmospheric pollution; Clouds and cloud types; El Niño and La Niña phenomena; Greenhouse gases and greenhouse effect; Isobars; Scientific data management in Earth Sciences; Weather balloon; Weather forecasting methods; Weather radar; Weather satellite; Weathering and weathering series; Wind chill; Wind shear

MID-ATLANTIC RIDGE • *see* MID-OCEAN RIDGES AND RIFTS

MID-OCEAN RIDGES AND RIFTS

The ocean floor is mountainous and uneven, much like Earth's surface. As oceanographers began mapping the ocean bottom, they discovered that the sea floor is full of vast rising slopes, or ridges, and dramatic open valleys, or rifts. During World War II, oceanographer William Maurice Ewing began mapping the complex ocean bottom with sophisticated instruments such as sonar depth finders and underwater cameras that helped trace the contours of the ocean bottom. Ewing set out to measure and record a massive chain of undersea mountains called the Midatlantic Ridge. When Ewing and his crew began mapping the massive ridge, they encountered a problem: the sonar beams were bouncing back. This problem led to another great discovery. They realized that there were frequent oceanic earthquakes occurring along the ridge. This was an exciting discovery because it opened up the possibility that oceanic earthquakes might be connected to ridges and rifts. Using data from other expeditions, Bruce Charles Heezen (b. 1924) more accurately measured the Midatlantic Ridge as he began mapping the ocean floor. The Ridge measured up to 1.9 miles (3 km) high and 45,954 miles (73,940 km) long. Interestingly, however, he detected a gully in the ridge that led to the Heezen-Ewing theory in 1958, which formally recognized the Midatlantic Ridge as containing a rift. Their discovery sparked interest in other scientists and explorers who questioned the existence of other rifts in ocean ridges.

In the late 1950s, American and Soviet oceanographic vessels began mapping the ocean floor so that their nuclear submarines could navigate deep underwater. The ensuing data provided maps that revealed extraordinary natural phenomena. Submerged peaks and undersea ridges form a continuous mountain chain that reaches up to 10,000 feet (3,048 m) and measures 40,000 miles (64,360 km). This mid-ocean ridge system circles the earth several times and is now known as one of Earth's dominant features, extending over an **area** greater than all the major land mountain ranges combined. Along a great deal of its length, the ridge system is sliced down its middle by a sharp gully, a rift that is the outlet of powerful heat flows. **Temperature** surveys demonstrate that heat seeps out of the earth in these mountainous regions of the middle Atlantic, adding to the complexity of the ocean floor. This evidence of heat emitting from Earth's giant cracks and faults helped reveal the existence of earthquakes and **volcanic eruptions** beneath the ocean. Most of this heat and movement take place in the Atlantic Ocean where the ridge is steeper and more jagged than in the Pacific or Indian **Oceans**.

In some of the most active volcanic areas another unusual natural phenomenon takes place, discovered by **Harry Hammond Hess**. Hess studied the isolated mountains rising from the ocean floor and discovered "sea-mounts," which he named **guyots** in honor of the Swiss-American geographer Arnold Henry Guyot (1807–1884). Hundreds of these strange undersea protrusions lie under the Pacific Ocean, all of which were probably sunken islands created from volcanic **lava**. Some of these guyots broke away and gradually wandered further away from the volcanoes. Before oceanographers studied the floor of the great oceans, there was little evidence to support the **continental drift theory**, which assumed that all the great landmasses were once joined in one supercontinent. Hess's discovery of guyots and other studies of seafloor movement helped reveal the spreading movement of the ocean floor. Hess proposed that hot **rock** swelled from deep within the earth, constantly forcing the ridges and rifts to part and spread. Later these discoveries of seabed movement helped build on the findings of Alfred L. Wegener's thoery of continental drift.

See also Sea-floor spreading

MID-PLATE EARTHQUAKES

Mid- or intra-plate earthquakes are those that occur within the boundaries of the major crustal plates. Most (over 90%) earthquakes occur along the tectonically active plate boundaries, where the major crustal plates are moving past and/or away from each other. However, large and damaging earthquakes can and have occurred within the plates, far from the boundaries. For example, the New Madrid earthquakes of 1811–1812 occurred well within the North American Plate. Both intra-plate and plate boundary earthquakes occur along faults, zones of crustal weakness that have experienced and/or continue to experience relative movement and deformation associated with tectonic activity. Intra-plate earthquakes are difficult to explain because they occur in the relatively stable interiors of plates. Seismic data, however, indicate that **faults and fractures** are very common in the upper **crust** of the earth. These act as zones of weakness that can be reactivated if the *in situ* stress field becomes favorable. Studies of the New Madrid events indicate that the earthquakes occurred along fault zones of an ancient rift system within the North American Plate.

In the winter of 1811–1812, one of the largest sequences of earthquakes in recorded history occurred within the North American Plate near the town of New Madrid, Missouri. The magnitudes of at least three of those earthquakes are estimated to have been greater than 8.0 on the **Richter scale**. The earthquakes were felt over an **area** of millions of square miles and as far away as Boston. During the quakes, large areas of land rose and large areas of land fell several feet. **Lakes** were created and islands disappeared. The Mississippi River was disrupted, waterfalls formed, and large waves swamped many boats. During one of the events, the Mississippi River gave the illusion of flowing backwards. In 1895, another significant but smaller **earthquake** occurred in this region near the town of Charleston, Missouri. That quake was felt in 23 states.

Earthquakes along plate boundaries are relatively understandable given the tectonic activity localized there. Intra-plate earthquakes are much more enigmatic. One explanation offered for the New Madrid earthquakes are the increased stresses that were induced in the region by the unloading of the **ice** sheets that covered much of the northern United States until about 20,000 years ago. Geological evidence indicates that large earthquakes have occurred in this region before 1811. Some modeling suggests that the stresses that caused the earthquakes may persist for thousands of years in the future.

See also Plate tectonics; Rifting and rift valleys

MIGMATITE

A migmatite, or "mixed rock" in Greek, is a banded, heterogenous **rock** composed of intermingled metamorphic and igneous components. Veins, contorted layers, and irregular pods of silica-rich **granite** occur within the structure of foliated **iron** and magnesium-rich metamorphic rocks like **gneiss**,

schist and amphibolite. Because metamorphic rocks form by recrystallizaton of **minerals** without **melting**, and **igneous rocks** like granite form by crystallization of minerals from molten **magma**, it is difficult to explain their coexistence in a single rock. It is clear, however, that migmatites form at the threshold between high-grade metamorphic recrystallization, and complete igneous melting. Migmatites were partially melted during formation.

Some migmatites appear to have formed by intrusion of liquid granitic melt into a preexisting banded **metamorphic rock**. In these examples, the granite inclusions have sharp contacts with the metamorphic bands, and cut across the metamorphic fabric in places. In other cases, the boundaries between metamorphic and igneous components are gradational, or indistinct, suggesting that at least some migmatites form during a single phase of **partial melting** and fractional recrystallization. Metamorphic and igneous petrologists have rigorously debated these two hypotheses regarding the formation of migmatites. As is sometimes the case, both hypotheses are probably correct, and some migmatites form in several phases of **metamorphism** and melting, while others from during a single phase.

Migmatites generally occur in plate tectonic settings where regional belts of continental **crust** have been subjected to very high temperatures and pressures. The metamorphic portion of most migmatites includes the minerals horneblende, **plagioclase feldspar**, and garnet. This mineral assemblage indicates so-called amphibolite-grade metamorphism typical of convergent plate tectonic boundaries where rocks are subjected to very high pressures, strong directional stresses, and high temperatures.

See also Plate tectonics

Milankovitch cycles

The Serbian astronomer Milutin Milankovitch (1879–1958) developed a theory that explained climatic variations in astrophysical terms. He was particularly concerned with the origin of an **ice** age during the Pleistocene. Through observations of the stars, Milankovitch found that the basic elements that govern the earth's orbit around the **Sun** are not constant. First, he noticed that the eccentricity of the elliptical path of the earth's **revolution** around the Sun changes with cycles of roughly 100,000 and 400,000 years. Second, he found that the obliquity, that is the angle of the earth's spin axis with the plane of its eccentric orbit changes with a frequency of roughly 41,000 years between 22 and 25 degrees. Third, he took into consideration that the earth's axis of revolution behaves like the spin axis of a top that is winding down. The spin axis traces a circle on the celestial sphere over a period of approximately 22,000 years. This motion, which is called the precession of the equinoxes, was probably detected by Hipparch of Nikaia (about 150 B.C.). It is the reason why a person then Aries-born, is born under the sign of Pisces today.

In 1920, Milankovitch calculated the effect of each of these cycles on the total summer insulation at a **latitude** of 65 degrees North. He reasoned that at this latitude, small insulation changes might have a big effect because a decrease of the summer insulation allowed the snow and ice of winter to persist through the summer months, and into the following winter. In this way, big ice sheets can develop and accelerate in a positive feed-back through enhancing the northern latitude albedo.

It required a few decades for the theory to have its break-through. Today, however, it is generally accepted that Milankovitch's theory describes the main causes for the waxing and waning of the Pleistocene ice sheets correctly. Proof for this theory is derived from cyclic variations of the chemical and paleontological composition of marine, lacustrine, and terrestrial sediments. Investigations into the temporal variation of the ratio of **oxygen** isotopes in particular have shown that Milankovitch cycles do indeed influence the **climate** greatly on time scales covering thousands to hundreds of thousands of years, although continental distribution, ocean patterns, the total solar irradiance, and other factors play an additional role. The Sun, for instance, is obviously a prominent factor for climatic variations on a time scale between years to a few thousand years, which is summarized in the so-called Athenian hypothesis.

See also Ice ages

Miller-Urey experiment

A classic experiment in molecular biology and genetics, the Miller-Urey experiment, established that the conditions that existed in Earth's primitive atmosphere were sufficient to produce amino acids, the subunits of proteins comprising and required by living organisms. In essence, the Miller-Urey experiment fundamentally established that Earth's primitive atmosphere was capable of producing the building blocks of life from inorganic materials.

In 1953, University of Chicago researchers Stanley L. Miller and Harold C. Urey set up an experimental investigation into the molecular origins of life. Their innovative experimental design consisted of the introduction of the molecules thought to exist in early Earth's primitive atmosphere into a closed chamber. Methane (CH_4), hydrogen (H_2), and ammonia (NH_3) gases were introduced into a moist environment above a water-containing flask. To simulate primitive **lightning** discharges, Miller supplied the system with electrical current.

After a few days, Miller observed that the flask contained organic compounds and that some of these compounds were the amino acids that serve as the essential building blocks of protein. Using chromatological analysis, Miller continued his experimental observations and confirmed the ready formation of amino acids, hydroxy acids, and other organic compounds.

Although the discovery of amino acid formation was of tremendous significance in establishing that the raw materials of proteins were easy to obtain in a primitive Earth environment, there remained a larger question as to the nature of the origin of genetic materials—in particular the origin of DNA and RNA molecules.

Stanley Miller working in the lab where he simulated atmospheric conditions similar to those on Earth 3.5 billion years ago and created organic compounds. ***© Bettmann/Corbis. Reproduced by permission.***

Continuing on the seminal work of Miller and Urey, in the early 1960s Juan Oro discovered that the nucleotide base adenine could also be synthesized under primitive Earth conditions. Oro used a mixture of ammonia and hydrogen cyanide (HCN) in a closed aqueous enviroment.

Oro's findings of adenine, one of the four nitrogenous bases that combine with a phosphate and a sugar (deoxyribose for DNA and ribose for RNA) to form the nucleotides represented by the genetic code: adenine (A), thymine (T), guanine (G), and cytosine (C). In RNA molecules, the nitrogenous base uracil (U) substitutes for thymine. Adenine is also a fundamental component of adenosine triphosphate (ATP), a molecule important in many genetic and cellular functions.

Subsequent research provided evidence of the formation of the other essential nitrogenous bases needed to construct DNA and RNA.

The Miller-Urey experiment remains the subject of scientific debate. Scientists continue to explore the nature and composition of Earth's primitive atmosphere and thus, continue to debate the relative closeness of the conditions of the Miller-Urey experiment (e.g., whether or not Miller's application of electrical current supplied relatively more electrical energy than did lightning in the primitive atmosphere. Subsequent experiments using alternative stimuli (e.g., ultraviolet light) also confirm the formation of amino acids from the gases present in the Miller-Urey experiment. During the 1970s and 1980s, astrobiologists and astrophyicists, including American physicist **Carl Sagan**, asserted that ultraviolet light bombarding the primitive atmosphere was far more energetic that even continual lightning discharges. Amino acid formation is greatly enhanced by the presence of an absorber of ultraviolet radiation such as the hydrogen sulfide molecules (H_2S) also thought to exist in the early Earth atmosphere.

Although the establishment of the availability of the fundamental units of DNA, RNA and proteins was a critical component to the investigation of the origin of biological molecules and life on Earth, the simple presence of these molecules is a long step from functioning cells. Scientists and evolutionary biologists propose a number of methods by which these molecules could concentrate into a crude cell surrounded by a primitive membrane.

See also Cosmology; Evolution, evidence of; Evolutionary mechanisms; Solar system

MINERALOGY

Mineralogy is the study of **minerals**. Rocks in the earth's **crust** are composed of one or more minerals. A mineral in the geologic sense is a naturally occurring, inorganic, crystalline solid. A particular mineral has a specific chemical composition. Each mineral has its own physical properties such as color, hardness, and density.

Most minerals are chemical compounds that are made of two or more different elements. The composition of a mineral is shown by its chemical formula, which states each of the **chemical elements** present in the mineral as well as the ratios of each element. For example, the mineral **quartz** has the chemical formula SiO_2. This means that quartz is made of the elements **silicon** (Si) and **oxygen** (O). The formula also shows that for every one silicon **atom**, two oxygen atoms are present. The mineral orthoclase has the chemical formula $KAlSi_3O_8$. A molecule of orthoclase contains one potassium (K) atom, one **aluminum** (Al) atom, three silicon atoms, and eight oxygen atoms. Some minerals always have the same chemical formula. Quartz always is composed of SiO_2 and halite is always made of sodium (Na) and chlorine (Cl), with the chemical formula NaCl. Some minerals can have more than one chemical formula, depending on their composition. Sometimes an element can substitute for another in a mineral. This occurs when the atoms of two elements are the same charge and close to the same size. For example, an **iron** (Fe) atom and a magnesium (Mg) atom are both about the same size, so they can substitute for each other. The chemical formula for the mineral **olivine** is $(Mg,Fe)_2SiO_4$. The (Mg,Fe) indicates that either magnesium, iron, or a combination of the two may be present in an olivine sample.

Native elements are minerals that are composed of only one element. These are the substances that dietitians call minerals. Examples of native elements include gold (Au), silver (Ag), and platinum (Pt). Two other native elements are **graphite** and **diamond**, both of which are entirely made of **carbon** (C).

All minerals are crystalline solids. A crystalline solid is a solid consisting of atoms arranged in an orderly three-dimensional matrix. This matrix is called a crystal lattice. A crystalline solid is composed of molecules with a large amount of order. The molecules in a crystalline solid occupy a specific place in the arrangement of the solid and do not move. The molecules not only occupy a certain place in the solid, but they are also oriented in a specific manner. The molecules in a crystalline solid vibrate a bit, but they maintain this highly ordered arrangement. The molecules in a crystalline solid can be thought of as balls connected with springs. The balls can vibrate due to the contractions and expansions of the springs between them, but overall they stay in the same place with the same orientation. It is not easy to deform a crystalline solid because of the strong attractive forces at work within the structure. Crystalline solids tend to be hard, highly ordered, and very stable.

Under ideal conditions, mineral crystals will grow and form perfect crystals. An ideal condition would be in a place where the crystals are allowed to grow slowly without disturbances, such as in a cavity. A perfect crystal has crystal faces (planar surfaces), sharp corners, and straight edges. The external crystal form is controlled by the internal structure. When the atoms in a crystal are arranged in a perfectly orderly fashion, the crystal will also be formed in a perfect orderly fashion. Even if a perfect crystal is not formed, the internal crystalline structure can be shown. Many minerals exhibit a property called cleavage. A mineral that has cleavage will break or split along planes. If the internal structure is formed in an orderly crystal arrangement, then the breaks will occur along the planes of the internal crystal structure.

There have been over 3,500 minerals identified and described. Only about two dozen of these are actually common. There are a limited number of minerals, mainly because there are only a certain number of chemical elements that can combine to form chemical compounds. Some combinations of elements are unstable, such as a potassium-sodium or a silicon-iron compound. In addition, only eight elements are found abundantly in the earth's crust, where minerals are formed. Oxygen and silicon alone account for more than 74% of the earth's crust. These factors place a limit on the number of possible minerals.

The minerals that have been discovered and studied can be placed into one of five groups. These groups are the silicate minerals, carbonate minerals, oxides, sulfides, and halides. The silicate minerals are those that contain silica, a combination of silicon and oxygen. Examples of silicates include quartz, orthoclase, and olivine. The silicate minerals are the most common, making up approximately one-third of all known minerals. They are composed of building blocks called the silica tetrahedron. A silica tetrahedron is one silicon atom and four oxygen atoms. The atoms are arranged in a four-faced pyramidal structure (the tetrahedron) with the silicon atom in the center. The silicon atom has a +4 charge, and each of the four oxygen atoms have a –2 charge. As a result, a silicon tetrahedron has a net charge of –4. Because of this charge, it does not occur in isolation in nature. A silica tetrahedron is always bound to other atoms or molecules.

There are two types of silicate minerals, the ferromagnesian and the nonferromagnesian silicates. Ferromagnesian silicates are those containing iron, magnesium, or both. These minerals tend to be dark colored and more dense than the nonferromagnesian silicates. An example of a ferromagnesian silicate is olivine. Nonferromagnesian silicates do not have iron and magnesium. These minerals are light colored and less dense. The most common nonferromagnesian silicates are the feldspars.

The carbonate minerals contain the carbonate ion, $(CO_3)^{2-}$. Calcite ($CaCO_3$), the main component of **limestone**, is an example of a carbonate mineral. The oxides are minerals that contain an element combined with oxygen. An example of an oxide is hematite, Fe_2O_3. The sulfides contain a cation combined with sulfur (S^{2-}). An example of a sulfide is galena (PbS), which is **lead** (Pb) combined with sulfur. The halides all contain halogen elements, such as chlorine and fluorine (F). Examples of halite minerals include halite (NaCl) and fluorite (CaF_2).

All minerals posses specific physical properties such as color, luster, crystal form, cleavage, fracture, hardness, and specific **gravity**. The physical characteristics of a mineral depend on its internal structure and chemical composition. The physical properties of minerals can be used for identification purposes by mineralogists. Color is the least reliable of the physical properties. Many minerals display a variety of colors due to impurities. Some generalizations can be made, however. Ferromagnesian silicates, for example, are usually black, brown, or dark green. The luster of a mineral refers to the way in which light is reflected off the mineral. Two types of luster can be displayed: metallic or nonmetallic.

The crystal form of a mineral is also a physical property specific for the type of mineral being observed. This property is most easily observed when the mineral has formed a perfect crystal. Cleavage is the tendency of a mineral to break or split along planes. There are different types of cleavage that correspond to the different internal crystal structures that make up individual minerals. Fracture occurs when a mineral does not break along smooth planes, rather along irregular surfaces. Some minerals display cleavage, others display fracture.

The hardness of a mineral is its resistance to being scratched. The Mohs hardness scale can be used to determine how hard a mineral is by determining what will scratch its surface. The specific gravity of a mineral is the ratio of its density to the density of **water**. For example, a mineral with a specific gravity of 4.0 is four times as dense as water, meaning that an certain volume of the mineral would weigh four times as much as an equal volume of water. The specific gravity of a mineral is determined by its composition and structure.

Mineralogy is an interesting science that studies the nature of minerals—naturally occurring, inorganic crystalline solids. There are many different minerals, each with its own properties determined by its chemical composition. Mineralogists continue to search for new, useful minerals in the earth's crust.

See also Chemical bonds and physical properties; Crystals and crystallography; Ferromagnetic; Mohs' scale

MINERALS

The term mineral is often used to denote any material that occurs naturally in the ground, including oil and **natural gas**. However, mineralogists and geologists restrict its use to naturally occurring solids having specific chemical compositions. For example, all solid forms of pure silica (SiO_2) are minerals, including natural **glass** and **quartz**, but **coal** is not a mineral because it has no definite and universal chemical composition.

Solids produced by living things—bones, shells, pearls, and the like—are a special case. Scientists usually consider these objects non-minerals even when they have definite a chemical composition, as do the calcium carbonate ($CaCO_3$) shells of marine animals. The distinction is more professional than physical; mineralogists study minerals, but biologists study shells and bones, so shells and bones must not be minerals. However, biological solids that have been completely rearranged at the atomic level are officially regarded as minerals. For example, **graphite** and **diamond** formed by metamorphosis of coal are minerals.

Because solidity is part of the definition of a mineral, substances may change from mineral to non-mineral or vice versa by **melting** or solidifying. Liquid **water** has a definite chemical composition (H_2O) but is not considered a mineral because it is not solid; **ice**, however, is a mineral. **Magma** or molten **lava** are not minerals because they have no definite, universal composition and are liquids; solidified, they become mixtures of specific minerals.

The atoms making up a mineral may be arranged either randomly, like mixed marbles in a bag, or in an orderly pattern, like squares on a chessboard. If a mineral's atoms show long-range organization, the mineral is termed crystalline. The objects commonly called crystals are crystalline minerals of relatively large size that happen to have developed smooth faces. Many crystals, however, are too small to see with the naked eye, and most have imperfectly developed faces or none at all. Most rocks consist of chunks of several crystalline minerals fused together. In some rocks, such as **granite**, these individual pieces are large enough to see, while in others, such as **slate**, they are too small.

If a mineral's atoms are randomly arranged it is termed an **amorphous** mineral or a mineraloid. The most common amorphous mineral is glass—the solid formed by cooling magma or molten lava so quickly that its atoms do not have time to organize into crystals. Molten lava quenched in air or water, or intrusive magma cooled rapidly by contact with **rock** form glasses. All glasses are metastable; that is, they tend to lapse into crystalline form, much as water molecules in cold vapor organize themselves into snowflakes. In the case of glasses, this spontaneous crystallization process is termed devitrification. The processes of devitrification causes glasses to be rare in proportion to their age. Most natural glasses date from the last 60 or 70 million years, a mere tenth of the time since the beginning of the **Cambrian Period**. The remainder have devitrified.

Because **oxygen**, **silicon**, and other elements may be present in any ratio in a glass, depending on the composition of the original melt, some mineralogists do not consider glasses minerals and restrict the term mineral to naturally occurring crystals. For the remainder of this article, the term mineral will be used in this restricted sense.

Earth's **crust** and mantle consist almost entirely of minerals, yet the number of known minerals is less than 3,000. Two factors limit the number of possible and actual minerals. First, a crystal's atoms must be arranged in some periodically repeating, three-dimensional pattern, but only a finite number of such patterns exists. Second, there are only a few score naturally occurring elements, many of which are rare and eight of which—oxygen, silicon, **aluminum**, **iron**, calcium, sodium, potassium, and magnesium, in order of decreasing common-

Display of minerals in Seitenstetten, Austria. © *Massimo Listri/Corbis. Reproduced by permission.*

ness—comprise 98.5% of Earth's crust by weight. Oxygen alone makes up approximately 47% of the crust by weight (over 90% by volume), and silicon makes up approximately another 27%. The number of minerals that can form is therefore finite, and many of those that could theoretically form do so rarely.

The atoms of the two most common elements on earth, silicon and oxygen, readily arrange themselves into tetrahedra (four-sided pyramids) having a silicon **atom** at the center and an oxygen atom at each point. This unit is the silicate radical, $(SiO_4)^{4-}$. Silicate radicals can link into sheets, chains, or three-dimensional frameworks by sharing oxygen atoms. If every oxygen atom participates in two tetrahedra, then the overall ratio of silicon to oxygen is 1:2, and the resulting chemical formula is that of silica, SiO_2. Minerals built mostly of silica are termed silicate minerals. The mineral quartz is pure crystalline silica; other silicate minerals result when atoms of elements other than silicon are introduced at regular intervals. For example, some of the tetrahedra in the silicate framework may be centered on aluminum atoms rather than silicon atoms. In this case, atoms of other elements (usually calcium, potassium, or barium) must be present to balance the ionic charges in the framework. The silicate minerals having this particular structure are the feldspars, which make up approximately 60% of the earth's crust by volume.

When atoms of elements other than silicon unite with oxygen to form the basic building block of a mineral, nonsilicate minerals result: carbonates from **carbon** (e.g., calcite [$CaCO_3$]), sulfates from sulfur (e.g., anhydrite [$CaSO_4$]), phosphates from phosphorus (e.g., apatite [$Ca_5(PO_4)_3F$]), and the oxide minerals, in which O^{2-} alternates with positively charged ions (e.g., spinel [$MgAl_2O_4$]). Other mineral groups do not involve oxygen at all, including the halides (e.g., salt [$NaCl$]), the sulfides (e.g., pyrite [FeS_2]), and the native elements (pure sulfur, carbon, gold, etc.).

Although for simplicity's sake chemical formulas have been identified with mineral species in the preceding paragraph, the identity and properties of a mineral depend not only on what kinds of atoms compose it but on the arrangement of these atoms in **space**. Diamond and graphite, for instance, both consist entirely of carbon atoms and so have the same chemical formula (C), but differ in structure. A mineral's structure, in turn, depends partly on its chemical formula and partly on its history, that is, on the changes in pressure, **temperature**, and chemical context through which it has passed in reaching its present state. A simple example of a mineral structure recording process is the production of glass by rapid cooling of molten silica. To hold a piece of glass is to know a small, specific piece of history; this silica must have cooled rapidly. The dependence of mineral formation on time and temperature is exactly analogous to cookery. Indeed, geologists routinely speak of how the formation of minerals in large bodies of cooling magma is influenced by the "baking" of the magma. Minerals are therefore studied not only for their directly useful properties but for what their very existence reveals about the history of the earth.

See also Crystals and crystallography; Minerology; Obsidian

MIOCENE EPOCH

Notable in the development of primates and human **evolution**, are fossilized remains of *Ardipithecus ramidus*, perhaps one of the earliest identifiable ancestors of man. Fossilized remains found in Ethiopia date to approximately six million years ago, near the end of the Miocene Epoch. Importantly, the fossilized bones found provide evidence that *Ardipithecus ramidus* could walk upright. Anthropologists assert that the ancestral line between apes and humans diverged six to eight million years ago from a common ancestor that lived during the Miocene Epoch.

In **geologic time**, the Miocene Epoch occurs during the **Tertiary Period** (65 million years ago to 2.6 million years ago—and is also sometimes divided or referred to in terms of a Paleogene Period from 65 million years ago to 23 million years ago) and a Neogene Period (23 million years ago to 2.6 million years ago) instead of a singular Tertiary Period—of the **Cenozoic Era** of the **Phanerozoic Eon**. The Miocene Epoch is the fourth epoch in the Tertiary Period (in the alternative, the earliest epoch in the Neogene Period).

The Miocene Epoch ranges from approximately 23 million years ago (mya) to 5 mya. The Miocene Epoch was preceded by the **Oligocene Epoch** and was followed by the **Pliocene Epoch**.

The Miocene Epoch is further subdivided into (from earliest to most recent) Aquitanian (23 mya to 21 mya), Burdigalian (21 mya to 16 mya), Langhian (16 mya to 14 mya), Serravallian (14 mya to 10 mya), Tortonian (10 mya to 7 mya), and Messinian (7 mya to 5 mya) stages.

Craters dating to the end of the Oligocene Epoch and start of the Miocene Epoch can be studied in Northwest Canada and in Logancha, Russia. Smaller impact craters dating to the end of the middle of the Miocene Epoch are evident in Russia and Germany.

Other notable finds in the **fossil record** that date to the Miocene Epoch include evidence of the continued extensive development of grasslands initiated during the preceding Eocene and Oligocene Epochs. The grassland development offered a chance for grazing animals to become well established. Many of the modern migratory patterns date to the Miocene Epoch. The fusion of the Arabian plate to the Eurasian plate provided a land bridge from **Africa** to **Asia** allowing migration of species and mixing of genetic traits among reproductively compatible sub-species.

The paleobotanical record provides evidence that kelp **forests** also became well developed during the Miocene Epoch as the climate cyclically warmed and cooled, but more generally became less humid.

See also Archean; Cambrian Period; Cretaceous Period; Dating methods; Devonian Period; Evolution; Evolution, evidence of; Evolutionary mechanisms; Fossils and fossilization; Historical geology; Holocene Epoch; Jurassic Period; Mesozoic Era; Miocene Epoch; Mississippian Period; Ordovician Period; Paleocene Epoch; Paleozoic Era; Pennsylvanian Period; Pleistocene Epoch; Precambrian; Proterozoic Eon; Quaternary Period; Silurian Period; Triassic Period

MISSISSIPPIAN PERIOD

Shallow, low-latitude **seas** and lush, terrestrial swamps covered the interior of the North American continent during the Mississippian Period of the **Paleozoic Era**, from about 360 to 320 million years ago. The Pennsylvanian and Mississippian Periods are uniquely American terms for the upper and lower sections of the Carboniferous, a geologic period defined by a sequence of **coal** and limestone-bearing strata delineated by European geologists in the early nineteenth century. In 1822, English geologists William Conybeare (1787–1857) and William Phillips (1775–1828) coined the term Carboniferous for the period of **geologic time** typified by the British Coal Measures, Millstone Grit, and Mountain **Limestone**, all important strata that appeared on Adam Smith's (1769–1839) famous map of the **geology** of England in 1815. (They also included the Old Red **Sandstone** that was later reassigned to the Devonian Period.)

Coal-bearing strata analogous to the British Coal Measures also exist in **North America**, especially in Pennsylvania. In 1870, Alexander Winchell (1824–1891), used the term Mississippian to describe a series of limestone beds exposed below the coal beds in the Mississippi River valley near St. Louis. After much confusion regarding mapping and **correlation** of American Carboniferous strata, United States Geological Survey (USGS) geologist Henry Shaler Williams (1847–1918), suggested the terms Pennsylvanian and Mississippian in 1891. While these sub-divisions of the Carboniferous are commonly used in North American, European geologists never adopted them. Furthermore, the Mississippian-Pennsylvanian boundary is younger than the Lower Carboniferous–Upper Carboniferous boundary, so the terms cannot be interchanged.

During the Mississippian, the North American and Eurasian continents were part of a northern supercontinent called Laurasia. The similarity between Carboniferous rocks of **Europe** and North America is thus not coincidental, as the two regions were connected at the time. **South America**, **Africa**, India, **Australia** and **Antarctica** were assembled into the southern supercontinent of Gondwana. Polar **glaciation** was minimal, and Laurasia was located near the equator. Tropical rainforests and swamps rich with vegetation that would later become coal beds grew on exposed land, and shallow, tropical seas covered large regions of the present-day American Midwest and South. Mississippian marbles and limestones, filled with **fossils** of flower-like invertebrates called crinoids, intricate corals, and other Paleozoic carbonate organisms, are exposed throughout the American Midwest.

See also Fossil record

MISTRAL • *see* SEASONAL WINDS

MOHOROVICIC, ANDRIJA (1857-1936)

Croatian seismologist and meteorologist

Croatian seismologist and meteorologist, Andrija Mohorovicic was the first one to suggest the existence of a boundary surface separating the **crust** of the earth from the underlying mantle. This layer, which is 5 mi (8 km) deep under the **oceans** and about 20 mi (32 km) deep under the continents in average, was later named the **Mohorovicic discontinuity**. In 1970, a large crater on the far side of the **Moon** was also named in Mohorovicic's honor.

Mohorovicic was born on in Volosko, now in Croatia. After spending his early school years in Volosko, then Rijeka, he studied **physics** and mathematics at the Faculty of Philosophy, in Prague. In 1882, he started his nine-year carrier in the Nautical School in Bakar teaching **meteorology**, and beginning his scientific work. A few years later, in 1887, he founded a meteorological station in Bakar.

In 1891, Mohorovicic transferred to Zagreb, and in 1892, he became the head of the Meteorological Observatory in Zagreb. Here he studied and wrote mainly about **clouds**, rainstorms, thunderstorms, tornadoes, whirlwinds, hail, and winds, focusing interest on these meteorological phenomena and their scientific interpretation, as well as studying the **climate** of Zagreb. In 1892, he started astronomical observations of stars. In 1893, he established a network of stations for thunderstorm observations. Still in 1893, he received his doctor of philosophy degree at the Zagreb University. In 1899, he founded hail stations, and the same year he started a research project about harnessing **wind** energy.

In 1901, Mohorovicic became director of the meteorological service of Croatia and Slovenia. In his meteorological research, he was still interested in the detection of tornadoes and thunderstorm tracking. Mohorovicic's last contribution to meteorology was in 1901, when he published his paper on vertically decreasing atmospheric **temperature**. After the turn of the twentieth century, Mohorovicic's scientific interest focused exclusively on the problems of **seismology**.

Mohorovicic gradually extended the activities of the observatory to other fields of geophysics: seismology, geomagnetism and gravitation, although as chief of the observatory, Mohorovicic was still responsible for recording all the meteorological data for Croatia and Slovenia. The **earthquake** in 1909 in Croatia directed his interest towards the examination of seismic waves, and in 1910, Mohorovicic published his findings. His plot (arrival time versus epicenter distance to recording station) used the data from 29 stations within 1,491 mi (2,400 km) of the epicenter. He concluded that at around 31 mi (50 km), there must be an abrupt change in the material in the interior of the earth, because he observed an abrupt change in the velocity of the earthquake waves. Although this conclusion was not accepted immediately, a few years later, in 1915, other researchers confirmed it. This discontinuity region under all the continents and oceans is today called the Mohorovicic discontinuity, or in short, the Moho. Although others later refined the study of crust and upper mantle with the application of new methods, Andrija Mohorovicic was clearly a pioneer of this **area**.

Mohorovicic also published a paper in 1909 on the effect of earthquakes on buildings that described periods of oscillation, which was considered by his contemporaries to be ahead of the times not only in his own country, but worldwide. In 1910, he became an associate university professor. From 1893 to 1917, he taught subjects in the fields of geophysics and **astronomy** at the Faculty of Philosophy in Zagreb. In 1893, Mohorovicic first became a corresponding member, then in 1898, a full member of the Academy of Sciences and Arts in Zagreb. Although at the end of 1921 he retired, he worked actively until the late 1920s. He died in 1936, and he is buried in Zagreb.

Because of his extensive work studying epicenters, seismographs, and travel-time curves, much of our knowledge of how earthquakes occur, as well as the current models of the earth's structure can be traced back to the work of Andrija Mohorovicic. Among his other achievements, Mohorovicic also found a procedure for identifying the unique location of earthquake epicenters and formulated an analytical expression for the increase of elastic wave velocity with depth, which was later named Mohorovicic's law. Mohorovicic's thoughts and ideas were original, and he focused his interest in more than one area: the effects of earthquakes on buildings, harnessing the energy of the bora, models of the earth, locating earthquake epicenters, seismographs, and many other subjects also in meteorology. Andrija Mohorovicic was an outstanding scientist and researcher, and his scientific work in the field of seismology rightfully gave him world recognition, making him one of the founders of modern seismology.

See also Earth, interior structure

Mohorovicic discontinuity (Moho)

The Mohorovicic discontinuity, sometimes referred to as "Moho," is the boundary where Earth's **crust** meets Earth's upper mantle (approximately 31 mi/50 km below the surface), and where seismic waves travel at a different and more rapid rate than the crust or mantle. The Moho is named after **Andrija Mohorovicic** (1857–1936), a Croatian meteorologist and seismologist who was fascinated with the faults and movements in the earth's infrastructure that result in earthquakes. The discovery of the Moho was most important because it helped scientists discover a second layer, or mantle, inside the earth. It also helped scientists to determine more accurately where this second layer was located in relation to Earth's surface, or crust.

Since the early 1900s, scientists were almost certain that Earth, like an onion, was made up of many layers, but they did not know exactly where the layers started and ended. In 1906, Mohorovicic studied Yugoslavian **earthquake** records, which revealed the existence of two different sets of earth shock waves from one earthquake. Because the second set of waves exactly mirrored the first set, Mohorovicic discovered that the additional set was actually the first bouncing back from a resistant surface, or a layer of different material inside the earth. This resistant surface, or discontinuity, allowed Mohorovicic to postulate the existence of a second stratum of material under the crust. He did this by gauging the time between the waves, which helped him determine how far this layer resided from the earth's surface.

Mohorovicic also noticed from these experiments that the waves, or tremors, traveled at different speeds depending on the thickness of the material inside Earth. This information helped scientists discover the different types of rocks in areas where drilling was impossible. For example, the lowest level of the crust is composed of basaltic **rock**, the material that rests next to the mantle. After the Moho was discovered, scientists were able to further plot seismic wave movements on sensitive shock recording devices called seismographs. From this information, we know that the outer crust of Earth is 20–25 mi (32–40 km) thick except under many places in the ocean, where it is only 3 mi (4.8 km) thick. The mantle is only the second interior layer. Deeper within the earth lies the most interior layer, Earth's core. We know from mountains and valleys that Earth's surface has changed and shifted with the ages. Similar to Earth's uneven crust, Earth's mantle is thought to be comparably uneven, mostly caused by enormous pressures inside Earth forcing the weaker areas of the rocky sub-layers out of alignment. When the weaker sub-layers, or plates, give way to pressure or stronger plates, earthquakes result. Ever since the existence of the mantle became certain, scientists sought to probe into the physical nature of the earth's inner layer. And because the Moho is located so much closer to the surface beneath the ocean, there were plans in the late 1950s to drill into the Moho from floating platforms out at sea. After a number of test drillings, and a drop in funding, the project—Project Mohole—was abandoned in the mid-1960s.

See also Crust; Earth (planet); Earth, interior structure

MOHS' SCALE

Mohs' hardness scale provides an index and relative measure of mineral hardness (i.e., resistance to abrasion). German geologist Frierich Mohs (1773–1839) devised a scale with specimen **minerals** that offered comparison of "hardness" qualities that allows the assignment of a Mohs hardness number to a mineral. Mohs' scale utilizes 10 specific representative materials that are arranged numerically from the softest (1) to the hardest (10). The reference minerals are (1) talc, (2) **gypsum**, (3) calcite, (4) fluorite, (5) apatite, (6) orthoclase **feldspar**, (7) **quartz**, (8) topaz, (9) corundum, and (10) **diamond**.

The softest mineral, talc, can be used in body powder. The hardest, diamond, is used in drill bits to cut through the most dense crustal materials. Mohs' scale is a relative index scale, meaning that a determination of Mohs' hardness number for a mineral is based upon scratch tests. For example, gypsum (Mohs' hardness number 2) will scratch talc (Mohs' hardness number 1). Talc, however, will not scratch gypsum. **Glass** is assigned a Mohs hardness number of 5.5 because it will scratch apatite (Mohs' hardness number 5) but will not scratch orthoclase feldspar (Mohs' hardness number 6).

Scratch tests are a common method used to identify mineral hardness relative to Mohs' scale. Streak tests are often carried out on streak plates. Mineral hardness is a fundamental property of minerals and can be used to identify unknown minerals. In the absence of comparative minerals, geologists often resort to common objects with a relatively well-established Mohs' hardness number. In addition to glass (5.5), copper pennies measure 3.5, and the average human fingernail averages a Mohs' hardness of 2.5.

The Mohs' scale is a comparative index rather than a linear scale. In fact, Mohs' scale has a near logarithmic relationship to absolute hardness. At the lower, softer end of the scale, the difference in hardness is close to linear, but at the extremes of harness, there are much greater increases in absolute hardness (e.g., a greater increase in the hardness between corundum and diamond than between quartz and topaz).

Hardness is a property of minerals derived from the nature and strength of chemical bonds in and between **crystals**. The number of atoms and the spatial density of bonds also influences mineral hardness. Softer minerals are held together by weak van der Waals bonds. The hardest minerals tend to be composed of dense arrays of atoms covalently bonded together.

Hardness characteristics—especially in calcite crystals—may vary as a property dependent upon the direction of the scratch (i.e., able show evidence of a particular Mohs' number if scratched along one face or direction as opposed to a different hardness number if scratched in a different direction).

See also Chemical bonds and physical properties; Field methods in geology; Mineralogy

MOLINA, MARIO (1943-)

Mexican-born American chemist

Mario Molina is an important figure in the development of a scientific understanding of the atmosphere. Molina earned national prominence by theorizing, with fellow chemist **F. Sherwood Rowland**, that chlorofluorocarbons (CFCs) deplete the earth's ozone layer. Molina and Sherwood shared the 1995 Nobel Prize in chemistry, along with the Dutch chemist Paul Crutzen, for their work on the depletion of **ozone** in the atmosphere. In his years as a researcher at the Jet Propulsion Lab at the California Institute for Technology (CalTech) and a professor at the Massachusetts Institute of Technology (MIT), Molina has continued his investigations into the effects of chemicals on the atmosphere.

Mario José Molina was born in Mexico City to Roberto Molina-Pasquel and Leonor Henriquez. Following his early schooling in Mexico, he graduated from the Universidad Nacional Autónoma de México in 1965 with a degree in chemical engineering. Immediately upon graduation, Molina went to West Germany to continue his studies at the University of Freiburg, acquiring the equivalent of his master's degree in polymerization kinetics in 1967. Molina then returned to Mexico to accept a position as assistant professor in the chemical engineering department at his alma mater, the Universidad Nacional Autónoma de México.

In 1968, Molina left Mexico to further his studies in physical chemistry at the University of California at Berkeley. He received his Ph.D. in 1972 and became a postdoctoral associate that same year. His primary **area** of postdoctoral work was the chemical laser measurements of vibrational energy distributions during certain chemical reactions. The following year, 1973, was a turning point in Molina's life. In addition to marrying a fellow chemist, Luisa Y. Tan, Molina left Berkeley to continue his postdoctoral work with physical chemist, Professor F. Sherwood Rowland, at the University of California at Irvine.

Both Molina and Rowland shared a common interest in the effects of chemicals on the atmosphere. Both were also well aware that every year millions of tons of industrial pollutants were bilged into the atmosphere. They also had questions about emissions of nitrogen compounds from supersonic aircraft. Molina and Rowland decided to conduct experiments to determine what happens to chemical pollutants that reach both the atmosphere directly above us but also at stratospheric levels, some 10–20 mi (16–32 km) above the earth. Both men knew that within the **stratosphere**, a thin, diffuse layer of ozone gas encircles the planet, which acts as a filter screening out much of the Sun's most damaging ultraviolet radiation. Without this ozone shield, life could not survive in its present incarnation.

They concentrated their research on the impact of a specific group of chemicals called chlorofluorocarbons, which are widely used in such industrial and consumer products as aerosol spray cans, pressurized containers, etc. They found that when CFCs are subjected to massive ultraviolet radiation they break down into their constituent chemicals: chlorine, fluorine, and **carbon**. It was the impact of chlorine on ozone that alarmed the two scientists. They found that each chlorine **atom** could destroy as many as 100,000 ozone molecules before becoming inactive. With the rapid production of CFCs for commercial and industrial use, millions of tons annually, Molina and Rowland were alarmed that the impact of CFCs on the delicate ozone layer within the stratosphere could be life-threatening.

Mario Molina published the results of his and Rowland's research in *Nature* magazine in 1974. Their findings had startling results. Molina was invited to testify before the United States House of Representative's Subcommittee on Public Health and Environment. Suddenly CFCs were a popular topic of conversation. Manufacturers began searching for alternative propellant gases for their products.

Over the next several years, Molina refined his work and, with Rowland, published additional data on CFCs and the destruction of the ozone layer in such publications as *Journal of Physical Chemistry*, *Geophysical Research Letter,* and in a detailed piece entitled "The Ozone Question" in *Science*. In 1976, Mario Molina was named to the National Science Foundation's Oversight Committee on Fluorocarbon Technology Assessment.

In 1982, Molina became a member of the technical staff at the Jet Propulsion Laboratory at CalTech; two years later he was named senior research scientist, a position he held for an additional five years. In 1989, Mario Molina left the West Coast to accept the dual position of professor of **atmospheric chemistry** at the MIT's department of Earth, atmosphere and planetary sciences, and professor in the department of chemistry. In 1990, he was one of 10 environmental scientists awarded grants of $150,000 from the Pew Charitable Trusts Scholars Program in Conservation and the Environment. In 1993, he was selected to be the first holder of a chair at MIT established by the Martin Foundation, Inc., "to support research and education activities related to the studies of the environment."

Molina has published more than fifty scientific papers, the majority dealing with his work on the ozone layer and the chemistry of the atmosphere. In 1992, Molina and his wife, Luisa, wrote a monograph entitled "Stratospheric Ozone" published in the book *The Science of Global Change: The Impact of Human Activities on the Environment* published by the American Chemical Society.

His later work has also focused on the atmosphere-biosphere interface which Molina believes is "critical to understanding global climate change processes." He is the recipient of more than a dozen awards including the 1987 American Chemical Society Esselen Award, the 1988 American Association for the Advancement of Science Newcomb-Cleveland Prize, the 1989 NASA Medal for Exceptional Scientific Advancement, and the 1989 United Nations Environmental Program Global 500 Award.

See also Atmospheric circulation; Atmospheric composition and structure; Atmospheric pollution

Monocline

Monoclines are **folds** consisting of two horizontal (or nearly so) limbs connected by a shorter inclined limb. They can be compared to anticlines, which consist of two inclined limbs dipping away from each other, and synclines, which consist of two inclined limbs dipping towards each other.

Folds such as monoclines, anticlines, and synclines are defined solely on the basis of their geometry, and the names therefore have no genetic connotations. Monoclines are, however, characteristic of regions in which **sedimentary rocks** have been deformed by **dip** slip movement along vertical or steeply dipping faults in older and deeper rocks, such as the Colorado Plateau of the southwestern United States. An excellent example of a Colorado Plateau monocline is the Waterpocket fold in Capitol Reef National Park. Most monoclines are classified as drape folds or forced folds because the sedimentary rocks are draped or forced as a result of movement along the underlying faults. Drape folds and forced folds are not necessarily monoclinal, though, so care must be taken to distinguish between geometric and genetic names.

The shape of a monoclinal drape fold is controlled in part by the nature of the underlying fault (normal, reverse, or vertical) and in part by the mechanical nature of the strata being folded. A sequence of relatively thin strata, for example, is less resistant to folding than a single **rock** unit of the same aggregate thickness. Thick rock units such as massive sandstones may fracture during monoclinal folding because they are too stiff to be bent without breaking, whereas rock units such as shales may be easily folded because they consist of innumerable and thin laminae that are weakly bonded to each other. Likewise, the rock type of the lowermost layer being folded has an influence on the form of the fold. A weak rock such as shale, salt, or **gypsum** can attenuate much of the movement along the underlying fault and reduce the amplitude of the resulting fold.

See also Faults and fractures; Syncline and anticline

Monsoon • *see* Seasonal winds

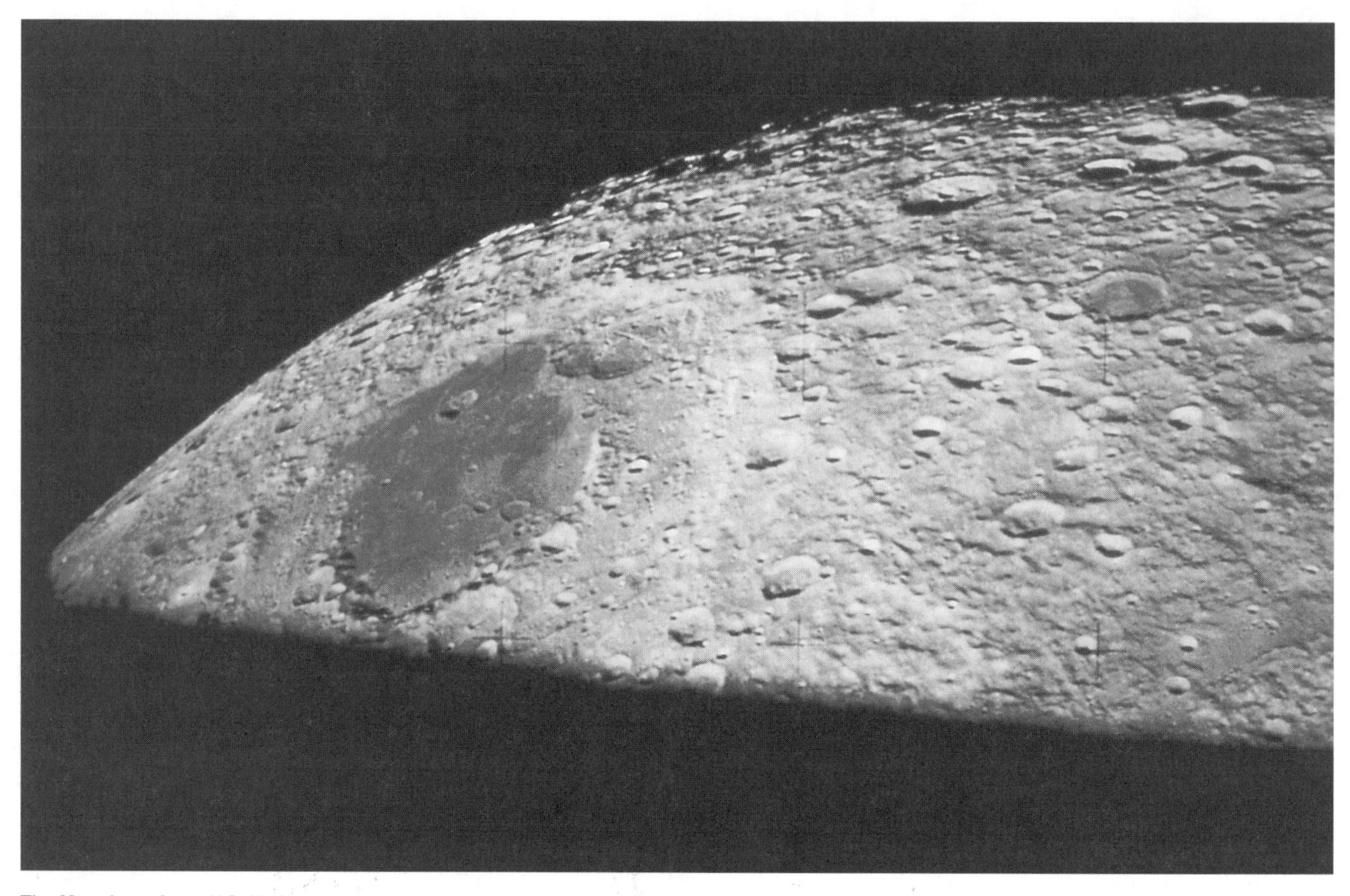

The Moon's surface. ***U.S. National Aeronautics and Space Administration (NASA).***

Moon

The Moon is Earth's only natural **satellite**. Reflecting light from the **Sun**, the Moon is often the brightest object in the night sky.

The Moon orbits Earth at an average distance of approximately 240,000 miles (385,000 km). With **revolution and rotation** periods of approximately 27.32 Earth days, the Moon is in synchronous orbit about the earth. This synchronous orbit maintains a "near side" and "far side" of the Moon. The "near side" faces Earth, while the far side is not visible from Earth. Although Russian **space** probes—and later many American probes—took the first pictures of the far side of the Moon years earlier, it was not until the flight of *Apollo 8* that United States astronauts became the first humans to directly view the far side of the Moon.

Orbital dynamics between the Sun, Moon, and Earth cause different patterns of illumination on the surface of the Moon as seen from Earth. As the Moon revolves about the earth, it appears to go through a series of illumination phases. The Sun constantly illuminates one-half of the lunar surface. The changing orientation in the three body system (Sun, Earth, and Moon), changes to what extent that solar illumination covers areas on the surface of the Moon that are visible from Earth.

Because the earth is revolving about the Sun, the displacement of the earth along it's orbital path establishes the time it takes to complete a cycle of lunar phases—a synodic month—and return the Sun, Earth, and Moon to the same starting alignment. This synodic month is approximately 29.5 days, and is longer than the 27.32-day sideral month.

A waxing moon is one where the **area** illuminated increases each night. A waning moon describes a decreasing area of illumination.

The Moon's phases are a cyclic repetition of illumination patterns described as: new moon, waxing crescent moon, waxing half moon, waxing gibbous moon, full moon, waning gibbous moon, waning half moon, waning crescent moon, followed by a return to the new moon phase.

A new moon occurs when the Moon's orbital path places it between the earth and the Sun. Only the side of the Moon not visible to Earth is illuminated and the Moon is lost in the bright sunlight. Occasionally when the Moon is also in the proper plane of alignment, it may provide a full or partial solar **eclipse** over portions of Earth's surface.

Relative to the Sun and starfield, the Moon appears to move eastward. Following the new moon, the next night, a small sliver or crescent becomes illuminated. The waxing crescent moon is low on the western horizon and is visible just after sunset (i.e., the Moon "sets" shortly after sunset).

As the orbital dynamics shift, the crescent grows larger—and the Moon sets later—each night following sunset. Approximately one week following the new moon, the Moon is one quarter of the way through it's orbital revolution of Earth, and one half of the lunar surface is illuminated as a waxing half moon. Depending upon **latitude**, the waxing half moon appears nearly directly overhead (at the zenith of the celestial meridian) at sunset. The waxing half moon will set about midnight local time. During the next week, the area of the Moon reflecting sunlight to Earth covers more than half of the visible lunar surface, and is described as a waxing gibbous moon.

Approximately two weeks after the new moon, the visible surface of the Moon becomes fully illuminated because the Moon is on the opposite side of Earth relative to the Sun. If the earth and Moon are in the proper plane, Earth may actually block the Sun's light over a portion of the lunar surface and cause a partial to full lunar eclipse. The full moon rises at sunset and sets at dawn.

Following the full moon, the Moon begin to progressively darken through waning gibbous phases until about a week following the full moon it forms a waning half moon. The waning half moon rises about midnight and sets about noon the next day. Continued darkening over the last week of the lunar cycle provides a waning crescent moon that finally returns full cycle to the new moon state, where the Moon and Sun, on the same side of Earth's orbit about the Sun, appear to rise and set together.

The phases of the Moon proved one of the most fundamental astronomical calendars for ancient peoples and the ancient Greek astronomers asserted that the Moon reflected the Sun's light. Phases of the Moon remain critical in determining the date and timing of many religious observances (e.g., Passover, Easter, Ramadan, Visakha Puja, etc.)

Because the earth is larger than the Moon and relatively close to the Moon, it casts a large shadow that causes lunar eclipses. Solar eclipses (where the Moon blocks the Sun) are less frequent and are only possible because, although the Sun is much larger than the Moon, the Moon is much closer to Earth. The present set of orbital dynamics and distances allow solar eclipses because the Sun and Moon have the same angular size (approximately 0.5°) when viewed from Earth. The average human thumb, held out at arm's length obscures approximately 0.5° degrees and will thus, block both the Sun and Moon. (*Warning: Direct viewing of the Sun may cause blindness or optic injury and should not be attempted. Solar observation requires special protective goggles that filter and reduce the intensity of sunlight.*)

The Moon appears to shift its position eastward on the celestial sphere by approximately 13° per night (i.e., appears to move 13° to the east from its prior position if observed at the same time on successive nights).

The Moon is nearly spherical with polar and equatorial radii varying by about a mile. The equatorial radius of the Moon is approximately 1,080 miles (1,738 km). The diurnal temperatures (the day/night temperatures) on the Moon range from approximately –280°F to +260°F (–173°C to +126°C). Contrary to popular belief, the Moon does have a thin atmosphere that consists of helium, argon, methane, minute amounts of **oxygen**, and other trace elements. The density of the lunar atmosphere is only approximately 2×10^5 particles/cm^3 and results in a lunar **atmospheric pressure** of only 8.86×10^{-14} inHg (3×10^{-12} mb) in contrast to Earth's average surface atmospheric pressure of 29.92 inHg (1,014 mb).

The thin and dry lunar atmosphere provides no substantial **weathering** agents (e.g., **wind**, **water**, etc.) and so erosional processes are greatly slowed—essentially reduced to heating, cooling, and slow geochemical changes. The thin atmosphere also offers no protection from meteor impacts and the combination of lack of protection and lack of Earth-like **erosion** produces a heavily cratered lunar landscape that preserves billions of years of accumulated impact craters.

Although the Moon is a quarter of Earth's size, it has only approximately 1.2% of Earth's mass. The gravitational attraction at the surface of the Moon is about one-sixth that of the gravitational attraction at Earth's surface. Accordingly, neglecting air friction (something easily accomplished on the Moon but not on Earth) an object in freefall near Earth's surface accelerates at 9.8 m/s^2, but near the lunar surface, the acceleration due to gravity is approximately 1.62 m/s^2.

See also Celestial sphere: The apparent movements of the Sun, Moon, planets, and stars; Diurnal cycles; Earth (planet); History of manned space exploration; Gravity and the gravitational field; Solar system

MORAINES

Moraines are glacial deposits of **till** (sediment) that are classified by their position relative to the glacial **ice** sheet. Moraines are classified as terminal moraines, lateral moraines, medial moraines, or ground moraines.

Moraines are formed of the sedimentary materials of varying sizes. Accordingly, they are poorly sorted and are essentially the sedimentary "dump" of material broken, dislodged, smashed, ground, and then deposited during the movement of **glaciers**.

End moraines form at the end of the glacial sheet. The end moraine that forms the maximum of glacial extent of glacial coverage (i.e., the farthest extent of "movement" of the glacier) is termed the terminal moraine.

Lateral moraines are deposits of till that accumulate along the lateral margins of the glacial sheet. As glacial sheets fuse, the corresponding fusion of lateral moraines forms a medial moraine.

Ground moraines form from deposits beneath the glacial sheet.

Moraines are geomorphologic features that can last long after glacial retreat (i.e., glacial **melting**) and moraines often form the base for varied and hilly landscapes in areas subject to periodic **glaciation**. Moraines can themselves be reshaped by subsequent glaciations and thus moraine patterns are often used to determine the extent and pattern of glaciations from which further information regarding climate changes can be derived.

See also Archeological mapping; Fjords; Ice ages; Ice heaving and ice wedging; Landforms; Landscape evolution; Topography and topographic maps

MORLEY, EDWARD WILLIAMS (1838-1923)

American physicist

Originally trained for the ministry, Edward Williams Morley decided instead in 1868 to pursue a career in science, the other great love of his life. Initially, Morley devoted himself primarily to teaching, but gradually became engaged in original research. His work can be divided into three major categories: the first two involved the determination of the **oxygen** content of the atmosphere and efforts to evaluate Prout's hypothesis. His third field of research involved experiments on the velocity of light, and it was this research that brought him scientific notoriety.

Morley was born in Newark, New Jersey, on January 29, 1838. His mother was the former Anna Clarissa Treat, a schoolteacher, and his father was Sardis Brewster Morley, a Congregational minister. According to a biographical sketch of Morley in the December 1987 issue of the *Physics Teacher*, the Morley family had come to the United States in early Colonial days and "was noted for its deep patriotism and religious devotion."

Morley's early education took place entirely at home, and he first entered a formal classroom at the age of nineteen when he was admitted to Williams College in Williamston, Massachusetts, as a sophomore. His plans were to study for the ministry and to follow his father in a religious vocation. He also took a variety of courses in science and mathematics, including **astronomy**, **chemistry**, calculus, and optics. His courses at Williams were a continuation of an interest in science that he had developed at home as a young boy.

Morley graduated from Williams as valedictorian of his class with a bachelor of arts degree in 1860. He then stayed on for a year to do astronomical research with Albert Hopkins. Morley's biographers allude to the careful and precise calculations required in this work as typical of the kind of research Morley most enjoyed doing.

After completing his work with Hopkins, Morley entered the Andover Theological Seminary to complete his preparation for the ministry, while concurrently earning his master's degree from Williams. Morley graduated from Andover in 1864, but rather than finding a church, he took a job at the Sanitary Commission at Fortress Monroe in Virginia. There he worked with Union soldiers wounded in the Civil War.

His work completed at Fortress Monroe, Morley returned to Andover for a year and then, failing to find a ministerial position, took a job teaching science at the South Berkshire Academy in Marlboro, Massachusetts. It was at Marlboro that Morley met his future wife, Isabella Birdsall. The couple was married in 1868. Morley had finally received an offer in September, 1868, to become minister at the Congregational church in Twinsburg, Ohio. He accepted the offer but, according to biographers David D. Skwire and Laurence J. Badar in the *Physics Teacher*, became disenchanted with "the low salary and rustic atmosphere" at Twinsburg and quickly made a crucial decision: he would leave the ministry and devote his life to science.

The opportunity to make such a change had presented itself shortly after Morley arrived in Twinsburg when he was offered a position teaching chemistry, botany, **geology**, and **mineralogy** at Western Reserve College in Hudson, Ohio. Morley accepted, and when Western Reserve was moved to Cleveland in 1882, Morley followed. While still in Hudson, Morley was assigned to a full teaching load, but still managed to carry out his first major research project. That project involved a test of the so-called Loomis hypothesis, which held that during periods of high **atmospheric pressure**, air is carried from upper parts of the atmosphere to the earth's surface. Morley made precise measurements of the oxygen content in air for 110 consecutive days, and his results appeared to confirm the theory.

In Cleveland, Morley became involved in two important research studies almost simultaneously. The first was an effort at obtaining a precise value for the atomic weight of oxygen, in order to evaluate a well-known hypothesis proposed by the English chemist William Prout in 1815. Prout had suggested that all atoms are constructed of various combinations of hydrogen atoms.

Morley (as well as many other scientists) reasoned that should this hypothesis by true, the atomic weight of oxygen (and other elements) must be some integral multiple of that of hydrogen. For more than a decade, Morley carried out very precise measurement of the ratios in which oxygen and hydrogen combine and of the densities of the two gases. He reported in 1895 that the atomic weight of oxygen was 15.897, a result that he contended invalidated Prout's hypothesis.

Even better known than his oxygen research, however, was a line of study carried out by Morley in collaboration with Albert A. Michelson, professor of **physics** at the Case School of Applied Science, adjacent to Western Reserve's new campus in Cleveland. Morley and Michelson designed and carried out a series of experiments on the velocity of light. The most famous of those experiments were designed to test the hypothesis that light travels with different velocities depending on the direction in which it moves, a hypothesis required by current theories regarding the way light is transmitted through **space**. A positive result for that experiment was expected and would have confirmed existing beliefs that the transmission of light is made possible by an invisible "ether" that permeates all of space.

In 1886, Michelson and Morley published their report of what has become known as the most famous of all negative experiments. They found no difference in the velocity with which light travels, no matter what direction the observation is made. That result caused a dramatic and fundamental rethinking of many basic concepts in physics and provided a critical piece of data for Albert Einstein's theory of relativity.

The research on oxygen and the velocity of light were the high points of Morley's scientific career. After many years of intense research, Morley's health began to deteriorate. To recover, he took a leave of absence from Western Reserve for a year in 1895 and traveled to **Europe** with his wife. When he returned to Cleveland, he found that his laboratory had been dismantled and some of his equipment had been destroyed. Although he remained at Western Reserve for another decade, he never again regained the enthusiasm for research that he had had before his vacation.

Morley died on February 24, 1923, in West Hartford, Connecticut, where he and Isabella had moved after his retirement in 1906; she predeceased him by only three weeks. Morley was nominated for the Nobel Prize in chemistry in 1902, and received a number of other honors including the Davy Medal of the Royal Society in 1907, the Elliot Cresson Medal of the Franklin Institute in 1912, and the Willard Gibbs Medal of the Chicago section of the American Chemical Society in 1917. He served as president of the American Association for the Advancement of Science in 1895 and of the American Chemical Society in 1899.

See also Atmospheric chemistry; Atmospheric composition and structure; Quantum theory and mechanics

MORLEY, LAWRENCE WHITAKER (1920-)

Canadian geologist

Lawrence Morley has had a long, successful career in geophysics, including work on **remote sensing**, aerophysics, palaeomagnetic research, and geophysical instrumentation and interpretation, publishing 65 scientific and technical papers. Born in Toronto, Ontario, Morley received his education at Collingwood, Owen Sound and Lakefield College School, Ontario. He then attended the University of Toronto, studying **Physics** and Geology, but his degree was interrupted by the Second World War, during which time he served four years with the Royal Navy as a radar officer. Morley graduated in 1946, after hostilities had ended. In 1949, Morley returned to the University of Toronto to complete a Masters and then a Ph.D. in geophysics, during which time he studied palaeomagnetism under Professor **J. Tuzo Wilson**.

Morley was the first to suggest the theory of the magnetic imprinting of rocks on the ocean floor in 1963, during a talk to the Royal Society of Canada. Previously in 1915, Alfred Wegner (1880–1930) had proposed that there had once been a super-continent, which he named Pangaea, that had slowly moved apart. However, Wegner's theory could not explain how such movement occurred, and his theory was largely ignored. In the early 1960s, Harry Hess (1906–1969) and Robert Dietz both independently hypothesized that that seafloor spreading was responsible for the motion of the continents. It had long been known that the earth had undergone a number of magnetic reversals in its long history. Morley suggested that new **lava** emerging from an ocean ridge would produce **rock** fixed in the current **magnetic field** of Earth. Rocks older than the last magnetic reversal would have an opposing polarity. These should appear as parallel stripes on both sides of an ocean ridge.

However, a paper submitted to the journal *Nature* was rejected, as was an article given to the *Journal of Geophysical Research*. By the time his ideas appeared in a special publication of the Royal Society of Canada, in 1964, Frederick J. Vine (1939–) and **Drummond Matthews** (1931–), working independently, had already published the same hypothesis. While the theory of magnetic striping in the earth's **crust** is sometimes referred to as the Vine-Matthews-Morley Hypothesis, just as often Morley's name is omitted. Surveys in the mid-1960s found the expected stripes at every ocean ridge. This evidence confirmed the ideas of Wegner, Dietz and Hess, and resulted in a revolution in **Earth science**, giving rise to the theory of **plate tectonics**.

Morley did pioneering work on aerial mineral and **petroleum** surveys using the airborne magnetometer in the 1940s, working in Venezuela, Columbia, and Canada. He was appointed to the Geological Survey of Canada in 1952, where he served for 17 years as Chief of the Geophysics Division to the Geological Survey. He promoted the Federal/Provincial Aeromagnetic Survey Program, which eventually covered the whole of Canada, producing more than 7,000 detailed maps. Much of his work was aimed at finding mineral and petroleum deposits, and he directed the surveying of Hudson's Bay and the Canadian offshore regions for oil reserves.

With Lee Godby, Morley helped establish the Canada Centre for Remote Sensing and was its founding Director-General from 1971 to 1980. He also served with the Canadian High Commission in London, England as its Science Counselor. Morley was consulted by York University to promote a University/Industry Institute for Space Research, becoming the founding Executive Director of the Institute for Space and Terrestrial Science in 1986. In the 1980s, Morley formed a consulting company specializing in remote sensing and geophysical exploration, and since 1990, this has been his main focus.

Morley is a fellow of the Royal Society of Canada and the Canadian Aeronaut and Space Institute, as well as being a member of the Society for Explorational Geophysics, the American Geophysics Union, the American Society of Photogrammetry and Remote Sensing, and the Canadian Institute of Surveys and Mapping. Morley has received a number of honors, including the McCurdy medal (1974), and the Tuzo Wilson Prize (1980).

See also Plate tectonics; Sea-floor spreading

MOSANDER, CARL GUSTAF (1797-1858)

Swedish chemist

In a large part, credit for unraveling the complex nature of the rare Earth elements goes to Carl Gustaf Mosander. Mosander was born in Kalmar, Sweden, on September 10, 1797. He was educated as a physician and pharmacist and served as an army surgeon for many years.

Mossander's most important professional association was with the eminent Swedish chemist J. J. Berzelius. Mosander lived with Professor and Mrs. Berzelius for many years and worked as Berzelius' assistant at the Stockholm Academy of Sciences. Eventually, Mosander became curator of **minerals** at the Academy and, in 1832, succeeded Berzelius as Permanent Secretary of the Academy. Mosander was also Professor of **Chemistry** and **Mineralogy** at the Caroline Institute for many years.

Mosander became interested in the rare Earth elements in the late 1830s. Fifty years earlier, a Swedish army officer, Carl Axel Arrhenius, had discovered a new mineral near the small town of Ytterby that he named ytterite. Chemists spent much of the next century trying to separate the mineral into its many chemically-similar parts.

The first breakthrough in this effort occurred in 1794 when Johan Gadolin (1760–1852) showed that ytterite contained a large fraction of a new oxide, which he called yttria. A decade later, M. H. Klaproth, Berzelius, and Wilhelm Hisinger (1766–1852) showed that ytterite also contained a second oxide, which they termed ceria.

Mosander first concentrated his efforts on the ceria component of ytterite. In 1839, he found that the ceria contained a new element, which he named lanthanum (for hidden). Mosander did not publish his results immediately, however, because he was convinced that yet more discoveries were to be made. He was not disappointed in these hopes. In 1841, he identified a second new component of ceria. He named the component didymium, for twin, because it was so closely related to lanthanum. Later research showed that didymium was not itself an element, but a complex mixture of other rare earth elements.

In 1843, Mosander turned his attention to the yttria component of ytterite. He was able to show that the yttria consisted of at least three components. He kept the name yttria for one and called the other two erbia and terbia. The last two of these components are now known by their modern names of erbium and terbium. Mosander is acknowledged as the discoverer, then, of three elements: lanthanum, erbium, and terbium. Mosander died in Ångsholm, Sweden, on October 15, 1858.

See also Chemical elements

MOUNTAIN CHAINS

Mountain chains are elongate, elevated areas of the earth's surface comprising several sub-parallel mountain ranges. Each mountain range is a connected series of mountain peaks (i.e., large **rock** masses that rise abruptly above the surrounding landscape). Mountain chains may be a thousand or more kilometers long and hundreds of kilometers wide. Mountain chains are formed by the interplay of endogenic and exogenic processes. Endogenic processes are those that originate within the earth, such as orogenesis and volcanism. Exogenic processes are external processes, such as **weathering** and **erosion** due to the action of **water**, **ice**, and **wind**.

Volcanism during subduction of oceanic **crust** beneath oceanic crust creates an island arc. **Island arcs** may comprise an arcuate alignment of volcanic island peaks (e.g., the Aleutian Islands) or a continuous land **area** comprising a central mountain chain formed by volcanic and tectonic processes (e.g., Japan). Subduction of oceanic crust beneath continental crust creates a Cordilleran or Andean-style mountain chain. The best-known example is the Andes in **South America**, where the oceanic Nazca Plate is subducted beneath the continental South American Plate. In the Barisan Mountains of Sumatra, Indonesia, oceanic **lithosphere** of the Indo-Australian Plate is obliquely subducted beneath continental crust of the Sunda Plate. A network of faults comprising a transcurrent fault system dissects the resulting volcanic mountain chain.

Collisional mountain chains result from the collision between two continental **lithospheric plates** or between a continental plate and an island arc. The continental lithosphere is greatly thickened in this process called orogenesis. The resulting belt of uplifted crust forms a mountain chain with a commensurate deep lithospheric keel or root. Such isostatic balance can be likened to an iceberg where only a proportion protrudes out of the water, with a large part of its mass being below water. When material in the mountain is removed by erosion or tectonically through extensional faulting, compensatory uplift will occur while the lithospheric root remains. Rocks previously at greater depths are, therefore, brought closer to the surface in a process called exhumation. Tens of kilometers of rock may be removed by erosion before a collisional mountain chain is eventually flattened. Eroded sediments are deposited in adjacent extensional or foreland basins. Extensive erosion plus or minus tectonic exhumation due to displacement along faults results in exhumation of high-grade rocks. Collisional mountain belts include:

- the European Alps, formed by collision of the European Plate with the Adriatic Plate following closure of part of the Tethyan Ocean,
- the Himalayas, formed by the collision between the Eurasian Plate with island arcs and the Indian Plate.

Folding of rock layers may also occur in the hinterland to a collisional orogenic belt or in a cover sequence where basement rocks slide past one another at the same time as undergoing regional shortening (i.e., a transpressional belt). Trains of **folds** can also form during gravitationally induced sliding on a weak basal décollement horizon without regional shortening. In all such folded areas, rock layers more resistant to erosion in regional-scale antiforms may define a continuous mountain range along each **anticline**. Such fold mountain chains generally lack the presence of a deep lithospheric root and will more rapidly disappear due to the effects of erosion over time.

Steep to moderately dipping normal faults formed during regional, horizontal extension during collapse of a collisional mountain belt or **rifting** may downthrow blocks of rock called graben, leaving elongate, fault-bounded high blocks (horsts). Block-faulted mountain chains are formed if displacements are great enough and erosion rates are low. Rock layers between parallel-dipping normal faults are tilted and this may result in asymmetrical mountain ranges with steep faces and

Mount Everest is the tallest mountain of the tallest, and youngest, mountain chain in the world, the Himalayas. ***Archive Photos. Reproduced by permission.***

long **dip** slopes parallel to layering. Valleys along intervening graben in which the eroded sediments are deposited separate the ridges. Mountain chains formed by extensional faulting occur in areas of widely distributed extension, such as the Basin and Range Province of Utah and Wyoming. Displacement can also take place along shallowly dipping extensional detachment faults that widen to form ductile **shear zones** at greater depth. The earth's lithosphere is thinned during regional extension. In order to compensate for this, the underlying **asthenosphere** is arched upward beneath the area of greatest lithospheric thinning. Extensional detachments are folded and deep crustal rocks exhumed, forming metamorphic core complexes. Metamorphic core complexes commonly produce elongate domal mountains as their metamorphic and possibly igneous core is likely to be more resistant to subsequent weathering than the surrounding low-grade rocks.

See also Orogeny; Plate tectonics; Subduction zone

MOUNTAIN FORMATION • *see* OROGENY

MT. PINATUBO VOLCANIC ERUPTION • *see* VOLCANIC ERUPTIONS

MT. ST. HELENS VOLCANIC ERUPTION • *see* VOLCANIC ERUPTIONS

MUD FLOW

The term mud flow, although not part of the classification system used by most **landslide** specialists, is a form of **mass movement** or **mass wasting** widely used in a manner that is synonymous with wet to very wet, rapid to extremely rapid earth flow. Mud itself is defined by most geologists as an unlithified mixture of silt, **clay**, and **water**; therefore, a mud flow is a flow consisting primarily of silt, clay, water, and other minor constituents such as **sand**, cobbles, boulders, trees, and other objects. A flow in which mud is a minor constituent relative to sand-size or coarser particles is by definition a **debris flow** or, if large pieces of **bedrock** are involved, a **rock** flow.

Because volcanic ash deposits commonly **weather** into clayey materials, mud flows are common on and around volcanoes as well as areas covered by deposits of fine-grained volcanic ash known as loess. A mudflow on a **volcano** can also be referred to as a fine-grained or muddy **lahar**.

Like debris flows, mud or earth flows can begin by mobilization from a landslide, incorporation of muddy sediments into flooding, or rapid **melting** of snow and **ice** during a volcanic eruption. Regardless of their mode of origin, mud or earth flows can be dangerous and destructive because of their great density (typically more than 50% solid material) and velocity. The density of debris and mudflows also allows them to transport unusually large boulders compared to **floods** consisting primarily of water. A typical debris or mudflow consisting of 60% solids and 40% water would have a density of about 125 lb/ft^3 (2,000 kg/m^3), or twice that of water. Thus, the buoyant force exerted on a boulder by a mud or debris flow would be about twice that exerted on the same boulder by water.

See also Floods

MUIR, JOHN (1838-1914)

Scottish-born American naturalist

John Muir—naturalist, conservationist, mountaineer, and chronicler of the American frontier—was born in Dunbar, Scotland on April 21, 1838. During his lifetime, Muir published more than 300 articles and 10 books recounting his travels, scientific observations, and opinions on nature conservation. His wanderlust led him on expeditions around the globe, but California's Sierra Nevadas were his home. In addition to his descriptive and inspirational nature writing, Muir advanced a number of scientific theories, including the now-accepted hypothesis that **glaciers** carved Yosemite Valley. His love of the Sierras, and his concern for their preservation, led him to become one of America's first environmental activists. Muir co-founded the Sierra Club in 1871, and he served as the club's first president until his death in 1914.

John Muir immigrated to Fountain Lake, Wisconsin in 1849 with his family at age 11. The Muir family's hard-working frontier life left John no time to continue the formal schooling he had begun in Scotland. He did, however, maintain his passion for reading and natural science, and excursions into the woods provided a welcome diversion from his father's strict discipline and grueling work schedule. John put his self-taught knowledge to use at the Muir homestead by inventing an assortment of machines, including a table saw and a machine that dumped him out of bed for morning chores.

In 1860, John Muir left home at age 22 to exhibit his inventions at the Wisconsin state fair in Madison. There he received his first public recognition in the form of a *Wisconsin State Journal* article describing his prize-winning whittled clocks. He also met one of the exhibit judges, Mrs. Jeanne Carr, and her husband, Dr. Ezra Carr, a professor at the University of Wisconsin, who would become his lifelong friends and mentors. Muir attended classes at the University of Wisconsin from 1861 until 1863 when a lack of funds and the Civil War draft led him to return home.

No letter came from the draft board, and Muir set out on a summer plant-collecting trip that became a four-year walking expedition into Canada. He financed his botanical studies with a series of factory jobs and contributing his inventions to improve production along the way. In spring of 1867, Muir suffered a blinding eye injury at a carriage factory in Indianapolis. When his sight returned after a month of painful recovery, he decided to devote his newly regained vision to observations of nature. After a visit home, Muir walked 1,000 mi (1,609 km) to the **Gulf of Mexico**, and boarded a ship to Cuba, New York, and finally Panama. He traveled across the Isthmus, and sailed on to California. John Muir was 30 when he arrived in San Francisco in March of 1868.

From San Francisco, Muir walked east across the San Joaquin Valley. He described his first impression of the Sierras in his book, *My First Summer in the Sierra*: "...from the eastern boundary of this vast golden flower-bed rose the mighty Sierra, miles in height, and so gloriously colored and so radiant, it seemed not clothed with light but wholly composed of it, like the wall of some celestial city.... Then it seemed to me that the Sierra should be called...the Range of Light." Muir spent the summer of 1869 herding sheep, or "hooved locusts" as he would later call them, at Tuolumne Meadows.

From 1869 until 1880, John Muir systematically explored the mountains of California from his cabin in Yosemite Valley. He traveled, unarmed, through the mountains carrying a tin cup, food, and a notebook. He observed active mountain glaciers, and hypothesized that the slow grinding of **ice** had carved Yosemite's soaring **granite** cliffs. His glacial theory, published in 1871 by the *New York Tribune,* gained him the respect of University of California geologist, Joseph LeConte, among others. His friends, the Carrs, moved to Oakland in 1869, and encouraged Muir to pursue his writing during this period. They also sent their influential academic friends to visit him in Yosemite, including Harvard botanist, Asa Gray, and, in May 1871, Ralph Waldo Emerson.

John Muir married Luisa Wanda Strentzel in 1880, and moved to Martinez, California to run the Strentzel's profitable fruit ranch, and help "Louie" raise their two daughters. Even during that 10-year period of relative domesticity, Muir continued to write and travel extensively, exploring Yellowstone, **Europe**, **Africa**, **Australia**, China, Japan, **South America**, and, of course, the Sierras. During the 1890s, he conducted a well-timed study of Alaska that coincided with the Klondike gold rush. His most popular book, *Stickeen,* is an account of a summer spent exploring Alaska's glaciers with a little black dog.

By the turn of the century, Muir had become a leading literary figure. His almost-spiritual descriptions of nature inspired influential and common people alike. Muir's articles in the *Century Magazine* gained him the attention and friendship of its like-minded editor, Robert Underwood Johnson. Their combined efforts led to an act of Congress that created Yosemite National Park in 1890. Muir and Johnson were subsequently involved in further conservation acts that resulted in the protection of Sequoia, Mount Rainier and Petrified Forest, and Grand **Canyon** National Parks. President Theodore Roosevelt visited Muir in Yosemite in 1901. Camping together in the shadow of El Capitan, they laid plans for the wilderness conservation programs that became Roosevelt's legacy.

In his last years, Muir turned his considerable energy to the preservation of wild lands. Muir, Johnson, and others formed the Sierra Club in 1892 to, as Muir wrote, "do something for wildness, and make the mountains glad." The fight to prevent erection of a dam in Hetch Hetchy valley was one of the Sierra Club's most dramatic early battles. Hetch Hetchy reservoir was filled in 1913, and Muir died, disappointed, on December 24, 1914 at the age of 76. His enduring legacy, however, were his books and essays that continue to inspire new generations of nature lovers and environmental activists. John Muir was America's first environmentalist, and was perhaps America's most influential naturalist.

See also Environmental pollution; Glacial landforms; History of exploration II (Age of exploration)

Murchison meteorite

The Murchison meteorite was a meteorite that entered Earth's atmosphere in September, 1969. The meteor fragmented before impact and remnants were recovered near Murchison, **Australia** (located about 60 mi [97 km] north of Melbourne). The fragments recovered dated to nearly five billion years ago—to the time greater than the estimated age of Earth. In addition to interest generated by the age of the meteorite, analysis of fragments revealed evidence of carbon-based compounds. The finds have fueled research into whether the organic compounds were formed from inorganic processes or are proof of extraterrestrial life dating to the time of Earth's creation.

In particular, it was the discovery of amino acids and the percentages of the differing types of amino acids found in the meteorite (e.g., the number of left handed amino acids vs. right handed amino acids), that made plausible the apparent evidence of extraterrestrial organic processes as opposed to biological contamination by terrestrial sources.

If the compounds prove to be from extraterrestrial life, this would constitute a profound discovery that would have far-reaching global scientific and social impact concerning prevailing hypotheses about the **origin of life**. For example, some scientists, notably one of the discoverers of the structure of DNA, Sir Francis Crick, assert that in the period from the formation of Earth to the time of the deposition of the earliest discovered fossilized remains, there was insufficient time for evolutionary process to bring forth life in the abundance and variety demonstrated in the **fossil record**. Crick and others propose that a form of organic molecular "seeding" by meteorites exemplified by the Murchison meteorite (meteorites rich in complex **carbon** compounds) greatly reduced the time needed to develop life on Earth.

In fact, the proportions of the amino acids found in the Murchison meteorite approximated the proportions proposed to exist in the primitive atmosphere modeled in the **Miller-Urey experiment**. First conducted in 1953, University of Chicago researchers Stanley L. Miller and Harold C. Urey developed an experiment to test possible mechanisms in Earth's primitive atmosphere that could have produced organic molecules from inorganic processes. Methane (CH_4), hydrogen (H_2), and ammonia (NH_3) gases were introduced into a moist environment above a water-containing flask. To simulate primitive **lightning** discharges, Miller supplied the system with electrical current. Within days organic compounds formed—including some amino acids. A classic experiment in molecular biology, the Miller-Urey experiment, established that the conditions that existed in Earth's primitive atmosphere were sufficient to produce amino acids, the subunits of proteins comprising and required by living organisms. It is possible, however, that extraterrestrial organic molecules could have accelerated the formation of terrestrial organic molecules by serving a molecular templates.

In 1997, NASA scientists announced evidence that the Murchison meteorite contained microfossils that resemble microorganisms. The microfossils were discovered in fresh breaks of meteorite material. The potential finding remains the subject of intense scientific study and debate.

University of Texas scientists Robert Folk and F. Leo Lynch also announced the observation of **fossils** of terrestrial nanobacteria in another carbonaceous chondrite meteorite named the Allende meteorite. Other research has demonstrated that the Murchison and Murray meteorites (a carbonaceous chondrite meteorite found in Kentucky) contain sugars critical for the development of life.

See also Cosmology; Evolution, evidence of; Evolutionary mechanisms

Murchison, Roderick (1792-1871)

Scottish geologist

Roderick Murchison was an amateur geologist of the Victorian age in Britain. Murchison's paramount achievement is his naming of the periods of the stratigraphic column.

Murchison was wealthy and traveled often; both qualities enabled him to make a significant contribution to geological research in England and Wales while still an amateur in the field. Born in Ross in Scotland, Murchison was destined for a military career and trained at Durham and Great Marlow Military College. Until 1818, he lived in Ross-shire at which time he moved to England. A colleague, chemist Humphrey Davy, along with Murchison's wife, persuaded Murchison to attend some lectures on **chemistry** and **geology**, and he then turned his vast energies to science.

Murchison named the Silurian System in 1839, after an **Iron** Age tribe in mid-Wales and the Permian in 1841, after the Perm region of Russia in the Ural Mountains. For his work on the Permian rocks of Russia, he was made an honorary member of the Academy and Natural History Society of St. Petersburg and Moscow. He named the **Devonian Period** with Adam Sedgwick after the British county of Devon. After great public arguments, however, Murchison and Sedgwick could not agree on the boundary between the Cambrian and the Silurian, and their friendship ended. Sedgwick had been mapping the Cambrian rocks of North Wales and the system is named after the Latin name for Wales. This famous controversy was not resolved until both men had gone to their graves as bitter enemies in the mid 1870s. Charles Lapworth resolved the crisis by naming the **Ordovician Period** (after another Iron Age Welsh tribe) between the Cambrian and Silurian.

In another scientific dispute, Murchison was against the theory of widespread **glaciation** as proposed by **Louis Agassiz** (1807–1873) and was one of the few geologists of his time who died not accepting it.

Murchison was knighted in 1846 following his presidency of the Geological Society of London in 1842 and his founding of the Royal Geographical Society (1830). He was subsequently four times the president of the later. When he died in 1871, he stipulated in his will that a medal and fund should be awarded each year by the Geological Society of London in his honor. It is awarded to outstanding Earth scientists who have shown a breadth and achievement over more than a narrow field. He was also a trustee of the British Museum. Murchison became the second Director-General of the Geological Survey of Great Britain in 1855, which gave him tremendous influence over the advancement of geology in Britain during the mid-Victorian age. He had little interest in economic geology such as hydrology, and this facet of geology did not develop at the survey during his time as Director-General. Historians debate that he actively prohibited research, but he certainly had an influence on slowing down work on the applied areas of geology. Murchison also founded a chair of geology and **mineralogy** at the University of Edinburgh.

See also Geologic time; Stratigraphy

N

NATURAL GAS

Natural gas is a mixture of **hydrocarbons** (molecules that contain only **carbon** and hydrogen) and gases (most notably methane, ethane, propane, and butane) that exist naturally in rocks beneath the surface of the earth. It is widely used as a heating source, and in some cases, specific portions of the natural gas are used as starting materials in industrial processes. Natural gas is the product of the decaying of living matter over millions of years (as is also true for **petroleum**). Specific conditions, including low **oxygen** levels, are necessary for this to occur. The hydrocarbon gases are trapped in geological formations known as anticlines. Each of the major hydrocarbon components of natural gas is used as a fuel source.

Natural gas has its origins in decayed living matter, most likely as the result of the action of bacteria upon dead animal and plant material. In order for most bacteria to effectively break down organic matter to hydrocarbons, there must be low levels of oxygen present. This would mean that the decaying matter was buried (most likely under **water**) before it could be completely degraded to **carbon dioxide** and water. Conditions such as this are likely to have been met in coastal areas where **sedimentary rocks** and marine bacteria are common. The actions of heat and pressure along with bacteria produced a mixture of hydrocarbons. The smaller molecules which exist as gases were then either trapped in porous rocks or in underground reservoirs where they formed sources of hydrocarbon **fuels**.

Natural gas, like petroleum, is a mixture of many organic substances. The exact composition of different sources of natural gas varies slightly, but in all cases, methane is by far the most common component, with other hydrocarbons also being very common. Other gases such as oxygen, argon, and carbon dioxide make up the rest of most natural gas sources. The largest sources of natural gas in the United States are found in Alaska, Texas, Oklahoma, western Pennsylvania, and Ohio. It is estimated that the supply of natural gas in this country may be sufficient to last for two centuries—although the more readily accessible sources have been used, meaning that it will be more expensive to obtain natural gas in the future.

Natural gas is believed to have been first discovered and used by the Chinese, perhaps as early as 1000 B.C. Shallow stores of natural gas were released from just beneath the ground and piped short distances to be used as a fuel source. Natural gas could provide a continuous source of energy for flames. These "eternal fires" were found in temples and also used as attractions for visitors. In the 1800s, natural gas began to be piped short distances as a light source. With the discovery of oil in the 1860s, natural gas was largely ignored as a fuel source. One of the early difficulties with natural gas was in transporting it from the source to other sites for use. The combination of electric lights and petroleum meant that containers of natural gas were used as heat sources for cooking in homes but for little else.

As the technology for piping gas from the source began to improve, it became possible to pipe natural gas over thousands of miles. This has meant that natural gas has become as convenient as petroleum and **coal** to use as a fuel source, and often with far less pollution. Natural gas burns with almost no byproducts except for carbon dioxide and water (as opposed to coal which often has large amounts of sulfur in it), and the heat released from the reaction (combustion of any of the hydrocarbon components of natural gas is an exothermic process). The combustion of methane, the most prevalent component of natural gas, is described by the reaction below:

$$CH_4 + O_2 \rightarrow CO_2 + H_2O + \text{heat energy}$$

Ethane is used less as a fuel source than as a starting material for the production of ethylene (acetylene), which is used in welding.

Both butane and propane are relatively easy to liquefy and store. Liquefied propane and butane are used in disposable lighters and as camping fuels.

Because gases take up large amounts of **space**, they can be inconvenient to transport and store. The ability to liquefy the components of natural gas (either as a mixture or in isolation) has made natural gas much more practical as an energy source. The liquefaction of natural gas takes advantage of the different boiling points of methane, ethane, and other gases as a way of purifying each substance. A combination of refrigeration and increased pressure allows the individual gases to be stored and transported conveniently. At one time, the natural gas that often accompanied petroleum in the ground was simply burned off as a means of getting rid of it. Recently, however, this gas has been collected, liquefied and used along with the petroleum.

See also Fuels and fuel chemistry; Petroleum extraction

NEPTUNE • *see* SOLAR SYSTEM

NESOSILICATES

The most abundant rock-forming **minerals** in the **crust** of the earth are the silicates. They are formed primarily of **silicon** and **oxygen**, together with various **metals**. The fundamental unit of these minerals is the silicon-oxygen tetrahedron. These tetrahedra have a pyramidal shape, with a relatively small, positively charged silicon cation (Si^{+4}) in the center and four larger, negatively charged oxygen anions (O^{-2}) at the corners, producing a net charge of –4. **Aluminum** cations (Al^{+3}) may substitute for silicon, and various anions such as hydroxyl (OH^{-}) or fluorine (F^{-}) may substitute for oxygen. In order to form stable minerals, the charges that exist between tetrahedra must be neutralized. This can be accomplished by the sharing of oxygen cations between tetrahedra, or by the binding together adjacent tetrahedra with various metal cations. This in turn creates characteristic silicate structures that can be used to classify silicate minerals into **cyclosilicates**, **inosilicates**, nesosilicates, **phyllosilicates**, **sorosilicates**, and **tectosilicates**.

The simplest silicates are the nesosilicates, formed by individual silicon-oxygen tetrahedra. There is some substitution of aluminum for silicon in the tetrahedra, but not as much as in other types of silicate minerals. The negatively charged, isolated tetrahedra in nesosilicates are held together by various metal cations. The garnet group of nesosilicates is commonly found in metamorphic rocks and more rarely in **igneous rocks**, and the metal cations that are typically found in garnet minerals include aluminum, calcium, chromium, magnesium, manganese, and **iron**. Garnet minerals include almandine ($Fe_3Al_2Si_3O_{12}$), grossularite ($Ca_3Al_2Si_3O_{12}$), and uvarovite ($Ca_3Cr_2Si_3O_{12}$). The minerals of the **olivine** group are nesosilicates commonly found in iron- and magnesium-rich igneous rocks. The chemical formula of olivine is given as $(Mg,Fe)_2SiO_4$, but a complete solid solution exists between the end-members forsterite (Mg_2SiO_4) and fayalite (Fe_2SiO_4). The nesosilicate mineral zircon ($ZrSiO_4$) commonly forms in igneous rocks, and is so chemically stable that it becomes a common accessory mineral in many sediments and **sedimentary rocks**.

See also Mineralogy

NEUTRONS • *see* ATOMIC THEORY

NEW MADRID EARTHQUAKE • *see* MID-PLATE EARTHQUAKES

NEWTON, SIR ISAAC (1642-1727)

English physicist

In 1687, English physicist Sir Isaac Newton published a law of universal gravitation in his important and profoundly influential work *Philosophiae Naturalis Principia Mathematica* (Mathematical principles of natural philosophy). Newton articulated a law of universal gravitation that states that bodies with mass attract each other with a force that varies directly as the product of their masses and inversely as the square of the distance between them. This mathematically elegant law, along with Newton's laws of motion, became the guiding models for the future development of physical law.

Newton admitted having no fundamental explanation for mechanism of gravity itself. In *Principia* Newton stated, "I have been unable to discover the cause of those properties of gravity from phenomena, and I feign no hypotheses" (regarding its mechanism). Moreover, Newton asserted, "To us it is enough that gravity does really exist, and act according to the laws which we have explained, and abundantly serves to account for the motions of the celestial bodies and our seas." Newton's law of gravitation proved to be a precise and effective tool. A truly universal law, it could be verified by the simplest fall of an apple or measured against the most detailed observations of celestial movements. Ultimately, Newton's law of universal gravitation would provide, in the twentieth century, evidence of the existence of black holes. The Newtonian methodology of simplifying mass to a point mass (i.e., with regard to gravitational fields, all of the mass of a body can be considered to lie in a center of mass without physical **space**) also proved a brilliant simplification that enabled the mathematical advancement of mechanics and electromagnetism.

Because Newton's law of universal gravitation was so mathematically simple and precise it strengthened the idea that all the laws describing the universe should be mathematical.

Born in Woolsthorpe, Lincolnshire, England, Newton was a premature baby who was not expected to live. His father had died three months before the birth, and his mother remarried three years later, leaving Newton in the care of his grandparents. Newton did not distinguish himself in school, and his mother removed him in the late 1650s to work on the family farm, but Newton proved a worse farmer than scholar. His uncle, however, encouraged the boy to go to Cambridge in 1660. Five years later Newton graduated, even though he had failed a scholarship exam in 1663 due to his lack of knowledge concerning geometry.

Newton returned to the farm shortly thereafter to escape the Bubonic plague, which at the time was decimating London. While at the farm in 1666, Newton first developed his law of universal gravitation. When Newton made calculations of what the rate of fall for the **Moon** should be, he came up short of what was actually observed and was quite disappointed. The problem was twofold; first, the radius of the earth was not known with precision and the size Newton used was too small. Second, he was not absolutely certain he was correct in making his calculations based on the gravitational force at the center of Earth, as opposed to the surface. Because of these issues, he set aside his work on gravity for 15 years.

During this same time, Newton began to experiment with light. Newton passed a beam of sunlight through a prism of **glass** and observed it was refracted into a spectrum. He passed the spectrum through a second prism, and the light was recombined into a white spot. In 1672, Newton was elected to the Royal Society.

During his life Newton became involved in often bitter disputes with other scholars, especially his dispute with German mathematician Gottfried Wilhelm von Leibniz (1646–1716) over credit for the development of calculus. Although the notations and nomenclature used in modern calculus most directly trace back to the work of Leibniz, Newton has received the most credit for the development of the calculus in textbooks. Modern historians of science generally conclude that the feud between Newton and Leibniz was essentially groundless. A modern analysis of the notes of Newton and Leibniz clearly established that Newton secretly developed calculus some years before Leibniz published his version but that Leibniz independently developed the calculus so often credited exclusively to Newton.

Newton died on March 20, 1727, at the age of 84, and his vast influence upon science continued, later rivaled only by that of Charles Darwin and **Albert Einstein**.

See also Gravitational constant; Gravity and the gravitational field; Relativity theory

NICHE

In geophysical and ecological terms, a niche designates the relationship between a species and its **area** of inhabitation. The term is specifically used to describe a species' unique position both in terms of physical area, and as a set of characteristics that relate the species' biological and ecological functions to its geophysical environment.

Although not the subject of this article, the term niche is also used to describe a type of glacier (e.g., niche glacier) that forms inside an irregular recess on or within a mountainside.

Four distinct stages of niche theory development in biological ecology can be identified: (1) Joseph Grinnell's original formulation of niche (in 1917 and 1928) as a geophysical spatial unit; (2) Charles Elton's formulation (in 1927) of niche as a functional unit; (3) Gause's (1934) competitive exclusion principle; and (4) E. Evelyn Hutchinson's concept of multidimensional niche in the 1950s.

Although Darwin understood the idea of niche and a few other biologists used the term earlier, Grinnell is credited with its formal development. To Grinnell, niche was a spatial unit that stood for the "concept of the ultimate distributional unit, within which each species is held by its structural and instinctive limitations." His conception of niche was "preinteractive"—that is, it referred to the entire area within which an organism could survive in the absence of other organisms. This is in contrast to the "post-interactive" niche, the actual place occupied by the organism in an environment after it has interacted with other organisms.

At about the same time, Charles Elton was developing the niche concept along somewhat different lines. Elton conceived of niche as a functional unit to describe the organism's "place in the biotic environment, relations to food and enemies." Although Elton presented niche as an organism's ecological position in a larger framework like a community or ecosystem, he then restricted its use to the food habits of an organism. Accordingly, Elton's niche is considered to be postinteractive.

Gause is credited with being the first investigator to perceive the connection between natural selection, competition, and niche and to see the interacting aspects of these concepts. Gause stated that "it is admitted that, as a result of competition, two similar species scarcely ever occupy similar niches, but displace each other in such a manner that each takes possession of certain peculiar kinds of food and modes of life in which it has an advantage over its competitor. Gause experimentally tested the general conclusions drawn from the Lotka-Volterra competitive equations, confirming and amplifying them. These conclusions are summarized in the "competitive exclusion principle," which states that two species cannot coexist at the same locality if they have identical ecological requirements. Gause based the principle on an Eltonian definition of niche.

The Eltonian niche dominated ecological theory during the period 1930–1950 and began to be referred to as an organism's "occupation" or "profession." Hutchinson responded to this rather limited idea of niche by incorporating selected features from both Grinnell's and Elton's niche definitions and redefining niche as an "n-dimensional hypervolume," an abstract multidimensional **space** defining the environmental limits within which an organism is able to survive and reproduce. Hutchinson's "fundamental niche" is preinteractive, composed of "close to innumerable" dimensions, each corresponding to some requisite for a species. By setting the number of defining dimensions at "close to innumerable," Hutchinson attempted to illustrate the complexity of the systems within which organisms exist and interact. He depicted it by plotting each identifiably important environmental variable along an axis to show the points below which and above which the given organism could not survive.

Hutchinson's "realized niche" usually corresponds to a smaller hypervolume because competition and other interactions serve to restrict organisms from some parts of their fundamental or potential niche. Although most current works in niche theory use some variation of Hutchinson's multidimensional niche, both the Eltonian and the Hutchinson niches are still found in contemporary ecology and are still useful. Any application of niche, however, is only an approximation of reality, because niche dimensions are too numerous to be counted.

See also Archeological mapping; Physical geography; Topography and topographic maps

NIMBOSTRATUS CLOUD • *see* CLOUDS AND CLOUD TYPES

NORTH AMERICA

The landmass occupied by the present-day countries of Canada, the United States, and the Republic of Mexico make up North America. Greenland (Kalaallit Nunaat), an island landmass to the northeast of Canada, is also included in North America, for it has been attached to Canada for almost two billion years.

Plate tectonics is the main force of nature responsible for the geologic history of North America. Over time, the plates have come together to form the continents, including North America. Other processes, such as **sedimentation** and **erosion**, modify the shape of the land that has been forged by plate tectonics.

North American geologic history includes several types of mountain ranges as a result of plate tectonics. When the edge of a plate of Earth's **crust** runs over another plate, forcing the lower plate deep into Earth's elastic interior, a long, curved mountain chain of volcanoes usually forms on the forward-moving edge of the upper plate. When this border between two plates forms in the middle of the ocean, the volcanic mountains form a string of islands, or archipelago, such as the Antilles and the Aleutians. This phenomenon is called an island arc.

When the upper plate is carrying a continent on its forward edge, a mountain chain, like the Cascades, forms right on the forward edge. This edge, heavily populated with volcanoes, is called a continental arc. The volcanic mountains on the plate border described above can run into a continent, shatter the collision **area** and stack up the pieces into a mountain range. This is how the Appalachians were formed.

When a continent-sized "layer cake" of **rock** is pushed, the upper layers move more readily than the lower layers. The layers separate from each other, and the upper few miles of rock move on ahead, floating on fluid pressure between the upper and lower sections of the crust like a fully loaded tractor trailer gliding effortlessly along an icy road. The flat surface where moving layers of crust slide along the top of the layers beneath it is called a thrust fault, and the mountains that are heaved up where the thrust fault reaches the surface are one kind of fault block mountains. The mountains of Glacier National Park slid along the Lewis thrust fault over younger rocks, and out onto the Great Plains.

Mountain ranges start being torn down by physical and chemical forces while they are still rising. North America has been criss-crossed by one immense range of mountains after another throughout its almost four-billion-year history.

A range of mountains may persist for hundreds of millions of years, like the Appalachians. On repeated occasions, the warped, folded rocks of the Appalachians were brought up out of the continent's basement and raised thousands of feet by tectonic forces. If mountains are not continuously uplifted, they are worn down by erosion in a few million years. In North America's geologic past, eroded particles from its mountains were carried by streams and dumped into the continent's inland **seas**, some of which were as large as the present-day Mediterranean. Those **rivers** and seas are gone from the continent, but the sediments that filled them remain, like dirt in a bathtub when the **water** is drained. The roots of all the mountain ranges that have ever stood in North America all still exist, and much of the **sand** and **clay** into which the mountains were transformed still exists also, as rock or **soil** formations.

Various parts of North America were formed all over the world, at various times over four billion years, and were brought together and assembled into one continent by the endless process of plate tectonics. What is now called North America began to form in the first two and one-half billion years of Earth's history, a period of time called the **Archean** Eon.

Some geologists speculate that the earth that created the oldest parts of North America barely resembled the middle-aged planet on which we live. The planet of four billion years ago had cooled enough to have a solid crust, and **oceans** of liquid water. But the crust may have included hundreds of small tectonic plates, moving perhaps 10 times faster than plates move today. These small plates, carrying what are now the most ancient rocks, scudded across the oceans of a frantic crazy-quilt planet. Active volcanoes and rifts played a role in rock formation on the Archean Earth. The oldest regions in North America were formed in this hyperactive world. These regions are in Greenland, Labrador, Minnesota, and Wyoming.

In the late Archean Eon, the plates of Earth's crust may have moved at a relatively high speed. Evidence of these wild times can be found in the ancient core of North America. The scars of tectonic events appear as rock outcrops throughout the part of northern North America called the Canadian Shield. One example of this kind of scar, a **greenstone belt**, may be the mangled remains of ancient **island arcs** or rifts within continents. Gold and chromium are found in the greenstone belts, and deposits of copper, zinc, and nickel. Formations of **iron** ore also began to form in the Archean Eon, and **fossils** of microscopic cyanobacteria-the first life on Earth—are found imbedded in them.

North America's little Archean continents slammed together in a series of mountain-building collisions. The core of the modern continent was formed 1,850 million years ago when five of these collisions occurred at once around northeastern Canada. This unified piece of ancient continental crust, called a **craton**, lies exposed at the surface in the Canadian Shield, and forms a solid foundation under much of the rest of the continent.

In the two billion years of the Proterozoic Eon (2,500–570 million years ago), North America's geologic setting became more like the world as we know it. The cores of the modern continents were assembled, and the first collections of continents, or **supercontinents**, appeared. Life, however, was limited to bacteria and algae, and unbreatheable gases filled the atmosphere. Rampant erosion filled the rivers with mud and sand, because no land plants protected Earth's barren surface from the action of rain, **wind**, heat, and cold.

Rich accumulations of both rare and common metallic elements make Proterozoic rocks a significant source of mineral wealth for North America, as on other continents. Chromium, nickel, copper, tin, titanium, vanadium, and platinum ores are found together in the onion-like layers of crystallized **igneous rocks** called layered intrusions. Greenstone belts are mined for copper, **lead**, and zinc, each of which is mixed with sulfur to form a sulfide mineral. Sulfide **minerals** of lead and zinc are found in limestones formed in shallow seas, while mines in the ancient continental river and **delta** sediments uncover buried vanadium, copper, and uranium ores.

During the middle to late Proterozoic Eon, continental collisions attached new pieces of continental crust to North America's southern, eastern, and western borders. Between 30% and 40% of North America joined the continent in the Proterozoic. The crust underlying the continental United States east of Nevada joined the craton, as well as the crust underlying the Sierra Madre Occidental of Sonora, Chihuahua, and Durango in Mexico. The Mazatzal Mountains, whose root outcrop is in the Grand Canyon's inner gorge, rose in these mountain-building times in southern and central North America.

North America experienced the sea washing over its boundaries many times during the three-billion-plus years of its Archean and Proterozoic history. Life had flourished in the shallow tidewater. Algae, a long-term resident of North America, was joined later by worms and other soft-bodied animals. Little is known of early soft-bodied organisms, because they left no skeletons to become fossils. Only a handful of good fossils remain from the entire world's immense **Precambrian** rock record.

Then, about 570 million years ago, several unrelated lineages of shell-bearing sea animals appeared. This was the beginning of the **Phanerozoic Eon** of earth history, which has lasted from 570 million years ago to the present day. Vast seas covered much of North America in the early Phanerozoic, their shorelines changing from one million-year interval to the next. The seas teemed with creatures whose bones and shells we have come to know in the **fossil record**. These oceanic events are memorialized in the layers of stone each sea left behind, lying flat in the continent's heartland, and folded and broken in the cordilleras. Geologists have surveyed the stacked sheets of stone left by ancient North American seas and have made maps of the deposits of each continental sea. The stacked layers are divided into sequences, each named for the sea that laid it down. Each sequence consisted of a slow and complex flooding of the continent. Sea level mountain uplift, the growth of deltas, and other factors continually changed the shape of the continental sea.

The eastern coast of North America was once part of an ancient "Ring of Fire" surrounding an ocean that has disappeared forever from Earth. From Greenland to Georgia, and through the Gulf Coast states into Mexico, the collision of continents raised mountains comparable to the Himalayas and Alps of today. Several ranges were raised up on the eastern border of North America between 480 and 230 million years ago.

Another collision about 450 million years ago created the Acadian mountain range, whose roots are exposed today in Newfoundland. These mountains began to be torn down by rain and wind, and by the time they had worn down to nothing, more than 63,000 cubic miles of sediment made from them had been dumped into the shallow continental sea between New York and Virginia—about the same amount of rock as the Sierra Nevadas of today. The bones of amphibians, the first land animals, are found in the rocks laid down by the streams of East Greenland.

The sleepless crust under North America's Pennsylvanian-age borders tossed and turned in complex ways. Three hundred million years ago, North America sat on the equator, its vast inland sea surrounded by rain **forests** whose fossilized remains are the **coal** deposits of the eastern United States. Small mountain ranges rose out of the sea that covered the center of the continent in Colorado, Oklahoma, and Texas. The Ouachitas stood in the Gulf Coast states, the last great mountain range to stand there. In the eastern United States, the Allegheny Mountain Range stood where the Acadian and Taconic Ranges had stood before.

The Ouachitas welded **South America** to the Gulf coast, at roughly the same time as the Alleghenies welded the East coast of North America to West **Africa**. The Ouachitas and Alleghenies stretched, unbroken, all the way around the eastern and southern coasts of North America. This joining of the world's continents formed Pangaea, the most recent supercontinent in geologic history. Pangaea's 150 million year history ended with the birth of the Atlantic Ocean and the separation of North and South America. As South America and Africa tore away from North America, Florida was left behind, attached to the intersection of the Allegheny and Ouachita Mountains. Another legacy of this cracking of Earth's crust is the New Madrid Fault, which runs through the North American Plate under the Mississippi Valley.

Around 340 million years ago, an offshore island arc, called the Antler Arc, struck the shores about where Nevada and Idaho now are (then the westernmost part of the continent), extending the shoreline of North America a hundred miles westward.

By 245 million years ago, the beginning of the age of dinosaurs, another island arc had run into the American West. The Golconda Arc added a Sumatra-sized piece of land to North America, and the continent bulged out to present-day northern California.

After the Golconda Arc piled onto the West Coast of that time, the crust broke beneath the continent's border, and the ocean's plate ran under North America's west coast. A continental arc was born around 230 million years ago in western North America, and its volcanoes have been erupting frequently from the dawn of the age of dinosaurs (the **Mesozoic Era**) until today.

Several more island arcs struck western North America since the middle **Jurassic Period**. The **granite** mountains of the Sierra Nevada are the roots of one of these island arcs. Landmasses created on the Pacific Plate have been scraped from it. This mechanism is the origin of the West Coast's ranges, the Cascades, and much of British Columbia and Alaska's southern coast.

A range of fault block mountains rose far inland as the continent was squeezed from west to east. The Sevier Mountains stood west of the Cretaceous Period's interior seaway, in what is now Montana, Idaho, Nevada, and Utah. The dinosaurs of that time (80–130 million years ago) left their tracks and remains in the mud and sand worn off these mountains.

In the same manner as large island chains were carried to North America on moving plates of oceanic crust, small pieces of land came to the coasts in this way as well. Numerous "exotic terrains," impacting on the western coasts during the Mesozoic and Cenozoic Eras, added large areas now covered by British Columbia, Washington, Oregon, California, and Mexico. These little rafts of continental crust were formed far from their present location, for the fossils in them are of creatures that lived halfway around the world—but never in North America. A sizeable piece of continental crust—southern Mexico as far south as the Isthmus of Tehuantepec—joined northern Mexico between 180–140 million years ago.

Starting 80 million years ago, new forces began to act on the inland west. Geologists do not know exactly what happened beneath the crust to cause these changes, but the features created on the surface by tectonic action underneath the crust are well known.

At the same time as the Sevier Mountains ceased to rise, a similar range, facing the opposite direction, began to move upward. Earth's upper crust beneath the Rocky Mountain states was shoved westward as the Laramides were forming, lifting the Rocky Mountains for the first time. These first modern Rocky Mountains drained the continent's last great shallow sea of inland North America as they rose. Huge mountains now stood in places where seas had rolled over Colorado, Wyoming, Utah, Idaho, Montana, and Alberta. In Mexico, the Laramides raised the Sierra Madre Occidental, and formed the mineral deposits that enrich Sonora, Chihuahua, Durango, and Zacatecas. In Colorado, Wyoming, and neighboring states, the Rocky Mountains began to erode away, and by 55 million years ago, the first Rockies had disappeared from the surface—the mountains' roots were buried in sediment from the eroded mountaintops. More recent uplift again exposed the Rockies, and Ice Age **glaciers** sculpted their tops into today's sharp peaks.

Twenty-five million years ago, after a quiet interlude, North America's western continental arc awoke, and its abundant volcanoes again added new rock to the continent from British Columbia to Texas and down the mountainous spine of Mexico. The only area in the Southwest in which volcanoes were uncommon was the Colorado Plateau, whose immunity to the tectonic forces around it is still a mystery. Around the borders of the Colorado Plateau's remarkably thick crust, one volcanic catastrophe after another covered the land. In this time, the San Juan Mountains were formed in Colorado. The Rocky Mountains began to slide westward and rose again on the thrust faults beneath them.

Ten million years ago, the Great Basin area of the United States was much shorter when measured east to west than it is today. It was then a mountainous highland. Some geologists propose that Nevada was an alpine plateau like Tibet is today—perhaps more than 10,000 ft (3,048 m) high. Starting then and continuing for five million years, this area began to be pulled apart. Long faults opened in the crust, and mountain-sized wedges slowly fell between ridges that were still standing on the unbroken basement rock miles below. Sediment from the erosion of these new ridges filled the valleys, enabling the valleys to become reservoirs of underground water, or aquifers. The low parts got so low that the area is indeed a basin; water does not flow out of it. Some geologists believe that the Basin and Range province stretches around the Colorado Plateau, into Texas, and extends down the Sierra Madre Occidental as far south as Oaxaca.

Another kind of pulling-apart of the continent happened in New Mexico's Rio Grande Rift. As at the Keewenaw Rift a billion years before, tectonic forces from beneath Earth's crust began pulling the surface apart just as east Africa is being pulled apart today. The broad rift's mountainous walls eroded, and the sediment from that erosion piled up in the ever-widening valley. A new ocean was about to be formed in the southwest. **Lava** poured from fissures in the crust near Taos, New Mexico, filling the valley floor. Also like the Keewenaw Rift, the Rio Grande Rift stopped growing after a few million years, as the tectonic processes ceased pulling the continent apart. The modern Rio Grande was born as a consequence of this rift, and still runs through the rift valley.

A cataclysmic volcanic event happened in Oregon and Washington 17 million years ago. For an unknown reason, perhaps a disturbance deep in Earth's mantle, or a meteor impact, lava began pouring out of cracks in Earth. So much lava poured onto the surface at once that it ran from southeastern Oregon down the Columbia River valley to the Pacific Ocean. Huge cracks in the ground called fissures flooded broad areas with **basalt** lava over about 500,000 years. This flood of basalt is called the Columbia River Plateau. A hot spot, or an upwelling of molten rock from Earth's mantle, appears to have caused the Columbia River Plateau. As the North American Plate moved westward between then and now, the hot spot stayed in one place, scorching holes in Earth's crust under Idaho and erupting the lava that makes the Snake River Plain a fertile farmland. This hot spot is assumed to be the heat source that powers the geysers of Yellowstone National Park.

In the early Jurassic Period, 200 million years ago, the northernmost edge of North America tore away from the continent and began rotating counterclockwise. This part of the continent came to rest to the northwest of North America, forming the original piece of Alaska—its northernmost mountains, the Brooks Range. In the late **Cretaceous Period**, the farthest part of this landmass from North America struck the edge of Siberia, and became the Chukotsk Peninsula. The remaining landmass of Alaska joined North America bit by bit, in the form of exotic terrains. The Aleutians, a classic island arc, formed in the **Tertiary Period**. The about 40 active volcanoes of the Aleutians have erupted numerous times in the twentieth century, including several eruptions in the last decade from Mt. Augustine, Pavlov, Shishaldin, and Mt. Redoubt.

National Bison Range, Montana. ***© Annie Griffiths Belt/Corbis. Reproduced by permission.***

For reasons that are not yet fully understood, Earth periodically enters a time of planet-wide cooling. Large areas of the land and seas are covered in ice sheets thousands of feet thick, which remain unmelted for thousands or hundreds of thousands of years. Today, only Greenland and **Antarctica** lie beneath continent-sized glaciers. But in the very recent geologic past, North America's northern regions, including the entire landmass of Canada, were ground and polished by an oceanic amount of water frozen into a single mass of ice. This ice began to accumulate as the planet's **weather** cooled, and began to stay frozen all year round. As it built up higher and higher, it began to move out from the piled-high center, flowing while still solid.

Vast amounts of Canadian soil and rock, called glacial **till**, rode on the ice sheets as they moved, or surfed slowly before the front of the ice wall. Some of the richest farmland in the United States Midwest and northeast arrived in its present location in this way-as well as boulders that must be removed from fields before plowing. In the unusual geographic conditions following the retreat of the ice sheets, barren soil lay on the landscape, no longer held down by the glacier. Windstorms moved tremendous amounts of this soil far from where the glacier left it, to settle out of the sky as a layer of fertile soil, called loess in German and English. Loess soils settled in the Mississippi and Missouri Valleys, and also Washington, Oregon, Oklahoma, and Texas.

This continental **glaciation** happened seven times over the last 2.2 million years. Warm intervals, some of them hundreds of thousands of years long, stretched between these planetary deep-freezes. Geologists do not agree whether the ice will return or not. Even if the present day is in a warm period between glaciations, tens or hundreds of thousands of years may elapse before the next advance of the ice sheets.

California lies between two different kinds of plate boundaries. To the south, the crust under California is growing; to the north it is shrinking. The part of California that sits on the Pacific Plate between these two forces is moved northward in sudden increments of a few feet which are felt as earthquakes. A few feet at a time, in earthquakes that happen every few decades, the part of California west of the San Andreas Fault will move northward along the coast.

Active faults also exist elsewhere in the United States, in the Midwest and in South Carolina. The last sizeable earthquakes in these regions occurred more than a hundred years ago, and geologists assume that earthquakes will probably occur within the next hundred years. The Pacific Northwest and Alaska, sitting atop active tectonic environments, will certainly be shaken by earthquakes for millions of years to come.

The Great Basin, the western Rocky Mountains, and the United States northeast are all considered tectonically active enough for earthquakes to be considered possible.

North America's volcanic mountain ranges, the Cascades, and the relatively recent Mexican Volcanic Belt, have erupted often in the recent geologic past. These mountains will certainly continue to erupt in the near geologic future.

See also Continental drift theory; Continental shelf; Earth (planet); Faults and fractures; Fossils and fossilization; Ice ages; Orogeny

NORTH POLE • *see* GEOGRAPHIC AND MAGNETIC POLES

NORTHERN LIGHTS • *see* AURORA BOREALIS AND AURORA AUSTRALIALIS

NUCLEAR FISSION

Nuclear fission is a process in which the nucleus of an **atom** splits, usually into two daughter nuclei. Spontaneous fission of uranium and other elements in Earth's interior provides an internal source of heat that drives **plate tectonics**. Fission tracks in mineral **crystals**, a result of spontaneous fission of uranium, can be used in radiometric dating of **rock** and rock layers.

Long before the internal construction of the atom was well understood in terms of protons, neutrons, and electrons, nuclear transformations that resulted in observable **radioactivity** were observed as early as 1896 by French physicist Henri Becquerel (1852–1908). In 1905, British physicist Ernst Rutherford (1871–1937) and American physicist Bertram Borden Boltwood (1870–1927) first used radioactive decay measurements to date **minerals**.

The fission reaction was discovered when a target of uranium was bombarded by neutrons. Fission fragments were shown to fly apart with a large release of energy. The fission reaction was the basis of the atomic bomb first developed by the United States during World War II. After the war, controlled energy release from fission was applied to the development of nuclear reactors. Reactors are utilized for production of **electricity** at nuclear power plants, for propulsion of ships and submarines, and for the creation of radioactive isotopes used in medicine and industry.

The fission reaction was first articulated by two German scientists, Otto Hahn (1879–1968) and Fritz Strassmann (1902–1980). In 1938, Hahn and Strassmann conducted a series of experiments in which they used neutrons to bombard various elements. Bombardment of copper, for example, produced a radioactive form of copper. Other elements became radioactive in the same way. When uranium was bombarded with neutrons, however, an entirely different reaction occurred. The uranium nucleus apparently underwent a major disruption. Accordingly, the initial evidence for the fission process came from chemical analysis. Hahn and Strassmann published a scientific paper showing that small amounts of barium (element 56) were produced when uranium (element 92) was bombarded with neutrons. Hahn and Strassmann questioned how a single neutron could transform element 92 into element 56.

Lise Meitner (1878–1968), a long-time colleague of Hahn who had left Germany due to Nazi persecution, suggested a helpful model for such a reaction. One can visualize the uranium nucleus to be like a liquid drop containing protons and neutrons. When an extra neutron enters, the drop begins to vibrate. If the vibration is violent enough, the drop can break into two pieces. Meitner named this process "fission" because it is similar to the process of cell division in biology. Moreover, it takes only a relatively small amount of energy to initiate nuclear instability.

Scientists in the United States and elsewhere quickly confirmed the idea of uranium fission, using other experimental procedures. For example, a cloud chamber is a device in which vapor trails of moving nuclear particles can be seen and photographed. In one experiment, a thin sheet of uranium was placed inside a cloud chamber. When it was irradiated by neutrons, photographs showed a pair of tracks going in opposite directions from a common starting point in the uranium. Clearly, a nucleus had been photographed in the act of fission.

Another experimental procedure used a Geiger counter, which is a small, cylindrical tube that produces electrical pulses when a radioactive particle passes through it. For this experiment, the inside of a modified Geiger tube was lined with a thin layer of uranium. When a neutron source was brought near it, large voltage pulses were observed, much larger than from ordinary radioactivity. When the neutron source was taken away, the large pulses stopped. A Geiger tube without the uranium lining did not generate large pulses. Evidently, the large pulses were due to uranium fission fragments. The size of the pulses showed that the fragments had a very large amount of energy.

To understand the high energy released in uranium fission, scientists made some theoretical calculations based on German-American physicist Albert Einstein's (1879–1955) famous equation $E=mc^2$. The Einstein equation states that mass (m) can be converted into energy (E) (and, conversely that energy can create mass). The conversion factor becomes c, the velocity of light squared. One can calculate that the total mass of the fission products remaining at the end of the reaction is slightly less than the mass of the uranium atom plus neutron at the start. This decrease of mass, multiplied by c, shows numerically why the fission fragments are so energetic.

Through fission, neutrons of low energy can trigger off a very large energy release. With the imminent threat of war in 1939, a number of scientists began to consider the possibility that a new and very powerful "atomic bomb" could be built from uranium. Also, they speculated that uranium perhaps could be harnessed to replace **coal** or oil as a fuel for industrial power plants.

Nuclear reactions in general are much more powerful than chemical reactions. A chemical change such as burning coal or even exploding TNT affects only the outer electrons of an atom. A nuclear process, on the other hand, causes changes among the protons and neutrons inside the nucleus. The energy of attraction between protons and neutrons is about a million times greater than the chemical binding energy between atoms. Therefore, a single fission bomb, using nuclear energy, might destroy a whole city. Alternatively, nuclear electric power plants theoretically could run for a whole year on just a few tons of fuel.

In order to release a substantial amount of energy, many millions of uranium nuclei must split apart. The fission process itself provides a mechanism for creating a so-called chain reaction. In addition to the two main fragments, each fission event produces two or three extra neutrons. Some of these can enter nearby uranium nuclei and cause them in turn to fission, releasing more neutrons, which causes more fission, and so forth. In a bomb explosion, neutrons have to increase very rapidly, in a fraction of a second. In a controlled reactor, however, the neutron population has to be kept in a steady state. Excess neutrons must be removed by some type of absorber material (e.g., neutron absorbing control rods).

In 1942, the first nuclear reactor with a self-sustaining chain reaction was built in the United States. The principal designer was Enrico Fermi (1901–1954), an Italian physicist and the 1938 Nobel Prize winner in physics. Fermi emigrated to the United States to escape from Benito Mussolini's fascism. Fermi's reactor design had three main components: lumps of uranium (the fuel), blocks of **carbon** (the moderator, which slows down the neutrons), and control rods made of cadmium (an excellent neutron absorber). Fermi and other scientists constructed the first nuclear reactor pile at the University of Chicago. When the pile of uranium and carbon blocks was about 10 ft (3 m) high and the cadmium control rods were pulled out far enough, Geiger counters showed that a steady-state chain reaction had been successfully accomplished. The power output was only about 200 watts, but it was enough to verify the basic principle of reactor operation. The power level of the chain reaction could be varied by moving the control rods in or out.

General Leslie R. Groves was put in charge of the project to convert the chain reaction experiment into a usable military weapon. Three major laboratories were built under wartime conditions of urgency and secrecy. Oak Ridge, Tennessee, became the site for purifying and separating uranium into bomb-grade material. At Hanford, Washington, four large reactors were built to produce another possible bomb material, plutonium. At Los Alamos, New Mexico, the actual work of bomb design was started in 1943 under the leadership of the physicist J. Robert Oppenheimer (1904–1967).

The fissionable uranium isotope, uranium-235, constitutes only about 1% of natural uranium, while the non-fissionable neutron absorber, uranium-238, makes up the other 99%. To produce bomb-grade, fissionable uranium-235, it was necessary to build a large isotope separation facility. Since the plant would require much electricity, the site was chosen to be in the region of the Tennessee Valley Authority (TVA). The technology of large-scale isotope separation involved solving many difficult, unprecedented problems. By early 1945, the Oak Ridge Laboratory was able to produce kilogram amounts of uranium-235 purified to better than 95%.

An alternate possible fuel for a fission bomb is plutonium-239. Plutonium does not exist in nature but results from radioactive decay of uranium-239. Fermi's chain reaction experiment had shown that uranium-239 could be made in a reactor. However, to produce several hundred kilograms of plutonium required a large increase from the power level of Fermi's original experiment. Plutonium production reactors were constructed at Hanford, Washington, located near the Columbia river to provide needed cooling **water**. A difficult technical problem was how to separate plutonium from the highly radioactive fuel rods after irradiation. This was accomplished by means of remote handling apparatus that was manipulated by technicians working behind thick protective **glass** windows.

With uranium-235 separation started at Oak Ridge and plutonium-239 production under way at Hanford, a third laboratory was set up at Los Alamos, New Mexico, to work on bomb design. In order to create an explosion, many nuclei would have to fission almost simultaneously. The key concept was to bring together several pieces of fissionable material into a so-called critical mass. In one design, two pieces of uranium-235 were shot toward each other from opposite ends of a cylindrical tube. A second design used a spherical shell of plutonium-239, to be detonated by an "implosion" toward the center of the sphere.

The first atomic bomb was tested at an isolated **desert** location in New Mexico on July 16, 1945. President Truman then issued an ultimatum to Japan that a powerful new weapon could soon be used against them. On August 8, a single atomic bomb destroyed the city of Hiroshima with over 80,000 casualties. On August 11, a second bomb was dropped on Nagasaki with a similar result. Japan surrendered three days later to end WWII.

The decision to use the atomic bomb has been vigorously debated over the years. It ended the war and avoided many casualties that a land invasion of Japan would have cost. However, the civilians who were killed by the bomb and the survivors who developed radiation sickness left an unforgettable legacy of fear. The horror of mass annihilation in a nuclear war is made vivid by the images of destruction at Hiroshima. The possibility of a ruthless dictator or a terrorist group obtaining nuclear weapons is a continuing threat to world peace. In late 2001, in the aftermath of the terrorist attacks on the World Trade Center in New York, evidence became public of terrorist attempts to acquire weapons-grade uranium and the other technology related to bomb production.

The first nuclear reactor designed for producing electricity was put into operation in 1957 at Shippingsport, Pennsylvania. From 1960 to 1990, more than 100 nuclear power plants were built in the United States. These plants now generate about 20% of the nation's electric power. Worldwide, there are over 400 nuclear power stations.

The most common reactor type is the pressurized water reactor (abbreviated PWR). The system operates like a coal-burning power plant, except that the firebox of the coal plant is replaced by a reactor. Nuclear energy from uranium is released in the two fission fragments. The fuel rod becomes very hot because of the cumulative energy of fissioning nuclei. A typical reactor core contains hundreds of these fuel rods.

The immense destructive power of the atom unleased in the atomic bomb, forming its distinctive mushroom-shaped cloud. © *UPI/Corbis-Bettmann. Reproduced by permission.*

Water is circulated through the core to remove the heat. The hot water is prevented from boiling by keeping the system under pressure (i.e., creating superheated steam).

The pressurized hot water goes to a heat exchanger where steam is produced. The steam then goes to a turbine, which has a series of fan blades that rotate rapidly when hit by the steam. The turbine is connected to the rotor of an electric generator. Its output goes to cross-country transmission lines that supply the electrical users in the region with electricity. The steam that made the turbine rotate is condensed back into water and is recycled to the heat exchanger.

Safety features at a nuclear power plant include automatic shutdown of the fission process by insertion of control rods, emergency water-cooling for the core in case of pipeline breakage, and a concrete containment shell. It is impossible for a reactor to have a nuclear explosion because the fuel enrichment in a reactor is intentionally limited to about 3% uranium-235, while almost 100% pure uranium-235 is required for a bomb. Regardless, nuclear power plants remain potential targets for terrorists who would seek to cause massive and lethal release of radioactivity by compromising the containment shell.

The fuel in the reactor core consists of several tons of uranium. As the reactor is operated, the uranium content gradually decreases because of fission, and the radioactive waste products (the fission fragments) build up. After about a year of operation, the reactor must be shut down for refueling. The old fuel rods are pulled out and replaced. These fuel rods, which are very radioactive, are stored under water at the power plant site. After five to 10 years, much of their radioactivity has decayed. Only those materials with a long radioactive lifetime remain, and eventually they will be stored in a suitable underground depository.

There are vehement arguments for and against nuclear power. As with other forms of producing electricity, nuclear power generation can have serious and unintended environmental impacts. The main objections to nuclear power plants are the fear of possible accidents, the unresolved problem of nuclear waste storage, and the possibility of plutonium diversion for weapons production by a terrorist group. The issue of waste storage becomes particularly emotional because leakage from a waste depository could contaminate ground water. Opponents of nuclear power often cite accidents at the Three Mile Island nuclear power plant in the United States and the massive leak at the Chernobyl nuclear plant in the USSR (now the Ukraine) as evidence that engineering or technical failures can have long lasting and devastating environmental and public health consequences.

The main advantage of nuclear power plants is that they do not cause **atmospheric pollution**. No smokestacks are needed because nothing is being burned. France initiated a large-scale nuclear program after the Arab oil embargo in 1973 and has been able to reduce its **acid rain** and **carbon dioxide** emissions by more than 40%. Nuclear power plants do not contribute to **global warming**. Shipments of fuel are minimal so the hazards of coal transportation and oil spills are avoided.

Environmentalists remain divided in their opinions of nuclear power. It is widely viewed as a hazardous technology but there is growing concern about atmospheric pollution and dwindling fossil fuel reserves that may make increased usage of nuclear power more widespread.

See also Atom; Atomic mass and weight; Atomic number; Atomic theory; Chemical elements; Chemistry; Dating methods; Energy transformations; Radioactive waste storage (geological considerations); Radioactivity

Nuclear fusion

Nuclear fusion is the process by which two light atomic nuclei combine to form one heavier atomic nucleus. As an example, a proton (the nucleus of a hydrogen **atom**) and a neutron will, under the proper circumstances, combine to form a deuteron (the nucleus of an atom of "heavy" hydrogen). In general, the mass of the heavier product nucleus is less than the total mass of the two lighter nuclei. Nuclear fusion is the initial driving process for the process of nucelosynthesis.

When a proton and neutron combine, the mass of the resulting deuteron is 0.00239 **atomic mass** units (amu) less than the total mass of the proton and neutron combined. This "loss" of mass is expressed in the form of 2.23 MeV (million electron volts) of kinetic energy of the deuteron and other particles and as other forms of energy produced during the reaction. Nuclear fusion reactions are like **nuclear fission** reactions, therefore, in that some quantity of mass is transformed into energy. This is the reason stars "shine" (i.e., radiate tremendous amounts of electromagnetic energy into **space**).

The particles most commonly involved in nuclear fusion reactions include the proton, neutron, deuteron, a triton (a proton combined with two neutrons), a helium-3 nucleus (two protons combined with a neutron), and a helium-4 nucleus (two protons combined with two neutrons). Except for the neutron, all of these particles carry at least one positive electrical charge. That means that fusion reactions always require very large amounts of energy in order to overcome the force of repulsion between two like-charged particles. For example, in order to fuse two protons with each other, enough energy must be provided to overcome the force of repulsion between the two positively charged particles.

As early as the 1930s, a number of physicists considered the possibility that nuclear fusion reactions might be the mechanism by which energy is generated in the stars. No familiar type of chemical reaction, such as combustion or oxidation, could possibly explain the vast amounts of energy released by even the smallest star. In 1939, the German-American physicist Hans Bethe worked out the mathematics of energy generation in which a proton first fuses with a **carbon** atom to form a nitrogen atom. The reaction then continues through a series of five more steps, the net result of which is that four protons are consumed in the generation of one helium atom.

Bethe chose this sequence of reactions because it requires less energy than does the direct fusion of four protons and, thus, is more likely to take place in a star. Bethe was able to show that the total amount of energy released by this sequence of reactions was comparable to that which is actually observed in stars.

The Bethe carbon-cycle is by no means the only nuclear fusion reaction. A more direct approach, for example, would be one in which two protons fuse to form a deuteron. That deuteron could then fuse with a third proton to form a helium-3 nucleus. Finally, the helium-3 nucleus could fuse with a fourth proton to form a helium-4 nucleus. The net result of this sequence of reactions would be the combining of four protons (hydrogen nuclei) to form a single helium-4 nucleus. The only net difference between this reaction and Bethe's carbon cycle is the amount of energy involved in the overall set of reactions.

Other fusion reactions include D-D and D-T reactions. The former stands for deuterium-deuterium and involves the combination of two deuterium nuclei to form a helium-3 nucleus and a free neutron. The second reaction stands for deuterium-tritium and involves the combination of a deuterium nucleus and a tritium nucleus to produce a helium-4 nucleus and a free neutron.

The term "less energy" used to describe Bethe's choice of nuclear reactions is relative, however, since huge amounts of energy must be provided in order to bring about any kind of fusion reaction. In fact, the reason that fusion reactions can occur in stars is that the temperatures in their interiors are great enough to provide the energy needed to bring about fusion. Since those temperatures generally amount to a few million degrees, fusion reactions are also known as thermonuclear (thermo = heat) reactions. The heat to drive a thermonuclear reaction is created during the conversion of mass to energy during other thermonuclear reactions.

The understanding that fusion reactions might be responsible for energy production in stars brought the accompanying realization that such reactions might be a very useful source of energy for human needs. The practical problems of building a fusion power plant are incredible, however, and scientists are still a long way from achieving a containment vessel or field in which controlled fusion reactions could take place. A much simpler challenge, however, is to construct a "fusion power plant" that does not need to be controlled, that is, a fusion bomb.

Scientists who worked on the first fission (atomic) bomb during World War II were aware of the potential for building an even more powerful bomb that operated on fusion principles. A fusion bomb uses a fission bomb as a trigger (a source of heat and pressure to create a fusion chain reaction. In the microseconds following a fission explosion, fusion begins to occur within the casing surrounding the fission bomb. Protons, deuterons, and tritons begin fusing with each other, releasing more energy, and initiating other fusion reactions among other hydrogen isotopes. The original explosion of the fission bomb would have ignited a small star-like reaction in the casing surrounding it.

From a military standpoint, the fusion bomb had one powerful advantage over the fission bomb. For technical reasons, there is a limit to the size one can make a fission bomb. However, there is no technical limit on the size of a fusion bomb—one simply makes the casing surrounding the fission bomb larger. On August 20, 1953, the Soviet Union announced the detonation of the world's first fusion bomb. It was about 1,000 times more powerful than was the fission bomb that had been dropped on Hiroshima less than a decade earlier. Since that date, both the Soviet Union (now Russia) and the United States have stockpiled thousands of fusion bombs and fusion missile warheads. The manufacture, maintenance, and destruction of these weapons remain a source of scientific and geopolitical debate.

With research on fusion weapons ongoing, attempts were also being made to develop peaceful uses for nuclear fusion. The containment vessel problems remain daunting because at the temperatures at which fusion occurs, known materials vaporize instantly. Traditionally, two general approaches have been developed to solve this problem: magnetic and inertial containment.

One way to control that plasma is with a **magnetic field**. One can design such a field so that a swirling hot mass of plasma within it can be held in a specified shape. Other proposed methods of control include the use of suspended microballoons that are then bombarded by the laser, electron, or atomic beam to cause implosion. During implosion, enough energy is produced to initiate fusion.

The production of useful nuclear fusion energy depends on three factors: **temperature**, containment time, and energy release. That is, it is first necessary to raise the temperature of the fuel (the hydrogen isotopes) to a temperature of about 100 million degrees. Then, it is necessary to keep the fuel suspended at that temperature long enough for fusion to begin. Finally, some method must be found for tapping off the energy produced by fusion.

In the late twentieth century, scientists began to explore approaches to fusion power that departed from magnetic and inertial confinement concepts. One such approach was called the PBFA process. In this machine, electric charge is allowed to accumulate in capacitors and then discharged in 40-nanosecond micropulses. Lithium ions are accelerated by means of these pulses and forced to collide with deuterium and tritium targets. Fusion among the lithium and hydrogen nuclei takes place, and energy is released. However, the PBFA approach to nuclear fusion has been no more successful than has that of more traditional methods.

In March of 1989, two University of Utah electrochemists, Stanley Pons and Martin Fleischmann, reported that they had obtained evidence for the occurrence of nuclear fusion at room temperatures (i.e., cold fusion). During the electrolysis of heavy **water** (deuterium oxide), it appeared that the fusion of deuterons was made possible by the presence of palladium electrodes used in the reaction. If such an observation could have been confirmed by other scientists, it would have been truly revolutionary. It would have meant that energy could be obtained from fusion reactions at moderate temperatures. The Pons-Fleischmann discovery was the subject of immediate and intense scrutiny by other scientists around the world. It soon became apparent, however, that evidence for cold fusion could not consistently be obtained by other researchers. A number of alternative explanations were developed by scientists for the apparent fusion results that Pons and Fleischmann believed they had obtained and most researchers now assert that Pons and Fleischmann's report of "cold fusion" was an error and that the results reported were due to other chemical reactions that take place during the electrolysis of the heavy water.

See also Atom; Atomic mass and weight; Atomic number; Atomic theory; Big Bang theory; Chemical elements; Chemistry; Electricity and magnetism; Energy transformations; Radioactive waste storage (geological considerations); Radioactivity

Nuclear winter

Nuclear winter is a theory estimating the global climatic consequences of a nuclear war: prolonged and worldwide cooling and darkening caused by sunlight-blocking smoke and soot entering the atmosphere. During the Cold War after World War II, the concern about nuclear weapons was increasing all over the world. Initially, only the danger of radioactive fallout was recognized, but later also the possible environmental effects of a nuclear war became the subject of several studies. The term nuclear winter was first defined and used by American astronomer **Carl Sagan** (1934–1996) and his group of colleagues in their 1983 article (later referred to as the TTAPS-article, from the initials of the authors' family names). This article was the first one to take into consideration not only the direct damage, but also the indirect effects of a nuclear war.

The basic assumption during a nuclear war is that the exploding nuclear warheads would create huge fires, resulting in smoke and soot from burning cities and **forests** being emitted into the **troposphere** in vast amounts. This would block the sun's incoming radiation from reaching the surface of Earth, causing cooling of the surface temperatures. The smoke and soot soon would rise because of their high **temperature**, allowing them to drift at high altitudes for weeks without being washed out. Finally, the particles would settle in the Northern Hemisphere mid-latitudes as a black particle cloud belt, blocking sunshine for several weeks. The darkness and cold, combined with nuclear fallout radiation, would kill most of Earth's vegetation and animal life, which would lead to starvation and diseases for the human population surviving the nuclear war itself. At the same time, the upper troposphere temperatures would rise because the smoke would absorb sunlight and warm it up, creating a temperature inversion, which would keep **smog** at the lower levels. Another possible consequence is that nuclear explosions would produce nitrogen oxides, which would damage the protective ozone layer in the **stratosphere**, thus allowing more ultraviolet radiation to reach the earth's surface.

Although the basic findings of the original TTAPS-article have been confirmed by later reports, some later studies report a lesser degree of cooling would occur, only around 25 degrees of temperature drop and only for weeks instead of the initially estimated months. According to different scenarios, depending on the number of nuclear explosions, their spatial distribution, targets, and many other factors, this cloud of soot and dust could remain for many months, reducing sunlight almost entirely, and decrease average temperatures to as low as –40°C in the Northern Hemisphere continents. There are other studies, that mention the possibility of a not so severe nuclear winter as originally estimated, hence it is named a nuclear fall. Other researchers even talk about nuclear summer, stating that a worldwide warming would follow a nuclear war because of the many small contributions to the **green-**

house effect from **carbon dioxide**, **water** vapor, **ozone**, and various aerosols entering the troposphere and stratosphere. What all scenarios agree on is that a nuclear war would have a significant effect on the atmosphere and **climate** of the earth and, consequently, many aspects of life such as food production or energy consumption would be drastically effected.

Opponents of the nuclear winter theory argue that there are many problems with the hypothesized scenarios either because of the model's incorrect assumptions (e.g., the results would be right only if exactly the assumed amount of dust would enter the atmosphere, or the model assumes uniformly distributed, constantly injected particles), or because important effects, processes and/or feedback mechanisms are not taken into consideration (e.g., the moderating effects of the **oceans**, or small-scale processes are not included, or the biological effects are not addressed), or simply because there are many uncertainties involved in the estimates. The topic even at present day remains controversial, because the exact level of damage, along with the extent and duration of the effects, cannot be agreed upon with full confidence.

See also Atmospheric circulation; Atmospheric composition and structure; Atmospheric inversion layers; Atmospheric lapse rate; Atmospheric pollution

Nucleate and nucleation

A nucleation event is the process of **condensation** or aggregation (gathering) that results in the formation of larger drops or **crystals** around a material that acts as a structural nucleus around which such condensation or aggregation proceeds. Moreover, the introduction of such structural nuclei can often induce the processes of condensation or crystal growth. Accordingly, nucleation is one of the ways that a phase transition can take place in a material.

In addition to an importance in explaining a wide variety of geophysical and geochemical phenomena—including crystal formation—the principles of nucleation were used in **cloud seeding weather** modification experiments where nuclei of inert materials were dispersed into **clouds** with the hopes of inducing condensation and rainfall.

During a phase transition, a material changes from one form to another. For example, **ice** melts to form liquid **water**, or a liquid boils to form a gas. Phase transitions occur due to changes in **temperature**. Certain transitions occur smoothly throughout the whole material, while others happen suddenly at different points in the material. When the transitions occur suddenly, a bubble forms at the point where the transition began, with the new phase inside the bubble and the old phase outside. The bubble expands, converting more and more of the material into the new phase. The creation of a bubble is called a nucleation event.

Phase transitions are grouped into two categories, known as first order transitions and second order transitions. Nucleation events happen in first order transitions. In this kind of transition, there is an obstacle to the transition occurring smoothly. A prime example is condensation of water vapor to form liquid water. Condensation requires that many water molecules collide and stick together almost simultaneously. This requirement for simultaneous collisions presents a temporary but measurable barrier to the formation of a bubble of liquid phase. Following formation, the bubble expands as more water molecules strike the surface of the bubble and are absorbed into the liquid phase. Because of the obstacle to the phase transition, a liquid may exist in its gaseous state even though the temperature is well below the boiling point.

A liquid in this state is said to be supercooled. Accordingly, in order for a liquid to be supercooled, it must be pure, because dust or other impurities act as nucleation centers. If the liquid is very pure, however, it may remain supercooled for a long time. A supercooled state is termed metastable due to its relatively long lifetime.

The other type of phase transition is called second order, and it proceeds simultaneously throughout the whole material. An example of a second order transition is the **melting** of a solid. As the temperature rises, the magnitude of the thermal vibrations of molecules causes the solid to break apart into a liquid form. As long as the solid is in thermal equilibrium and the melting occurs slowly, the transition takes place at the same time everywhere in the solid, rather than taking place through nucleation events at isolated points.

See also Bowen's reaction series; Chemical bonds and physical properties; Chemical elements; Chemistry; Crystals and crystallography; Minerals; Precipitation

Nuee ardent

A nuee ardent, or "glowing cloud," is a type of explosive volcanic eruption characterized by a dense, very hot mass of ash, gasses, and volcanic material traveling down a volcanic slope at high velocity. A nuee ardent, also called a pyroclastic flow, can reach speeds of 450 mi/hr (720 km/hr) and temperatures of 1,500°F (830°C). They can travel as much as 124 mi (200 km) from the source and cover areas as large as 12,000 mi^2 (20,000 km^2). The volume of transported material can be as large as 250 mi^3 (1,000 km^3) or more.

Nuees ardentes can be destructive and deadly. An eruption of Mount Pelee, Martinique, in 1920 produced a nuee ardent that, within minutes, killed 30,000 inhabitants of a nearby town. In 1982, eruptions and pyroclastic flows from El Chichòn **Volcano**, southeastern Mexico, killed 2000 people, in villages as far as 5 mi (8 km) from the source.

Nuees ardentes are typically composed of two parts; a basal or lower part that hugs the ground and contains the larger volcanic fragments, and an upper part composed of a turbulent mix of hot ash and gas. When cooled, the gas and ash can form **pumice**, a very porous and lightweight volcanic **rock**.

Pyroclastic flows commonly are produced either by the downslope movement of fragments ejected upwards as they fall back, or by explosive eruptions that pulverize existing rock and throw the rock, ash, and gas in a more horizontal direction. In many cases, pyroclastic flows result from a combination of mechanisms. The sequence of events leading up to and during the 1980 eruption of Mount St. Helens illustrates some of the complexity of the process. An increase in seismic activity beneath the volcano in March marked the beginning phase of the eruption. **Magma** injected into the cone of the volcano at high pressure created a bulge on the north flank that grew outward at rates as high as 8.2 ft (2.5 m) per day. For the next two months only minor eruptions occurred. On May 18, a 5.1 magnitude **earthquake** triggered a series of massive landslides of material over the bulge. Remove of this material destabilized the north slope, which failed explosively releasing a pyroclastic flow horizontally. The eruptions flattened trees and killed wildlife in a 212 mi^2 (550 km^2) **area**. The eruption removed the upper 1,300 ft (400 m) of the cone and left a crater 2,000 ft (625 m) deep, 1.7 mi (2.7 km) long, and 1.3 mi (2.0 km) wide.

O

Obsidian

Obsidian is volcanic **glass** with a chemical composition similar to that of **rhyolite** or **granite**. It is most commonly black, although greenish to reddish and banded varieties also occur. In addition to its dark color and glassy luster, obsidian is characterized by bowl-shaped or concave conchoidal fractures.

In the hands of a skilled practitioner, obsidian can be worked into razor sharp stone tools such as knives and spear points. Thus, obsidian tools are often found during archaeological excavations. Obsidian is also used for scalpel blades by some modern surgeons because obsidian blades can be sharper and thinner than their steel counterparts.

Obsidian is formed from silica-rich **lava** that cools too quickly for **crystals** to grow. The result is a glass that does not possess the regular crystal structure of **minerals** in rhyolite and granite, even though all three have the same general chemical composition. The lack of crystal structure is also responsible for the conchoidal fractures characteristic of obsidian. The dark color of obsidian is due to the presence of **iron** oxide minerals distributed throughout the otherwise clear glass. Magnetite produces black obsidian, whereas more highly oxidized hematite produces reddish varieties. Because its silica content makes it extremely viscous, the rhyolitic lava that freezes into obsidian is extremely resistant to flow and obsidian is most commonly found in small dome-like bodies very close to volcanic vents.

Obsidian artifacts and outcrops can be dated using the thickness of a hydration rind that forms when **water** vapor diffuses into a freshly broken surface. The age of the fresh surface is estimated by comparing the rind thickness with the results of a mathematical model of water vapor diffusion through obsidian. Rind thickness was originally estimated using optical microscopes, but recently developed techniques such as secondary ionization mass spectrometry provide more accurate measurements of the rind thickness. When applied to artifacts, the method gives the age of fractures produced when a human made the tool; when applied to **rock** outcrops, it gives the age of fractures presumably produced during or shortly after the eruption.

See also Dating methods; Minerals; Pumice; Volcanic eruptions; Volcanic vent; Volcano

Ocean circulation and currents

Ocean circulation is the large, connected system of **water** movements in the **oceans**, including not only the surface movement, but also the slow, deep-water circulations. While the surface currents are caused by the winds and a geographically uneven **solar energy** distribution, the deep ocean currents are the result of sinking and upwelling water, and termohaline (**temperature** and salinity) differences. The atmospheric and oceanic circulations are linked together, and this global ocean circulation system transfers heat from low to higher latitudes, making the oceans responsible for about 40% of the global heat transport. Although there are similarities between the atmospheric and oceanic circulations, ocean currents move slower than winds, with a speed of about several kilometers per day to a few kilometers per hour.

The system of surface currents resembles the major **wind** patterns of the atmosphere. However, the surface currents do not move exactly in the same direction, but at a 45-degree angle to wind direction, with a water speed of less than 3% of the original wind speed. It turns to right on the Northern Hemisphere, and to left on the Southern Hemisphere. Going down to a deeper layer, it tends to turn even further compared to the surface, and going even deeper in the water (until about 109 yd [100 m], where the depth of no motion is reached), eventually, the angle between the surface winds and the deep movements reaches 90 degrees. This is called the Ekman spiral.

The surface currents in each ocean can be described with a simplified scheme (although the actual currents differ in depth, size or exact location). The dominant features are the subtropical gyres (semi-closed circles of the currents), and the boundary currents, both in the east and west ocean basins. The gyres are centered around 30 degrees **latitude** in the major ocean basins, and rotate clockwise in the Northern Hemisphere, and counterclockwise in the Southern Hemisphere because of the earth's **rotation** and the change in wind direction with latitude. The basic ocean circulation looks like a westward flow near the equator, which moves warm water to higher latitudes at the western ocean basins. At about 35 degrees of both North and South latitudes, the major currents turn in an eastern direction, to carry warm water to higher latitudes. To balance this poleward movement at the eastern ocean basins, cold water has to return in the currents. In the Southern Hemisphere, the gyre turns in the opposite direction, and because there is no land between 35 and 60 degrees latitude, a strong, cold current around **Antarctica** completes the system.

The clockwise rotating gyre in the northern part of the Atlantic Ocean consists of the warm North Equatorial Current flowing near the equator, joining the western boundary of the warm **Gulf Stream**, which turns towards east and becomes the warm North Atlantic Drift, while the cold Labrador Current, West Greenland Drift, and East Greenland Drift return cold water from the North. The eastern boundary, the southward Canary Current close to West **Africa** also brings cold water towards the equator, closing the North Atlantic Ocean gyre. In the southern part of the Atlantic Ocean, in the counterclockwise gyre, the warm South Equatorial Current joins the warm Brazil Current going south towards Antarctica at the east coast of **South America**, meeting the cold Falkland Current, and the cold, West Wind Drift going east around Antarctica. The gyre is closed by the cold Benguela Current going north at the southwestern coast of Africa, while the eastward flowing warm Equatorial Countercurrent connects the gyres from the north and South Atlantic Oceans.

The system of currents in the Pacific Ocean is very similar to the Atlantic currents. The clockwise rotating gyre in the northern part of the Pacific Ocean consists of the warm North Equatorial Current flowing westward north of the equator, joining the western boundary warm Kuroshio Current, which turns towards the East and becomes the warm North Pacific Drift, while the cold Oyashio Current returns cold water from the North. At the coast of Alaska loops the warm Alaska Current. The eastern boundary California Current, close to the coast of California, also brings cold water towards the equator, closing the North Atlantic Ocean gyre. This is where upwelling, the rising of cold water replacing surface water that drifts away due to the winds, occurs. Although it brings cold water, low **clouds**, and sometimes **fog** in summer, the nutrient-rich, cold water helps the fishing industry. South of this gyre but still in the Northern Hemisphere, the eastward flowing warm North Equatorial Countercurrent can be found. In the southern part of the Pacific Ocean, the warm, westward South Equatorial Current flows opposite of the eastward warm South Equatorial Countercurrent, and continues towards the South at the east coast of **Australia**. While the West Wind Drift is heading east around the Antarctica, the cold Peru or Humboldt Current going northward at the southwestern coast of South America closes this gyre.

The system of surface ocean currents is simpler in the Indian Ocean, because instead of the "8-shaped" two gyres, only one is present, influenced by seasonally changing winds. It consists of the warm North Equatorial Current, the South Equatorial Countercurrent, the South Equatorial Current, and the eastward, circumpolar current around Antarctica, the cold West Wind Drift.

See also Gulf stream; Oceans and seas

OCEAN TRENCHES

A deep-sea trench is a narrow, elongate, v-shaped depression in the ocean floor. Trenches are the deepest parts of the ocean, and the lowest points on Earth, reaching depths of nearly 7 mi (10 km) below sea level. These long, narrow, curving depressions can be thousands of miles in length, yet as little as 5 mi (8 km) in width. Deep-sea trenches are part of a system of tectonic processes termed subduction. Subduction zones are one type of **convergent plate boundary** where either an oceanic or a continental plate overrides an oceanic plate. A trench is formed where the oceanic plate dives below (is subducted by) the (less dense) overriding plate. They are associated with a certain type of volcanic chain called an island arc and with zones of high **earthquake** activity. The trenches can extend for thousands of kilometers parallel to the volcanoes of the **island arcs** located on the overriding plate. Examples include the Aleutian islands, an arc bordered to the south by the Aleutian trench, and the Marianas, bordered by the Mariana trench, the deepest in the world. Along the western coast of **South America**, the Peru-Chile trench marks where the Nazca plate is being subducted beneath the South American plate. The volcanic activity and uplift of the Andes mountains are a result of the subduction process.

Trenches and active subduction zones are found along much of the Ring of Fire, a zone of volcanism and earthquake activity that borders the Pacific ocean. The tectonic processes of subduction form the trenches and island arcs and are also responsible for the earthquake activity. Major earthquakes occur along the plunging boundary between the subducting and overriding plates. This boundary is called a **Benioff zone**. Many scientists believe that the volcanic island arcs are formed from magmas produced by the **partial melting** of the descending and/or the overriding plate. Considerable volcanic activity worldwide is the result of subduction.

Most geologists argue that the size of the earth has not changed significantly in the past several hundred million years. According to tectonic theory, new **crust** is generated along the divergent plate boundaries such as the mid-Atlantic ridge at rates on the order of a few centimeters a year. As the new crust is created, an equal amount must be destroyed at roughly the same rate for the size of Earth to remain unchanged. Subduction along the trenches of the convergent plate boundaries appears to maintain that balance.

See also Plate tectonics

OCEANIC CRUST • *see* CRUST

OCEANOGRAPHY

Oceanography, the study of the **oceans**, is a combination of the sciences of biology, **chemistry**, **geology**, **physics**, and **meteorology**.

Ancient explorers of the ocean were sailors and fishermen who learned about marine biology by observing the sea life and discovering when it was most plentiful. They observed the effects of **wind**, currents, and **tides**, and learned how to use them to their advantage or to avoid them. These early humans discovered that salt could be retrieved from seaweed and grasses.

Polynesians combined what they knew about the **weather**, winds, and currents to investigate the Pacific Ocean, while the Phoenicians, Greeks, and Arabs explored the Mediterranean Sea. The early Greeks in general and Herodotus (484–428 B.C.) in particular believed that the world was round. Herodotus performed studies of the Mediterranean, which helped sailors of his time. He was able to take depth measurements of the sea floor by using the fathom as a unit of measure, which was the length of a man's outstretched arms. Today the fathom has been standardized to measure 6 ft (1.8 m) in length.

Aristotle (384–322 B.C.) also studied marine life. One of his contemporaries, a geographer by the name of Poseidarius, studied the tides and their relationship to the phases of the **Moon**.

Pliny the Elder (A.D. 23–79) was a Roman naturalist who discovered, by studying marine biology, that some organisms had medicinal uses. One of his predecessors, Seneca (4 B.C.–A.D. 65) predicted that interest in the oceans would fade and "a huge land would be revealed," foreshadowing the age of exploration and discovery of the New World. A period of about 1,000 years followed when no new studies were done until the fifteenth century. Christopher Columbus performed oceanographic studies on his voyages.

Captain **James Cook**, the explorer, was one of the first scientists to study the oceans' natural history. A surge in scientific studies took place in the seventeenth century, during which scientists tried for the first time to combine the **scientific method** with sailors' knowledge.

U.S. Navy lieutenant Matthew Fontaine Maury (1806–1873) is considered the father of modern oceanography. It was during the nineteenth century that the name was given to the science.

In December 1872, the British ship HMS *Challenger* began a four-year journey, which lasted until May of 1876. This was the first major study of the ocean approached from a scientific viewpoint, and since that time significant strides have been made. The advent of submersible vehicles allowed for first-hand study of the ocean floor and the **water** above it. In 1900, Prince Albert of Monaco established two institutes to study oceanography.

Two areas of focus within oceanography today are physical and chemical oceanography. Physical oceanography is the study of ocean basin structures, water and sediment transportation, and the interplay between ocean water, air and sediments and how this relationship affects processes such as tides, upwellings, **temperature**, and salinity. Findings aid oceanic engineers, coastal planners, and military defense strategists. Current areas of research also include oceanic circulation, especially ocean currents and their role in predicting weather-related events, and changes in sea level and climate.

Chemical oceanography investigates the chemical make-up of the oceans. Many studies in this **area** are geared to understanding how to use the oceans' resources to produce food for a growing population.

Even though the study of the oceans has entered the technological age, there is much still unexplored and unknown. Oceanographers of the 1990s use satellites to study changes in salt levels, temperature, currents, biological events, and transportation of sediments. Deep-sea studies are underway using unmanned robotic submersible craft to study ocean floor hydrothermal vents, **sea-floor spreading**, and subduction zones that lie beneath the ocean floor. As scientists develop new technologies, the future will open new doors to the study of oceanography.

See also Bathymetric mapping; Beach and shoreline dynamics; Continental drift theory; Continental shelf; Convergent plate boundary; Coral reefs and corals; Delta; Depositional environments; Desalination; Douglas sea scale; Dunes; Earth (planet); Earth science; Earth, interior structure; El Nino and La Nina phenomena; Geothermal deep ocean vents; Global warming; Gulf of Mexico; Gulf stream; Guyots and atolls; Hawaiian Island formation; History of exploration II (Age of exploration); Hydrogeology; Hydrologic cycle; Hydrothermal processes; Icebergs; Land and sea breeze; Latitude and longitude; Marine transgression and marine regression; Mid-ocean ridges and rifts; Ocean circulation and currents; Ocean trenches; Oceans and seas; Offshore bars; Petroleum; RADAR and SONAR; Red tide; Saltwater encroachment; Seawalls and beach erosion; Tides; Wave motions

OCEANS AND SEAS

Oceans are large bodies of **saltwater** connected together, unevenly covering 70% of the earth, and containing about 97% of Earth's **water** supply as salt water. The five major oceans, in descending order of size by **area**, are the Pacific Ocean (64 million square miles, about 165 million square kilometers), Atlantic Ocean (33 million square miles, about 85 million square kilometers), Indian Ocean (28 million square miles, about 70 million square kilometers), Southern Ocean (almost 8 million square miles, about 20 million square kilometers), and Arctic Ocean (5,000,000 square miles, almost 13 million square kilometers).

The Pacific Ocean is the largest ocean, covering about one third of the earth's surface, which is a bigger area than all of the continents combined. The Pacific Ocean is not only the home of the highest mountain on Earth (Mauna Kea, 33,476 ft, or 10,203 m), but also has the deepest trench (Mariana trench, 36,198 ft, or 11,033 m deep). The Atlantic Ocean is the second largest, covering 20% of Earth's surface. It is also the youngest among the five oceans, and it is the ocean where the most shipping occurs. The third largest ocean is the Indian Ocean, which provides an important trade route between **Africa** and **Asia**, and is home to the most expressed monsoon system. The next in size is the Southern Ocean, which surrounds **Antarctica**, and was officially named in the year 2000 by the International Hydrographic Organization. Finally, the Arctic Ocean is the smallest ocean of all five, but contains the widest **continental shelf**.

Seas are the smaller bodies of salt water connecting the oceans, which can be partially or entirely enclosed by land. Examples of seas adjacent to the oceans include the South China Sea (the largest sea), Caribbean Sea, Mediterranean Sea, Bering Sea, **Gulf of Mexico**, Hudson Bay, Gulf of California, Sea of Japan, and Persian Gulf. Examples of inland seas (or **lakes**) are the Caspian Sea, Sea of Galilee, or Dead Sea.

Earth's oceans and seas are unique in that there is no other planet in our **Solar System** that has liquid water on its surface. Life on Earth began in the oceans, and today they still are the home to some of the most spectacular wildlife in the world. Most of the oceans' wildlife is located in the upper ocean layers, which contain about two percent of the oceans' volume. The oceans also play a significant role in the earth's water cycle. Oceans are a large source of water vapor for the atmosphere, which is important in heat transportation in the atmosphere in the form of latent heat. Additionally, the oceans gather water at the end of the water cycle not only from **precipitation**, but also from surface **runoff** and return flow from **rivers** and as **groundwater** flows from land. The oceans are major reservoirs in the **carbon** cycle. In order to double the current atmospheric **carbon dioxide**, it would be necessary to release only 2% of the carbon currently stored in the oceans.

Oceans produce important and widespread effects on Earth's atmosphere, **weather**, and **climate**. Oceans and land exhibit different heating and cooling properties; **solar energy** penetrates deeper into water than into land, and water can circulate that absorbed heat easily into deeper layers. Because the specific heat of water (the amount of heat required to raise the **temperature** of one gram of water by one degree Celsius) is higher than that of land, it takes about five times more energy to warm up water by one degree Celsius than to equally heat a **rock**. Consequently, oceans not only warm more slowly than land, but also cool more slowly. Oceans, therefore, act as a giant heat reservoir, which heats the land during winter and cools it during summer, moderating the climate of the land located next to it.

Oceans not only moderate the climate of adjacent areas by absorbing and storing solar energy, they also distribute heat between lower and higher latitudes by a global, interconnected system of ocean currents. An example of the climatic effects of oceans on lands is the **Gulf Stream**, which is not only part of the heat redistributing process by carrying warm waters towards higher latitudes, but also brings mild air to the British Isles and Northwest **Europe**, causing a significantly milder climate than it would normally have according to its latitudes.

Interesting examples of the interaction between the oceans and the atmosphere are the **El Niño and La Niña phenomena** patterns. Along with the Southern Oscillation, El Niño and La Niña influence not only nearby areas in the Pacific Ocean, but effect the entire global climate system, along with the ecologies and economies of many countries worldwide, from New Zealand to the United States.

As the oceans and seas are a significant part of the atmosphere and the climate system with many interactions and feedback mechanisms, there is a recent debate about their role in anthropogenic climate change, and also about the possible consequences of climate change for the oceans. Rising sea levels could occur due to both the thermal expansion of the oceans, and from the **partial melting** of the **polar ice** because of **global warming**. Although there is a debate between scientists about the possibility and the intensity of this prediction, if rising sea levels happen at any level, it could be potentially devastating for many coastal cities around the world.

See also El Niño and La Niña phenomena; Hydrologic cycle; Ocean circulation and currents

ODÉN, SVANTE N. F. (1924-1986)

Swedish soil scientist and chemist

Svante Odén is known because of his efforts in the 1960s to publicize the problem of acid rain and connect it with the deterioration of **forests** and fisheries. **Acid rain** was known to European scientists in the seventeenth century, named by Robert Angus Smith (1817–1884) in 1872, and accurately described by Eville Gorham (b. 1925) in the 1950s, but only became a popular environmental concern after the Swedish National Science Research Council published Odén's findings in 1968 as "The Acidification of Air and Precipitation and its Consequences in the Natural Environment" in *Ecology Community Bulletin No. 1.*

Odén was born on April 29, 1924 in Oscar's Parish, Stockholm, Sweden. After passing his upper secondary school examination in 1943, he earned a master of science degree in agriculture in 1954 and a licentiate in agriculture in 1957. He taught **soil** science and ecological **chemistry** at the Swedish University of Agricultural Sciences, Uppsala, for the rest of his career. He disappeared and was presumed dead in 1986.

In 1961, Odén began collecting field data about the relationship between bodies of fresh **water** and surrounding soils. This research, following that of two other Swedes, Carl Gustav Arvid Rossby (1898–1957) and Erik Eriksson at the **Meteorology** Institute of Stockholm University, led him toward analyzing the chemistry of rain and eventually to the retroactive monitoring of air pollutants emitted from several parts of **Europe**. He soon correlated these findings with meteorological evidence and concluded that industrial air pollution, especially sulfur dioxide, from Britain and Germany adversely affected the rain that fell in Sweden.

Comparing his data with that of fisheries inspector Ulf Lunden, Odén became alarmed that the unchecked continuation of this trend could destroy Swedish fish populations and reduce Swedish crop yields. Accordingly, he took the unusual step of publishing his results, not first in a scientific journal, but in a newspaper, the October 24, 1967, issue of *Dagens Nyheter*. The Swedish government and people immediately mobilized behind Odén, and in 1972 the United Nations Conference on the Human Environment, held in Stockholm, reported extensive fish kills, forest and farm damage, and human health problems in Sweden, all related to acid rain.

Sweden was the first country where the study of acid rain was taken seriously. Since much of the Swedish economy centers around fish—and since acid rain can decrease fish populations—Odén's publications in both the scientific and the popular press galvanized public opinion and rallied scientists and politicians to defend and protect Swedish interests. Odén remained a tireless advocate for policy reform regarding acid rain, but sometimes overdramatized the case, for example by claiming that Western Europe was waging "chemical warfare" against Scandinavia. He conducted a lecture tour of the United States in 1971 and spoke at the Nineteenth International Limnological Congress in Winnipeg, Manitoba, in 1974. These trips inspired many American and Canadian scientists to begin studying acid rain.

See also Atmospheric pollution; Forests and deforestation; Freshwater; Groundwater; Hydrologic cycle; Lakes; Precipitation; Rivers; Water pollution; Wind

OFFSHORE BARS

Bars are elongate ridges and mounds of **sand** or gravel deposited beyond a shoreline by currents and waves. The term offshore bar has been used to describe both submerged bars, and emergent islands separated from a shoreline by a lagoon, features more correctly identified as **barrier islands**. Submerged bars are only exposed at low tide, if ever, while barrier islands remain at least partially exposed, even at high tide. Because of this ambiguity, the term offshore bar is no longer used as a descriptor in coastal geomorphology.

Longshore, tidal, and fluvial currents construct submerged bars in shallow **water** coastal environments. The amount of unconsolidated sediment available in a shore-zone system, called its sand budget, determines the number of bars and other depositional features that form along the coastline. A shore-zone system's dominant mode of sediment transport controls the shapes and orientations of its depositional forms, including the types of submerged bars.

Along wave-dominated shorelines, longshore currents carry sand along the shoreface and deposit it in submerged bars parallel to the shore. Because waves almost never approach a shoreline perpendicularly, there is an angle between approaching and retreating waves, and sediment grains move down the beach in a zigzag pattern called **longshore drift**. Wave refraction and surf-zone interaction also create strong longshore currents that transport and deposit sediment parallel to the shore in deeper water. Along coastlines where **tides** are the dominant sediment transport mechanism, bidirectional tidal currents build submerged tidal bars perpendicular to the coastline. Tides also transport sand into and out of coastal lagoons through barrier island inlets, forming submerged mounds called ebb-tidal and flood-tidal deltas on either side of the inlets. Submerged bars also form where **rivers** enter the ocean. When sediment-laden fresh water from a confined channel discharges into an unconfined salt-water ocean basin, the current slows and deposits its coarse-grained sediment, or bed load, at the river mouth. The resulting submerged mound of sand and gravel is called a channel mouth bar.

Wave action, tidal currents and fluvial input each influence a shore-zone to some extent, and the different types of submerged bars—longshore bars, tidal bars, and channel mouth bars—usually coexist in a single depositional environment. Combinations of shore-parallel and shore-perpendicular processes create bars with intermediate curved or obliquely oriented morphologies. All types of submerged bars typically obstruct natural and man-made outlets into the ocean, and are well-known navigational hazards.

See also Beach and shoreline dynamics; Bedforms

OIL • *see* PETROLEUM

OLIGOCENE EPOCH

In **geologic time**, the Oligocene Epoch occurs during the **Tertiary Period** (also sometimes divided or referred to in terms of a Paleogene Period and a Neogene Period) of the **Cenozoic Era** of the **Phanerozoic Eon**. The Oligocene Epoch is the third epoch in the Tertiary Period (in the alternative, the latest (most recent) epoch in the Paleogene Period).

The Oligocene Epoch lasts from approximately 34 million years ago (mya) to 23 mya.

The Oligocene Epoch is further subdivided into (from earliest to most recent) Rupelian (34 mya to 29 mya) and Chattian (29 mya to 23 mya) stages. The Oligocene Epoch was preceded by the **Eocene Epoch** and was followed by the **Miocene Epoch**.

Large impact craters dating to the end of the Eocene Epoch and the start of the start of the Oligocene Epoch are evident in Russia (Popigal crater) and in the Chesapeake Bay of the United States. Craters dating to the end of the Oligocene Epoch and start of the Miocene Epoch can be studied in Northwest Canada and in Logancha, Russia. Volcanic activity also increased during the Oligocene Epoch.

The Oligocene Epoch **climate** was warmer than the modern climate. Evidence of the start of a generalized cooling trend is, however, in accord with the rise of warm-blooded mammals as the dominant land species. The Oligocene Epoch continued to present the slow climatic changes that allowed continued development and diversification of mammals.

Notable finds in the **fossil record** that date to the Oligocene Epoch include Branisella monkeys. The first **fossils** of Australian marsupials date to Oligocene Epoch fossil beds. Roses and orchids appeared by the end of the Oligocene Epoch.

See also Archean; Cambrian Period; Cretaceous Period; Dating methods; Devonian Period; Evolution, evidence of; Fossils and fossilization; Historical geology; Holocene Epoch; Jurassic Period; Mesozoic Era; Mississippian Period; Ordovician Period; Paleocene Epoch; Paleozoic Era; Pennsylvanian Period; Pleistocene Epoch; Pliocene Epoch; Precambrian; Proterozoic Era; Quaternary Period; Silurian Period; Supercontinents; Triassic Period

OLIVINE

The olivines are a class of common silicate **minerals** named for their greenish or olive color. They are glassy, fracture conchoidally (i.e., along curving cleavage surfaces), and are often found in meteorites and in **mafic igneous rocks** such as **basalt**, dunite, gabbro, and **peridotite**.

Like the feldspars, the olivines consist of a **silicon** (Si) and **oxygen** (O) framework interspersed with atoms of a metallic additive, usually magnesium (Mg) or **iron** (Fe) but sometimes calcium (Ca). Forsterite (Mg_2SiO_4) is olivine containing no additive but magnesium, while fayalite (Fe_2SiO_4) is olivine containing no additive but iron. Between these two minerals there is a continuum of olivines containing varying percentages of forsterite and fayalite in solid solution. Olivine with 10–30% fayalite is defined as chrysolite; 30–50%, hyalosiderite; 50–70%, hortonolite; and 70–90%, ferrohortonite. The remainder in all cases is forsterite. An olivine with less than 10% fayalite is classified simply as forsterite, while one with less than 10% forsterite is classified simply as fayalite. Confusingly, the term chrysolite is also sometimes used as a synonym for olivine in general.

Magnesium-rich olivine is the majority ingredient of the **rock** peridotite, the main component of Earth's upper mantle. The interface between the underside of the **crust** and the olivine-rich peridotite of the mantle, is called the **Mohorovicic discontinuity** or Moho for short, and is of great importance in **seismology**. The Moho is generally located at a depth of 3.7 mi (6 km) beneath the **oceans** and 19 mi (30 km) beneath the continents.

Compression of olivine's atomic structure to its spinel phases under extreme pressure causes a second seismic discontinuity at approximately 250 mi (400 km) and a third at approximately 420 mi (670 km). These olivine–spinel phase transitions affect the mechanical properties of the whole mantle, which in turn determine the convective flow processes that drive **plate tectonics** and thus much of the geological history Earth.

Implosive collapse of olivine to spinel in slabs of oceanic crust being subducted rapidly into the mantle may be violent enough to generate deep-focus earthquakes.

Olivine is readily altered to the mineral serpentine by the hydration of its crystal structure by hot (400–800°C) **water**. This process occurs along **mid-ocean ridges** and other places where mafic rock is exposed to superheated water.

Peridot (pronounced PER-ih-do) is a transparent variety of olivine valued as a gemstone. Olivine is used industrially as a lining in furnaces due to its heat resistance.

See also Bowen's reaction series; Earth, interior structure; Mantle plumes

OPHIOLITE SUITES

Since the late 1970s, the term ophiolite has been used to describe sections of oceanic **crust** and upper mantle, along with **sedimentary rocks** deposited on the sea floor, emplaced as thrust slices onto continental **lithosphere**. This process, called obduction, results from continent-continent collision following subduction of oceanic crust and the closure of **oceans** or back-arc basins. A typical ophiolite suite comprises (from base to top):

- Harzburgite, dunite, **peridotite,** and pyroxenite (ultramafic **igneous rocks** composed of varying amounts of **olivine** and pyroxene) representing upper, oceanic mantle. They are commonly altered to the slippery, shiny green-black **rock** serpentinite. Serpentinite is named after its resemblance to the skin of a snake. Indeed, the word ophiolite is derived from the Greek words *ophis*, meaning snake, and *lithos*, meaning stone, because of the presence of serpentinites.
- Iron-titanium and magnesium gabbros. Layered, cumulate-textured gabbros dominate the basal section of oceanic crust. Higher-level gabbros tend to be more massive and associated with plagiogranites.
- Sheeted **mafic** dykes, feeders to overlying volcanics.
- Pillow basalts, formed as **lava** flows on the sea floor. The characteristic pillow shapes result from rapid cooling when lava contacts seawater.
- Radiolarian **chert** and **limestone**, graywacke and mudstone or their metamorphic equivalents **marble**, quartzite and mica schist.

Oceanic crust in ophiolites is produced at spreading centers, above zones of subduction in extensional or transtensional arcs (supra-subduction zone ophiolites), and along some leaky **transform faults**. Ophiolites that lack ultramafic, upper mantle rocks may represent obducted slices of seamounts or oceanic plateaus.

Many ophiolites have undergone high pressure-low **temperature**, blueschist **metamorphism** in subduction zones prior to their emplacement. Blueschists are named after blue-colored glaucophane and other sodium-rich amphiboles formed in rocks of appropriate composition. Eclogite facies metamophism occurs when rocks are subducted to greater depth. In eclogites, pyroxene, olivine, and **plagioclase** recrystallize to sodium-rich pyroxene and garnet. During collision, blueschists plus or minus eclogites are thrust as a series of imbricate slices onto lower-grade, continental rocks. Ophiolites may be overprinted by greenschist facies metamorphic assemblages and exhumed during collapse of a thrust-thickened orogen.

Ophiolites in orogenic (mountain building) belts represent sutures between two continental plates. Their recognition is therefore important in tectonic reconstructions. Ophiolites host a range of mineral deposits. Ultramafic and gabbroic rocks may contain deposits of chromium or platinum-group elements. Chrysotile asbestos occurs in serpentinites. Copper, zinc, cobalt and nickel sulfides (marine exhalatives) may occur in economic amounts. Some ophiolites host shear controlled epithermal or mesothermal gold mineralization.

See also Plate tectonics; Subduction zone

Ordovician Period

In **geologic time**, the Ordovician Period, the second period of the **Paleozoic Era**, covers the time roughly 505 million years ago (mya) until 438 mya. The name Ordovician derives from that of the Ordovices, an ancient British tribe.

The Ordovician Period spans three epochs. The Lower Ordovician Epoch is the most ancient, followed in sequence by the Middle Ordovician Epoch, and the Upper Ordovician Epoch. The Ordovician Period is divided chronologically (from the most ancient to the most recent) into the Tremadocian, Arenigian, Llanvirnian, Llandeilian, Caradocian and Ashgillian stages.

Much of the continental **crust** that exists now had already been formed by the time of the Ordovician Period and the forces driving **plate tectonics** actively shaped the fusing continental landmasses. Near the margins of the continental landmasses, extensive **orogeny** (mountain building) allowed the development of **mountain chains**.

The **fossil record** provides evidence to support the demarcation of the preceding **Cambrian Period** from the Ordovician Period. Drastic changes of sea levels resulted in massive extinctions among marine organisms. In accord with a mass extinction, many **fossils** dated to the Cambrian Period are not found in Ordovician Period formations.

The fossil record establishes that vertebrates existed during the Ordovician Period. As with the Cambrian Period, the Ordovician Period ended with a mass extinction of nearly a third of all species. This mass extinction, approximately 438 mya, marked the end of the Ordovician Period and the start of the **Silurian Period**.

Although there is no evidence of an occurrence equivalent to the **K-T event**, it is possible that an impact from a large meteorite may have been responsible for the mass extinction marking the end of the Cambrian Period and start of the Ordovician Period. Impact craters dating to the Ordovician Period have been identified in **Australia**.

See also Archean; Cenozoic Era; Cretaceous Period; Dating methods; Devonian Period; Eocene Epoch; Evolution, evidence of; Fossils and fossilization; Historical geology; Holocene Epoch; Jurassic Period; Mesozoic Era; Miocene Epoch; Mississippian Period; Oligocene Epoch; Paleocene Epoch; Pennsylvanian Period; Phanerozoic Eon; Pleistocene Epoch; Pliocene Epoch; Proterozoic Era; Quaternary Period; Tertiary Period; Triassic Period

Orientation of strata

Strata are layers of **rock**, whether of sedimentary (e.g., **sandstone** or **limestone**) or of extrusive igneous (e.g., **lava** flow) origin. Sedimentary strata are formed when Earth's **gravity** acts upon particles being transported by **wind**, **water**, or **ice** and pulls them down to the earth's surface, where they form a layer. Sedimentary strata also may form from debris flows and viscous mud flows that move according to gravity. Extrusive igneous strata are formed when Earth's gravity acts upon particles within viscous molten rock and pulls them into a sheet-like or tabular mass called a lava flow. Extrusive igneous strata can also form when pyroclastic material is blown out of a **volcano** and falls to Earth, forming a layer of volcanic debris. All such layers obey the laws of **superposition**, original horizontality, and lateral continuity. Of these laws, original horizontality is most pertinent to this discussion because this law predicts the original orientation of rock strata (horizontal). Horizontal is the original orientation of essentially all rock strata.

For this reason, if rock strata is found with some orientation other than horizontal, a force has acted upon the rock strata to re-orient it (change it from its original state). Re-orientation of rock strata occurs principally due to tectonic forces acting within Earth's **crust**.

Orientation of rock strata is defined as the attitude of layers of rock in three-dimensional **space**. In order to measure the orientation of rock strata, geologists use a system of measure consisting of two different compass bearings and an angular measurement. The first compass bearing is that of **strike**. Strike is defined as the compass bearing, relative to north, of the line of intersection between an imaginary horizontal plane and a dipping rock stratum. The second compass bearing is that of **dip** direction, which is the direction of maximum inclination down from strike (dip direction is always perpendicular to strike). The angular measurement is called dip magnitude, which is the smaller of two angles formed by the intersection of an imaginary horizontal plane and a dipping rock stratum. The compass bearings and angular measurement are made in the field by using a hand-held device called a Brunton compass. In subsurface strata, e.g., layers drilled during **petroleum** exploration, orientation of strata (i.e., **strike and dip** measurements) are made by electrical sensor devices lowered into the drill hole on a cable.

Studies of the orientation of rock strata are useful for helping understand the origin of crustal deformation in general and mountain building in particular. Orientation of rock strata can help geologists deduce the direction and type of stresses in Earth's crust that produced the observed deformation. Orientation of rock strata is studied also for purely practical purposes, e.g., in predicting the distribution of **hydrocarbons**, mineral deposits, or **groundwater** within dipping strata. For example, groundwater will flow down the dip direction within an inclined **aquifer** stratum.

Orientation of strata has a strong effect upon surficial **landforms** developed by **weathering** of stratified rocks. For example, flat-topped hills and mountains (e.g., butte, mesa, and pinnacle) have a flat-lying (i.e., horizontally oriented) layer of relatively insoluble rock on top, which protects weaker, underlying strata from **erosion**. Another example of a landform created by rock strata is the flatiron, which is an asymmetrical hill formed by the upturned edges of a dipping layer (rock stratum).

Orientation of rock strata can pose problems for engineering geologists because contacts between steeply dipping strata (i.e., strata with high dip-magnitude angles) can act as planes of weakness. Such planes can promote rockslides and rock falls, both of which can take lives and damage property due to sudden movement of rock material along detachments formed by weak contacts between rock strata.

The orientation of rock strata can be found on most geological maps. The standard symbol for orientation of rock strata is the strike-dip symbol, which consists of a long bar, parallel to strike, and a short spike, parallel to dip direction. This symbol is two to three millimeters in size on most maps and is printed directly upon the mapped layer or stratum possessing the measured dip and strike.

See also Folds; Plate tectonics

Origin of Life

The origin of life has been a subject of speculation in all known cultures and indeed, all have some sort of creation idea that rationalizes how life arose. In the modern era, this question has been considered in terms of a scientific framework, meaning that it is approached in a manner subject to experimental verification as far as that is possible. Geological formations contain a wealth of information concerning the origin of life on Earth and provide abundant evidence of the relationships between physical and biological evolutionary processes.

Radioactive dating provides evidence that that Earth formed at least 4.6 billion years ago. Yet, the earliest known **fossils** of microorganisms, similar to modern bacteria, are only about 3.5–3.8 billion years old. The earlier prebiotic era (i.e., before life began) left no direct record, and so it cannot be determined from the geologic record exactly how life arose. It is possible, however, to at least demonstrate the kinds of abiotic reactions that may have led to the formation of living systems through laboratory experimentation. It is generally accepted that the development of life occupied three stages: First, chemical evolution, in which simple geologically occurring molecules reacted to form complex organic polymers. Second, collections of these polymers self organized to form replicating entities. At some point in this process, the transition from a lifeless collection of reacting molecules to a living system probably occurred. The third process following organization into simple living systems was biological evolution, which ultimately produced the complex web of modern life.

The underlying biochemical and genetic unity of organisms suggests that life arose only once, or if it arose more than once, the other life forms must have become rapidly extinct. All organisms are made of chemicals rich in the same kinds of carbon-containing, organic compounds. The predominance of **carbon** in living matter is a result of its tremendous chemical versatility compared with all the other elements. Carbon has the unique ability to form a very large number of compounds as a result of its capacity to make as many as four highly stable covalent bonds (including single, double, triple bonds) combined with its ability to form covalently linked carbon-carbon (C—C) chains of unlimited length. The same 20 carbon and nitrogen containing compounds called amino acids combine to make up the enormous diversity of proteins occurring in living things. Moreover, all organisms have their genetic blueprint encoded in nucleic acids, either DNA or RNA. Nucleic acids contain the information needed to synthesize specific proteins from their amino acid components. Enzymes, catalytic proteins, which increase the speed of specific chemical reactions, regulate the activity of nucleic acids and other biochemical functions essential to life, while other proteins provide the structural framework of cells. These two types of molecules, nucleic acids and proteins, are essential enough to all organisms that they, or closely related compounds, must also have been present in the first life forms.

Scientists suspect that the primordial Earth's atmosphere was very different from what it is today. The modern atmosphere with its 79% nitrogen, 20% **oxygen**, and trace quantities of other gases is an oxidizing atmosphere. The primordial atmosphere is generally believed not to have contained significant quantities of oxygen, having instead rather small amounts of gases such as carbon monoxide, methane, ammonia and sulphate in addition to the **water**, nitrogen and **carbon dioxide** that it still contains today. With these combinations of gases, the atmosphere at that time would have been a reducing atmosphere providing the hydrogen atoms for the synthesis of compounds needed to create life. In the 1920s, the Soviet scientist Aleksander Oparin (1894–1980) and the British scientist J.B.S. Haldane (1892–1964) independently suggested that ultraviolet (UV) light, which today is largely absorbed by the ozone layer in the higher atmosphere, or violent **lightning** discharges, caused molecules of the primordial reducing atmosphere to react and form simple organic compounds (e.g., amino acids, nucleic acids and sugars). The possibility of such a process was demonstrated in 1953 by Stanley Millar and **Harold Urey**, who simulated the effects of lightning storms in a primordial atmosphere by subjecting a refluxing mixture of water, methane, ammonia and hydrogen to an electric discharge for about a week. The resulting solution contained significant amounts of water-soluble organic compounds including amino acids.

The American scientist, Norman H. Horowitz proposed several criteria for living systems, saying that they all must exhibit replication, catalysis and mutability. One of the chief features of living organisms is their ability to replicate. The primordial self-replicating systems are widely believed to have been nucleic acids, like DNA and RNA, because they could direct the synthesis of molecules complementary to themselves. One hypothesis for the evolution of self-replicating systems is that they initially consisted entirely of RNA. This idea is based on the observation that certain species of ribosomal RNA exhibit enzyme-like catalytic properties and also all nucleic acids are prone to mutation. Thus RNA can demonstrate the three Horowitz criteria and the primordial world may well have been an "RNA world." A cooperative relationship between RNA and protein could have arisen when these self-replicating protoribosomes evolved the ability to influence the synthesis of proteins that increased the efficiency and accuracy of RNA synthesis. All these ideas suggest that RNA was the primary substance of life and the later participation of DNA and proteins were later refinements that increased the survival potential of an already self-replicating living system. Such a primordial pond where all these reactions were evolving eventually generated compartmentalization amongst its components. How such cell boundaries formed is not known, though one plausible theory holds that membranes first arose as empty vesicles whose exteriors served as attachment sites for entities such as enzymes and chromosomes in ways that facilitated their function.

See also Atmospheric chemistry; Cambrian Period; Carbon dating; Earth (planet); Evolution, evidence of; Evolutionary mechanisms; Evolution; Geologic time; Miller-Urey experiment; Precambrian; Uniformitarianism

OROGENY

The terms orogeny and orogenesis are synonymous for tectonic processes that result in the formation of **mountain chains**. Orogeny and orogenesis are derived from the Greek words *oros*, meaning mountain, and *geneia*, meaning born. Orogeny can also have a time connotation when used in naming periods of intense tectonic activity and mountain building, whilst orogenesis is only used to describe the process. For example, "Grenvillian Orogeny" is used to refer to the period of orogenesis in many parts of the world approximately one billion years ago, synchronous with the collision between Laurentia and Baltica in the Grenville Province of **North America**. Areas in which mountain building have occurred in the past, although no mountains may remain today, are called orogens or orogenic belts.

Orogeny or orogenesis most commonly involves the collision between two continental **lithospheric plates** or the collision between a continental plate and an island arc. When a continent collides with an island arc, arc rocks are thrust over continental **crust**. In some orogens (including the Grenville Province), collision between a continental plate and one or more **island arcs** has occurred prior to final continent-continent collision. Collision follows subduction of oceanic **lithosphere** beneath one continent (or possibly even both continents), or a continent and an island arc. Subduction progressively closes the ocean that previously separated them. Continental crust may be partially subducted following ocean closure. Evidence for this comes from deep reflection seismic profiles and seismic tomographic images. High-pressure metamorphic **minerals** such as coesite and glaucophane in exhumed continental crust formed during subduction. In continent-continent collisions, oceanic upper mantle and crust, along with sediments deposited on the seafloor and in the trench above the **subduction zone**, are thrust over continental crust as a series of imbricate slices in a process called obduction. The former oceanic material (collectively called an ophiolite) marks the suture between the two plates. Ophiolitic rocks may be overthrust by continental crust of the colliding plate. Continued shortening (resulting from protracted convergence following plate collision) further deforms the orogen. Early-formed thrust slices are folded and thrusts developed with both the same and opposite sense of tectonic transport to earlier thrusts. In transpressional orogens (e.g., the Mesoproterozoic Albany Mobile Belt of Western **Australia** or the Tertiary Spitsbergen fold and thrust belt, Svalbard, Norway), lateral displacement between plates occurs in addition to folding and thrusting. Transpressional orogens are created by the oblique convergence between two continental plates and do not necessitate closure of an ocean.

Both upper and exhumed lower crustal rocks in collisional and transpressional orogens frequently display evidence for late orogenic extension, or cyclic changes between regional shortening and extension. Uplifted, upper crust may collapse along normal faults and extensional detachments as a result of gravitational instability. Gravitational collapse of upper crustal rocks may take place even though regional shortening continues at deeper crustal levels. Faults formed during gravitational collapse may trend parallel, orthogonal or oblique to earlier-formed thrust faults. Extension in the middle to lower crust in a thrust-thickened orogen may result from convective removal or peeling away (delamination) of part of a lithospheric root. Extension may also be induced by gravitational instabilities within the lower crust. Where the thickened crustal root comprises buoyant material, lateral and upward **gravity** spreading of the lower crust (root rebound) results in horizontal flattening and formation of shallow extensional ductile **shear zones**. Late- to post-orogenic granitoids that intrude as large batholiths or as sheet-like bodies along older shear zones result from **partial melting** of the extended crust. In **Precambrian** orogens, AMCG-suite rocks (anorthosite-mangerite-charnockite-granodiorite) are also commonly the product of late-orogenic extension.

See also Mountain chains; Plate tectonics

ORTELIUS, ABRAHAM (1527-1598)

Belgian cartographer, geographer, and archaeologist

Commercially the most successful cartographer of his time, Ortelius satisfied the ever increasing demand for more and better maps during the Age of Exploration and pushed accurate mapmaking toward the status of fine art.

Ortelius is variously known as "Oertel," "Wortel," "Wortels," "Ortel," "Ortels," "Ortello," "Ortellius," and even "Portello." Born on either April 14 or 4, 1527, in Antwerp,

Belgium, the son of a rich merchant from Augsburg, Germany, he began selling maps as a young boy and joined the chart colorers guild of St. Luke when he was 20. Some historians speculate that he may have had to work to help support his family after his father died in 1535, but there is little evidence for that speculation. It was common then for children of the merchant class to begin learning a trade at a very early age.

Ortelius's mercantile shrewdness, artistic talent, and technical mapmaking skill made him successful by the 1550s. He contracted as an engraver for one of the most important early printers, Christophe Plantin (1520–1589). He traveled widely, conducting business throughout Germany and the Low Countries. In 1559 and 1560, he toured France with his new friend **Gerhard Mercator** (1512–1594), who encouraged him to make original maps, rather than just copies. Inspired by Mercator and urged by map aficionado and merchant Gillis Hooftman (1520?–1581) and Hooftman's protégé, Johan Radermacher de Oude (1538–1617), Ortelius envisioned greater works and began to create them.

In 1570, Plantin published Ortelius's major work, *Theatrum orbis terrarum* (Theatre of the lands of the world), which contained 70 maps on 53 sheets with accompanying text. A book of maps was not called an "atlas" until 1585, when Mercator coined the term for that purpose, but the *Theatrum* was the first modern atlas of the world. Subsequent editions had more maps and it was in print long into the next century, edited by the Flemish engraver Joan Babtista de Vrients (1552–1612) after Ortelius's death. Among the most useful features of the *Theatrum* is Ortelius's historical synopsis of eighty-seven of his predecessors. In many cases, his words are all that is now known about these early cartographers.

In 1575, King Philip II of Spain rewarded Ortelius for the *Theatrum* by appointing him royal geographer, but only after the influential Spanish Benedictine monk, Benedictus Arias Montanus (1527–1598), had assured Philip that Ortelius was a Roman Catholic. Philip's patronage made Ortelius rich.

During Ortelius's lifetime, his atlases sold better than Mercator's. He was not an innovative mathematical cartographer like Mercator, but was a better artist, and an expert at editing, updating, and accurately presenting the data of explorers, geographers, and previous cartographers as far back as **Ptolemy** (fl. ca. 130). A broadly learned man, he kept company with scholars and corresponded with the Catholic humanist Justus Lipsius (1547–1606). In 1577, he toured the British Isles and met the most prominent British and Irish geographers. Among his other publications was the *Thesaurus geographicus* (Geographical treasury) in 1587. He died in Antwerp on either July 4 or June 28, 1598.

See also Cartography; History of exploration II (Age of exploration); Mapping techniques

OWEN, SIR RICHARD (1804-1892)

English biologist

Sir Richard Owen was a comparative anatomist, paleontologist, and zoologist who originated the term "dinosaurus." After insisting that a group of **fossils** he observed belonged to a separate taxonomic order of extinct reptiles unrecognized at the time, Owen named the animal by combining the Greek words "deinos" for terrible and awe-inspiring with "sauros" meaning lizard. Owen noticed that the dinosaur sacral vertebrae were fused, thereby allowing the animal exceptional strength.

Owen was the Hunterian Professor of Comparative Anatomy and Physiology at the Royal College of Surgeons, London, from 1836 to 1856. He then became the superintendent of the Natural History Section of the British Museum in London in 1849, and was superintendent of the entire museum from 1856 to 1883. He is best known as an influential paleontologist during an exciting time in the nineteenth century, when the fossils of extinct dinosaurs were first discovered and their significance in chronicling Earth's biological history began to be understood. In coining the word "dinosaur," Owen and was largely responsible for kindling the dinosaur mania that began in the mid-nineteenth century, and continues today. His first, great popularizing event was the erection of a series of life-sized models of dinosaurs and other extinct creatures at the Crystal Palace in London in 1854, which created an absolute sensation among the Victorian population. Remarkably, a formal dinner partly was held within the body of one of the giant dinosaurs, a model of Iguanodon, as it was nearing completion. Owen sat at the "head" of the table, in the head of the dinosaur.

Along with his extensive work on extinct species of vertebrates, Owen also conducted some important studies of living animals. One of his works involved the confirmation of the earlier observations of James Paget, that the deadly parasite *Trichina spiralis* was the cause of trichinosis in humans, and was transmitted by eating inadequately cooked pork.

Owen was a strong opponent of the theory of Charles Darwin concerning natural selection as a critical force of **evolution**. Throughout his life, Owen refused to accept Darwinian evolution, but modified his anti-evolution views by the mid-1840s. Because of his extensive observations in comparative vertebrate anatomy, Owen eventually asserted that all vertebrate animals evolved from the same archetype, or prototype, that was inspired and created by God. Darwin's most outspoken ally, Thomas Henry Huxley, sparked a 20-year debate with Owen on the principles of evolution that exceeded scientific circles, capturing the attention of Victorian writers, artists, philosophers, and the public at large.

See also Evolution, evidence of; Evolutionary mechanisms; Fossil record; Fossils and fossilization

OUTER CORE • *see* EARTH, INTERIOR STRUCTURE

OXBOW LAKES • *see* CHANNEL PATTERNS

Oxidation-reduction reaction

Oxidation-reduction reactions are significant to many geochemical reactions (e.g., the production of **natural gas**). In addition, oxidation-reduction reactions are critical in many carbon-based biological processes.

The term oxidation was originally used to describe reactions in which an element combines with **oxygen**. In contrast, reduction meant the removal of oxygen. By the turn of this century, it became apparent that oxidation always seemed to involve the loss of electrons and did not always involve oxygen. In general, oxidation-reduction reactions involve the exchange of electrons between two species.

An oxidation reaction is defined as the loss of electrons, while a reduction reaction is defined as the gain of electrons. The two reactions always occur together and in chemically equivalent quantities. Thus, the number of electrons lost by one chemical species (a variation of an element or chemical compound) is always equal to the number of electrons gain by another chemical species. The combination of the two reactions is known as a redox reaction. Chemical species that participate in redox reactions are described as either reducing or oxidizing agents. An oxidizing agent is a chemical species that causes the oxidation of another chemical species. The oxidizing agent accomplishes this by accepting electrons in a reaction. A reducing agent causes the reduction of another chemical species by donating electrons to the reaction.

In general, a strong oxidizing agent is a species that has an attraction for electrons and can oxidize another chemical species. The standard voltage reduction of an oxidizing agent is a measure of the strength of the oxidizing agent. The more positive the chemical species' standard reduction potential, the stronger the chemical species is as an oxidizing agent.

In reactions where the reactants and products are not ionic, there is still a transfer of electrons between chemical species. Chemists have devised a way to keep track of electrons during chemical reactions where the charge on the atoms is not readily apparent. Charges on atoms within compounds are assigned oxidation states (or oxidation numbers). An oxidation number is defined by a set of rules that describes how to divide up electrons shared within compounds. Oxidation is defined as an increase in oxidation state, while reduction is defined as a decrease in oxidation state. Because an oxidizing agent accepts electrons from another chemical species, a component **atom** of the oxidizing agent will decrease in oxidation number during the redox reaction.

There are many examples of oxidation-reduction reactions in the world. Important processes that involve oxidation-reduction reactions include combustion reactions that convert energy stored in **fuels** into thermal energy, the **corrosion** of **metals**, and metabolic reactions.

Oxidation-reduction reactions occur in both physical and biological settings (where carbon-containing compounds such as carbohydrates are oxidized). The burning of natural gas is an oxidation-reduction reaction that releases energy [$CH_4(g) + 2O_2(g) \rightarrow CO_2(g) + 2H_2O(g)$ + energy]. In many organisms, including humans, redox reactions burn carbohydrates that provide energy [$C_6H_{12}O_6(aq) + 6O_2(g) \rightarrow 6CO_2(g) + 6H_2O(l)$]. In both examples, the carbon-containing compound is oxidized, and the oxygen is reduced.

See also Chemical bonds and physical properties; Chemical elements; Chemistry

Oxygen

Oxygen is the simplest group VIA element and is, under normal atmospheric conditions, usually found as a colorless, odorless, and tasteless gas. Oxygen has an **atomic number** of 8 and an **atomic mass** of 16.0 amu. The liquid and solid forms, which are strongly paramagnetic, are a pale blue color. Oxygen has a boiling point of –297°F (–182.8°C) and a **melting** point of –368.7°F (–222.6°C).

Oxygen is the third most abundant element found in the **Sun**, after hydrogen and helium, and plays an important role in the carbon-nitrogen cycle. Oxygen composes 21% of Earth's atmosphere by volume and is vital to the existence of carbon-based life forms.

Although English chemist Joseph Priestley (1733–1804) is generally credited with the discovery of oxygen in 1774, many science historians contend that Swedish chemist Carl Scheele (1742–1786) probably discovered oxygen a few years prior to Priestly. French chemist Antoine Lavoisier's (1743–1794) contributions to the study of the important reactions, combustion and oxidation, were spurred by the discovery of oxygen. Lavoisier noticed that something was absorbed when combustion took place and that it was obtained from the surrounding air. Lavoisier noted that the increase in the weight of the substance burned was equal to the decrease in the weight of the air used. His studies lead to Lavoisier's oxidation theory, which eventually superseded the phlogistonists' theory (i.e., that every combustible substance was thought to contain a phlogiston, or inherent principal of fire, liberated through burning, along with a residue) that was widely accepted at that time. Lavoisier eventually named the gas he studied oxygen from the Greek *oxys* meaning acid or sharp, and *geinomial* meaning forming. Lavoisier named the gas oxygen because he noted that the burned materials were converted into acids.

Although oxygen has nine isotopes, natural oxygen is a mixture of only three of these. The most abundant isotope, oxygen-18, is stable and available commercially. The most common use for commercial oxygen gas is in enrichment of steel blast furnaces and for medical purposes. Large quantities are also used in making synthetic ammonia gas, methanol and ethylene oxide. Oxygen is also consumed in oxy-acetylene welding. Most commercial oxygen is produced in air separation plants. It is estimated that the United States consumes 20 million tons of oxygen in commercial use per year and the demand is expected to increase dramatically.

When oxygen is exposed to ultraviolet light, as from the Sun, or an electrical discharge, as from lightening, **ozone** (O_3) is formed. Although ozone is toxic to breathe, the 0.12 in (3 mm) thick layer of ozone in the earth's atmosphere provides a shield from harmful **ultraviolet rays** from the Sun. The ozone layer has recently been the subject of intense scientific interest to determine whether, and to what extent, it may be deteriorating, mainly from pollutants in the atmosphere. Unlike pure oxygen gas, ozone has a bluish color and its liquid and solid forms are bluish black to violet-black.

See also Atmospheric chemistry; Atmospheric composition and structure; Global warming; Greenhouse gases and greenhouse effect; Ozone layer and hole dynamics; Ozone layer depletion

OZONE

The name ozone comes from the Greek *Ozon* meaning smell. At atmospheric temperatures, ozone is a colorless gas with an odor similar to chlorine that can usually be detected at a level of about 0.01 parts per million.

High in the atmosphere, ozone plays an important protective role by diminishing the amount of potentially damaging ultraviolet radiation reaching Earth. In sufficient concentration, however, ozone is a poison that at lower atmospheric levels, is a pollutant that can be damaging to health. Ozone is also a strong oxidizing agent used in many industrial processes for bleaching and sterilization. Although ozone is often used in **water** treatment, the largest commercial application of ozone is in the production of pharmaceuticals, synthetic lubricants, and other commercially useful organic compounds.

In the atmosphere, ozone is formed predominantly by electric discharges (e.g., **lightning**). In the laboratory, ozone can be extracted form a mixture of **oxygen** and ozone by fractionation.

Ozone can also be formed by ultraviolet light. Ultraviolet light is energetic, and when it strikes the atmosphere it can break down some oxygen molecules producing highly energized oxygen atoms (free radicals). These free radicals can then react with molecular oxygen to produce ozone. The absorption of energetic light radiation also triggers the decomposition of ozone. As a result, ozone is an unstable molecule that exists in a dynamic equilibrium of formation and destruction. Consequently, the protective ozone layer is also in dynamic equilibrium.

The **area** where ozone is formed at the fastest rate is in the atmosphere at a height of approximately 164,042 ft (50 km). At this height, the number of free radicals made by ultraviolet light and electric discharge is balanced by the concentration of diatomic oxygen, which is sufficiently high to ensure that reactive collisions occur.

The protective ozone layer is found in the upper reaches of the atmosphere (between 98,000–295,000 ft [30–90 km]) where it absorbs ultraviolet radiation that, in excess, can be harmful to biological organisms. The potential detrimental effects of increased exposure to ultraviolet light due to a lessening of atmospheric ozone are of great concern. Holes in the ozone layer, or a global breakdown of stratospheric ozone would lead to increasing doses of ultraviolet radiation at Earth's surface. Scientists fear that significant increases exposure to ultraviolet light will increase risks of cancer in animal skin, eyes, and immune systems. Studies have shown that high ultraviolet radiation doses can supply the needed energy for chemical reactions that produce highly reactive radicals that have the potential to damage DNA and other cell regulating chemicals and structures.

There are several atmospheric trace elements, including ozone, that are important in the regulation of the global **climate**. Although the atmosphere consists of mainly of nitrogen and oxygen, approximately one percent of Earth's atmosphere is made of small amounts of other gases. Trace gases include water vapor, **carbon dioxide**, nitrous oxide, methane, chlorofluorocarbons (CFCs), and ozone. Because the amount of trace gases in the atmosphere is small, human activities can significantly affect the proportions of atmospheric trace gases.

Chloroflourocarbons (CFCs) easily react with ozone, which has the effect of breaking down an already unstable molecule. Until recently, CFCs were commonly used in refrigeration and in aerosol propellants (a pressurized gas used to propel substances out of a container). After evidence indicating that the use of CFCs was tipping the ozone equilibrium toward overall **ozone layer depletion**, many industrialized countries opted to enforce restrictions on the use of CFCs. Consumer aerosol products in the United States have not used ozone-depleting substances such as CFCs since the late 1970s. Federal regulations, including the Clean Air Act and Environmental Protection Agency (EPA) regulations restrict the use of ozone-depleting substances.

Ozone played a critical role in the development of life on Earth. Once primitive plants evolved, oxygen started to accumulate in the atmosphere. Some of this oxygen was converted into ozone and the developing ozone layer gave needed protection from disruptively energetic ultraviolet radiation. As a consequence, complex organic molecules which would otherwise have been destroyed began to accumulate.

As well as being found high in the atmosphere, ozone can be found at ground level. At these locations it is regarded as a pollutant. Ozone at ground level can be manufactured as part of photochemical **smog**. This is brought about by the disassociation of oxides of nitrogen that produce oxygen free radicals. These free radicals can react with diatomic oxygen to produce ozone. Pollutant ozone can also be a by-product of the action of photocopiers and computer printers. Low level ozone is usually found at a concentration of less than 0.01 parts per million, whereas in photochemical smog, it can be encountered at levels as high as 0.5 parts per million. Levels of ozone exposure between 0.1 and 1 part per million cause headaches, burning eyes, and irritation to the respiratory passages in humans. Elderly people, asthma sufferers, and those exercising in photochemical smog suffer the greatest adverse effects.

Some plant species (e.g., the tobacco plant) are particularly sensitive to low-lying ozone. The presence of excessive ozone causes a characteristic spotting of the leaves. High ozone levels are also known to damage structural material such as rubber.

Replacing more dangerous chlorine gas, ozone is used in many waste treatment facilities to purify water. Ozone is responsible for disinfecting the water and the efficient removal of trace elements such as pesticides. Ozone kills bacteria and other small life forms and it reacts with organic compounds. During the process, the ozone is transformed to molecular oxygen.

See also Atmospheric pollution; Ozone layer and ozone hole dynamics

Ozone layer and ozone hole dynamics

In 1985, atmospheric scientists discovered that stratospheric **ozone** over **Antarctica** had been reduced to half its natural level. This local loss, termed the Antarctic ozone hole, was traced to destruction of stratospheric ozone by human-made chemicals, especially chlorofluorocarbons (CFCs; artificial compounds consisting of chlorine, fluorine, and **carbon** and widely used as refrigerants and aerosol spray propellants). Other evidence indicates that ozone levels potentially declining over other regions, though nowhere as drastically as over Antarctica.

The ozone hole covers an **area** over the Antarctic continent, the surrounding ocean, the southern tip of **South America** in which stratospheric ozone begins to diminish every August (at the beginning of the Southern hemisphere's spring season), reaches a minimum of less than 50% of its natural value in October, and returns to normal levels by the beginning of December.

Essential to the formation of the Antarctic ozone hole is the polar vortex, which forms every winter over the South Pole. The pole is in 24-hour darkness in midwinter, so the air above it becomes very cold. Cooling air lowers its pressure. Air nearer the equator, warmer and therefore at higher pressure, is sucked toward the pole by the low pressure there. As this warm air moves southward it is twirled into a circular **wind** by the spin of the earth. This circular wind, the polar vortex, sits over the South Pole like a halo, isolating the air over the pole and allowing it to become even colder. Intermittently, the **stratosphere** over the pole becomes cold enough to form **clouds**. The droplets and **ice crystals** in these stratospheric clouds accelerate the breakdown of ozone by chlorine, essentially eliminating ozone from the lower stratosphere and allowing twice the usual amount of UV-B to reach the surface.

No ozone hole forms at the North Pole because the north-polar winter vortex is smaller and warmer than the southern one. There is nevertheless a 30% decline in north-polar ozone every March. Ozone levels have also declined by 3–6% over the inhabited (middle) latitudes, allowing more UV-B to reach the surface and increasing skin cancer rates.

The ozone layer protects the earth by absorbing UV-B, which can cause skin cancer and eye damage. Low-altitude ozone, however, blocks little UV-B and is toxic to plant and animal life.

Ozone (O_3) is a trace ingredient of the atmosphere that stops most solar radiation in the 280–315-nm ultraviolet (UV-B) band from reaching the ground. Ozone is produced in the stratosphere by the breakup of molecular **oxygen** (O_2) by solar radiation. It is also produced artificially in the lower atmosphere (**troposphere**) by the burning of **coal** and gasoline. Ninety percent of the atmosphere's ozone is concentrated in the lower stratosphere, about 6–30 miles (10–50 km) up; this concentration of ozone is the ozonosphere or ozone layer.

Ozone is formed in the stratosphere when an O_2 molecule is split by a photon in the 175–242-nm ultraviolet band (1 nm[nanometer] = 10^{-9} m.) Each O then joins with an O_2 to form an O_3 (ozone) molecule. Ozone converts the energy it gains from absorbing ultraviolet (and infrared) photons into heat, supplying an average 15 watts of power to every square meter of the stratosphere. This ozone-driven heating defines the temperature-versus-altitude structure of the stratosphere.

Because ozone is created by sunlight it forms more rapidly over the tropics, where there is more sunlight per square meter. Some ozone created at tropical latitudes circulates through the upper stratosphere to the polar regions, but natural polar ozone levels remain lower than tropical levels. This contributes to the greater vulnerability of the polar regions to ozone depletion by CFCs and other chemicals, discussed below.

Ozone is destroyed primarily by the ClO (chlorine oxide) radical that is produced by the breakdown by sunlight of more complex chlorine-bearing molecules. ClO facilitates the reaction, participating as a catalyst. ClO radicals are free to facilitate reactions again and again. This catalytic persistence explains how minute concentrations of a human-made substance can alter the **chemistry** of an entire layer of the atmosphere: ozone is a million or so times more abundant in the stratosphere than ClO, but each ClO radical destroys thousands of ozone molecules.

Not all chlorine-containing compounds threaten the ozone layer, because not all are capable of reaching the stratosphere. Only non-water-soluble compounds such as CFCs, carbon tetrachloride (CCl_4), and methyl chloroform (CH_3CCl_3) can evade **water** capture in the troposphere and eventually circulate to the ozone layer. There they last anywhere from 5 years (methyl chloroform) to 100 years (CFC-12). CFC-F11 (CCl_3F), the primary contributor to stratospheric chlorine and therefore to ozone loss, has a lifetime of 45 years in the stratosphere.

CFCs are not the only compounds that affect stratospheric ozone; nitrous oxide (N_2O), the bromine-containing compounds termed halons, and methane (CH_4), also do so. Sulfur dioxide (SO_2) injected into the stratosphere by violent **volcanic eruptions**, such as that of Mt. Pinatubo in 1991, can cause significant, albeit temporary, drops in global stratospheric ozone.

In 1987, over 100 nations signed an international agreement to reduce emissions of CFCs and other ozone-depleting chemicals, the Montreal Protocol. Later amendments to the Protocol greatly increased its effectiveness, and today scientists estimate that with strict observance of the Protocol, and barring unforeseen side effects of global **climate** change, stratospheric ozone will cease to decline at some point in the next 10–20 years and recover to 1980 levels by about 2050.

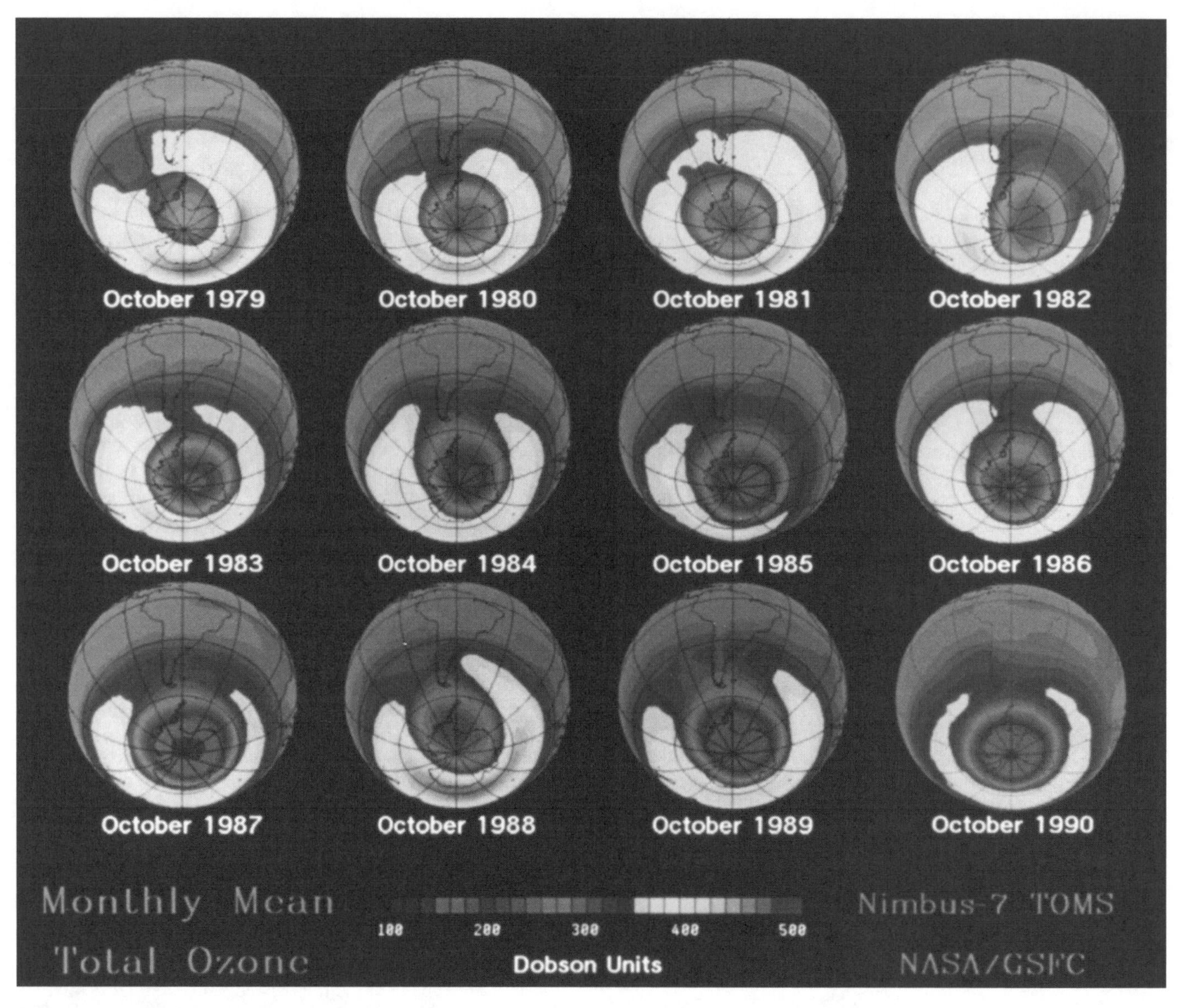

False color images showing ozone depletion from October 1979 to October 1990. ***U.S. National Aeronautics and Space Administration (NASA).***

See also Atmospheric chemistry; Atmospheric circulation; Atmospheric composition and structure; Atmospheric pollution; Atmospheric pressure

OZONE LAYER DEPLETION

The ozone layer is a part of the atmosphere between 18.6 mi and 55.8 mi (30 and 90 km) above the ground. The **ozone** present is responsible for blocking potentially harmful ultraviolet radiation reaching the surface of the earth. During the last twenty years, evidence has accumulated that human activity may be the cause of a generalized depletion of the ozone layer. This phenomena is global and distinct from the natural factors that induce annual ozone layer hole formation over **Antarctica**.

Ozone is constantly created and destroyed in natural processes (manufactured by the action of **lightning** on **oxygen** and destroyed by the action of ultraviolet radiation), however the amounts normally balance each other out so there is no net increase or decrease due to natural processes. In 1970, Paul Crutzen showed that naturally occurring oxides of nitrogen can catalytically destroy ozone. In 1974, **F. Sherwood Rowland** and **Mario Molina** demonstrated that chlorofluorcarbons (CFCs) could also destroy ozone. In 1995, all three were jointly awarded the Nobel Prize for chemistry.

The CFCs that were observed as being damaging included Freon 11 ($CFCl_3$) and Freon 12 (CF_2Cl_2). These chemicals are widely used in industry and the home. They have uses as propellants in aerosol spray cans, refrigerant gases, and foaming agents for blown plastics. One problem associated with these gases is their relative lack of reactivity.

When released there is very little that will break them down and, as they are not soluble in **water**, they are not removed from the atmosphere by rain. As a consequence, once released they tend to concentrate in the upper regions of the atmosphere. It is estimated that some several million tons of CFCs are present in the atmosphere.

Once in the upper atmosphere the CFCs are exposed to high energy radiation that can cause disassociation of the molecule, producing free chlorine atoms. This atomic chlorine reacts readily with ozone to produce chlorine monoxide and molecular oxygen. The chlorine monoxide can further react to produce molecular oxygen and more atomic chlorine. This all accelerates the destruction of ozone beyond its natural ability to regenerate. Overall, there is a net reduction in the amount of ozone present in the upper atmosphere. This has led to a thinning of the ozone layer. The majority of this loss is at an altitude between 7.44 mi and 18.6 mi (12 and 30 km) and in the late 1990s evidence was seen that suggested losses were also occurring at other altitudes. In addition to the annual holes in the ozone layer now detected over Antarctica, in the late 1990s, holes were detected over **Australia** and atmospheric sampling indicated a dramatic thinning of the ozone layer in the Northern Hemisphere during the winter months. In the Northern Hemisphere losses of some 30% have been recorded at an altitude of 12.4 mi (20 km).

In 1987, the Montreal Protocol was signed with the appropriate countries agreeing to reduce CFC production. By 1996, more than 100 countries agreed to cease widespread commercial use of CFCs and to stop or curtail production of CFCs.

In the absence of the ozone layer, harmful ultraviolet radiation is able to reach the surface of the earth in higher doses. This can lead to increases in skin cancers.

See also Atmospheric chemistry; Greenhouse gases and greenhouse effect; Ozone layer and hole dynamics; Ultraviolet rays and radiation

P

Pacific Ocean • *see* Oceans and seas

Pahoehoe flow • *see* Lava

Paleocene Epoch

In **geologic time**, the Paleocene Epoch occurs during the **Tertiary Period** (also sometimes divided or referred to in terms of a Paleogene Period and a Neogene Period instead of a Tertiary Period) of the **Cenozoic Era** of the **Phanerozoic Eon**. The Paleocene Epoch is the earliest epoch in the Tertiary Period (in the alternative, the earliest epoch in the Paleogene Period).

The Paleocene Epoch spans the time between roughly 65 million years ago (mya) and 55 mya.

The Paleocene Epoch is further subdivided into (from earliest to most recent) Danian (65 mya to 61 mya) and Thanetian (61 mya to 55 mya) stages. The **Eocene Epoch** followed the Paleocene Epoch.

The onset of the Paleocene Epoch is marked by the K-T boundary or **K-T event**, a large mass extinction. Most scientists argue that the K-T extinction resulted from—or was initiated by—a large asteroid impact in a submerged **area** off the Yucatan Peninsula of Mexico termed the Chicxulub crater. The impact caused widespread primary damage due to blast impact and firestorms. Post-impact damage to Earth's ecosystem occurred as dust, soot, and debris from the collision occluded the atmosphere to sunlight. The global darkening was sufficient to slow photosynthesis and the resulting climatic changes and widespread starvation resulted in extinction of the largest life forms with the greatest metabolic energy needs (e.g., the dinosaurs).

Other impact craters that date within the Paleocene Epoch include sites in Alberta, Canada and Marquez, Texas.

The Paleocene Epoch marks the rise of mammals as the dominant land species. During the Paleocene Epoch, climatic moderations reduced the evolutionary pressure of extreme swings in climate. Although a diversity of mammals had evolved and widely populated the changing continental landmasses long before the Paleocene Epoch, the reduction in predator species allowed land mammals to dominate and thrive—eventually setting the stage from the **evolution** of homo sapiens (humans). Pine trees appeared during the Paleocene Epoch and avian species flourished and diversified.

See also Archean; Cambrian Period; Cretaceous Period; Dating methods; Devonian Period; Evolution, evidence of; Fossil record; Fossils and fossilization; Historical geology; Jurassic Period; Mesozoic Era; Miocene Epoch; Mississippian Period; Oligocene Epoch; Ordovician Period; Paleozoic Era; Pennsylvanian Period; Pleistocene Epoch; Pliocene Epoch; Precambrian; Proterozoic Era; Quaternary Period; Silurian Period; Supercontinents; Triassic Period

Paleoclimate

Paleoclimate is the variation of the **climate** in past geologic times, prior to instrumental measurements. Paleoclimate is expressed by its parameters—paleotemperature, **precipitation** in the past, circulation, sea surface **temperature** (SST), and sea level.

The general state of the earth's climate is dependent upon the amount of energy the earth receives from the solar radiation, and the amount of energy the earth releases back to **space** in the form of infrared heat energy. Causes of climate change involve any process that can alter the global energy balance (climate forcing). Climate forcing processes can be divided into internal and external types. External processes include variations in Earth's orbit around the **Sun**. These variations change the amount of energy received from the Sun, and also cause variations of the distribution of sunlight reaching the earth's surface. Long periods of solar luminosity variations cause variations of the global climate, although lower intensity variations of luminosity may not produce detectable changes in local climate if circulation patterns modulate them. Internal processes operate within the earth's climate system, and include changes in **ocean circulation** and changes in the composition of the atmosphere. Other climate forcing processes include the impacts of large **volcanic eruptions**, and collisions with **comets** or meteorites.

Over much of the earth's geologic history, the global climate has been warmer and wetter than at present. Global temperatures early in Earth's history were 46–59°F (8–15°C) warmer than today. Polar regions were free of **ice** until periods of glaciations occurred 2,300 million years ago. For about the past one billion years, Earth's dominant climate pattern has been one of tropical regions, cool poles, and periodic **ice ages**. Most recent **glaciers** reached their maximum thickness and extent about 18,000 years ago, and then **glaciation** ended abruptly about 10,000 years ago. Since the last glacial period, sea levels changed from –132.2 yd (–120 m) during glaciation to +10.9 yd (+10 m) during interglacial maximum, due to ice sheet **melting**.

The Medieval Climatic Optimum occurred around A.D. 1000–1250. The Northern Hemisphere experienced a warm and dry climate. Most of Greenland was ice free and, therefore, was named Greenland. The Little Ice Age was a period of rapid cooling, which began after the end of the Medieval Warm period and lasted nearly until the end of the eighteenth century, reaching its peak from 1460 to 1705. During this period, the average global temperature dropped 33.8–35.6°F (1–2°C). Solar activity may have lead to climatic changes like the Little Ice Age and Medieval Climatic Optimum. The Little Ice Age coincides with a period of absence of aurora from 1460 to 1550 called the Spoerer Solar Minimum, and an absence of sunspots from 1645 to 1715 called the Maunder Minimum. The number of sunspots has been related to solar output and the emission of the radiant heat from the Sun.

Reconstructions of Paleoclimate are made by use of records of different proxies (models) of different climatic parameters. These models can be divided to quantitative, qualitative, and indirect from the point of view of the precision of the reconstruction of past climates they provide. Quantitative models are able to reconstruct the exact values of the temperature, annual precipitation or sea level, and to estimate its error. Qualitative models are able to reconstruct only their principal variations expressed through the variation of the model. Indirect models do not express variations of the climate, but of something dependent on climate through a complicated mechanism such as distribution of a certain plant type.

Speleothems (stalagmites, **stalactites** and flowstones) are producing a tremendous range of reconstructions of different types of paleoclimatic parameters, including many quantitative records. Calcite speleothems display luminescence, which is produced by calcium salts of humic and fulvic acids derived from soils above the **cave**. The luminescence of speleothems depends exponentially on the solar **insolation** (if **soil** surface is heated directly by the Sun) or on the air temperature (if the cave is covered by forest or bush). Therefore, luminescence records represent solar insolation or temperature in the past. Luminescence of many speleothems is exhibited by annual bands much like tree rings. Distance between them is a quantitative proxy of annual precipitation in the past.

Changes in the thickness of the tree rings records temperature changes if derived from temperature-sensitive tree rings, or records precipitation if derived from precipitation-sensitive tree rings. These records are modulated to some degree by the other climatic parameters.

The stable isotope records of past glaciations are preserved in glacier ice and in sea cores. These records are primarily a measure of changing volume of glacier ice. The ratio of stable isotopes in **water** is temperature dependent, and is altered whenever water undergoes a phase change. Now, the volume of land ice is relatively small, but during glacial periods, much isotopically light water was removed from **oceans** and stored in glaciers on land. This caused slight enrichment of seawater, while glacier ice had lower values of the ratio. Sea cores do not allow for a better resolution than 1,000 years, and cannot be dated precisely. Corals and speleothems often allow measurements with minor time increments. Plants and animals adapt to the climatic changes, so may be used as indirect paleoclimatic indicators. Fossil evidence provides a good record of the advancing and retreating of ice sheets, while various pollen types indicate advances and retreats of northern **forests**.

See also Insolation and total solar irradiation; Marine transgression and marine regression; Milankovitch cycles; Stalactites and stalagmites

PALEOMAGNETICS

The first ever treatise on experimental science by thirteenth century scholar Petrus Peregrinus of Marincourt dealt with **magnetism** ("Epistola de Magnete"). However, direct observations of the geomagnetic field were not recorded until the late sixteenth century, when the magnetic compass became a widespread tool for navigation. In order to understand nature and origin of Earth's **magnetic field**, however, much longer records are necessary. Paleomagnetic research draws this information from rocks that acquire a remanent magnetization upon formation.

The natural magnetization of a **rock** is parallel to the ambient magnetic field. It is carried by minute amounts of ferrimagnetic **minerals** and can be stable over geological time scales. Precise snap-shots of the past geomagnetic field are recorded by volcanic rocks, while **sedimentary rocks** retain smoothed records acquired over discrete intervals of time. Sequences of rocks can thus act like a magnetic tape, which records a piece of music. Unfortunately, the original record is usually altered secondarily through time and various **weathering** processes. Paleomagnetic methods have to be employed to remove this magnetic noise and extract a true primary magnetization.

Paleomagnetic research has shown that Earth's magnetic field has been a dipole field for more than 99.9% of Earth's history. Its shape resembles that of the field of a bar-magnet. The field lines emerge at one pole and re-enter at the other pole. The earth's magnetic field however is not caused by a huge mass of **iron** with a remanent magnetization, but its origin lies in the outer fluid core where convective motion generates the magnetic field in a self-sustaining dynamo action. This dynamic origin of the geomagnetic field is the main reason why its shape and orientation are not constant but subject to temporal variations on time scales that range from millions of years to days. Recently, for example, the dipole axis is inclined by about 11 (against the spin axis). Averaged over time spans greater than 100,000 years, the dipole axis is parallel with the earth's spin axis.

The earth's magnetic field can characteristically reverse its polarity, meaning that the magnetic poles can switch position. Other second order phenomena are termed the secular variation of Earth's magnetic field.

The temporal variations of Earth's magnetic field are widely used in geosciences. The understanding that Earth's magnetic field can truly reverse its polarity had a huge impact on our view of Earth, because the idea was crucial for the development and break-through of **plate tectonics**. This view of Earth as a dynamic system which was put forward in 1915 by German geophysicist **Alfred Wegener** (1880–1930), but was not commonly accepted until the 1960s. Only then was the cause for the characteristic pattern of the oceanic magnetic anomalies understood. They are characterized by alternating stripes of normally and inversely magnetized rocks parallel to the **mid-ocean ridges** and are caused by the continuous addition of newly formed rocks, adding new layers and pushing the rims away from the ridge, while the geomagnetic field frequently reversed.

The fact that Earth's magnetic field never fundamentally changed its shape through its history allowed paleomagnetists to investigate the movement of plates by calculating the position of the North magnetic pole from the magnetization of rocks. Assuming the earth's field is a dipole over large intervals of time, one can calculate the geographic **latitude** of a rock at the time when its remanent magnetization was acquired. By investigating rocks from subsequent time intervals, it is then possible to reconstruct the path of a plate relative to the magnetic pole, or vice versa. It allows tracking of the former distribution of plates and continents through time.

Another important application of paleomagnetism to geoscience is the opportunity to use a sequence of reversals for dating and correlating sedimentary sequences on a global scale. Magnetostratigraphy uses the globally simultaneous occurrence of dated polarity changes. This dating method can at best resolve an average of 100,000 years. However, for rocks younger than approximately 10,000 years it is possible to use calibration curves of the paleosecular variation for dating with accuracy better than a few hundred years.

Recently, another branch of paleomagnetism has become a method in its own right. Rock **magnetism** was developed as a tool to judge the reliability of the paleomagnetic record. Today it is widely used in environmental and paleoclimatic research.

See also Earth (planet); Ferromagnetic; Plate tectonics; Polar axis and tilt

PALEOZOIC ERA

In **geologic time**, the Paleozoic Era, the first era in the **Phanerozoic Eon**, covers the time between roughly 544 million years ago (mya) and until 245 mya.

The Paleozoic Era spans six geologic time periods including the **Cambrian Period** (544 to 500 mya); **Ordovician Period** (500 mya to 440 mya); Silurian (440 mya to 410 mya); Devonian (410 mya to 360 mya); and the Carboniferous Period (360 mya to 286 mya) (in many modern geological texts, especially those in the United States, the time of Carboniferous Period is covered by two alternate geologic periods, the **Mississippian Period** [360 mya to 325 mya] and the **Pennsylvanian Period** [325 mya to 286 mya]. The final geologic time period in the Paleozoic Era is the **Permian Period** (286 mya to 245 mya).

The onset of the Paleozoic Era is marked by the "Cambrian explosion," the sudden appearance of numerous **fossils**. Although life certainly started in **Precambrian** time, The start of the Paleozoic Era marks the point at which life developed to a variety or organisms capable of leaving fossils. Although **fossilization** is difficult under any circumstances, organisms with structures such as shells have a much greater chance of leaving fossilized remains than did single celled microorganisms.

The Paleozoic Era spanned that period of geologic time during which the **evolution** of the first invertebrates, vertebrates, terrestrial (land-based) plants, bony fish, reptiles, insects, etc. took place. The end of the Paleozoic Era (approximately 245 mya) marks the largest mass extinction of species in Earth's history. During this mass extinction an estimated 90% of all Earth's marine species suddenly became extinct.

Six major continental landmasses developed during the Paleozoic Era. Although not located in their present global positions, parts of the modern continents can be traced to these landmasses. For example, continental **crust** now located in the North American continent was located near the equator during the Paleozoic Era. The forces of **plate tectonics** were active, and not only moved the continents but helped shape the continental margins though uplift and subduction. Enormous changes on sea state relative to the continents meant extensive flooding, **marine transgression**, and marine regression that added large sedimentary deposits (e.g., large **limestone** deposits). The abundance of organic life provided the start for abundant **coal** formation during the Carboniferous Period (Pennsylvanian Period and Mississippian Period).

See also Archean; Cenozoic Era; Cretaceous Period; Dating methods; Eocene Epoch; Evolution, evidence of; Fossil record; Fossils and fossilization; Historical geology; Holocene Epoch; Jurassic Period; Mesozoic Era; Miocene Epoch; Oligocene Epoch; Paleocene Epoch; Pennsylvanian Period; Pleistocene Epoch; Pliocene Epoch; Quaternary Period; Tertiary Period; Triassic Period

PANGAEA • *see* SUPERCONTINENTS

PARALLELS • *see* LATITUDE AND LONGITUDE

PARTIAL MELTING

A process known as partial **melting** generates the molten **rock**, known as **magma**, that cools to form crystalline rocks in the earth's outer compositional layer, or its **crust**. The terms "partial melting," "partial fusion," and "anatexis" refer to processes that create a magmatic melt from a portion of a solid rock less than the whole. Because most crystalline, or igneous, rocks in the earth's crust are composed of a number of silicate **minerals** that melt at different temperatures, and of minerals with heterogeneous crystal lattices, almost all magmas are generated by partial melting.

Incongruent melting occurs over a range of temperatures; the mineral components with the lowest melting temperatures melt first, and the minerals with the highest melting temperatures melt last. Partial melts are thus enriched in the chemical components of minerals with lower melting temperatures, and the remaining unmelted portion of the rock is composed of minerals with the highest melting temperatures. There are two end member types of partial melting. In equilibrium fusion, the liquid melt continuously reacts with the residual **crystals**, changing composition until the whole rock has melted. In fractional fusion, the melted material is separated from the remaining solid rock as it is produced. Fractional fusion leads to differentiation of chemical components in the melt, and to creation of different rock types from the same magmatic source.

Earthquake wave velocities and travel paths through the earth's interior suggest that the outer core is the only fully liquid layer of our planet. However, the outer core is composed entirely of **iron**, and is not a possible source of siliceous magma. Magmatic source areas are thus confined to areas of the upper mantle and lower crust that seismic shear waves indicate to be almost entirely solid. Only a very small portion ($<5\%$) of the rock in magmatic source areas is thought to be liquid. Partial melts migrate upward from their source areas to intermediate staging areas, or magma chambers, in the middle and upper crust before erupting from volcanoes, or cooling to form intrusive igneous plutons. Magmas are generated by partial melting in a number of present-day plate tectonic settings, including subduction zones, **mid-ocean ridges,** and hot spots. The granitic continental interiors, called continental shields or cratons, probably formed above ancient subduction zones, or by melting at the base of the crust during the **Precambrian** and Paleozoic Eras when more heat was escaping from the inner earth.

See also Geothermal gradient; Phase state changes

PEBBLE • *see* ROCK

PEGMATITE

A pegmatite is an intrusive igneous body of highly variable grain size that often includes coarse crystal growth. A pegmatite may be a segregation within an associated plutonic **rock** or a **dike** or vein that intrudes the surrounding **country rock.**

The composition range of pegmatites is similar to that of other intrusive **igneous rocks** and is indicated by using modifier, e.g., **granite** pegmatite or gabbro pegmatite. However, pegmatites occur most commonly in granites and the term applied alone usually refers to a granitic composition. The **mineralogy** of pegmatites can be simple or exotic. A simple granite pegmatite may contain only **quartz**, **feldspar**, and mica. More complex pegmatites are often zoned and can contain **minerals** like tourmaline, garnet, beryl, fluorite, lepidolite, spodumene, apatite, and topaz.

Pegmatites are formed as part of the cooling and crystallization process of intrusive rocks. As the parent body begins to cool, a sequential crystallization process occurs that concentrates many volatile constituents such as H_2O, boron, fluorine, chlorine, and phosphorous in a residual **magma**. In simple cases, the presence of residual **water** has simply allowed the magma to cool slowly enough to permit coarse crystal growth. More complex pegmatites are the result of the presence of numerous exotic volatiles that are eventually incorporated into rare minerals.

The most distinguishing characteristic of pegmatites is the unusually large crystal size of the minerals, which ranges from less than an inch to several feet. Single **crystals** of spodumene from the Black Hills have reached 40 ft (12 m) in length. A Maine pegmatite contained a beryl crystal 27 ft (8 m) long and 6 ft (1.8 m) wide. These exceptionally large crystals are not free-growing, rather they are intergrown with the rest of the pegmatite. However, pegmatites do produce large and beautiful individual crystals of many different minerals that are highly prized by gem and mineral collectors.

Pegmatites are also valued for the suite of rare elements that tend to be concentrated in the residual magmas. For example, beryllium is obtained from beryl, lithium from spodumene and lepidolite, and boron from tourmaline. Other rare elements obtained from pegmatites include tin, tantalum, and niobium.

See also Intrusive cooling; Pluton and plutonic bodies

PENNSYLVANIAN PERIOD

The Pennsylvanian Period lasted from 320 to 286 million years ago. During the Pennsylvanian Period, widespread swamps laid down the thick beds of dead plant material that today constitute most of the world's **coal**. The term Pennsylvanian is a U.S. coinage based on the frequency of rocks of this period in the state of Pennsylvania; internationally, the terms late Carboniferous Period or Silesian Period are preferred.

Although most artist's conceptions of the Pennsylvanian Period emphasize its prolific swamps, these were characteristic only of the equatorial regions. The

Southern Hemisphere, which was dominated by the huge continent Gondwana, underwent a series of **ice ages** during this period. These **ice** ages sequestered **water** in times of ice growth and released it in times of **melting**, causing the ocean to cyclically regress (uncover coastal lands) and transgress (cover coastal lands) around the world. Repeating sequences of sedimentary **rock** layers record these changes in sea level.

From the bottom up, a typical sequence is **sandstone**, shale, coal, **limestone**, and sandstone again. Each such unit is termed a cyclothem and was formed as follows: (1) As ice melted in Gondwana, **seas** rose globally. **Rivers** and streams deposited **sand** and gravel in the coastal lowlands as they sought new equilibrium profiles (i.e., stable altitude-vs.-distance cross-sections). This sand layer eventually became sandstone. Although the coastal zones where sandstone deposition was taking place at any one time were narrow, larges areas were blanketed by these sediments as the seas rose and coastlines swept slowly inland. (2) As the rising sea neared a given location, a lush coastal swamp developed. This deposited a thick layer of dead leaves, tree trunks, and other organic material rich in **carbon** that would eventually form coal. (3) When the sea finally submerged the swamp, a shallow marine environment appeared. The remains of shelly marine animals built up on the sea floor and eventually became limestone. (4) Ice began to build again in Gondwana, and sea levels began to drop in a new phase of regression. (5) **Erosion** of re-exposed coastal lands scraped off the topmost sediments left by the last transgression, including some of the limestone layer. (6) Ice began to melt again in Gondwana, triggering a fresh cycle of transgression.

As many as 90 cyclothems have been found in one place, one on top of the other. Each such cylothem records a complete climatic cycle like the one described above.

The first reptiles evolved during the Pennsylvanian Period. These were small (about a foot long) and outnumbered by the amphibians, which were prosperous, diverse, and achieved lengths of up to 15 ft (4.6 m). Insects also throve; dragonflies with 2.5-ft (0.76 m) wingspans were common. Over 1,000 species of Pennsylvanian cockroach have been identified, giving this period the alternative, informal title of the "age of cockroaches."

See also Archean; Cambrian Period; Carbon dating; Cenozoic Era; Continental drift theory; Cretaceous Period; Dating methods; Devonian Period; Eocene Epoch; Evolution, evidence of; Fossil record; Fossils and fossilization; Geologic time; Historical geology; Holocene Epoch; Jurassic Period; Mesozoic Era; Miocene Epoch; Mississippian Period; Oligocene Epoch; Ordovician Period; Paleocene Epoch; Paleozoic Era; Phanerozoic Eon; Pleistocene Epoch; Pliocene Epoch; Precambrian; Proterozoic Era; Quaternary Period; Silurian Period; Tertiary Period

PENZIAS, ARNO (1933-)

German-born American astrophysicist

Arno Penzias shared the Nobel Prize for physics in 1978 with Robert Wilson for a discovery that supported the **big bang theory** of the universe. The two radio astronomers at what was then American Telephone & Telegraph's (AT&T) Bell Telephone Laboratories were using a 20-ft (6.1-m) horn reflector antenna that year to measure the intensity of radio waves emitted by the halo of gas surrounding the galaxy. The two scientists were bothered by a persistent noise that they could not explain. At first they pinned it on two pigeons that were nesting in the antenna throat. But even after they evicted the birds, the noise continued. Eventually the scientists were able to conclude that the noise came from cosmic background, or microwave, radiation. This came to be widely considered as remnant microwave radiation from the "big bang" in which the universe was created billions of years ago. The Penzias-Wilson discovery came to be considered a major finding in astrophysics.

Arno Allan Penzias was born in Munich, Germany, to Jewish parents Karl and Justine (Eisenreich) Penzias. Hitler's campaign to wipe out the Jews of **Europe** was well underway when the family escaped in 1940. Arriving in New York, Penzias had to acclimate to a new culture and language and suffer through hard times for his family. Naturalized in 1946, he demonstrated scientific acumen at Brooklyn Technical High School and went on to obtain his B.S. at City College in New York in 1954. He married Anne Pearl Barras that same year; the union produced three children. After a two-year stint in the U.S. Army Signal Corps, Penzias obtained both his master's and Ph.D. degrees at Columbia University. He has said he chose to study physics because he asked a professor if he could make a living in the field and was told, "Well, you can do the same things engineers can do and do them better."

In 1961, Penzias was hired at Bell Labs in Holmdel, New Jersey. AT&T was a telecommunications monopoly at that time and Bell Labs was its research center, attracting the best and brightest scientific minds. In this context, Penzias demonstrated his capabilities early on. Asked to join a committee of older scientists who were trying to devise how to calculate the precise positions of communication satellites by triangulation, young Penzias suggested they use radio stars, which emit characteristic frequencies from fixed positions, as reference points. For his abilities, Penzias rose through the Bell ranks to become director of the facility's Radio Research Laboratory in 1976, and executive director of the Communications Sciences Research Division in 1979. He also took part in the pioneering Echo and Telstar communications **satellite** experiments of the 1970s.

It was astronomer George Gamow who in 1942 first calculated the conditions of **temperature** and density that would have been required for a fireball explosion or "big bang" origin of the universe 15 billion years ago. Astronomers Ralph Alpher and Robert Herman later concluded that cosmic radiation would have resulted from this event. This theory was confirmed by Penzias and Wilson. According to the theory, the background radiation resulting from the big bang would have lost energy; it would have essentially "cooled." Gamow and Alpher calculated in 1948 that the radiation should now be characteristic of a perfectly emitting body—or black body—with a temperature of about 5 Kelvin, or –268° C. The scientists said this radiation should lie in the microwave region of the spectrum; their calculations were verified by physicists Robert Dicke and P. J. E. Peebles.

Penzias's and Wilson's contribution to the issue began with a 20-ft (6.1-m) directional radio antenna, the same kind of radio antenna designed for satellite communication. Investigating an irritating noise emitted by the antenna, the two men realized in May of 1964 that what they heard was not instrumental noise but microwave radiation coming from all directions uniformly. Penzias and Wilson calculated the radiation's temperature as about 3.5 K. Dicke and Peebles, who had made the earlier calculations, got reinvolved from nearby Princeton University with a scientific explanation of the Penzias-Wilson discovery. More experiments followed, confirming that the radiation was unchanging when measured from any direction. Even after the duo received the Nobel Prize in 1978 (also awarded that year to Pyotr Kapitsa for unrelated work in physics) they continued to collaborate on research into intergalactic hydrogen, galactic radiation and interstellar abundances of the isotopes.

At the time of the federal lawsuit which led to the breakup of AT&T in 1984, Penzias, who had become vice president of research in 1981, predicted that without the operating companies as a base, Bell Labs would become "a sinking ship." That did not happen. Instead, in September of 1990, Penzias presided over the realignment of Bell Labs into a facility whose research is streamlined and oriented towards the activities of its business units.

While rearranging Bell Labs, Penzias has kept an eye on the outside world, writing *Ideas and Information: Managing in a High-Tech World* in 1989 and staying involved in the national dialogue regarding the growth of computer technology and international competition. He told *Forbes* magazine in March of 1989, "You go into Sears, the best cordless telephone you can buy is an AT&T phone. It works better. You try it."

In his personal life, Penzias, who is the proud grandfather of three, is also an avid skier, swimmer, and runner with an interest in kinetic sculpture and writing limericks. Penzias is a member of the National Academy of Sciences and the National Academy of Engineering, as well as the vice-chairman of the Committee of Concerned Scientists, devoted to political freedom for scientists internationally. He has written over 100 articles and collected more than 20 honorary degrees. Penzias holds that technology can be liberating. As he wrote in *Fortune* magazine in March of 1990: "Everybody is overstressed.... We've got to stop going to meetings and have them electronically instead.... How far away are we from realizing this dream? My guess is that by the time *Fortune* marks its 100th anniversary, a lot of this will have happened. In fact, long before I retire, I hope to have at least a multimedia terminal in my office so that I can integrate voice, data, high-definition video, conference video, document access, and shared software... who's going to do all this? I hope it's AT&T. But it could be IBM, Apple—it could be anybody." Penzias officially retired from Bell Labs in the spring of 1998, but continues to serve in a technical advisory role.

See also Cosmic microwave background radiation

Peridotite

Peridotite is a dark-colored, coarse-grained igneous **rock** believed by many scientists to be the primary rock of the exterior of Earth's mantle. The rock typically forms in volcanic pipes and is forced to the surface from great depths during a volcanic eruption.

Peridotite consists of a dense **iron** and magnesia mineral called **olivine**, as well as pyroxenes and a small amount of **feldspar**. It is a pistachio-green color when fresh, but **weathering** creates iron oxides that turn it a medium brown. The iron and magnesia-rich rock is the most common host for naturally occurring diamonds. South African diamonds are obtained from a mica-rich form of peridotite called kimberlite. Periodite is also a source of valuable ores and **minerals** including chromite, platinum, nickel, and precious garnet. In rare instances, individual olivine **crystals** in periodite are large enough and pure enough to be designated a gem. The resulting gem is a semi-precious mineral called peridot.

Periodite is found worldwide, but particularly in New Zealand.

See also Gemstones; Igneous rocks; Plagioclase

Periodic table (Predicting the structure and properties of the elements)

An element is defined by the number of protons in the nucleus of its atoms, but its chemical reactivity is determined by the number of electrons in its outer shell—a property fundamental to the organization of the periodic table of the elements.

In the second half of the nineteenth century, data from laboratories in France, England, Germany, and Italy were assembled into a pamphlet by Stanislao Cannizzaro (1826–1910), a teacher in what is now northern Italy. In this pamphlet, Cannizzaro demonstrated a way to determine a consistent set of atomic weights, one weight for each of the elements then known. Cannizzaro distributed his pamphlet and explained his ideas at an 1860 international meeting held in Karlsruhe, Germany, that was organized to discuss new ideas about the theory of atoms. When Russian chemist and physicist Dmitri Mendeleyev (1834–1907) returned from the meeting to St. Petersburg, Russia, he pondered Cannizzaro's list of atomic weights along with an immense amount of information he had gathered about the properties of elements. Mendeleyev found that when he arranged the elements in order of increasing atomic weight, similar properties were repeated at regular intervals—they displayed periodicity. Mendeleyev used the periodic repetition of chemical and physical properties to construct a chart much like the Periodic Table we currently use.

Early in the twentieth century work initiated by Joseph John Thomson (1856–1940) in England led to the discovery of the electron and, later, the proton. In 1932, James Chadwick (1891–1974), also in England, proved the existence of the neutron in the atomic nucleus. The discovery of these elementary particles and the experimental determination of their actual weights led scientists to conclude that different atoms have different weights because they contain different numbers of protons and neutrons. However, it was not yet clear how many subatomic particles were present in any but the simplest atoms, such as hydrogen, helium and lithium.

In 1913, a third British scientist, Henry G. J. Moseley (1887–1915), determined the frequency and wavelength of x rays emitted by a large number of elements. By this time, the number of protons in the nucleus of some of the lighter elements had been determined. Moseley found the wavelength of the most energetic x ray of an element decreased systematically as the number of protons in the nucleus increased. Moseley then hypothesized the idea could be turned around: he could use the wavelengths of x rays emitted from heavier elements to determine how many protons they had in their nuclei. His work set the stage for a new interpretation of the Periodic Table.

Moseley's results led to the conclusion that the order of elements in the periodic chart was based on some fundamental principle of atomic structure. As a result of Moseley's work, scientists were convinced that the periodic nature of the properties of elements is due to differences in the numbers of subatomic particles. As each succeeding element is added across a row on the periodic chart, one proton and one electron are added. The number of neutrons added is unpredictable but can be determined from the total weight of the **atom**.

When Mendeleyev placed elements in his Periodic Table, he had all elements arranged in order of increasing relative atomic weight. However, in the modern Periodic Table, the elements are placed in order of the number of protons in the nucleus. As atomic weight determinations became more precise, discrepancies were found. The first case of a heavier element preceding a lighter one in the modern Periodic Table occurs for cobalt and nickel (58.93 and 58.69, respectively). In Mendeleyev's time, both atomic weights had been determined to be 59. Mendeleyev grouped both elements together, along with **iron** and copper. From the work of Moseley and others, the number of protons in the nuclei of the elements cobalt and nickel had been found to be 27 and 28, respectively. Therefore, the order of these elements, and the reason for their similar behavior with respect to other members of their chemical families, arises because nickel has one more proton and one more electron than cobalt.

Whereas Mendeleyev based his order of elements on mass and chemical and physical properties, the arrangement of the table now arises from the numbers of subatomic particles in the atoms of each element. The stage was now set for examining how the number of subatomic particles affects the **chemistry** of the elements. The role of the electrons in determining chemical and physical properties was obscure early in the twentieth century, but that would soon change with the pivotal work of American chemist Gilbert N. Lewis (1875–1946).

After Moseley's work, the idea that the periodic patterns in chemical reactivity might actually be due to the number of electrons and protons in atoms intrigued many chemists. Among the most notable was Lewis, then at the University of California at Berkeley. Lewis explored the relationship between the number of electrons in an atom and its chemical properties, the kinds of substances formed when elements reacted together to form compounds, and the ratios of atoms in the formulas for these compounds. Lewis concluded that chemical properties change gradually from metallic to nonmetallic until a certain "stable" number of electrons is reached.

An atom with this stable set of electrons is a very unreactive species. But if one more electron is added to this stable set of electrons, the properties and chemical reactivities of this new atom change dramatically: the element is again metallic, with the properties like elements of Group 1. Properties of subsequent elements change gradually until the next stable set of electrons is reached and another very unreactive element completes the row.

A stable number of electrons is defined as the number of electrons found in an unreactive or "noble" gas. Lewis suggested electrons occupied specific areas around the atom, called shells. The noble gas atoms have a complete octet of electrons in the outermost shell.

The observation that each element starting a new row has just one electron in a new shell opens the door to relating chemical properties to the number of electrons in a shell. Mendeleyev put elements together in a family because they had similar reactivities and properties; Lewis proposed that elements have similar properties because they have the same number of electrons in their outer shells.

Many observations of the chemical behavior of elements are consistent with this idea: the number of electrons in the outer shell of an atom (the valence electrons) determines the chemical properties of an element. Lewis extended his ideas about the importance of the number of valence electrons from the properties of elements to the bonding of atoms together to form compounds. He proposed that atoms bond with each other either by sharing electrons to form covalent bonds or by transferring electrons from one atom to another to form ionic bonds. Each atom forms stable compounds with other atoms when all atoms achieve complete shells. An atom can achieve a complete shell by sharing electrons, by giving them away completely to another atom, or by accepting electrons from another atom.

Many important compounds are formed from the elements in rows two and three in the Periodic Table. Lewis predicted these elements would form compounds in which the number of electrons about each atom would be a full shell, like the noble gases. The noble gases of rows two and three, neon or argon, each have eight electrons in the outermost valence shell. Thus, Lewis's rule has become known as the octet rule and simply states that there should be eight electrons in the outer shell of an atom in a compound. An important exception to this is hydrogen for which a full shell consists of only two electrons.

The octet rule is followed in so many compounds it is a useful guide. However, it is not a fundamental law of chemistry. Many exceptions are known, but the octet rule is a good starting point for learning how chemists view compounds and how the periodic chart can be used to make predictions about the likely existence, formulas and reactivities of chemical substances.

Elements in a vertical column of the Periodic Table typically have many properties in common. After all, Mendeleyev used similarities in properties to construct a periodic table in the first place. Because they show common characteristics, elements in a column are known as a family. Sometimes a family had one very important characteristic many chemists knew about: that characteristic became the family name. Four important chemical family names of elements still widely used are the alkali **metals**, the alkaline earths, the halogens, and the noble gases. The alkali metals are the elements in Group 1, excluding hydrogen, which is a special case. These elements—lithium, sodium, potassium, rubidium, cesium and francium—all react with **water** to give solutions that change the color of a vegetable dye from red to blue. These solutions were said to be highly alkaline or basic; hence the name alkali metals was given to these elements.

The elements of Group 2 are also metals. They combine with **oxygen** to form oxides, formerly called "earths," and these oxides produce alkaline solutions when they are dissolved in water. Hence, the elements are called alkaline earths.

The name for Group 17, the halogens, means salt former because these elements all react with metals to form salts.

The name of Group 18, the noble gases, has changed several times. These elements have been known as the rare gases, but some of them are not especially rare. In fact, argon is the third most prevalent gas in the atmosphere, making up nearly 1% of it. Helium is the second most abundant element in the universe—only hydrogen is more abundant. Another name used for the Group 18 family is the inert gases. However, Neil Bartlett, while at the University of British Columbia in Vancouver, Canada, showed over 30 years ago that several of these gases could form well-defined compounds. The members of Group 18 are now known as noble gases. They do not generally react with the common elements but do on occasion, especially if the common element is as reactive as fluorine.

Knowing the chemistry of four families of the periodic table—groups 1, 2, 17, and 18, the alkali metals, the alkaline earths, the halogens and the noble gases—enables chemists to divide the elements in the Periodic Chart into other general categories: metals and nonmetals. Metals are hard but ductile substances that conduct **electricity**. Groups 1 (excluding hydrogen) and 2 are families of metallic elements. Groups 17 and 18 contain elements with very different properties perhaps best described by what they are not—they are not metals, and hence are called nonmetals. Between Groups 1 and 2, and Groups 17 and 18 is a dividing line between these two types of elements. Most periodic charts have a heavy line cutting between **aluminum** and **silicon** and descending downward and to the right in a stair-step fashion. Elements to the left of the line are metallic; those to the right, nonmetallic. The boundary is somewhat fuzzy, however, because the properties of elements change gradually as one moves across and down the chart, and some of the elements touching that border have a blend of characteristics of metals and nonmetals; they are frequently called semi-metals or metalloids.

The elements in the center region of the table, consisting of dozens of metallic elements in Groups 3–12, including the lanthanide and actinide elements, are called the transition elements or transition metals. The other elements, Groups 1,2, and 13–18, are called the representative elements.

There is a **correlation** among the representative elements between the number of valence electrons in an atom and the tendency of the element to act as a metal, nonmetal, or metalloid. Among the representative elements, the metals are located at the left and have few valence electrons. The nonmetals are at the right and have nearly a full shell of electrons. The metalloids have an intermediate number of valence electrons.

The structure and bonding of a compound determine its chemical and physical properties. Lewis's idea of stable, filled electron shells can be used to predict what atom is bonded to what other atom in a molecule. In many cases, Lewis's octet rule is followed by taking one or more electrons from one atom to form a cation and donating the electron or electrons to another atom to form an anion. Metallic elements on the far left of the Periodic Table can lose electrons and elements on the far right can readily accept electrons. When these elements combine, ionic bonds result. An example of an ionic compound is sodium chloride. The sodium cation, Na^+, forms an ionic bond with chloride anion, Cl^-.

In covalent bonds, electrons are shared between atoms. Lewis defined a covalent bond as a union between two atoms resulting from the sharing of two electrons. Thus, a covalent bond must be considered a pair of electrons shared by two atoms. Elemental bromine, Br_2, is an example of a covalent compound. Each bromine atom has seven electrons in its outer shell and requires one electron to achieve a noble gas configuration. Each can pick up the needed electron by sharing one with the other bromine atom.

Water, the solvent of life and an important agent in many geochemical processes, is a compound formed by the combination of atoms of two nonmetallic elements, hydrogen and oxygen. Each hydrogen atom requires just one electron to fill its shell because the first shell (the number of electrons of the noble gas helium) holds only two electrons. Oxygen lacks two electrons compared with neon, the nearest noble gas. If each hydrogen can obtain one electron by sharing electrons from the oxygen atom and the oxygen atom can share one electron from each of the two hydrogen atoms, every atom will have a full shell of electrons, and two covalent bonds will be formed as a result of sharing two pairs of electrons.

One of the most important properties of an element that can be used to predict bonding characteristics is whether the element is metallic or nonmetallic.

Pure metals are typically shiny and malleable. Chemists have found metals also have common chemical properties. Metals combine in similar ways with other elements and form compounds with common characteristics. Metals combine with nonmetals to form salts. In salts, the metals tend to be cations. Salts conduct electricity well when melted or when dissolved in water or some other solvents but not when they are solid.

Most pure metals, when freshly cut to expose a new surface, are lustrous, but most lose this luster quickly by combining with oxygen, **carbon dioxide,** or hydrogen sulfide to form oxides, carbonates or sulfides. Only a few metals such as gold, silver, and copper are found pure in nature, uncombined with other elements.

Nonmetals in their elemental form are usually gases or solids. A few are shiny solids, but instead of being metallic gray they are typically black (boron, **carbon** as **graphite**), colorless (carbon as **diamond**), or highly colored (violet iodine, yellow sulfur). At room **temperature**, only one of them is a liquid (bromine).

Nonmetallic elements combine with metallic elements to form salts. In salts, the nonmetallic elements tend to be anions. Non-metals accept electrons in forming anions while metals donate electrons to form cations. This reflects a periodic property of elements: as one moves from left to right across a row on the periodic chart, on the left are the atoms of metals which tend to give up electrons relatively easily and on the right side are nonmetals which do not readily give up electrons in forming **chemical bonds**. At the start of the next row, the trend is repeated. This periodic property is referred to as electronegativity. The more readily atoms accept electrons in forming a bond, the higher their electronegativity. Metals are characterized by low electronegativities; nonmetals, by high electronegativity. Electronegativity increases across a row on the periodic chart.

Nonmetallic elements combine with each other to form compounds. Although some nonmetallic elements form solutions when mixed with other nonmetallic elements, most react with other nonmetals to form new substances. For example, at the high temperatures and pressures of an internal combustion engine, nitrogen and oxygen gases from the atmosphere react to form nitrogen oxides such as nitric oxide, NO, and nitrogen dioxide, NO_2. Nonmetallic elements form covalent bonds with each other by sharing electron pairs. This tendency to bond by sharing electrons reflects the periodic trend described above: elements on the right side of the periodic chart do not give up electrons easily when forming bonds; their electronegativity is high. They tend to either accept electrons from metals to form salts or share electrons with other nonmetals to form covalent compounds.

Metalloids typically show physical characteristics (e.g., electrical conductivity) intermediate between the metals and nonmetals. Metalloids typically act more like nonmetals than metals in their chemistry. They more often combine with nonmetals to form covalent compounds rather than salts, but they can do both. This reflects their intermediate position on the Periodic Table. They can form alloys with metals and with the other metalloids. Semiconductors are typically made from combinations of two metalloids. The minor constituent, for example germanium, is said to be "doped" into the major constituent, which is often silicon.

The boundaries between metals, nonmetals, and metalloids are arbitrary. The changes in properties as one moves from element to element on the chart are gradual.

Earth's atmosphere contains slightly more than 20% oxygen. Because oxygen is quite reactive, most elements can be found in nature as oxides. The alkali metals (Group 1) and alkaline earths (Group 2) were so named because the metallic oxides formed when the metals reacted with oxygen produced basic solutions when dissolved in water. Metallic oxides are known as basic anhydrides (anhydrous, meaning without water), because basic solutions are formed when they are added to water.

Nonmetallic elements combine with oxygen to form oxides, many of which, such as carbon dioxide, sulfur dioxide and nitrogen dioxide, are gases. When oxides of nonmetallic elements are dissolved in water, they tend to form acidic solutions or neutral solutions. Nonmetal oxides that form acidic solutions when dissolved in water are called acid anhydrides.

Transition metals react with oxygen to form a wide variety of oxides, some of which are basic and some acidic. A few transition metals are relatively unreactive and may be found in nature as pure elements.

See also Atmospheric chemistry; Atomic mass and weight; Atomic number; Atomic theory; Atoms; Chemical bonds and physical properties; Chemical elements

PERMAFROST

About 20% of Earth's surface is covered by permafrost, land that is frozen year-round. Permafrost occurs at high latitudes or at very high altitudes—anywhere the mean annual **soil temperature** is below **freezing**. About half of Canada and Russia, much of northern China, most of Greenland and Alaska, and probably all of **Antarctica** are underlain by permafrost. Areas underlain by permafrost are classified as belonging to either the continuous zone or the discontinuous zone. Permafrost occurs everywhere within the continuous zone, except under large bodies of **water**, and underlies the discontinuous zone in irregular zones of varying size. Fairbanks, Alaska, lies within the discontinuous zone, while Greenland is in the continuous zone.

The surface layer of soil in a permafrost zone may thaw during the warmer months, and the upper layer of the frozen zone is known as the permafrost table. Like the **water table**, it may rise and fall according to environmental conditions. When the surface layer thaws, it often becomes waterlogged because the meltwater can only permeate slowly, or not at all, into the frozen layer below. **Partial melting** coupled with irregular drainage leads to the creation of hummocky **topography**. Walking on permafrost is extremely difficult, because the surface is spongy, irregular, and often wet. Waterlogging of the surface layer also causes slopes in permafrost areas to be unstable and prone to failure.

Permafrost provides a stable base for construction only if the ground remains frozen. Unfortunately, construction often warms the ground, thawing the upper layers. Special care must be taken when building in permafrost regions, and structures are often elevated above the land surface on stilts. The Trans-Alaska Pipeline, along much of its length, is elevated on artificially cooled posts, and communities in permafrost regions often must place pipes and wires in above-ground conduits rather than burying them. Even roads can contribute to warming and thawing of permafrost, and are generally built atop a thick bed of gravel and dirt.

See also Creep

PERMEABILITY • *see* POROSITY AND PERMEABILITY

PERMIAN PERIOD

In **geologic time**, the Permian Period, the last period of the **Paleozoic Era**, covers the time roughly 286 million years ago (mya) until 245 mya.

The Permian Period spans two epochs. The Early Permian Epoch is the most ancient, followed by the Late Permian Epoch.

The Early Permian Epoch is divided chronologically (from the most ancient to the most recent) into the Asselian, Sakmarian, and Artinskian stages. The Late Permian Epoch is divided chronologically (from the most ancient to the most recent) into the Kungurian, Kazanian, and Tatarian stages.

In terms of paleogeography (the study of the **evolution** of the continents from **supercontinents** and the establishment of geologic features), the Permian Period was dominated by the movements of the supercontinent Pangaea, that during the Permian Period was located along the equator. Plate tectonic activity along the western border of Pangaea formed an extensive **subduction zone** that survives today as a large number of volcanoes located around the Pacific rim (i.e., the Pacific "Ring of Fire").

Differentiated by fossil remains and continental movements, the Carboniferous Period (360 mya to 286 mya) preceded the Permian Period. In many modern geological texts, especially those in the United States, the time of Carboniferous Period is covered by two alternate geologic periods, the **Mississippian Period** (360 mya to 325 mya) and the **Pennsylvanian Period** (325 mya to 286 mya). The Permian Period is followed in geologic time by start of the **Triassic Period** of the **Mesozoic Era**. The largest mass extinction in Earth's history—a catastrophic extinction of marine life—marks the close of both the Permian Period and the Paleozoic Era. Accordingly, many **fossils** dated to the Permian Period are not found in Mesozoic Era formations.

The **fossil record** indicates that more than 95% of all Permian species became extinct at the close of the Permian Period. Alternative hypotheses integrate differently the effects of loss of marine habitat due to the continued fusion of continents into Pangaea.

There were a number of major impacts from large meteorites during the Permian Period. Although no crater has been specifically identified with the impact possibly associated with the mass extinction of species, indirect evidence in the form of catastrophically fused **quartz** crystals (shocked quartz) in **area** of **Antarctica** indicates that the crater measured approximately 300 mi (450 km) in diameter. Other but smaller impact craters dating to the Permian Period have been identified in modern Florida, Quebec, and Brazil.

Because of the fusion and confluence of continental land masses in Pangaea, locations as diverse as Texas (Glass Mountains), Nova Scotia (Brule Trackways), and Germany share a similar fossil record dating to the Permian Period.

See also Archean; Cambrian Period; Cenozoic Era; Cretaceous Period; Dating methods; Devonian Period; Eocene Epoch; Evolution, evidence of; Fossils and fossilization; Historical geology; Holocene Epoch; Jurassic Period; Miocene Epoch; Oligocene Epoch; Ordovician Period; Paleocene Epoch; Phanerozoic Era; Pleistocene Epoch; Pliocene Epoch; Precambrian; Proterozoic Era; Quaternary Period; Silurian Period; Tertiary Period

PERUTZ, MAX (1914-2002)

English crystallographer, molecular biologist, and biochemist

Max Perutz transformed a fascination of geological processes and crystal structure into one of the fundamental techniques upon which modern molecular biology was founded. Ultimately, Perutz pioneered the use of x-ray crystallography to determine the atomic structure of proteins by combining two lines of scientific investigation—the physiology of hemoglobin and the **physics** of x-ray crystallography. His efforts resulted in his sharing the 1962 Nobel Prize in chemistry with his colleague, biochemist John Kendrew. A passionate mountaineer and skier, Perutz also applied his expertise in x-ray crystallography to the study of glacier structure and flow.

Perutz's work in deciphering the diffraction patterns of protein **crystals** opened the door for molecular biologists to study the structure and function of enzymes—specific proteins that are the catalysts for biochemical reactions in cells. Known for his impeccable laboratory skills, Perutz produced the best early pictures of protein crystals and used this ability to determine the structure of hemoglobin and the molecular mechanism by which it transports **oxygen** from the lungs to tissue.

Perutz was born in Vienna, Austria, on May 19, 1914. His parents were Hugo Perutz, a textile manufacturer, and Adele Goldschmidt Perutz. In 1932, Perutz entered the University of Vienna, where he studied organic chemistry. In 1936, Perutz landed a position as research student in the Cambridge laboratory of Desmond Bernal, who was pioneering the use of x-ray crystallography in the field of biology. Perutz, however, was disappointed again when he was assigned to research **minerals** while Bernal closely guarded his crystallography work, discussing it only with a few colleagues and never with students.

Perutz's received excellent training in the promising field of x-ray crystallography, albeit in the classical mode of mineral crystallography.

In the early 1930s, crystallography had been successfully used only in determining the structures of simple crystals of **metals**, minerals, and salts. However, proteins such as hemoglobin are thousands of times more complex in atomic structure. Physicists William Bragg and Lawrence Bragg, the only father and son to share a Nobel Prize, were pioneers of x-ray crystallography. Focusing on minerals, the Braggs found that as x rays pass through crystals, they are buffeted by atoms and emerge as groups of weaker beams which, when photographed, produce a discernible pattern of spots. The Braggs discovered that these spots were a manifestation of Fourier synthesis, a method developed in the nineteenth century by French physicist Jean Baptiste Fourier to represent regular signals as a series of sine waves. These waves reflect the distribution of atoms in the crystal.

The Braggs successfully determined the amplitude of the waves but were unable to determine their phases, which would provide more detailed information about crystal structure. Although amplitude was sufficient to guide scientists through a series of trial and error experiments for studying simple crystals, proteins were much too complex to be studied with such a haphazard and time consuming approach.

Initial attempts at applying x-ray crystallography to the study of proteins failed, and scientists soon began to wonder whether proteins in fact produce x-ray diffraction patterns. However, in 1934, Desmond Bernal and chemist Dorothy Crowfoot Hodgkin at the Cavendish laboratory in Cambridge discovered that by keeping protein crystals wet, specifically with the liquid from which they precipitated, they could be made to give sharply defined x-ray diffraction patterns. Still, it would take 23 years before scientists could construct the first model of a protein molecule.

Perutz and his family, like many other Europeans in the 1930s, tended to underestimate the seriousness of the growing Nazi regime in Germany. While Perutz himself was safe in England as Germany began to invade its neighboring countries, his parents fled from Vienna to Prague in 1938. That same summer, they again fled to Switzerland from Czechoslovakia, which would soon face the onslaught of the approaching German army. Perutz was shaken by his new classification as a refugee and the clear indication by some people that he might not be welcome in England any longer. He also realized that his father's financial support would certainly dwindle and die out.

As a result, in order to vacation in Switzerland in the summer of 1938, Perutz sought a travel grant to apply his expertise in crystallography to the study of glacier structures and flow. His research on **glaciers** involved crystallographic studies of snow transforming into **ice**, and he eventually became the first to measure the velocity distributions of a glacier, proving that glaciers flow faster at the surface and slower at the glacier's bed.

Finally, in 1940, the same year Perutz received his Ph.D., his work was put to an abrupt halt by the German invasions of Holland and Belgium. Growing increasingly wary of foreigners, the British government arrested all enemy aliens, including Perutz. Transported from camp to camp, Perutz ended up near Quebec, Canada, where many other scientists and intellectuals were imprisoned, including physicists Herman Bondi and Tom Gold. Always active, Perutz began a camp university, employing the resident academicians to teach courses in their specialties. It didn't take the British government long, however, to realize that they were wasting valuable intellectual resources and, by 1941, Perutz followed many of his colleagues back to his home in England and resumed his work with crystals.

Perutz, however, wanted to contribute to the war effort. After repeated requests, he was assigned to work on the mysterious and improbable task of developing an aircraft carrier made of ice. The goal of this project was to tow the carrier to the middle of the Atlantic Ocean, where it would serve as a stopping post for aircrafts flying from the United States to Great Britain. Although supported both by then British Prime Minister Winston Churchill and the chief of the British Royal Navy, Lord Louis Mountbatten, the ill-fated project was terminated upon the discovery that the amount of steel needed to construct and support the ice carrier would cost more than constructing it entirely of steel.

Perutz married Gisela Clara Peiser in 1942; the couple later had a son and a daughter. After the war, in 1945, Perutz was finally able to devote himself entirely to the study of hemoglobin crystals. He returned to Cambridge, and was soon joined by John Kendrew. In 1946 Perutz and Kendrew founded the Medical Research Council Unit for Molecular Biology, and Perutz became its director. Many advances in molecular biology would take place there, including the discovery of the structure of deoxyribonucleic acid (DNA).

Over the next years, Perutz refined the x-ray crystallography technology. Often bogged down by tedious mathematical calculations, the development of computers hastened the process tremendously.

By 1957, Kendrew had delineated the first protein structure through crystallography, again working with myoglobin. In 1962, Perutz and Kendrew were awarded the Nobel Prize in chemistry for their codiscoveries in x-ray crystallography and the structures of hemoglobin and myoglobin, respectively. The same year, Perutz left his post as director of the Unit for Molecular Biology and became chair of its laboratory.

Perutz was a Fellow of the Royal Society. He died on February 6, 2002.

See also Atomic theory; Crystals and crystallography

PETROLEUM

Petroleum is a term that includes a wide variety of liquid **hydrocarbons**. Many scientists also include **natural gas** in their definition of petroleum. The most familiar types of petroleum are tar, oil, and natural gas. Petroleum forms through the accumulation, burial, and transformation of organic material—such as the remains of plants and animals—by chemical reactions over long periods of time. After petroleum has been generated, it migrates upward through the earth, seeping out at the surface of the earth if it is not trapped below the surface. Petroleum accumulates when it migrates into a porous **rock** called a reservoir that has a non-porous seal or cap rock that prevents the oil from migrating farther. To fully understand how petroleum forms and accumulates requires considerable knowledge of **geology**, including **sedimentary rocks**, geological structures (faults and domes, for example), and forms of life that have been fossilized or transformed into petroleum throughout the earth's long history.

Tremendous petroleum reserves have been produced from areas all over the world. In the United States, the states of Alaska, California, Louisiana, Michigan, Oklahoma, Texas, and Wyoming are among the most important sources of petroleum. Other countries that produce great amounts of petroleum include Saudi Arabia, Iran, Iraq, Kuwait, Algeria, Libya, Nigeria, Indonesia, the former Soviet Union, Mexico, and Venezuela.

Petroleum products have been in use for many years. Primitive man might have used torches made from pieces of wood dipped in oil for lighting as early as 20,000 B.C. At around 5,000 B.C., the Chinese apparently found oil when they were digging underground. Widespread use of petroleum probably began in the Middle East by the Mesopotamians, perhaps by 3,000 B.C., and probably in other areas where oil seeps were visible at the surface of the earth. Exploration for petroleum in the United States began in 1853, when George Bissell, a lawyer, recognized the potential use of oil as a source of lamp fuel. Bissell also recognized that boring or drilling into the earth, as was done to recover salt, might provide access to greater supplies of petroleum than surface seeps. In 1857, Bissell hired Edwin Drake—often called "Colonel" Drake despite having worked as a railroad conductor—to begin drilling the first successful oil well. The well was drilled in 1859 in Titusville, Pennsylvania. Once the usefulness of oil as a fuel was widely recognized, exploration for oil increased. By 1885, oil was discovered in Sumatra, Indonesia. The famous "gusher" in the Spindletop field in eastern Texas was drilled in 1901. The discoveries of giant oil fields in the Middle East began in 1908 when the company now known as British Petroleum drilled a well in Persia (now Iran). During World Wars I and II, oil became a critical factor in the ability to successfully wage war.

Currently, petroleum is among our most important natural resources. We use gasoline, jet fuel, and diesel fuel to run cars, trucks, aircraft, ships, and other vehicles. Home heat sources include oil, natural gas, and **electricity**, which in many areas is generated by burning natural gas. Petroleum and petroleum-based chemicals are important in manufacturing plastic, wax, fertilizers, lubricants, and many other goods. Thus, petroleum is an important part of many human activities.

Petroleum, including liquid oil and natural gas, consists of substances known as hydrocarbons. Hydrocarbons, as their name suggests, comprise hydrogen and **carbon**, with small amounts of impurities such as nitrogen, **oxygen**, and sulfur. The molecules of hydrocarbons can be as simple as that of methane, which consists of a carbon **atom** surrounded by four hydrogen atoms, abbreviated as CH_4. More complex hydrocarbons, such as naphthenes, include rings of carbon atoms (and attached hydrogen atoms) linked together. Differences in the number of hydrogen and carbon atoms in molecules as well as their molecular structure (carbon atoms arranged in a ring structure, chain, or tetrahedron, for example) produce numerous types of petroleum.

Different types of petroleum can be used in different ways. Jet fuel differs from the gasoline that automobiles consume, for example. Refineries separate different petroleum products by heating petroleum to the point that heavy hydrocarbon molecules separate from lighter hydrocarbons so that each product can be used for a specific purpose. Refining reduces the waste associated with using limited supplies of more expensive petroleum products in cases in which a cheaper, more plentiful type of petroleum would suffice. Thus, tar or asphalt, the dense, nearly solid hydrocarbons, can be used for road surfaces and roofing materials, waxy substances called paraffins can be used to make candles and other products, and less dense, liquid hydrocarbons can be used for engine **fuels**.

Petroleum is typically found beneath the surface of the earth in accumulations known as fields. Fields can contain oil, gas, tar, **water**, and other substances, but oil, gas, and water are the most common. In order for a field to form, there must be some sort of structure to trap the petroleum, a seal on the trap that prohibits leakage of the petroleum, and a reservoir rock that has adequate pore space, or void space, to hold the petroleum. To find these features together in an **area** in which petroleum has been generated by chemical reactions affecting organic remains requires many coincidences of timing of natural processes.

Petroleum generation occurs over long periods of time—millions of years. In order for petroleum generation to occur, organic matter such as dead plants or animals must accumulate in large quantities. The organic matter can be deposited along with sediments and later buried as more sediments accumulate on top. The sediments and organic material that accumulate are called source rock. After burial, chemical activity in the absence of oxygen allows the organic material in the source rock to change into petroleum without the organic matter simply rotting. A good petroleum source rock is a sedimentary rock such as shale or **limestone** that contains between 1% and 5% organic carbon. Rich source rocks occur in many environments, including **lakes**, deep areas of the **seas** and **oceans**, and swamps. The source rocks must be buried deep enough below the surface of the earth to heat up the organic material, but not so deep that the rocks metamorphose or that the organic material changes to **graphite** or materials other than hydrocarbons. Temperatures less than 302°F (150°C) are typical for petroleum generation.

Once a source rock generates and expels petroleum, the petroleum migrates from the source rock to a rock that can store the petroleum. A rock capable of storing petroleum in its pore spaces, the void spaces between the grains of sediment in a rock, is known as a reservoir rock. Rocks that have sufficient pore space through which petroleum can move include **sandstone**, limestone, and rocks that have many fractures. A good reservoir rock might have pore space that exceeds 30% of the rock volume. Poor quality reservoir rocks have less than 10% void space capable of storing petroleum. Rocks that lack pore space tend to lack **permeability**, the property of rock that allows fluid to pass through the pore spaces of the rock. With very few pores, it is not likely that the pores are connected and less likely that fluid will flow through the rock than in a rock with larger or more abundant pore spaces. Highly porous rocks tend to have better permeability because the greater number of pores and larger pore sizes tend to allow fluids to move through the reservoir more easily. The property of permeability is critical to producing petroleum: if fluids can not migrate through a reservoir rock to a petroleum production

Barrels of motor oil are one product of the distillation, or cracking, of crude oil. ***© Vince Streano/Corbis. Reproduced by permission.***

well, the well will not produce much petroleum and the money spent to drill the well has been wasted.

In order for a reservoir to contain petroleum, the reservoir must be shaped and sealed like a container. Good petroleum reservoirs are sealed by a less porous and permeable rock known as a seal or cap rock. The seal prevents the petroleum from migrating further. Rocks like shale and salt provide excellent seals for reservoir rocks because they do not allow fluids to pass through them easily. Seal-forming rocks tend to be made of small particles of sediment that fit closely together so that pore spaces are small and poorly connected. The permeability of a seal must be virtually zero in order to retain petroleum in a reservoir rock for millions to hundreds of millions of years, the time span between formation of petroleum to the discovery and production of many petroleum fields. Likewise, the seal must not be subject to forces within the earth that might cause fractures or other breaks in the seal to form.

Reservoir rocks and seals work together to form a trap for petroleum. Typical traps for petroleum include hills shaped similar to upside-down bowls below the surface of the earth, known as anticlines, or traps formed by faults. Abrupt changes in rock type can form good traps, such as sandstone deposits next to shale deposits, especially if a **sand** deposit is encased in a rock that is sufficiently rich in organic matter to act as a petroleum source and endowed with the properties of a good seal.

An important aspect of the formation of petroleum accumulations is timing. The reservoir must have been deposited prior to petroleum migrating from the source rock to the reservoir rock. The seal and trap must have been developed prior to petroleum accumulating in the reservoir, or else the petroleum would have migrated farther. The source rock must have been exposed to the appropriate **temperature** and pressure conditions over long periods of time to change the organic matter to petroleum. The necessary coincidence of several conditions is difficult to achieve in nature.

Petroleum exploration and production activities are performed primarily by geologists, geophysicists, and engineers. Geologists look for areas of the earth where sediments accumulate. They then examine the area of interest more closely to determine whether or not source rocks and reservoir rocks exist there. They examine the rocks at the surface of the earth and information from wells drilled in the area. Geologists also examine **satellite** images of large or remote areas to evaluate the rocks more quickly.

Geophysicists examine seismic data, data derived from recording waves of energy introduced into the rock layers of the earth through dynamite explosions or other means, to determine the shape of the rock layers beneath the surface and whether or not traps such as faults or anticlines exist.

Once the geologist or geophysicist has gathered evidence of potential for a petroleum accumulation, called a prospect, an engineer assists in determining how to drill a well or multiple wells to assess the prospect. Drilling a well to explore for petroleum can cost as little as $100,000 and as much as $30,000,000 or more, depending on how deep the well must be drilled, what types of rocks are present, and how remote the well location is. Thus, the scientists must evaluate how much the well might cost, how big the prospect might be, and how likely the scientific predictions are to be correct. In general, approximately 15% of exploration wells are successful.

Once a successful exploration well has been drilled, the oil and/or gas flow are pumped to the surface of the earth through the well. At the surface, the petroleum either moves through a pipeline or is stored in a tank or on a ship until it can be sold.

Estimates of the amount of recoverable oil and natural gas in the United States are 113 billion barrels of oil and 1,074 trillion cubic feet of natural gas. Worldwide estimates of recoverable oil and natural gas are 1 trillion barrels of oil and 5 quadrillion cubic feet of natural gas. These worldwide reserves are expected to supply 45 years of fuel at current production rates with expected increases in demand. However, such estimates do not take into account reserves added through new discoveries or through the development of new technology that would allow more oil and natural gas to be recovered from existing oil and natural gas fields.

Daily consumption of oil in the United States exceeds 17 million barrels of oil per day, of which approximately 7 million barrels are in the form of gasoline for vehicles. Over half the petroleum consumed in the United States is imported from other countries. (Assuming oil costs $20 per barrel and 8.5 million barrels per day are imported, over one billion dollars per week are spent on oil imports). While the United States has tremendous reserves of petroleum, the undiscovered fields that remain tend to be smaller than the fields currently producing petroleum outside of the United States. Thus, less expensive foreign reserves are imported to the United States. When foreign petroleum increases in price, more exploration occurs in the United States as it becomes more profitable to drill wells in order to exploit smaller reservoirs.

Current research in petroleum includes many different activities. Within companies that explore for and produce petroleum, scientists and engineers try to determine where they should explore for petroleum, how they might recover more petroleum from a given field, and what types of tools can be lowered into wells in order to enhance our understanding of whether or not that individual well might have penetrated an oil or gas field. They also examine fundamental aspects of how the earth behaves, such as how rocks form and what forms of life have existed at various times in the earth's history. The United States Geological Survey continues to evaluate petroleum reserves and new technology to produce oil and gas. The federal government operates several facilities called Strategic Petroleum Reserves that store large quantities of petroleum for use in times of supply crisis.

Petroleum exploration specialists are using a type of geophysical data known as three-dimensional seismic data to study the structures and rock types below the surface of the earth in order to determine where exploration wells might successfully produce petroleum. Geochemists are assessing the results of studies of the **chemistry** of the surface of the earth and whether or not these results can improve the predictions of scientists prior to drilling expensive exploratory wells.

Significant recent discoveries of petroleum have been made in many areas of the world: Algeria, Brazil, China, Egypt, Indonesia, the Ivory Coast, Malaysia, Papua New Guinea, Thailand, the United Kingdom, and Vietnam, among others. In the United States, the **Gulf of Mexico**, Gulf Coast states, California, and Alaska continue to attract the interest of explorers.

See also Fossils and fossilization; Fuels and fuel chemistry; Geochemistry; Petroleum detection; Petroleum, economic uses of; Petroleum extraction; Petroleum, history of exploration; Sedimentation; Syncline and anticline

Petroleum Detection

Four main issues control the occurrence and distribution of oil and gas: source, reservoir, seal, and trap.

A source is a fine-grained **rock** unit containing sufficient organic matter so that when it is heated and/or placed under pressure (maturation), **hydrocarbons** are generated. If the organic matter is of marine algal origin, then the source rock is most likely to generate oil under optimum maturation conditions, whereas rocks dominated by land plant matter will tend to create gaseous hydrocarbons. The hydrocarbons are of lower **gravity** than the surrounding **groundwater** and, therefore, move away from and generally upwards (migration) from the source rock until they are trapped in a reservoir.

A reservoir is a rock unit that acts as a storage device for the hydrocarbons that migrate from the source rock. Hydrocarbons are retained within the reservoir because these rocks contain numerous pores (essentially microscopic-sized holes) between the mineral grains making up the fabric of the reservoir. In good quality reservoirs the **porosity** is frequently over 20% of the rock volume. However, the pores need to be interlinked in such a manner that the fluids can move into (and out of, if we are to exploit the oil and gas) the reservoirs over geological time. This is known as **permeability**. There are two main rock types that make up the giant reservoirs around the world—sandstones that are made up of **sand** grains (**quartz** and **feldspar** in the majority) and carbonates that are made up of organically created calcium carbonate grains (corals, algae and shells) or mud. In order to stop the upward movement of hydrocarbons and constrain them to one zone of the subsurface (trap), there must be a barrier to prevent fluid migration. Generally this mechanism or seal consists of rocks that are impermeable to fluid flow. The most effective of these seals are mudstones or shales, very fine-grained rocks containing abundant **clay minerals**. Occasionally the impermeable layers are dense **igneous rocks** and in rare situations, there may be significant rock and fluid pressure differences in a region that prevents fluid flow and acts as a seal.

Equally important is the presence of a trapping mechanism. They are either of structural form where the reservoir rock unit is contorted to produce a zone where fluids naturally accumulate against a seal, or of stratigraphic nature where a reservoir rock unit changes laterally into an impermeable unit reflecting changes in depositional environment along one bed. Upwarping of rocks (anticlinal **folds**) are particularly good at trapping hydrocarbons along with faults where permeable titled reservoir strata are moved up against impermeable strata.

Detection of hydrocarbons in the subsurface during exploration takes a number of forms: direct identification of hydrocarbons at the surface, direct hydrocarbon indicators (DHI) in the subsurface, and indirect indicators both at the surface and in the subsurface. Traditionally, oil exploration was primarily conducted by recognizing seeps of hydrocarbons at the surface. The Chinese, for example, used oil (mostly bitumen) obtained from seeps for use in medication, waterproofing, and warfare several thousand years ago. The ancient Chinese frequently dug shallow pits or horizontal tunnels at the seep locations in order to recover the oil. In Baku, Azerbaijan, there are still gas and oil seeps that are permanently alight and have been used to light caravanserai since the times of Marco Polo and the Silk route. With the dawning of the modern era in Oil Creek, Pennsylvania, Colonel Edwin Drake drilled the first well to intentionally look for oil in the subsurface in 1859. Again, this was based on direct identification of seeped hydrocarbons at the surface. Initially the oil was used to provide kerosene for lamps but the later invention of automobiles drove up demand and ushered in modern methods of oil exploration.

Around the turn of the century and up until the 1950s, the main exploration tool used for finding oil was the use of intensive and detailed geological mapping. This was frequently in terrain that was remote and inhospitable. The early pioneers working their way through the jungles of Burma, the deserts of Iraq, or the mountains of Iran, would conduct detailed evaluations of the nature and distribution of rock units that could represent potential reservoirs, seals, and source units as well as frequency, orientation, and geological history of folds or faults that could act as traps for the migrating hydrocarbons. If all four of the features required for oil or gas to be created and trapped can be recognized in a region, then a variety of play concepts can be generated. Detailed local study might identify a suitable target (prospect) and then a shallow well would be drilled to test the features.

One of the most important recent discoveries in **petroleum** studies has been **plate tectonics**. Not only has this revolutionized the earth sciences, but also it has provided a conceptual setting for oil exploration. The movement of plates around the surface of Earth creates large-scale depressions into which substantial quantities of sediments eroded from the surrounding high ground may accumulate. These accumulations can exceed thicknesses of several thousand kilometers and are referred to as sedimentary basins. By comparison of basins around the world and by analogy to existing producing hydrocarbon regions, an exploration team can say which basins are worth looking at in more detail. Then explorationists will spend time ensuring that within the basin there are present all the key elements that control the presence of hydrocarbons. Assuming that all the needed features are present, the team would agree that the basin contained a viable petroleum system and prospect generation can proceed.

In modern exploration programs, the mapping of gravity and magnetic anomalies would normally be the first two methods to be applied to a new basin or region being evaluated. These techniques would be used to identify large-scale changes in the structure of the basement and sedimentary basins together with major differences in rock density such as the influx of dense igneous rocks or light salt into a sedimentary sequence. These techniques are large scale, can be applied over both land and **water** and can even be collected remotely from plane or **satellite**.

At the same time, **remote sensing** of onshore areas initially based on large scale photogeological surveys and, after the 1970s, by satellite imaging, can identify areas with anticlinal and faulted structural features, seeps or salt domes frequently associated with oil occurrences. Offshore remote sensing of the sea surface can lead to the identification of slicks associated with the seepage of oil (both natural and man-made) into the water column. A coarse two-dimensional grid of seismic data is then collected to obtain a picture of the subsurface in the **area** to be targeted. Seismic data collection involves the generation of a seismic wave using an energy source such as an air-gun in water, dynamite in drill holes inland or a truck with a plate that is thumped down onto the road/soil surface (vibroseis). The wave travels through the earth's rock layers and reflects back off key surfaces. The time taken for the waves to be received back at the surface along with their strength is recorded via geophones and displayed on a seismic section. Processing the two-dimensional seismic sections using highly sophisticated software reveals the detailed structure of the subsurface and in certain circumstances shows the presence of direct hydrocarbon indicators such as bright spots associated with gas/water differences. Primarily, though, seismic is used to indicate the nature of folded and faulted structures that could prove to be suitable hydrocarbon traps. These are frequently referred to as leads.

The objective of seismic acquisition and processing is to acoustically image the subsurface in a geologically accurate manner with as high resolution as possible. For a detailed analysis of a small area representing a field or prospect, a high density and calibrated three-dimensional seismic is collected. Modern technology also allows scientists to accurately map changes in fluid movements through time (repeat multiple 3-D seismic surveys, known as 4-D seismic) and this technique is now particularly important in monitoring production performance of the reservoir.

Ultimately, however, the only way of confirming the presence or absence of hydrocarbons at depth is by drilling the prospect. In certain areas of the world where drilling is cheap and the subsurface has been explored extensively, such as certain onshore basins of the United States, drilling is commonly preferred to extensive and expensive seismic acquisition. Wells are then analyzed using electric, sonic, and radioactive logging techniques that measure characteristics of the rocks and fluids. These methods can identify the presence of oil and gas, which can then be tested to see if they occur at commercially viable production levels. On the other hand, at a cost of over ten million dollars per offshore exploration well, the oil companies are also likely to employ the sophisticated battery of direct and indirect detection techniques first before resorting to drilling in these areas.

See also Petroleum, economic uses of; Petroleum extraction; Petroleum, history of exploration

PETROLEUM, ECONOMIC USES OF

From the dawn of time up through the late 1800s, economic development depended largely on the strength of man, animal, and to limited use **water**, **wind**, and steam. Economic conditions progressed from clans of primitive gatherers to reasonably advanced agricultural societies. As the industrialized age was in its early stages, the primary sources of energy were wood, **coal**, and whale oil. The environmental impact of utilizing these energy sources was extreme and growing worse.

Petroleum has been a known commodity through areas of natural seepage to the surface since early man, but was generally inaccessible to the masses. Its full potential had not yet begun to be realized. Population growth placed great economic stress on traditional **fuels**, and rising prices encouraged the search for alternatives.

In 1854, Canadian Abraham Gesner discovered an alternative to whale oil for use in lighting lamps by distilling kerosene from coal and oil. Edwin Laurentine Drake drilled the first successful oil well in 1859 in Titusville, Pennsylvania, and created an industry that would go on to make petroleum the most significant single economic factor to date over the entire history of the world. Industries not possible without petroleum and its derivatives now dominate world economy.

Petroleum exists within the substrata of the earth in a number of forms depending on the hydrocarbon source, maturation process, elemental exposure, and the **temperature** and pressure of the reservoir. Crude oil is the most common form of petroleum and may range in specific **gravity** from being as light as 0.73 to being as heavy as over 1.07. Under standard surface conditions, the lightest crude oil will be a thin liquid of a brown or brownish-blue/green color, while the heaviest will be a black solid tar-like substance. The corresponding physical properties and chemical compositions also vary widely and determine which products may be derived from each specific crude oil and what refining processes will be most efficient in doing so.

Early refining techniques yielded barely 4.5 gal (17 L) of gasoline per standard barrel of crude oil (42 gal, or 159 L) and much of the remaining raw materials were underutilized. Continuing technical advances in a wide range of chemical processes, however, has significantly improved the efficient conversion of a barrel of crude oil to an ever-widening range of products that contribute to almost every facet of modern life. The typical barrel now yields 21 gal (80 L) of gasoline, 3 gal (11 L) of jet fuel, 9 gal (34 L) of distillates and petrochemical feedstock, 4 gal (15 L) of lubricants, and 3 gal (11 L) of heavy residue.

Gasoline is the primary fuel used to power internal combustion engines widely used in vehicles and machines. Jet fuel is used to power the extremely powerful engines that drive high performance aircraft.

Distillates are used to produce lower grade fuels such as kerosene for use as a heating fuel and diesel fuel for use in powerful vehicles such as trucks, ships and industrial machinery. Other even lower grade fuels are used to provide energy to industrial processes not requiring the same combustion quality required by higher speed engines. Distillates also yield a wide variety of waxes that are turned into products used for lining milk cartons, as water repellant coatings, cosmetics, electrical insulators, sealants, medicinal tablet coatings, crayons, candles, and many other everyday items.

Petrochemical feedstock is processed into supplying an ever growing assortment of products such as anti-freeze, bases for paints, cleaning agents, detergents, dyes, explosives, fertilizers, industrial resins, plastics, synthetic fibers (nylon, polyester, rayon), synthetic rubber, solvents, thinners, and varnishes. Though all of these products have helped improve how people live, the impact of plastics is among the most consequential petroleum products in the civilized world.

Lubricants help overcome friction and are produced in an assortment of greases and oils used to lubricate moving parts in machinery; pull electrical wire through insulating conduit; lubricate sewing needles, sliding doors, heavy loads, and surgical medical equipment; and to reduce drag on surf boards as they pass through water.

Finally, the heavy residue left over is in the form of tar, pitch, and asphalt. Tar and pitch were first discovered by early man laying in surface seeps or pools, having been cooked off from oil deposits deep within Earth's surface and was used to seal boats and preserve wood. Today, more refined forms of these heavy residues are used in much the same way and also to pave roadways.

See also Fuels and fuel chemistry; Geochemistry; Petroleum detection; Petroleum extraction; Petroleum, history of exploration

PETROLEUM EXTRACTION

After an exploration effort has successfully discovered **petroleum** within an acceptable range of reserve potential, the challenge becomes how to best optimize extraction of recoverable reserves in a manner yielding an acceptable economic return on total cash expenditures required over the life of the project. Surface and subsurface conditions of a discovery have considerable impact on the extraction process, its related costs, and ultimate project success or failure. Technical success is one thing; economic success is another. Real world experience has shown that economic success is by far the more difficult accomplishment, as it is dependent on factors well beyond the means of science and technology.

Petroleum reserves exist as oil or gas within trapping sections of reservoir **rock** formed by structural and or stratigraphic geologic features. **Water** is the predominant fluid found in the **permeability** and **porosity** of subsurface strata within the earth's **crust**. Both oil and gas have a low specific **gravity** relative to water and will thus, float through the more porous sections of reservoir rock from their source **area** to the surface unless restrained by a trap. Typically, reservoir rock consists of **sand**, **sandstone**, **limestone**, or **dolomite**. A trap is a reservoir that is overlain by a dense cap rock or a zone of very low or no porosity that restrains migrating hydrocarbon. Petroleum bearing reservoirs can exist from surface seeps to subsurface depths over 4 mi (6.4 km) below sea level. Reservoirs vary from being quite small to covering several thousands of acres, and range in thickness from a few inches to hundreds of feet or more.

The process of evaluating how to best optimize extraction of recoverable reserves begins with a development plan. The development plan considers all available geologic and engineering data to make an initial estimate of reserves in-place, to project recovery efficiencies and optimal recoverable reserve levels under various producing scenarios, and to evaluate development plan alternatives. Development alternatives will include the number of wells to be drilled and completed for production or injection, well spacing and pattern, processing facility requirements, product transportation options, cost projections, project schedules, depletion plans, operational programs, and logistics and economic studies.

In general, petroleum is extracted by drilling wells from an appropriate surface configuration into the hydrocarbon-bearing reservoir or reservoirs. Wells are designed to contain and control all fluid flow at all times throughout drilling and producing operations. The number of wells required is dependent on a combination of technical and economic factors used to determine the most likely range of recoverable reserves relative to a range of potential investment alternatives.

The complexity and cost of drilling wells and installing all necessary equipment to produce reserves can vary significantly. The development of an onshore shallow gas reservoir located among other established fields may be comparatively low cost and nominally complex. A deep oil or gas reservoir located in 4,000+ ft (1,219+ m) of water depth located miles away from other existing producing fields will push the limits of emerging technology at extreme costs. Individual wells in deepwater can and have cost in excess of 50 million dollars to drill, complete, and connect to a producing system. Onshore developments may permit the phasing of facility investments as wells are drilled and production established to minimize economic risk. However, offshore projects may require 65% or more of the total planned investments to be made before production start up, and impose significant economic risk.

Once production begins, the performance of each well and reservoir is monitored and a variety of engineering techniques are used to progressively refine reserve recovery estimates over the producing life of the field. The total recoverable reserves are not known with complete certainty until the field has produced to depletion or its economic limit and abandonment.

The ultimate recovery of original in-place volumes may be as high as one-third for oil and 80% or more for gas. There are three phases of recovering reserves. Primary recovery occurs as wells produce because of natural energy from expansion of gas and water within the producing formation, pushing fluids into the well bore and lifting them to the surface. Secondary recovery occurs as artificial energy is applied to lift fluids to surface. This may be accomplished by injecting gas down a hole to lift fluids to the surface, installation of a subsurface pump, or injecting gas or water into the formation itself. Secondary recovery is done when well, reservoir, facility, and economic conditions permit. Tertiary recovery occurs when means of increasing fluid mobility in oil reservoirs within the reservoir are introduced in addition to secondary techniques. This may be accomplished by introducing additional heat into the formation to lower the viscosity (thin the oil) and improve its ability to flow to the well bore. Heat may be introduced by either injecting steam in a "steam flood" or injecting **oxygen** to enable the ignition and combustion of oil within the reservoir in a "fire flood." Such methods are undertaken only in a few unique situations where technical, environmental and economic conditions permit. Most gas reserves are produced during the primary recovery phase. Secondary recovery has significantly contributed to increasing oil recoveries.

Many technical assumptions become better understood and more certain with the evaluation of performance data over the producing life of a field. One of the most critical assumptions, however, remains uncertain and holds project success at risk to the very end—the oil and gas price forecast.

See also Petroleum, detection; Petroleum, economic uses of; Petroleum, history of exploration

PETROLEUM, HISTORY OF EXPLORATION

Exploration for **hydrocarbons** (oil, gas, and condensate) is commonly acknowledged to have begun with the discovery at Oil Creek, Pennsylvania, by "Colonel" Edwin Drake in 1859. However, this was only the start of the modern global era of technology-driven advances in exploration. Traditionally, oil exploration was conducted by recognizing seeps of hydrocarbons at the surface. The Chinese, for example, used oil (mostly bitumen) obtained from seeps in medication, waterproofing, and warfare several thousand years ago. They frequently dug shallow pits or horizontal tunnels at seep locations but also, as early as 200 B.C., drilled down as much as 3,500 ft (1,067 m) using rudimentary bamboo poles (making Drake's 69.5 ft [21.2 m] over 2,000 years later seem puny by comparison). In Baku, Azerbaijan, there are still gas and oil seeps that are permanently on fire and have been used to light caravanserai since the times of Marco Polo and the Silk Route. Similarly, seeps were recognized and exploited in the Caucasus (Groznyy region of Chechnya), Ploesti in Romania, Digboi in Assam, Sanga Sanga in eastern Borneo and Talara in Peru.

The first oil well, drilled by Colonel Edwin Drake near Titusville, Pennsylvania. ***AP/Wide World. Reproduced by permission.***

Even Drake's well, the first to intentionally look for oil in the subsurface, was based on direct identification of seeped hydrocarbons at the surface. Initially, the oil produced was used to provide kerosene for lamps, but the later invention of automobiles drove up demand and ushered in modern methods of oil exploration. In fact, most oil until the turn of the twentieth century was in one form or another related to seep identification. However, one theory developed during this time was to have a profound impact on exploration. In the mid 1800s, William Logan, first Director of the Geological Survey of Canada, recognized oil seeps associated with the crests of convex-upward folded rocks and employed a geologist, Thomas Hunt, to formalize his "anticlinal theory." This idea, however, was only recognized as a viable tool for exploration when Spindletop was discovered on the Gulf Coast of Texas in 1901. For the next 30 years, the anticlinal theory dominated exploration, to the extent that many believed that there were no other types of hydrocarbon accumulation. As a result, geologists became critical to understand the structural configurations of **rock** sequences which, when combined with seep occurrences, proved to be the keys to discovering the main oil-producing provinces of the United States, Mexico, and Venezuela. For a period of time before World War I, Oklahoma, Texas, and California were the World's leading production areas.

Around the turn of the century and up until the 1950s, the main exploration tool used for finding oil was the use of intensive and detailed geological mapping. This was frequently in terrain that was remote and inhospitable. The early pioneers working their way through the jungles of Burmah, India (Burmah oil company, now part of British **Petroleum**), and Borneo (Shell), the deserts of Iraq or the mountains of Iran (the Anglo-Persian Oil Company that became British Petroleum), would conduct detailed evaluations of the nature and distribution of rock units. These rock units represented potential reservoirs, seals, and source units, as well as frequency, orientation, and geological history of **folds** or faults that could act as traps for the migrating hydrocarbons.

It took until the 1920s for explorers to realize that hydrocarbons could occur in situations where no **anticline** was preserved. For example, it was noted as far back as 1880 that oil was trapped in the Venango Sands of Pennsylvania, not in the form of an anticlinal structure, but by the lithologies occurring in a moving palaeoshoreline. In fact, oil trapped by **stratigraphy** was discovered more often by chance rather than design even until the 1970s. By the 1920s, mapping of surface features was complimented by the development of seismic refraction, **gravity**, and magnetic geophysical methods. In particular, gravity and seismic methods proved effective in locating oil trapped against buried salt domes in the onshore **Gulf of Mexico**. At this time, another significant advance in exploration of the subsurface took place with the application of geophysical techniques by the Schlumberger brothers to measuring properties of rocks and fluids encountered whilst drilling for hydrocarbons. In France in 1927, they initially measured the resistivity of the rocks in shallow wells (drilled primarily for **water** distribution), but later went on to add other electric, sonic, and radioactive logging tools. It is now even possible to log **porosity**, **permeability**, **mineralogy**, and fluids and image the structures and rock types downhole. Ultimately, these developments have been one of the main reasons why Schlumberger has become one of the largest electronics companies in the world.

Aerial **remote sensing** for features favored for hydrocarbon accumulation became an important and effective technique, particularly in areas of sparse vegetation cover following World War II when low-cost, rapid reconnaissance of large areas became feasible. Large-scale features such as faults and folds could be identified and targeted for detailed seismic acquisition. In the 1970s, this capability was improved dramatically by the use of **satellite** remote sensing technologies (LANDSAT).

From the 1940s to the 1960s, there were important developments in the understanding of the controls on lateral and vertical variations within reservoir sequences. In particular, the new discipline of sedimentology used modern depositional analogues from around the world to understand the nature, distribution and controls over ancient reservoir sequences. There was also much interest generated over the discovery of carbonate oil-bearing reservoirs in West Texas and Canada (Leduc Reef), and recognition that modern intertidal carbonate-evaporite sequences in the UAE had equivalents in ancient reservoirs. These developments lead to the discovery of many super giant carbonate oil fields in the United States (Yates Field), Mexico (Posa Rica), Middle East (Kirkuk), and Russia (a number of Siberian oil fields).

Other tools such as **geochemistry**, developed during this period, have helped to quantify the level of maturity and the nature and distribution of source potential in a region. Micropalaeontology was developed in Tertiary Basins such as Trinidad and the Caucasus for horizon identification and **correlation** using planktonic foraminefera, but spread rapidly to the United States Gulf Coast. Now, geochemistry and biostratigraphy, including palynology (the study of spores and other organic matter), have become standard tools in the explorationist's armory.

Also beginning in the 1970s, there was a significant advance in the power and reduction in size and cost of computers that has lead directly to a dramatic increase in the ability of geophysicists to acquire, process, and interpret large quantities of seismic data. Initially, this was in the form of 2-D reflection seismic onshore, but this trend has continued to the present day and now oil companies regularly undertake, mostly offshore, 3-D seismic surveys and even 4-D field surveys. Three-dimensional surveys are repeated over the same **area** every few years to monitor fluid movement within reservoirs and thereby optimally manage hydrocarbon recovery. Highly complex three-dimensional models of the subsurface can be displayed on sophisticated workstations or in the form of a fully enclosed room where staff can be totally immersed in the data using special glasses and can "walk through" the reservoirs to, for example, choose the optimal location and direction of wells.

Exploration for oil and gas has progressed dramatically in the last 30 years, driven forward by the ever-increasing power and capabilities of the computer. As a result, it now takes only a fraction of the time required 20 years ago to find and develop oil fields. However, technology in itself does not find oil or gas fields; it frequently requires a flash of inspiration that is the mark of a true explorer to discover some of the major new exploration plays in such areas as Equatorial Guinea, Angola, Nigeria, Trinidad, the Gulf of Mexico, and the northern Canadian Rockies.

See also Fuels and fuel chemistry; Petroleum detection; Petroleum, economic uses of; Petroleum extraction

Petroleum microbiology

Microorganisms play an important role in the formation, recovery, and uses of **petroleum**. Petroleum is broadly considered to encompass both oil and **natural gas**. The microorganisms of concern include bacteria and fungi.

Much of the experimental underpinnings of petroleum microbiology are a result of the pioneering work of Claude ZoBell. Beginning in the 1930s and extending through the late 1970s, ZoBell's research established that bacteria are important in a number of petroleum related processes.

Bacterial degradation can consume organic compounds in the ground, which is a prerequisite to the formation of petroleum.

Some bacteria can be used to improve the recovery of petroleum. For example, experiments have shown that starved bacteria, which become very small, can be pumped down into an oilfield, and then resuscitated. The resuscitated bacteria plug up the very porous areas of the oilfield. When **water** is subsequently pumped down into the field, the water will be forced to penetrate into less porous areas, and can push oil from those regions out into spaces where the oil can be pumped to the surface.

Alternatively, the flow of oil can be promoted by the use of chemicals that are known as surfactants. A variety of bacteria produce surfactants, which act to reduce the surface tension of oil-water mixtures, leading to the easier movement of the more viscous oil portion.

In a reverse application, extra-bacterial polymers, such as glycocalyx and xanthan gum, have been used to make water more gel-like. When this gel is injected down into an oil formation, the gel pushes the oil ahead of it.

A third **area** of bacterial involvement involves the modification of petroleum **hydrocarbons**, either before or after collection of the petroleum. Finally, bacteria have proved very useful in the remediation of sites that are contaminated with petroleum or petroleum by-products.

The bioremediation aspect of petroleum microbiology has grown in importance in the latter decades of the twentieth century. In the 1980s, the massive spill of unprocessed (crude) oil off the coast of Alaska from the tanker Exxon *Valdez* demonstrated the usefulness of bacteria in the degradation of oil that was contaminating both seawater and land. Since then, researchers have identified many species of bacteria and fungi that are capable of utilizing the hydrocarbon compounds that comprise oil. The hydrocarbons can be broken down by bacteria to yield **carbon dioxide** and water. Furthermore, the bacteria often act as a consortium, with the degradation waste products generated by one microorganism being used as a food source by another bacterium, and so on.

A vibrant industry has been spawned around the use of bacteria as petroleum remediation agents and enhancers of oil recovery. The use of bacteria involves more than just applying an unspecified bacterial population to the spill or the oilfield. Rather, the bacterial population that will be effective depends on factors such as the nature of the contaminant, **pH**, **temperature**, and even the size of the spaces between the rocks (i.e., **permeability**) in the oilfield.

Not all petroleum microbiology is concerned with the beneficial aspects of microorganisms. Bacteria such as *Desulfovibrio hydrocarbonoclasticus* utilize sulfate in the generation of energy. While originally proposed as a means of improving the recovery of oil, the activity of such sulfate reducing bacteria (SRBs) actually causes the formation of acidic compounds that "sour" the petroleum formation. SRBs can also contribute to dissolution of pipeline linings that lead to the burst pipelines, and plug the spaces in the **rock** through which the oil normally would flow on its way to the surface. The growth of bacteria in oil pipelines is such a problem that the lines must regularly be scoured clean in a process that is termed "pigging," in order to prevent pipeline blowouts. Indeed, the formation of acid-generating adherent populations of bacteria has been shown to be capable of dissolving through a steel pipeline up to one-half an inch thick within a year.

See also Biosphere; Fuels and fuel chemistry; Petroleum detection; Petroleum, economic uses of; Petroleum extraction; Petroleum, history of exploration

PH

pH is a measure of the acidity or alkalinity of a solution. The variability of pH can have a dramatic effect on geochemical processes (e.g., **weathering** processes).

The pH scale was developed by Danish chemist Søren Peter Lauritz Sørensen (1868–1939) in 1909 and is generally presented as ranging from 0 to 14, although there are no theoretical limits on the range of the scale (there are substances with negative pH's and with pH's greater than 14, although for most substances the range of 0–14 suffices). A solution with a pH of less than 7 is acidic and a solution with a pH of greater than 7 is basic (alkaline). The midpoint of the scale, 7, is neutral. The lower the pH of a solution, the more acidic the solution is and the higher the pH, the more basic it is. Mathematically, the potential hydronium ion concentration (pH) is equal to the negative logarithm of the hydronium ion concentration: $pH = -\log [H_30^+]$, where H_3O^+ represents the hydronium ion.

Litmus paper is specially treated to turn either blue or pink if exposed to acidic or basic compounds. ***© Yoav Levy/Phototake NYC. Reproduced by permission.***

Essentially, the hydronium ion can be thought of as a **water** molecule with a proton attached. The square brackets indicate the concentration of, in moles per liter. Thus, $[H_3O^+]$ indicates the concentration of hydronium ions in moles per liter.

The hydronium ion is an important participant in the chemical reactions that take place in aqueous (water, H_20) solutions.

Through a process termed self-ionization, a small number of water molecules in pure water dissociate (separate) in a reversible reaction to form a positively charged H^+ ion and a negatively charged OH^- ion. In aqueous solution, as one water molecule dissociates, another is nearby to pick up the loose, positively charged, hydrogen proton to form a positively charged hydronium ion (H_3O^+).

Water molecules have the ability to attract protons and form hydronium ions because water is a polar molecule. **Oxygen** is more electronegative than hydrogen. As a result, the electrons in each of water's two oxygen-hydrogen bonds to spend more time near the oxygen **atom**. Because the electrons are not shared equally—and because the bond angles of the water molecule do not cancel out this imbalance—the oxygen atom carries a partial negative charge that can attract positively charged protons donated by other molecules.

In a sample of pure water, the concentration of hydronium ions is equal to 1×10^{-7} moles per liter (0.0000001 M). The water molecule that lost the hydrogen proton—but that kept the hydrogen electron—becomes a negatively charged hydroxide ion (OH^-).

The equilibrium (balance) between hydronium and hydroxide ions that results from self-ionization of water can be disturbed if other substances that can donate protons are put into solution with water.

The pH of solutions may be measured electronically with a pH meter (better pH meters can measure to 0.001 pH units) or by using acid base indicators, chemicals that change color in solutions of different pH.

See also Acid rain; Geochemistry; Weathering and weathering series

PHANEROZOIC EON

The Phanerozoic Eon represents **geologic time** from the end of **Precambrian** time, approximately 544 to 570 million years ago (mya), until the present day. As such, the Phanerozoic Eon includes the **Paleozoic Era**, the **Mesozoic Era**, and the current **Cenozoic Era**. The Phanerozoic Eon and constituent eras are then further divided into 12 geologic periods.

The Phanerozoic Eon derives its name from *phaneros*, meaning visible or evident, and *zoon*, meaning life. Although early life existed in Precambrian time—including prokaryotes (e.g., bacteria) and eukaryotes (organisms with a true nucleus containing DNA)—the onset of the Phanerozoic Eon marks the start of complex life (e.g., invertebrates) found in the **Cambrian Period**.

In terms of the **fossil record**, the Phanerozoic Eon represents not the **origin of life**, but of life capable of leaving extensive fossil remains (e.g., organisms with shells, etc.). Fossilization refers to the series of postmortem (after death) changes that lead to replacement of **minerals** in the original hard parts (shell, skeleton, teeth, horn, scale) with different minerals, a process known as remineralization. Infrequently, soft parts may also be mineralized and preserved as **fossils**. Fossils of soft bodied Precambrian time fossils have been found but, as expected, they are rare and present an incomplete evolutionary record.

See also Archean; Cretaceous Period; Dating methods; Devonian Period; Eocene Epoch; Evolution, evidence of; Fossil record; Fossils and fossilization; Historical geology; Holocene Epoch; Jurassic Period; Miocene Epoch; Mississippian Period; Oligocene Epoch; Ordovician Period; Paleocene Epoch; Pennsylvanian Period; Pleistocene Epoch; Pliocene Epoch; Proterozoic Era; Quaternary Period; Silurian Period; Tertiary Period; Triassic Period

PHASE STATE CHANGES

A change of state occurs when matter is converted from one physical state to another. For example, when **water** is heated, it changes from a liquid to a gas—when cooled water will eventually freeze into a solid: **ice**. A change of state is usually accompanied by a change in **temperature** and/or pressure.

Matter commonly exists in one of three forms, or states: solid, liquid, or gas. One fundamental way in which these three states differ from each other is the energy of the particles of which they are made. The particles in a solid contain relatively little energy and move slowly. The particles in a liquid are at a higher energy level and move more rapidly. The particles in a gas are at an even higher energy level and move most rapidly.

The state in which matter occurs can be changed by changing the energy state of the matter or the system surrounding matter that has the capacity to come into equilibrium with that system. When water is heated, molecules begin to move more rapidly. Eventually, they are moving fast enough to change to the gaseous, or vapor, state. The term vapor is used to describe the gaseous state of a substance that is normally a liquid at room temperature.

Imagine a block of ice at 14°F (–10°C). The molecules of water in the ice are vibrating in a crystalline array. As heat is added to the ice, the molecules begin vibrate more rapidly. At some point, they vibrate rapidly enough to break the lattice array and move freely in a liquid state. The point at which this occurs is the **melting** point. The melting point is the temperature at which a solid changes to a liquid. The melting point of ice is 32°F (0°C).

If additional heat is added to the liquid water, water molecules move even faster. The increase in speed with which they move is measured as an increase in temperature. The temperature of the liquid water increases from 32°F (0°C) to 212°F (100°C). At a temperature of 212°F (100°C), the water molecules are moving fast enough to change to a vapor, called steam. The temperature at which a liquid changes to a gas is called its boiling point. The temperature is a function of **atmospheric pressure**. The lower the pressure (e.g., lower pressures found at altitude) the lower the boiling point.

If the steam formed in this process is heated further, its temperature continues to increase.

Changes of state occur also when a material is cooled. Suppose the steam in this example is cooled below 212°F (100°C). When that happens, the water reverts to a liquid. The steam is said to condense to a liquid. The **condensation** point is the temperature at which a gas or vapor changes to a liquid. It is the same as the boiling point of the liquid.

Under the proper conditions, at some point, the liquid cools sufficiently to change to a solid. At this point, the liquid becomes frozen. The **freezing** point of a liquid is the temperature at which the liquid changes to a solid. The freezing point is the same as the melting point of the solid.

Some materials behave differently from water when they are heated. They may pass directly from the solid state to the gaseous state. Iodine is an example. When solid iodine is heated, it does not melt. Instead, it changes directly into a vapor. Substances that behave in this way are said to sublime. The sublimation point of a substance is the temperature at which it changes directly from a solid to a vapor.

Dry ice (solid CO_2), which rapidly undergoes sublimation from solid to vapor at room temperatures, is often used to create **fog** on stage and movie sets. A white, opaque solid, it is

Dry ice, or frozen carbon dioxide, goes from a solid state to a gaseous state at normal pressure and temperature, unlike water ice, which goes from a solid state to a liquid state. This process is called sublimation. Please note that dry ice should never be touched with bare hands, as it is cold enough to cause severe damage to exposed skin. Proper precautions should always be taken when handling dry ice. ***R. Fowell. National Audubon Society Collection/Photo Researchers, Inc. Reproduced by permission.***

also widely used as a cryogenic agent in industry to reduce bacteria growth and maintain low temperatures. At atmospheric pressures found on Earth, dry ice undergoes sublimation at –109.3°F (–78.5°C). Special care must be taken when working with dry ice to avoid frostbite. In addition, dry ice must be used in a well-ventilated environment because, as it undergoes sublimation, dry ice will reduce the percentage of available **oxygen**.

See also Atmospheric chemistry; Atoms; Hydrothermal processes; Ice heaving and ice wedging; Rate factors in geologic processes

PHYLLITE

Phyllite is an intermediate-grade, foliated **metamorphic rock** type that resembles its sedimentary parent **rock**, shale, and its lower-grade metamorphic counterpart, **slate**. Like slate, phyllite can be distinguished from shale by its foliation, called slaty cleavage, and its brittleness, or fissility. Both slate and phyllite are generally dark-colored; their most common color is dark gray-blue, but dark red and green varieties also exist. Unlike slate, phyllite has a characteristic glossy sheen, its foliation is usually slightly contorted, and it rarely retains traces of the original sedimentary **bedding**. Phyllite also lacks the large, visible mica **crystals** and high-grade index **minerals** diagnostic of **schist**, its higher-grade metamorphic cousin.

Heating and compression of clay-rich, bedded **sedimentary rocks** called shales creates a series of rock types of increasing metamorphic grade: slate, phyllite, schist, and **gneiss**. During **metamorphism** of shales, and occasionally volcanic ash layers, metamorphism transforms platy **clay** minerals into small sheets of mica. As the intensity of heating and compression, the so-called metamorphic grade, increases, the mica sheets align themselves perpendicular to the direction of stress, and they grow larger. In phyllite, the crystals of sheet-silicate minerals like chlorite, biotite, and muscovite are large enough to give the rock its distinctive satin sheen and slaty cleavage, but not large enough to be visible to the unaided eye. The amount of heat and pressure required to transform shale to phyllite is generally sufficient to destroy any original sedimentary layering. Additional metamorphism transforms phyllite to schist; all the original clay and small mica crystals transform into large mica crystals, any remaining organic material is destroyed, and high-grade metamorphic index minerals like garnet and staurolite grow in the micaceous matrix.

Slates and phyllites typically form along the edges of regional metamorphic belts where clay-rich, marine sedimentary rocks have been caught between colliding continental plates, or scraped off the seafloor into an accretionary wedge above a **subduction zone**. Slates and phyllites may also form in sedimentary basins where marine muds have been extremely deeply buried. The assemblage of minerals usually present in phyllite is referred to as greenschist facies, and includes chlorite, muscovite, sodium-rich **plagioclase feldspar**, and a small amount of **quartz**. Greenschist metamorphism of shales requires moderate amounts of both heating and compression, consistent with the conditions present in accretionary wedges, shallow continental fold belts, and very deep sedimentary basins.

See also Greenstone belt

PHYLLOSILICATES

Phyllosilicates are a group of minerals that are fundamentally composed of extended flat sheets of linked silicon-oxygen tetrahedra. Included in the phyllosilicate family are micas and clays. The name is derived from the Greek word *phyllos,* meaning leaf. As the name implies, phyllosilicates often display a platy or flaky crystal habit or perfect planar cleavage.

The phyllosilicate structure consists of sheets of hydrated SiO_4 rings (they are hydrated because there is one hydroxyl ion [OH^-] in the middle of each ring) and an octahedral sheet consisting of AlO or MgO. The simplest phyllosilicate is a sole octahedral layer: brucite in the case of MgO, gibbsite if the octahedral layer is AlO. However, most phyllosilicates are composed of interlocking tetrahedral and octahedral layers combined in the ratio of 1:1 or 2:1. One tetrahedral layer linked to one octahedral layer yields the 1:1 structure. This forms kaolinite when an AlO octahedral layer is used, antigorite if the octahedral layer consists of MgO. Other **clay** minerals such as talc and pyrophyllite are composed of a 2:1 structure: two tetrahedral layers sandwiching one octahedral layer. In both structures, 1:1 and 2:1, a weak residual interlayer charge exists. This interlayer charge is the source of the van der Waals forces that hold each interlocked layer to one another. These weak interlayer bonds are easily broken and allow the sheets to slip, resulting in the low hardness and greasy feel that is typical of clays.

Other phyllosilicate minerals are derived from further complications of the interlocking sheet system. For example, the micas are formed when atoms of **aluminum** substitute for **silicon** in the tetrahedral layer. Because these two elements carry different charges, the net charge of the 2:1 layers is increased. The result is an interlayer bond that is stronger than a van der Waals forces and produces the perfect planar cleavage and slightly higher hardness of micas.

See also Clay; Phyllite

Physical geography

Physical geography is a scientific discipline that addresses the distribution of natural features and processes within a spatial, or geographical, reference frame. This subdiscipline of geography is an interdisciplinary amalgam of such diverse subjects as **geology**, ecology, environmental science, computer science, and aerospace engineering. When examined, the narrow definition of physical geography as the study and creation of physical maps expands into a broad array of topics from **satellite remote sensing** to computer-aided mapping known as geographic information systems (**GIS**), to the study of surficial geological processes. The basic work of physical geology lies in determining how natural phenomena are spatially ordered, and in illuminating these geographic patterns using maps and images; the fundamental question behind physical geographic studies is why these patterns exist in nature.

The history of physical geography spans nearly four thousand years. Archeologists have discovered maps created by ancient Chinese, Phoenician, and Egyptian explorers, including a Babylonian map carved in a **clay** tablet dated at about 2300 B.C. Aristotle (384–322 B.C.) suggested that the earth is a sphere based on his observations of lunar cycles. Eratosthenes (circa 276–194 B.C.) accurately calculated the circumference of the earth using a geometric proof. **Ptolemy** (circa A.D. 100–170) developed a number of map **projection** schemes, as well as the coordinate system using **latitude and longitude**. Most of these early Greek and Roman geographical insights were forgotten during the Middle Ages (600–1400), especially in **Europe**. In fact, the idea of a spherical Earth did not resurface until the Renaissance when European navigators, including Christopher Columbus, **Ferdinand Magellan**, and Sir Francis Drake, explored the **oceans** and the Americas. Increasingly detailed physical and cultural geographic studies accompanied the rapid population growth, European colonization, and exploration of the American frontier that took place from the late 1700s the early 1900s.

Physical geography underwent a quantitative revolution beginning in the 1950s, when geographic investigations became more scientifically rigorous. This revolution continues today with ever-improving geographical methods and tools like satellite-aided navigation using the Global Positioning System (**GPS**), and images of the earth collected from **space**. Another change in the science of physical geography since the 1950s has been an increasing focus on the ways that humans affect their natural environment. Precise geographical information that documents or explains anthropogenic changes in the natural world is valuable to decision-makers across the ideological spectrum, from environmentalists, to government resource managers, to insurance actuaries.

See also History of exploration I (Ancient and classical); History of exploration II (Age of exploration); Mapping techniques; Projections

Physics

Physics and **astronomy**, from which all other sciences derive their foundation, are attempts to provide a rational explanation for the structure and workings of the Universe. The creation of the earliest civilizations and of mankind's religious beliefs was profoundly influenced by the movements of the **Sun**, **Moon**, and stars across the sky. As our most ancient ancestors instinctively sought to fashion tools through which they gained mechanical advantage beyond the strength of their limbs, they also sought to understand the mechanisms and patterns of the natural world. From this quest for understanding evolved the science of physics. Although these ancient civilizations were not mathematically sophisticated by contemporary standards, their early attempts at physics set mankind on the road toward the quantification of nature.

In Ancient Greece, in a natural world largely explained by the whim of gods, the earliest scientists and philosophers of record dared to offer explanations of the natural world based on their observations and reasoning. Pythagoras (582–500 B.C.) argued about the nature of numbers, Leucippus (c. 440 B.C.), Democritus (c. 420 B.C.), and Epicurus (342–270 B.C.) asserted matter was composed of extremely small particles called atoms.

Many of the most cherished arguments of ancient science ultimately proved erroneous. For example, in Aristotle's (384–322 B.C.) physics, for example, a moving body of any mass had to be in contact with a "mover," and for all things there had to be a "prime mover." Errant models of the universe made by **Ptolemy** (c. A.D. 100–170) were destined to dominate the Western intellectual tradition for more than a millennium. Midst these misguided concepts, however, were brilliant insights into natural phenomena. More then 1700 years before the Copernican revolution, Aristarchus of Samos (310–230 B.C.) proposed that the earth rotated around the Sun and **Eratosthenes Of Cyrene** (276–194 B.C.), while working at

Researchers at the Fermi National Accelerator Laboratory, or Fermilab, are expanding the limits of our knowledge of particle physics. *© Michael S. Yamashita/Corbis. Reproduced by permission.*

the great library at Alexandria, deduced a reasonable estimate of the circumference of the earth.

Until the collapse of the Western Roman civilization there were constant refinements to physical concepts of matter and form. Yet, for all its glory and technological achievements, the science of ancient Greece and Rome was essentially nothing more than a branch of philosophy. Experimentation would wait almost another two thousand years for injecting its vigor into science. Although there were technological advances and more progress in civilization than commonly credited, during the Dark and Medieval Ages in **Europe** science slumbered. In other parts of the world, however, Arab scientists preserved the classical arguments as they developed accurate astronomical instruments and compiled new works on mathematics and optics.

At the start of the Renaissance in Western Europe, the invention of the printing press and a rediscovery of classical mathematics provided a foundation for the rise of empiricism during the subsequent Scientific Revolution. Early in the sixteenth century, Polish astronomer Nicolaus Copernicus's (1473–1543) reassertion of heliocentric theory sparked an intense interest in broad quantification of nature that eventually allowed German astronomer and mathematician **Johannes Kepler** (1571–1630) to develop laws of planetary motion. In addition to his fundamental astronomical discoveries, Italian astronomer and physicist **Galileo Galilei** (1564–1642) made concerted studies of the motion of bodies that subsequently inspired seventeenth century English physicist and mathematician Sir Isaac Newton's (1642–1727) development of the laws of motion and gravitation in his influential 1687 work, *Philosophiae Naturalis Principia Mathematica (Mathematical Principles of Natural Philosophy).*

Following the *Principia*, scientists embraced empiricism during an Age of Enlightenment. Practical advances spurred by the beginning of an industrial revolution resulted in technological advances and increasingly sophisticated instrumentation that allowed scientists to make exquisite and delicate calculations regarding physical phenomena. Concurrent advances in mathematics, allowed development of sophisticated and quantifiable models of nature. More tantalizingly for physicists, many of these mathematical insights ultimately pointed toward a physical reality not necessarily limited to three dimensions and not necessarily absolute in time and **space**.

Nineteenth century experimentation culminated in the formulation of Scottish physicist James Clerk Maxwell's (1831–1879) unification of concepts regarding **electricity**, **magnetism**, and light in his four famous equations describing electromagnetic waves.

During the first half of the twentieth century, these insights found full expression in the advancement of quantum and **relativity theory**. Scientists, mathematicians, and philosophers united to examine and explain the innermost workings of the universe—both on the scale of the very small subatomic world and on the grandest of cosmic scales.

By the dawn of the twentieth century more than two centuries had elapsed since Newton's *Principia* set forth the foundations of classical physics. In 1905, in one grand and sweeping Special Theory of Relativity, German-American physicist **Albert Einstein** (1879–1955) provided an explanation for seemingly conflicting and counter-intuitive experimental determinations of the constancy of the speed of light, length contraction, time dilation, and mass enlargements. A scant decade later, Einstein once again revolutionized concepts of space, time and **gravity** with his General Theory of Relativity.

Prior to Einstein's revelations, German physicist Maxwell Planck (1858–1947) proposed that atoms absorb or emit electromagnetic radiation in discrete units of energy termed quanta. Although Planck's quantum concept seemed counter-intuitive to well-established Newtonian physics, quantum mechanics accurately described the relationships between energy and matter on atomic and subatomic scale and provided a unifying basis to explain the properties of the elements.

Concepts regarding the stability of matter also proved ripe for revolution. Far from the initial assumption of the indivisibility of atoms, advancements in the discovery and understanding of **radioactivity** culminated in renewed quest to find the most elemental and fundamental particles of nature. In 1913, Danish physicist **Niels Bohr** (1885–1962) published a model of the hydrogen **atom** that, by incorporating **quantum theory**, dramatically improved existing classical Copernican-like atomic models. The quantum leaps of electrons between orbits proposed by the **Bohr model** accounted for Planck's observations and also explained many important properties of the photoelectric effect described by Einstein.

More mathematically complex atomic models were to follow based on the work of the French physicist Louis Victor de Broglie (1892–1987), Austrian physicist **Erwin Schrödinger** (1887–1961), German physicist Max Born (1882–1970) and English physicist P.A.M Dirac (1902–1984). More than simple refinements of the Bohr model, however, these scientists made fundamental advances in defining the properties of matter—especially the wave nature of subatomic particles. By 1950, the articulation of the elementary constituents of atoms grew dramatically in numbers and complexity and matter itself was ultimately to be understood as a synthesis of wave and particle properties.

Against a maddeningly complex backdrop of politics and fanaticism that resulted in two World Wars within the first half of the twentieth century, science knowledge and skill became more than a strategic advantage. The deliberate misuse of science scattered poisonous gases across World War I battlefields at the same time that advances in physical science (e.g., x-ray diagnostics) provided new ways to save lives. The dark abyss of WWII gave birth to the atomic age. In one blinding flash, the Manhattan Project created the most terrifying of weapons that could—in an blinding flash—forever change the course of history for all peoples of the earth.

The insights of relativity theory and quantum theory also stretched the methodology of science. No longer would science be mainly exercise in inductively applying the results of experimental data. Experimentation, instead of being only a genesis for theory, became a testing ground to test the apparent truths unveiled by increasingly mathematical models of the universe. Moreover, with the formulation of quantum mechanics, physical phenomena could no longer be explained in terms of deterministic causality, that is, as a result of at least a theoretically measurable chain causes and effects. Instead, physical phenomena were described as the result of fundamentally statistical, unreadable, indeterminant (unpredictable) processes.

The development of quantum theory, especially the delineation of Planck's constant and the articulation of the Heisenburg uncertainty principle carried profound philosophical implications regarding limits on knowledge. Modern cosmological theory (i.e., theories regarding the nature and formation of the universe) provided insight into the evolutionary stages of stars (e.g., neutron stars, pulsars, black holes, etc.) that carried with it an understanding of nucleosythesis (the formation of elements) that forever linked mankind to the lives of the very stars that had once sparked the intellectual journey towards an understanding of nature based upon physical laws.

With specific regard to **geology**, the twentieth century development of geophysics and advances in sensing technology made possible the revolutionary development of plate tectonic theory.

See also Earth Science; History of exploration I (Ancient and classical); History of exploration II (Age of exploration); History of exploration III (Modern era); History of geoscience: Women in the history of geoscience; History of manned space exploration

Piccard, Jacques Ernest-Jean (1922-)

Swiss oceanic engineer, physicist, and economist

Jacques Piccard is a Swiss oceanic engineer best noted for making the deepest ocean dive (with Lt. Don Walsh) in the bathyscaph *Trieste*, a submersible vessel he helped build with his father, Auguste Piccard.

Jacques Ernest-Jean Piccard was born in Brussels, Belgium in 1922, where his Swiss-born father taught at the city's university. He attended the École Nouvelle de Suisse Romande in Lausanne, Switzerland, and in 1943, enrolled at the University of Geneva where he studied economics, history and **physics**. Piccard put his education on hold for a year in 1944 to serve with the French First Army. Upon leaving the service, he resumed studies and went on to receive his licentiate in 1946.

In the 1950s, Piccard joined his father in designing new and improved deep ships or "bathyscaphes." Their 50-ft (15-m) long navigable diving vessel, the *Trieste*, consisted of a heavier-than-water steel cabin that could resist sea pressure and a float filled with gasoline to provide lift. On August 1, 1953, the pair launched the *Trieste* to a depth of 10,168 ft (3,099 m) off the coast of Ponza, Italy.

Jacques Piccard then took the idea to the United States government in 1956; two years later the U.S. Navy acquired the *Trieste* and redesigned the cabin so the vehicle could descend to deeper **ocean trenches**. Subsequently, the Navy asked Piccard to serve as their consultant.

It was during this time that Piccard performed what many believe to be his most noteworthy accomplishment. On January 23, 1960, Piccard, along with U.S. Navy Lieutenant Don Walsh, made a dive in the *Trieste* to the deepest known point on Earth. The team descended 35,810 ft (10,916 m) in **area** known as the Challenger Deep in Pacific's Mariana Trench. Piccard and Walsh sat in a 6-ft-diameter (1.8-m) steel capsule at the base of the ship while the vessel made the nearly five-hour dive to the ocean floor. Just shy of 7 mi (11 km), the dive set a new submarine depth record. In the decades since the feat, no one has come within 10,000 ft (3,048 m) of Piccard and Walsh's record.

In later years, the *Trieste* helped locate the sunken nuclear submarine U.S.S. *Thresher*, and documented information on another sunken sub, U.S.S. *Scorpion*. The original *Trieste* now sits in the Navy Museum in Washington, D.C.

In the 1960s, Piccard continued to work with his father and designed the first tourist submarine. The mesoscaphe, as they called it, was an underwater observation vehicle capable of carrying 40 tourists.

For most of his life, Piccard has continued the oceanographic work inspired by his father. He has served as a consultant for a number of private deep-sea research organizations, including the Grumman Aircraft Engineering Corporation. His son, Bertrand, has kept the Piccard spirit of adventure alive, but by traveling up instead of down. In 1999, Bertrand Piccard and his teammate, Brian Jones, became the first to circumnavigate the globe in a hot-air balloon.

See also Deep sea exploration

PILLOW LAVA • *see* LAVA

PIPE, VOLCANIC

The central conduit by which **magma** rises through a **volcano** is termed a volcanic pipe. A volcanic pipe may be anywhere from a few yards to about 0.5 mi (0.8 km) in width. When a volcano ceases to erupt, its pipe generally becomes plugged by a column of solidified magma mixed with angular fragments of **rock** ripped from the walls of the pipe. This solid column is also termed a pipe (or neck, or plug). **Erosion** may strip the cone from around such a plug to create a free-standing pillar.

A pipe forms when magma from a deep reservoir drills or blasts upward. One mechanism by which this is occurs involves convection, that is, vertical circulation driven by the density difference between hotter and cooler magmas: hotter magma rises, cooler magma sinks. Magma in a narrow vertical pipe quickly loses heat to surrounding rocks, and the magma thus cooled sinks along the sides of the pipe while hot, fresh magma ascends in the pipe's center. This central fountain erodes chunks of rock from the pipe's roof, extending the pipe upward. These chunks are transported by down-convecting magma to the reservoir below, where they are melted down and assimilated. This process enables a pipe to rise through many miles of rock without having to push rocks aside.

Magma containing large amounts of dissolved gas can widen a pipe explosively by a mechanism resembling that of an erupting **geyser**. If magma reaches the surface via a relatively narrow pipe and encounters substantial **groundwater**, a large steam explosion may occur: the pipe explodes at the top. This suddenly removes weight from the magma column in the pipe, reducing the pressure on magma deeper down. Gas dissolved in this deeper magma boils out explosively, blowing still more material out of the top of the pipe and further reducing the pressure on magma still deeper down. A series of explosive eruptions can thus propagate downward to great depth. The rubble-choked pipe left after such an eruption is termed a diatreme.

If an ascending pipe full of hot magma encounters a layer of groundwater but conditions are not right for a downward-propagating explosion, a simple steam explosion at the surface may result that excavates a large, shallow crater or maar. Maars closely resemble meteor impact craters because they do not rise above the terrain surrounding them.

See also Crater, volcanic; Fissure; Magma chamber; Volcanic eruptions; Volcanic vent

PLAGIOCLASE

Plagioclase is a form of the mineral **feldspar**. All feldspars are **crystals** of **aluminum**, **oxygen**, and **silicon**, plus a major additive. In the case of plagioclase, the additive is calcium, sodium, or a blend of both. Miscellaneous impurities may also be present.

Like other feldspars, plagioclases have a vitreous (glassy) luster, are translucent or transparent, and are typically pink, gray, or white in color. Plagioclase cleaves along two planes that intersect at about 86 degrees, hence its name (from the Greek *plagio*, slanted, and *clase*, breaking). About 29% of the earth's **crust** consists of plagioclase.

There are two pure or extreme forms of plagioclase. Albite ($NaAlSi_3O_8$) contains sodium but no calcium, and anorthite $CaAl_2Si_2O_8$ contains calcium but no sodium. Many plagioclases consist of both **minerals** microscopically blended to form what is termed a solid solution. Plagioclase blends can occur along a linear continuum from albite to anorthite, intermediate minerals being mixtures of both. Plagioclase with 10–30% anorthite is defined as oligoclase; 30–50%, andesine; 50–70%, labradorite; and 70–90%, bytownite. The remainder in all cases is albite. A plagioclase with less than 10% anorthite is classified simply as albite, while one with less than 10% albite is classified simply as anorthite. A continuous, evenly represented series of plagioclases is not actually found in the field; oligoclases with anorthite percentages in the low twenties are most common, while some other points on the continuum are scarcely represented at all. The plagioclases are difficult to distinguish from each other without laboratory tests.

Many plagioclases are zoned. Zoning is the concentric or onion-skin structuring of an individual plagioclase crystal (often only a few millimeters long) with layers of varying sodium–calcium content. For example, a plagioclase crystal may consist of an andesine core (30–50% anorthite) surrounded by thin shells or zones of labradorite (50–70% anorthite) alternating with zones of andesine. Zoning records the chemical and thermal environment in which a plagioclase crystal formed.

Plagioclases are used for ceramics and glassmaking, and several gemstone varieties exist.

See also Bowen's reaction series; Mineralogy

PLANCK, MAX (1858-1947)

German physicist

Max Planck is best known as one of the founders of the **quantum theory** of **physics**. As a result of his research on heat radiation, Planck concluded that energy can sometimes be described as consisting of discrete units, later given the name *quanta*. This discovery was important because it made possible, for the first time, the use of matter-related concepts in an analysis of phenomena involving energy. Planck also made important contributions in the fields of thermodynamics, relativity, and the philosophy of science. He was awarded the 1918 Nobel Prize in physics for his discovery of the quantum effect.

Max Karl Ernst Ludwig Planck was born in Kiel, Germany. His parents were Johann Julius Wilhelm von Planck, originally of Göttingen, and Emma Patzig, of Griefswald. Max was the couple's fourth child.

Johann von Planck was descended from a long line of lawyers, clergyman, and public servants and was himself Professor of Civil Law at the University of Kiel. Young Max began school in Kiel, but moved at the age of nine with his family to München. There he attended the Königliche Maximillian Gymnasium until his graduation in 1874.

Planck entered the University of München in 1874 with plans to major in mathematics. He soon changed his mind, however, when he realized that he was more interested in practical problems of the natural world than in the abstract concepts of pure mathematics. Although his course work at München emphasized the practical and experimental aspects of physics, Planck eventually found himself drawn to the investigation of theoretical problems. It was, biographer Hans Kango points out in *Dictionary of Scientific Biography*, "the only time in [his] life when he carried out experiments."

Planck's tenure at München was interrupted by illness in 1875. After a long period of recovery, he transferred to the University of Berlin for two semesters in 1877 and 1878. At Berlin, he studied under a number of notable physicists, including Hermann Helmholtz and Gustav Kirchhoff. By the fall of 1878, Planck was healthy enough to return to München and his studies. In October of that year, he passed the state examination for higher-level teaching in math and physics. He taught briefly at his alma mater, the Maximillian Gymnasium, before devoting his efforts full time to preparing for his doctoral dissertation. He presented that dissertation on the second law of thermodynamics in early 1879 and was granted a Ph.D. by the University of München in July of that year.

Planck's earliest field of research involved thermodynamics, an **area** of physics dealing with heat energy. He was very much influenced by the work of Rudolf Clausius, whose work he studied by himself while in Berlin. He discussed and analyzed some of Clausius's concepts in his own doctoral dissertation. Between 1880 and 1892, Planck carried out a systematic study of thermodynamic principles, especially as they related to chemical phenomena such as osmotic pressure, boiling and **freezing** points of solutions, and the dissociation of gases. He brought together the papers published during this period in his first major book, *Vorlesungen über Thermodynamik*, published in 1897.

During the early part of this period, Planck held the position of Privat-Dozent at the University of München. In 1885, he received his first university appointment as extraordinary professor at the University of Kiel. His annual salary of 2,000 marks was enough to allow him to live comfortably and to marry his childhood sweetheart from München, Marie Merck. They eventually had three children.

Planck's research on thermodynamics at Kiel soon earned him recognition within the scientific field. Thus, when Kirchhoff died in 1887, Planck was considered a worthy successor to his former teacher at the University of Berlin. Planck was appointed to the position of assistant professor at Berlin in 1888 and assumed his new post the following spring. In addition to his regular appointment at the university, Planck was also chosen to head the Institute for Theoretical Physics, a facility that had been created especially for him. In 1892, Planck was promoted to the highest professorial rank, ordinary professor, a post he held until 1926.

Once installed at Berlin, Planck turned his attention to an issue that had long interested his predecessor, the problem of black body radiation. A black body is defined as any object that absorbs all frequencies of radiation when heated and then gives off all frequencies as it cools. For more than a decade, physicists had been trying to find a mathematical law that would describe the way in which a black body radiates heat.

The problem was unusually challenging because black bodies do not give off heat in the way that scientists had predicted that they would. Among the many theories that had been proposed to explain this inconsistency was one by the German physicist Wilhelm Wien and one by the English physicist John Rayleigh. Wien's explanation worked reasonably well for high frequency black body radiation, and Rayleigh's appeared to be satisfactory for low frequency radiation. But no one theory was able to describe black body radiation across the whole spectrum of frequencies. Planck began working on the problem of black body radiation in 1896, and by 1900, had found a solution to the problem. That solution depended on a revolutionary assumption, namely that the energy radiated by a black body is carried away in discrete "packages" that were later given the name *quanta* (from the Latin, *quantum*, for "how much"). The concept was revolutionary because physicists had long believed that energy is always transmitted in some continuous form, such as a wave. The wave, like a line in geometry, was thought to be infinitely divisible.

Planck's suggestion was that the heat energy radiated by a black body be thought of as a stream of "energy bundles," the magnitude of which is a function of the wavelength of the radiation. His mathematical expression of that concept is relatively simple: E = h 6,50 - υ, where E is the energy of the quantum, υ is the wavelength of the radiation, and h 6,50 - is a constant of proportionality, now known as *Planck's constant*. Planck found that by making this assumption about the nature of radiated energy, he could accurately describe the experimentally observed relationship between wavelength and energy radiated from a black body. The problem had been solved.

The numerical value of Planck's constant, h 6,50 -, can be expressed as 6.62 x 10^{-27} erg second, an expression that is engraved on Planck's headstone in his final resting place at the Stadtfriedhof Cemetery in Göttingen. Today, Planck's constant is considered to be a fundamental constant of nature, much like the speed of light and the **gravitational constant**. Although Planck was himself a modest man, he recognized the significance of his discovery. Robert L. Weber in *Pioneers of Science: Nobel Prize Winners in Physics* writes that Planck remarked to his son Erwin during a walk shortly after the discovery of the quantum concept, "Today I have made a discovery which is as important as Newton's discovery." That boast

Max Planck. ***Library of Congress.***

has surely been confirmed. The science of physics today can be subdivided into two great eras, classical physics, involving concepts worked out before Planck's discovery of the quantum, and modern physics, ideas that have been developed since 1900, often as a result of that discovery. In recognition of this accomplishment, Planck was awarded the 1918 Nobel Prize in physics.

After completing his study of black body radiation, Planck turned his attention to another new and important field of physics: relativity. Albert Einstein's famous paper on the theory of general relativity, published in 1905, stimulated Planck to look for ways on incorporating his quantum concept into the new concepts proposed by Einstein. He was somewhat successful, especially in extending Einstein's arguments from the field of electromagnetism to that of mechanics. Planck's work in this respect is somewhat ironic in that it had been Einstein who, in another 1905 paper, had made the first productive use of the quantum concept in his solution of the photoelectric problem.

Throughout his life, Planck was interested in general philosophical issues that extended beyond specific research questions. As early as 1891, he had written about the importance of finding large, general themes in physics that could be used to integrate specific phenomena. His book *Philosophy of Physics*, published in 1959, addressed some of these issues. He also looked beyond science itself to ask how his own discipline might relate to philosophy, religion, and society as a whole. Some of his thoughts on the **correlation** of science, art, and religion are presented in his 1935 book, *Die Physik im Kampf um die Weltanschauung*.

Planck remained a devout Christian throughout his life, often attempting to integrate his scientific and religious views. Like Einstein, he was never able to accept some of the fundamental concepts of the modern physics that he had helped to create. For example, he clung to the notion of causality in physical phenomena, rejecting the principles of uncertainty proposed by Heisenberg and others. He maintained his belief in God, although his descriptions of the Deity were not anthropomorphic but more akin to natural law itself.

By the time Planck retired from his position at Berlin in 1926, he had become the second most highly respected scientific figure in **Europe**, if not the world, behind Einstein. Four years after retirement, he was invited to become president of the Kaiser Wilhelm Society in Berlin, an institution that was then renamed the Max Planck Society in his honor. Planck's own prestige allowed him to speak out against the rise of Nazism in Germany in the 1930s, but his enemies eventually managed to have him removed from his position at the Max Planck Society in 1937. During an air raid on Berlin in 1945, Planck's home was destroyed with all of his books and papers. During the last two and a half years of his life, Planck lived with his grandniece in Göttingen, where he died at the age of 89.

See also Quantum electrodynamics (QED)

PLANKTIC FORAMINIFERA • *see* CALCAREOUS OOZE

PLATE TECTONICS

Plate tectonics is the theory explaining geologic changes that result from the movement of **lithospheric plates** over the **asthenosphere** (the molten, ductile, upper portion of the earth's mantle). The visible continents, a part of the lithospheric plates upon which they ride, shift slowly over time as a result of the forces driving plate tectonics. Moreover, plate tectonic theory is so robust in its ability to explain and predict geological processes that it is equivalent in many regards to the fundamental and unifying principles of **evolution** in biology, and nucleosynthesis in **physics** and **chemistry**.

Based upon centuries of cartographic depictions that allowed a good fit between the western coast of **Africa** and the eastern coast of **South America**, in 1858, French geographer Antonio Snider-Pellegrini, published a work asserting that the two continents had once been part of larger single continent ruptured by the creation and intervention of the Atlantic Ocean. In the 1920s, German geophysicist Alfred Wegener's writings advanced the hypothesis of continental drift depicting the movement of continents through an underlying oceanic **crust**. Wegner's hypothesis met with wide skepticism but found support and development in the work and writings of

South African geologist Alexander Du Toit, who discovered a similarity in the **fossils** found on the coasts of Africa and South Americas that derived from a common source.

The technological advances necessitated by the Second World War made possible the accumulation of significant evidence now underlying modern plate tectonic theory.

Plate tectonic theory asserts that Earth is divided into core, mantle, and crust. The crust is subdivided into oceanic and continental crust. The oceanic crust is thin (3–4.3 mi [5–7 km]), basaltic (<50% SiO_2), dense, and young (<250 million years old). In contrast, the continental crust is thick (18.6–40 mi [30–65 km]), granitic (<60% SiO_2), light, and old (250–3,700 million years old). The outer crust is further subdivided by the subdivision of the lithospheric plates, of which it is a part, into 13 major plates. These lithospheric plates, composed of crust and the outer layer of the mantle, contain a varying combination of oceanic and continental crust. The lithospheric plates move on top of mantle's asthenosphere.

Boundaries are adjacent areas where plates meet. Divergent boundaries are areas under tension where plates are pushed apart by **magma** upwelling from the mantle. Collision boundaries are sites of compression either resulting in subduction (where lithospheric plates are driven down and destroyed in the molten mantle) or in crustal uplifting that results in **orogeny** (mountain building). At transform boundaries, exemplified by the San Andreas fault, the continents create a shearing force as they move laterally past one another.

New oceanic crust is created at divergent boundaries that are sites of **sea-floor spreading**. Because Earth remains roughly the same size, there must be a concurrent destruction or uplifting of crust so that the net **area** of crust remains the same. Accordingly, as crust is created at divergent boundaries, oceanic crust must be destroyed in areas of subduction underneath the lighter continental crust. The net area is also preserved by continental crust uplift that occurs when less dense continental crusts collide. Because both continental crusts resist subduction, the momentum of collision causes an uplift of crust, forming **mountain chains**. A vivid example of this type of collision is found in the ongoing collision of India with **Asia** that has resulted in the Himalayan Mountains that continue to increase in height each year. This dynamic theory of plate tectonics also explained the formation of **island arcs** formed by rising material at sites where oceanic crust subducts under oceanic crust, the formation of mountain chains where oceanic crust subducts under continental crust (e.g., Andes mountains), and volcanic arcs in the Pacific. The evidence for deep, hot, convective currents combined with plate movement (and concurrent continental drift) also explained the mid-plate "hot spot" formation of volcanic island chains (e.g., Hawaiian Islands) and the formation of rift valleys (e.g., Rift Valley of Africa). **Mid-plate earthquakes**, such as the powerful New Madrid **earthquake** in the United States in 1811, are explained by interplate pressures that bend plates much like a piece of sheet metal pressed from opposite sides.

As with **continental drift theory** two of the proofs of plate tectonics are based upon the geometric fit of the displaced continents and the similarity of **rock** ages and Paleozoic fossils in corresponding bands or zones in adjacent or corresponding geographic areas (e.g., between West Africa and the eastern coast of South America).

Modern understanding of the structure of Earth is derived in large part from the interpretation of seismic studies that measure the reflection of seismic waves off features in Earth's interior. Different materials transmit and reflect seismic shock waves in different ways, and of particular importance to the theory of plate tectonics is the fact that liquid does not transmit a particular form of seismic wave known as an S-wave. Because the mantle transmits S-waves, it was long thought to be a cooling solid mass. Geologists later discovered that radioactive decay provided a heat source within Earth's interior that made the asthenosphere plasticine (semi-solid). Although solid-like with regard to transmission of seismic S-waves, the asthenosphere contains very low velocity (inches per year) currents of **mafic** (magma-like) molten materials.

Another line of evidence in support of plate tectonics came from the long-known existence of ophiolte suites (slivers of oceanic floor with fossils) found in upper levels of mountain chains. The existence of ophiolte suites are consistent with the uplift of crust in collision zones predicted by plate tectonic theory.

As methods of dating improved, one of the most conclusive lines of evidence in support of plate tectonics derived from the dating of rock samples. Highly supportive of the theory of sea floor spreading (the creation of oceanic crust at a **divergent plate boundary** (e.g., Mid-Atlantic Ridge) was evidence that rock ages are similar in equidistant bands symmetrically centered on the divergent boundary. More importantly, dating studies show that the age of the rocks increases as their distance from the divergent boundary increases. Accordingly, rocks of similar ages are found at similar distances from divergent boundaries, and the rocks near the divergent boundary where crust is being created are younger than the rocks more distant from the boundary. Eventually, radioisotope studies offering improved accuracy and precision in rock dating also showed that rock specimens taken from geographically corresponding areas of South America and Africa showed a very high degree of correspondence, providing strong evidence that at one time these rock formations had once coexisted in an area subsequently separated by movement of lithospheric plates.

Similar to the age of rocks, studies of fossils found in once adjacent geological formations showed a high degree of correspondence. Identical fossils are found in bands and zones equidistant from divergent boundaries. Accordingly, the **fossil record** provides evidence that a particular band of crust shared a similar history as its corresponding band of crust located on the other side of the divergent boundary.

The line of evidence, however, that firmly convinced modern geologists to accept the arguments in support of plate tectonics derived from studies of the magnetic signatures or magnetic orientations of rocks found on either side of divergent boundaries. Just as similar age and fossil bands exist on either side of a divergent boundary, studies of the magnetic orientations of rocks reveal bands of similar magnetic orientation that were equidistant and on both sides of divergent boundaries. Tremendously persuasive evidence of plate tectonics is also derived from **correlation** of studies of the magnetic orientation

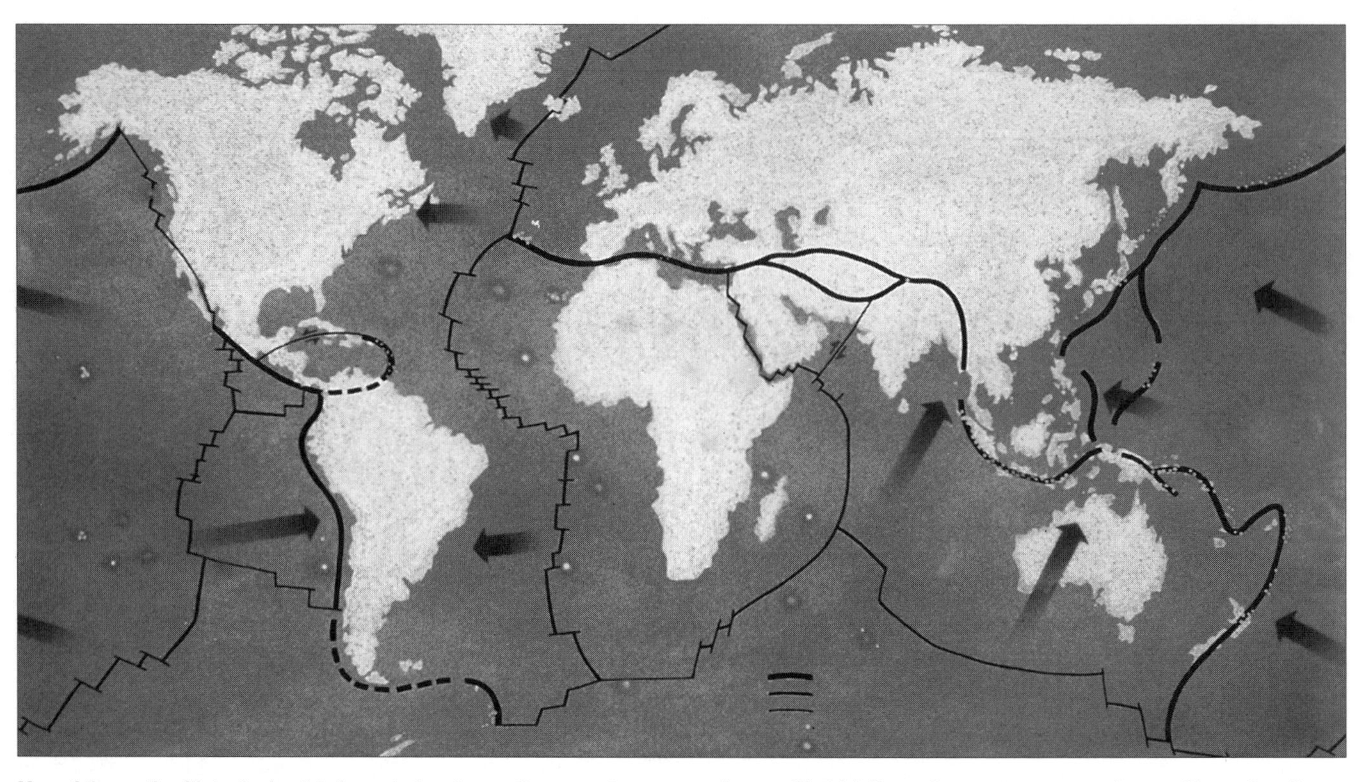

Map of the earth with tectonic plate boundaries shown. Convergent zones are shown with thick lines, divergent zones are shown with medium lines, and transform boundaries are shown with thin lines. Arrows indicate the direction of plate motion. Large orange dots represent "hot spots" and small orange dots represent volcanoes. *David Hardy/Science Photo Library. Reproduced by permission.*

of the rocks to known changes in Earth's **magnetic field** as predicted by electromagnetic theory. Paleomagnetic studies and discovery of polar wandering, a magnetic orientation of rocks to the historical location and polarity of the magnetic poles as opposed to the present location and polarity, provided a coherent map of continental movement that fit well with the present distribution of the continents.

Paleomagnetic studies are based upon the fact that some hot **igneous rocks** (formed from volcanic magma) contain varying amounts of **ferromagnetic minerals** (e.g., Fe_3O_4) that magnetically orient to the prevailing magnetic field of Earth at the time they cool. Geophysical and electromagnetic theory provides clear and convincing evidence of multiple polar reversals or polar flips throughout the course of Earth's history. Where rock formations are uniform—i.e., not grossly disrupted by other geological processes—the magnetic orientation of magnetite-bearing rocks can also be used to determine the approximate **latitude** the rocks were at when they cooled and took on their particular magnetic orientation. Rocks with a different orientation to the current orientation of Earth's magnetic field also produce disturbances or unexpected readings (anomalies) when scientists attempt to measure the magnetic field over a particular area.

This overwhelming support for plate tectonics came in the 1960s in the wake of the demonstration of the existence of symmetrical, equidistant magnetic anomalies centered on the Mid-Atlantic Ridge. Geologists were comfortable in accepting these magnetic anomalies located on the sea floor as evidence of sea floor spreading because they were able to correlate these anomalies with equidistant radially distributed magnetic anomalies associated with outflows of **lava** from land-based volcanoes.

Additional evidence continued to support a growing acceptance of tectonic theory. In addition to increased energy demands requiring enhanced exploration, during the 1950s there was an extensive effort, partly for military reasons related to what was to become an increasing reliance on submarines as a nuclear deterrent force, to map the ocean floor. These studies revealed the prominent undersea ridges with undersea rift valleys that ultimately were understood to be divergent plate boundaries. An ever-growing network of seismic reporting stations, also spurred by the Cold War need to monitor atomic testing, provided substantial data that these areas of divergence were tectonically active sites highly prone to earthquakes. Maps of the global distribution of earthquakes readily identified stressed plate boundaries. Improved mapping also made it possible to view the retrofit of continents in terms of the fit between the true extent of the continental crust instead of the current coastlines that are much variable to influences of **weather** and ocean levels.

In his important 1960 publication, *History of Ocean Basins*, geologist and U.S. Navy Admiral Harry Hess (1906–1969) provided the missing explanatory mechanism for plate tectonic theory by suggesting that the thermal convection currents in the asthenosphere provided the driving force behind plate movements. Subsequent to Hess's book, geologists

Drummond Matthews (1931–1997) and Fred Vine (1939–1988) at Cambridge University used magnetometer readings previously collected to correlate the paired bands of varying **magnetism** and anomalies located on either side of divergent boundaries. Vine and Matthews realized that magnetic data revealing strips of polar reversals symmetrically displaced about a divergent boundary confirmed Hess's assertions regarding seafloor spreading.

See also Dating methods; Earth, interior structure; Fossil record; Fossils and fossilization; Geologic time; Hawaiian Island formation; Lithospheric plates; Mantle plumes; Mapping techniques; Mid-ocean ridges and rifts; Mohorovicic discontinuity (Moho); Ocean trenches; Rifting and rift valleys; Subduction zone

PLATEAU • *see* LANDFORMS

PLEISTOCENE EPOCH

In **geologic time**, the Pleistocene Epoch represents the first epoch in current **Quaternary Period** (also termed the Anthropogene Period) of the **Cenozoic Era** of the **Phanerozoic Eon**. The Pleistocene Epoch spans the time between roughly 2.6 million years ago (mya) and onset of the current **Holocene Epoch** 10,000 to 11,000 years ago.

The Quaternary Period contains two geologic epochs. The earliest epoch, the Pleistocene Epoch is further subdivided into (from earliest to most recent) Gelasian and Calabrian stages. The Calabrian stage is also frequently replaced by a series of geologic stages, from earliest to most recent, including the Danau, Donau-Günz, Günzian, Günz-Mindel, Mindelian, Mindel-Riss, Rissian, Riss-Würm, and Würmian stages.

During the Pleistocene Epoch, Earth's continents almost completely assumed their modern configuration.

Glaciation cycles dominated the major climatic changes of the Pleistocene Epoch. There were at least four distinct glacial advances and recessions. In addition to tremendous **landscape evolution**, climatic cooling contributed to mass extinction in selective areas of the world, but not nearly on the scale as earlier mass extinctions.

The size of land mammals generally increased throughout the Pleistocene Epoch and the **fossil record** established that during the Pleistocene Epoch, hominid (human-like) species became established and evolved into humans (*Homo sapiens*).

Near the start of the Pleistocene Epoch, a number of related species (e.g., *Australopithecus afarensis*) lived and became extinct before modern humankind (*Homo sapiens*) appeared. Early in the Pleistocene Epoch, *Homo habilis* and *Homo rudolfensis* lived and became extinct. Their extinctions are dated to approximately the appearance of *Homo ergaster*, a species some anthropologists argue is one of the earliest identifiable direct ancestors of *Homo erectus*. Although often confused with *Homo erectus*, many scientists assert that *Homo ergaster* is a common ancestor that lead more directly to the subsequent development of *Homo heidelbergensis*, *Homo neanderthalensis*, and humans (*Homo sapiens*).

The last major **impact crater** with a diameter over 31 mi (50 km) struck Earth near what is now Kara-Kul, Tajikistan, at the start of the Pleistocene Epoch. The last major impacts producing craters greater than 6.2 mi (10 km) in diameter occurred during the Pleistocene Epoch about 1.2 million years ago in what are now Kazakhastan and Ghana.

See also Archean; Cambrian Period; Cretaceous Period; Dating methods; Devonian Period; Eocene Epoch; Evolution, evidence of; Fossils and fossilization; Glacial landforms; Glaciers; Historical geology; Ice ages; Jurassic Period; Mesozoic Era; Miocene Epoch; Mississippian Period; Oligocene Epoch; Ordovician Period; Paleocene Epoch; Paleozoic Era; Pennsylvanian Period; Precambrian; Proterozoic Era; Silurian Period; Triassic Period

PLIOCENE EPOCH

In **geologic time**, the Pliocene Epoch occurs during the **Tertiary Period** (65 million years ago [mya] to 2.6 mya) of the **Cenozoic Era** of the **Phanerozoic Eon**. The Tertiary Period is sometimes divided into—or referred to in terms of—a Paleogene Period (65 mya to 23 mya) and a Neogene Period (23 mya to 2.6 mya). The Pliocene Epoch is the last epoch on the Tertiary Period or, in the alternative, the last epoch in the Neogene Period.

The Pliocene Epoch spans the time 5 mya to 2.6 mya.

The Pliocene Epoch is further subdivided into Zanclian (5 mya to 3.9 mya) and Placenzian (3.9 mya to 2.6 mya) stages.

By the end Pliocene Epoch, Earth's continents assumed their modern configuration. The Pacific Ocean separated **Asia** and **Australia** from **North America** and **South America**; the Atlantic Ocean separated North and South America from **Europe** (Eurasian plate) and **Africa**. The Indian Ocean filled the basin between Africa, India, Asia, and Australia. The Indian plate driving against and under the Eurasian plate uplifted both and resulted in rapid mountain building. As a result of the ongoing collision, ancient oceanic **crust** bearing marine **fossils** was uplifted into the Himalayan chain. The collision between the Indian and Eurasian plate continues. The reemergence of the land bridge between North America and South America at the isthmus of Panama about 3 mya allowed migration of species and mixing of gene pools in subspecies.

Climatic cooling increased during Pliocene Epoch, and grasslands continued the rapid development found in the **Miocene Epoch**. Eventually, **glaciation** became well established and a general glacier advance started that continued into the subsequent **Pleistocene Epoch** of the **Quaternary Period**.

The Pliocene Epoch spanned that period of geologic time during which the **evolution** of humans becomes increasingly well documented in the **fossil record**. Notable in the development of primates and human evolution, are fossilized remains of *Ardipithecus ramidus*, *Australopithecus anamensis*, *Australopithecus afarensis*, *Australopithecus garhi*, and *Australopithecus africanus* that date to the Pliocene Epoch.

Although these species became extinct during the Pliocene Epoch, they at a minimum co-existed with the ancestors of humans (*Homo sapiens*); analysis of remains indicate that these species walked upright. Anthropologists argue that apes and humans diverged six to eight mya from a common ancestor that lived during the Miocene Epoch. By the end of the Pliocene Epoch, the subsequent extinctions of *Homo habilis* and *Homo rudolfensis* were almost contemporaneous with the appearance of *Homo ergaster*, a species some anthropologists argue is one of the earliest identifiable direct ancestors of *Homo sapiens*.

The last major **impact crater** with a diameter over 31 mi (50 km) struck Earth near what is now Kara-Kul, Tajikistan at the Pliocene Epoch and Pleistocene Epoch geologic time boundary.

See also Archean; Cambrian Period; Cretaceous Period; Dating methods; Devonian Period; Eocene Epoch; Evolution, evidence of; Evolutionary mechanisms; Fossils and fossilization; Historical geology; Holocene Epoch; Jurassic Period; Mesozoic Era; Mississippian Period; Oligocene Epoch; Ordovician Period; Paleocene Epoch; Paleozoic Era; Pennsylvanian Period; Precambrian; Proterozoic Era; Silurian Period; Triassic Period

PLUTO • *see* SOLAR SYSTEM

PLUTON AND PLUTONIC BODIES

Plutons or plutonic bodies are masses of intrusive igneous **rock** that have solidified underground, as opposed to volcanic (extrusive) rocks that solidify only after erupting onto the surface. Plutonic rocks are characterized by a coarse crystalline texture in which individual **crystals** can be easily seen by the naked eye. The word plutonic is derived from the name of Pluto, the Greek god of the underworld.

The composition of plutonic rocks ranges from **mafic** (gabbro and diorite) to **felsic** (granodiorite and **granite**), and the coarse crystalline texture develops because plutonic rocks are insulated by the surrounding **country rock** and cool very slowly. Plutons, however, are classified according to their size, shape, and relationship to the surrounding rock rather than the kind of rock composing the pluton. Of the many varieties of plutons, four types are described below.

Batholiths are large plutons with more than about 38.6 mi^2 (100 km^2) of surface exposure in map view, such as the Sierra Nevada **batholith** that forms the core of the Sierra Nevada mountain range. Batholiths are discordant plutons, meaning that they cut across the layering of the rocks that they have intruded, and generally do not have an identifiable bottom. Detailed studies have shown that batholiths are most commonly composed of many different igneous intrusions with chemical compositions that vary in **space** and time.

Laccoliths are concordant plutons that follow the existing rock layers and push up overlying strata to form an intrusion that is mushroom-shaped in cross section, as exemplified by the Henry Mountains in Utah. Laccoliths tend to be circular, or nearly so, in map view and less than approximately 5 mi (8 km) in diameter. The thickness of laccoliths can range from a few meters near the edges to several hundred meters near the center of the intrusion. Laccoliths are understood to form when **magma** rises through a feeder **dike**, then begins to spread laterally along a plane of weakness such as a **bedding** plane separating different layers of sedimentary rock.

Two relatively common kinds of small plutons, both tabular in shape, are dikes and sills. Dikes are discordant whereas sills are concordant. In both cases, the thickness of the pluton is very small compared to its lateral extent.

See also Dike; Granite; Igneous rocks; Intrusive cooling; Sill

POINTBARS • *see* CHANNEL PATTERNS

POLAR AXIS AND TILT

The polar axis is an imaginary line that extends through the north and south geographic poles. Earth rotates on its axis as it revolves around the **Sun**. Earth's axis is tilted approximately 23.5 degrees to the plane of the ecliptic (the plane of planetary orbits about the Sun or the apparent path of the Sun across in imaginary celestial sphere). The tilt of the polar axis is principally responsible for variations in solar illumination (**insolation**) that result in the cyclic progressions of the **seasons**.

Earth rotates about the polar axis at approximately 15 angular degrees per hour and makes a complete **rotation** in 23.9 hours. The length of day has changed throughout Earth's history and as rotation slows, the time to complete one rotation about the polar axis will continue to increase. Rate of rotation is a function of planet's mass and orbital position. As Earth rotates on its polar axis, it makes a slightly elliptical orbital revolution about the Sun in 365.26 days. The rates of rotation and revolution are functions of a planet's mass and orbital position.

Rotation about the polar axis results in a diurnal cycle of night and day, and causes the apparent motion of the Sun across the imaginary celestial sphere. The celestial sphere is an imaginary **projection** of the Sun, stars, planets, and all astronomical bodies upon an imaginary sphere surrounding Earth. The celestial sphere is a useful conceptual and tracking remnant of the geocentric theory of the ancient Greek astronomer **Ptolemy**.

During revolution about the Sun, Earth's polar axis exhibits parallelism to Polaris (also known as the North Star). Although observing parallelism, the orientation of Earth's polar axis exhibits precession—a circular wobbling exhibited by gyroscopes—that results in a 28,000-year-long precessional cycle. Currently, Earth's polar axis points roughly in the direction of Polaris (the North Star). As a result of precession, over the next 11,00 years, Earth's axis will precess or wobble so that it assumes an orientation toward the star Vega.

Precession also affects the dates of solstice. At the summer solstice (currently occurring about June 21), the north polar axis points in a direction 23.5 degrees from vertical—

relative to the plane of the ecliptic—toward the Sun. At the winter solstice (currently occurring about December 21) the north polar axis points away from the Sun. At equinox neither pole is tilted toward the Sun but rather in a 23.5 degree tilt from vertical oriented at right angles to line between an imaginary line drawn between the Sun and Earth.

Annual changes in the orientation of the polar axis relative to the Sun result in the apparent movement of the path of the Sun (the ecliptic) across the celestial sphere. The maximum variation in the altitude of the ecliptic above the horizon is two times the polar axial tilt (i.e., 47 degrees).

Milankovitch cycles attempt to integrate and relate changes in Earth's orbital eccentricity, polar axial tilt, and polar axis precession to changes in **climate** (e.g., **glaciation** cycles).

The International **Latitude** Service was established in 1899 to collect data concerning polar axial motion. In 1962, the International Polar Motion Service assumed data collection. In 1988, the International Polar Motion Service, the Earth Rotation Division of the Bureau International de l'Heure, combined operations to form the International Earth Rotation Service. A number of geodetic measuring techniques—including VLBI (Very Long Baseline Interferometry), lunar laser ranging, and the current Global Positioning System (**GPS**) contribute to accurate measurements concerning polar wobbling and rotational rates.

See also Astrolabe; Astronomy; Celestial sphere: The apparent movements of the Sun, Moon, planets, and stars; Revolution and rotation; Solar illumination: Seasonal and diurnal patterns

POLAR ICE

The polar **ice** caps cover the North and South Poles and their surrounding territory, including the entire continent of **Antarctica** in the south, the Arctic Ocean, the northern part of Greenland, parts of northern Canada, and bits of Siberia and Scandinavia also in the north. Polar ice caps are dome-shaped sheets of ice that feed ice to other glacial formations, such as ice sheets, ice fields, and ice islands. They remain frozen year-round, and they serve as sources for **glaciers** that feed ice into the polar **seas** in the form of **icebergs**. Because the polar ice caps are very cold (temperatures in Antarctica have been measured to –126.8°F [–88°C]) and exist for a long time, the caps serve as deep-freezes for geologic information that can be studied by scientists. Ice cores drawn from these regions contain important data for both geologists and historians about paleoclimatology and give clues about the effects human activities are currently having on the world.

Polar ice caps also serve as reservoirs for huge amounts of the earth's **water**. Geologists suggest that three-quarters of the world's fresh water is frozen at the North and South Pole. Most of this **freshwater** ice is in the Southern Hemisphere. The Antarctic ice cap alone contains over 90% of the world's glacial ice, sometimes in huge sheets over 2.5 mi (4 km) deep and averaging 1.5 mi (2 km) deep across the continent. It has been estimated that enough water is locked up in Antarctica to raise sea levels around the globe over 200 ft (61 m), drowning most of the world's major cities, destroying much of the world's food-producing capacity, and ending civilization.

Although the polar ice caps have been in existence for millions of years, scientists disagree over exactly how long they have survived in their present form. It is generally agreed that the polar cap north of the Arctic Circle, which covers the Arctic Ocean, has undergone contraction and expansion through some 26 different glaciations in just the past few million years. Parts of the Arctic have been covered by the polar ice cap for at least the last five million years, with estimates ranging up to 15 million. The Antarctic ice cap is more controversial; although many scientists believe extensive ice has existed there for 15 million years, others suggest that volcanic activity on the western half of the continent it covers causes the ice to decay, and the current south polar ice cap is therefore no more than about three million years old.

At least five times since the formation of the earth, because of changes in global **climate**, the polar ice has expanded north and south toward the equator and has stayed there for at least a million years. The earliest of these known **ice ages** was some two billion years ago, during the Huronian Epoch of the **Precambrian** Era. The most recent ice age began about 1.7 million years ago in the **Pleistocene Epoch**. It was characterized by a number of fluctuations in North polar ice, some of which expanded over much of modern **North America** and **Europe**, covered up to half of the existing continents, and measured as much as 1.8 mi (3 km) deep in some places. These glacial expansions locked up even more water, dropping sea levels worldwide by more than 30 ft (100 m). Animal species that had adapted to cold **weather**, like the mammoth, thrived in the polar conditions of the Pleistocene glaciations, and their ranges stretched south into what is now the southern United States.

The glaciers completed their retreat and settled in their present positions about 10,000–12,000 years ago. There have been other fluctuations in global temperatures on a smaller scale, however, that have sometimes been known popularly as ice ages. The 400-year period between the fourteenth and the eighteenth centuries is sometimes called the Little Ice Age. Contemporaries noted that the Baltic Sea froze over twice in the first decade of the 1300s. Temperatures in Europe fell enough to shorten the growing season, and the production of grain in Scandinavia dropped precipitously as a result. The Norse communities in Greenland could no longer be maintained and were abandoned by the end of the fifteenth century. Scientists believe that we are currently in an interglacial period, and that North polar ice will again move south some time in the next 23,000 years.

Scientists believe the growth of polar ice caps can be triggered by a combination of several global climactic factors. The major element is a small drop (perhaps no more than 15°F [9°C]) in average world temperatures. The factors that cause this drop can be very complex. They include fluctuations in atmospheric and oceanic **carbon dioxide** levels, increased amounts of dust in the atmosphere, heightened winds—especially in equatorial areas—and changes in surface oceanic currents. The Milankovitch theory of glacial cycles also cites as factors small variations in Earth's orbital path around the **Sun**, which in the long term could influence the expansion and contraction of the polar ice caps. Computer models based on the Milankovitch theory correlate fairly closely with observed behavior of **glaciation** over the past 600 million years.

Scientists use material preserved in the polar ice caps to chart these changes in global glaciation. By measuring the relationship of different **oxygen** isotopes preserved in ice cores, they have determined both the mean **temperature** and the amount of dust in the atmosphere in these latitudes during the recent ice ages. Single events, such as **volcanic eruptions** and variations in solar activity and sea level, are also recorded in polar ice. These records are valuable not only for the information they provide about past glacial periods; they serve as a standard to compare against the records of more modern periods.

The process of **global warming**, which has been documented by scientific investigation, is also reflected in the ice caps. Should global warming continue unchecked, scientists warn, it could have a drastic effect on polar ice. Small variations over a short period of time could shrink the caps and raise world sea levels. Even a small rise in sea level could affect a large percentage of the world's population, and it could effectively destroy major cities like New York. Ironically, global warming could also delay or offset the effects of the coming ice age.

See also Glacial landforms

POLYMORPH

A polymorph is a chemical composition that can crystallize into more than one type of structure. This results in different **minerals** with identical compositions and distinguished by their crystallography.

Some common examples of polymorphs are calcite and aragonite. The composition of these two minerals is $CaCO_3$, but calcite is rhombohedral while aragonite is orthorhombic. **Diamond** and **graphite**, both of which are pure **carbon**, are also polymorphs. Diamond, however, is cubic while graphite is hexagonal. Pyrite is the cubic form of FeS_2, marcasite, the orthorhombic version.

A single chemical composition that can form polymorphs does so as a response to varying conditions of formation. The **temperature**, pressure, and chemical environment all affect the crystallization process and can determine the resulting polymorph. For example, diamond requires very high pressure to crystallize, while graphite forms at lower pressures. For the composition $CaCO_3$, calcite is the high temperature-low pressure polymorph while aragonite forms at higher pressures and lower temperatures.

Many polymorphs are only stable within a certain range of conditions and solid-state transitions from one polymorph to another are possible. When low-quartz, which is rhombohedral, is heated to above 1063°F (573°C), it instantaneously goes through an internal structural displacement, or shift, to form hexagonal high-quartz. This type of polymorphic transition is reversible if the temperature is lowered. Other polymorphic transitions involve extensive internal rearrangement and reconstruction of the crystal and subsequently require significantly more energy. The examples of diamond-graphite, pyrite-marcasite, and calcite-aragonite are all known as reconstructive transitions. The large amounts of energy required to effect these polymorphic changes makes the resulting mineral more stable and the process less reversible than with a displacive transition.

See also Crystals and crystallography; Mineralogy

PONNAMPERUMA, CYRIL (1923-1994)

Sri Lankan-born American chemist

Cyril Ponnamperuma, an eminent researcher in the field of chemical evolution, rose through several National Aeronautics and Space Administration (NASA) divisions as a research chemist to head the Laboratory of Chemical Evolution at the University of Maryland, College Park. His career focused on explorations into the **origin of life** and the "primordial soup" that contained the precursors of life. In this search, Ponnamperuma took advantage of discoveries in such diverse fields as molecular biology and astrophysics.

Born in Galle, Ceylon (now Sri Lanka) on October 16, 1923, Cyril Andres Ponnamperuma was educated at the University of Madras (where he received a B.A. in Philosophy, 1948), the University of London (B.Sc., 1959), and the University of California at Berkeley (Ph.D., 1962). His interest in the origin of life began to take clear shape at the Birkbeck College of the University of London, where he studied with J. D. Bernal, a well-known crystallographer. In addition to his studies, Ponnamperuma also worked in London as a research chemist and radiochemist. He became a research associate at the Lawrence Radiation Laboratory at Berkeley, where he studied with Melvin Calvin, a Nobel laureate and experimenter in chemical evolution.

After receiving his Ph.D. in 1962, Ponnamperuma was awarded a fellowship from the National Academy of Sciences, and he spent one year in residence at NASA's Ames Research Center in Moffet Field, California. After the end of his associate year, he was hired as a research scientist at the center and became head of the chemical evolution branch in 1965.

During these years, Ponnamperuma began to develop his ideas about chemical evolution, which he explained in an article published in *Nature.* Chemical evolution, he explained, is a logical outgrowth of centuries of studies both in **chemistry** and biology, culminating in the groundbreaking 1953 discovery of the structure of deoxyribonucleic acid (DNA) by James Watson and Francis Crick. Evolutionist Charles Darwin's studies affirming the idea of the "unity of all life" for biology could be extended, logically, to a similar notion for chemistry: protein and nucleic acid, the essential elements of biological life, were, after all, chemical.

In the same year that Watson and Crick discovered DNA, two researchers from the University of Chicago, Stanley Lloyd Miller and **Harold Urey**, experimented with a primordial soup concocted of the elements thought to have made up Earth's early atmosphere—methane, ammonia, hydrogen, and **water**. They sent electrical sparks through the mixture, simulating a **lightning** storm, and discovered trace amounts of amino acids.

During the early 1960s, Ponnamperuma began to delve into this primordial soup and set up variations of Miller and Urey's original experiment. Having changed the proportions of the elements from the original Miller-Urey specifications slightly, Ponnamperuma and his team sent first high-energy electrons, then ultraviolet light through the mixture, attempting to recreate the original conditions of the earth before life. They succeeded in creating large amounts of adenosine triphosphate (ATP), an amino acid that **fuels** cells. In later experiments with the same concoction of primordial soup, the team was able to create the nucleotides that make up nucleic acid—the building blocks of DNA and ribonucleic acid (RNA).

In addition to his work in prebiotic chemistry, Ponnamperuma became active in another growing field: exobiology, or the study of extraterrestrial life. Supported in this effort by NASA, he was able to conduct research on the possibility of the evolution of life on other planets. Theorizing that life evolved from the interactions of chemicals present elsewhere in the universe, he saw the research possibilities of spaceflight. He experimented with lunar **soil** taken by the *Apollo 12* space mission in 1969. As a NASA investigator, he also studied information sent back from Mars by the unmanned Viking, Pioneer, and Voyager probes in the 1970s. These studies suggested to Ponnamperuma that Earth is the only place in the **solar system** where there is life.

In 1969, a meteorite fell to Earth in Muchison, **Australia**. It was retrieved still warm, providing scientists with fresh, uncontaminated material from space for study. Ponnamperuma and other scientists examined pieces of the meteorite for its chemical make-up, discovering numerous amino acids. Most important, among those discovered were the five chemical bases that make up the nucleic acid found in living organisms. Further interesting findings provided tantalizing but puzzling clues about chemical evolution, including the observation that light reflects both to the left and to the right when beamed through a solution of the meteorite's amino acids, whereas light reflects only to the left when beamed through the amino acids of living matter on Earth.

Ponnamperuma's association with NASA continued as he entered academia. In 1979, he became a professor of chemistry at the University of Maryland and director of the Laboratory of Chemical Evolution—established and supported in part by the National Science Foundation and by NASA. He continued active research and experimentation on meteorite material. In 1983, an article in the science section of the *New York Times* explained Ponnamperuma's chemical evolution theory and his findings from the Muchison meteorite experiments. He reported the creation of all five chemical bases of living matter in a single experiment that consisted of bombarding a primordial soup mixture with **electricity**.

Ponnamperuma's contributions to scholarship include hundreds of articles. He wrote or edited numerous books, some in collaboration with other chemists or exobiologists, including annual collections of papers delivered at the College Park Colloquium on Chemical Evolution. He edited two journals, *Molecular Evolution* (from 1970 to 1972) and *Origins of Life* (from 1973 to 1983). In addition to traditional texts in the field of chemical evolution, he also co-authored a software program entitled "Origin of Life," a simulation model intended to introduce biology students to basic concepts of chemical evolution.

Although Ponnamperuma became an American citizen in 1967, he maintained close ties to his native Sri Lanka, even becoming an official governmental science advisor. His professional life has included several international appointments. He was a visiting professor of the Indian Atomic Energy Commission (1967); a member of the science faculty at the Sorbonne (1969); and director of the UNESCO Institute for Early Evolution in Ceylon (1970). His international work included the directorship of the Arthur C. Clarke center, founded by the science fiction writer, a Sri Lankan resident.

Ponnamperuma was a member of the Indian National Science Academy, the American Association for the Advancement of Science, the American Chemical Society, the Royal Society of Chemists, and the International Society for the Study of the Origin of Life, which awarded him the A. I. Oparin Gold Medal in 1980. In 1991, Ponnamperuma received a high French honor—he was made a Chevalier des Arts et des Lettres. Two years later, the Russian Academy of Creative Arts awarded him the first Harold Urey Prize. In October 1994, he was appointed to the Pontifical Academy of Sciences in Rome. He married Valli Pal in 1955; they had one child. Ponnamperuma died on December 20, 1994.

See also Evolutionary mechanisms; Miller-Urey experiment

Porosity and permeability

Porosity and permeability are two of the primary factors that control the movement and storage of fluids in rocks and sediments. They are intrinsic characteristics of these geologic materials. The exploitation of natural resources, such as **groundwater** and **petroleum**, is partly dependent on the properties of porosity and permeability.

Porosity is the ratio of the volume of openings (voids) to the total volume of material. Porosity represents the storage capacity of the geologic material. The primary porosity of a sediment or **rock** consists of the spaces between the grains that make up that material. The more tightly packed the grains are, the lower the porosity. Using a box of marbles as an example, the internal dimensions of the box would represent the volume of the sample. The **space** surrounding each of the spherical marbles represents the void space. The porosity of the box of marbles would be determined by dividing the total void space by the total volume of the sample and expressed as a percentage.

The primary porosity of unconsolidated sediments is determined by the shape of the grains and the range of grain sizes present. In poorly sorted sediments, those with a larger range of grain sizes, the finer grains tend to fill the spaces between the larger grains, resulting in lower porosity. Primary porosity can range from less than one percent in crystalline rocks like **granite** to over 55% in some soils. The porosity of some rock is increased through fractures or solution of the material itself. This is known as secondary porosity.

Permeability is a measure of the ease with which fluids will flow though a porous rock, sediment, or **soil**. Just as with porosity, the packing, shape, and sorting of granular materials control their permeability. Although a rock may be highly porous, if the voids are not interconnected, then fluids within the closed, isolated pores cannot move. The degree to which pores within the material are interconnected is known as effective porosity. Rocks such as **pumice** and shale can have high porosity, yet can be nearly impermeable due to the poorly interconnected voids. In contrast, well-sorted **sandstone** closely replicates the example of a box of marbles cited above. The rounded **sand** grains provide ample, unrestricted void spaces that are free from smaller grains and are very well linked. Consequently, sandstones of this type have both high porosity and high permeability.

The range of values for permeability in geologic materials is extremely large. The most conductive materials have permeability values that are millions of times greater than the least permeable. Permeability is often directional in nature. The characteristics of the interstices of certain materials may cause the permeability to be significantly greater in one direction. Secondary porosity features, like fractures, frequently have significant impact on the permeability of the material. In addition to the characteristics of the host material, the viscosity and pressure of the fluid also affect the rate at which the fluid will flow.

See also Aquifer; Hydrogeology

POTENTIAL ENERGY • *see* ENERGY TRANSFORMATIONS

PRAIRIE

The term prairie is an ecological term used to describe a geologic plain covered by mostly grass. Prairies have been subdivided into smaller, more specific categories by the type of vegetation they support. Short grass and long grass prairies historically covered most of the central portion of the United States. However, the grasses have been replaced by urbanization and agriculture, but the plain still exists.

The Great Plains of the United States support one of the most famous prairies in the world. As with all prairies, the **area** is supported underneath by a firm **bedrock**. In this case, the bedrock is composed of **limestone** deposited by a relatively continuous series of ancient **seas** that advanced and retreated across America and Canada for millions of years. **Dolomite**, containing high levels of magnesium, is the primary building block of the **bedforms**.

Overlying the bedrock are massive and extensive fossil **coral reefs**. These ancient reefs are quite impressive. They began to form in the warm shallow seas after the Silurian about 400 million years ago. Growth was intermittent as the seas transgressed (grew) and regressed (receded) in a cyclic pattern.

During the last 1.9 million years, the entire upper North American continent was covered with **ice**. The ice sheets grew and shrank according to global climate fluctuations. The grinding of the massive ice sheets produced a fine sediment called glacial **till**. The meltwaters of the **glaciers** moved the till away from the sheets and out onto the dolomitic plain in a process known as glacial drift. The drift formed many distinctive structures including eskers, **moraines**, and kettles. The sequences of advancing ice are recorded in the layers of the sediments. What was once considered a distinct pattern of a few ice advances is now understood to be a complicated chronology of at least 29 different episodes. The last identifiable age of ice deposition is called the Wisconsin **glaciation**. This event stripped much of the prairie of high physiographic features while depositing the characteristic soils of the current prairie.

The soils that are high in carbonates are not very rich for growing vegetation. Trees are exposed to extremes of temperatures and varying **precipitation**. They do not fare as well as the hardier grasses. Consequently, the **evolution** of grasses has been intimately tied with the development of the prairie. The prairies have been threatened by the introduction of foreign grass types. Some areas of the prairie have been set aside as national grasslands and non-native grasses are sought out and removed. The majesty of the prairie still exists in these historic places.

See also Glacial landforms; Ice ages; Marine transgression and marine regression; Soil and soil horizons

PRECAMBRIAN

In **geologic time**, Precambrian time encompasses the time from Earth's formation, approximately 4.5 billion years ago, until the start of the Cambrian approximately 540 million years ago (mya). Because the Precambrian is not a true geologic eon, era, period, or epoch, geologists often refer to it as Precambrian time (or simply, Precambrian). Precambrian time represents the vast bulk of Earth's geologic history and covers nearly 90% of Earth's history.

Although scientists do not yet know all the exact steps by which the earth formed, cooled, and took on its approximate shape and physical characteristics, a good deal of reliable evidence can be inferred from studies that concentrate on the formation of landmass, **oceans**, and atmosphere. Astrophysical data—and theories of **physics** that explain the **evolution** of physical law and nucelosynthesis—make these studies of Earth's formation both possible and reliable because the same laws of physics and **chemistry** that exist now operated during the formation of Earth's **solar system**.

Radiological dating provides overwhelming evidence that dates known terrestrial (Earth origin) **rock** specimens to more than 3.6 billion years old. Earth and lunar meteorites date to 4.5 billion years.

Precambrian time is subdivided into Hadean time (4.5 billion years ago to 3.8 billion years ago); **Archean** time (3.8 billion years ago to 2.5 billion years ago; Paleoproterozoic time

(2.5 billion years ago to 1.6 billion years ago); Mesoproterozoic time (1.6 billion years ago to 900 million years ago); and Neoproterozoic time (900 million years ago to 540 mya).

Hadean time represents the time during which the solar system formed. During the subsequent course of Precambrian time, Earth's **lithospheric plates** formed and the mechanisms of geologic change described by modern plate tectonic theory began to occur. During Precambrian time, life arose on Earth. The oldest known fossil evidence (fig tree group fossils in what is now **Africa**) dates to early in Archaean time. During the Paleoproterozoic, Earth's primitive atmosphere made a transition to an **oxygen** rich atmosphere. Soon thereafter in geologic time, i.e. within a few hundred million years, there is evidence of the earliest appearance of eukaryotes (organisms with a true nucleus containing DNA). Evidence of the oldest fossilized animal remains dates to the end of Neoproterozic time.

The extensive debris field that existed in the early solar system assured frequent bombardment of Earth's primitive atmosphere by **asteroids** and **comets**. Despite the consuming effects of geological **weathering** and **erosion**, evidence of Precambrian time impacts dating almost 2.0 billion years ago have been found in what are now South Africa and Canada.

See also Cambrian Period; Cenozoic Era; Cretaceous Period; Dating methods; Devonian Period; Eocene Epoch; Evolution, evidence of; Fossil record; Fossils and fossilization; Historical geology; Holocene Epoch; Jurassic Period; Mesozoic Era; Miller-Urey experiment; Miocene Epoch; Mississippian Period; Oligocene Epoch; Ordovician Period; Paleocene Epoch; Paleozoic Era; Pennsylvanian Period; Phanerozoic Eon; Pleistocene Epoch; Pliocene Epoch; Precambrian; Proterozoic Era; Quaternary Period; Silurian Period; Tertiary Period; Triassic Period

PRECIOUS METALS

Gold, silver, and platinum have historically been valued for their beauty and rarity. They are the precious **metals**. Platinum usually costs slightly more than gold, and both metals are about 80 times more costly than silver. Precious metal weights are given in Troy ounces (named for Troyes, France, known for its fairs during the Middle Ages) a unit approximately 10% larger than 1 oz (28.35 g).

The ancients considered gold and silver to be of noble birth compared to the more abundant metals. Chemists have retained the term noble to indicate the resistance these metals have to **corrosion**, and their natural reluctance to combine with other elements.

The legends of King Midas and Jason's search for the golden fleece hint at prehistoric mankind's early fascination with precious metals. The proof comes in the gold and silver treasure found in ancient Egyptian tombs and even older Mesopotamian burial sites.

The course of recorded history also shows twists and turns influenced to a large degree by precious metals. It was Greek silver that gave Athens its Golden Age, Spanish gold and silver that powered the Roman Empire's expansion, and the desire for gold that motivated Columbus to sail west across the Atlantic. The exploration of Latin America was driven in large part by the search for gold, and the Jamestown settlers in **North America** had barely gotten their "land legs" before they began searching for gold. Small amounts of gold found in North Carolina, Georgia, and Alabama played a role in the 1838 decision to remove the Cherokee Indians to Oklahoma. The California gold rush of 1849 made California a state in 1850, and California gold fueled northern industry and backed up union currency, two major factors in the outcome of the Civil War.

Since ancient times, gold has been associated with the **Sun**. Its name is believed to be derived from a Sanskrit word meaning "to shine," and its chemical symbol (Au) comes from *aurum*, Latin for "glowing dawn." Pure gold has an attractive, deep yellow color and a specific **gravity** of 19.3. Gold is soft enough to scratch with a fingernail, and the most malleable of metals. A block of gold about the size of a sugar cube can be beaten into a translucent film some 27 ft (8 m) on a side. Gold's purity is expressed either as fineness (parts per 1,000) or in karats (parts per 24). An **alloy** containing 50% gold is 500 fine or 12 karat gold. Gold resists corrosion by air and most chemicals but can be dissolved in a mixture of nitric and hydrochloric acids, a solution called *aqua regia* because it dissolves the "king of metals".

Gold is so rare that one ton of average **rock** contains only about eight pennies worth of gold. Gold ore occurs where geologic processes have concentrated gold to at least 250 times the value found in average rock. At that concentration, there is still one million times more rock than gold and the gold is rarely seen. Ore with visible gold is fabulously rich.

Gold most commonly occurs as a pure metal called native gold or as a natural alloy with silver called electrum. Gold and silver combined with tellurium are of local importance. Gold and silver tellurides are found, for example, in the mountains around the old mining boom-town of Telluride, Colorado. Gold is found in a wide variety of geologic settings, but placer gold and gold veins are the most economically important.

Placer gold is derived from gold-bearing rock from which the metal has been freed by **weathering**. Gravity and running **water** then combine to separate the dense grains of gold from the much lighter rock fragments. Rich concentrations of gold can develop above deeply weathered gold veins as the lighter rock is washed away. The "Welcome Stranger" from the gold fields of Victoria, **Australia**, is a spectacular 158–16 (71.5-kg) example of this type of occurrence.

Gold washed into mountain streams also forms placer deposits where the stream's velocity diminishes enough to deposit gold. Stream placers form behind boulders and other obstructions in the streambed, and where a tributary stream merges with a more slowly moving river. Placer gold is also found in gravel bars where it is deposited along with much larger rocky fragments.

The discovery of placer gold set off the California gold rush of 1849 and the rush to the Klondike in 1897. The largest river placers known are in Siberia, Russia. Gold-rich sands there are removed with jets of water, a process known as hydraulic mining. A fascinating byproduct of Russia's hydraulic mining is the unearthing of thousands of woolly mammoths, many with flesh intact, locked since the **Ice** Age in frozen tundra gravel.

Stream placer deposits have their giant ancient counterparts in paleoplacers, and the Witwatersrand district in South **Africa** outproduces all others combined. Gold was reported from the Witwatersrand (White Waters Ridge) as early as 1834, but it was not until 1886 that the main deposit was discovered. From that time until today, it has occupied the paramount position in gold mining history. Witwatersrand gold was deposited between 2.9 and 2.6 billion years ago in six major fields, each produced by an ancient river system.

Placer and paleoplacers are actually secondary gold deposits, their gold having been derived from older deposits in the mountains above. The California 49ers looked upstream hoping to find the mother lode, and that's exactly what they called the system of gold veins they discovered.

Vein gold is deposited by hot subterranean water known as a hydrothermal fluid. Hydrothermal fluids circulate through rock to leach small amounts of gold from large volumes of rock and then deposit it in fractures to form veins. Major U.S. gold vein deposits have been discovered at **Lead** in the Black Hills of South Dakota and at Cripple Creek on the slopes of Pike's Peak, Colorado. Important vein deposits are also found in Canada and Australia. All these important deposits were located following the discovery of placer gold in nearby streams.

Gold's virtual indestructibility means that almost all gold ever mined could still be in use today. Today, gold is being mined in ever-increasing amounts from increasingly lower-grade deposits. It is estimated that 70% of all gold recovered has been mined in this century. Each year nearly 2,000 tons are added to the total. Nevada currently leads the nation in gold production, and the Republic of South Africa is the world's leading gold-producing nation.

Gold has traditionally been used for coinage, bullion, jewelry, and other decorative uses. Gold's chemical inertness means that gold jewelry is hypoallergenic and remains tarnish-free indefinitely.

Silver is a brilliant white metal and the best metal in terms of thermal and electrical conductivity. Its chemical symbol, Ag, is derived from its Latin name, *argentum*, meaning "shining white." Silver is not nearly as precious, dense, or noble as gold or platinum. The ease with which silverware tarnishes is an example of its chemical reactivity. Although native silver is found in nature, it most commonly occurs as compounds with other elements, especially sulfur.

Hydrothermal veins constitute the most important source of silver. The Comstock Lode, a silver bonanza 15 mi (24 km) southeast of Reno, Nevada, is a well-known example. Hydrothermal silver veins are formed in the same manner as gold veins, and the two metals commonly occur together. Silver, however, being more reactive than gold, can be leached from surface rocks and carried downward in solution. This process, called supergene enrichment, can concentrate silver into exceedingly rich deposits at depth.

Mexico has traditionally been the world's leading silver producing country, but the United States, Canada, and Peru each contribute significant amounts. Although silver has historically been considered a precious metal, industrial uses now predominate. Significant quantities are still used in jewelry, silver ware, and coinage; but even larger amounts are consumed by the photographic and electronics industries.

Platinum, like silver, is a silver-white metal. Its chemical symbol is Pt and its name comes from the Spanish world for silver (*plata*), with which it was originally confused. Its specific gravity of 21.45 exceeds that of gold, and, like gold, it is found in pure metallic chunks in stream placers. The average crustal abundance of platinum is comparable to that of gold. The **melting** point of platinum is 3,219°F (1,769°C), unusually high for a metal, and platinum is chemically inert even at high **temperature**. In addition, platinum is a catalyst for chemical reactions that produce a wide range of important commodities.

Platinum commonly occurs with five similar metals known as the platinum group metals. The group includes osmium, iridium, rhodium, palladium, and ruthenium. All were discovered in the residue left when platinum ore was dissolved in aqua regia. All are rare, expensive, and classified chemically as noble metals.

Platinum is found as native metal, in natural alloys, and in compounds with sulfur and arsenic. Platinum ore deposits are rare, highly scattered, and one deposit dominates all others much as South Africa's Witwatersrand dominates world gold production. That platinum deposit is also in the Republic of South Africa.

Placer platinum was discovered in South Africa in 1924 and subsequently traced to a distinctively layered igneous rock known as the Bushveld Complex. Although the complex is enormous, the bulk of the platinum is found in a thin layer scarcely more than three feet thick. Nearly half of the world's historic production of platinum has come from this remarkable layer.

The Stillwater complex in the Beartooth mountains of southwestern Montana also contains a layer rich in platinum group metals. Palladium is the layer's dominant metal, but platinum is also found. The layer was discovered during the 1970s, and production commenced in 1987.

Platinum is used mostly in catalytic converters for vehicular pollution control. Low-voltage electrical contracts form the second most common use for platinum, followed closely by dental and medical applications, including dental crowns, and a variety of pins and plates used internally to secure human bones. Platinum is also used as a catalyst in the manufacture of explosives, fertilizer, gasoline, insecticides, paint, plastics, and pharmaceuticals. Platinum crucibles are used to melt high-quality optical **glass** and to grow **crystals** for computer chips and lasers. Hot glass fibers for insulation and nylon fibers for textiles are extruded through platinum sieves.

Because of their rarity and unique properties, the demand for gold and platinum are expected to continue to increase. Silver is more closely tied to industry, and the demand for silver is expected to rise and fall with economic conditions.

PRECIPITATION

Precipitation is **water** in either solid or liquid form that falls from Earth's atmosphere. Major forms of precipitation include rain, snow, and hail. When air is lifted in the atmosphere, it expands and cools. Cool air cannot hold as much water in vapor form as warm air, and the **condensation** of vapor into droplets or **ice crystals** may eventually occur. If these droplets or crystals continue to grow to large sizes, they will eventually be heavy enough to fall to the earth's surface.

Precipitation in liquid form includes drizzle and raindrops. Raindrops are on the order of a millimeter (one thousandth of a meter) in radius, while drizzle drops are approximately a tenth of this size. Important solid forms of precipitation include snowflakes and hailstones. Snowflakes are formed by aggregation of solid ice crystals within a cloud, while hailstones involve supercooled water droplets and ice pellets. They are denser and more spherical than snowflakes. Other forms of solid precipitation include graupel and sleet (ice pellets). Solid precipitation may reach Earth's surface as rain if it melts as it falls. Virga is precipitation that evaporates before reaching the ground.

Precipitation forms differently depending on whether it is generated by warm or cold **clouds**. Warm clouds are defined as those that do not extend to levels where temperatures are below 32°F (0°C), while cold clouds exist at least in part at temperatures below 32°F (0°C). **Temperature** decreases with height in the lower atmosphere at a moist adiabatic rate of about 3.3°F per 3,281 ft (6°C per 1,000 m), on average. High clouds, such as cirrus, are therefore colder and more likely to contain ice. As discussed below, however, temperature is not the only important factor in the formation of precipitation.

Even the cleanest air contains aerosol particles (solid or liquid particles suspended in the air). Some of these particles are called cloud condensation nuclei, or CCN, because they provide favorable sites on which water vapor can condense. Air is defined to be fully saturated, or have a relative **humidity** of 100%, when there is no net transfer of vapor molecules between the air and a plane (flat) surface of water at the same temperature. As air cools, its relative humidity will rise to 100% or more, and molecules of water vapor will bond together, or condense, on particles suspended in the atmosphere. Condensation will preferentially occur on particles that contain water soluble (hygroscopic) material. Types of particles that commonly act as CCN include sea-salt and particles containing sulfate or nitrate ions; they are typically about 0.0000039 in (0.0001 mm) in radius. If relative humidity remains sufficiently high, CCN will grow into cloud droplets 0.00039 in (0.01 mm) or more in size. Further growth to precipitation size in warm clouds occurs as larger cloud droplets collide and coalesce (merge) with smaller ones.

Although large quantities of liquid water will freeze as the temperature drops below 32°F (0°C), cloud droplets sometimes are supercooled; that is, they may exist in liquid form at lower temperatures down to about –40°F (–40°C). At temperatures below –40°F (–40°C), even very small droplets freeze readily, but at intermediate temperatures (between –40 and 32°F or –40 and 0°C), particles called ice nuclei initiate the **freezing** of droplets. An ice nucleus may already be present within a droplet, may contact the outside of a droplet and cause it to freeze, or may aid in ice formation directly from the vapor phase. Ice nuclei are considerably more rare than cloud condensation nuclei and are not as well understood.

Once initiated, ice crystals will generally grow rapidly because air that is saturated with respect to water is supersaturated with respect to ice; i.e., water vapor will condense on an ice surface more readily than on a liquid surface. The habit, or shape, of an ice crystal is hexagonal and may be plate-like, column-like, or dendritic (similar to the snowflakes cut from paper by children). Habit depends primarily on the temperature of an ice crystal's formation. If an ice crystal grows large enough to fall through air of varying temperatures, its shape can become quite intricate. Ice crystals can also grow to large sizes by aggregation (clumping) with other types of ice crystals that are falling at different speeds. Snowflakes are formed in this way.

Clouds that contain both liquid water and ice are called mixed clouds. Supercooled water will freeze when it strikes another object. If a supercooled droplet collides with an ice crystal, it will attach itself to the crystal and freeze. Supercooled water that freezes immediately will sometimes trap air, forming opaque (rime) ice. Supercooled water that freezes slowly will form a more transparent substance called clear ice. As droplets continue to collide with ice, eventually the shape of the original crystal will be obscured beneath a dense coating of ice; this is how a hailstone is formed. Hailstones may even contain some liquid water in addition to ice. Thunderstorms are dramatic examples of vigorous mixed clouds that can produce high precipitation rates. The electrical charging of precipitation particles in thunderstorms can eventually cause **lightning** discharges.

Precipitation reaching the ground is measured in terms of precipitation rate or precipitation intensity. Precipitation intensity is the depth of precipitation reaching the ground per hour, while precipitation rate may be expressed for different time periods. Typical precipitation rates for the northeastern United States are 2–3 in (50–80 mm) per month, but in Hilo, Hawaii, 49.9 in (127 cm) of rain fell in March 1980. Average annual precipitation exceeds 80 in (200 cm) in many locations. Because snow is less compact than rain, the mass of snow in a certain depth may be equivalent to the mass of rain in only about one-tenth that depth (i.e., one inch of rain contains as much water as about 10 in [25 cm] of snow). Certain characteristics of precipitation are also measured by radar and satellites.

The earth is unique in our **solar system** in that it contains water, which is necessary to sustain life as we know it. Water that falls to the ground as precipitation is critically important to the **hydrologic cycle**, the sequence of events that moves water from the atmosphere to the earth's surface and back again. Some precipitation falls directly into the **oceans**, but precipitation that falls on land can be transported to the oceans through **rivers** or underground in aquifers. Water stored in this permeable **rock** can take thousands of years to reach the sea. Water is also contained in reservoirs such as **lakes** and the **polar ice** caps, but about 97% of the earth's water is contained in the oceans. The sun's energy heats and evaporates water from the ocean surface. On average, **evaporation** exceeds precipitation over the oceans, while precipitation exceeds evaporation over land masses. Horizontal air motions can transfer evaporated water to areas where clouds and precipitation subsequently form, completing the circle which can then begin again.

The distribution of precipitation is not uniform across the earth's surface, and varies with time of day, season and year. The lifting and cooling that produces precipitation can be caused by solar heating of the earth's surface, or by forced lifting of air over obstacles or when two different air masses converge. For these reasons, precipitation is generally heavy in the tropics and on the upwind side of tall mountain ranges. Precipitation over the oceans is heaviest at about 7°N **latitude** (the intertropical convergence zone), where the tradewinds converge and large thunderstorms frequently occur. While summer is the "wet season" for most of **Asia** and northern **Europe**, winter is the wettest time of year for Mediterranean regions and western **North America**. Precipitation is frequently associated with large-scale low-pressure systems (cyclones) at mid-latitudes.

Precipitation is obviously important to humankind as a source of drinking water and for agriculture. It cleanses the air and maintains the levels of lakes, rivers, and oceans, which are sources of food and recreation. Interestingly, human activity may influence precipitation in a number of ways, some of which are intentional, and some of which are quite unintentional. These are discussed below.

The irregular and frequently unpredictable nature of precipitation has led to a number of direct attempts to either stimulate or hinder the precipitation process for the benefit of humans. In warm clouds, large hygroscopic particles have been deliberately introduced into clouds in order to increase droplet size and the likelihood of collision and coalescence to form raindrops. In cold clouds, ice nuclei have been introduced in small quantities in order to stimulate precipitation by encouraging the growth of large ice crystals; conversely, large concentrations of ice nuclei have been used to try to reduce numbers of supercooled droplets and thereby inhibit precipitation formation. Silver iodide, which has a crystalline structure similar to that of ice, is frequently used as an ice nucleus in these "cloud seeding" experiments. Although certain of these experiments have shown promising results, the exact conditions and extent over which **cloud seeding** works and whether apparent successes are statistically significant is still a matter of debate.

Acid rain is a phenomenon that occurs when acidic pollutants are incorporated into precipitation. It has been observed extensively in the eastern United States and northern Europe. Sulfur dioxide, a gas emitted by power plants and other industries, can be converted to acidic sulfate compounds within cloud droplets. In the atmosphere, it can also be directly converted to acidic particles, which can subsequently act as CCN or be collected by falling raindrops. About 70 megatons of sulfur is emitted as a result of human activity each year across the planet. (This is comparable to the amount emitted naturally.) Also, nitrogen oxides are emitted by motor vehicles, converted to nitric acid vapor, and incorporated into clouds in the atmosphere.

Acidity is measured in terms of **pH**, the negative logarithm of the hydrogen ion concentration; the lower the pH, the greater the acidity. Water exposed to atmospheric **carbon dioxide** is naturally slightly acidic, with a pH of about 5.6. The pH of rainwater in remote areas may be as low as about 5.0 due to the presence of natural sulfate compounds in the atmosphere. Additional sulfur and nitrogen containing acids introduced by anthropogenic (human-induced) activity can increase rainwater acidity to levels that are damaging to aquatic life. Recent reductions in emissions of sulfur dioxide in the United Kingdom have resulted in partial recovery of some affected lakes.

Recent increases in anthropogenic emissions of trace gases (for example, **carbon** dioxide, methane, and chlorofluorocarbons) have resulted in concern over the so-called **greenhouse effect**. These trace gases allow energy in the form of sunlight to reach the earth's surface, but "trap" or absorb the infrared energy (heat) that is emitted by the earth. The heat absorbed by the atmosphere is partially re-radiated back to the earth's surface, resulting in warming. Trends in the concentrations of these **greenhouse gases** have been used in **climate** models (computer simulations) to predict that the global average surface temperature of the earth will warm by 3.6–10.8°F (2–6°C) within the next century. For comparison, the difference in average surface temperature between the Ice Age 18,000 years ago and present day is about 9°F (5°C).

Greenhouse warming due to anthropogenic activity is predicted to have other associated consequences, including rising sea levels and changes in cloud cover and precipitation patterns around the world. For example, a reduction in summertime precipitation in the Great Plains states is predicted by many models and could adversely affect crop production. Other regions may actually receive higher amounts of precipitation than they do currently. The level of uncertainty in these model simulations is fairly high, however, due to approximations that are made. This is especially true of calculations related to aerosol particles and clouds. Also, the natural variability of the atmosphere makes verification of any current or future trends extremely difficult unless actual changes are quite large.

As discussed above, gas-phase pollutants such as sulfur dioxide can be converted into water-soluble particles in the atmosphere. Many of these particles can then act as nuclei of cloud droplet formation. Increasing the number of CCN in the atmosphere is expected to change the characteristics of clouds. For example, ships' emissions have been observed to cause an increase in the number of droplets in the marine stratus clouds above them. If a constant amount of liquid water is present in the cloud, the average droplet size will be smaller. Higher concentrations of smaller droplets reflect more sunlight, so if pollution-derived particles alter clouds over a large enough region, climate can be affected. Precipitation rates may also decrease, since droplets in these clouds are not likely to grow large enough to precipitate.

See also Air masses and fronts; Atmospheric chemistry; Atmospheric circulation; Atmospheric composition and structure; Atmospheric pollution; Atmospheric pressure; Blizzards and lake effect snows; Clouds and cloud types; Greenhouse gases and greenhouse effect; Rainbow; Seasonal winds; Tropical cyclone; Water pollution and biological purification; Weather forecasting methods; Weather forecasting

One method of projection. ***© First Light/Corbis. Reproduced by permission.***

PROJECTION

Because the earth is a sphere, all flat maps of its surface contain inherent distortions. Map projections represent a curved land surface in two dimensions while minimizing these unavoidable errors of shape, distance, azimuth, scale and **area**. Most projections accurately portray one type of geographical information at the expense of another type, and cartographers choose a projection based on a map's intended use. A conformal projection, for example, shows relatively undistorted shapes, but inaccurate areas, while an equal-area projection makes the opposite choice. The Mercator projection is accurate at the equator but becomes progressively more distorted toward the poles, while polar stereographic maps preserve high-latitude coordinates at the expense of equatorial regions. The Mercator map of the world is responsible for the mistaken impression that Greenland covers almost as large an area as **Africa**.

The method of projecting a sphere onto a two-dimensional surface defines three classes of map projections: cylindrical, conic, and azimuthal. The alignment of the projection cylinder, cone, or plane relative to the globe further divides these classes into subtypes. Cylindrical equal-area, Mercator, Miller cylindrical, oblique Mercator, and transverse Mercator are all cylindrical map projections. Mercator maps have straight, evenly spaced lines of **latitude and longitude** that intersect at right angles, and are undistorted in scale at the equator, or at two lines of **latitude** equidistant to the equator. Mercator maps are useful for marine navigation because straight lines drawn on the map are true headings. Transverse Mercator maps are created by projecting the global sphere onto a cylinder tangent to a line of **longitude**, or meridian. The British National Grid System (BNG), used by the British Ordnance Survey, and the Universal Tranverse Mercator projection (UTM) are widely-used transverse Mercator **mapping techniques**.

Conic and azimuthal projections are less common than cylindrical projections. In a number of specific cases, however, projection of the globe onto a cone or a plane presents the most suitable map scheme. Albers equal area, equidistant conic, Lambert conformal equal area, and polyconic are all conic projections used in maps of **North America**. Most United States Geographical Survey (USGS) topographic quadrangles use a polyconic projection. Azimuthal projections are variously used for aeronautical navigation (azimuthal equidistant), maps of the ocean basins (Lambert azimuthal equal area), maps of the hemispheres (orthographic), and polar navigation (stereographic).

Proterozoic Era

The Proterozoic Era, also termed the Algonkian, is the second of the two eras into which the **Precambrian** has traditionally been divided. The Precambrian includes over four fifths of Earth's history: the 4.5 billion years from the formation of Earth to the start of the **Cambrian Period** some 570 million years ago. The first half of the Precambrian is known as the **Archean** Era and the second half as the Proterozoic Era.

Eukaryotic cells (cells with nuclei) first appeared in the early Proterozoic, about 2.5 billion years ago. Until that time only prokaryotic cells (cells without nuclei) existed. Bacteria and marine algae also evolved during the Proterozoic, and, late in the era, the first multicellular life appeared. During the Proterozoic, photosynthetic bacteria and algae liberated enough **oxygen** (O_2) from **carbon dioxide** (CO_2) to change Earth's atmosphere from oxygen-free to oxygen-rich. This chemical transformation made the Cambrian explosion of multicellular life possible.

Significant geological changes also took place during the late Archean Era and early Proterozoic Era. The continents first began to form wide, stable continental shelves at this time and to be moved about by plate-tectonic processes. On the continents, which were still devoid of plant life, **erosion** and deposition proceeded rapidly. Numerous extremely thick beds of pure **quartz sandstone** formed—some, kilometers thick. In contrast, more recently formed beds of this type are, usually, at most, 109 yards (100 meters) thick.

Throughout both the Archean Era and the Proterozoic Era, beds of the banded **iron** formation were formed. This type of banded formation consists of alternating thin layers of quartz and iron oxide and were not formed during any later period. Today they are the world's major source of iron ore.

For decades, some geologists have disputed the usefulness of the term Proterozoic (from the Greek *protero*, earlier, and *zoic*, life). The Archean-Proterozoic distinction was first devised to describe the striking unconformity (change in **rock** type with depth) that runs horizontally through the Canadian shield, a vast **area** of Precambrian rock that rings Hudson Bay and includes Greenland. However, this dramatic division has not been found globally in Precambrian rocks. Furthermore, some geologists argue that it is misleading to lump 4.5 billion years of various geological history into just two compartments. Therefore, vaguer terms—early, middle, and late (or lower, middle, and upper) Precambrian—are often used.

See also Cenozoic Era; Cretaceous Period; Dating methods; Devonian Period; Eocene Epoch; Evolution, evidence of; Fossil record; Fossils and fossilization; Geologic time; Historical geology; Holocene Epoch; Jurassic Period; Mesozoic Era; Miocene Epoch; Mississippian Period; Oligocene Epoch; Ordovician Period; Origin of life; Paleocene Epoch; Paleozoic Era; Pennsylvanian Period; Phanerozoic Eon; Pleistocene Epoch; Pliocene Epoch; Quaternary Period; Silurian Period; Tertiary Period

Protons • *see* Atomic theory

Pteropodes • *see* Calcareous ooze

Ptolemy (ca. 100-170)

Greek astronomer

Very little is known about Ptolemy's early life. Born in Alexandria, Egypt, as Ptolemais Hermii, his name was later latinized as Claudius Ptolemaeus, and later Ptolemy.

Ptolemy's chief contribution to science is a series of books in which he compiled the knowledge of the ancient Greeks, his primary source being Hipparchus (fl. second century B.C.). Because most of Hipparchus' writings have not survived from antiquity, many of the ideas he espoused about the universe have become known as the Ptolemaic system.

Ptolemy's system placed Earth directly at the center of the universe. The **Sun**, **Moon** and planets all orbited Earth. However, since such a scheme did not match the observed motions of the planets, Ptolemy added small orbits to the planets called *epicycles*, and introduced other mathematical devices to make a better fit.

Despite its errors and complications, the Ptolemaic system was adequate enough to make predictions of planetary positions, and it influenced thinking for 1,400 years. It was not until 1543 that Polish astronomer Nicolaus Copernicus (1473–1543) published his book refuting the Ptolemaic system. After Danish astronomer Tycho Brahe's (1546–1601) exceptionally accurate measurements of the positions of the planets showed Ptolemy's system was inaccurate, it fell upon German astronomer and mathematician **Johannes Kepler** (1571–1630) to devise a better explanation of planetary orbits.

Hipparchus had made a catalogue of stars, which were grouped into 48 constellations. Ptolemy placed them in his book and gave these patterns the names that are still in use today. He also included Hipparchus' work on trigonometry, his estimate of the distance between Earth and the Moon, which was fairly accurate, as well as Aristarchus' (third century B.C.) incorrect estimate of Earth's distance from the Sun.

Ptolemy's book was entitled *Mega (mathematike) syntaxis* ("Great [mathematical] compilation") although *Mega* was sometimes replaced by *Megiste* ("Greatest"). When the Arabs adopted the work, they called it *Al-majisti* ("The

Greatest"), which it is known as today. It was translated into Latin in 1175 (as "Almagesti" or "Almagestum") and dominated European thinking for four centuries.

In the field of optics, Ptolemy wrote about the reflection and refraction of light. He lists tables for the refraction of light as it passes into **water** at different angles. Another book, *Tetrabiblos*, is a serious treatment of astrology.

Ptolemy also wrote a treatise that dealt with geography and included maps as well as tables of **latitude and longitude**. It explained how those lines could be mathematically determined, but only a few latitudes were calculated. He had accepted Poseidonius' (ca. 135–51 B.C.) erroneously small estimate of the size of Earth, instead of Eratosthenes' (ca. 276–194 B.C.) more accurate figure, and Ptolemy unwittingly may have altered the history of the world. After his geography had been translated into Latin, it eventually came to the attention of Christopher Columbus (1451–1506), who accepted the incorrect size and concluded that his search for a short-cut to **Asia** was possible.

See also History of exploration I (Ancient and classical); History of exploration II (Age of exploration)

PUMICE

Pumice is a vesicular volcanic **rock** that is commonly light enough to float in **water**. It typically has a chemical composition similar to **rhyolite** (or its plutonic counterpart, **granite**), although **magma** of virtually any composition can form pumice. The term vesicular refers to the presence of vesicles, or irregularly shaped cavities, that produce a sponge-like or bubbly texture and very low density in volcanic rocks.

Pumice can be thought of as a volcanic foam that forms when dissolved gases expand rapidly as magma rises towards the surface and confining pressure decreases. This process is similar to the foaming that occurs when a bottle of carbonated water or soda is opened. Upon eruption, the magma surrounding the gas bubbles quickly freezes into a delicate **glass** framework that produces the distinctive vesicular texture and light weight of pumice. Pumice will float if most vesicle walls remain intact and form air-filled chambers.

Reticulite is a type of pumice formed from basaltic magma in which most of the vesicle walls have burst to form a honeycomb-like structure of glassy threads. Because very few of the vesicle walls remain intact, reticulite will not float in water. Scoria, which is darker and heavier than but otherwise superficially similar to pumice, forms as a vesicular **crust** atop basaltic and andesitic **lava** flows. Close examination usually shows that scoria is much more crystalline than pumice—indicating a slower rate of cooling—and is composed of dark ferromagnesian **minerals**. It is too heavy to float in water.

The liberation of dissolved gases that produces pumice is also responsible for explosive pyroclastic eruptions. Thus, pumice fragments are commonly found within deposits of volcanic ejecta known collectively as tephra, and ash-flow deposits known as tuffs.

Pumice has a several commercial uses and is obtained from strip mines or open pit mines in volcanic rocks located throughout the western United States and elsewhere. It is most commonly used for garment softening (principally stone washed denim), as aggregate in lightweight cinder blocks and prefabricated concrete panels, as landscaping rock, as an abrasive, and as an inert filter material.

See also Andesite; Basalt; Glass; Igneous rocks; Volcanic eruptions; Volcano

Q

Quantum electrodynamics (QED)

Quantum electrodynamics (QED) is a scientific theory that is also known as the **quantum theory** of light. QED describes the quantum properties (properties that are conserved and that occur in discrete amounts called quanta) and mechanics associated with the interaction of light (i.e., electromagnetic radiation) with matter. The practical value of QED rests upon its ability, as a set of equations, to allow calculations related to the absorption and emission of light by atoms and to allow scientists to make very accurate predictions regarding the result of the interactions between photons and charged atomic particles such as electrons. QED is a fundamentally important scientific theory because it accounts for all observed physical phenomena except those associated with aspects of **relativity theory** and radioactive decay.

QED is a complex and highly mathematical theory that paints a picture of light that is counter-intuitive to everyday human experience. According to QED theory, light exists in a duality consisting of both particle and wave-like properties. More specifically, QED asserts that electromagnetism results from the quantum behavior of the photon, the fundamental "particle" responsible for the transmission electromagnetic radiation. According to QED theory, a seeming particle vacuum actually consists of electron-positron fields. An electron-positron pair (positrons are the positively charged antiparticle to electrons) comes into existence when photons interact with these fields. In turn, QED also accounts for the subsequent interactions of these electrons, positrons, and photons.

Photons, unlike other "solid" particles, are thought to be "virtual particles" constantly exchanged between charged particles such as electrons. Indeed, according to QED theory the forces of **electricity and magnetism** (i.e., the fundamental electromagnetic force) stem from the common exchange of virtual photons between particles and only under special circumstances do photons become observable as light.

According to QED theory, "virtual photons" are more like the wavelike disturbances on the surface of **water** after it is touched. The virtual photons are passed back and forth between the charged particles much like basketball players might pass a ball between them as they run down the court. As virtual particles, photons cannot be observed because they would violate the laws regarding the conservation of energy and momentum. Only in their veiled or hidden state do photons act as mediators of force between particles. The "force" caused by the exchange of virtual photons causes charged particles to change their velocity (speed and/or direction of travel) as they absorb or emit virtual photons.

Only under limited conditions do the photons escape the charged particles and thereby become observable as electromagnetic radiation. Observable photons are created by perturbations (i.e., wave-like disruptions) of electrons and other charged particles. According to QED theory, the process also works in reverse as photons can create a particle and its antiparticle (e.g., an electron and its oppositely charged antiparticle, a positron).

In QED dynamics, the simplest interactions involve only two charged particles. The application of QED is, however, not limited to these simple systems; interactions involving an infinite number of photons are described by increasingly complex processes termed second-order (or higher) processes. Although QED can account for an infinite number of processes (i.e., an infinite number of interactions) the theory also dictates that more interactions also become increasingly rare as they become increasingly complex.

The genesis of QED was the need for physicists to reconcile theories initially advanced by British physicist **James Clerk Maxwell** regarding electromagnetism in the later half of the nineteenth century (i.e., that **electricity** and **magnetism** are two aspects of a single force) with quantum theory developed during the early decades of the twentieth century. Prior to WWII, British physicist Paul Dirac, German physicist Werner Heisenberg, and Austrian-born American physicist Wolfgang Pauli all made significant contributions to the mathematical foundations related to QED. Even for these experienced physicists, however, working with QED posed formidable obstacles because of the presence of "infinities" (infinite values) in the mathematical calculations (e.g., for emission rates or determinations of mass). It was often difficult to make predictions match observed phenomena and early attempts at using QED theory often gave physicists wrong or incomprehensible answers.

The calculations used to define QED were made more accessible and reliable by a process termed renormalization, independently developed by American physicist **Richard Feynman** (1918–1988), American physicist Julian Schwinger (1918–1994), and Japanese physicist Shin'ichiro Tomonaga (1906–1979). In essence the work of these three renowned scientists concentrated on making the needed corrections to Dirac's infinity problems and his advancement of QED theory, which helped reconcile quantum mechanics with Einstein's special theory of relativity. Their "renormalization" allowed positive infinities to cancel out negative infinities and thus, allowed measured values of mass and charge to be used in QED calculations.

The use of renormalization initially allowed QED predictions to accurately predict the observed interactions of electrons and photons. During the later half of the twentieth century, based principally on the work of Feynman, Schwinger, Tomonaga and another influential physicist Freeman Dyson, QED became an important model used to explain the structure, properties and reactions of quarks, gluons and other subatomic particles. Although Feynman, Schwinger, and Tomonaga each worked separately on the refinement of different aspects of QED theory, in 1965, these physicists jointly shared the Nobel Prize for their work.

Because QED is compatible with special relativity theory, and special relativity equations are part of QED equations, QED is termed a relativistic theory. QED is also termed a gauge-invariant theory, meaning that it makes accurate predictions regardless of where applied in **space** or time. Like **gravity**, QED mathematically describes a force that becomes weaker as the distance between charged particles increases, reducing in strength as the inverse square of the distance between particles. Although the photons themselves are electrically neutral, the predictions of interactions made possible by QED would not be possible between uncharged or electrically neutral particles. Accordingly, in QED theory there are two values for electric charge on particles, positive and negative.

QED theory was revolutionary in **physics**. In contrast to theories that strove to explain natural phenomena in terms of direct causes and effects, the development of QED stemmed from a growing awareness of the limitations on scientist's ability to make predictions regarding the subatomic realm. In fact, QED was unique precisely because QED did not always make specific predictions. QED relied instead on developing an understanding of the properties and behavior of subatomic particles characterized by probabilities rather than by traditional cause-and-effect certainties. Instead of allowing scientists to make specific predictions regarding the outcome of certain interactions—Tomonaga's predictions were often mystifyingly incompatible with human experience (e.g., that an electron could be in two places at once)—QED allowed the calculation of probabilities regarding outcomes (e.g., the probability that an electron would take one path as opposed to another).

In particular, Feynman's work, teaching, and contributions to QED theory reached near legendary status within the physics community. In 1986, Feynman published *QED: The Strange Theory of Light and Matter*. In his book, Feynman attempted to explain QED theory in much the same manner as Einstein's writing on relativity theory a half century earlier. In fact, although Feynman's profound contributions to QED theory were well beyond the understanding of the general public, no other physicist since Einstein and Oppenheimer had so captured the attention of the lay public. In addition, Feynman also became somewhat of a celebrity for chronicles relating to his life and studies.

Feynman's work redefined QED theory, quantum mechanics, and electrodynamics, and Feynman's writings remain the definitive explanation of QED theory. With regard to QED theory, Feynman is perhaps best remembered for his invention of simple diagrams, now widely known among physicists as "Feynman diagrams," to portray the complex interactions of atomic particles. The diagrams allow visual representation of the ways in which particles can interact by the exchange of virtual photons. In addition to providing a tangible picture of processes outside the human capacity for observation, Feynman's diagrams precisely portray the interactions of variables used in the complex QED mathematical calculations.

Schwinger and Tomonaga also refined the mathematical methodology of QED theory so that predictions became increasingly consistent with predictions of phenomena made by the special theory of relativity. Tomonaga also solved a perplexing inconsistency that vexed Dirac's work (e.g., that an electron could, inconceivably, and not in accord with observations, be calculated to have a seemingly infinite amount of energy). Tomonaga's mathematical improvements, along with refinements made by Schwinger and Feynman, resolved this incompatibility and allowed for the calculation of finite energies for electrons. In a master-stroke, Tomonaga renormalized and made more accurate the prediction of particle properties (e.g., magnetic properties) and the process of radiation.

QED went on to become, arguably, the best tested theory in science history. Most atomic interactions are electromagnetic in nature and, no matter how accurate the equipment yet devised, the predictions made by renormalized QED theory hold true. Some tests of QED—for example, predictions of the mass of some subatomic particles—offer results accurate to six significant figures or more. Even with the improvements made by the renormalization of QED, however, the calculations often remain difficult. Although some predictions can be made using one Feynman diagram and a few pages of calculations, others may take hundreds of Feynman diagrams and the access to supercomputing facilities to complete the necessary calculations.

The development of QED theory allowed scientists to predict how subatomic particles are created or destroyed. Just as Feynman, Schwinger and Tomonaga's renormalization of QED allowed for calculation of finite properties relating to mass, energy, and charge-related properties of electrons, physicists hope that such improvements offer a model to improve other gauge theories (i.e., theories which explain how forces, such as the electroweak force, arise from underlying symmetries). The concept of forces such as electromagnetism arising from the exchange of virtual particles has intriguing ramifications for the advancement of theories regarding the working mechanisms underlying the strong, weak, and gravitational forces.

Many scientists assert that if a unified theory can be found, it will rest on the foundations established during the development of QED theory. Without speculation, however, is the fact that the development of QED theory was, and remains today, an essential element in the verification and development of quantum field theory.

See also Atomic structure; Quantum theory and mechanics

QUANTUM THEORY AND MECHANICS

Quantum mechanics describes the relationships between energy and matter on the atomic and subatomic scale. At the beginning of the twentieth century, German physicist Maxwell Planck (1858–1947) proposed that atoms absorb or emit electromagnetic radiation in bundles of energy termed quanta. This quantum concept seemed counter-intuitive to well-established Newtonian **physics**. Advancements associated with quantum mechanics (e.g., the uncertainty principle) also had profound implications with regard to the philosophical scientific arguments regarding the limitations of human knowledge.

The classical model of the **atom** evolved during the last decade of the nineteenth century and early years of the twentieth century was similar to the Copernican model of the **solar system** where, just as planets orbit the **Sun**, electrically negative electrons moved in orbits about a relatively massive, positively charged nucleus. Most importantly, in accord with Newtonian theory, the classical models allowed electrons to orbit at any distance from the nucleus. Problems with these models, however, continued to vex the leading physicist of the time. The classical models predicted that when, for example, a hydrogen atom was heated it should produce a continuous spectrum of colors as it cooled. Nineteenth century spectroscopic experiments, however, showed that hydrogen atoms produced only a portion of the spectrum. Moreover, physicist James Clerk Maxwell's (1831–1879) studies on electromagnetic radiation predicted that an electron orbiting around the nucleus according to Newton's laws would continuously lose energy and eventually fall into the nucleus.

Planck proposed that atoms absorb or emit electromagnetic radiation only in certain units or bundles of energy termed quanta. The concept that energy existed only in discrete and defined units seemed counter-intuitive, that is, outside the human experience with nature. Regardless, Planck's quantum theory, that also asserted that the energy of light was directly proportional to its frequency, proved a powerful theory that accounted for a wide range of physical phenomena. Planck's constant relates the energy of a photon with the frequency of light. Along with the constant for the speed of light, Planck's constant (h = 6.626 ù 10^{-34} Joule-second) is a fundamental constant of nature.

Prior to Planck's work, electromagnetic radiation (light) was thought to travel in waves with an infinite number of available frequencies and wavelengths. Planck's work focused on attempting to explain the limited spectrum of light emitted by hot objects and to explain the absence of what was termed the "violet catastrophe" predicted by nineteenth century theories developed by Prussian physicist Wilhelm Wien (1864–1928) and English physicist Baron (John William Strutt) Rayleigh (1842–1919).

Danish physicist **Niels Bohr** (1885–1962) studied Planck's quantum theory of radiation and worked in England with physicists J.J. Thomson (1856–1940), and Ernest Rutherford (1871–1937) improving their classical models of the atom by incorporating quantum theory. During this time, Bohr developed his model of atomic structure. To account for the observed properties of hydrogen, Bohr proposed that electrons existed only in certain orbits and that, instead of traveling between orbits, electrons made instantaneous quantum leaps or jumps between allowed orbits. According to the **Bohr model**, when an electron is excited by energy it jumps from its ground state to an excited state (i.e., a higher energy orbital). The excited atom can then emit energy only in certain (quantized) amounts as its electrons jump back to lower energy orbits located closer to the nucleus. This excess energy is emitted in quanta of electromagnetic radiation (photons of light) that have exactly same energy as the difference in energy between the orbits jumped by the electron.

The electron quantum leaps between orbits proposed by the Bohr model accounted for Plank's observations that atoms emit or absorb electromagnetic radiation in quanta. Bohr's model also explained many important properties of the photoelectric effect described by **Albert Einstein** (1879–1955).

The development of quantum mechanics during the first half of the twentieth century replaced classical Copernican-like atomic models of the atom. Using probability theory, and allowing for a wave-particle duality, quantum mechanics also replaced classical mechanics as the method by which to describe interactions between subatomic particles. Quantum mechanics replaced electron "orbitals" of classical atomic models with allowable values for angular momentum (angular velocity multiplied by mass) and depicted electrons position in terms of probability "clouds" and regions.

When Planck started his studies in physics, Newtonian or classical physics seemed fully explained. In fact, Planck's graduate advisor once claimed that there was essentially nothing new to discover in physics. By 1918, however, the importance of the quantum mechanics was recognized and Planck received the Nobel Prize in physics. The philosophical implications to quantum theory seemed so staggering, however, that Planck himself admitted that he did not fully understand the theory. In fact, Planck initially regarded the development of quantum mechanics as a mathematical aberration or temporary answer to be used only until a more intuitive or commonsense model was developed.

Despite Planck's reservations, however, Einstein's subsequent Nobel Prize winning work on the photoelectric effect was heavily based on Planck's theory. Expanding on Planck's explanation of blackbody radiation, Einstein assumed that light was transmitted in as a stream of particles termed photons. By extending the well-known wave properties of light to include a treatment of light as a stream of photons, Einstein was able to explain the photoelectric effect.

The Bohr model of atomic structure was published in 1913, and Bohr's work earned a Nobel Prize in 1922. Bohr's model of the hydrogen atom proved to be insufficiently complex to account for the fine detail of the observed spectral lines. Prussian physicist Arnold (Johannes Wilhelm) Sommerfeld's (1868–1951) refinements (e.g., the application of elliptical, multi-angular orbits), however, explained the fine structure of the observed spectral lines.

Later in the 1920s, the concept of quantization and its application to physical phenomena was further advanced by more mathematically complex models based on the work of the French physicist Louis Victor de Broglie (1892–1987) and Austrian physicist **Erwin Schrödinger** (1887–1961) that depicted the particle and wave nature of electrons. De Broglie showed that the electron was not merely a particle but a waveform. This proposal led Schrödinger to publish his wave equation in 1926. Schrödinger's work described electrons as "standing wave" surrounding the nucleus and his system of quantum mechanics is called wave mechanics. German physicist Max Born (1882–1970) and English physicist P. A. M. Dirac (1902–1984) made further advances in defining the subatomic particles (principally the electron) as a wave rather than as a particle and in reconciling portions of quantum theory with **relativity theory**.

Working at about the same time, German physicist Werner Heisenberg (1901–1976) formulated the first complete and self-consistent theory of quantum mechanics. Matrix mathematics was well established by the 1920s, and Heisenberg applied this powerful tool to quantum mechanics. In 1926, Heisenberg put forward his uncertainty principle that states that two complementary properties of a system, such as position and momentum, can never both be known exactly. This proposition helped cement the dual nature of particles (e.g., light can be described as having both wave and particle characteristics). Electromagnetic radiation (one region of the spectrum of which comprises visible light) is now understood as having both particle and wave-like properties.

In 1925, Austrian-born physicist Wolfgang Pauli (1900–1958) published the Pauli exclusion principle that states that no two electrons in an atom can simultaneously occupy the same quantum state (i.e., energy state). Pauli's specification of spin (+½ or –½) on an electron gave the two electrons in any suborbital differing quantum numbers (a system used to describe the quantum state) and made completely understandable the structure of the **periodic table** in terms of electron configurations (i.e., the energy related arrangement of electrons in energy shells and suborbitals).

In 1931, American chemist Linus Pauling published a paper that used quantum mechanics to explain how two electrons, from two different atoms, are shared to make a covalent bond between the two atoms. Pauling's work provided the connection needed in order to fully apply the new quantum theory to chemical reactions.

Quantum mechanics posed profound questions for scientists and philosophers. The concept that particles such as electrons making quantum leaps from one orbit to another, as opposed to simply moving between orbits, seems counter-intuitive. Like much of quantum theory, the proofs of how nature works at the atomic level are mathematical. Bohr himself remarked, "Anyone who is not shocked by quantum theory has not understood it."

The rise of the importance and power of quantum mechanics carried important philosophical consequences. When misapplied to larger systems—as in the famous paradox of Schrödinger's cat—quantum mechanics was often misinterpreted to make bizarre predictions (i.e., a cat that is simultaneously dead and alive). On the other hand, quantum mechanics made possible important advances in cosmological theory.

Quantum and relativity theories strengthened philosophical concepts of complementarity, wherein phenomena can be looked upon in mutually exclusive yet equally valid perspectives. In addition, because of the complexity of quantum relationships, the rise of quantum mechanics fueled a holistic approach to explanations of physical phenomena. Following the advent of quantum mechanics, the universe could no longer be explained in terms of Newtonian causality but only in terms of statistical, mathematical constructs.

In particular, Heisenberg's uncertainty principle asserts that knowledge of natural phenomena is fundamentally limited—to know one part allows another to move beyond recognition. Quantum mechanics, particularly in the work of Heisenberg and Schrödinger, also asserted a indeterminist (no preferred frame of reference) epistemology (study of the nature and limits of human knowledge).

Fundamental contradictions to long-accepted Newtonian causal and deterministic theories made even the leading scientists of the day resistant to the philosophical implications of quantum theory. Einstein argued against the seeming randomness of quantum mechanics by asserting, "God does not play dice!" Bohr and others defended quantum theory with the gentle rebuttal that one should not "prescribe to God how He should run the world."

See also Atomic structure; Quantum electrodynamics (QED)

Quartz

Quartz (SiO_2), a common mineral, is the product of the two most prevalent elements in the earth's **crust**: **silicon** and **oxygen**. Quartz can be found as giant **crystals** or small grains, and is the main component of most types of **sand**. It is the hardest common mineral, and for this reason is often used in the making of sandpaper, grindstones, polishers, and industrial cleaners. Though quartz is clear and glassy in its large crystal form, called **rock** quartz, it also can be found in several shades of coloration, the most familiar being rose quartz (pink), smoky quartz (brown), and amethyst (purple).

Rose quartz. ***U.S. National Aeronautics and Space Administration (NASA).***

Quartz has a variety of scientific and industrial uses, chiefly because it possesses piezoelectricity. Discovered by the French physicist and chemist **Pierre Curie** (1859–1906), the piezoelectric effect is a phenomenon demonstrated by certain crystals: when squeezed or stretched, a voltage is produced across the crystal's face. This effect is reversible as well, for when a voltage is applied to a piezoelectric crystal it will stretch; if the polarity of the voltage is alternated, the crystal will rapidly expand and contract, producing a vibration. It is this vibration that makes quartz especially useful. Every kind of piezoelectric crystal has a natural vibration frequency that is determined by its thickness—the thinner the crystal, the higher the frequency. When a crystal is made to vibrate at its natural frequency by the application of a voltage, the system is said to be in resonance. A crystal in resonance will maintain a constant, unfaltering frequency. When coupled with vacuum tubes or transistors, this constant frequency can be changed into a radio signal. Such was the design of the quartz radio, used primarily during World War II. Another common use of quartz is in timekeeping. All clocks rely upon some form of oscillator to keep regular time; for example, mechanical clocks sometimes use a pendulum to regulate the motion of their hands. In a quartz timepiece, a small ring-shaped piece of crystal is made to vibrate at its natural frequency. A microchip reads how many times the quartz vibrates each second and uses that information to keep accurate time. Because the crystal's vibration is unfaltering, quartz clocks are among the most precise timekeeping devices, losing less than one hundred thousandth of a second each day. Quartz crystals can be used to regulate both digital and analog clocks and watches.

Because of the many applications for quartz, the demand for clear, flawless rock crystal is often greater than the supply. Shortly after World War II, scientists developed a process by which quartz can be "grown" in the laboratory. Scientists begin with a small piece of natural crystal called a seed. Placing the seed within an alkaline solution, along with a supply of silica, they apply heat and pressure to the mixture. Slowly, the silica bonds with the seeds, eventually forming large, near-perfect crystals. Another type of man-made quartz, called fused quartz, is made by **melting** down many pieces of natural quartz and reforming it into almost any shape. Fused quartz displays many useful properties not found in natural quartz. First, because it neither expands nor contracts with changing temperatures, it makes an ideal component of precise scientific equipment, such as **telescope** and microscope lenses. It also is an unsurpassed conductor of heat, light, and **ultraviolet rays**, and in many cases it can be used to direct light rays through bends and angles. Additionally, fused quartz, which is nearly impervious to acids and other chemicals, is often used to make test tubes and other chemical containers.

See also Industrial minerals

QUASARS

Quasi-stellar radio sources (quasars) are the most distant cosmic objects observed by astronomers. Although not visible to the naked eye, quasars are also among the most energetic of cosmic phenomena. Even though some quasars may be physically smaller in size than our own **solar system**, some quasars are calculated to be brighter than hundreds of galaxies combined. Quasars and active galaxies appear to be related phenomena, each associated with massive rotating black holes in their central region. As a type of active galaxy, the enormous energy output of quasars can be explained using the theory of general relativity.

The great distance of quasars means that the light observed coming from them was produced when the universe was very young. Because of the finite speed of light, large cosmic distances translate to looking back in time. The observation of quasars at large distances and of their nearby scarcity argues that quasars were much more common in the early universe. Correspondingly, quasars may also represent the earliest stages of galactic **evolution**. This change in the universe over time (e.g., specifically the rate of quasar formation) contradicted steady-state cosmological models that relied on a universe that was the same in all directions (when averaged

Quasar PKS. Quasars are highly energetic objects billions of light years away, and some of the oldest objects in the known universe. ***U.S. National Aeronautics and Space Administration (NASA).***

over a large span of **space**) and at all times. Along with the discovery of ubiquitous cosmic background radiation, the discovery of quasars tilted the cosmological argument in favor of Big Bang based cosmological models.

In 1932, American engineer Karl Jansky (1905–1945) discovered the existence of radio waves emanating from beyond the solar system. By the mid-1950s, an increasing number of astronomers using radio telescopes sought explanations for mysterious radio emissions from optically dim stellar sources.

In 1962, British radio astronomer Cyril Hazard used the **moon** as an occultive shield to discover strong radio emissions traceable to the constellation Virgo. Optical telescopes pinpointed a faint star-like object (subsequently designated quasar 3C273—3rd Cambridge Catalog, 273rd radio source) as the source of the emissions. Of greater interest was an unusual emission spectrum found associated with 3C273. In 1963, American astronomer Marten Schmidt explained the abnormal spectrum from 3C273 as evidence of a highly redshifted spectrum. Redshift describes the Doppler-like shift of spectral emission lines toward longer (hence, redder) wavelengths in objects moving away from an observer. Observers measure the light coming from objects moving away from them as redshifted (i.e., at longer wavelengths and at a lower frequency when the light was emitted). Conversely, observers measure the light coming from objects moving toward them as blueshifted (i.e., at shorter wavelengths and at a higher frequency when the light was emitted). Most importantly, the determination of the amount of an object's redshift allows the calculation of a recession velocity. Moreover, because the recession rate increases with distance, the recession velocity is a function (known as the Hubble relation) of the distance to the receding object. After 3C273, many other quasars were discovered with similarly redshifted spectra.

Schmidt's calculation of the redshift of the 3C273 spectrum meant that 3C273 was approximately three billion light-years away from Earth. It became immediately apparent that, if 3C273 was so distant, it had to be many thousands of times more luminous than a normal galaxy for the light to appear as bright as it did from such a great distance. Refined calculations involving the luminosity of 3C273 indicate that, although dim to optical astronomers, the quasar is actually five trillion times as bright as the **Sun**. The high redshift of 3C273 also implied a great velocity of recession measuring one-tenth the speed of light.

Astronomers now assert that quasars represent a class of galaxies with extremely energetic centers. Large radio emissions seem most likely associated with massive black holes with great amounts of matter available to enter the accretion disk. In fact, prior to more direct observations late in the twentieth century, the discovery of quasars provided at least tacit proof of the existence of black holes. Black holes form around a singularity (the remnant of a collapsed massive star) with a gravitational field so intense that not even light can escape. Located outside the black hole is the accretion disk, an **area** of intense radiation emitted as matter heats and accelerates toward the black hole's event horizon (the boundary past which nothing can escape). Further, as electrons in the accretion disk are accelerated to near light speed, they are influenced by a strong **magnetic field** to emit quasar-like radio waves in a process termed synchrotron radiation. Electromagnetic waves similar to the electromagnetic waves emanating from quasars are observed on Earth when physicists pass high-energy electrons through synchrotron particle accelerators. Studies of Quasar 3C273 and other quasars identified jets of radiation blasting tens of thousands of light-years into space.

In addition to radio and visible light emissions, some quasars emit light in other regions of the **electromagnetic spectrum** including ultraviolet, infrared, x ray, and gamma-ray regions. In 1979, an x-ray quasar was found to have a redshift of 3.2, indicating a recession velocity equaling 97% the speed of light.

Not all quasars or active galaxies are alike. Although they seem optically similar to energetic quasars, at least 90% of active galaxies appear to be radio quiet. Accordingly, Seyfert galaxies or quasi-stellar objects (QSO) may be radio silent or emit electromagnetic radiation at greatly reduced levels. More than 1,500 quasars have now been identified as distant QSO. One hypothesis accounts for these quiet quasars by linking them to smaller black holes, or to black holes in regions of space with less matter available for consumption.

The limitations of ground-based telescopes and the need to study quasars was officially cited as one of the principal reasons to build the **Hubble Space Telescope** launched by the United States in 1990. In addition to direct studies of quasars, astronomers use quasars as an electromagnetic backdrop that can be used to study the primitive gas **clouds** found in the early universe.

See also Big bang theory; Stellar life cycle

Quaternary Period

In **geologic time**, the Quaternary Period (also termed the Anthropogene Period), the second geologic period in the **Cenozoic Era**, spans the time between roughly 2.6 million years ago (mya) and present day. On the geologic time scale, Earth is currently in the Quaternary Period of the Cenezoic Era of the **Phanerozoic Eon**.

The Quaternary Period contains two geologic epochs. The earliest epoch, the **Pleistocene Epoch** ranges from approximately 2.6 mya to 10,000 years ago. The Pleistocene Epoch is further subdivided into (from earliest to most recent) Gelasian and Calabrian stages. The Calabrian stage is also frequently replaced by a series of geologic stages, from earliest to most recent, including the Danau, Donau-Günz, Günzian, Günz-Mindel, Mindelian, Mindel-Riss, Rissian, Riss-Würm, and Würmian stages. The latest, most recent, and current epoch, the **Holocene Epoch** ranges from approximately 10,000 years ago until present day. According to geologic time, Earth is currently in the Holocene Epoch.

During the Quaternary Period, Earth's continents assumed their modern configuration. The Pacific Ocean separated **Asia** and **Australia** from **North America** and **South America**; the Atlantic Ocean separated North and South America from **Europe** (Euro-Asia) and **Africa**. The Indian Ocean filled the basin between Africa, India, Asia, and Australia. The Arabian Plate wedged between the Eurasian and African plates continues to provide high levels of tectonic activity (e.g., earthquakes) in the **area** of modern day Turkey. The Indian plate driving against and under the Eurasian plate uplifts both in rapid mountain building. As a result of the ongoing collision, ancient oceanic **crust** bearing marine fossils was uplifted into the Himalayan chain. The collision between the Indian and Eurasian plate continues with a resulting slow—but measurable—increase in the altitude of the highest Himalayan mountains (e.g., Mt. Everest) each year.

Glaciation (e.g., **ice ages**), and fluctuating climatic conditions—possibly at least partially explainable by Milankovitch cycles—during both the Tertiary and Quaternary Periods brought about sweeping changes in the landscape evident in modern topographical features.

The **fossil record** provides evidence that by the end of the **Tertiary Period** (also known as the Neogene period), the species *Ardipithecus ramidus* walked upright in an area now encompassing modern Ethiopa. Near the start of the Quaternary Period, a number of species lived and became extinct before modern humankind (*Homo sapiens*) appeared. Many of these species, including *Australopithecus anamensis*, *Australopithecus afarensis*, *Australopithecus garhi*, and *Australopithecus africanus* were only collateral rungs on the ladder of **evolution** to *Homo sapiens,* and do not provide a direct evolutionary link to humans. Although these species became extinct near the start of the Quaternary Period, they at least co-existed with the direct ancestors of humans. Early in the Quaternary Period *Homo habilis* and *Homo rudolfensis* lived and became extinct. Their extinctions are dated to approximately the appearance of *Homo ergaster*, a species some anthropologists argue is one of the earliest identifiable direct ancestors of *Homo erectus*, *Homo heidelbergensis*, *Homo neanderthalensis*, and *Homo sapiens*.

The last major **impact crater** with a diameter over 31 mi (50 km) struck Earth near what is now Kara-Kul, Tajikistan at the **Pliocene Epoch** and Pleistocene Epoch geologic time boundary that established the start of the Quaternary Period.

See also Archean; Cambrian Period; Cretaceous Period; Dating methods; Devonian Period; Eocene Epoch; Evolution, evidence of; Fossil record; Fossils and fossilization; Historical geology; Jurassic Period; Mesozoic Era; Miocene Epoch; Mississippian Period; Oligocene Epoch; Ordovician Period; Paleocene Epoch; Paleozoic Era; Pennsylvanian Period; Precambrian; Proterozoic Era; Silurian Period; Supercontinents; Triassic Period

R

RADAR AND SONAR

Although they rely on two fundamentally different types of wave transmission, Radio Detection and Ranging (RADAR) and Sound Navigation and Ranging (SONAR) both are **remote sensing** systems with important military, scientific and commercial applications. RADAR sends out electromagnetic waves, while active SONAR transmits acoustic (i.e., sound) waves. In both systems, these waves return echoes from certain features or targets that allow the determination of important properties and attributes of the target (i.e., shape, size, speed, distance, etc.). Because electromagnetic waves are strongly attenuated (diminished) in **water**, RADAR signals are mostly used for ground or atmospheric observations. Because SONAR signals easily penetrate water, they are ideal for navigation and measurement under water.

The threat of submarine warfare during World War I made urgent the development of SONAR and other means of echo detection. The development of the acoustic transducer that converted electrical energy to sound waves enabled the rapid advances in SONAR design and technology during the last years of the war. Although active SONAR was developed too late to be useful during World War I, the push for its development reaped enormous technological dividends. Not all of the advances, however, were restricted to military use. After the war, echo sounding devices were placed aboard many large French ocean-liners.

During the early battles of World War II, the British Anti-Submarine Detection and Investigation Committee (its acronym, ASDIC, became a name commonly applied to British SONAR systems) made efforts to outfit every ship in the British fleet with advanced detection devices. The use of ASDIC proved pivotal in the British effort to repel damaging attacks by German submarines upon both British warships and merchant ships keeping the island nation supplied with munitions and food.

While early twentieth century SONAR developments proceeded, another system of remote sensing was developed based upon the improved understanding of the nature and propagation of electromagnetic radiation achieved by Scottish physicist **James Clerk Maxwell** (1831–1879) during the 19th century. Scottish physicist and meteorologist Sir Robert Alexander Watson-Watt (1892–1973) successfully used short-wave radio transmissions to detect the direction of approaching thunderstorms. Another technique used by Watson-Watt and his colleagues at the British Radio Research Station measured the altitude of the **ionosphere** (a layer in the upper atmosphere that can act as a radio reflector) by sending brief pulses of radio waves upward and then measuring the time it took for the signals to return to the station. Because the speed of radio waves was well established, the measurements provided very accurate determinations of the height of the reflective layer. In 1935, Watson-Watt had the ingenious idea of combining these direction and range finding techniques and, in so doing, he invented RADAR. Watson-Watt built his first practical RADAR device at Ditton Park.

Shortly thereafter, without benefit of a test run, Watson-Watt and Ministry scientists conducted an experiment to test the viability of RADAR. Watson-Watt's apparatus was found able to illuminate (i.e., detect) aircraft at a distance of up to eight miles. Within a year, Watson-Watt improved his RADAR systems so that it could detect aircraft at distances up to seventy miles. Pre-war Britain quickly put Watson-Watt's invention to military use and by the end of 1938, primitive RADAR systems dotted the English coast. These stations, able to detect aircraft regardless of ground fogs or **clouds**, were to play an important role in the detection of approaching Nazi aircraft during World War II. By the end of the war, the British and American forces had developed a number of RADAR types and applications including air interception (AI), air-to-surface vessel (ASV), ground controlled interception (GCI), and various gun sighting and tracking RADARs.

Regardless of their application, both RADAR and SONAR targets scatter, deflect, and reflect incoming waves. This scattering is, however, not uniform, and in most cases a strong echo of the image is propagated back to the signal

RADAR and SONAR allow ships to safely operate in conditions where vision is either not possible or feasible. ***AP/Wide World. Reproduced by permission.***

transmitter in much the same way as a smooth mirror can reflect light back in the specular direction. The strength of the return signal is also characteristic of the target and the environment in which the systems are operating. Because they are electromagnetic radiations, RADAR waves travel through the atmosphere at the speed of light (in air). SONAR waves (compression waves) travel through water at much slower pace, the speed of sound. By measuring the time it takes for the signals to travel to the target and to return echoes, both RADAR and SONAR systems are capable of accurately determining the distance to their targets.

Within their respective domains, both RADAR and SONAR can operate reliably under a wide variety of adverse conditions to extend human sensing capabilities.

RADAR technology also had a dramatic impact on the fledgling science of radio **astronomy**. During the Second World War, British officer, J.S. Hey correctly determined that the **Sun** was a powerful source of radio transmissions. Hay discovered this while investigating the causes of system wide jamming of the British RADAR net that could not be attributed to enemy activity (Hey attributed the radio emission to increased solar flare activity). Although kept secret during the war, British RADAR installations and technology became the forerunners of modern radio telescopes as they recorded celestial background noise while listening for the telltale signs of enemy activity.

Remote sensing tools such as RADAR and SONAR also allow scientists, geologists, and archaeologists to map **topography** and subsurface features on Earth and on objects within the **solar system**. SONAR readings led to advances in underwater seismography that allowed the mapping of the ocean floors and the identification of mineral and energy resources.

See also Sound transmission

RADIATION FOG • *see* FOG

Radioactive decay • *see* Dating methods

Radioactive waste storage (geologic considerations)

In the 1960s, nuclear power gained popularity as a means of producing power for civilian use. During the next two decades, several nuclear power plants were built, but there was little consensus about how to best dispose of radioactive waste. Waste from plants, as well as from military and defense operations, was usually stored on site or in nearby storage facilities. Low-level waste, such as that from hospitals, research labs, and power plants is generally placed into containment facilities on-site. However, the disposal of high-level waste, materials that are highly radioactive, remains more problematic. Spent nuclear **fuels** from power plants are sometimes shipped to containment facilities, and sometimes stored in specially constructed containment pools on-site. Radioactive waste is thus, stored in various locations, governed by federal regulations. Forty-three states in the United States, and several Canadian provinces, currently have nuclear waste storage facilities. In the late 1990s, the government proposed plans for a central storage facility for high-level waste at Yucca Mountain, Nevada. In May, 2002, the United States House of Representatives approved a measure that would establish the site at Yucca Mountain, and in July 2002, the United States Senate approved the site for development, following 20 years of debate in Congress. The Yucca Mountain site has sparked ongoing controversy over the environmental impact of nuclear waste storage, much of which focuses on the unique geological and environmental conditions of the region.

When looking for a site for permanent storage of high level waste, engineers and geologists took several factors into consideration, including: **water table**, geological stability, **rock** composition, seismic (**earthquake**) activity, and proximity to population areas. Furthermore, the site must have a high probability of remaining undisturbed for tens of thousands of years, or as long as the materials in storage are radioactive. Yucca Mountain is located in a rural region, with sparse population. Las Vegas, 100 mi (161 km) from the site, is the nearest metropolitan **area**. Within a 100-mi radius of the proposed site, there are approximately 35,000 inhabitants. Thus, Yucca Mountain is relatively secluded.

Yucca Mountain itself has a **desert climate**, receiving less than six inches of rain per year. The lack of rain means that **cave** systems within the mountain are dry, and that there is minute seepage from the surface of the mountain to the deep **water** table 2000 ft (670 m) below ground. This ensures that waste stored in the mountain would have fewer chances of polluting ground water if specially engineered storage containers ever rupture. The deep location of the water table at the site also means that the cavity, or storage room, would lie equidistant from the surface of the mountain to ground water stores—about 1000 ft, or 304 m. This isolates the waste, and removes the chance of accidental disturbance from future drilling or other means of exploration.

Some aspects of the geological composition of the mountain itself further makes Yucca Mountain a candidate for a nuclear waste repository. Dense volcanic rock, as well as thick and nearly impenetrable **bedrock** mean the Yucca Mountain's interior is relatively stable, not very porous, and resistant to water and heat. Under the most extreme conditions, this deep and solid rock could help contain minor seepage, as well as insulate the repository—possibly making it as safe as a band of untapped uranium ore.

Yucca Mountain's unique **geology** and environment is unequaled by that of any of the nation's other current nuclear waste repositories, many of which pose a greater potential threat to cities, drinking water, and their local environments. Centralization could potentially lead to tighter regulation of waste, better handling, and less environmental damage.

While Yucca Mountain does meet much of the criteria for a safe storage site, it is not a perfect location. The region around Yucca Mountain contains several **faults and fractures** (cracks in the Earth's **crust** where movement causes earthquakes), and is considered seismically active. Earthquakes could change the patterns of water flow inside the mountain, and well as endanger the integrity of the storage cavities within the mountain. Increased hydrothermal activity could promote seepage and water contamination.

Researchers also explored the possibility of the storage cavity filling with water, thus exposing the **aquifer** and **groundwater** to radioactive contaminants. Geologists studied core samples and cave linings to determine the extent to which **minerals** permeated the walls of the cavities. The scientists found that there were only scant traces of opal and calcite, telltale signs of flooding and water seepage, at the lower levels of the mountain. Thus, the cavities did not have a history of filling with water. A corresponding study of the geological history of the mountain further confirmed the relative stability of the site's water table, drainage, and seepage.

However, under Yucca Mountain is a deep aquifer. In the desert region, the aquifer provides drinking and irrigation water. As metropolitan centers, such as Las Vegas, continue to grow, the aquifer might play a significant role as a water resource for the region. The nuclear storage site would have to remain stable and well sealed for tens of thousands of years in order to ensure the continued safety of the aquifer.

Part of the problem in designing high-level waste storage facilities is the time span for which these sites must remain secure and safe. Lab tests are inadequate to insure the stability of the mountain, the fortitude of containers and casks, and the security of the site from accidental intrusion for the tens of thousands of years necessary for radioactive waste to be rendered harmless. Project planners face not only design difficulties such as preventing accidents and mitigating environmental impact, but also how to document the site in ways that will ensure that people 10,000 years from now will recognize the hidden danger of the mountain storage facility. People today have only scant artifacts and generalized understanding of civilizations and people that lived ten thousand years ago.

Geologists and other scientists disagree on the possible effect that the waste could have on the behavior of the mountain itself. Some predict that heat generated by the waste could alter the mountain's geological and hydrological behavior, causing rocks to crack and water to seep into and out of the storage cavity in ways that we cannot predict. Some raise concerns over the unpredictable nature of seismic activity in the area. Other scientists assert that the stable pattern of geological processes at Yucca Mountain will remain unchanged, and that the site is predictably stable. Geologists have to account for not only the mountain's history, but also predict its future in order to insure the safety of the site for future generations.

While much of the scientific community's assessment of the safety of the Yucca Mountain project centers on geology, public concerns focus on technology. Though waste is currently stored in forty-three states, little of the nation's spent nuclear materials travel long distances. The creation of the Yucca Mountain site would require that waste be shipped by truck and rail to the central storage facility. Engineers and researchers have developed safe casks, or storage bins, which are impervious to accidents, water, and fire specifically for shipping high-level waste, but may people are discomforted simply by the perceived risk (the threat that people feel is associated with a given project, not the statistical risk) of shipping nuclear materials.

The controversy surrounding the Yucca Mountain waste repository is both political and scientific. The perceived threat of nuclear materials heavily influences public opinion, and environmentalists are reticent to trade many smaller environmental problems for a large potential hazard. Some people cite the Yucca Mountain facility as a means of centralizing the problem of nuclear waste. Project proponents claim that the repository will lessen environmental risk and keep volatile, dangerous materials secure and controlled. The scientific community is also in discord over the geological impact of the project, thus exposing the many unknown variables that **Earth science** still has not defined.

See also Radioactivity; Water pollution and biological purification; Water table

RADIOACTIVITY

Radioactivity originates from extraterrestrial sources and terrestrial geologic sources. All elements with more than 83 protons (i.e., an **atomic number** greater than 83) are radioactive. Some radioactive isotopes also occur in elements with lower atomic numbers.

Atoms that are radioactive emit radioactivity during spontaneous transformation from an unstable isotope to a more stable one. Natural radioactive decay provides a source of heating in Earth's interior that drives mantle dynamics and **plate tectonics**. Both natural and man-made sources of radioactivity at certain levels may represent a significant health risk to humans and other organisms. Radioactive materials must be isolated from the environment until their radiation level has decreased to a safe level, a process which requires thousands of years for some materials.

Radiation is classified as being ionizing or nonionizing. Both types can be harmful to humans and other organisms. Nonionizing radiation is relatively long-wavelength electromagnetic radiation, such as radio waves, microwaves, visible radiation, ultraviolet radiation, and very low-energy electromagnetic fields. Nonionizing radiation is generally considered less dangerous than ionizing radiation. However, some forms of nonionizing radiation, such as ultraviolet radiation, can damage biological molecules and cause health problems. Scientists do not yet fully understand the longer-term health effects of some forms of nonionizing radiation, such as that from very low-level electromagnetic fields (e.g., high-voltage power lines), although the evidence to date suggests that the risks are extremely small.

Ionizing radiation is the short wavelength radiation or particulate radiation emitted by certain unstable isotopes during radioactive decay. There are about 70 radioactive isotopes, all of which emit some form of ionizing radiation as they decay. A radioactive isotope typically decays through a series of intermediate isotopes until it reaches a stable isotope state. As indicated by its name, ionizing radiation can ionize the atoms or molecules with which it interacts. In other words, ionizing radiation can cause other atoms to release their electrons. These free electrons can damage many biochemicals, such as proteins, lipids, and nucleic acids (including DNA). In intense radioactivity, this damage can cause severe human health problems, including cancers, and death.

Ionizing radiation can be either short-wavelength electromagnetic radiation or particulate radiation. Gamma radiation and x radiation are short-wavelength electromagnetic radiation. Alpha particles, beta particles, neutrons, and protons are particulate radiation. Alpha particles, beta particles, and gamma rays are the most commonly encountered forms of radioactive pollution. Alpha particles are simply ionized helium nuclei, and consist of two protons and two neutrons. Beta particles are electrons, which have a negative charge. Gamma radiation is high-energy electromagnetic radiation.

Scientists have devised various units for measuring radioactivity. A Curie (Ci) represents the rate of radioactive decay. One Curie is 3.7×10^{10} radioactive disintegrations per second. A rad is a unit representing the absorbed dose of radioactivity. One rad is equal to an absorbed energy dose of 100 ergs per gram of radiated medium. A rem is a unit that measures the effectiveness of radioactivity in causing biological damage. One rem is equal to one rad times a biological weighting factor. The weighting factor is 1.0 for gamma radiation and beta particles, and it is 20 for alpha particles. The radioactive **half-life** is a measure of the persistence of radioactive material. The half-life is the time required for one-half of an initial quantity of atoms of a radioactive isotope to decay to a different isotope.

In the United States, people are typically exposed to about 350 millirems of ionizing radiation per year. On average, 82% of this radiation comes from natural sources and 18% from anthropogenic sources (i.e., those associated with human activities). The major natural source of radiation is radon gas, which accounts for about 55% of the total radiation dose. The principal anthropogenic sources of radioactivity are medical x rays and nuclear medicine. Radioactivity from the fallout of nuclear weapons testing and from nuclear power plants make up less than 0.5% of the total radiation dose, i.e., less than 2 millirems. Although the contribution to the total human radiation dose is extremely small, radioactive isotopes released during previous atmospheric testing of nuclear weapons will remain in the atmosphere at detectable levels for the next 100 to 1000 years.

People who live in certain regions are exposed to higher doses of radiation. For example, residents of the Rocky Mountains of Colorado receive about 30 millirems more cosmic radiation than people living at sea level. This is because the atmosphere is thinner at higher elevations, and therefore less effective at shielding the surface from cosmic radiation. Exposure to cosmic radiation is also high while people are flying in an airplane, so pilots and flight attendants have an enhanced, occupational exposure. In addition, residents of certain regions receive higher doses of radiation from radon-222, due to local geological anomalies. Radon-222 is a colorless and odorless gas that results from the decay of naturally occurring, radioactive isotopes of uranium. Radon-222 typically enters buildings from their ground level.

Personal lifestyle also influences the amount of radioactivity to which people are exposed. For example, miners, who spend a lot of time underground, are exposed to relatively high doses of radon-222 and consequently have relatively high rates of lung cancer. Cigarette smokers expose their lungs to high levels of radiation, because tobacco plants contain trace quantities of polonium-210, lead-210, and radon-222. These radioactive isotopes come from the small amount of uranium present in fertilizers used to promote tobacco growth. Consequently, the lungs of a cigarette smoker are exposed to thousands of additional millirems of radioactivity, although any associated hazards are much less than those of tar and nicotine.

The U.S. Nuclear Regulatory Commission has strict requirements regarding the amount of radioactivity that can be released from a nuclear power reactor. In particular, a nuclear reactor can expose an individual who lives on the fence line of the power plant to no more than 10 millirems of radiation per year. Actual measurements at U.S. nuclear power plants have shown that a person who lived at the fence line would actually be exposed to much less than 10 millirems.

Thus, for a typical person who is exposed to about 350 millirems of radiation per year from all other sources, much of which is natural background, the proportion of radiation from nuclear power plants is extremely small. In fact, **coal-** and oil-fired power plants, which release small amounts of radioactivity contained in their **fuels**, are responsible for more airborne radioactive pollution in the United States than are nuclear power plants.

By far, the worst nuclear reactor accident occurred in 1986 in Chernobyl, Ukraine. An uncontrolled build-up of heat resulted in a meltdown of the reactor core and combustion of **graphite** moderator material in one of the several generating units at Chernobyl, releasing more than 50 million Curies of radioactivity to the ambient environment. The disaster killed 31 workers and resulted in the hospitalization of more than 500 other people from radiation sickness. According to Ukrainian authorities, during the decade following the Chernobyl disaster an estimated 10,000 people in Belarus, Russia, and Ukraine died from cancers and other radiation-related diseases caused by the accident. In addition to these relatively local effects, the atmosphere transported radioactive fallout from Chernobyl into **Europe** and throughout the Northern Hemisphere.

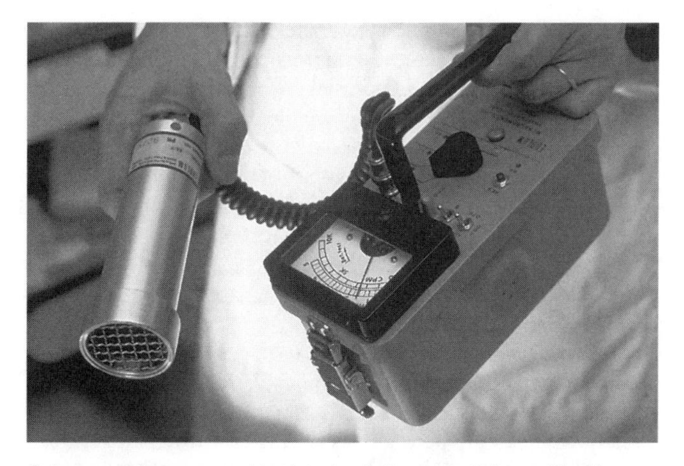

Geiger counters are used to detect subatomic particles emitted by radioactive substances. ***Hank Morgan. National Audubon Society Collection/Photo Researchers, Inc. Reproduced by permission.***

The large amount of radioactive waste generated by nuclear power plants is another important problem. This waste will remain radioactive for many thousands of years, so technologists must design systems for extremely long-term storage. One obvious problem is that the long-term reliability of the storage systems cannot be fully assured, because they cannot be directly tested for the length of time they will be used (i.e., for thousands of years). Another problem with nuclear waste is that it will remain extremely dangerous for much longer than the expected lifetimes of existing governments and social institutions. Thus, future societies of the following millennia, however they may be structured, will be responsible for the safe storage of nuclear waste that is being generated today.

See also Atmospheric chemistry; Atmospheric pollution; Atomic mass and weight; Atomic theory; Atoms; Atomic theory; Carbon dating; Cosmic microwave background radiation; Environmental pollution; Geochemistry; Radioactive waste storage (geological considerations); Radon production, detection and elimination; Ultraviolet rays and radiation

RADIOMETRIC DATING • *see* DATING METHODS

RADON PRODUCTION, DETECTION, AND ELIMINATION

Radon (usually in the form of the Radon-222 isotope) is a colorless and odorless radioactive gas formed from radioactive decay. The most common geologic source of radon derives from the decay of uranium. Radon is commonly found at low levels in widely dispersed crustal formations, **soil**, and **water** samples. To some extent, radon can be detected throughout the United States. Specific geologic formations, however, frequently present elevated concentration of radon that may pose a significant health risk. The Surgeon General of the United States and the Environmental Protection Agency identify radon exposure as the second leading cause of lung cancer in the United States. Cancer risk rates are based upon magnitude and duration of exposure.

Produced underground, radon moves toward the surface and eventually diffuses into the atmosphere or in **groundwater**. Because radon has a **half-life** of approximately four days, half of any size sample deteriorates during that time. Regardless, because radon can be continually supplied, dangerous levels can accumulate in poorly ventilated spaces (e.g., underneath homes, buildings, etc.). Moreover, the deterioration of radon produces alpha particle radiation and radioactive decay products that can exhibit high surface adherence to dust particles. Radon detection tests are designed to detect radon gas in picocuries per liter of air (pCi/L). The picocurie is used to measure the magnitude of radiation in terms of disintegrations per minute. One pCi, one trillionth of a Curie, translates to 2.2 disintegrations per minute. EPA guidelines recommend remedial action (e.g., improved ventilation) if long term radon concentrations exceed 4 pCi/L.

Working level units (WL) are used to measure radon decay product levels. The working level unit is used to measure combined alpha radiation from all radon decay products. Commercial test kits designed for use by the general public are widely available. The most common forms include the use of charcoal canisters, alpha track detectors, liquid scintillation detectors, and ion chamber detectors. In most cases, these devices are allowed to measure cumulative radon and byproduct concentrations over a specific period of time (e.g., 60–90 days) that depends on the type of test and geographic radon risk levels. The tests are usually designed to be returned to a qualified laboratory for analysis.

The EPA estimates that nearly one out 15 homes in the United States has elevated radon levels.

Radon can be kept at low concentration levels by ventilation and the use of impermeable sheeting to prevent radon seepage into enclosed spaces. Radon in water does not pose nearly the health risk as does breathable radon gas. Regardless, radon removal protocols are increasingly a part of water treatment programs. Radon is removed from water by aeration or **carbon** filtration systems.

Exposure to radon is cumulative. Researchers are presently conducting extensive research into better profiling the mutagenic risks of long term, low-level radiation exposure.

See also Atmospheric pollution; Atomic theory; Atoms; Cosmic microwave background radiation; Environmental pollution; Radioactivity

RAIN • *see* PRECIPITATION

RAINBOW

Water droplets and light form the basis of all rainbows, which are circular arcs of color with a common center. Because only water and light are required for rainbows, one will see them in rain, spray, or even **fog**.

A raindrop acts like a prism and separates sunlight into its individual color components through refraction, as light will do when it passes from one medium to another. When the white light of the **Sun** strikes the surface of the raindrop, the light waves are bent to varying degrees depending on their wavelength. These wavelengths are reflected on the far surface of the water drop and will bend again as they exit. If the light reflects off the droplet only once, a single rainbow occurs. If the rays bounce inside and reflect twice, two rainbows will appear: a primary and a secondary. The second one will appear fainter because there is less light energy present. It will also occur at a higher angle.

Not all the light that enters the raindrop will form a rainbow. Some of the light, that which hits the droplet directly at its center, will simply pass through the other side. The rays that strike the extreme lower portions of the drop will produce the secondary bow, and those that enter at the top will produce the primary bow.

The formation of the arc was first discussed by Rene Descartes in 1637. He calculated the deviation for a ray of red light to be about 138 degrees. Although light rays may exit the drop in more than one direction, a concentration of rays emerges near the minimum deviation from the direction of the incoming rays. Therefore the viewer sees the highest intensity looking at the rays that have minimum deviation, which form a cone with the vertex in the observer's eye and with the axis passing through the Sun.

The color sequence of the rainbow is also due to refraction. It was **Sir Isaac Newton**, however, 30 years after Descartes, who discovered that white light was made up of different wavelengths. Red light, with the longest wavelength, bends the least, while violet, being the shortest wavelength, bends the most. The vertical angle above the horizon will be a little less than 41° for the violet (about 40°) and a little more for the red (about 42°). The secondary rainbow has an angular radius of about 50° and its color sequence is reversed from the primary. It is universally accepted that there are seven rainbow colors, which appear in the order: red, orange, yellow, green, blue, indigo, and violet. However, the rainbow is a whole continuum of colors from red to violet and even beyond the colors that the eye can see.

Supernumerary rainbows, faintly colored rings just inside of the primary bow, occur due to interference effects on the light rays emerging from the water droplet after one internal reflection.

No two people will see the same rainbow. If one imagines herself or himself standing at the center of a cone cut in half lengthwise and laid on the ground flatside down, the raindrops that bend and reflect the sunlight that reach the person's eye as a rainbow are located on the surface of the cone. A viewer standing next to the first sees a rainbow generated by a different set of raindrops along the surface of a different imaging cone.

A double rainbow over the desert. ***FMA. Reproduced by permission.***

Using the concept of an imaginary cone again, a viewer could predict where a rainbow will appear by standing with his back to the sun and holding the cone to his eye so that the extension of the axis of the cone intersects the Sun. The rainbow will appear along the surface of the cone as the circular arc of the rainbow is always in the direction opposite to that of the Sun.

A rainbow lasts only about a half-hour because the conditions that create it rarely stay steady much longer than this. In many locations, spring is the prime rainbow-viewing month. Rainfall is usually more localized in the spring, and brief showers over limited areas are a regular feature of atmospheric behavior. This change is a result of the higher springtime sun warming the ground more effectively than it did throughout the previous winter months. This process produces local convection. These brief, irregular periods of **precipitation** followed by sunshine are ideal rainbow conditions. Also, the Sun is low enough for much of the day to allow a rainbow to appear above the horizon—the lower the sun, the higher the top of a rainbow.

The "purity" or brightness of the colors of the rainbow depends on the size of the raindrops. Large drops or those with diameters of a few millimeters create bright rainbows with well-defined colors; small droplets with diameters of about 0.01 mm produce rainbows of overlapping colors that appear nearly white.

For refraction to occur, the light must intersect the raindrops at an angle. Therefore no rainbows are seen at noon when the sun is directly overhead. Rainbows are more frequently seen in the afternoon because most showers occur in midday rather than morning. Because the horizon blocks the other half of a rainbow, a full 360° rainbow can only be viewed from an airplane.

The sky inside the arc will appear brighter than that surrounding it because of the number of rays emerging from a raindrop at angles smaller that those that are visible. But there is essentially no light from single internal reflections at angles greater than those of the rainbow rays. In addition to the fact that there is a great deal of light directed within the arc of the bow and very little beyond it, this light is white because it is a mixture of all the wavelengths that entered the raindrop. This is just the opposite in the case of a secondary rainbow, where the rainbow ray is the smallest angle and there are many rays that emerge at angles greater than this one. A dark band forms where the primary and secondary bows combine. This is known as the Alexander's Dark Band, in honor of Alexander of Aphrodisias who discovered this around 200 B.C.

If a viewer had a pair of polarizing sunglasses, he or she would see that light from the rainbow is polarized. Light vibrating horizontally at the top of the bow is much more intense than the light vibrating perpendicularly to it across the bow and it may be as much as 20 times as strong.

Although rare, a full **moon** can produce a lunar rainbow when it is bright enough to have its light refracted by raindrops just as is the case for the Sun.

See also Electromagnetic spectrum

RAINFOREST • *see* FORESTS AND DEFORESTATION

RAPIDS AND WATERFALLS

Rapids are stream sections with extremely strong currents, numerous obstacles, and steps in their streambeds. A waterfall is a vertical drop in a streambed. Both are sites of vigorous **erosion**. Rapids often form where resistant **bedrock** confines a stream to a narrow channel, and forces an increase in **water** velocity. Fast-moving water, laden with abrasive **sand** and gravel, cuts into the bedrock, forming cliffs on either side of the cataract. Large boulders fall from the cliffs, creating obstacles in the streambed, and increasing water turbulence. Rapids are navigational hazards that have hampered exploration, travel, and trade on the world's **rivers** throughout human history. Today, adventurers explore remote natural areas, and test their athletic abilities, by kayaking, canoeing, and rafting along these treacherous stretches of rivers.

Waterfalls form where fast-flowing water traverses a geologic contact between more resistant and less resistant **rock** layers, or across a fault that has juxtaposed different rock types. In other words, waterfalls often form at the end of a

Victoria Falls, bordering with Zimbabwe and Zambia. ***Cynthia Bassett. Reproduced by permission.***

series of rapids. (The classic movie scene in which the protagonists survive a trip through rapids only to be carried over a waterfall contains an element of reality.) Turbulent, sediment-laden water quickly erodes the less-resistant rocks, creating a vertical step in the streambed. Falling water erodes the soft rock even more quickly, and the waterfall grows taller. Turbulence at the base of the waterfall undercuts the newly-formed cliff, and it moves upstream. Niagara Falls, for example, retreats upstream about 3.3 ft (1 m) each year. Waterfalls also form where streams flow across pre-existing cliffs. They are common where streams reach the ocean along eroding or tectonically uplifting coastlines. For example, many streams end at spectacular waterfalls along the Scandinavian **fjords**. Yosemite Valley's famous waterfalls occur where small streams flow over the rim of the valley. Angel Falls, in Venezuela, is Earth's tallest waterfall at 3212 ft (979 m). Victoria Falls, on the Zambizi River along the border of Zambia and Zimbabwe, is one of the natural wonders of world; its beauty and mythical history are legendary.

See also Bed or traction load; Stream valleys, channels and floodplains

Rate factors in geologic processes

The rates at which geologic processes occur range from imperceptibly slow to exceptionally fast.

At the slow end of the spectrum, mountain ranges rise, basins subside, and tectonic plates move over time periods that span many millions of years. Recent research using repeated **GPS** (global positioning **satellite**) measurements of crustal deformation has shown, for example, that the Los Angeles basin is being shortened at a rate that is on the order of a few millimeters per year as the result of movement along faults throughout the basin. Likewise, the rate at which the earth's **crust** rebounds after the retreat of continental **glaciers** is generally on the order of a few millimeters per year and decreases with time. Measurements of offset strata cut by faults, combined with radiometric age dating, also suggest that crustal deformation rates are on the order of millimeters per year.

At the opposite end of the spectrum, acoustic waves traveling through **rock** (for example, from earthquakes, blasting, or seismic exploration surveys) typically travel at velocities of one to ten thousand meters per second. Thus, an observer standing 62 mi (100 km) from the epicenter of a large

earthquake may have to wait 10 or 20 seconds before ground shaking begins. The catastrophic debris **avalanche** that occurred in conjunction with the 1980 eruption of Mount St. Helens traveled at a velocity greater than 149 mph (240 kph), which is fast in terms of human perception but is on the order of one-millionth of the velocity of typical seismic waves.

The rate at which any given geologic process occurs can vary significantly. Different kinds of landslides, for example, can move at rates as slow as a few centimeters per year to rates as rapid as tens of meters per second, a range that extends over eight orders of magnitude. Similarly, the flow of **groundwater** through porous aquifers is controlled by a combination of the hydraulic gradient and the **permeability** of the **aquifer**, which can vary over many orders of magnitude. In general, however, groundwater and other subsurface fluids generally flow at velocities so low that their kinetic energy (proportional to the square of the velocity) can safely be ignored in calculations. The viscosity, or resistance to flow, of fluids such as **lava** and crude oil depend strongly on the **temperature** and chemical composition of the fluid. For example, the viscosity of **basalt** lava falls by a factor of 100 to 1,000 as its temperature increases from 2,100°F to 2,550°F (1,150°C to 1,400°C). Therefore basalt lava, which is typical of Hawaiian volcanoes, flows as easily as honey (honey at room temperature has the same viscosity as basalt at 2,550°F/1,400°C) when it first erupts but slows and eventually solidifies in place as it cools. The rates of chemical reactions, for example those associated with **metamorphism** and ore deposit formation, are also strongly dependent upon pressure and temperature.

The rates at which rocks are subjected to stress can also control their response. A rock struck sharply with a hammer will behave as a brittle substance, deforming elastically and, if the blow is strong enough, breaking into pieces. Earthquakes are an example of naturally occurring elastic deformation of rocks. Over the lengths of time required to build mountain ranges, however, rocks appear to deform as very viscous fluids via a process known as slow creeping flow. The viscosity of deforming rocks is also influenced by pressure and temperature conditions during mountain building episodes.

Geologic process rates must also be viewed in relationship to the rates at which society functions. In many parts of the world, for example, groundwater pumping rates exceed those at which aquifers are naturally replenished, resulting in land subsidence that can occur at the rate of tens of centimeters per year. Humans have also become significant geologic agents and are responsible for the movement of 37 billion tons of **soil** and rock per year throughout the world. Likewise, the rates at which mineral and energy resources are consumed by society need to be balanced by the rates at which humans discover new deposits and develop new technologies for extraction, as well as the extremely slow rates at which mineral and energy resources accumulate over **geologic time**.

See also Catastrophism; Half-life; Historical geology; Uniformitarianism

RAY, JOHN (1627-1705)

English naturalist

A predecessor of Carl Linnaeus, John Ray was the first naturalist to use the idea of species to distinguish different organisms from each other. Focusing primarily on the classification of plants and basing his system on the work of Aristotle, Ray divided plants into two groups: the moncotyledons and the dicotyledons. Both are still recognized today. In 1693, Ray published the final volume of *Histora Plantarum*, a complete classification of plants and one of the first natural systems of classification that was based on physical characteristics rather than origin and perceived use.

John Ray was born in Black Notley, Essex, England, to Roger Ray, a blacksmith, and Elizabeth Ray, an amateur herbalist and medical practitioner. He attended Trinity College at Cambridge from 1644–1651, receiving both a bachelor and masters degree. After graduation, he continued at Trinity as an appointed fellow of the college. He taught a number of courses, including Greek, mathematics, and humanities. Ray left his post at Trinity during the Reformation, when he refused to sign an oath required by the Act of Uniformity in 1662. It was at this time that his contribution to taxonomy flourished.

Without employment, Ray relied on the patronage of former students. One such patron was Francis Willughby, a wealthy contemporary from Cambridge. With the support of Willughby, Ray was able to expand his classification of plants from a part-time endeavor restricted to the indigenous species of Chambridgeshire to the whole of the British Isles and beyond. Willughby accompanied Ray on his many expeditions, and his interest in animals complimented Ray's own interests in plants. Ray's collaboration with Willughby ended in 1672, with the death of Willughby. That same year, Ray married Margaret Oakeley. They settled in Black Notley, where Ray continued his scientific endeavors.

As part of his work, Ray was able to convincingly show that **fossils** represented extinct species. At the time, the link between fossils and extinct species was not an accepted model; however, Ray's evidence provided the basis for the formation of a more thorough system of paleontology. Such a view was unusual for a naturalist at this time, particularly considering Ray's strong religious beliefs.

John Ray never lost his love and wonder for nature and had no problem reconciling his views of the world with his views of religion. As well as publishing extensively on natural history, Ray also published many theological works, including *The Wisdom of God*, and he was only stopped from taking priestly orders by the English civil war and Reformation. According to Ray, the study of nature was a way to reveal the omnipotence of God and to be a naturalist was a way to work within divinity.

See also Fossil record; Fossils and fossilization; History of exploration II (Age of exploration)

Red tides are caused by a population explosion of microorganisms called dinoflagellates. The seawater becomes toxic to most life, and kills fish in the area. ***AP/Wide World. Reproduced by permission.***

RED TIDE

Red tide is a condition in which a huge **area** of seawater turns to a reddish-brown hue. This rusty-red discoloration is caused by an exploding population of tiny single-celled microorganisms called dinoflagellates, which are usually found in ocean **water**, but occasionally in **lakes** and **rivers** as well. Red **tides** have occurred naturally since **oceans** were formed, but today they are becoming more common because of human influence. During summer months, the warm **Sun** and an abundance of food in the water create optimal conditions for the breeding of dinoflagellates, which are a type of phytoplankton. This multiplication, or bloom, happens rapidly, and the seawater becomes extremely dense with dinoflagellates; sometimes their numbers can reach many millions per cup of seawater. Even though most red tides are harmless, many of them are toxic and extremely dangerous to fish, shellfish, birds, and even humans. Certain species of dinoflagellates are capable of producing highly-toxic substances.

When these toxic red tides appear in warm coastal places like Texas and Florida, people are warned not to swim, fish, or eat locally-caught fish. Clams, oysters, mussels, and other shellfish are especially dangerous because they feed on the dinoflagellates and retain the toxins. If ingested by humans, contaminated shellfish can cause nausea and diarrhea or worse. In severe cases, the poisons attack human muscle fibers and can cause partial paralysis or even death. In addition to being warned not to eat or catch fish, people are generally advised to stay away from coastal areas during red tide. Decaying bodies of dead fish and birds can create foul smells in the air. Moreover, when people inhale the air around windblown red tide, their lungs can become irritated.

Today, red tides are increasingly common in the **Gulf of Mexico**. Many rivers, including the Mississippi, empty into the Gulf, depositing sewage, industrial waste, and chemicals into the ocean. These pollutants contain phosphorous and nitrogen which then serve as food for the dinoflagellate algae. As the algae organisms consume the nitrogen and phosphorous, they spread their color across the water, cutting off sunlight and **oxygen** to other marine life. The severity of red tide is unpredictable because of such factors as the **weather**, water composition, marine life, and pollution levels. Red tides can last for a few hours or up to several months. The size can range from less than a few square yards to more than 1,000 miles (1613 km).

See also Environmental pollution; Oceans and seas; Water pollution

REGOLITH

Regolith is a layer of loose or unlithified **soil** and **rock** debris that overlies and blankets **bedrock**. It is derived from the Greek roots *rhegos*, for blanket, and *lithos*, for stone. The term is descriptive and non-genetic, meaning that regolith refers to any blanket of unlithified material regardless of its origin. Lunar regolith (also known as lunar soil) is the loose material ranging in size from dust to boulders on the surface of the **Moon**, and is believed to have been formed as a result of meteorite impact and fragmentation.

Regolith can be transported by a variety of geologic processes or formed in place by the chemical and physical **weathering** of bedrock. Transported regolith can include volcanic ash, glacial deposits, alluvium, colluvium (debris accumulating near the base of a slope due to **creep** or related processes), talus, loess (windblown silt), eolian deposits, **landslide** and **debris flow** deposits, and soil. Regolith resting on slopes can also be the source material for landslides, debris flows, and rock falls. Because regolith can be the product of many different geological processes, its physical characteristics (for example, density, **permeability**, **porosity**, and strength) can be highly variable over short distances. Geologic maps typically do not portray regolith unless it is so thick that it completely obscures bedrock or the purpose of the map is specifically to show unlithified surficial deposits.

In the fields of engineering **geology**, geotechnical engineering, and soil mechanics, regolith is often referred to as soil even though it may not have been altered by biological and chemical soil forming processes. These conflicting definitions of soil can cause confusion, so it is always important to understand the context in which the word is being used.

Characterization of the physical properties of regolith is of paramount importance in many engineering and environmental projects. In some cases, regolith must be removed in order to construct foundations on bedrock. In other cases, it is the regolith itself in which structures are anchored or which constitutes the source of material for embankments and other earthworks. Regolith can also be a locally significant **aquifer** that yields **groundwater** relatively easily and inexpensively. Because regolith lies at the earth's surface, however, aquifers in it are much more easily polluted than those in deeper **artesian** aquifers.

See also Alluvial system; Eolian processes; Glacial landforms; Landscape evolution; Soil and soil horizons; Stream valleys, channels, and floodplains; Talus pile or Talus slope

RELATIVE AGE • *see* DATING METHODS

RELATIVE HUMIDITY • *see* HUMIDITY

RELATIVITY THEORY

Relativity theory, a general term encompassing special and general relativity theories, sets forth a specific set of laws relating motion to mass, **space**, time, and **gravity**. Relativity theory allows calculations of the differences in mass, space, and time as measured in different reference frames.

At the start of the twentieth century the classical laws of **physics** contained in Sir Isaac Newton's (1642–1727) 1687 work, *Philosophiae Naturalis Principia Mathematica (Mathematical Principles of Natural Philosophy)* adequately described the phenomena of everyday existence. In accord with these laws, more than century of experimental and mathematical work in **electricity and magnetism** resulted in Scottish physicist James Clerk Maxwell's (1831–1879) four equations describing light as an electromagnetic wave. Prior to Maxwell's equations it was thought that all waves required a medium or ether for propagation. Such an ether would also serve as an absolute reference frame against which absolute motion, space and time could be measured. Ironically, although Maxwell's equations established that electromagnetic waves do not require such a medium, Maxwell and others remained unconvinced and the search for an elusive ether continued. For more than three decades, the lack of definable or demonstrable ether was explained away as simply a problem of experimental accuracy. The absence of a need for an ether for the propagation of electromagnetic radiation was demonstrated in late nineteenth century experiments conducted by Albert Michelson (1852–1931) and Edward Morley (1838–1923).

In 1904, French mathematician Jules-Henri Poincaré (1854–1912) pointed out important problems with concepts of simultaneity by asserting that observers in different reference frames must measure time differently. In 1905, a German-born clerk in the Swiss patent office named **Albert Einstein** (1879–1955) published a theory of light that incorporated implications of Maxwell's equations, demonstrated the lack of need for an ether, explained FitzGerald-Lorentz contractions, and explained Poincaré's reservations concerning differential time measurement. Both Einstein and his special relativity theory went to revolutionize modern physics.

In formulating his special theory of relativity, Einstein assumed that the laws of physics are the same in all inertial (moving) reference frames and that the speed of light was measured as a constant regardless of its direction of propagation. Moreover, the measured speed of light was independent of the velocity of the observer.

Einstein's special theory of relativity also related mass and energy. Einstein published a formula relating mass and energy, $E=mc^2$ (Energy=mass times speed of light squared). Einstein's equation implied that tremendous energies were contained in small masses. Along with advancements in **atomic theory**, Einstein's insights ultimately allowed the development of atomic weapons during World War II and the dawn of the nuclear age.

Special relativity also gave rise to a number of counterintuitive paradoxes dealing with the passage of time (e.g., the twin paradox) and with problems dependent upon an assumption of simultaneity. According to the postulates of special relativity, under certain conditions it would be impossible to determine when one event happened in relation to another event.

Although Einstein's special theory was limited to special cases dealing with systems in uniform nonaccelerated motion, the theory did away with the need for an absolute frame of reference. In addition, the implications of special relativity on the equivalence of mass and energy revolutionized classical laws regarding conservation of mass and energy. A more complete understanding of the conservation of mass and energy now relies upon mass-energy systems. Einstein's special relativity theory was also important in the development of **quantum theory**. German Physicist **Max Planck** (1858–1947) and others who were in the process of developing quantum theory, set out to reconcile (often unsuccessfully) relativity theory with quantum theory.

In 1915, a decade after publishing his special relativity theory, Einstein published his general theory of relativity that soon came to supplant well-understood Newtonian concepts of gravity. Although Newtonian theories of gravity hold valid for most objects, there were small but noticeable errors in calculations regarding the motion of bodies at high velocities or for description of motion in massive gravitational fields. These small errors were completely corrected by the general theory of relativity that described nonuniform, or accelerated, motion.

General relativity's impact on calculations regarding gravity sparked dramatic revisions in **cosmology** that continue today. The conceptual fusion of traditional three-dimensional space with time to create space-time also made observers more integral to measurement of phenomena.

In a sophisticated elaboration of Newton's laws of motion, in general relativity theory the motion of bodies is explained by the assertion that in the vicinity of mass, space-time curves. The more massive the body the greater the curvature of space-time and, consequently, the greater the force of gravitational attraction.

It may be fairly argued that the most stunning philosophical consequence of general relativity was that space-time is a creation of the universe itself. Under general relativity, the universe is not simply expanding into preexisting space and time, but rather creating space-time as a consequence of expansion. In this regard, general relativity theory set the stage for the subsequent development of **Big Bang theory**.

Unlike the esoteric proofs of special relativity, the proofs of general relativity could be measured by conventional experimentation. General relativity's assertion that gravitational fields would bend light was confirmed during a 1919 solar **eclipse**. Other predictions regarding shifts in the perihelion of Mercury and in redshifted spectra also found confirmation. Using relativity based equations, German physicist Karl Schwarzschild (1873–1916) mathematically described the gravitational field near massive compact objects. Schwarzschild's work subsequently enabled the predication and discovery of the evolutionary stages in massive stars (e.g., neutron stars, black holes, etc.).

Relativity theory was quickly accepted by the general scientific community, and its implications on general philosophical thought were profound. Although Newtonian physics still enjoys widespread utility, along with quantum physics, relativity-based theories have replaced Newtonian cosmological concepts.

See also Astronomy; Cosmic microwave background radiation; Cosmology

RELIC DUNES • *see* DUNES

RELIEF

Relief is the difference in altitude between the highest and lowest point of a defined **area** (Relief = highest point – lowest point).

Although to humans the earth is composed of towering mountains and deep **ocean trenches**, Earth's relief, when compared to its overall size, is very small. From a not too distant point in **space**, the earth appears essentially smooth.

For example, using sea level as a base, in 1999 Mt. Everest—the highest point on Earth—measured slightly over 29,000 ft (8,850 m) above sea level. The Marinas Trench, at an estimated depth of 37,000 ft (11,300 m) below sea level (approximately 7 mi, or 11.2 km), is the lowest point on Earth. Using these approximate figures, the relief of Earth is then calculated to be an estimated 66,000 ft (20,117 m). [66,000= 29,000 ft – –37,000 ft (minus 37,000 ft because the reference point of 0 is assigned to sea level)]

The "smooth" character of the earth is fairly argued when comparing the scant 12.5-mi (20.1-km) relief of Earth's surface with Earth's approximate 7,900-mile (12,714-km) diameter. The relief measures less than two-tenths of one percent of the overall size of the earth.

Topographic maps depict elevation and contours (lines of equal elevation) show the progression of surface altitude changes. Relief is a critical component when defining certain area geographic features. For example, a plateau is a broad area with steep sided uplifts but with low relief on the surface. Correspondingly, a basin is often described as a low-lying area with low relief.

Although relief generally changes with geologic slowness (e.g., the uplift of Mt. Everest), relief in some **desert** areas—highly exposed to **wind** forces—often shows dramatic and rapid changes.

See also Cartography; GIS; Landforms; Landscape evolution; Topography and topographic maps

REMOTE SENSING

Remote sensing is the science and art of obtaining and interpreting information about an object, **area**, or phenomenon through the analysis of data acquired by a sensor that is not in contact with the object, area, or phenomenon being observed. There are four major characteristics of a remote sensing system, namely, an electromagnetic energy source, transmission path, target, and sensor.

The **Sun** is a common source of electromagnetic energy. It radiates **solar energy** in all directions. Earth reflects the energy from the Sun and emits some energy in the form of heat.

Based on the energy source, remote sensing systems can be grouped into two types, passive and active systems. Passive remote sensing systems detect radiation that is reflected and/or emitted from the surface features of Earth. Examples are the Landsat and European SPOT **satellite** systems. Active remote sensing systems provide their own energy source. For example, the Radarsat-1 synthetic aperture radar (SAR) system has an antenna that beams pulses of electromagnetic energy towards the target.

The transmission path is the **space** between the electromagnetic energy source and the target, and back to the sensor. In the case of Earth observation, the transmission path is usually the atmosphere of Earth. While passing through Earth's atmosphere, the electromagnetic energy can be scattered by minute particles or absorbed by gases such that its strength and spectral characteristics are modified before being detected by the sensor.

The target could be a particular object, an area, or phenomenon. For example, it could be a ship, city, forest cover, mineralized zone, and **water** body contaminated by oil slick, a forest fire, or a combination thereof.

Electromagnetic energy that hits a target, called incident radiation, interacts with matter or the target in several ways. The energy could be reflected, absorbed, or transmitted. When incident radiation hits a smooth surface, it is reflected or bounced in the opposite direction like light bouncing off a mirror. If it hits a relatively rough surface, it could be scattered in all directions in a diffuse manner. When incident radiation is absorbed, it loses its energy largely to heating the matter. A portion of the energy may be emitted by the heated substance, usually at longer wavelengths. When incident radiation is transmitted, it passes through the substance such as from air into water.

The sensor is a device that detects reflected and/or emitted energy. Passive remote sensing systems carry optical sensors that detect energy in the visible, infrared, and thermal infrared regions of the **electromagnetic spectrum**. Common sensors used are cameras and charge-coupled detectors (**CCD**) mounted on either airborne or space-borne platforms. In active remote sensing systems, the same antenna that sends out energy pulses detects the return pulse.

Present applications of remote sensing are numerous and varied. They include land cover mapping and analysis, land use mapping, agricultural plant health monitoring and harvest forecast, water resources, wildlife ecology, archeological investigations, snow and **ice** monitoring, disaster management, geologic and **soil** mapping, mineral exploration, coastal resource management, military surveillance, and many more.

One main advantage of a remote sensing system is its ability to provide a synoptic view of a wide area in a single frame. The width of a single frame, or swath width, could be 37 mi x 37 mi (60 km x 60 km) in the case of the European

SPOT satellite, or as wide as 115 mi x 115 mi (185 km x 185 km) in the case of Landsat. Remote sensing systems can provide data and information in areas where access is difficult as rendered by terrain, **weather**, or military security. The towering Himalayas and the bitterly cold Antarctic regions provide good examples of these harsh environments. Active remote sensing systems provide cloud-free images that are available in all weather conditions, day or night. Such systems are particularly useful in tropical countries where constant cloud cover may obscure the target area. In 2002, the United States military initiatives in Afghanistan used remote sensing systems to monitor troops and vehicle convoy movements at spatial resolutions of less than one meter to a few meters. Spatial resolution or ground resolution is a measure of how small an object on Earth's surface can be "seen" by a sensor as separate from its surroundings.

The greater advantage of remote sensing systems is the capability of integrating multiple, interrelated data sources and analysis procedures. This could be a multistage sensing wherein data on a particular site is collected from the multiple sources at different altitudes like from a low altitude aircraft, a high altitude craft, a **space shuttle** and a satellite. It could also be a multispectral sensing wherein data on the same site are acquired in different spectral bands. Landsat-5, for example, acquires data simultaneously in seven wavelength ranges of the electromagnetic spectrum. Or, it could be a multitemporal sensing whereby data are collected on the same site at different dates. For example, data may be collected on rice-growing land at various stages of the crop's growth, or on a **volcano** before and after a volcanic eruption.

Two satellite systems in use today are the Landsat and Radarsat remote sensing systems. Landsat is the series of Earth observation satellites launched by the U.S. National Aeronautics and Space Administration (NASA) under the Landsat Program in 1972 to the present. The first satellite, originally named Earth Resources Technology Satellite-1 (ERTS-1), was launched on July 22, 1972. In 1975, NASA renamed the "ERTS" Program the "Landsat" Program and the name ERTS-1 was changed to Landsat-1. All following satellites carried the appellation of Landsat. As of 2002, there are seven Landsat satellites launched. The latest, Landsat-7 was launched on July 15, 1999.

Landsat-7 carries the Enhanced Thematic Mapper Plus (ETM+) sensor. The primary features of Landsat-7 include a panchromatic band with 49 ft (15 m) spatial resolution and a thermal infrared channel (Band 6) with 197 ft (60 m) spatial resolution. Like its predecessors the Landsat-4 and -5, Landsat-7 ETM+ includes the spectral bands 1,2,3,4,5,6 and 7. The spatial resolution remains at 98 ft (30 m), except for band 6 in which the resolution is increased from 394 ft (120 m) to 197 ft (60 m). Landsat-7 orbits Earth at an altitude of 438 mi (705 km). It has a repeat cycle of 16 days, meaning it returns to the same location every 16 days.

Radarsat is the series of space-borne SAR systems developed by Canada. Radarsat-1, launched on November 4, 1995 by NASA, carries a C-band 2.2 in (5.6 cm wavelength) antenna that looks to the right side of the platform. The antenna transmits at 5.3 GHz with an HH polarization (Horizontally transmitted, Horizontally received). It can be steered from 10 to 59 degrees. The swath width can be varied to cover an area from 31 mi (50 km) in fine mode to 311 mi (500 km) in ScanSAR Wide mode. Radarsat-1 orbits Earth at an altitude of 496 mi (798 km) and has a repeat cycle of 24 days.

Several space-borne remote sensing systems planned for launch in the near future include the Radarsat-2 and the Advanced Land Observing Satellite (ALOS) in 2003, and the Landsat-8 in 2005.

See also Archeological mapping; Earth, interior structure; Mapping techniques; Petroleum, history of exploration; RADAR and SONAR; Seismograph

REVOLUTION AND ROTATION

Although often confused, there is a distinct and important difference in the concepts of revolution and rotation. Earth rotates on its axis as it revolves around the **Sun**.

Earth rotates about its axis at approximately 15 angular degrees per hour. Rotation dictates the length of the diurnal cycle (i.e, the day/night cycle), creates "time zones" with differing local noons, and also causes the apparent movement of the **Moon**, stars, and planets across the "celestial sphere". The rotation of Earth is eastward (from west to east) making the apparent rotation of the celestial sphere from east to west.

The rates of rotation and revolution are functions of a planet's mass and orbital position. For example, the mass of Jupiter is approximately 317.5 times Earth's mass and the rotation time (the time for Jupiter to revolve once about its axis) is approximately nine hours.

Earth takes approximately 365.25 days to complete one revolution around the Sun in a slightly elliptical orbit with the Sun at one focal point of the ellipse. Ranging between the extremes of perihelion (closest approach) in January and aphelion (most distant orbital position) in July, Earth's orbital distance from the Sun ranges from approximately 91.5 to approximately 94 million miles (147–151 million km), respectively. Although these distances seem counterintuitive to residents of the Northern Hemisphere who experience summer in July and winter in January—the **seasons** are not nearly as greatly affected by distance as they are by changes in solar illumination caused by the fact that Earth's **polar axis** is inclined 23.5 degrees from the perpendicular to the ecliptic (the plane of the **solar system** through or near which most of the planet's orbits travel) and because the Earth exhibits parallelism (currently toward Polaris, the North Star) as it revolves about the Sun.

At the extreme of the solar system, Pluto, usually the most distant planet (i.e., at certain times Neptune's orbit actually extends farther than Pluto's orbit) takes approximately 247 Earth years (the time it takes the Earth to revolve about the Sun) to complete one orbital revolution about the Sun.

Rotation, revolution, polar tilt, parallelism, and Earth's oblate spheroid shape combine to produce an unequal distribution of **solar energy**, the changing of seasons, the changing lengths of day and night, and influence the circulation of the atmosphere and **oceans**.

In addition to Earth's rotation about the Sun, the solar system is both moving with the Milky Way galaxy and revolving around the galactic core.

See also Celestial sphere: The apparent movements of the Sun, Moon, planets, and stars

RHYOLITE

Rhyolite is an **aphanitic** volcanic **rock** with the equivalent mineralogical composition of **granite**. Rhyolite contains less than 5% phenocrysts, or mineral grains visible without magnification. The rest of the more than 95% of the rock consists of a ground mass too fine to discern without magnification. This texture is the result of the rapid cooling of extruded **lava**, which does not allow sufficient time for larger **crystals** to grow.

The mineralogical composition of rhyolite is defined as containing mostly **quartz** and **feldspar** with a total silica content of more than 68%. Quartz in rhyolite may be as low as 10% but is usually present in amounts of 25% to 30%. Feldspars often comprise 50% to 70% of rhyolite, with potassium feldspar present in at least twice the amount of **plagioclase** feldspar. Ferromagnesian, or dark, **minerals** are rare as phenocrysts, being mostly biotite when present. Trace accessory minerals may also include muscovite, pyroxenes, amphiboles, and oxides.

Rhyolite often appears very uniform in texture, although lava flow structures may be evident. They range in color from white to gray to pink. Due to the fine grained nature, the differentiation of rhyolite from aphanitic rocks of differing composition is not always conclusive based on color alone, but any light colored volcanic aphanitic rock is likely to be a rhyolite.

The high silica content of rhyolite creates a high viscosity lava, or one that is strongly resistance to flow. The viscous lava tends to build up volcanic gases instead of allowing them to escape. Eruptions of rhyolite can be highly explosive due to the spontaneous release of large amounts of trapped gases. This accounts for some of the very quickly cooled textural variations of rhyolite. For example, **obsidian** is a pure volcanic **glass** of rhyolitic composition and **pumice** is rhyolite glass that has cooled in the form of gas bubbles.

See also Extrusive cooling

RICHARDSON, LEWIS FRY (1881-1953)

English physicist and meteorologist

Lewis Fry Richardson was an English physicist with a penchant for trying to solve a wide range of scientific problems using mathematics. During his career as a scientist and educator, Richardson explored mathematical solutions to predict **weather**, to explain the flow of **water** through peat, and to identify the origins of war.

Richardson was the youngest of seven children born to David Richardson, a tanner, and his wife, Catherine Fry, who came from a family of corn merchants. Richardson was born on October 11, 1881, in Newcastle upon Tyne. After completing his high school education in 1898, Richardson studied science at Durham College in Newcastle for two years before entering King's College at Cambridge, where he ultimately earned a doctorate in **physics** and then later returned to study and receive a degree in psychology. After graduating from King's College, Richardson held a number of positions in the years leading up to World War I. These included working as a scientist for a tungsten lamp factory, the National Peat Industries, Ltd., and serving four years as superintendent of the Eskdalemuir Observatory operated by the National Meteorological Office.

Richardson, who was born into a Quaker family, served with the French army as a member of the Friends' Ambulance Unit during the war from 1916 to 1919. Following the end of hostilities, Richardson returned to England, where he combined his scientific inquiry with teaching. In 1920, he accepted a position as director of the physics department at Westminster Training College. This was followed by an appointment as principal of Paisley Technical College in 1929, a post that he held until his retirement in 1940. Retirement allowed Richardson to continue his primary love, research.

Richardson began his research looking at practical problems, such as examining the flow of water through peat while he worked for the National Peat Industries, Ltd. Using differential equations, Richardson came up with ways to determine water flow that were far more accurate than other methods. His work eventually led to attempts at developing a system of weather prediction based on newly understood knowledge of the upper atmosphere and the roles played by radiation and eddies, or atmospheric currents which move contrary to main air flow. Richardson's work led to the publication of his book, *Weather Prediction by Numerical Process*, in 1922.

Richardson's experiences in France during the First World War also inspired him to probe the causes of human conflict using mathematics, and he published a paper in 1919 on the mathematical psychology of war. Eventually, he enlarged upon this early work in the book *Arms and Insecurity* and went on to complete a mathematical study of the world's wars. This work, which resulted in *Statistics of Deadly Quarrels*, examined the causes and magnitude of these conflicts. In his research, Richardson tried to define the relations between countries in terms of mathematical equations.

Richardson's pioneering use of mathematics resulted in him being elected a fellow in the Royal Society in 1926. Richardson died on September 30, 1953.

See also Atmospheric composition and structure; Weather forecasting methods; Weather forecasting

RICHTER, CHARLES F. (1900-1985)

American seismologist

Charles F. Richter is remembered every time an **earthquake** happens. With German-born seismologist Beno Gutenberg,

Richter developed the scale that bears his name and measures the magnitude of earthquakes. Richter was a pioneer in seismological research at a time when data on the size and location of earthquakes were scarce. He authored two textbooks that are still used as references in the field and are regarded by many scientists as his greatest contribution, exceeding the more popular **Richter scale**. Devoted to his work all his life, Richter at one time had a **seismograph** installed in his living room, and he welcomed queries about earthquakes at all hours.

Charles Francis Richter was born on a farm near Hamilton, Ohio, north of Cincinnati. His parents were divorced when he was very young. He grew up with his maternal grandfather, who moved the family to Los Angeles in 1909. Richter went to a preparatory school associated with the University of Southern California, where he spent his freshman year in college. He then transferred to Stanford University, where he earned an A.B. degree in **physics** in 1920.

Richter received his Ph.D. in theoretical physics from the California Institute of Technology (Cal Tech) in 1928. That same year he married Lillian Brand of Los Angeles, a creative writing teacher. Robert A. Millikan, a Nobel Prize-winning physicist and president of Cal Tech, had already offered Richter a job at the newly established Seismological Laboratory in Pasadena, then managed by the Carnegie Institution of Washington. Thus Richter started applying his physics background to the study of the earth.

As a young research assistant, Richter made his name early when he began a decades-long collaboration with Beno Gutenberg, who was then the director of the laboratory. In the early 1930s, the pair was one of several groups of scientists around the world who were trying to establish a standard way to measure and compare earthquakes. The seismological laboratory at Cal Tech was planning to issue regular reports on southern California earthquakes, so the Gutenberg-Richter study was especially important. They needed to be able to catalog several hundred quakes a year with an objective and reliable scale.

At the time, the only way to rate shocks was a scale developed in 1902 by the Italian priest and geologist Giuseppe Mercalli. The Mercalli scale classified earthquakes from 1 to 12, depending on how buildings and people responded to the tremor. A shock that set chandeliers swinging might rate as a 1 or 2 on this scale, while one that destroyed huge buildings and created panic in a crowded city might count as a 10. The obvious problem with the Mercalli scale was that it relied on subjective measures of how well a building had been constructed and how used to these sorts of crises the population was. The Mercalli scale also made it difficult to rate earthquakes that happened in remote, sparsely populated areas.

The scale developed by Richter and Gutenberg, which became known by Richter's name only, was instead an absolute measure of an earthquake's intensity. Richter used a seismograph—an instrument generally consisting of a constantly unwinding roll of paper, anchored to a fixed place, and a pendulum or magnet suspended with a marking device above the roll—to record actual earth motion during an earthquake. The scale takes into account the instrument's distance from the epicenter, or the point on the ground that is directly

Charles Richter. ***AP/Wide World. Reproduced by permission.***

above the earthquake's origin. Richter chose to use the term "magnitude" to describe an earthquake's strength because of his early interest in **astronomy**; stargazers use the word to describe the brightness of stars. Gutenberg suggested that the scale be logarithmic, so that a quake of magnitude 7 would be ten times stronger than a 6, a hundred times stronger than a 5, and a thousand times stronger than a 4. (The 1989 Loma Prieta earthquake that shook San Francisco was magnitude 7.1.)

The Richter scale was published in 1935, and immediately became the standard measure of earthquake intensity. Richter did not seem concerned that Gutenberg's name was not included at first; but in later years, after Gutenberg was already dead, Richter began to insist that his colleague be recognized for expanding the scale to apply to earthquakes all over the globe, not just in southern California. Since 1935, several other magnitude scales have been developed. Depending on what data is available, different ones are used, but all are popularly known by Richter's name.

For several decades, Richter and Gutenberg worked together to monitor seismic activity around the world. In the late 1930s they applied their scale to deep earthquakes, ones that originate more than 185 miles below the ground, which rank particularly high on the Richter scale—8 or greater. In 1941, they published a textbook, *Seismicity of the Earth*, which in its revised edition became a standard reference book in the field. They worked on locating the epicenters of all the major earthquakes and classifying them into geographical groups. All his life, however, Richter warned that seismological records only reflect what people have measured in populated areas and are not a true representative sample of what shocks have actually occurred. He long remained skeptical of some scientists' claims that they could predict earthquakes.

Richter remained at Cal Tech for his entire career, except for a visit to the University of Tokyo from 1959 to 1960 as a Fulbright scholar. He became involved in promoting good earthquake building codes, while at the same time discouraging the overestimation of the dangers of an earthquake in a populated **area** like Los Angeles. He pointed out that statistics reveal freeway driving to be much more dangerous than living in an earthquake zone. He often lectured on how loss of life and property damage were largely avoidable during an earthquake, with proper training and building codes—he opposed building anything higher than thirty stories, for example. In the early 1960s, the city of Los Angeles listened to Richter and began to remove extraneous, but potentially dangerous, ornaments and cornices from its buildings. Los Angeles suffered a major quake in February of 1971, and city officials credited Richter with saving many lives. Richter was also instrumental in establishing the Southern California Seismic Array, a network of instruments that has helped scientists track the origin and intensity of earthquakes, as well as map their frequency much more accurately. His diligent study resulted in what has been called one of the most accurate and complete catalogs of earthquake activity, the Cal Tech catalog of California earthquakes.

Later in his career, Richter would recall several major earthquakes. The 1933 Long Beach earthquake was one, which he felt while working late at Cal Tech one night. That quake caused the death of 120 people in the then sparsely populated southern California town; it cost the Depression-era equivalent of $150 million in damages. Nobel Prize-winning physicist **Albert Einstein** was in town for a seminar when the earthquake struck, according to a March 8, 1981 story in the *San Francisco Chronicle*. Einstein and a colleague of Richter's were crossing the campus at the time of the quake, so engrossed in discussion that they were oblivious to the swaying trees. Richter also remembered the three great quakes that struck in 1906, when he was a six-year-old on the Ohio farm. That year, San Francisco suffered an 8.3 quake, Colombia and Ecuador had an 8.9, and Chile had an 8.6.

In 1958, Richter published his text *Elementary Seismology*, which was derived from the lectures he faithfully taught to Cal Tech undergraduates as well as decades of earthquake study. Many scientists consider this textbook to be Richter's greatest contribution, since he never published many scientific papers in professional journals. *Elementary Seismology* contained descriptions of major historical earthquakes, tables and charts, and subjects ranging from the nature of earthquake motion to earthquake insurance and building construction. Richter's colleagues maintained that he put everything he knew into it. The book was used in many countries.

In the 1960s, Richter had a seismograph installed in his living room so that he could monitor quakes at any time. He draped the seismographic records—long rolls of paper covered with squiggly lines—over the backs of the living room chairs. (His wife, Richter maintained, considered the seismograph a conversation piece.) He would answer press queries at any hour of the night and never seemed tired of talking about his work. Sometimes he grew obsessive about speaking to the press; when a tremor happened during Cal Tech working hours, Richter made sure he would be the one answering calls—he put the lab's phone in his lap.

Richter devoted his entire life to **seismology**. He even learned Russian, Italian, French, Spanish, and German, as well as a little Japanese, in order to read scientific papers in their original languages. His dedication to his work was complete; in fact, he became enraged at any slight on it. For instance, at his retirement party from Cal Tech in 1970, some laboratory researchers sang a clever parody about the Richter scale. Richter was furious at the implication that his work could be considered a joke. During his lifetime he enjoyed a good deal of public and professional recognition, including membership in the American Academy of Arts and Sciences and a stint as president of the Seismological Society of America, but he was never elected to the National Academy of Sciences. After his retirement, Richter helped start a seismic consulting firm that evaluated buildings for the government, for public utilities such as the Los Angeles Department of **Water** and Power, and for private businesses.

Richter enjoyed listening to classical music, reading science fiction, and watching the television series *Star Trek*. One of his great pleasures, ever since he grew up walking in the southern California mountains, was taking long solitary hikes. Richter died in Pasadena at the age of 85.

See also Faults and fractures; Folds

Richter scale

The earliest **earthquake** measurements were simple descriptions called intensity ratings. These results were unreliable depending on the distance between the quake's source (epicenter), and the people evaluating the event.

A more systematic approach was developed by an Italian seismologist, Guiseppe Mercalli in 1902. He gauged earthquake intensity by measuring the damage done to buildings. The United States Coast and Geodetic Survey adapted his method, which they called the modified Mercalli Scale, dividing the measurements into 12 categories: level II was "felt by persons at rest," but at level VII it was "difficult to stand." Level X caused most buildings to collapse, and level XII, the most intense, combined ground fissures with tsunamis (tidal waves) and almost total destruction. Despite the specific detail of descriptions, this method, like the intensity ratings, was influenced by the measurement's distance from the earthquake's epicenter. Seismologists needed a way to determine the size, or magnitude, of an earthquake. They needed a quantitative, numerical measurement that would compare the strength of earthquakes in a meaningful way, not merely catalog damage or record perceptions as Mercalli's qualitative method did. This critical factor was finally determined in 1935 by American seismologist **Charles F. Richter**, a professor of **seismology** at the California Institute of Technology. His system of measurement, called the Richter scale, was based on his studies of earthquakes in southern California. It has become the most widely used assessment of earthquake severity in the world.

Richter measured ground movement with a **seismograph**, compared the reading to others taken at various distances from the epicenter, then calculated an average magnitude from all reports. The results are plotted on a logarithmic scale, in whole numbers and tenths, from 1 to 9. Each whole number increase means that the magnitude of the quake is ten times greater than the previous whole number. Thus, an earthquake with a magnitude of 6.5 has ten times the force of one with a magnitude of 5.5; an earthquake of 7.5 has 100 times the intensity of the 5.5 earthquake. An 8.5 measurement is 1,000 times stronger, and so on.

The amount of energy an earthquake releases is calculated in a different manner. Instead of tenfold jumps with each increase in magnitude, energy released is measured in roughly thirtyfold increments. Thus, an earthquake with a value of 7 releases 30 times the amount of energy as an earthquake measured at 6, while an earthquake of 8 would have 900 times the energy as one valued at 6.

Today the modified Mercalli scale is often used in combination with the Richter scale because both methods are helpful in gauging the total impact of an earthquake.

See also Seismology

RIDE, SALLY (1951-)

American astronaut

Sally Ride is best known as the first American woman sent into outer **space**. She also served the National Aeronautics and Space Administration (NASA) in an advisory capacity, and was the only astronaut chosen for President Ronald Reagan's Rogers Commission investigating the mid-launch explosion of the **space shuttle** *Challenger* in January, 1986, writing official recommendation reports and creating NASA's Office of Exploration. Both scientist and professor, she has served as a fellow at the Stanford University Center for International Security and Arms Control, a member of the board of directors at Apple Computer Inc., and a space institute director and **physics** professor at the University of California at San Diego. Ride has chosen to write primarily for children about space travel and exploration. Her commitment to educating the young earned her the Jefferson Award for Public Service from the American Institute for Public Service in 1984, in addition to her National Spaceflight Medals recognizing her two groundbreaking shuttle missions in 1983 and 1984.

Sally Kristen Ride is the older daughter of Dale Burdell and Carol Joyce (Anderson) Ride of Encino, California, and was born May 26, 1951. As author Karen O'Connor describes tomboy Ride in her young reader's book, *Sally Ride and the New Astronauts,* Sally would race her dad for the sports section of the newspaper when she was only five years old. An active, adventurous, yet also scholarly family, the Rides traveled throughout **Europe** for a year when Sally was nine and her

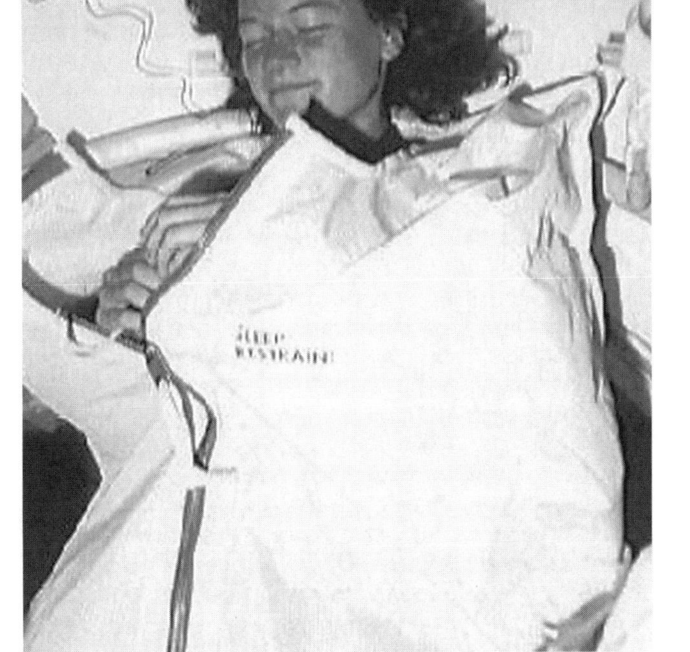

Astronaut Sally Ride suspended in a shuttle sleeping bag. *U.S. National Aeronautics and Space Administration (NASA).*

sister Karen was seven after Dale took a sabbatical from his political science professorship at Santa Monica Community College. While sister Karen was inspired to become a minister, in the spirit of her parents who were elders in their Presbyterian church, Sally Ride's own developing taste for exploration would eventually lead her to apply to the space program almost on a whim. "I don't know why I wanted to do it," she confessed to *Newsweek* prior to embarking on her first spaceflight.

From her earliest years in school, Ride was so proficient and efficient at once, she proved to be an outright annoyance to some of her teachers. Though she was a straight-A student she was easily bored, and her intellect only came to the fore in high school when she was introduced to the world of science by her physiology teacher. The impact of this mentor, Dr. Elizabeth Mommaerts, was so profound that Ride would later dedicate her first book primarily to her, as well as the fallen crew of the *Challenger*. While she was adept at all forms of sport, playing tennis was Ride's most outstanding talent, which she had developed since the age of ten. Under the tutelage of a four-time U.S. Open champion, Ride eventually ranked eighteenth nationally on the junior circuit. Her ability won her a partial scholarship to Westlake School for Girls, a prep school in Los Angeles. After graduating from there in 1968, Ride preferred to work on her game full time instead of the physics program at Swarthmore College, Pennsylvania, where she had originally enrolled. It was only after Ride had fully tested her dedication to the game that she decided against a professional career, even though tennis pro Billie Jean King had once told her it was within her grasp. Back in California as an undergraduate student at Stanford University, Ride followed her burgeoning love for Shakespeare to a double major, receiving B.S. and B.A. degrees in tandem by 1973. She narrowed her focus to physics for her masters, also from Stanford, awarded in 1975. Work toward her dissertation continued at Stanford; she submitted "The Interaction of X Rays with the Interstellar Medium" in 1978.

Ride was just finishing her Ph.D. candidacy in physics, **astronomy**, and astrophysics at Stanford working as a research assistant when she got the call from NASA. She became one of 35 chosen from an original field of applicants numbering 8,000 for the spaceflight training of 1978. "Why I was selected remains a complete mystery," she later admitted to John Grossmann in a 1985 interview in *Health*. "None of us has ever been told." Even after three years of studying x-ray astrophysics, Ride had to go back to the classroom to gain skills to be part of a team of astronauts. The program included basic science and math, **meteorology**, guidance, navigation, and computers as well as flight training on a T-38 jet trainer and other operational simulations. Ride was selected as part of the ground-support crew for the second (November, 1981) and third (March, 1982) shuttle flights, her duties including the role of "capcom," or capsule communicator, relaying commands from the ground to the shuttle crew. These experiences prepared her to be an astronaut.

Ride would subsequently become, at 31, the youngest person sent into orbit as well as the first American woman in space, the first American woman to make two spaceflights, and, coincidentally, the first astronaut to marry another astronaut in active duty. She and Steven Alan Hawley were married in 1982. Hawley, a Ph.D. from the University of California, had joined NASA with a background in astronomy and astrophysics. When asked during a hearing by Congressman Larry Winn, Jr. of the House Committee on Science and Technology how she would feel when Hawley was in space while she remained earthbound, Ride replied, "I am going to be a very interested observer." Eventually, the couple divorced.

Ride points to her fellow female astronauts Anna Fisher, Shannon Lucid, Judith Resnik, Margaret Seddon, and Kathryn Sullivan with pride. Since these women were chosen for training, Ride's own experience could not be dismissed as tokenism, which had been the unfortunate fate of the first woman in orbit, the Soviet Union's **Valentina Tereshkova**, a textile worker. Ride expressed her concern to *Newsweek* reporter Pamela Abramson in the week before her initial shuttle trip: "It's important to me that people don't think I was picked for the flight because I am a woman and it's time for NASA to send one."

From June 18 to June 24, 1983, flight STS-7 of the space shuttle *Challenger* launched from Kennedy Space Center in Florida, orbited the earth for six days, returned to Earth, and landed at Edwards Air Force Base in California. Among the shuttle team's missions were the deployment of international satellites and numerous research experiments supplied by a range of groups, from a naval research lab to various high school students. With Ride operating the shuttle's robot arm in cooperation with Colonel John M. Fabian of the U.S. Air Force, the first **satellite** deployment and retrieval using such an arm was successfully performed in space during the flight.

Ride was also chosen for *Challenger* flight STS-41G, which transpired between October 5 and October 13, 1984. This time the robot arm was put to some unusual applications, including "ice-busting" on the shuttle's exterior and readjusting a radar antenna. According to Henry S. F. Cooper, Jr., in his book *Before Lift-off,* fellow team member Ted Browder felt that because Ride was so resourceful and willing to take the initiative, less experienced astronauts on the flight might come to depend upon her rather than develop their own skills, but this mission also met with great success. Objectives during this longer period in orbit covered scientific observations of the earth, demonstrations of potential satellite refueling techniques, and deployment of a satellite. As STS-7 had been, STS-41G was led by Captain Robert L. Crippen of the U.S. Navy to a smooth landing, this time in Florida.

As leader of a task force on the future of the space program, Ride wrote *Leadership and America's Future in Space.* According to *Aviation Week and Space Technology,* this status report initiated a proposal to redefine NASA goals as a means to prevent the "space race" mentality that might pressure management and personnel into taking untoward risks. "A single goal is not a panacea," the work stated in its preface. "The problems facing the space program must be met head-on, not oversimplified." The overall thrust of NASA's agenda, Ride suggested, should take environmental and international research goals into consideration. A pledge to inform the public and capture the interest of youngsters should be taken as a given. Ride cited a 1986 work decrying the lack of math and science proficiency among American high school graduates, a mere 6% of whom are fluent in these fields, compared to up to 90% in other nations.

While with NASA, Ride traveled with fellow corps members to speak to high school and college students on a monthly basis. Speaking at Smith College in 1985, she announced that encouraging women to enter math and science disciplines was her "personal crusade." Ride noted in *Publishers Weekly* the next year that her ambition to write children's books had been met with some dismay by publishing houses more in the mood to read an autobiography targeted for an adult audience. Her youth-oriented books were both written with childhood friends. Susan Okie, coauthor of *To Space and Back,* eventually became a journalist with the *Washington Post. Voyager* coauthor Tam O'Shaughnessy, once a fellow competition tennis player, grew up to develop workshops on scientific teaching skills.

Ride left NASA in 1987 for Stanford's Center for International Security and Arms Control, and two years later she became director of the California Space Institute and physics professor at the University of California at San Diego. She has flown Grumman Tiger aircraft in her spare time since getting her pilot's license. The former astronaut keeps in shape when not teaching or fulfilling the duties of her various professional posts by running and engaging in other sports, although she once told *Health* magazine she often winds up eating junk food. Ride admitted not liking to run but added, "I like being in shape."

See also Space physiology; Spacecraft, manned

Rifting and rift valleys

Rifting is the process in which continental **crust** is extended and thinned, forming extensional sedimentary basins and/or **mafic** dyke-swarms. Rifts commence as intracratonic, down-thrown blocks dominated by normal or oblique-extensional (transtensional) faults (e.g., the Rhine Graben in Germany and the East African Rift). Rift flanks may be uplifted. Continued rifting results in the break-up of continental plates and creation of oceanic crust (typically 20 to 60 million years after the onset of rifting, but ranging from 7 to 280 million years). Outpouring of flood basalts (also called traps, e.g., the Deccan Traps in western India) can occur over large areas prior to break-up. Marine **sedimentary rocks** are deposited over the rift sequence during the ensuing phase of post-rift, thermal subsidence.

Rifts commonly develop above upwelling convection cells in the **asthenosphere**, such as over a mantle plume. Extensional stresses are induced or enhanced by shear-traction on the base of the **lithosphere** by outwards asthenospheric flow from zones of upwelling. Continents may be split along rifts linking two or more **mantle plumes** (e.g., **South America** and **Africa** were separated along rifts linking the St. Helena, Tristan and Bouvet plumes). Rifting may also represent the far-field reactivation of pre-existing crustal weaknesses, such as old orogenic (mobile) belts, during collision at a distant convergent plate margin. For example, Permo-Triassic 'Gondwanan' rifts in southern Africa, India, and **Australia** represent the orthogonal or oblique extensional reactivation of Proterozoic orogenic belts on the margins of Archaean cratons during collision on the Paleopacific margin of Gondwanaland. Rifts develop in back-arc settings where extensional stresses can be induced by decreasing rates of plate convergence and/or roll-back of a **subduction zone**. Small oceanic basins, such as the Sea of Japan, may form by back-arc rifting. Small rifts may also form by the stepping of transcurrent faults, such as the Salton Sea **area** of the San Andreas fault system in California.

An understanding of rift architecture and structural styles is important as rifts contain major hydrocarbon provinces and mineral deposits. The two main styles of rifting are pure-shear and simple-shear. Many rifts, however, exhibit different elements of these two end-member styles. In pure-shear rifting, steep to moderately dipping normal faults form symmetrically either side of the rift axis. Rift valleys (such as along the East African rift) are developed in the early stages of rifting. Rift valleys are elongate, wide, and typically flat-bottomed topographic depressions along down-thrown blocks. As rift valleys are bounded by normal faults, their sides tend to be steep. Continued crustal extension results in further subsidence and formation of a sedimentary basin along the rift axis. The asthenosphere is bowed upward as an isostatic response to lithospheric thinning. Mafic dykes may be intruded along fractures in the overlying crust. The area of thinnest crust corresponds to the shallowest asthensophere.

Other, highly asymmetrical rifts display a different pattern of structures. In early simple-shear rift models, a through-going shear zone was proposed to extend from the upper crust to the upper mantle. Brittle deformation along the upper part of the extensional detachment was thought to progressively change to ductile shearing over a broader zone at greater depth. It is now thought more likely that a zone of ductile flow in the lower to middle crust separates and decouples displacement along a shallowly dipping extensional detachment in the upper to middle crust from a normal shear zone cutting lithospheric mantle (possibly reactivating a former suture) or bowed up lithospheric mantle. In simple shear rifts, areas of greatest crustal thinning may be offset from areas of greatest asthenospheric uplift. Greatest heat flow and hence volcanic activity and dyke intrusion may be offset from a zone of highly extended crust. This zone comprises highly rotated blocks between imbricate, curved (listric) normal faults. Extensional detachments may be folded by regional antiforms during asthenospheric uplift. The opposite margins of two continents rifted apart may therefore be quite dissimilar. The lower plate margin may show a broad zone of shallowly dipping listric normal faults draped by shallowly dipping sedimentary strata deposited during the post-rift, thermal subsidence phase. Dykes and volcanic rocks may be absent to sparse. In contrast, the opposite, upper plate margin may comprise a narrow zone of steeply dipping faults and contain volcanic rocks and dykes.

See also Continental drift theory; Earth, interior structure; Faults and fractures; Orogeny; Plate tectonics

Rip current

A rip current is a narrow, river-like channel of **water** moving away from the surf (breaking) zone and back toward sea. Rip currents can travel up to 3 mph (4.8 kph) and stretch 100 ft (30.5 m) wide. Some cease just past the breaking surf; others extend a thousand feet offshore. In some regions, the rip current, or rip tide, is a permanent feature of the sea. In other areas, one can appear suddenly or intensify after a storm or a breach in an offshore sandbar.

Rip currents are fed by long shore currents, or feeders, which flow parallel to the beach inside the **surf zone**. In addition to the feeder, each current consists of a neck (main channel) and a head. The neck is the point where feeder currents converge and move back out to sea through a weak spot in the breakers. The head is the widest part of the rip current. Rip currents are typically found near jetties of irregular beaches and along straight, uninterrupted beaches. Although often mistakenly called one, a rip current is not an undertow.

Telltale signs of a rip current include murkier or darker waters, changes in wave formation (large, choppier waves inside the current, calmer ones up front), and foam moving seaward.

Rip currents are dangerous to swimmers, who may suddenly be pulled far out to sea or underwater. ***Courtesy of Kelly A. Quin.***

The intense currents can pull even the most experienced swimmer into dangerously deep ocean water. Attempting to swim to safety against the current can result in exhaustion, panic, and sometimes drowning. An estimated 80% of United States lifeguard rescues are due to rip currents. Experts advise swimming parallel to shore until you surpass the current, then head toward land. Inexperienced swimmers should tread water and call for help.

See also Ocean circulation and currents; Surf zone

Ritter, Carl (1779-1859)

German geographer

Along with his countryman and mentor **Alexander von Humboldt** (1769–1859), Carl Ritter is recognized as one of the two founders of modern geography.

Ritter was born the son of a physician on August 7, 1779, in Quedlinburg, Germany. After his father's premature death in 1784, his mother enrolled him at the age of five in Schnepfenthal, the experimental school of Christian Gotthilf Salzmann (1744–1811), where he acquired an amazing breadth of basic education. Based on the humanistic pedagogical theories of Jean Jacques Rousseau (1712–1778), Salzmann's system emphasized empirical science, practical living, natural law, history, philosophy, theology, art, and modern languages, but not classical languages. His geography teacher was Johann Christoph Friedrich GutsMuths (1759–1839), who also taught history and French. Upon completing this curriculum, Ritter was hired as private tutor for the children of Bethmann Hollweg, a wealthy banker in Frankfurt. From 1798 to 1814, he worked for Hollweg, who financed his university education, first at Frankfurt, then at Göttingen from 1813 to 1819. During this time he also taught himself Latin and Greek.

Ritter's first geographical publication appeared in 1804. By 1816, he was a well-established scholar, with many articles and a two-volume textbook of European geography. He had traveled throughout **Europe**, but not beyond it. On the strength of his excellent reputation, he became professor of history at the University of Frankfurt in 1819 then in 1820 professor of geography at the University of Berlin, holding Germany's first endowed professorship of geography. He spent the rest of his career there, enjoying honors as a popular lecturer and prolific writer. Among his students at Berlin was the geologist **Arnold Henri Guyot** (1807–1884). While in Berlin, he also taught at a military academy and, in 1828, co-founded the Berlin Geographical Society. He died in Berlin on September 28, 1859.

A child of the Enlightenment, Ritter developed a strong affinity for the progressive ideas of Swiss educator Johann Heinrich Pestalozzi (1746–1827), German philosopher Johann Gottfried von Herder (1744–1803), and Humboldt. He integrated the studies of history and geography, perhaps sometimes to the disadvantage of traditional **physical geography**. He envisioned his life's work as a comprehensive geographical treatise of the entire world. With a long title but commonly known as simply *Erdkunde* (Geography), the first volume appeared in 1817. Eventually it ran to nineteen volumes, but when Ritter died it was still incomplete, covering only **Asia** and **Africa**. Immediately successful, it defined the discipline of geography as the study of the relation between humans and their various environments throughout the world. The human aspect of the study was Ritter's innovation, based on GutsMuths's principles. Geography was no longer "just maps."

See also Earth science

Rivers

Rivers and streams are bodies of flowing surface **water** that transport sediment from continental highlands to **lakes**, alluvial fans, and ultimately the ocean. Streams are the main agent of **erosion** of the earth's continental **crust**, and they play a major role in shaping the landscape. Streams are also a focus of humans' interaction with our environment. Human agriculture, industry, and essential biology require fresh, accessible water. Ancient human civilizations first arose in the fertile val-

The Nile River as viewed from the space shuttle. The Nile is the longest river in the world. ***AP/Wide World. Reproduced by permission.***

leys of some of the world's greatest rivers: the Yangtze and Yellow Rivers in China, the Tigris and Euphrates Rivers in the Middle East, and the Nile in Egypt. The distribution of the earth's river systems has influenced human population patterns, commerce, and conquest since then, and the availability of uncontaminated surface water for irrigation, industrial and municipal uses remains a pressing geopolitical issue.

Streamflow is a gravity-driven process that acts to level continental **topography**. Stream erosion balances uplift at plate tectonic boundaries by mechanically and chemically eroding upland rocks, and transporting the resulting siliclastic sediments and dissolved ions and molecules toward the ocean. Current velocity determines a stream's capacity to transport a given volume of suspended and bedload sediment. Sediment transport is intermittent, and individual grains are deposited and re-entrained by turbulent streamflow many times before final deposition in deltas and alluvial fans.

Stream erosion and deposition act in dynamic equilibrium to maintain a concave longitudinal stream profile, called a graded profile, with steep headwaters to low-gradient downstream portions. The elevation where a stream enters another body of water, called base level, controls the downstream end of a stream profile, and the elevation of the headwaters determines the upstream end. Streams cannot erode below base level. Sea level is the ultimate base level for most river systems, and a sea-level change creates a string of compensatory adjustments throughout a stream system. Base level for an individual tributary, however, is controlled by the elevation of the next body of water it enters. If base level falls, or uplift occurs, current velocity increases, and the stream erodes downward. If base level rises, or subsidence occurs, a stream slows down and deposits sediment.

Streams flow in valleys that encompass an **area** between uplands. Some rivers carve their own valleys, and some flow in preexisting valleys created by other geologic processes like **rifting** or glacial erosion. The stream channel that contains flow during non-flood times runs through the stream valley flanked by its overspill areas called **floodplains**. Over time, a stream fills its valley with its own deposits; the **stratigraphy** of a river basin thus shows the depositional history of the stream. Most streams have a valley, a channel, and a floodplain, but their morphology varies between three end-member types—straight, meandering, and braided—depending on the stream gradient, the rate of sediment supply, and the sediment grain size.

Straight streams develop in regions where uplift and/or base level fall force rapid regrading by channel incision. Meandering streams develop at the low-gradient, downstream ends of stream profiles. Because they cannot erode below base level, streams near base level maintain their profile by moving horizontally across the stream valley, eroding and depositing sediment with little effect on the overall sediment flux. Meandering streams develop an organized pattern of fluvial **landforms** and deposits: coarse-grained point bars, gravel channel lags, sandy natural levees, abandoned meanders called oxbow lakes, and fine-grained flood deposits. Braided streams form in mountainous and glaciated areas where rapid currents, voluminous sediment supply, and coarse-grained sediment prevent a stream from forming an orderly pattern of channels and bars. Braided streams have many interlaced channels separated by longitudinal gravel bars that shift over time.

Stream systems are organized into **drainage basins** with small tributary streams that feed into larger trunk streams, and finally into a major river that lets out into the ocean. Drainage divides are topographic highs that separate drainage basins. Drainage basins and divides vary in scale from small hillside watersheds separated by ridges, to the two halves of the North American continent separated by the **continental divide** along the spine of the Rocky Mountains. The outcrop pattern of underlying geologic strata determines the geometry of a stream system. Tree-shaped, or dendritic, drainage patterns form when water flows randomly downhill without encountering geologic obstacles or conduits. Dendritic drainages are the most common and form when **bedrock** layers are horizontal. Trellis-shaped drainages develop in continental fold belts. Rectangular patterns are common in areas of fractured crystalline rocks, and streams flow down the sides of volcanoes in a radial pattern.

See also Alluvial systems; Drainage basins and drainage patterns; Canyon; Estuary; Hydrogeology; Sedimentation; Stream capacity and competence; Stream piracy; Stream valleys, channels, and floodplains

ROCHE MONTONEU • *see* GLACIAL LANDFORMS

ROCK

To the geologist, the term rock means a naturally occurring aggregate of **minerals** that may include some organic solids (e.g., **fossils**) and/or **glass**. Rocks are generally subdivided into three large classes: igneous, sedimentary, and metamorphic. These classes relate to common origin, or genesis. **Igneous rocks** form from cooling liquid rock or related volcanic eruptive processes. **Sedimentary rocks** form from compaction and cementation of sediments. Metamorphic rocks develop due to solid-state, chemical and physical changes in pre-existing rock because of elevated **temperature**, pressure, or chemically active fluids.

With igneous rocks, the aggregate of minerals comprising these rocks forms upon cooling and crystallization of liquid rock. As **crystals** form in the liquid rock, they become interconnected to one another like jigsaw puzzle pieces. After total crystallization of the liquid, a hard, dense igneous rock is the result. Also, some volcanic lavas, when extruded on the surface and cooled instantaneously, will form a natural glass. Glass is a mass of disordered atoms, which are frozen in place due to sudden cooling, and is not a crystalline material like a mineral. Glass composes part of many extrusive igneous rocks (e.g., **lava** flows) and pyroclastic igneous rocks. Alternatively, some igneous rocks are formed from volcanic processes, such as violent volcanic eruption. Violent eruptions eject molten, partially molten, and non-molten igneous rock, which then falls in the vicinity of the eruption. The fallen material may solidify into a hard mass, called pyroclastic igneous rock. The texture of igneous rocks (defined as the size of crystals in the rock) is strongly related to cooling rate of the original liquid. Rapid cooling of liquid rock promotes formation of small crystals, usually too small to see with the unaided eye. Rocks with this cooling history are called fine-textured igneous rocks. Slow cooling (which usually occurs deep underground) promotes formation of large crystals. Rocks with this cooling history are referred to as coarse-textured igneous rocks.

The mineral composition of igneous rocks falls roughly into four groups: **silicic**, intermediate, **mafic**, and ultramafic. These groups are distinguished by the amount of silica (SiO_4), **iron** (Fe), and magnesium (Mg) in the constituent minerals. Mineral composition of liquid rock is related to place of origin within the body of the earth. Generally speaking, liquids from greater depths within the earth contain more Fe and Mg and less SiO_4 than those from shallow depths.

In sedimentary rocks, the type of sediment that is compacted and cemented together determines the rock's main characteristics. Sedimentary rocks composed of sediment that has been broken into pieces (i.e., clastic sediment), such as gravel, **sand**, silt, and **clay**, are clastic sedimentary rocks (e.g., conglomerate, **sandstone**, siltstone, and shale, respectively). Sedimentary rocks composed of sediment that is chemically derived (i.e., chemical sediment), such as dissolved elements like calcium (Ca), sodium (Na), iron (Fe), and **silicon** (Si), are chemical sedimentary rocks. Examples of chemical sedimentary rocks are **limestone** (composed of calcium carbonate), rock salt (composed of sodium chloride), rock **gypsum** (composed of calcium sulfate), ironstones (composed of iron oxides), and **chert** (composed of hydrated silica). Biochemical sedimentary rocks are a special kind of chemical sedimentary rock wherein the constituent particles were formed by organisms (typically as organic hard parts, such as shells), which then became sedimentary particles. Examples of this special kind of sedimentary rock include chalk, fossiliferous limestone, and coquina. Sedimentary rocks are formed from sediment in two stages: compaction and cementation. Compaction occurs when sediments pile up to sufficient thickness that overlying mass squeezes out **water** and closes much open **space**. Cementation occurs when water flowing through the

Pebbles on lake shore. *© Paul A. Souders/Corbis. Reproduced by permission.*

compacted sediment deposits mineral crystals upon particles thus binding them together. The main cement minerals are calcite ($CaCO_3$), hematite (Fe_2O_3), and **quartz** (SiO_2).

With metamorphic rocks, the nature of the pre-existing rock (protolith) determines in large part the characteristics of the ultimate **metamorphic rock**. Regardless of protolith, however, almost all metamorphic rocks are harder and more dense than their protoliths. A protolith with flat or elongate mineral crystals (e.g., micas or amphiboles) will yield a metamorphic rock with preferentially aligned minerals (due to directed pressure). Such metamorphic rocks are called foliated metamorphic rocks (e.g., **slate** and **schist**). Non-foliated metamorphic rocks (e.g., **marble** and quartzite) come from protoliths that have mainly equidimensional mineral crystals (e.g., calcite and quartz, respectively). For example, a protolith shale will yield a foliated metamorphic rock, and a protolith limestone will yield marble, a non-foliated metamorphic rock. Metamorphic rocks possess distinctive grades or levels of metamorphic change from minimal to a maximum near total **melting**. Low-grade metamorphic rocks generally have fine-textured crystals and low-temperature indicator minerals like the mica chlorite. High-grade metamorphic rocks generally have coarse-textured crystals and very distinctive foliation, plus high-temperature indicator minerals like the silicate mineral staurolite.

Rock is a brittle natural solid found mainly in the outer reaches of Earth's **crust** and upper mantle. Material that would be brittle rock at such shallow depths becomes to one degree or another rather plastic within the body of the earth. The term "rock" is not generally applied to such non-brittle internal Earth materials. Therefore, rock is a concept related to the outer shell of the earth. The term rock may also be properly applied to brittle natural solids found on the surfaces of other planets and satellites in our **solar system**. Meteorites are rock. Naturally occurring **ice** (e.g., brittle water ice in a glacier, H_2O) is also a rock, although we do not normally think of ice this way.

Rock has been an important natural resource for people from early in human **evolution**. Rocks' properties are the key to their specific usefulness, now as in the past. Hard, dense rocks that could be chipped into implements and weapons were among the first useful possessions of people. Fine-textured and glassy rocks were particularly handy for these applications. Later on, rock as building stone and pavement material became very important, and this continues today in our modern world. All of Earth's natural mineral wealth, fossil energy resources, and most **groundwater** are contained within rocks of the earth's crust.

See also Lithification; Metamorphism

ROCK AGE • *see* DATING METHODS

ROCKETS • *see* SPACECRAFT, MANNED

ROCKFALL

Rockfall is a form of **mass movement** or **mass wasting** in which pieces of **rock** travel downward through some combination of falling, bouncing, and rolling after they are initially separated from the slope. The sizes of rockfall blocks can range from cubic centimeters to tens of thousand of cubic meters. Although some sliding may occur as the rock is becoming detached, sliding is a minor component of the process. Free fall typically occurs on slopes steeper than 76 degrees, bouncing on slopes between 45 and 76 degrees, and rolling on slopes below 45 degrees. Because slopes are commonly irregular, a rock may alternate between the three modes during its downslope movement. Talus slopes along the bases of cliffs are the products of uncounted rockfalls over thousands of years.

The size of rockfall blocks is controlled by **bedding** planes, joints, and fractures that form mechanical discontinuities and allow the blocks to become detached from the slope. Fracture lengths, and therefore rockfall volumes, tend to follow power law or fractal distributions, meaning that their numbers decrease exponentially as fracture length or rockfall volume increases. Field studies have also shown that **freezing and melting** of **ice** within cracks may control rockfall timing in some areas, although the falls generally seem to be more closely related to **melting** rather than ice wedging.

Extremely large rockfalls can constitute **catastrophic mass movements** because their weight and vertical fall distance (which combine to define the potential energy of the rock before it falls) produce potentially destructive kinetic energy. A large rockfall in Yosemite National Park during the summer of 1996, for example, involved two large rock slabs with a total volume of between 30,000 and 49,700 yd^3 (23,000 and 38,000 m^3) that became detached from a cliff face and became airborne 2,181 ft (665 m) above the valley floor. The estimated velocity of the slabs at impact was approximately 246–268 mph (396–431 kph) and the event was recorded on seismographs as far as 124 mi (200 km) away. The impact produced an air blast with estimated velocities as high as 246 mph, which snapped and uprooted trees, generated a dense cloud of dust that plunged the **area** into darkness, and killed a hiker. The catastrophic rock avalanches, such as those that destroyed the towns of Elm, Austria and Frank, Alberta probably also involved significant amounts of rockfall that evolved into rock avalanches as they moved downslope.

See also Landslide; Talus pile or Talus slope

RODINIA • *see* SUPERCONTINENTS

ROTATION • *see* REVOLUTION AND ROTATION

ROWLAND, F. SHERWOOD (1927-)

American atmospheric chemist

In 1974, F. Sherwood Rowland and his research associate, **Mario Molina**, first sounded the alarm about the harmful effects of chlorofluorocarbons, or CFCs, on the earth's ozone layer. CFCs, which have been used in air conditioners, refrigerators, and aerosol sprays, release chlorine atoms into the upper atmosphere when the sun's ultraviolet light hits them. The chlorine then breaks down atmospheric **ozone** molecules, destroying a shield that protects the earth from damaging **ultraviolet rays**. In the mid–1980s, a National Aeronautics and Space Administration (NASA) **satellite** actually confirmed the existence of a continent-sized hole in the ozone layer over **Antarctica**. By the early 1990s, National Oceanographic and Atmospheric Administration (NOAA) scientists warned that yet another ozone hole, this one over the Arctic, could imperil Canada, Russia, **Europe**, and, in the United States, New England. This news might have been gratifying affirmation for Rowland, a professor of **chemistry** at the University of California at Irvine, but rather than rest on his laurels he continued to steadfastly and soberly warn the world of the ozone danger. His efforts have won him worldwide renown and prestigious awards, including the Charles A. Dana Award for Pioneering Achievement in Health in 1987, the Peter Debye Award of the American Chemical Society in 1993, the Roger Revelle Medal from the American Geophysical Union for 1994, and the Japan Prize in Environmental Science and Technology, presented to Rowland by the Japanese emperor in 1989.

Frank Sherwood Rowland was born in Delaware, Ohio, the son of a math professor, Sidney A. Rowland, and his wife, Latin teacher Margaret Drake Rowland. In an interview with Joan Oleck, Rowland claimed that math always came easy for him. "I always liked solving puzzles and problems," he said. "I think the rule we had in our family that applied even to my own children was you had your choice in school as to what order you took biology, chemistry, and **physics**, but not whether."

Sidetracked by World War II, Rowland was still in boot camp when peace arrived. In 1948, he received his B.A. degree from Ohio Wesleyan University; after three years—the summers of which he spent playing semiprofessional baseball—he obtained his master's from the University of Chicago. His Ph.D. came a year later, in 1952. That same year he married Joan E. Lundberg; the couple would eventually have a son and daughter.

Also in 1952, Rowland obtained his first academic job, an instructorship in chemistry at Princeton University, where he would remain for four years. In 1956, Rowland moved his family west to the University of Kansas, where he was a professor for eight years, and then farther west still, to Irvine, California, where he took over as chemistry department chairman at the University of California in Irvine in 1964. He has stayed at Irvine ever since, enjoying stints as Daniel G. Aldrich, Jr., Professor of Chemistry from 1985 to 1989, and as Donald Bren Professor of Chemistry since 1989.

At Chicago, Rowland's mentor had been **Willard F. Libby**, winner of the Nobel Prize for his invention of carbon–14 dating, a way to determine the age of an object by measuring how much of a radioactive form of **carbon** it contains. The **radioactivity** research Rowland conducted with Libby led the young scientist eventually to **atmospheric chemistry**. Realizing, as he told Oleck, that "if you're going to be a professional scientist, one of the things you're going to do is stay out ahead of the pack," Rowland looked for new avenues to explore. In the 1970s, Rowland was inspired by his daughter's dedication to the then-fledgling environmental movement and by the tenor of the times: 1970 was the year of the first Earth Day. In 1971, the chemist helped allay fears about high levels of mercury in swordfish and tuna by showing that preserved museum fish from a hundred years earlier contained about the same amount of the element as modern fish.

Later events pushed him further in the direction of environmental concerns. At a meeting in Salzburg, Austria, Rowland met an Atomic Energy Commission (AEC) staffer who was trying to get chemists and meteorologists into closer partnerships. Sharing a train compartment with the AEC man to Vienna, Rowland was invited to another professional meeting. And it was there, in 1972, that he first began to think about chlorofluorocarbons in the atmosphere.

In those days, production of CFCs for household and industrial propellants was doubling every five years. A million tons of CFCs were being produced each year alone, but scientists were not particularly alarmed; it was believed they were inert in the atmosphere. Rowland, however, wanted to know more about their ultimate fate. Ozone, a form of **oxygen**, helps make up the **stratosphere**, the atmospheric layer located between eight and 30 miles above the earth. Ozone screens out dangerous ultraviolet rays, which have been linked to skin cancer, malfunctions in the immune system, and eye problems such as cataracts. Performing lab experiments with Molina, Rowland reported in 1974 that the same chemical stability that makes CFCs useful also allows them to drift up to the stratosphere intact. Once they rise thorough the ozone shield, Rowland and Molina warned, they pose a significant threat to ozone: each chlorine **atom** released when CFCs meet ultraviolet light can destroy up to one hundred thousand ozone molecules.

Sounding the alarm in the journal *Nature* in June of 1974 and in a subsequent presentation to the American Chemical Society that September, Rowland attracted attention: a federal task force found reason for serious concern; the National Academy of Sciences (NAS) confirmed the scientists' findings; and by 1978, the Environmental Protection Agency (EPA) had banned nearly all uses of CFC propellants. There were setbacks: In the 1980s President Ronald Reagan's EPA administrator, Anne Gorsuch, dismissed ozone depletion as a scare tactic. And Rowland himself discovered that the whole matter was more complex than originally thought, that another chemical reaction in the air was affecting calculations of ozone loss. The NAS's assessment of the problem was similarly vague, generalizing future global ozone losses as somewhere between 2%–20%.

Then came a startling revelation. In the mid–1980s a hole in the ozone shield over Antarctica the size of a continent was discovered; NASA satellite photos confirmed it in 1985. The fall in ozone levels in the **area** was drastic, as high as 67%. These events led to increased concern by the international community. In 1987 the United States and other CFC producers signed the Montreal Protocol, pledging to cut production by 50% by the end of the millennium. Later, in the United States, President George Bush announced a U.S. plan to speed up the timetable to 1995.

There were more accelerations to come: Du Pont, a major producer, announced plans to end its CFC production by late 1994, and the European Community set a 1996 deadline. And producers of automobile air-conditioning and seat cushions—two industries still using CFCs—began looking for alternatives. These goals only became more urgent in the face of the 1992 discovery of another potential ozone hole, this one over the Arctic. Scientists have attributed the extreme depletion of ozone over the poles to **weather** patterns and seasonal **sun** that promote an unusually rapid cycle of chlorine-ozone chain reactions.

In addition to the development of holes in the ozone layer, the atmosphere is further threatened because of the time delay before CFCs reach the stratosphere. Even after a complete ban on CFC production is achieved, CFCs would continue to rise through the atmosphere, and were predicted to reach peak concentrations in the late 1990s. Some remained skeptical of the dangers, however. In the early 1990s a kind of "ozone backlash" occurred, with a scientist as prominent as Nobel Prize-winning chemist Derek Barton joining those who called for a repeal of the CFC phase-out pact.

Meanwhile Rowland continued his examination of the atmosphere. Every three months, his assistants have fanned out around the Pacific Ocean to collect air samples from New Zealand to Alaska. The news from his research has been sobering, turning up airborne compounds that originated from the burning of rain **forests** in Brazil and the aerial pollution of oil fields in the Caucasus mountains. "The major atmospheric problems readily cross all national boundaries and therefore can affect everyone's security," Rowland said in his President's Lecture before the American Association for the Advancement of Science (AAAS) in 1993. "You can no longer depend upon the 12-mi (19.3-km) offshore limit when the problem is being carried by the winds." An instructive reminder of the international nature of such insecurity was given by the arrival only two weeks later in Irvine, California, of trace amounts of the radioactive fission products released by the 1986 Chernobyl nuclear reactor accident in the former Soviet Union.

Rowland has said in interviews that he's pleased with the progress he's helped set in motion. "One of the messages is that it is possible for mankind to influence his environment negatively," Rowland told Oleck. "On the other side there's the recognition on an international basis that we can act in unison. We have the [Montreal] agreement, people are following it, and it's not only that they have said they will do these things but they *are* doing them because the measurements in the atmosphere show it. People have worked together to solve the problem."

See also Atmospheric circulation; Atmospheric composition and structure; Atmospheric pollution; Ozone layer and hole dynamics

RUNCORN, S.K. (1922-1995)

English geophysicist

Geophysicist S.K. Runcorn made significant contributions to the understanding of several areas within his field, including Earth's **magnetism** and the theory of continental drift. During the 1950s, he helped establish the discipline of paleomagnetism—the study of the intensity and direction of residual magnetization found in ancient rocks. Later, his research encompassed lunar **magnetism**. Runcorn published prolifically, with the publication of more than 180 papers and editing over two dozen books.

Stanley Keith Runcorn was born on November 19, 1922 in Southport, England. He was the eldest of two children born to William Henry Runcorn, a businessman, and Lily Idina Roberts Runcorn. As Runcorn related to contributor Linda Wasmer Smith in a letter, "My interest in science as a child was certainly stimulated... by excellent maths and **physics** teaching in my grammar school." In 1941, Runcorn began studies at Gonville and Caius College of Cambridge University. He passed the tripos, or final honors examination, in mechanical sciences two years later. Runcorn earned a B.A. degree from Cambridge in 1944 and an M.A. in 1948, before transferring to Manchester University to obtain a Ph.D. in 1949. Later, he returned to Cambridge, where he received an Sc.D. degree in 1963.

Runcorn's early years at college coincided with World War II. From 1943 through 1946, he worked on radar research for the ministry of supply at Malvern. For three years afterward, he was a lecturer in physics at Manchester University. His department head there was **Patrick Maynard Stuart Blackett**, who won the 1948 Nobel Prize in physics. Under Blackett's leadership, Runcorn first began a long line of investigations into geomagnetism, which extended well past his move back to Cambridge as assistant director of geophysics research in 1950.

At the time, the idea was rapidly gaining currency in England that many rocks contain within them a fossilized record of the magnetic conditions under which they were formed. This is the basic assumption behind paleomagnetic research. Runcorn compared the results of tests done on rocks from Great Britain and the United States. His analysis seemed to support the hypothesis that over hundreds of millions of years the earth's magnetic poles had undergone large-scale movement, or polar wandering. However, the polar migration routes were different depending on whether the tested rocks came from **Europe** or **North America**. This suggested that the continents themselves had actually moved. Thus, Runcorn became a proponent of the theory called continental drift. Although this idea had first been put forth in 1912, it had not up to that point won widespread acceptance. It was not until the mid-1950s that Runcorn and his colleagues published convincing evidence for its existence.

Advocates of continental drift argued that the direction of magnetization within rocks from different continents would align if only the land masses were oriented differently. However, this suggestion was not immediately embraced by most scientists, partly because a physical mechanism to explain continental drift had yet to be found. By the early 1960s, though, Runcorn had proposed that, under very high **temperature** and pressure, rocks beneath Earth's cold, outer shell—the lithosphere—might gradually "creep," or flow. The resulting upward transfer of heat by convection currents could be the force that moved continents. This idea contributed to the modern theory of **plate tectonics**, which posits that the earth's shell is divided into a number of rigid plates floating on a viscous underlayer.

In 1956, Runcorn accepted a post as professor and head of the physics department at King's College, part of the University of Durham. Seven years later, King's College became the University of Newcastle upon Tyne, and Runcorn was appointed head of the school of physics there, a position he held until 1988. During this period, Runcorn was also a visiting scientist or professor at several institutions around the world, including the University of California, Los Angeles and Berkeley; Dominion Observatory, Ottawa; the California Institute of Technology; the University of Miami; the Lunar Science Institute, Houston; the University of Texas, Galveston; and the University of Queensland, **Australia**.

By the late 1960s, Runcorn's attention had turned toward the **Moon**. At the time, the Moon was generally presumed to be a dead body. As early as 1962, though, Runcorn had suggested that the Moon, too, might be subject to the forces of convection—an idea that was initially rejected by most scientists. However, examination of lunar samples brought back by the Apollo missions showed that some of them were magnetized, which implied that they had been exposed to magnetic fields while they were forming. Runcorn and his colleagues concluded that the Moon had probably once possessed its own strong **magnetic field**, generated within an **iron** core.

This magnetic field seemed to have pointed in different directions at different times in lunar history. When Runcorn and his co-workers calculated the strengths and directions of this ancient magnetism, they found evidence of polar wandering. Runcorn subsequently proposed that the wandering could have been caused by the same impacts that created large basins near the Moon's equator. According to this hypothesis, the force of the impacts could have shifted the Moon's entire surface, so that regions once near the poles might have been relocated closer to the equator. However, attempts to confirm this notion have so far proved inconclusive.

Runcorn's remarkable skill as a theorist was widely recognized. In 1989, he assumed an endowed chair in the natural sciences at the University of Alaska in Fairbanks, a position he held until his death. He also received honorary degrees from universities in Utrecht, Netherlands; Ghent, Belgium; Paris; and Bergen, Norway. Among the many prestigious awards he received are the Napier Shaw Prize of Britain's Royal Meteorological Society in 1959, the John Adams Fleming Medal of the American Geophysical Union in 1983, the Gold Medal of Britain's Royal Astronomical Society in 1984, and the Wegener Medal of the European Union of Geosciences in 1987.

In addition, Runcorn was elected a fellow or member of such respected associations as the British Royal Society, the American Physical Society, the European Geophysical Society, the Royal Netherlands Academy of Science, the Indian National Science Academy, the Royal Norwegian Academy of Science and Letters, the Pontifical Academy of Science, and Academia Europaea.

Runcorn, who never married, was an aficionado of sports and the arts. Among his favorite pastimes was rugby, which he enjoyed as a participant until he was past fifty and as a spectator thereafter. In a letter to Wasmer Smith, Runcorn described himself also as an avid fan of "squash rackets and swimming,... visiting art galleries, seeing opera and ballet, reading history and politics, hiking in the country, and seeing architecture in my travels." Runcorn died at the age of 73 during a lecture tour stop in San Diego, California, where he was robbed and murdered in his hotel room.

See also Continental drift theory; Magnetic field; Paleomagnetics

RUNOFF

Runoff is the component of the **hydrologic cycle** through which **water** is returned to the ocean by overland flow. The term runoff is considered synonymous with streamflow and comprises surface runoff resulting from **precipitation** and that portion of the streamflow that is contributed by **groundwater** flow entering the stream channel.

Surface runoff consists of that portion of the precipitation reaching the surface that neither infiltrates into the ground nor is retained on the surface. The quantity of surface runoff is controlled by a complex variety of factors. Included among these are precipitation intensity and duration, **permeability** of the ground surface, vegetation type and density, **area** of drainage basin, distribution of precipitation, stream-channel geometry, depth to **water table**, and topographic slope.

In the early stages of a storm, much of the precipitation may be intercepted by vegetation or captured in surface depressions. Water held in this manner often presents a large surface area and is likely to be evaporated. Any water reaching the surface at this stage is more likely to infiltrate before the upper layer of the ground becomes saturated. Thus, storms of light intensity or short duration may produce little or no surface runoff. As storm intensity or duration increases, interception becomes less effective, infiltration capacity of the **soil** decreases, and surface depressions fill. The result is increasing surface runoff leading to greater flow rates within local stream channels.

Variations in permeability within the soil may cause a portion of the water that infiltrates into the soil to migrate laterally as interflow. Some of the remaining infiltrate will percolate downward to the water table and flow with the groundwater. Ultimately, both of these sources may intercept a stream channel and contribute to the total runoff.

During a particular storm event, the contribution of runoff to a stream varies significantly through time. Inflow to the stream begins with direct channel precipitation followed by overland surface runoff when the appropriate conditions exist. Lateral interflow and groundwater contributions typically move more slowly and impact the stream level later. The groundwater portion of the runoff frequently supports the flow of a stream both during and between storm events.

See also Evaporation

S

SAGAN, CARL (1934-1996)

American astronomer

One of the first scientists to take an active interest in the possibility that life exists elsewhere in the Universe, and an astronomer that was both a best-selling author and a popular television figure, Carl Sagan was one of the best-known scientists in the world. Sagan made important contributions to studies of Venus and Mars, and he was extensively involved in planning NASA's *Mariner* missions. Regular appearances on the *Tonight Show* with Johnny Carson began a television career that culminated in the series Sagan hosted on public television called *Cosmos*, seen in sixty countries by over 400 million people. He was also one of the authors of a paper that predicted drastic global cooling after a nuclear war; the concept of "nuclear winter" affected not only the scientific community, but also national and international policy, as well as public opinion about nuclear weapons and the arms race. Although some scientists considered Sagan too speculative and insufficiently committed to detailed scientific inquiry, many recognized his talent for explaining science, and acknowledged the importance of the publicity generated by Sagan's enthusiasm.

Sagan was born in Brooklyn, New York, the son of Samuel Sagan, a Russian emigrant and a cutter in a clothing factory, and Rachel Gruber Sagan. He became fascinated with the stars as a young child, and was an avid reader of science fiction, particularly the novels by Edgar Rice Burroughs about the exploration of Mars. By the age of five, Sagan was sure he wanted to be an astronomer, but, as he told Henry S.F. Cooper, Jr., of the *New Yorker*, he sadly assumed it was not a paying job; he expected he would have to work at "some job I was temperamentally unsuited for, like door-to-door salesman." When he found out a few years later that astronomers actually got paid, he was ecstatic. "That was a splendid day," he told Cooper.

Sagan's degrees, all of which he earned at the University of Chicago, include an A.B. in 1954, a B.S. in 1955, an M.S. in **physics** in 1956, and a doctorate in **astronomy** and astrophysics in 1960. As a graduate student, Sagan was deeply interested in the possibility of life on other planets, a discipline known as exobiology. Although this interest then was considered beyond the realm of responsible scientific investigation, he received important early support from scientists such as Nobel laureates Hermann Joseph Muller and Joshua Lederberg. He also worked with Harold C. Urey, who had won the 1934 Nobel Prize in chemistry and had been Stanley Lloyd Miller's thesis adviser when he conducted his famous experiment on the **origin of life**. Sagan wrote his doctoral dissertation, "Physical Studies of the Planets," under Gerard Peter Kuiper, one of the few astronomers at that time who was also a planetologist. It was during his graduate student days that Sagan met Lynn Margulis, a biologist, who became his wife in 1957. The couple had two sons before divorcing in 1963.

From graduate school, Sagan moved to the University of California at Berkeley, where he was the Miller residential fellow in astronomy from 1960 to 1962. He then accepted a position at Harvard as an assistant professor from 1962 to 1968. On April 6, Sagan married the painter Linda Salzman; Sagan's second marriage, which ended in a divorce, produced a son. From Harvard, Sagan went to Cornell University, where he was first an associate professor of astronomy at the Center for Radiophysics and **Space** Research. He was then promoted to professor and associate director at the center, serving in that capacity until 1977, when he became the David Duncan Professor of Astronomy and Space Science.

Sagan's first important contributions to the understanding of Mars and Venus began as insights while he was still a graduate student. Color variations had long been observed on the planet Mars, and some believed these variations indicated the seasonal changes of some form of Martian plant life. Sagan, working at times with James Pollack, postulated that the changing colors were instead caused by Martian dust, shifting through the action of **wind** storms; this interpretation was confirmed by *Mariner 9* in the early 1970s. Sagan also suggested that the surface of Venus was incredibly hot, since

Carl Sagan. ***Library of Congress.***

the Venusian atmosphere of **carbon dioxide** and **water** vapor held in the sun's heat, thus creating a version of the "greenhouse effect." This theory was also confirmed by an exploring spacecraft, the Soviet probe *Venera IV*, which transmitted data about the atmosphere of Venus back to Earth in 1967. Sagan also performed experiments based on the work of Stanley Lloyd Miller, studying the production of organic molecules in an artificial atmosphere meant to simulate that of a primitive Earth or contemporary Jupiter. This work eventually earned him a patent for a technique that used gaseous mixtures to produce amino acids.

Sagan first became involved with spaceflight in 1959, when Lederberg suggested he join a committee on the Space Science Board of the National Academy of Sciences. He became increasingly involved with NASA (National Aeronautics and Space Administration) during the 1960s, and participated in many of their most important robotic missions. He developed experiments for the Mariner Venus mission and worked as a designer on the *Mariner 9* and the *Viking* missions to Mars, as well as on the *Pioneer 10*, the *Pioneer 11*, and the *Voyager* spacecrafts. Both the *Pioneer* and the *Voyager* spacecrafts have left our **solar system** carrying plaques which Sagan designed with Frank Drake as messages to any extraterrestrials that find them. The plaques contain pictures of two humans, a man and a woman, as well as various astronomical information. The nude man and woman were drawn by Sagan's second wife, Linda Salzman, and they provoked many letters to Sagan denouncing him for sending "smut" into space. During this project Sagan met the writer Ann Druyan, the project's creative director, who eventually became his wife. Sagan and Druyan had two children.

Sagan continued his involvement in space exploration in the 1980s and 1990s. The expertise he developed in biology and genetics while working with Muller, Lederberg, Urey, and others, is unusual for an astronomer, and he extensively researched the possibility that Saturn's **moon**, Titan, which has an atmosphere, might also have some form of life. Sagan was also involved in less direct searches for life beyond Earth. He was one of the prime movers behind NASA's establishment of a radio astronomy search program that Sagan called CETI, for Communication with Extra-Terrestrial Intelligence.

A colleague of Sagan's working on the Viking mission explained to Cooper of the *New Yorker* that this desire to find extraterrestrial life is the focus of all of Sagan's various scientific works. "Sagan desperately wants to find life someplace, anyplace—on Mars, on Titan, in the solar system or outside it. I don't know why, but if you read his papers or listen to his speeches, even though they are on a wide variety of seemingly unrelated topics, there is always the question 'Is this or that phenomenon related to life?' People say, 'What a varied career he has had,' but everything he has done has had this one underlying purpose." When Cooper asked Sagan why this was so, the scientist had a ready answer. "I think it's because human beings love to be alive, and we have an emotional resonance with something else alive, rather than with a molybdenum atom."

During the early 1970s Sagan began to make a number of brief appearances on television talk shows and news programs; Johnny Carson invited him on the *Tonight Show* for the first time in 1972, and Sagan soon was almost a regular there, returning to discuss science two or three times a year. However, it was *Cosmos*, which Public Television began broadcasting in 1980, that made him into a media sensation. Sagan narrated the series, which he wrote with Ann Druyan and Steven Soter, and they used special effects to illustrate a wide range of astronomical phenomena such as black holes. In addition to being extremely popular, the series was widely praised both for its showmanship and its content, although some reviewers had reservations about Sagan's speculations as well as his tendency to claim as fact what most scientists considered only hypotheses.

Sagan was actively involved in politics; as a graduate student, he was arrested in Wisconsin for soliciting funds for the Democratic Party, and he was also involved in protests against the Vietnam War. In December, 1983, he published, with Richard Turco, Brian Toon, Thomas Ackerman, and James Pollack, an article discussing the possible consequences of nuclear war. They proposed that even a limited number of nuclear explosions could drastically change the world's **climate** by starting thousands of intense fires that would throw hundreds of thousands of tons of smoke and ash into the atmosphere, lowering the average **temperature** ten to twenty degrees and bringing on what they called a "nuclear winter."

The authors happened upon this insight accidentally a few years earlier, while they were observing how **dust storms** on the planet Mars cooled the Martian surface and heated up the atmosphere. Their warning provoked a storm of controversy at first; their article was then followed by a number of studies on the effects of war and other human interventions on the world's climate. Sagan and his colleagues stressed that their predictions were only preliminary and based on certain assumptions about nuclear weapons and large-scale fires, and that their computations had been done on complex computer models of the imperfectly understood atmospheric system. However, despite numerous attempts to minimize the concept of a **nuclear winter**, the possibility that even a limited nuclear war might well lead to catastrophic environmental changes was supported by later research.

The idea of nuclear winter not only led to the reconsideration of the implications of nuclear war by many countries, institutions, and individuals, but it also produced great advances in research on Earth's atmosphere. In 1991, when the oil fields in Kuwait were burning after the Persian Gulf War, Sagan and others made a similar prediction about the effect the smoke from these fires would have on the climate. Based on the nuclear winter hypothesis and the recorded effects of certain **volcanic eruptions**, these predictions turned out to be inaccurate, although the smoke from the oil fires represented about 1% of the volume of smoke that would be created by a full-scale nuclear war.

In 1994, Sagan was diagnosed with myelodysplasia, a serious bone-marrow disease. Despite his illness, Sagan kept working on his numerous projects. His last book, *The Demon-Haunted World: Science as a Candle in the Dark*, was published in 1995. At the time of his death, Sagan was co-producing a film version of his novel *Contact*. His partner in this project was his wife, Ann Druyan, who had co-authored *Comet*. Released in 1997, the film received popular and critical acclaim as a testimony to Sagan's enthusiasm for the search for extraterrestrial life. Sagan, who lived in Ithaca, New York, died at the Fred Hutchinson Cancer Research Center in Seattle in 1996.

Carl Sagan won a Pulitzer Prize in 1978 for his book on **evolution** called *The Dragons of Eden*. He also won the A. Calvert Smith Prize (1964), NASA's Apollo Achievement Award (1969), NASA's Exceptional Scientific Achievement Medal (1972), NASA's Medal for Distinguished Public Service (twice), the International Astronaut Prize (1973), the John W. Campbell Memorial Award (1974), the Joseph Priestly Award (1975), the Newcomb Cleveland Prize (1977), the Rittenhouse Medal (1980), the Ralph Coats Roe Medal from the American Society of Mechanical Engineers (1981), the Tsiolkovsky Medal of the Soviet Cosmonautics Federation (1987), the Kennan Peace Award from SANE/Freeze (1988), the Oersted Medal of the American Association of Physics Teachers (1990), the UCLA Medal (1991), and the Mazursky Award from the American Astronomical Association (1991). Sagan was a fellow of the American Association for the Advancement of Science, the American Academy of Arts and Sciences, the American Institute for Aeronautics and Astronautics, and the American Geophysical Union. Sagan was also the chairman of the Division for Planetary Sciences of the American Astronomical Society (from 1975 to 1976) and for twelve years was editor-in-chief of *Icarus*, a journal of planetary studies.

Salt wedging

Salt wedging in an **estuary** is the process by which a distinct layer of **saltwater** forms below a layer of **freshwater** due to differences in density. Salt wedging is the result of weak tidal currents that cannot mix the saltwater with the freshwater, thus creating a halocline. Because freshwater is less dense than saltwater, it will float on top of the saltwater. A halocline is a zone in the **water** column where an abrupt alteration in the salinity creates a sharp freshwater-saltwater interface. Salt wedging typically occurs in an estuary along a salinity gradient when a fresh body of water such as a river meets, but does not mix with saltwater from an ocean or sea.

The rate of freshwater **runoff** from a river into an estuary is a major determinant of salt wedge formation. Salt wedging occurs when there is continuous flow of freshwater running into an estuary that opens into an ocean or sea with small tidal currents. Additionally, **evaporation** must be significantly lower than the freshwater runoff in order for salt wedging to occur.

Conversely, if there is no runoff of freshwater into the estuary, or if the runoff of freshwater is less than its evaporation, the salt water flowing in from the ocean will become more diluted within the estuary. Because the rate of evaporation is higher than the freshwater runoff, the top layer of water where evaporation occurs, will have a higher salinity than the other layers of the estuary.

Typically, the weak bottom current of a salt wedge estuary flows toward land while the dominant, turbulent current on the surface of the estuary flows toward sea. The horizontal layer where the two opposing currents meet creates internal waves that grow and eventually break as they move out toward sea, causing water to flow upward. The breaks in the internal waves shift diminutive volumes of saltwater from the bottom layer into the surface layer, causing the bottom layer to invariably flow towards land.

See also Hydrogeology; Tides

Saltation

Saltation is the transportation of **sand** grains in small jumps by **wind** or flowing **water**. The term does not refer to salt, but is derived from the Latin *saltare*, to dance.

Certain conditions are necessary for saltation. First, a bed of sand grains must be covered by flowing air or water, as in a streambed or windy **desert**. Second, this flow must be turbulent. In turbulent flow, a fluid swirls and mixes chaotically—and virtually all natural flows of water and air are turbulent. Third, some of the eddies in the turbulence must be strong enough to lift individual sand grains from the bottom.

Fourth, the turbulence must not be so strong that grains cannot settle out again once suspended.

An individual saltating grain spends most of its time lying at rest on the bottom. Eventually an eddy happens to apply enough suction to the upper surface of the grain to overcome its weight, lifting it into the current. The grain is carried for a short distance, and then allowed to settle to the bottom again by the ever-shifting turbulence. After a random waiting period, the grain is lifted, carried, and dropped again, always farther downstream.

Grains too small to settle once suspended are carried indefinitely by the current; intermediate-size grains saltate; and grains too large to saltate either remain unmoved or move by sliding or rolling. Turbulent flow thus tends to sort grains by size.

See also Bed or traction load; Bedforms (ripples and dunes)

SALTWATER

Saltwater, or salt **water**, is a geological term that refers to naturally occurring solutions containing large concentrations of dissolved, inorganic ions. In addition, this term is often used as an adjective in biology, usually to refer to marine organisms, as in saltwater fish.

Saltwater most commonly refers to oceanic waters, in which the total concentration of ionic solutes is typically about 35 grams per liter (also expressed as 3.5%, or 35 parts per thousand). As a result of these large concentrations of dissolved ions, the density of saltwater (1.028 g/L at 4° C) is slightly greater than that of **freshwater** (1.00 g/L). Therefore, freshwater floats above saltwater in poorly mixed situations where the two types meet, as in estuaries and some underground reservoirs.

The ions with the largest concentrations in marine waters are sodium, chloride, sulfate, magnesium, calcium, potassium, and carbonate. In oceanic waters, sodium and chloride are the most important ions, having concentrations of 10.8 g/L and 19.4 g/L, respectively. Other important ions are sulfate (2.7 g/L), magnesium (1.3 g/L), and calcium and potassium (both 0.4 g/L). However, in inland saline waters, the concentrations and relative proportions of these and other ions can vary widely.

Other natural waters can also be salty, sometimes containing much larger concentrations of salt than the **oceans**. Some **lakes** and ponds, known as salt or brine surface waters, can have very large concentrations of dissolved, ionic solutes. These water bodies typically occur in a closed basin, with inflows of water but no outflow except by **evaporation**, which leaves salts behind. Consequently, the salt concentration of their contained water increases progressively over time. For example, the Great Salt Lake of Utah and the Dead Sea in Israel have salt concentrations exceeding 20%, as do smaller, saline ponds in Westphalia, Germany, and elsewhere in the world.

Underground waters can also be extremely salty. Underground saltwaters are commonly encountered in **petroleum** and gas well-fields, especially after the hydrocarbon resource has been exhausted by mining.

Both surface and underground saltwaters are sometimes "mined" for their contents of economically useful **minerals**.

Saltwater intrusions can be an important environmental problem, which can degrade water supplies required for drinking or irrigation. Saltwater intrusions are caused in places near the ocean where there are excessive withdrawals of underground supplies of fresh waters. This allows underground saltwaters to migrate inland, and spoil the quality of the **aquifer** for most uses. Saltwater intrusions are usually caused by excessive usage of ground water for irrigation in agriculture, or by excessive demands on freshwaters to supply drinking water to large cities.

See also Chemical bonds and physical properties; Hydrogeology; Water table; Weathering and weathering series

SALTWATER ENCROACHMENT

Saltwater encroachment or intrusion is the movement of saltwater into subsurface aquifers previously occupied by **freshwater**. Frequently, this occurs as freshwater from coastal aquifers is displaced as a result of the shoreward movement of seawater. Encroachment also occurs as the upconing of saltwater beneath pumping wells in areas where **groundwater** aquifers are underlain by more saline layers. In fact, the latter of these forms is quite common because two-thirds of the freshwater aquifers used for **water** supply in the United States are underlain by highly saline aquifers.

Because salt water contains dissolved **minerals**, its density is greater than that of freshwater. The lower density of fresh groundwater allows it to override or float on saltwater. As **precipitation** falls onto and infiltrates the land surface, freshwater accumulates in the ground above the saltwater in the shape of a lens. As more water is added through precipitation, the height of the **water table** above sea level is increased and the thickness of the freshwater lens also increases. Under certain conditions, the depth below sea level of the freshwater lens can extend to approximately 40 times the height of the water table above sea level. Along the lower curved boundary of the lens, some mixing of fresh and saltwater occurs due to movement of the interface by **tides** and other natural phenomena.

Saltwater encroachment occurs as a result of the withdrawal of fresh groundwater in the vicinity of saltwater. As freshwater is withdrawn from the ground, the pressure imposed on the saltwater by the overlying freshwater is reduced. This allows the saltwater/freshwater interface to migrate toward the point of withdrawals. The interface moves to a point where the pressure balance is restored. Lateral migration of the interface is most common in coastal locations whereas encroachment in areas overlying saline aquifers typically takes the form of an upward-pointing cone of saltwater directly beneath the pumping well.

Several methods have been examined for the control of saltwater encroachment. These techniques have included reduction of groundwater withdrawals, repositioning of withdrawal locations, utilization of recharge basins or injection wells to artificially maintain freshwater pressure, interception of intruding saltwater through a line of pumping wells parallel to the coastline, and emplacement of a subsurface groundwater barrier between the coastline and pumping wells. Reduction of groundwater withdrawals and relocation of pumping wells are the techniques found to be most effective and economically feasible in the control of saltwater encroachment.

See also Hydrologic cycle

SAN ANDREAS FAULT • *see* FAULTS AND FRACTURES

SAND

Sand is any material composed of loose, stony grains between 1/16 mm and 2 mm in diameter. Larger particles are categorized as gravel, smaller particles are categorized as silt or **clay**. Sands are usually created by the breakdown of rocks, and are transported by **wind** and **water**, before depositing to form soils, beaches, **dunes**, and underwater fans or deltas. Deposits of sand are often cemented together over time to form sandstones.

The most common sand-forming process is **weathering**, especially of **granite**. Granite consists of distinct **crystals** of **quartz**, **feldspar**, and other **minerals**. When exposed to water, some of these minerals (e.g., feldspar) decay chemically faster than others (especially quartz), allowing the granite to crumble into fragments. Sand formed by weathering is termed epiclastic.

Where fragmentation is rapid, granite crumbles before its feldspar has fully decayed and the resulting sand contains more feldspar. If fragmentation is slow, the resulting sand contains less feldspar. Fragmentation of **rock** is enhanced by exposure to fast-running water, so steep mountains are often source areas for feldspar-rich sands and gentler terrains are often source areas for feldspar-poor sands. Epiclastic sands and the sandstones formed from them thus record information about the environments that produce them. A sedimentologist can deduce the existence of whole mountain ranges long ago eroded, and of mountain-building episodes that occurred millions of years ago from sandstones rich in relatively unstable minerals like feldspar.

The behavior of sand carried by flowing water can inscribe even more detailed information about the environment in sand deposits. When water is flowing rapidly over a horizontal surface, any sudden vertical drop in that surface splits the current into two layers, (1) an upper layer that continues to flow downstream and (2) a slower backflow that curls under in the lee of the dropoff. Suspended sand tends to settle out in the backflow zone, building a slope called a "slip face" that tilts downhill from the dropoff. The backflow zone adds continually to the slip face, growing it downstream, and as the slip face grows downstream its top edge continues to create a backflow zone. The result is the deposition of a lengthening bed of sand. Typically, periodic avalanches of large grains down the slip face (or other processes) coat it with thin layers of distinctive material. These closely-spaced laminations are called "cross-bedding" because they angle across the main bed. Cross-bedding in **sandstone** records the direction of the current that deposited the bed, enabling geologists to map currents that flowed millions of years ago (paleocurrents).

Evidence of grain size, bed thickness, and cross-bedding angle, allows geologists to determine how deep and fast a paleocurrent was, and thus how steep the land was over which it flowed.

Ripples and dunes—probably the most familiar forms created by wind- or waterborne sand—involve similar processes. However, ripples and dunes are more typical of flow systems to which little or no sand is being added. The downstream slip faces of ripples and dunes are built from grains plucked from their upstream sides, so these structures can migrate without growing. When water or wind entering the system (e.g., water descending rapidly from a mountainous region) imports large quantities of sand, the result is net deposition rather than the mere migration of sandforms.

Grain shape, too, records history. All epiclastic grains of sand start out angular and become more rounded as they are polished by abrasion during transport by wind or water. Quartz grains, however, resist wear. One trip down a river is not enough to thoroughly round an angular grain of quartz; even a long sojourn on a beach, where grains are repeatedly tumbled by waves, does not suffice. The well-rounded state of many quartz sands can be accounted for only by crustal recycling. Quartz grains can survive many cycles of **erosion**, burial, cementation into sandstone, uplift, and re-erosion. Recycling time is on the order of 200 million years, so a quartz grain first weathered from granite 2.4 billion years ago may have gone through 10 or 12 cycles of burial and re-erosion to reach its present day state. An individual quartz grain's degree of roundness is thus an index of its antiquity. Feldspar grains can also survive recycling, but not as well, so sand that has been recycled a few times consists mostly of quartz.

Sand can be formed not only by weathering but by explosive volcanism, the breaking up of shells by waves, the cementing into pellets of finer-grained materials (pelletization), and the **precipitation** of dissolved chemicals (e.g., calcium carbonate) from solution.

Pure quartz sands are mined to make **glass** and the extremely pure **silicon** employed in microchips and other electronic components.

See also Beach and shoreline dynamics; Bed or traction load; Bedding; Bedforms (ripples and dunes); Desert and desertification; Dune fields; Sedimentary rocks; Sedimentation

Sandstone

Any sedimentary **rock** composed of stony grains between 1/16 mm and 2 mm in diameter that are cemented together is a sandstone.

Sandstone forms from beds of **sand** laid down under the sea or in low-lying areas on the continents. As a bed of sand subsides into the earth's **crust**, usually pressed down by overlying sediments, it is heated and compressed. Hot **water** flows slowly through the spaces between the sand grains, importing dissolved **minerals** such as **quartz**, calcium carbonate, and **iron** oxide. These minerals crystallize around the sand grains and cement them together into a sandstone. Spaces remain between the grains, resulting in a porous, spongelike matrix through which liquids can flow.

Petroleum and **natural gas** are often found in sandstones. They do not form there, but seek to float to the surface by percolating through water-saturated sandstones. Sandstone layers shaped into domes by folding or other processes (and overlaid by non-porous rock) act as traps for migrating oil and gas, that ascend into them but then have no way out. Such traps are much sought after by oil companies; indeed, most sandstone sedimentologists work for the petroleum industry.

Another useful feature of sandstones is that they tend to record the surface conditions that prevailed when their sands were created and deposited. For example, the diagonal laminations often seen running across sandstone beds (cross-bedding) record the direction and speed of the water or **wind** that deposited their original sand. Furthermore, the ratio of **feldspar** to quartz in a sandstone reveals whether its sand was produced by rapid **erosion**, such as occurs in young, steep mountain ranges, or more slowly, such as occurs in flatter terrain. Since sand beds are often deposited rapidly by wind or water, tracks of reptiles—and even the pocks made by individual raindrops—may be preserved as **fossils** in sandstone.

A sandstone may be uplifted to the surface and broken down by **weather** into sand. This sand may be deposited in a bed that subsides, turns to sandstone, returns to the surface, breaks down into sand again, and so on. Some individual grains of sand have participated in more than 10 such cycles, each of which lasts on the order of 200 million years.

See also Bedding; Petroleum detection; Petroleum extraction; Sedimentary rocks; Sedimentation

Santa Anna winds • *see* Seasonal winds

Satellite

In **astronomy**, a satellite refers to any object that is orbiting another larger, more massive object under the influence of their mutual gravitational force.

Thus, any planetary **moon** (e.g., the Moon revolving about Earth) is properly described as a satellite of that planet. Because the word is used to describe a single object, it is not used to designate rings of material orbiting a planet, even though such rings might be described as being made up of millions of satellites. In those rare instances where the mass of the satellite approaches that of the object around which it orbits, the system is sometimes referred to as a binary system. This is the reason that some people refer to Pluto and its moon Charon as a binary planet. This description is even more appropriate for some recently discovered **asteroids** which are composed of two similar sized objects orbiting each other.

In this century, scientific probes and commercial devices have been launched into Earth orbit or into orbits about the **Sun** or another planet. A tradition has developed to refer to these objects as man-made satellites to distinguish them from naturally occurring satellites. Surveillance satellites orbiting Earth have been used to measure everything from aspects of the planet's **weather** to movements of ships. Communications satellites revolve about Earth in geostationary orbits 25,000 mi (40,225 km) above the surface and a recent generation of navigation satellites and global positioning satellites (**GPS**) enables receiving stations on the surface of Earth to be determined with errors measured within a few meters.

Surveillance satellites have been placed in orbit about the Moon, Mars, and Venus to provide detailed maps of their surfaces and measure properties of their surrounding environment. A number of probes have at least temporarily entered the orbits of Jupiter, Saturn, or moons of these Jovian worlds.

Spacecraft missions to other planets in the **solar system** have revealed the existence of numerous previously unknown natural satellites and data from the **Hubble Space telescope** continue to reveal satellite objects that explain discrepancies in orbital paths and **rotation** rates of celestial bodies.

See also History of manned space exploration; Terra satellite and Earth Observing Systems (EOS); Weather forecasting methods

Satellite imagery • *see* Mapping techniques

Saturated zone

The saturated zone encompasses the **area** below ground in which all interconnected openings within the geologic medium are completely filled with **water**. Many hydrogeologists separate this zone into two subzones: the phreatic zone and the capillary fringe.

The phreatic zone is the area in which the interstitial water will freely flow from pores in the geologic material. Water in the pores of the phreatic zone is at a pressure greater than **atmospheric pressure**. Lying above, and separated from the phreatic zone by the **water table**, is the capillary fringe. Capillary action within the voids of the geologic medium causes water to be drawn upward from the top of the phreatic zone or captured as it percolates downward from the overlying **unsaturated zone**. Unlike the phreatic zone, however, the capillary action causes the water in the pores to have a pressure that is lower than atmospheric pressure. While the pores of both subzones are saturated, the different pressures in each cause the water to behave differently. Water within the phreatic zone will readily flow out of the pores while the negative pressures within the capillary fringe tightly hold the water in place. It is water from the phreatic zone that is collected and pumped from wells and flows into streams and **springs**.

Water within the phreatic portion of the saturated zone moves through the interconnected pores of the geologic material in response to the influences of **gravity** and pressure from overlying water. Rates of **groundwater** movement within the saturated zone ranges from a few feet per year to several feet per day depending upon local conditions. Only in larger fractures or karst systems do velocities approach those seen in surface flows.

The saturated zone extends downward from the capillary fringe to the depth where **rock** densities increase to the point that migration of fluids is impossible. In deep sedimentary basins, this may occur at depths of approximately 50,000 feet. At these extreme depths, the voids are no longer interconnected or not present.

Localized saturated zones can occur within the unsaturated zone when heterogeneities within the geologic medium cause differential downward percolation of water. Specifically, layers or lenses of low **permeability**, such as **clay** or shale, can retard the movement of water in the unsaturated zone and cause it to pool above the layer. This forms a perched zone of saturation.

See also Hydrogeology; Karst topography; Porosity and permeability

SATURN • *see* SOLAR SYSTEM

SCHIST

Schist is a **metamorphic rock** consisting of mineral grains that are more or less aligned in layers. Because of this structure, schist tends to cleave into flakes or slabs.

The parent **rock** of a schist may be igneous (e.g., **basalt**, **granite**, syenite) or sedimentary (e.g., **sandstone**, mudstone, impure **limestone**). The metamorphic grade of a schist depends on how thoroughly melted and recrystallized its parent rock has been; higher temperatures produce lower **water** content, coarser crystallization, more distinct layering, and reduced schistosity. At the high end of this scale, the schists blur into the gneisses.

The directional mineral structure of schists and gneisses arises during crystallization under anisotropic stress. Anisotropic stress is stress that does not point equally in all directions, such as would be produced by placing a block of any material on a table and leaning on it at an angle. During the formation of a schist or **gneiss**, the parent rock is heated sufficiently to mobilize its atoms. As it cools under anisotropic stress, its atoms assume the most stable, low-energy arrangements available to them: anisotropic **crystals** (plates or elongated shapes pointing in a common direction). Anisotropic crystal structure gives the schists their characteristic cleavage properties.

There are many varieties of schist. Two categories of particular importance are the greenschists and the blueschists. These have similar parent rocks but are formed under different pressure (P) and **temperature** (T) conditions. What **minerals** will crystallize during metamorphosis depend on both P and T. Greenschists form under high P and high T such as are found far below Earth's **crust**; blueschists form under high P and relatively *low* T. Both greenschists and blueschists are found in regionally metamorphosed landscapes—that is, land masses that have been submerged entirely in Earth's interior and metamorphosed in bulk. Regional metamorphosis often occurs at subduction zones, those places where one tectonic plate is being driven edgewise into the mantle beneath another. A large, cool chunk of Earth's crust takes a long time to reach ambient T when plunged into the mantle, but is raised to high P at once; a subducted mass that stays down long enough to achieve both high T and high P produces greenschist, while one that returns to the surface relatively quickly produces blueschist. One of the great unresolved puzzles of modern **geology** is that blueschist formation seems to have become globally more common in the last 300 million years, while greenschist formation has remained constant throughout Earth's history.

See also Metamorphism; Partial melting; Plate tectonics

SCHRÖDINGER, ERWIN (1887-1961)

Austrian physicist

Erwin Schrödinger shared the 1933 Nobel Prize for physics with English physicist Paul Dirac in recognition of his development of a wave equation describing the behavior of an electron in an **atom**. His theory was a consequence of French theoretical physicist Louis Victor Broglie's hypothesis that particles of matter might have properties that can be described by using wave functions. Schrödinger's wave equation provided a sound theoretical basis for the existence of electron orbitals (energy levels), which had been postulated on empirical grounds by Danish physicist **Niels Bohr** in 1913.

Schrödinger was born in Vienna, Austria. His father, Rudolf Schrödinger, enjoyed a wide range of interests, including painting and botany, and owned a successful oil cloth factory. Schrödinger's mother was the daughter of Alexander Bauer, a professor at the Technische Hochschule. For the first eleven years of his life, Schrödinger was taught at home. Though a tutor came on a regular basis, Schrödinger's most important instructor was his father, whom he described as a "friend, teacher, and tireless partner in conversation," as Armin Hermann quoted in *Dictionary of Scientific Biography*. From his father, Schrödinger also developed a wide range of academic interests, including not only mathematics and science but also grammar and poetry. In 1898, he entered the Akademische Gymnasium in Vienna to complete his pre-college studies.

Erwin Schrödinger. *Library of Congress.*

Having graduated from the Gymnasium in 1906, Schrödinger entered the University of Vienna. By all accounts, the most powerful influence on him there was Friedrich Hasenöhrl, a brilliant young physicist who was killed in World War I a decade later. Schrödinger was an avid student of Hasenöhrl's for the full five years he was enrolled at Vienna. He held his teacher in such high esteem that he was later to remark at the 1933 Nobel Prize ceremonies that, if Hasenöhrl had not been killed in the war, it would have been Hasenöhrl, not Schrödinger, being honored in Stockholm.

Schrödinger was awarded his Ph.D. in physics in 1910, and was immediately offered a position at the University's Second Physics Institute, where he carried out research on a number of problems involving, among other topics, **magnetism** and dielectrics. He held this post until the outbreak of World War I, at which time he became an artillery officer assigned to the Italian front. As the War drew to a close, Schrödinger looked forward to an appointment as professor of theoretical physics at the University of Czernowitz, located in modern-day Ukraine. However, those plans were foiled with the disintegration of the Austro-Hungarian Empire, and Schrödinger was forced to return to the Second Physics Institute.

During his second tenure at the Institute, on April 6, 1920, Schrödinger married Annemarie Bertel, whom he had met prior to the War. Not long after his marriage, Schrödinger accepted an appointment as assistant to Max Wien in Jena, but remained there only four months. He then moved on to the Technische Hochschule in Stuttgart. Once again, he stayed only briefly—a single semester—before resigning his post and going on to the University of Breslau. He received yet another opportunity to move after being at the University for only a short time: he was offered the chair in theoretical physics at the University of Zürich in late 1921.

The six years that Schrödinger spend at Zürich were probably the most productive of his scientific career. At first, his work dealt with fairly traditional topics; one paper of particular practical interest reported his studies on the relationship between red-green and blue-yellow color blindness. Schrödinger's first interest in the problem of wave mechanics did not arise until 1925. A year earlier, de Broglie had announced his hypothesis of the existence of matter waves, a concept that few physicists were ready to accept. Schrödinger read about de Broglie's hypothesis in a footnote to a paper by American physicist **Albert Einstein**, one of the few scientists who did believe in de Broglie's ideas.

Schrödinger began to consider the possibility of expressing the movement of an electron in an atom in terms of a wave. He adopted the premise that an electron can travel around the nucleus only in a standing wave (that is, in a pattern described by a whole number of wavelengths). He looked for a mathematical equation that would describe the position of such "permitted" orbits. By January of 1926, he was ready to publish the first of four papers describing the results of this research. He had found a second order partial differential equation that met the conditions of his initial assumptions. The equation specified certain orbitals (energy levels) outside the nucleus where an electron wave with a whole number of wavelengths could be found. These orbitals corresponded precisely to the orbitals that Bohr had proposed on purely empirical grounds thirteen years earlier. The wave equation provided a sound theoretical basis for an atomic model that had originally been derived purely on the basis of experimental observations. In addition, the wave equation allowed the theoretical calculation of energy changes that occur when an electron moves from one permitted orbital to a higher or lower one. These energy changes conformed to those actually observed in spectroscopic measurements. The equation also explained why electrons cannot exist in regions between Bohr orbitals since only non-whole number wavelengths (and, therefore, non-permitted waves) can exist there.

After producing unsatisfactory results using relativistic corrections in his computations, Schrödinger decided to work with non-relativistic electron waves in his derivations. The results he obtained in this way agreed with experimental observations and he announced them in his early 1926 papers. The equation he published in these papers became known as "the Schrödinger wave equation" or simply "the wave equation." The wave equation was the second theoretical mechanism proposed for describing electrons in an atom, the first being German physicist Werner Karl Heisenberg's matrix mechanics. For most physicists, Schrödinger's approach was preferable since it lent itself to a physical, rather than strictly mathematical, interpretation. As it turned out, Schrödinger was soon able to show that wave mechanics and matrix mechanics are mathematically identical.

In 1927, Schrödinger was presented with a difficult career choice. He was offered the prestigious chair of theoretical physics at the University of Berlin left open by German physicist Max Planck's retirement. The position was arguably the most desirable in all of theoretical physics, at least in the German-speaking world; Berlin was the center of the newest and most exciting research in the field. Though Schrödinger disliked the hurried environment of a large city, preferring the peacefulness of his native Austrian Alps, he did accept the position.

Hermann quoted Schrödinger as calling the next six years a "very beautiful teaching and learning period." That period came to an ugly conclusion, however, with the rise of National Socialism in Germany. Having witnessed the dismissal of outstanding colleagues by the new regime, Schrödinger decided to leave Germany and accept an appointment at Magdalene College, Oxford, in England. In the same week he took up his new post he was notified that he had been awarded the 1933 Nobel Prize for physics with Dirac.

Schrödinger's stay at Oxford lasted only three years; then, he decided to take an opportunity to return to his native Austria and accept a position at the University of Graz. Unfortunately, he was dismissed from the University shortly after German leader Adolf Hitler's invasion of Austria in 1938, but Eamon de Valera, the Prime Minister of Eire and a mathematician, was able to have the University of Dublin establish a new Institute for Advanced Studies and secure an appointment for Schrödinger there.

In September, 1939, Schrödinger left Austria with few belongings and no money and immigrated to Ireland. He remained in Dublin for the next seventeen years, during which time he turned to philosophical questions such as the theoretical foundations of physics and the relationship between the physical and biological sciences. During this period, he wrote one of the most influential books in twentieth-century science, *What Is Life?* In this book, Schrödinger argued that the fundamental nature of living organisms could probably be studied and understood in terms of physical principles, particularly those of quantum mechanics. The book was later to be read by and become a powerful influence on the thought of the founders of modern molecular biology.

After World War II, Austria attempted to lure Schrödinger home. As long as the nation was under Soviet occupation, however, he resisted offers to return. Finally, in 1956, he accepted a special chair position at the University of Vienna and returned to the city of his birth. He became ill about a year after he settled in Vienna, however, and never fully recovered his health. He died in 1961, in the Alpine town of Alpbach, Austria, where he is buried.

Schrödinger received a number of honors and awards during his lifetime, including election into the Royal Society, the Prussian Academy of Sciences, the Austrian Academy of Sciences, and the Pontifical Academy of Sciences. He also retained his love for the arts throughout his life, becoming proficient in four modern languages in addition to Greek and Latin. He published a book of poetry and became skilled as a sculptor.

See also Quantum electrodynamics (QED); Quantum theory and mechanics

Scientific data management in Earth sciences

Data constitute the raw material of scientific understanding. They are distinguished in analytical data (i.e. numbers with units) and meta-information (i.e. context describing analytical data). Data management is the control of data handling operations such as acquisition, analysis, quality check, processing, storage, retrieval, distribution, and sharing of data. However, it is not necessarily the generation and use of data. Data management ensures integrity of research, confidentiality, compliance with sponsor's requirements, and protects intellectual property. Scientific data management in Earth Sciences covers at least the four major fields of scientific activities, namely the geosphere, hydrosphere, atmosphere, and **biosphere**. Data cover time scales ranging from seconds to millions of millennia and provide baseline information for research in many disciplines, among them monitoring environmental changes—gradual or sudden, foreseen or unexpected, natural or man made.

Scientific data gathering has a long history, and evolved from descriptive cataloguing to a relational digital record of expertise. The Chinese chronicled information about solar and aurora activity in past millennia. In the eighteenth and nineteenth centuries, geological data were recorded in expedition reports. Since the middle of the twentieth century, inconceivably huge and heterogeneous numeric data loads came up during large-scale marine projects. At this time, a data management strategy termed 'the box of floppies' approach was developed. The data sets were supplied to the data center as discrete entities (usually on floppy disks) where they were checked, catalogued and stored. On demand, clients were supplied with the data sets necessary to satisfy their requirements. Data management philosophy in these scenarios was firmly focused on data archival. Today, the challenge of scientific data management is to provide standardized import and export routines to support the scientific community with comfortable and uniform retrieval functions and efficient tools for the graphical visualization of their analytical and meta-data through computers.

The unique requirement of data management in Earth Sciences compared to other (natural) sciences is that any datum has to be specified by a space-time geo-code, i.e., geographical (**latitude**, **longitude**, sample depth) and time dimensions (date/time, period of time, age [model]). Together with the key parameters sample compartment (e.g., **water** column, sediment), variable and unit, and principal investigator (i.e. the owner of a data series) any data collection can be mapped however heterogeneous it may be. To describe this so-called n-dimensional parameter catalogue, a meta-information catalogue was invented that comprises project fact, campaign information, station data, **scientific method**, public access status, and reference where the data were published first. Both parameter and meta-information catalogues itemize the analytical value and serve as unambiguous identifier. Validation and verification of data are the two most critical components in scientific data management. Even if scientific data are supposed to be correct, the definition of what is correct is far from straightforward. It can quite often be a matter of opinion, and opinions are subject to change as scientific knowledge changes. For example, the CLIMAP Project Members referred to the Last Glacial Maximum as "18k" (i.e., 18,000 years ago), whereas Bard revised this concept some 25 years later to "21,000 calendar years ago." Each datum reflects current scientific opinion at its time; however, it became subject to change with altered scientific knowledge. It is not essential to have only excellent quality data sets but it is important that exact information on the quality is provided. The user of a specific data set must be able to verify data by reading the reference publication and thus make a decision about the usefulness of retrieved data. Since (yet) unpublished data are even more sensitive than published data, the data management group is obliged to ensure that data are not accessed from outside a project until data are formally placed in public domain.

Consequently, a data management profile in Earth sciences seeks an information system that represents the n-dimensional parameter-catalogue and the accompanying meta-information catalogue by a suitable data model and archives the data collection in a way that any datum is described at any stage thoroughly and is traced back to its origin in order to protect copyright. Simultaneously, all interfaces are administered independently during data flow, i.e., from the scientific community to the data management, from the data management to the database, from the database to the data application (e.g., numerical model), from the application through the data management back to the scientific community. Finally, the archived data may be retrieved and presented as raw data and graphics. However, data format can be different each time. A popular conceptual construct of ideas applied to such an approach is the multidimensional view of data. This concept or data model, respectively, formally serves as a basis to inductively generate hypotheses with a search algorithm on specific data sets, which commonly is called data mining.

In practice, the conversion of multidimensional data model and data mining tool in Earth sciences is carried out by the International Council of Scientific Unions' World Data Center system (WDC). It works to guarantee access to any data in all fields of Earth sciences on a long-term basis. The categories of World Data Centers read like a *Who's Who* in Earth sciences: Air glow (Tokyo, Japan), **astronomy** (Beijing, China), atmospheric trace gases (Oak Ridge, United States), aurora (Tokyo, Japan), cosmic Rays (Toyokawa, Japan), Earth **tides** (Brussels, Belgium), geology (Beijing, China), geomagnetism (Copenhagen, Denmark; Edinburgh, United Kingdom; Kyoto, Japan; Mumbai, India), glaciology (Boulder, United States; Cambridge, United Kingdom; Lanzhou, China), human interactions in the environment (Palisades, United States), **ionosphere** (Tokyo, Japan), marine environmental sciences (Bremen, Germany), marine geology and geophysics (Boulder, United States; Moscow, Russia), **meteorology** (Asheville, United States; Beijing, China; Obninsk, Russia), nuclear radiation (Tokyo, Japan), **oceanography** (Obninsk, Russia; Silver Spring, United States; Tianjin, China), paleoclimatology (Boulder, United States), recent crustal movements (Ondrejov, Czech Republic), remotely sensed land data (Sioux Falls, United States), renewable resources and environment (Beijing, China), rockets and satellites (Obninsk, Russia), **rotation** of the Earth (Obninsk, Russia; Washington, United States), **satellite** information (Greenbelt, United States), **seismology** (Denver, United States; Beijing, China), soils (Wageningen, The Netherlands), solar activity (Meudon, France), solar radio emissions (Nagano, Japan), solar terrestrial **physics** (Boulder, United States; Didcot Oxon, United Kingdom; Moscow, Russia; Haymarket, **Australia**), solid Earth geophysics (Beijing, China; Boulder, United States; Moscow, Russia), **space** science (Beijing, China; Sagamihara, Japan), sunspot index (Brussels, Belgium).

Since the early beginnings of modern scientific data management in Earth sciences, the gathering and exchange of data has been transformed by rapid technological advances, such as the replacement of analog with digital instruments, the networking of digital instruments to simplify collection and exchange of data, unmanned automatic observatories etc. Personal computers and compact disc readers are ubiquitous. Many World Data Centers publish collections of digital data sets on compact discs for easy distribution. Digital communication networks make it possible to transfer large data files by electronic mail. Environmental disciplines make use of map-based data through Geographical Information Systems. The collaboration of international scientific bodies ensures the continuation of long-term monitoring of the Earth system, the permanent preservation of the data acquired for the mutual benefit of the international scientific community, and the dissemination mechanisms through publications, workshops, exhibitions, and other means.

See also GIS; Ice ages; International Council of Scientific Unions World Data Center System

Scientific method

Scientific thought aims to make correct predictions about events in nature. Although the predictive nature of scientific thought may not at first always be apparent, a little reflection usually reveals the predictive nature of any scientific activity. Just as the engineer who designs a bridge ensures that it will withstand the forces of nature, so the scientist considers the ability of any new scientific model to hold up under scientific scrutiny as new scientific data become available.

It is often said that the scientist attempts to understand nature. Ultimately, understanding something means being able to predict its behavior. Scientists, therefore, usually agree that events are not understandable unless they are predictable. Although the word "science" describes many activities, the notion of prediction or predictability is always implied when the word is used.

Until the seventeenth century, scientific prediction simply amounted to observing the changing events of the world, noting any regularities, and making predictions based upon those regularities. The Irish philosopher and bishop George Berkeley (1685–1753) was the first to rethink this notion of predictability.

Berkeley noted that each person experiences directly only the signals of his or her five senses. An individual can infer that a natural world exists as the source of his sensations, but he or she can never know the natural world directly. One can only know it through one's senses. In everyday life, people tend to forget that their knowledge of the external world comes to them through their five senses.

The physicists of the nineteenth century described the **atom** as though they could see it directly. Their descriptions changed constantly as new data arrived, and these physicists had to remind themselves that they were only working with a mental picture built with fragmentary information.

In 1913, **Niels Bohr** used the term *model* for his published description of the hydrogen atom. This term is now used to characterize theories developed long before Bohr's time. Essentially, a model implies some correspondence between the model itself and its object. A single correspondence is often enough to provide a very useful model, but it should never be forgotten that the intent of creating the model is to make predictions.

There are many types of models. A conceptual model refers to a mental picture of a model that is introspectively present when one thinks about it. A geometrical model refers to diagrams or drawings that are used to describe a model. A mathematical model refers to equations or other relationships that provide quantitative predictions.

New models are not constructed from observations of facts and previous models; they are postulated. That is to say, the statements that describe a model are assumed and predictions are made from them. The predictions are checked against the measurements or observations of actual events in nature. If the predictions prove accurate, the model is said to be validated. If the predictions fail, the model is discarded or adjusted until it can make accurate predictions.

The formulation of the scientific model is subject to no limitations in technique; the scientist is at liberty to use any method he can come up with, conscious or unconscious, to develop a model. Validation of the model, however, follows a single, recurrent pattern. Note that this pattern does not constitute a method for making new discoveries in science; rather it provides a way of validating new models after they have been postulated. This method is called the scientific method.

The scientific method 1) postulates a model consistent with existing experimental observations; 2) checks the predictions of this model against further observations or measurements; 3) adjusts or discards the model to agree with new observations or measurements.

The third step leads back to the second, so, in principle, the process continues without end. (Such a process is said to be recursive.) No assumptions are made about the reality of the model. The model that ultimately prevails may be the simplest, most convenient, or most satisfying model; but it will certainly be the one that best explains those problems that scientists have come to regard as most acute.

Paradigms are models that are unprecedented to attract an enduring group of adherents away from competing scientific models. A paradigm must be sufficiently open-ended to leave many problems for its adherents to solve. The paradigm is thus a theory from which springs a coherent tradition of scientific research. Examples of such traditions include Ptolemaic **astronomy**, Copernican astronomy, Aristotelian dynamics, Newtonian dynamics, etc.

To be accepted as a paradigm, a model must be better than its competitors, but it need not and cannot explain all the facts with which it is confronted. Paradigms acquire status because they are more successful than their competitors in solving a few problems that scientists have come to regard as acute. Normal science consists of extending the knowledge of those facts that are key to understanding the paradigm, and in further articulating the paradigm itself.

Scientific thought should in principle be cumulative; a new model should be capable of explaining everything the old model did. In some sense, the old model may appear to be a special case of the new model.

The descriptive phase of normal science involves the acquisition of experimental data. Much of science involves classification of these facts. Classification systems constitute abstract models, and it is often the case that examples are found that do not precisely fit in classification schemes. Whether these anomalies warrant reconstruction of the classification system depends on the consensus of the scientists involved.

Predictions that do not include numbers are called qualitative predictions. Only qualitative predictions can be made from qualitative observations. Predictions that include numbers are called quantitative predictions. Quantitative predictions are often expressed in terms of probabilities, and may contain estimates of the accuracy of the prediction.

The Greeks constructed a model in which the stars were lights fastened to the inside of a large, hollow sphere (the sky), and the sphere rotated about the earth as a center. This model predicts that all of the stars will remain fixed in position relative to each other. However, certain bright stars were found to wander about the sky. These stars were called planets (from the Greek word for wanderer). The model had to be modified to account for motion of the planets. In Ptolemy's (A.D. 100–170) model of the **solar system**, each planet moves in a small circular orbit, and the center of the small circle moves in a large circle around the earth as center.

Copernicus (1473–1543) assumed the **Sun** was near the center of a system of circular orbits in which the earth and planets moved with fair regularity. Like many new scientific ideas, Copernicus' idea was initially greeted as nonsense, but over time, it eventually took hold. One of the factors that led astronomers to accept Copernicus' model was that Ptolemaic astronomy could not explain a number of astronomical discoveries.

In the case of Copernicus, the problems of calendar design and astrology evoked questions among contemporary scientists. In fact, Copernicus's theory did not lead directly to any improvement in the calendar. Copernicus's theory suggested that the planets should be like the earth, that Venus should show phases, and that the universe should be vastly larger than previously supposed. Sixty years after Copernicus's death, when the **telescope** suddenly displayed mountains on the **moon**, the phases of Venus, and an immense number of previously unsuspected stars, the new theory received a great many converts, particularly from non-astronomers.

The change from the Ptolemaic model to the Copernican model is a particularly famous case of a paradigm change. As the Ptolemaic system evolved between 200 B.C. and 200 A.D., it eventually became highly successful in predicting changing positions of the stars and planets. No other ancient system had performed as well. In fact, the Ptolemaic astronomy is still used today as an engineering approximation. Ptolemy's predictions for the planets were as good as Copernicus's predictions. With respect to planetary position and precession of the equinoxes, however, the predictions made with Ptolemy's model were not quite consistent with the best available observations. Given a particular inconsistency, astronomers for many centuries were satisfied to make minor adjustments in the Ptolemaic model to account for it. Eventually, it became apparent that the web of complexity resulting from the minor adjustments was increasing more rapidly than the accuracy, and a discrepancy corrected in one place was likely to show up in another place.

Tycho Brahe (1546–1601) made a lifelong study of the planets. In the course of doing so, he acquired the data needed to demonstrate certain shortcomings in Copernicus's model. But it was left to **Johannes Kepler** (1571–1630), using Brahe's data after the latter's death, to come up with a set of laws consistent with the data. It is worth noting that the quantitative superiority of Kepler's astronomical tables to those computed from the Ptolemaic theory was a major factor in the conversion of many astronomers to the Copernican theory.

In fact, simple quantitative telescopic observations indicate that the planets do not quite obey Kepler's laws, and Isaac Newton (1642–1727) proposed a theory that shows why they should not. To redefine Kepler's laws, Newton had to neglect all gravitational attraction except that between individual planets and the sun. Since planets also attract each other, only approximate agreement between Kepler's laws and telescopic observation could be expected.

Newton thus generalized Kepler's laws in the sense that they could now describe the motion of any object moving in any sort of path. It is now known that objects moving almost as fast as the speed of light require a modification of Newton's laws, but such objects were unknown in Newton's day.

Newton's first law says that a body at rest remains at rest unless acted upon by an external force. His second law states quantitatively what happens when a force is applied to an object. The third law states that if a body A exerts a force F on body B, then body B exerts on body A a force that is equal in magnitude but opposite in direction to force F. Newton's fourth law is his law of gravitational attraction.

Newton's success in predicting quantitative astronomical observations was probably the single most important factor leading to acceptance of his theory over more reasonable but uniformly qualitative competitors.

It is often pointed out that Newton's model includes Kepler's laws as a special case. This permits scientists to say they understand Kepler's model as a special case of Newton's model. But when one considers the case of Newton's laws and relativistic theory, the special case argument does not hold up. Newton's laws can only be derived from Albert Einstein's (1876–1955) relativistic theory if the laws are reinterpreted in a way that would have only been possible after Einstein's work.

The variables and parameters that in Einstein's theory represent spatial position, time, mass, etc. appear in Newton's theory, and there still represent **space**, time, and mass. But the physical natures of the Einsteinian concepts differ from those of the Newtonian model. In Newtonian theory, mass is conserved; in Einstein's theory, mass is convertible with energy. The two ideas converge only at low velocities, but even then they are not exactly the same.

Scientific theories are often felt to be better than their predecessors because they are better instruments for solving puzzles and problems, but also for their superior abilities to represent what nature is really like. In this sense, it is often felt that successive theories come ever closer to representing truth, or what is "really there." Thomas Kuhn, the historian of science whose writings include the seminal book *The Structure of Scientific Revolution* (1962), found this idea implausible. He pointed out that although Newton's mechanics improve on Ptolemy's mechanics, and Einstein's mechanics improve on Newton's as instruments for puzzle solving, there does not appear to be any coherent direction of development. In some important respects, Professor Kuhn has argued, Einstein's general theory of relativity is closer to early Greek ideas than relativistic or ancient Greek ideas are to Newton's.

See also Historical geology; History of exploration I (Ancient and classical); History of exploration II (Age of exploration); History of exploration III (Modern era)

SCILLA, AGOSTINO (1629-1700)

Italian painter, paleontologist, and geologist

Agostino Scilla inaugurated the modern scientific study of **fossils**. Born the son of a minor government official in Messina, Sicily, he studied art in Messina under Antonio Ricci Barbalunga, who arranged for him to study in Rome for five years under Andrea Sacchi (1599–1661). Upon his return to Messina, Scilla associated with the Accademia della Fucina and established himself throughout eastern Sicily as a painter of religious scenes for church interiors, including some decorations for the cathedral in Syracuse. A gentleman of broad humanistic learning, with particular interest in ancient local culture, he became an expert on the history of Sicilian coins. During the 1650s or 1660s he began to study natural history, especially the fossils he found in the Sicilian hills. His expeditions were sometimes in company with the botanist Paolo Boccone (1633–1704). Scilla's training as a painter enhanced his skill at observation in general. He was intrigued by how the petrified forms of what looked like marine life could have come to rest at such high elevations so far from the sea.

Scilla's investigations of fossils culminated in the publication of his only scientific work, *La vana speculazione disingannata dal senso* (Vain Speculation Undeceived by Sense, 1670). In it, he famously opposed Francesco Stelluti (1577–1646) and Athanasius Kircher (1602–1680) on the question of why marine fossils are discovered inland. Stelluti, Kircher, and their allies believed that such fossils were "sports of nature," ruses of God to test our faith, or accidents explicable only through astrology, alchemy, or other fantastic means. To Scilla that was all nonsense. He wrote in plain language that he had no idea how the remains of corals, shells, shark teeth, and fish bones ended up in the hills, that he did not know of any method to try to learn how they got there, and that to speculate about their origin would be fanciful, unwarranted, and pointless. Scilla rejected the authority of ancient authors and medieval theologians, relying instead upon naturalistic observation, skeptical empiricism, and common sense. His style and degree of skepticism anticipated that of David Hume (1711–1776).

Prior to the work of Fabio Colonna (1567–1640), **John Ray** (1627–1705), Robert Hooke (1635–1703), Nicolaus Steno (1638–1686), and Scilla, there was no consensus that fossils were the remains of organic life. The *glossopetrae* ("tongue stones") commonly found throughout **Europe** were believed to have magical properties and mystic origins, by either the actions of lunar eclipses or the miracles of St. Paul. Unknown to each other, Steno and Scilla each positively identified *glossopetrae* as shark teeth. Their analysis of *glossopetrae* effectively undermined most earlier theories and superstitions about fossils.

In 1678, having participated in an unsuccessful Sicilian revolt against Spanish rule, Scilla was exiled. He went first to Turin, then, in 1679, to Rome, where he spent the rest of his life, making a living as a painter and becoming a prominent member of the Accademia di San Luca.

See also Fossil record; Fossils and fossilization; Marine transgression and marine recession; Sedimentary rocks; Sedimentation

SEAS • *see* OCEANS AND SEAS

SEA-FLOOR SPREADING

Earth's surface is composed of two kinds of **crust**, continental and oceanic. Most continental crust is over 3 billion years old, while virtually all oceanic crust is less than 180 million years old. Oceanic crust is young because it is continually destroyed in some places and created in others. Subduction is the process that destroys oceanic crust, and sea-floor spreading is the process that creates oceanic crust.

Sea-floor spreading is driven by crust formation along the **mid-ocean ridges**, meandering undersea mountain ranges that span Earth like the seams of a baseball. Oceanic crust is continually produced by **magma** welling up along the centerlines of the mid-ocean ridges. This new crust flows away from each ridgeline in two symmetric sheets, one on each side. The rate of sea-floor spreading resulting from this process is from 0.5 to 8 inches per year (1–20 cm/yr), depending on the particular mid-ocean ridge.

The Mid-Atlantic Ridge offers a particularly clear case of sea-floor spreading. About 165 million years ago, the Americas were matched to **Africa** and **Europe** like the pieces of a puzzle. Then, magma upwelling at the Mid-Atlantic Ridge began to produce oceanic crust, parting the continents to form the Atlantic Ocean. Today the Mid-Atlantic Ridge snakes down the center of the Atlantic all the way from Iceland to the Antarctic Plate and remains an active site of sea-floor spreading.

A dramatic proof of sea-floor spreading was discovered in the mid 1960s when data revealed alternating stripes of magnetic orientation on the sea floor, parallel to the mid-ocean ridges and symmetric across them—that is, a thick or thin stripe on one side of the ridge is always matched by a similar stripe at a similar distance on the other side. This mirror-image magnetic orientation pattern is created by steady sea-floor spreading combined with recurrent reversals of Earth's **magnetic field**. **Iron** atoms in liquid **rock** welling up along a mid-ocean ridge align with Earth's magnetic field. When this magma solidifies into crust, its iron atoms lock into position. This solid crust flows away from the mid-ocean ridge in both directions, carrying its original magnetic orientation with it. Eventually Earth's magnetic field reverses. Previously solidified crust retains its original field state, but crust just forming along the ridge is locked into the new orientation. As crust feeds steadily and symmetrically away from the ridgeline and Earth's magnetic field reverses over and over again, a symmetric striped pattern of **magnetism** is created.

See also Geographic and magnetic poles; Lithospheric plates; Magnetic field; Mantle plumes; Mapping techniques; Ocean trenches; Paleomagnetics; Plate tectonics

SEASONAL WINDS

A **wind** in low-latitude climates that seasonally changes direction between winter and summer is called a monsoon, and is a typical example of seasonal winds. Monsoons usually blow from the land in winter (called the dry phase, because it carries cool, dry air), and to the land in summer (called the wet phase, because it carries warm, moist air), causing a drastic change in the **precipitation** and **temperature** patterns of the **area**.

The word "monsoon" originates from the Arabic *mauzim,* meaning season. It was first used to depict the winds in the Arabian Sea, but later it was extended for seasonally changing wind systems all over the world. The main reason for monsoons is the difference in the heating of land and **water** surfaces, which results in land-ocean pressure differences. On a small scale, heat is transferred by land-sea breezes, to maintain the energy balance between land and water. On a larger scale, in winter when the air over the continents is colder than over the **oceans**, a large, high-pressure area builds up over

Siberia, resulting in air motion over the Indian Ocean and South China, causing dry, clear skies for East and South **Asia**. This is the winter monsoon. The opposite of this happens in summer; the air over the continents is much warmer than over the ocean, leading to moisture-carrying wind moving from the ocean towards the continent. When the humid air unites with relatively drier west airflow and crosses over mountains, it rises, reaches its saturation point, and thunderstorms and heavy showers develop. This is the summer monsoon in Southwest Asia—wind blowing from the ocean to the continent with wet, rainy **weather** patterns.

Although the most pronounced monsoon system is in eastern and southern Asia, monsoons can also be observed in West **Africa**, **Australia**, or the Pacific Ocean. Even in the southwestern United States, a smaller scale monsoonal circulation system exists (called North American Monsoon, Mexican monsoon, or Arizona monsoon). The North American Monsoon is a regional-scale circulation over southwest **North America** between July and September, bringing dramatic increases in rainfall in a normally arid region of Arizona, New Mexico, and northwestern Mexico. It is a monsoonal circulation because of its similarities to the original Southwest Asian monsoon—the west or northwest winds turn more south or southeast, bringing moisture from the Pacific Ocean, Gulf of California and **Gulf of Mexico**. As the moist air moves in, it is lifted up due to the mountains, which, combined with daytime heating from the **Sun**, causes thunderstorms.

The monsoon is an important feature of **atmospheric circulation**, because large areas in the tropics and subtropics are under the influence of monsoons, bringing humid air from over the oceans to produce rain over the land. In highly populated areas (e.g., Asia or India), this precipitation is essential for agriculture and food crop production. Sometimes a strong monsoon circulation can also bring flooding. Or, if the monsoon is late in a certain year, it can cause droughts.

A similar phenomenon to the monsoon also occurs in a smaller spatial and temporal scale, the mountain and valley breezes. The main reason they occur is also the difference in heating of the areas: during the day, the valley and the air around it warm and, because it is less dense, the air rises, and thus, a gentle upslope wind occurs. This wind is called the valley breeze. If the upslope valley winds carry sufficient moisture in the air, showers, even thunderstorms can develop in the early afternoon, during the warmest part of the day. The opposite happens at night, when the slopes cool down quickly, causing the surrounding air also to cool and glide down from the mountain to the valley, forming a mountain breeze (also called **gravity** winds or drainage winds). Although technically any kind of downslope wind is called a katabatic (or fall) wind, this term is usually used for a significantly stronger wind than a mountain breeze.

For katabatic winds carrying cold air, their ideal circumstances are mountains with steep downhill slopes and an elevated plateau. If winter snow accumulates on the plateau, it makes the surrounding air very cold, which then starts to move down as a cold, moderate breeze and can become a destructive, fast wind if it passes through a narrow **canyon** or channel. These katabatic winds have different names in different areas of the world. The bora is a northeast cold wind with speeds of sometimes more than 115 miles per hour (100 knots), blowing along the northern coast of the Adriatic Sea, when polar cold air from Russia moves down from a high plateau, reaching the lowlands. The mistral is a similar, although less violent cold wind in France, which moves down from the western mountains into the Rhone Valley then out to the Mediterranean Sea, often causing frost damage to vineyards. Even in Greenland and **Antarctica** there are occasional cold, strong katabatic winds.

Among the katabatic winds carrying warm air, the chinook wind is a dry warm wind, moving down the eastern slope of the Rocky Mountains in a narrow area between northeast New Mexico and Canada. When strong westerly winds blow over a north-south mountain, it produces low pressure on the eastward side of the mountain, forcing the air downhill, and causing a compressional heating. The chinook causes the temperature to rise over an area sharply, resulting in a sharp drop in the relative **humidity**. If chinooks move over heavy snow cover, they can even melt and evaporate a foot of snow in less than a day. The chinook is important because it can bring relief from a strong winter, uncovering grass, which can be fed to the livestock. A similar wind in the Alps is called foehn, a dry, warm wind descending the mountain slope then flowing across flat lands below. A warm and dry wind in South California blowing from the east or northeast is called the Santa Ana wind (named from the Santa Ana Canyon). Because this air originates in the **desert**, it is dry, and becomes even drier as it is heated. Brush fires and dried vegetation can follow the Santa Ana wind.

See also Air masses and fronts; Land and sea breeze

SEASONS

Seasons, which generally coincide with annual changes in **weather** patterns, are most pronounced in *temperate zones*. These zones extend from 23.5° north (and south) **latitude** to 66.5° north (and south) latitude. Within these latitudes, nature generally exhibits four seasons; spring, summer, autumn (or fall) and winter. Each season is characterized by differences in **temperature**, amounts of **precipitation**, and the length of daylight.

Seasonal observations have been noted in the earliest known written records of history. In fact, seasonal changes have affected the course of history in the outcomes of battles or movements of peoples in search of longer growing seasons has often been greatly influenced by seasonal changes. Spring comes from an Old English word meaning to rise; summer originated as a Sanskrit word meaning half year or season. Autumn comes originally from an Etruscan word for maturing. Winter comes from an Old English word meaning wet or **water**. The equatorial regions or torrid zones have no noticeable seasonal changes and one generally finds only a wet season and a dry season in these zones. In the polar regions the seasons are closely related to the amount of sunlight received, resulting in a light season and a dark season.

Seasons are tied to the apparent movements of the **Sun** and stars across the celestial sphere. In the Northern Hemisphere, spring begins at the vernal equinox (around March 21) when sunlight is directly incident on the equator with equal distribution of light to the Northern and Southern Hemispheres. Summer begins at the summer solstice (approximately June 21) when the Sun is at its apparent maximum declination. Autumn begins at the autumnal equinox around September 23, and winter at the winter solstice (minimum declination in the Northern Hemisphere) that occurs approximately December 21. Because every fourth year is a leap year and February then has 29 days, the dates of these seasonal starting points change slightly. In the Southern Hemisphere, the seasons are reversed with spring beginning in September, summer in December, fall in March, and winter in June. Seasons in the Southern Hemisphere are generally milder due to the moderating presence of larger amounts of ocean surface as compared to the Northern Hemisphere.

Changes in the seasons are caused by Earth's movement around the Sun. Because Earth orbits the Sun at varying distances, many people assume that the seasons result from the changes in the Earth-Sun distance. This belief is incorrect. In fact, Earth is actually closer to the Sun in January compared to June by approximately three million miles.

Earth makes one complete **revolution** about the Sun each year. The major reason that the seasons occur is that the axis of Earth's **rotation** is tilted with respect to the plane of its orbit. This tilt, called the obliquity of Earth's axis, is 23.5 degrees from a line drawn perpendicular to the plane of Earth's orbit. As Earth orbits the Sun, there are times of the year when the North Pole is alternately tilted toward the Sun (during northern hemispheric summer) or tilted away from the Sun (during northern hemispheric winter). At other times, the axis is generally perpendicular to the incoming Sun's rays. During summer, two effects contribute to produce warmer weather. First, the Sun's rays fall more directly on Earth's surface and this results in a stronger heating effect. The second reason for the seasonal temperature differences results from the differences in the amount of daylight hours versus nighttime hours. The Sun's rays warm Earth during daylight hours; Earth cools at night by re-radiating heat back into **space**. This is the major reason for the warmer days of summer and cooler days of winter. The orientation of Earth's axis during summer results in longer periods of daylight and shorter periods of darkness at this time of year. At the mid-northerly latitudes, summer days have about 16 hours of warming daylight and only eight hours of cooling nights. During mid-winter the pattern is reversed, resulting in longer nights and shorter days. To demonstrate that it is the daylight versus darkness ratio that produces climates that make growing seasons possible, one should note that even in regions only 30° from the poles one finds plants such as wheat, corn, and potatoes growing. In these regions the Sun is never very high in the sky but because of the orientation of Earth's axis, the Sun remains above the horizon for periods for over 20 hours a day from late spring to late summer.

Astronomers have assigned names to the dates at which the official seasons begin. When the axis of Earth is perpendicular to the incoming Sun's rays in spring the Sun stands directly over the equator at noon. As a result, daylight hours equal nighttime hours everywhere on Earth. This gives rise to the name given to this date, the vernal equinox. Vernal refers to spring and the word equinox means *equal night*. On the first day of fall, the autumnal equinox also produces 12 hours of daylight and 12 hours of darkness everywhere on Earth.

The name given for the first day of summer results from the observation that as the days get longer during the spring, the Sun's height over its noon horizon increases until it reaches June 21. Then on successive days, it dips lower in the sky as Earth moves toward the autumn and winter seasons. This gives rise to the name for that date, the summer solstice, because it is as though the Sun "stands still" in its noon height above the horizon. The winter solstice is likewise named because on December 21 the Sun reaches the lowest noon time height and appears to "stand still" on that date as well.

In the past, early humans celebrated the changes in the seasons on some of these cardinal dates. The vernal equinox was a day of celebration for the early Celtic tribes in ancient Britain, France, and Ireland. Other northern European tribes also marked the return of warmer weather on this date. Even the winter solstice was a time to celebrate, as it marked the lengthening days that would lead to spring. The ancient Romans celebrated the Feast of Saturnalia on the winter solstice. And even though there are no historical records to support the choice of a late December date for the birth of Christ, Christians in the fourth century A.D. chose to celebrate Christmas near the winter solstice.

See also Atmospheric circulation; Celestial sphere: The apparent movements of the Sun, Moon, planets, and stars; Latitude and longitude; Seasonal winds; Solar illumination: Seasonal and diurnal patterns

SEAWALLS AND BEACH EROSION

Beaches are one of the most important economic and environmental zones along coasts everywhere. In the United States alone, there are over 19,000 miles of beaches, 500 miles of which are within designated National Parks. Beaches are the boundaries between land and sea. A large part of the human population lives within driving distance of a beach. One of the recent environmental problems of importance is the increasing loss of beaches due to **erosion** of the shorelines.

Sediments, transported from inland resources, stock the rich supply of sands on beaches. **Rivers** drain inland valleys and plains and carry eroded particles from their source to their final deposition at sea. An enormous amount of **sand** and mud is carried by rivers such as the Nile and Mississippi. These rivers nourish farmers' lands and provide transport for economic goods. Because of their value, they are not often dammed, and their flow remains relatively unimpeded. However, unlike their famous counterparts, other rivers that feed coastal areas have been dammed in more than one place. The dams can provide everything from hydroelectric power to reservoirs for recreational and agricultural use.

The sediment load effect of large dams is often not observable until the river reaches the shore. The dams act as brakes for flowing **water**. When velocities are reduced, so is the carrying capacity of the river. Sediment loads are dumped on the upriver side of the dam. For local operators, the problem is one of continuously having to dredge the dams so the water can run through the turbines with the same force. For beaches, it is the reduction and even loss of sediment supply. The usual supply of sand is drastically reduced. Sand starvation occurs near the river mouths.

Another contributing factor to beach erosion is the loss of vegetation. The roots of plants keep sands in place. Shoreline plants are unique in that they have adapted to saltier conditions than their **freshwater** counterparts. When homes and business encroach on this delicate habitat the function they serve is lost. With increasing urbanization, more and more beach ecosystems are destroyed.

The Atlantic seaboard is an excellent example of a region that is battling beach erosion. Along the coast, a longshore current displaces the sand from the river source and down the coast. This current is referred to as littoral drift. Waves refract at often-steep angles off the shore and move sand from its original place to one a bit farther down the coast. Over time, the transport of sediment moves sand far away from the river mouths. It is deposited on beaches where it builds up into the shoreline. When sand supplies are decreased and the sediment loads are reduced, the waves continue to work away at the beach. They move sand in a daily cycle of littoral drift. Unfortunately, when the sand is reduced, the shorelines are eaten away by the longshore transport. For many this means the loss of sandy beaches and even the loss of homes.

Many types of remedies are attempted to reduce beach erosion. Some have had limited success while others appear to slow the erosion of sand. Some of the more long-lasting remedies are sea walls and jetties. Sea walls were designed to slow or reduce the impact of waves on the beach. They are built on top of and parallel to the beach. As waves come toward shore they first strike the sea wall and their energy is dissipated. When the waves finally reach the shore they are so weak that they carry hardly any sand away.

Another structure, the jetty, is built perpendicular to the beach. Jetties are walls of rocks or cement that jut out into the water. They are often constructed in groups. Their purpose is to control the flow of water along the shore and to sometimes block the movement of sand and sediment. Technically, a jetty keeps sand from moving to a certain location, such as a harbor, while a groin is a jetty that simply keeps sand from moving down the shore. When constructed in groups, groins are called groin fields. The fields buffer sand removal along the beach.

While certainly effective, none of these structures will ultimately end beach erosion. Increased awareness of the problem has teamed geologists with environmental scientists. Many alternate plans for saving beaches are being considered. Resuming the sediment flow is the ultimate solution, and remains an economic and technological challenge.

See also Beach and shoreline dynamics; Ocean circulation and currents; Sedimentation; Wave motions

SEDGWICK, REVEREND ADAM (1785-1873)

English geologist and Anglican priest

Reverend Adam Sedgwick contributed to the entire scope of **geology**, but mainly toward defining the Cambrian stratum of the **fossil record** and trying to show precisely when life originated in **geologic time**. As an accomplished and popular teacher, speaker, and writer, he successfully encouraged many young British scientists and intellectuals to pursue geologic inquiry, and thus set the course of British geology for over a century. The Sedgwick Museum of Geology at the University of Cambridge is named in his honor.

Sedgwick was born the son of the Anglican vicar in Dent, Yorkshire, England. His childhood hobby of **rock** collecting on the moors grew into his career as a geologist. After his secondary education at Sedbergh School, he matriculated at Trinity College, Cambridge, where he received his baccalaureate in mathematics in 1808. He became a Fellow of Trinity College in 1810, was ordained in the Church of England in 1817, assumed the Woodwardian Chair of Geology at Cambridge in 1818, served as president of the Geological Society of London from 1829 to 1831, and after 1845 held a variety of high administrative posts at Cambridge. Liberal in both politics and theology, he led the fight to allow non-Anglicans to study at Trinity and was among the first professors at Cambridge to allow women into his classes. He never married, but spent the rest of his life as a faculty member at Trinity, and died in Cambridge.

Sedgwick's research centered on the **Paleozoic Era**. Influenced by William Conybeare (1787–1857), he studied British **limestone** and **sandstone** deposits, concentrating especially on the Devonian, and became an expert on the fossil record throughout England and Wales. He established the contemporary agenda for **stratigraphy** and frequently challenged the findings of other geologists, notably Roderick Impey Murchison (1792–1871) with regard to the Silurian.

In 1831, one of Sedgwick's students was Charles Darwin (1809–1882). They maintained a lifelong friendship, even after 1859, when the clergyman expressed his strong objections to Darwin's recently published *Origin of Species*. Sedgwick did not accept a literalist or creationist interpretation of the *Bible*, but, consistent with his opposition to Darwinian **evolution**, believed in the cataclysmic or catastrophic theory of the history of Earth, whereby occasional dramatic events cause mass extinctions and radically alter further geologic development. In this he supported **Georges Cuvier** (1769–1832) and opposed **Charles Lyell** (1797–1875), a major proponent of a theory of gradual, steady, predictable geologic change.

See also Evolution, evidence of; Evolutionary mechanisms; Fossils and fossilization

SEDIMENTARY ROCKS

Sedimentary rocks form at or near the earth's surface from the weathered remains of pre-existing rocks or organic debris. The term sedimentary **rock** applies both to consolidated, or lithified sediments (bound together, or cemented) and unconsolidated sediments (loose, like **sand**). Although there is some overlap, most sedimentary rocks belong to one of the following groups: clastic, chemical, or organic.

Mechanical **weathering** breaks up rocks, while chemical weathering dissolves and decomposes rocks. Weathering of igneous, metamorphic, and sedimentary rocks produces rock fragments, or clastic sediments, and mineral-rich **water**, or mineral solutions. After transport and laying down, or deposition, of sediments by **wind**, water, or **ice**, compaction occurs due to the weight of overlying sediments that accumulate later. Finally, **minerals** from mineral-rich solutions may crystallize, or precipitate, between the grains and act as cement. Cementation of the unconsolidated sediments forms a consolidated rock. Clastic rocks are classified based on their grain size. The most common clastic sedimentary rocks are shale (grains less than 1/256 mm in diameter), siltstone (1/256–1/16 mm), **sandstone** (1/16–2 mm), and conglomerate (greater than 2 mm).

Chemical or crystalline sedimentary rocks form from mineral solutions. Under the right conditions, minerals precipitate out of mineral-rich water to form layers of one or more minerals, or chemical sediments. For example, suppose ocean water is evaporating from an enclosed **area**, such as a bay, faster than water is flowing in from the open ocean. Salt deposits will form on the bottom of the bay as the concentration of dissolved minerals in the bay water increases. This is similar to putting salt water into a **glass** and letting the water evaporate; a layer of interlocking salt **crystals** will precipitate on the bottom of the glass. Due to their interlocking crystals, chemical sediments always form consolidated sedimentary rocks. Chemical rocks are classified based on their mineral composition. Rock salt (composed of the mineral halite, or table salt), rock **gypsum** (composed of gypsum), and crystalline **limestone** (composed of calcite) are common chemical sedimentary rocks.

Organic sedimentary rocks form from organically derived sediments. These organic sediments come from either animals or plants and usually consist of body parts. For example, many limestones are composed of abundant marine **fossils**, so these limestones are of organic rather than chemical origin. **Coal** is an organic rock composed of the remains of plants deposited in coastal swamps. The sediments in some organic rocks (for example, fossiliferous limestone) undergo cementation; other sediments may only be compacted together (for example, coal). Geologists classify organic rocks by their composition.

The origin (clastic, chemical, or organic) and composition of a sedimentary rock provide geologists with many insights into the environment where it was deposited. Geologists use this information to interpret the geologic history of an area, and to search for economically important rocks and minerals.

See also Depositional environments; Lithification; Mineralogy; Sedimentation; Stratigraphy

SEDIMENTATION

Sediments are loose Earth materials such as **sand** that accumulate on the land surface, in river and lakebeds, and on the ocean floor. Sediments form by **weathering** of **rock**. They then erode from the site of weathering and are transported by **wind**, **water**, **ice**, and **mass wasting**, all operating under the influence of **gravity**. Eventually sediment settles out and accumulates after transport; this process is known as deposition. Sedimentation is a general term for the processes of **erosion**, transport, and deposition. Sedimentology is the study of sediments and sedimentation.

There are three basic types of sediment: rock fragments, or clastic sediments; mineral deposits, or chemical sediments; and rock fragments and organic matter, or organic sediments. Dissolved **minerals** form by weathering rocks exposed at the earth's surface. Organic matter is derived from the decaying remains of plants and animals.

Clastic and chemical sediments form during weathering of **bedrock** or pre-existing sediment by both physical and chemical processes. Organic sediments are also produced by a combination of physical and chemical weathering. Physical (or mechanical) weathering—the disintegration of Earth materials—is generally caused by abrasion or fracturing, such as the striking of one pebble against another in a river or stream bed, or the cracking of a rock by expanding ice. Physical weathering produces clastic and organic sediment.

Chemical weathering, or the decay and dissolution of Earth materials, is caused by a variety of processes. However, it results primarily from various interactions between water and rock material. Chemical weathering may alter the mineral content of a rock by either adding or removing certain chemical components. Some mineral by-products of chemical weathering are dissolved by water and transported below ground or to an ocean or lake in solution. Later, these dissolved minerals may precipitate out, forming deposits on the roof of a **cave** (as **stalactites**), or the ocean floor. Chemical weathering produces clastic, chemical, and organic sediments.

Erosion and transport of sediments from the site of weathering are caused by one or more of the following agents: gravity, wind, water, or ice. When gravity acts alone to move a body of sediment or rock, this is known as mass wasting. When the forces of wind, water, or ice act to erode sediment, they always do so under the influence of gravity.

Large volumes of sediment, ranging in size from mud to boulders, can move downslope due to gravity, a process called mass wasting. Rock falls, landslides, and mudflows are common types of mass wasting. If you have ever seen large boulders on a roadway you have seen the results of a rock fall. Rock falls occur when rocks in a cliff face are loosened by weathering, break loose, and roll and bounce downslope. Landslides consist of rapid downslope movement of a mass of rock or **soil**, and require that little or no water be present. Mudflows occur when a hillside composed of fine-grained material becomes nearly saturated by heavy rainfall. The water helps lubricate the sediment, and a lobe of mud quickly moves downslope. Other types of mass wasting include **slump**, **creep**, and subsidence.

Water is the most effective agent of transport, even in the **desert**. When you think of water erosion, you probably think of erosion mainly by stream water, which is channelized. However, water also erodes when it flows over a lawn or down the street, in what is known as sheet flow. Even when water simply falls from the sky and hits the ground in droplets, it erodes the surface. The less vegetation that is present, the more water erodes - as droplets, in sheets, or as channelized flow.

Wind is an important agent of erosion only where little or no vegetation is present. For this reason, deserts are well known for their wind erosion. However, as mentioned above, even in the desert, infrequent, but powerful rainstorms are still the most important agent of erosion. This is because relatively few areas of the world have strong prevailing winds with little vegetation, and because wind can rarely move particles larger than sand or small pebbles.

Ice in **glaciers** is very effective at eroding and transporting material of all sizes. Glaciers can move boulders as large as a house hundreds of miles.

Generally, erosive agents remove sediments from the site of weathering in one of three ways: impact of the agent, abrasion (both types of mechanical erosion, or corrasion), or **corrosion** (chemical erosion). The mere impact of wind, water, and ice erodes sediments; for example, flowing water exerts a force on sediments causing them to be swept away. The eroded sediments may already be loose, or they may be torn away from the rock surface by the force of the water. If the flow is strong enough, **clay**, silt, sand, and even gravel, can be eroded in this way.

Sediments come in all shapes and sizes. Sediment sizes are classified by separating them into a number of groups, based on metric measurements, and naming them using common terms and size modifiers. The terms, in order of decreasing size, are boulder (>256 mm), cobble (256–64 mm), pebble (64–2 mm), sand (2-1/16 mm), silt (1/16–1/256 mm), and clay (<1/256 mm). The modifiers in decreasing size order, are very coarse, coarse, medium, fine, and very fine. For example, sand is sediment that ranges in size from 2 millimeters to 1/16 mm. Very coarse sand ranges from 2 mm to 1 mm; coarse from 1 mm to 1/2 mm; medium from 1/2 mm to 1/4 mm; fine from 1/4 mm to 1/8 mm; and very fine from 1/8 mm to 1/16 mm. Unfortunately, the entire classification is not as consistent as the terminology for sand - not every group includes size modifiers.

When particles are eroded and transported by wind, water, or ice, they become part of the transport medium's sediment load. There are three categories of load that may be transported by an erosion agent: dissolved load, **suspended load**, and bedload. Wind is not capable of dissolving minerals, and so it does not transport any dissolved load. The dissolved load in water and ice is not visible; to be deposited, it must be chemically precipitated.

Sediment can be suspended in wind, water, or ice. Suspended sediment is what makes stream water look dirty after a rainstorm and what makes a windstorm dusty. Suspended sediment is sediment that is not continuously in contact with the underlying surface (a stream bed or the desert floor) and so is suspended within the medium of transport. Generally, the smallest particles of sediment are likely to be suspended; occasionally sand is suspended by powerful winds and pebbles are suspended by floodwaters. However, because ice is a solid, virtually any size sediment can be part of the suspended sediment load of a glacier.

Bedload consists of the larger sediment that is only sporadically transported. Bedload remains in almost continuous contact with the bottom, and moves by rolling, skipping, or sliding along the bottom. Pebbles on a riverbed or beach are examples of bedload. Wind, water, and ice can all transport bedload, however, the size of sediment in the bedload varies greatly among these three transport agents.

Because of the low density of air, wind only rarely moves bedload coarser than fine sand. Some streams transport pebbles and coarser sediment only during **floods**, while other streams may, on a daily basis, transport all but boulders with ease.

Floodwater greatly increases the power of streams. For example, many streams can move boulders during flooding. Flooding also may cause large sections of a riverbank to be washed into the water and become part of its load. Bank erosion during flood events by a combination of abrasion, hydraulic impact, and mass wasting is often a significant source of a stream's load. Ice in glaciers, because it is a solid, can transport virtually any size material if the ice is sufficiently thick and the slope is steep.

For a particular agent of transport, its ability to move coarse sediments as either bedload or suspended load is dependant on its velocity. The higher the velocity, the coarser the load.

Transport of sediments causes them to become rounder as their irregular edges are removed by both abrasion and corrosion. Beach sand becomes highly rounded due to its endless rolling and bouncing in the surf. Of the agents of transport, wind is most effective at mechanically rounding (abrading) clastic sediments, or clasts. Its low density does not provide much of a "cushion" between the grains as they strike one another.

Sorting, or separation of clasts into similar sizes, also happens during sediment transport. Sorting occurs because the size of the grains that a medium of transport can move is limited by the medium's velocity and density. For example, in a stream on a particular day, water flow may only be strong enough to transport grains that are finer than medium-grained sand. So all clasts on the surface of the streambed that are equal to or larger than medium sand will be left behind. The sediment, therefore, becomes sorted. The easiest place to recognize this phenomenon is at the beach. Beach sand is very well sorted because coarser grains are only rarely transported up the beach face by the approaching waves, and finer material is suspended and carried away by the surf.

Ice is the poorest sorter of sediment. Glaciers can transport almost any size sediment easily, and when ice flow slows down or stops the sediment is not deposited due to the density of the ice. As a result, sediments deposited directly by ice when it melts are usually very poorly sorted. Significant sorting only occurs in glacial sediments that are subsequently transported by meltwater from the glacier. Wind, on the other hand, is the best sorter of sediment because it can usually only transport sediment that ranges in size from sand to clay. Occasional variation in wind speed during transport serves to further sort out these sediment sizes.

When the velocity (force) of the transport medium is insufficient to move a clastic (or organic) sediment particle it is deposited. When velocity decreases in wind or water, larger sediments are deposited first. Sediments that were part of the suspended load will drop out and become part of the bedload. If velocity continues to drop, nearly all bedload movement will cease, and only clay and the finest silt will be left suspended. In still water, even the clay will be deposited, over the next day or so, based on size—from largest clay particles to the smallest.

During its trip from outcrop to ocean, a typical sediment grain may be deposited, temporarily, thousands of times. However, when the transport medium's velocity increases again, these deposits will again be eroded and transported. Surprisingly, when compacted fine-grained clay deposits are subjected to stream erosion, they are nearly as difficult to erode as pebbles and boulders. Because the tiny clay particles are electrostatically attracted to one another, they resist erosion as well as much coarser grains. This is significant, for example, when comparing the erodibility of stream bank materials—clay soils in a river bank are fairly resistant to erosion, whereas sandy soils are not.

Eventually the sediment will reach a final resting place where it remains long enough to be buried by other sediments. This is known as the sediment's depositional environment.

Unlike clastic and organic sediment, chemical sediment cannot simply be deposited by a decrease in water velocity. Chemical sediment must crystallize from the solution; that is, it must be precipitated. A common way for **precipitation** to occur is by **evaporation**. As water evaporates from the surface, if it is not replaced by water from another source (rainfall or a stream) any dissolved minerals in the water will become more concentrated until they begin to precipitate out of the water and accumulate on the bottom. This often occurs in the desert in what are known as saltpans or **lakes**. It may also occur along the sea coast in a salt marsh.

Another mechanism that triggers mineral precipitation is a change in water **temperature**. When ocean waters with different temperatures mix, the end result may be seawater in which the concentration of dissolved minerals is higher than can be held in solution at that water temperature, and minerals will precipitate. For most minerals, their tendency to precipitate increases with decreasing water temperature. However, for some minerals, calcite (calcium carbonate) for example, the reverse is true.

Minerals may also be forced to precipitate by the biological activity of certain organisms. For example, when algae remove **carbon dioxide** from water the acidity of the water decreases, promoting the precipitation of calcite. Some marine organisms use this reaction, or similar chemical reactions, to promote mineral precipitation and use the minerals to form their skeletons. Clams, snails, hard corals, sea urchins, and a large variety of other marine organisms form their exoskeletons by manipulating water **chemistry** in this way.

Landscapes form and constantly change due to weathering and sedimentation. The **area** where sediment accumulates and is later buried by other sediments is known as its depositional environment. There are many large-scale, or regional, environments of deposition, as well as hundreds of smaller subenvironments within these regions. For example, **rivers** are regional **depositional environments**. Some span distances of hundreds of miles and contain a large number of subenvironments, such as channels, backswamps, **floodplains**, abandoned channels, and sand bars. These depositional subenvironments can also be thought of as depositional **landforms**, that is, landforms produced by deposition rather than erosion.

Erosion, weathering, and sedimentation constantly work together to reshape the earth's surface. These are natural processes that sometimes require us to adapt and adjust to changes in our environment. However, too many people and too much disturbance of the land surface can drastically increase sedimentation rates, leading to significant increases in the frequency and severity of certain natural disasters. For example, disturbance by construction and related land development is sometimes a contributing factor in the mudflows and landslides that occur in certain areas of California. The resulting damage can be costly both in terms of money and lives.

The world's rivers carry as much as 24 million tons of sediment to the ocean each year. About two-thirds of this may be directly related to human activity, which greatly accelerates the natural rate of erosion. This causes rapid loss of fertile topsoil, which leads to decreased crop productivity.

Increased sedimentation also causes increased size and frequency of flooding. As stream channels are filled in, the capacity of the channel decreases. As a result, streams flood more rapidly during a rainstorm, as well as more often, and they drain less quickly after flooding. Likewise, sedimentation can become a major problem on dammed rivers. Sediment accumulates in the lake created by the dam rather than moving farther downstream and accumulating in a **delta**. Over time, trapped sediment reduces the size of the lake and the useful life of the dam. In areas that are forested, lakes formed by dams are not as susceptible to this problem. Sedimentation is not as great due to interception of rainfall by the trees and underbrush.

Vegetative cover also prevents soil from washing into streams by holding the soil in place. Without vegetation, erosion rates can increase significantly. Human activity that disturbs the natural landscape and increases sediment loads to streams also disturbs aquatic ecosystems. Many state and local governments are now developing regulations concerning erosion and sedimentation resulting from private and commercial development.

Seeing

Astronomical seeing refers to the ability to view celestial objects through the obscurations of the earth's atmosphere. These obscurations include opacity, scattering, turbulence, atmospheric and thermal emission, and ionization.

Opacity refers to the fact that Earth's atmosphere is transparent only to relatively narrow wavelength ranges of light. These include visual light, the near infrared, microwaves and radio waves with wavelengths between about 0.35 mm and 1 m. The atmosphere is almost completely opaque to ultraviolet light, x rays, gamma rays, and radio waves with

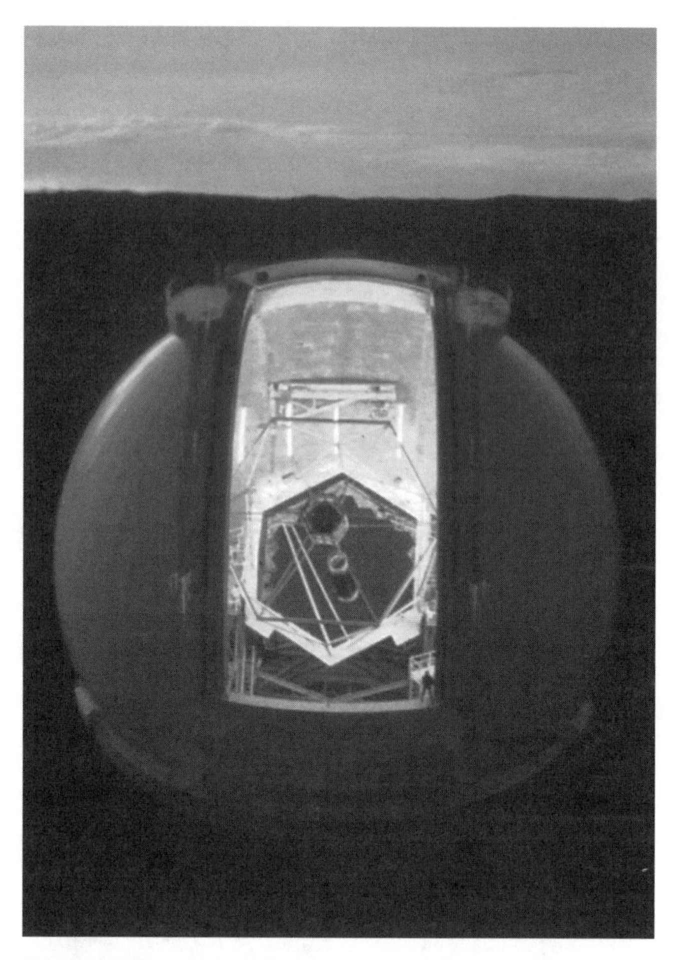

The Keck I Observatory in Hawaii is one of the most powerful ground-based telescopes in the world. ***© Roger Ressmeyer/Starlight/for the W. M. Keck Observatory, courtesy of California Associate for Research and Astronomy***

wavelengths greater than 1 m. The need to observe heavenly bodies outside of these narrow wavelength windows, along with the desirability to avoid the degrading affects of the atmosphere are among the main reasons for the development of space-based telescopes.

Ultraviolet photons are absorbed by electron transitions in **oxygen** and **ozone** atoms in the upper atmosphere. Because the amount of ozone varies greatly with location and seasonal time of year (e.g., the hole in the ozone layer over **Antarctica** during the winter) so does the amount of ultraviolet light reaching the surface of Earth. Ultraviolet light is mainly absorbed by molecular nitrogen. Infrared light is absorbed mainly by **water** vapor and **carbon dioxide** in the earth's atmosphere. Millimeter wavelengths are absorbed by rotational bands of water and molecular oxygen, while long radio waves are absorbed by ions high in the earth's atmosphere. Because 50% of the water vapor lies within three kilometers of the earth's surface, to some degree the infrared spectrum may be observed with instruments located on high mountains, or mounted on airplanes or balloons. Wavelength regions at which the atmosphere is transparent are called atmospheric windows.

Scattering of light particles degrades seeing. The mechanism by which molecules scatter visible light is called Rayleigh scattering, and the degree to which light is scattered is inversely proportional to the fourth power of the wavelength and proportional to the density of atmosphere. Thus, blue light is scattered more strongly than red light, accounting for the blue color of the sky. Red sunsets are an optical illusion caused by the intense scattering of blue light as the light rays travel through the horizon-level thickest regions of the atmosphere.

Atmospheric turbulence resulting from thermal currents, or **wind**, creates small changes in the density of pockets of air that cause the direction of light rays from point sources such as stars to be changed by refraction. In effect, the position of the star seems to shift slightly, and the star appears to twinkle. On a photographic plate, turbulence results in a smearing of the stellar image. Details of planetary features are often highly obscured and degraded by turbulence. The human eye, which processes light almost instantly, often sees a much sharper image than may be obtained with photographs.

Atmospheric emission of the night sky, also called airglow, is caused by the recombination of electrons with atoms that were ionized during the day by photochemical dissociation. This fluorescent light arises from neutral oxygen atoms and molecules, sodium, hydrogen, and hydroxide molecules, and is emitted about 100 km above the earth's surface. The total airglow over one arcsecond, roughly the apparent size of a star with a large **telescope**, corresponds to a visual magnitude of 22. Thus, stars dimmer than this, or galaxies with a surface **area** brightness less than this, are difficult to detect.

Thermal emission of the night sky is also a factor in the near infrared part of the spectrum. Any warm object in thermal equilibrium will emit black body radiation (e.g., **iron** heated to several hundred degrees will glow red or white). The earth's atmosphere below about 50 km emits a faint light by this mechanism, equivalent to a few Janskys (10^{-26} Watts/m^{2}Hz).

Ionization in the earth's **ionosphere** degrades radio waves. Turbulence in this layer of the atmosphere causes small fluctuations in the density of free electrons, which alter the direction of radio waves and cause dispersion in the frequencies of the radio waves. A radio wave emitted from a star of a particular wavelength or frequency will be dispersed into a small range of frequencies by ionization.

See also Astronomy; Atmospheric chemistry; Atmospheric circulation; Atmospheric composition and structure; Atmospheric inversion layers; Atmospheric pollution; Space and planetary geology

SEISMOGRAPH

A seismograph is an instrument used to measure and record ground vibration caused by explosions and **earthquake** shock waves. In the late 1800s, John Milne (1850–1913), an English mining engineer, developed the first precise seismometer, the sensor in a seismograph that detects and measures motion. Since then, seismograms, the data recorded by a seismograph, have helped seismologists predict much more than Earth movement. These devices have also led to discoveries about the nature of the earth's core.

The process of using a tool to detect ground motion dates back to the ancient Han Dynasty when Chang Heng, a first-century Chinese astronomer and mathematician, invented the first seismometer. He used a pendulum connected to an eight-spoked wheel in which each spoke was connected to a mounted dragon's head with moveable jaws. When the pendulum moved during an earthquake, a bronze ball in each of the heads would pop out if hit by the spoke of the pendulum wheel. While this did not lend clues about the force of an earthquake, it gave the ancient scientist an idea of the direction of the shock waves and their source. Since that time, Heng's concept has been refined considerably. Later seismographs employed a heavy pendulum with a *stylus*, or needle, suspended above a revolving drum. The drum contained a device on which the etchings from the needle could be recorded. During an earthquake, the pendulum and needle remained motionless while the drum on the base moved, recording the earth's movement. As much as these later pendulum seismographs improved upon the ancient Chinese method, they still fell short of providing answers to the many questions that arose with more precise readings. For example, once a strong motion set off a seismograph's pendulum, the pendulum would swing indefinitely, failing to record aftershocks that followed the initial disturbance. Additionally, the seismographs of the late 1800s recorded only a limited range of wave sizes and numbers. The inverted pendulum, invented by German seismologist Emil Weichert in 1899, helped overcome many of these limitations. Weichert employed a system of mechanical levers that linked the pendulum movement more closely to the earth's vibrations.

In 1906, Boris Golitsyn (1862–1916), a Soviet physicist and seismologist, devised the first electromagnetic seismograph; for the first time, a seismograph could be operated without mechanical levers. Although many of the modern seismographs are complicated technical devices, these instruments contain five basic parts. The clock records the exact time that the event takes place and marks the arrival time of each specific wave. The support structure, which is always securely attached to the ground, withstands the earth's vibrations during the earthquake or explosion. The inertial mass is a surface **area** that does not move although the earth and the support structure oscillate around it. The pivot holds the inertial mass in place, enabling it to record the earth's vibration. The vibrations are registered through the recording device: essentially a pen attached to the inertial mass and a roll of paper. The paper moves along with the earth's vibrations while the pen remains stationary. This shows the pattern of shock waves by recording thin, wavy lines, revealing the strength of the various waves as well as the frequency with which they occurred.

After the first modern seismograph was installed in the United States at the University of California at Berkeley, it recorded the 1906 earthquake that devastated San Francisco. Not long before that, Weichert and fellow scientist Richard Oldham (1858–1936) were finally able to determine the existence of the earth's core through precise recordings of seismic waves. In 1909, the use of a seismograph helped Yugoslavian seismologist and meteorologist Andrija Mohorovičic (1857–1936) discover the location at which the earth's **crust** meets the upper mantle. That discovery was followed in 1914 by Inge Lehmann's discovery of the boundary between the earth's outer and inner core. These important findings finally secured knowledge about the existence of boundaries for all of the earth's major layers: the inner core, the outer core, the mantle, and the crust. Seismographs also help miners determine the amount of dynamite needed for quarry blasts. Seismographs detect the force of atomic blasts and nuclear explosions, and are also used to detect the speed of seismic waves traveling in the earth. This data provides valuable information about the substances of which the earth is comprised, such as the natural resources oil and **coal**.

See also Earth, interior structure; Faults and fractures; Mohorovicic discontinuity (Moho); Richter scale

SEISMOLOGY

Seismology is the science that studies earthquakes and phenomena connected with them. Seismology is a branch of geophysics.

Seismology attempts to explain the origin of earthquakes, where, when, and why they occur, what accompanies them, and how to forecast them. Earthquakes were mentioned in written historical documents as early several thousands of years ago, but their serious study began only in the nineteenth century. As a rough guide, the **earthquake** is a vibration of the ground tangible in a definite place; the stronger these vibrations are, the more damage an earthquake can cause.

Two variables are usually used for describing the earthquake power: magnitude and intensity. Magnitude is an objective parameter, which is connected with the ground displacement at the point of its measurement; the bigger the displacement, the stronger is the earthquake. Earthquakes with magnitudes bigger than 5–6 are considered powerful ones. Intensity is a parameter that is not measured by a device. Different factors are taken into account to determine the intensity of an earthquake, and its value varies relative to locations accessed. The Modified Mercalli Intensity Scale determines earthquake intensity in the United States by gathering information, including witness accounts and building damage, and assigning a numeral scale from I (low intensity, no earth movement felt) to XII (visible earth movement seen, severe building damage).

The earth's **crust**, the upper layer of Earth's surface, consists of a solid medium with different values of parameters in different points, and is exposed to permanent action of different forces, which are also irregular. Action of these forces can lead to a situation in which some parts of the crust can occur under the condition of very high tension (for example, like a rod that is curved). If the tension remains too high, the crust is damaged in some points (the rod is broken). In this case, a very big amount of energy becomes free, and this energy transfers into elastic waves of different kinds, which can spread to great distances from the damage point. This illustrates a simplified model of an earthquake.

It is possible to distinguish the areas where earthquakes occur more often than in other places. Usually, these are mountain areas, and areas circumambient the Pacific (Japan, for example). Seismology studies these areas together with **geology**, in which during the late nineteenth century arose the special theory (**plate tectonics**) which explains a set of phenomena, in particular the occurrence of active seismic zones and non-active zones (platform regions). Earthquakes are practically nonexistent in non-active zones (for example, the Russian platform can be referred to as one of these regions).

Both seismology and geology use the achievements of **physics** (such as elasticity theory and hydrodynamics) in the creation of theories such as plate tectonics. In studying earthquakes, the oscillation theory and the theory of wave propagation in elastic media are used. The main devices used for earthquake study are seismometers, which record media oscillations at the point of the device location. Nowadays, such devices are located in many points on the earth's surface, on the ocean bottom, in shafts—often they are joined in special nets. The analysis of seismometers (seismograms), which take into account wave propagation theory, performs another important role; it permits scientists to understand deep-Earth structure. Even with improved theory, calculation methods, and equipment used in seismology, reliable forecasts of earthquakes still cannot be achieved.

Another important aspect of seismology is that its applied methods permit scientists to search useful **minerals**, especially oil. Seismologic methods also give the most precise results in underground nuclear tests control.

See also Earth, interior structure; Faults and fractures; Mid-ocean ridges and rifts; Mid-plate earthquakes; Petroleum detection; Richter scale; Subduction zone

SETI

SETI (The Search for Extraterrestrial Intelligence) is a term that encompasses several different groups and their efforts to seek out intelligent extraterrestrial life. The driving force behind these groups is the ancient human desire to understand the origin and distribution of life throughout the Universe. As technology progresses, SETI efforts move from the study of extraterrestrial rocks and meteors towards scanning the skies for a variety of signal types.

Cornell University professor Frank Drake founded the first SETI program in late 1959. Drake reinforced his idea of scanning the skies with his famous Drake Equation. The Drake Equation predicts the abundance of intelligent life within a certain galaxy.

The second major development of SETI took shape in the late 1960s when NASA joined the program. NASA was minimally involved the project, but spawned many SETI related programs. These programs included the Microwave Observing Project, Project Orion, the High Resolution Microwave Survey, Toward Other Planetary Systems, and more. One of the most intensive SETI related programs NASA would initiate began in 1992, but Congress cut funding for the program within a year. SETI projects now must rely on private funding, and SETI operates through the SETI Institute, a non-profit corporation.

The Very Large Array (VLA) is a collection of radio telescopes set up in a series, which turns the entire array into one vast radio telescope. ***JLM Visuals. Reproduced by permission.***

Historically, scientists used several different methods for searching for extraterrestrial intelligence. The earliest method, and still most commonly used in present research, is the scanning of electromagnetic emissions. Radio waves are picked up by an array of radio telescopes and scanned for non-random patterns. More modern methods expand the search to other regions of the **electromagnetic spectrum**, including the infrared spectrum.

As of 2002, the University of California at Berkeley hosts the most widespread SETI effort in history. Berkeley projects include SETI@Home, SERENDIP, Optical SETI, and Southern SERENDIP. SETI@Home collects its data in the background of the Arecibo Radio Observatory and relays it back to the lab in Berkeley. The data is then divided into workunits and sent out to the personal computers of volunteers throughout the world. These personal computers scan the data for candidate signals. If a candidate signal appears, it is relayed back to Berkeley, where the signal is checked for data integrity. Finally, the lab removes radio interference and scans the data for final candidates. The Berkeley faction of SETI will be expanding their efforts with the Allen **Telescope** Array (formerly known as the One Hectare Telescope) designated specifically for this research.

Project Phoenix, also run by the SETI Institute, concentrates on obtaining signals from targeted areas within our galaxy. The focused Phoenix receiver can amass radio energy for longer periods of time and with greater sensitivity than previous SETI radio telescopes, allowing for faster and more precise analysis.

Although only a small fraction of the sky has been scanned, so far, SETI initiatives have not confirmed a signal from an extraterrestrial source that is conclusive proof of an extraterrestrial intelligence. A few strong and unexplained signals have intrigued SETI scientists; the most well known was received in 1977 at the Ohio State Radio Observatory. None of the signals have ever repeated.

SHARDS • *see* TUFF

SHEAR ZONES

Shear zones are microscopic to regional scale domains across which displacement has occurred. Brittle, brittle-ductile, or ductile deformation processes occur in shear zones at shallow, intermediate, or deep levels in the **crust**, respectively.

In brittle shear zones (equivalent to fault zones), displacement occurs on discrete fracture surfaces. In brittle-ductile shear zones, all or portions of the zone may undergo both ductile and brittle deformation. Displacement across brittle-ductile shear zones can be accommodated by oblique, en échelon stepping extensional veins (tension fractures) and/or shear fractures in addition to through-going shears parallel to zone boundaries. Extensional veins may be subsequently deformed into sigmoidal shapes (S shapes) and cut by younger veins. The sense of stepping of veins and their sigmoidal shape is used to determine the sense of displacement. **Folds** within transcurrent brittle-ductile shear zones (i.e., where displacement across steeply dipping shear zones is sub-horizontal) are en échelon stepping and doubly plunging. Fold axial surfaces initiated at approximately 45° to zone boundaries progressively rotate towards parallelism with the shear zone in areas of greatest deformation commonly overlying a crustal-scale structure.

Layers offset by ductile shear zones are thinned and progressively bent into parallelism with the zone. Grain size reduction in shear zones produces rocks called mylonites. The word mylonite is derived from the Greek word *mulon* or *mulos* for mill and the suffix *–ite,* for product of. Despite the origin of the word mylonite, only harder **minerals** such as **feldspar** are fractured and mechanically ground. Minerals such as **quartz** deform in a plastic manner instead, and are smeared out to form quartz ribbons. The term ultramylonite is used where grain size reduction has been extreme. The precursor **rock** type may be difficult to distinguish, e.g., ultramylonitic **granite**, **felsic** volcanic, and metasedimentary rock may all appear almost identical. Where the sense of offset of markers is not apparent, as is often the case in regional-scale ductile shear zones, the sense of displacement may be determined from observation of sections perpendicular to the foliation and parallel to the displacement direction (given by the orientation of mineral elongation lineations). This often requires microscopic, thin section examination. Shear criteria include:

- The relative orientations of flattening (S) and shear (C) foliations. S develops at about 45° to C, and is bent towards parallelism with C with increasing strain.
- Synthetic (C' or ecc1) shears at approximately 15–30° to C displace the shear foliation in mylonitic rocks with the same sense as the bulk shear sense. Antithetic shears (ecc2) with the opposite sense of displacement form at a large angle to the foliation.
- The asymmetry of foliations and strain (pressure) shadows around hard mineral grains or competent clasts. Low strain areas caused by ductile flow around rigid objects are sites where minerals (e.g., quartz, calcite, amphiboles, and biotite) crystallize. Mineral growths in strain shadows and surrounding foliations may be deformed when rigid objects rotate during shearing.
- Synthetic or antithetic slip on mineral cleavages. Slip and separation may occur in minerals such as pyroxene with pronounced mineral cleavage planes. Antithetic domino or bookshelf style slip occurs on planes inclined against the sense of shear at a large angle to the shear foliation. Cleavages inclined at small angles towards the sense of shear show synthetic slip.
- Mica fish (fish-shaped mica **crystals** whose extremities are asymmetrically bent into shear planes).
- X-ray, laser, or optical measurements of crystallographic axes provide shear criteria in quartz-rich mylonites.

Folds develop within ductile shear zones where layers are inclined to shear zone boundaries or are adjacent to irregularities that perturb ductile flow. Folds in ductile shear zones generally initiate with axes at a large angle to the displacement direction and are overturned in a sense consistent with the sense of shear. Care must, however, be taken in using fold asymmetry to establish the sense of shear. Folds in overturned layers at high strain within shear zones, and folds formed by the back-rotation of layers between two shear zones, may be overturned in a sense opposite to the sense of shear displacement. Fold axes progressively rotate towards the shear direction with progressive deformation. Folds with curved axes and sheath-like folds may form at high strains, producing complex, refolded folds within a single deformation event.

See also Faults and fractures; Plate tectonics

SHEPARD, ALAN B. (1923-1998)

American astronaut

One of the original seven American astronauts, Alan B. Shepard, Jr. became the first American to venture into **space** in a suborbital flight aboard the Mercury capsule, *Freedom 7.* His achievements—including his landmark *Freedom 7* flight on May 5, 1961—symbolized the beginning of a technological **revolution** in the 1960s and marked the onset of "new frontiers" in space. A decade later, he commanded the *Apollo 14* lunar mission, becoming the fifth man to step on the Moon's surface and the only one of the original astronauts to make a flight to the **Moon**. In addition to his space flight accomplishments, Shepard served as Chief of the Astronaut Office and participated in the overall astronaut-training program. He received the National Aeronautics Space Administration

(NASA) Distinguished Service Medal from President John F. Kennedy for his Mercury flight. In 1971, appointed by President Nixon, he served as a delegate to the 26th United Nations General Assembly. He was promoted to rear admiral by the Navy in 1971, the first astronaut to achieve flag rank. In 1979, President Jimmy Carter awarded him the Medal of Honor for gallantry in the astronaut corps.

Alan Bartlett Shepard, Jr. was born in East Derry, New Hampshire, and spent most of his formative years in this New England setting. The son of a career military man—his father was an Army colonel—Shepard showed a strong interest at an early age for mechanical things, disassembling motors and engines and building model airplanes. He attended primary school in East Derry and received his secondary education from Pinkerton Academy in Derry in 1940. During high school, he did odd jobs at the local airport hangar in exchange for a chance to take airplane rides. There was little doubt in the family that Shepard would pursue a military career, and after completing a year's study at Admiral Farragut Academy in New Jersey, he entered the U.S. Naval Academy, graduating in 1944 with a B.S. in science. Shepard married Louise Brewer of Kennett Square, Pennsylvania; the couple eventually had two daughters.

Shepard's flying career began in 1947 after he served aboard the destroyer U.S.S. *Cogzwell* in the Pacific during the last year of World War II. He received flight training at both Corpus Christi, Texas, and Pensacola, Florida, receiving his wings in 1947. Between the years 1947 and 1950, he served with Fighter Squadron 42 at bases in Virginia and Florida, completing two cruises aboard carriers in the Mediterranean. In 1950, as a lieutenant, junior grade, he was selected to attend U.S. Navy Test Pilot School at Patuxent River in Maryland, serving two years in flight test work at that station. During those tours, he participated in high-altitude tests and experiments in the development of the Navy's in-flight refueling system. He was project test pilot on the F.S.D. *Skylancer* and was involved in testing the first angled deck on a U.S. Navy carrier. During his second tour to Patuxent for flight test work, the navy sent him to the Naval War College in Newport, Rhode Island. Upon graduation he became a staff officer at Atlantic Fleet Headquarters in Norfolk, in charge of aircraft readiness for the fleet. Being skipper of an aircraft squadron was a goal of any career pilot in the Navy and one that was of interest to Shepard. About this same time, NASA was developing Project Mercury and was seeking astronauts for America's space program.

Knowing that he met the required qualifications of NASA's advertised program, Shepard eagerly applied for a chance to serve his country and meet the challenge of the race to space. On April 27, 1959, NASA announced that Shepard and six other astronauts were selected as the first class of astronauts. A rigorous and intensive training program followed as preparations were being made for the first manned space flight. With the Russian space program forging ahead, it was imperative that a U.S. astronaut follow cosmonaut Yury Gagarin into space as soon as possible. Three astronauts—Shepard, Virgil "Gus" Grissom, and John Glenn—were selected to make three sub-orbital "up-and-down" missions to ready Mercury for orbital flight. Interest in the first manned American space flight was keen, forcing NASA to keep Shepard's identity secret until three days before the launch. At 9:45 A.M. on May 5th, 1961, Shepard, enclosed in the tiny bell-shaped Mercury capsule named *Freedom 7*, was thrust into space by a Redstone rocket to a distance of 2300 miles and a height of 113 miles above the surface of the Earth. The flight lasted only 15 minutes and 22 seconds and traveled at a speed of 5,180 mph. According to *Space Almanac*, Shepard, reporting from space that everything was "AOK," was in free-fall just five minutes before splashdown in the Atlantic Ocean. The U.S.S. *Lake Champlain* spotted his orange and white parachute 297 miles downrange from Cape Canaveral. Just before landing, the heat shield was dropped 4 feet, pulling out a rubberized landing bag designed to reduce shock. Shepard exclaimed "Boy what a ride," according to Tim Furniss in *Manned Spaceflight Log*, and with that successful, text-book perfect launch, NASA's space program gained support from the government and from people around the world.

Shepard's performance also showed the world the tradition of engineering excellence, professionalism and dedication that was evident in the subsequent missions. About ten weeks after this historic flight another Mercury-Redstone blasted Virgil Grissom's spacecraft for a similar flight. Shepard continued his training and space preparation and was selected for one of the early Gemini flights, but in early 1964, his career was sharply changed by an inner–ear ailment called Meniere's syndrome, which causes an imbalance and a gradual degradation of hearing. The Navy doctors would not let Shepard fly solo in jet planes, which forced NASA to ground him. The offer of a job as Chief of the Astronaut office with NASA came along about this time, and it helped allay some of the intense disappointment that Shepard experienced. As Chief, Shepard was in charge of all phases of the astronaut-training program and played an influential role in the selection of crews for upcoming missions. Periodic checks on his condition during this time showed a continued loss of hearing on the left side, and in May 1968, he submitted to an experimental operation to insert a plastic tube to relieve the pressure in his inner ear. After waiting six months for the final results of the operation, Shepard was declared by NASA officials and doctors fully fit to fly and to resume his role in the space flight program.

Shepard worked hard and long to ready himself for his next space endeavor, *Apollo 14*, which would last nine days and send a crew of three to the moon. The crew for this flight, Shepard as mission commander, Stuart Roosa as Command module pilot, and Edgar Mitchell as pilot of the lunar excursion module, was chosen in August 1969, just after the successful moon landing by *Apollo 11*. The mission was tentatively scheduled for an October 1970, launch date, but the explosion of the **oxygen** tank aboard *Apollo 13* called for several alterations in the *Apollo 14* spacecraft. One of the goals of this flight was to explore the Fra Mauro region of the Moon, and Shepard and Mitchell each spent more than 300 hours walking in **desert** areas and using simulators that resembled the lunar surface. A Saturn V rocket launched the

Apollo 14 capsule at 4:03 p.m. on January 3l, 1971. The astronauts had chosen the name *Kitty Hawk* as a tribute to the first manned powered flight in 1903, and named the lunar lander *Antares* for the star on which it would orient itself just before descending to the Fra Mauro landing site. *Apollo 14* entered lunar orbit on February 4, with touchdown in the uplands of the cone crater scheduled for the next day. Shepard and Mitchell departed from *Kitty Hawk* in *Antares* and descended smoothly to the surface, coming to rest on an 8 degree slope. Shepard descended the lander's ladder, stepping on the moon at 9:53 A.M., February 5, becoming the fifth man to walk on the moon. With much emotion, he reported to Houston, "I'm on the surface. It's been a long way, and I'm here," as quoted by Anthony J. Cipriano in *America's Journeys into Space*. He and Mitchell then collected 43 pounds of lunar samples and deployed TV, communications and scientific equipment in their first extra vehicular activity (EVA), which lasted 4 hours, 49 minutes. Their second EVA lasted 4 hours, 35 minutes, and the two astronauts used a Modularized Equipment Transporter for this landing. It was a rickshaw-like device in which they pulled their tools, cameras and samples with them across the moon. Shepard and Mitchell set off the first two moonquakes to be read by seismic monitors planted by earlier *Apollo* moonwalkers. As they prepared to leave the lunar surface in *Antares*, Shepard, an avid golfer, surprised his audience by making the first golf shot on the Moon, rigging a 6-iron club head to the end of a digging tool and hitting a ball hundreds of yards. On February 6, *Kitty Hawk* rocketed out of lunar orbit and headed for Earth. After nearly three days of coasting flight, *Apollo 14* splashed down in the Pacific Ocean, 4.6 miles from the recovery vessel *New Orleans*, on February 9—216 hours, 42 minutes after launch.

At Shepard's retirement from NASA and the U.S. Navy on August 1, 1974, Dr. James C. Fletcher, NASA Administrator, praised the astronaut's dedication and determination in a *NASA News* bulletin. "Al Shepard was the first American to make a space flight and his determination to overcome a physical ailment after his suborbital mission carried him to a highly successful manned lunar landing mission." Shepard joined the private sector as partner and chairman of the Marathon Construction Company of Houston, Texas. He became a successful businessman in Houston, pursuing interests as a commercial property developer, a venture capital group partner, and director of mutual fund companies. He also chaired the board of the Mercury Seven Foundation, created by the six living Mercury Seven astronauts and Grissom's widow to raise money for science and engineering scholarships. The Mercury capsule *Freedom 7* is on display at the National Air and Space Museum in Washington, D.C., and the *Apollo 14* command module *Kitty Hawk* is displayed at the Los Angeles County Museum in California. Shepard died in Houston at the age of 74.

See also Spacecraft, manned

SHOCK METAMORPHISM

Except for certain laboratory experiments and outdoor detonations of high explosives (including nuclear weapons), evidence of shock metamorphic conditions of extreme pressure and heat on Earth exist only within and around impact craters. Only hypervelocity impact between objects of substantial size moving at cosmic velocity (at least several kilometers per second) can produce these conditions. Meteorites larger than approximately 246 ft (75 m) in diameter, **asteroids**, and **comets** can pass through Earth's atmosphere and yet retain a very high percentage of their original velocity. In such events, this energy of motion is converted into seismic and heat energy almost instantaneously. On planetary bodies with no atmosphere, even smaller impacting bodies (even micrometeorites) can produce shock metamorphic effects. Meteorites recovered on Earth, which are fragments of larger bodies shattered by impact elsewhere in the **solar system**, also show shock metamorphic features and effects.

Shock **metamorphism** involves changes wrought by instantaneously applied extreme pressure and heat. This contrasts sharply with metamorphic changes accompanying development of most of Earth's metamorphic crustal rocks via long-term contact, cataclastic, and regional metamorphic conditions. Typical shock metamorphic pressures range from 72,519 to 14,503,774 psi (0.5 to 100 Gpa). Usually shock temperatures range from a few hundred to a few thousand degrees Fahrenheit or Celsius.

Shock metamorphism manifests itself through unique physical changes in mineral characteristics. There are five recognized shock stages, which are numbered 0, Ia, Ib, II, and III. **Quartz** (SiO_2), a very common mineral on Earth, displays a range of shock effects that make it one of the most studied **minerals** in shock metamorphism. At shock stage 0, Ia, and Ib, quartz displays progressively greater numbers of planar features (PFs), numbers of planar deformation features (PDFs), and extent of mosaicism. PFs form at threshold shock temperatures and pressures. PFs are microscopically thin fissures, spaced at about 20 microns or more, which are parallel to selected atomic planes within the quartz crystal. At higher temperatures and pressures, PFs give way to PDFs and mosaicism. PDFs are microscopic, parallel zones within the quartz cystal, spaced at 2–10 microns. PDFs are strongly planar and are arranged in specific crystallographic orientations. The number of PDFs and the specific orientation of PFs and PDFs are diagnostic of approximate levels of shock pressure. PFs and PDFs are found mainly in quartz, but also occur in some other silicate minerals as well. Mosaicism is a microscopic type of shock metamorphism in quartz and some other silicate minerals. It is an irregular, mottled pattern, as revealed under polarized light. Mosaicism is shock-induced expansion of crystal volume that results in multiple crystal development within the original (pre-shock) crystal.

At shock stages II and III, high-pressure forms (polymorphs) of pre-existing minerals and shock-produced melts can form. For quartz, the high-pressure polymorphs are stishovite and coesite, both SiO_2 like quartz, but with different internal structures of atoms (and higher densities). Other common minerals and their high-pressure polymorphs (in parentheses) are: **olivine** (ringwoodite); **plagioclase** (jadite); pyroxene (majorite); and **graphite** (lonsdaleite, a polycrystalline, impure **diamond**). For pressures over 60 gigapaschals, rocks can undergo complete **melting** and thus form impact melts. These melts may have very high temperatures (due to shock-wave passage); temperatures tend to be much higher than normal crustal processes or volcanic activity would produce. These extreme temperatures generate high-temperature polymorphs of common minerals such a lechatelierite (SiO_2, like quartz), which forms at temperatures over 3,092°F (1,700°C), and baddeleyite ($ZrSiO_4$, like zircon), which forms at temperatures over 3,452°F (1,900°C). Lechatelierite is not found in any other natural material, except fulgarites (fused **soil** or **sand** from **lightning** strikes). Impact melt **temperature** may exceed 18,000°F (10,000°C) and thus, pre-existing **rock** like **limestone**, which has a very high melting temperature, may become liquid. Rapid cooling of such melts produces various kinds of impact glassses.

See also Impact crater; Metamorphic rock

SIDEROPHILE • *see* CHALCOPHILES, LITHOPHILES, SIDEROPHILES, AND ATMOPHILES

SILICIC

Igneous rocks are classified by geologists using various schemes. One of the several schemes based on chemical composition divides igneous rocks into four categories according to silica (**silicon** dioxide, SiO_2) content: (1) Rocks containing more than 66% silica are silicic. (2) Rocks containing 52–66% silica are classified as intermediate. (3) Rocks containing 45–52% silica are **mafic**. (4) Rocks containing less than 45% silica are ultramafic.

Because silicon is less dense than the elements that often replace it in rock—especially **iron** and magnesium—silicic rocks are less dense than mafic rocks and tend to rise above them. The continents float on the mantle because they consist primarily of silicic rocks. Mafic and ultramafic rocks predominate in the mantle and oceanic **crust**.

All rocks high in **quartz**, which is simply crystalline silica, are silicic. **Feldspar** is a silicic mineral, and is a major component of many igneous rocks. Some common silicic rocks are **granite**, granodiorite, and **rhyolite**.

The terms silicic and siliceous (suh-LISH-us) are used synonymously by some geologists; others reserve silicic for high-silica igneous rocks, siliceous for high-silica **sedimentary rocks**. The term acid is sometimes used as a synonym for silicic and the terms basic and ferromagnesian as synonyms for mafic.

See also Isostasy

SILICON

Silicon (Si, element 14) is a nonmetallic chemical found in group IV, the **carbon** family, on the **Periodic table**. Swedish chemist Jons Jacob Berzelius first isolated and described the element in 1824.

In nature, silicon is always paired with another substance; it combines with **oxygen** to form **quartz** and **sand** (silicon dioxide, SiO_2) or with oxygen and a metal to form silicates, which are used to make **glass**, pottery, china, and other ceramics. The relatively inactive element occurs in nearly all rocks, as well as in **soil**, sand, and clays. It is the second most abundant element found in the earth's **crust**, surpassed only by oxygen.

Scientists create pure silicon by heating sand and coke in an electric furnace to remove oxide (oxygen) from the element. Pure silicon is colored dark gray and has a crystalline structure similar to **diamond**. The **crystals** are extremely hard and demonstrate remarkable insulating and semiconducting properties, which has made silicon an invaluable resource for the computing and electronics industries. A single purified silicon crystal contains millions of atoms accompanied by loosely attached electrons that break free upon the introduction of energy, such as light or heat. The flowing electrons conduct **electricity**, hence the term semiconductor. Today, silicon is the backbone of computer chips, transistors and many other electronic components.

Silicones, a chain of alternating silicon and oxygen atoms, are chemically inert and stable in the presence of high heat. The compounds are often used as lubricants, waterproofing materials and varnishes and enamels. Silicone gels have long been used as implants in the human body.

Silicon has an atomic weight of 28.086, a **melting** point of 2,570°F (1,410°C) and a boiling point of 4,270°F (2,355°C). Only three stable isotopes of silicon are known to exist: silicon-28, silicon-29 and silicon-30.

See also Earth (planet)

SILL

A sill is a formation of igneous **rock** found in features such as mesas, hogbacks, and cuestas. Although sills can become exposed, sills are formed underground and are thus composed of plutonic **igneous rocks**. Sills are an intrusive rock formation. Intrusive formations such as dikes and sills are formed from **magma** that solidifies beneath the Earth's surface and then intrudes into the overlying host or **country rock**.

Sills are distinguishable from the similarly formed dikes and dome forming laccoliths because sills are horizontal in orientation to the surface. Accordingly, sills are a horizontal intrusion.

Sills are sheet-like or tabular because the magma intrusion moves or intrudes horizontally before solidifying. Sills are characterized as concordant intrusive contacts. Because sills are an intrusive rock formation in contact with the host or country rock in the horizontal plane, they are often parallel to foliation or **bedding** planes (e.g., parallel to underlying sedimentary bedding planes). Sills form as rising magma encounters vertical resistance from host rock. The upwelling magma then spreads out in the horizontal plane into **area** of lower resistance to form sheet-like layers of rock.

Sill texture is a function of the time it takes for the magma to cool and solidify. Sills range from **aphanitic** to phaneritic in texture. In general, the longer the time to cool, the greater the extent and size of crystal formation. If a sill cools quickly, the texture is usually smooth and mineral **crystals** are not visible to the naked eye (aphanitic in texture). If conditions in the surrounding host rock are such that the magma cools over a long period of time, large visible crystals form a phaneritic texture.

When sills come to overlie sedimentary basins, the sills can act as horizontal obstructions that cap traps or reservoirs containing hydrocarbon **fuels**.

See also Dike; Petroleum detection; Pluton and plutonic bodies; Stratigraphy

SILT • *see* ROCK

SILURIAN PERIOD

In **geologic time**, the Silurian Period, the third period of the **Paleozoic Era**, covers the time from roughly 440 million years ago (mya) until 410 mya. The name, Silurian, derives from the Silures, an ancient British tribe. The Silurian Period spans two epochs. The Early Silurian Epoch is the most ancient, followed by the Late Silurian Epoch.

The Early Silurian Epoch is divided chronologically (from the most ancient to the most recent) into the Llandoverian and Wenlockian stages. The Late Silurian Epoch is divided chronologically (from the most ancient to the most recent) into the Ludlovian, and Pridolian stages.

In terms of paleogeography (the study of the **evolution** of the continents from **supercontinents** and the establishment of geologic features), the Silurian Period featured cleavage of some supercontinent landmass and fusion of plates into Laurussia. Collision with remnants of other continents later formed the supercontinent Laurasia and eventually the supercontinent Pangaea.

The **fossil record** establishes that the preceding **Ordovician Period** ended with a mass extinction. This mass extinction, approximately 440 mya, marked the end of the Ordovician Period and the start of the Silurian Period. In accord with a mass extinction, many fossils dated to the Ordovician Period are not found in Silurian Period formations. Differentiated by fossil remains and continental movements, the **Devonian Period** followed the Silurian Period.

The Silurian Period marked a geologically active period for volcanic activity. The accompanying ash deposits and **lava** flows are clearly evident in Silurian Period strata.

The fossil record indicates that it was during the Silurian Period that marine species made the evolutionary transition to terrestrial (land-based species). The first true insect fossils date to this period, as do fossils of jawed fish. Atmospheric changes, driven by increasingly diverse plant life, allowed the further development of the protective ozone layer, which filters out harmful ultraviolet radiation.

See also Archean; Cambrian Period; Cenozoic Era; Cretaceous Period; Dating methods; Eocene Epoch; Evolution, evidence of; Fossils and fossilization; Historical geology; Holocene Epoch; Jurassic Period; Mesozoic Era; Miocene Epoch; Mississippian Period; Oligocene Epoch; Paleocene Epoch; Pennsylvanian Period; Phanerozoic Eon; Pleistocene Epoch; Pliocene Epoch; Precambrian; Proterozoic Era; Quaternary Period; Tertiary Period; Triassic Period

SINKHOLES

Sinkholes are cavities that form when **water** erodes easily dissolved, or soluble, **rock** located beneath the ground surface. Water moves along joints, or fractures, enlarging them to form a channel that drains sediment and water into the subsurface. As the rock erodes, materials above subside into the openings. At the surface, sinkholes often appear as bowl-shaped depressions. If the drain becomes clogged with rock and **soil**, the sinkhole may fill with water. Many ponds and small **lakes** form via sinkholes.

Abundant sinkholes as well as caves, disappearing streams, and **springs**, characterize a type of landscape known as **karst topography**. Karst **topography** forms where **groundwater** erodes subsurface carbonate rock, such as **limestone** and **dolomite**, or evaporite rock, such as **gypsum** and halite (salt). **Carbon dioxide** (CO_2), when combined with the water in air and soil, acidifies the water. The slight acidity intensifies the corrosive ability of the water percolating into the soil and moving through fractured rock.

Geologists classify sinkholes mainly by their means of development. Collapse sinkholes are often funnel shaped. They form when soil or rock material collapses into a **cave**. Collapse may be sudden and damage is often significant; cars and homes may be swallowed by sinkholes.

Solution sinkholes form in rock with multiple vertical joints. Water passing along these joints expands them, allowing cover material to move into the openings. Solution sinkholes usually form slowly and minor damage occurs, such as cracking of building foundations.

Alluvial sinkholes are previously exposed sinkholes that, over time, partly or completely filled with Earth material. They can be hard to recognize and some are relatively stable.

Rejuvenated sinkholes are alluvial sinkholes in which the cover material once again begins to subside, producing a growing depression.

Uvalas are large sinkholes formed by the joining of several smaller sinkholes. Cockpits are extremely large sink-

Sinkholes can appear without warning, swallowing roads, cars, or in this case, a lake. ***© Fletcher W.K., 1982/Photo Researchers, Inc. Reproduced by permission.***

holes formed in thick limestone; some are more than a kilometer in diameter.

Sinkholes occur naturally, but are also induced by human activities. Pumping water from a well can trigger sinkhole collapse by lowering the **water table** and removing support for a cave's roof. Construction over sinkholes can also cause collapse. Sinkhole development may damage buildings, pipelines and roadways. Damage from the Winter Park sinkhole in Florida is estimated at greater than $2 million. Sinkholes may also serve as routes for the spread of contamination to groundwater when people use them as refuse dumps.

In areas where evaporite rock is common, human activities play an especially significant role in the formation of sinkholes. Evaporites dissolve in water much more easily than carbonate rocks. Salt mining and drilling into evaporite deposits allows water that is not already saturated with salt to easily dissolve the rock. These activities have caused the formation of several large sinkholes.

Sinkholes occur worldwide, and in the United States are common in southern Indiana, southwestern Illinois, Missouri, Kentucky, Tennessee, and Florida. In areas with known karst topography, subsurface drilling or geophysical **remote sensing** may be used to pinpoint the location of sinkholes.

See also Hydrogeology; Hydrologic cycle; Landscape evolution; Weathering and weathering series

SLATE

Slate is a hard, fine-grained **metamorphic rock** that forms when **sedimentary rocks**, such as shale and mudstone, are subjected to relatively low **temperature** and pressure. It occurs chiefly among older rocks. Millions of years of geological compression force the flaky **minerals** (mica, chlorite, **quartz**) within **clay** sediments to shift perpendicular to the pressure. This pushing alters the material's fundamental structure and creates a new feature known as slaty cleavage. True slate splits easily along this plane into thin, but durable, sheets.

While slate's characteristic color is gray-blue, varieties range from dark gray to black. Organic materials present in the parent **rock** can create different tinges. **Iron** oxide creates a reddish purple tinge; chlorite turns slate green. The rock also varies greatly in surface texture and luster; some slates have a dull, matte finish while others can be as shiny as mica.

Better grades of the rock are widely used for roofing, flooring and sidewall cladding. Slate is also used to make blackboards and pool tables. Pennsylvania and Vermont serve as the principal slate producers for the United States, although slate mines can also be found in Maine, Georgia, Lake Superior, and the Rocky Mountains.

See also Bedding; Metamorphism; Sedimentation

SLEET • *see* PRECIPITATION

SLUMP

The word slump is most commonly used as a colloquial description of a **landslide** with a markedly curved and concave-upward slip surface, which results in rotational movement of the mass above the slip surface. This stands in contrast to landslides with more nearly planar slip surfaces, above which the sliding motion is predominantly translational rather than rotational. Most landslides exhibit both kinds of motion, so the distinction is based on the predominant type of motion. Rotational slides tend to occur in slopes that are, at least mechanically speaking, relatively homogeneous whereas translational slides tend to occur in slopes that contain mechanical discontinuities such as steeply dipping **bedding** or soil-bedrock contacts that can evolve into slip surfaces.

Like the term mudslide, slump is frequently used but is not defined in the classification system used by most landslide specialists. Depending on the type of earth material involved, a slump can be properly described as a **rock** slide, debris slide, or earth slide with predominantly rotational movement.

The curved slip surface and resulting **rotation** of the material in a slump cause strata within a slump mass to be tilted backwards relative to undeformed strata beneath the slip surface. This back-tilting can produce a topographic depression that collects **water** and sediment, which can be used as a criterion to identify old slumps in the field. The material within a slump mass is in most cases remarkably undeformed, albeit rotated, which is a characteristic that distinguishes slides from flows.

A less common use of the term slump is in reference to the downslope flow of unlithified submarine sediments, which frequently occurs along topographic concavities such as submarine canyons. In this case the phenomenon would in most cases be properly described as a submarine debris or earth flow rather than a slide.

As is the case for all landslides, slumping begins when there is an imbalance between resisting and driving forces in a potentially unstable slope. If the slip surface is very nearly circular (as opposed to curved but non-circular), which is an idealized situation that does not often occur, stability can be assessed by comparing resisting and driving moments acting about a center of rotation. The resisting force (or moment) is a function of the shear strength of the **soil** or rock integrated over the **area** of the potential slip surface as well as any engineered structures put in place to increase slope stability. The primary effect of water within the slope is to reduce the normal stress acting across the potential slip surface, thereby reducing the shear strength along the surface. The driving force (or moment) is due primarily to the component of the slump block weight acting parallel to the potential slip surface, the weight of imprudently designed or constructed structures built on the slope, and seismic shaking. Movement can be triggered if the ratio of driving to resisting forces (or moments) is altered by adding water to the slope or by changing its geometry during construction projects.

See also Debris flow; Mass movement; Mass wasting

SMITH, WILLIAM (1769-1839)

English geologist and cartographer

William Smith is often called the founder of English **geology**, and the founder of stratigraphical geology. His interests in **fossils** and the countryside led to a method to identify **rock** strata, along with the first large-scale geological maps of any country. Smith contributed many practical innovations to the embryonic science of geology, and rose from humble beginnings to become a well known and respected scientific figure.

Smith was the eldest son of a village blacksmith, in Churchill, Oxfordshire. His father died when he was still young, and he was sent to live on his Uncle's nearby farm. He attended the small local school, receiving a limited education, but his interest in mathematics was encouraged by friends and relatives, who gave him further tuition. Smith's local reputation as an intelligent boy led him to become employed as an assistant to the surveyor Edward Webb. Webb initially employed Smith to take notes, hold the chain, and other trivial tasks, but was impressed enough to take the eighteen-year-old Smith into his home in Stow and give him an apprenticeship.

Surveying took Smith across much of England, and it was while in Somerset, just outside of Bath, that Smith began to formulate some key ideas. The **area** had many **coal** mines, and Smith was allowed to go into many to observe the rock strata. In 1795, he was employed by local landowners to survey a coal transportation canal, and this work offered Smith further observations of the local rocks. He had a keen interest in fossils, taking many samples in the course of his work. Smith began to speculate that there was a link between the types of fossils and the rock they were found in. He also began to make his first maps of the local rock structures.

After his work on the Bath canal was finished, in 1799, Smith traveled widely, performing a number of small engineering and surveying jobs, in which he observed much more of the English rock strata. While Smith discussed his ideas with many influential men, it was not until 1802 that his ideas began to be widely appreciated in the English scientific community. In that year he met Sir Joseph Banks, president of the Royal Society. Banks encouraged Smith to produce a book containing his ideas and maps. However, more than ten years were to pass before Smith produced any geological work. The pressures of work and some financial worries forced Smith to postpone and delay his writing.

In 1815, Smith, with the help of map engraver John Cary, finished his first of many geological maps, *A Delineation of the Strata of England and Wales, With part of Scotland.* Aside from being the first geological map of an entire country, it was also notable for the innovative use of colored contours to make differentiation clear. The map was well received, being exhibited to the Royal Institution, and Smith received an endowment from the Society for the Encouragement of the Arts, Manufacture and Commerce of fifty pounds.

A year later, over 250 copies of the map had been printed, and while they sold for the hefty price of five guineas (five and a quarter pounds) there were high printing costs. Smith found himself in grave financial difficulties at this time, mainly from a bad investment in a poor rock quarry, and was forced to sell his vast collection of fossils to the British Museum. He also encountered resistance to his rise in social status, in particular from the Geological Society of London president George Greenough. Greenough blocked Smith's membership, and produced a competing map which was cheaper.

Encouraged by his printer, Smith began to publish many more writings and smaller maps of the English counties, and finally he published a work on fossils, the four volume *Strata Identified by Organized Fossils* (1816–19). This presented Smith's observation that fossils of the same type always appeared in the same rock strata, and so could be used to identify the rocks. Smith also began to give lectures on his ideas in the North of England, where he was accompanied by his nephew John Phillips, who later became professor of Geology at Oxford University.

Smith began to suffer from arthritis, and became quite deaf, and was forced to give up lecturing. He became the land steward of Sir John Johnstone of Hackness, in Yorkshire. This gave him the opportunity to study the area in fine detail, and in 1832 he produced a map of the district to the scale of six and a half inches to the mile.

The Geological Society of London, under a new president, awarded Smith the first Wollaston Medal for his work in 1831, and he was given a number of other awards and degrees in recognition of his contributions to geology. He was also given a government pension, and finally achieved a degree of economic security, if somewhat late in life. He was selected as a member of the group to select the stone for the new Houses of Parliament, but once again he had bad luck with quarries, and the stone failed to withstand the detailed carvings of the ornately decorated buildings. However, Smith actually died before the final selection of stone was made, and his notes suggest he had some reservations about the quality of the stone.

Smith died in 1839, after catching a chill on his way to a meeting of the British Association in Birmingham. His work was mainly practical, and he stressed the commercial benefits that could be gained from his work. His mapping of strata enabled others to deduce areas of likely coal sources, and his many county maps remained in use for decades after his death. Some historians have commented that he was lucky that England has such 'well behaved' rock strata, as opposed to continental **Europe** where alpine folding made interpretation difficult. However, Smith should still be given credit for recognizing details others did not, and for making his ideas public.

See also Fossil record; Geologic map; Stratigraphy

SMOG

Smog refers to an atmospheric condition of atmospheric instability, poor visibility, and large concentrations of gaseous and particulate air pollutants. The word "smog" is an amalgam of the words "smoke" and "fog." There are two types of smog: reducing smog characterized by sulfur dioxide and particulates, and photochemical smog characterized by **ozone** and other oxidants.

Reducing smog refers to air pollution episodes characterized by high concentrations of sulfur dioxide and smoke (or particulate aerosols). Reducing smog is also sometimes called London-type smog, because of famous incidents that occurred in that city during the 1950s.

Reducing smogs first became common when industrialization and the associated burning of **coal** caused severe air pollution by sulfur dioxide and soot in European cities. This air pollution problem first became intense in the nineteenth century, when it was first observed to damage human health, buildings, and vegetation.

There have been a number of incidents of substantial increases in human illness and mortality caused by reducing smog, especially among higher-risk people with chronic respiratory or heart diseases. These toxic pollution events usually occurred during prolonged episodes of calm atmospheric conditions, which prevented the dispersion of emitted gases and particulates. These circumstances resulted in the accumulation of large atmospheric concentrations of sulfur dioxide and particulates, sometimes accompanied by a natural **fog**, which became blackened by soot. The term smog was originally coined as a label for these coincident occurrences of **atmospheric pollution** by sulfur dioxide and particulates.

Coal smoke, in particular, has been recognized as a pollution problem in England and elsewhere in **Europe** for centuries, since at least 1500. Dirty, pollution-laden fogs occurred especially often in London, where they were called "peasoupers." The first convincing linkage of a substantial increase in human mortality and an event of air pollution was in Glasgow in 1909, when about 1,000 deaths were attributed to noxious smog during an episode of atmospheric stagnation. A North American example occurred in 1948 in Donora, Pennsylvania, an industrial town located in a valley near Pittsburgh. In that case, a persistent fog and stagnant air during a four-day period coupled with large emissions of sulfur dioxide and particulates from heavy industries to cause severe air pollution. A large increase in the rate of human mortality

Wildfires in Indonesia caused smog thick enough that these girls need to wear masks on their way to school. ***© Michael S. Yamashita/Corbis. Reproduced by permission.***

was associated with this smog; 20 deaths were caused in a population of only 14,100. An additional 43% of the population was made ill in Donora, 10% severely so.

The most famous episode of reducing smog was the so-called "killer smog" that afflicted London in the early winter of 1952. In this case, an extensive atmospheric stability was accompanied by a natural, white fog. In London, these conditions transformed into a noxious "black fog" with almost zero visibility, as the concentrations of sulfur dioxide and particulates progressively built up. The most important sources of emissions of these pollutants were the use of coal for the generation of **electricity**, for other industrial purposes, and to heat homes because of the cold temperatures. In total, this smog caused 18 days of greater-than-usual mortality, and 3,900 deaths were attributed to the deadly episode, mostly of elderly or very young persons, and those with pre-existing respiratory or coronary diseases.

Smogs like the above were common in industrialized cities of Europe and **North America**, and they were mostly caused by the uncontrolled burning of coal. More recently, the implementation of clean-air policies in many countries has resulted in large improvements of air quality in cities, so that severe reducing smogs no longer occur there. Once the severe effects of reducing smogs on people, buildings, vegetation, and other resources and values became recognized, mitigative actions were developed and implemented.

However, there are still substantial problems with reducing smogs in rapidly industrializing regions of eastern Europe, the former Soviet Union, China, India, and elsewhere. In these places, the social priority is to achieve rapid economic growth, even if environmental quality is compromised. As a result, control of the emissions of pollutants is not very stringent, and reducing smogs are still a common problem.

To a large degree, oxidizing or Los Angeles-type smogs have supplanted reducing smog in importance in most industrialized countries. Oxidizing smogs are common in sunny places where there are large emissions of nitric oxide and **hydrocarbons** to the atmosphere, and where the atmospheric conditions are frequently stable. Oxidizing smogs form when those emitted (or primary) pollutants are transformed through photochemical reactions into secondary pollutants, the most important of which are the strong oxidant gases, ozone and peroxyacetyl nitrate. These secondary gases are the major components of oxidizing smog that are harmful to people and vegetation.

Typically, the concentrations of these various chemicals vary predictably during the day, depending on their rates of emission, the intensity of sunlight, and atmospheric stability. In the vicinity of Los Angeles, for example, ozone concentrations are largest in the early-to-mid afternoon, after which these gases are diluted by fresh air blowing inland from the Pacific Ocean. These winds blow the polluted smog further inland, where pine **forests** are affected on the windward slopes of nearby mountains. The light-driven photochemical reactions also cease at night. This sort of daily diurnal cycle is typical of places that experience oxidizing smog.

Humans are sensitive to ozone, which causes irritation and damage to membranes of the respiratory system and eyes, and induces asthma. People vary greatly in their sensitivity to ozone, but hypersensitive individuals can suffer considerable discomfort from exposure to oxidizing smog.

See also Atmospheric circulation; Atmospheric composition and structure; Atmospheric inversion layers; Biosphere; Ultraviolet rays and radiation

SNOW • *see* PRECIPITATION

SOIL AND SOIL HORIZONS

Soil is found in the top layers of **regolith**, the unconsolidated (uncompacted) matter comprised of soil, sediment, and portions of **bedrock** that form the outer crustal layer of the Earth's surface. Soil includes varying amounts of organic matter mainly derived from plants and animal decay.

Because soil is a superficial layer, it can be highly variable and has a composition that can be readily modified by **weather** (e.g., rainfall).

Soil is usually found in stratified layers (i.e. a layer of black soil over subsurface layers of **sand** and/or **clay**. Although different for each geographic **area**, geologists use a generic soil model from which to describe unique area differences.

A soil horizon is a coherent layer of soil—similar in characteristics such as composition, texture, and color—that define the horizon from other soil types. Geologists construct soil profiles of an area by describing the various soil horizons (horizontal layers) that exist within a vertical column of soil. Soil profiles, descriptive of the type and relation of soil horizons, are unique to different geologic and climatic areas.

The outermost (most superficial) soil horizon in a soil profile is termed the "A" horizon. Accordingly, "B," "C," and "D" horizons indicate successively deeper layers. A typical soil profile might then consist of vegetation (not strictly a part of the soil horizon), overlying an "A" horizon of humus (zone of leeching), overlying a "B" horizon of regolith (Zone of accumulation) that was superficial to a layer of bedrock.

The zone of leeching is so defined because **water** is able to percolate through the horizon.

Within the United States, in areas of the Eastern United States with a temperate, humid **climate** with adequate rainfall (generally defined as greater than 24 inches of rain per year) the soil profile typically consists of forest vegetation growing on the "A" horizon zone of leeching that consists of thick humus and sand. The underlying "B" horizon comprising the zone of accumulation is often clay that is rich in **iron** or **aluminum**. The "C" horizon is regolith and the "D" horizon is bedrock. A unique feature of this soil profile is that the mineral calcite ($CaCO_3$) often leeches out of the soil. The loss of this natural buffer allows the soil to become acidic.

In the Western United States—with a more arid climate—a much sparser layer of vegetation (an eventual contributor of organic material) covers an "A" zone of leeching horizon of thin humus and unaltered silicates. The "B" horizon (zone of accumulation) is composed of **caliche**. As with the Eastern profile the "C" and "D" horizons are regolith and bedrock. Caliche is rich in calcite because the calcium carbonate in the upper soil becomes briefly dissolved in the sparse rains that then wash the calcium carbonate down to the caliche layer.

A tropical soil profile features lush vegetation overlying a "B" horizon layer of accumulation that is rich in bauxites and iron oxide. In the tropical soil profile, the "A" horizon may be missing or just a few centimeters thick. In this thin layer, there is a rapid turnover of organic decay and decomposition. All the **quartz** or clay elements are generally "weathered out" and the loss of calcite severe.

See also Leaching; Mass wasting; Porosity and permeability; Rate factors in geologic processes; Runoff; Sedimentation; Stratigraphy

SOLAR ENERGY

Earth's surface receives energy from processes in Earth's interior and from the **Sun**. Heat from the interior comes from radioactive elements in the mantle and core, tidal kneading by the **Moon** and Sun, and residual heat from the earth's formation. This interior heat is radiated through the surface at a global rate of 3×10^{13} watts (W)—about .07 W per square yard (.06 W/m^2). The Sun, in contrast, provides 1.73×10^{17} W, 5,700 times more power than Earth radiates from within and about 30,000 times more than is released by all human activity. **Clouds**, air, land, and sea absorb 69% of the energy arriving from the Sun and reflect the rest back into **space**. The ocean, which covers about 70% of the earth's surface, does about 70% of the absorbing of solar energy.

Between its absorption as heat and its final return to space as infrared radiation, solar energy takes many forms, including kinetic energy in flowing air and **water** or latent heat in evaporated water. Solar energy keeps the **oceans** and atmosphere from **freezing** and drives all winds and currents. A small fraction of Earth's solar energy income is intercepted by green plants, providing the flow of food energy that sustains most

Solar energy is becoming a viable alternate source of power for many people. ***Library of Congress.***

earthly life. Only a few organisms, including thermophilic bacteria infiltrating the **crust** and organisms specialized to live in the vicinity of hydothermal deep-sea vents, derive their energy from Earth's interior rather than from the Sun.

Regional variations in solar input contribute to **weather** patterns and seasonal changes. On average Earth's surface is more nearly at a right angle to the Sun's rays near the equator, so the tropics absorb more solar energy than the higher latitudes. This creates an energy imbalance between the equator and the poles, an imbalance that the circulation of the atmosphere and oceans redress by transporting energy away from the equator. During each half of the year the daylight side of each hemisphere is tilted at a steeper angle to the sun than during the other half, and so intercepts less solar energy; this results in seasonal climatic changes.

Solar energy is also of technological importance. Utilization of the Sun as an energy source has been routine on spacecraft for decades and is becoming more frequent on the ground. Electromagnetic radiation from the Sun, unlike the major conventional power sources, produces no smokestack emissions, **greenhouse gases**, or radioactive wastes; and its production cannot be manipulated for profit or political leverage. On the down side, sunlight is a diffuse or spread-out energy source compared to any fuel and is directly available only during the day. Yet, even at high latitudes in **Europe** and **North America**, where most of the world's energy is consumed, the ground receives from the Sun a long-term average of 83.6 W per square yard (100 W/m^2). This average is inclusive of "dark" hours. Both indirect and direct harvesting of this energy income is possible. Indirect solar schemes, including **wind** power, wood heat, and the burning of alcohol, methane, or hydrogen, run on energy derived at second hand from sunlight. Direct schemes use sunlight as such to heat buildings or water, generate **electricity**, or supply high-temperature process heat to industrial systems.

Because conventional electricity generation is expensive and polluting, much effort has been devoted to solar electricity generation. Electricity can be generated from sunlight either thermally or photovoltaically. Thermal methods focus the Sun's rays on looped pipes through which molten salt, hot air, or steam flows. This hot fluid is then used either at first or second hand to run generators, much as heat from **coal** or nuclear fuel is used in conventional power plants. Photovoltaic electrical generation depends on flat, specially designed transistors (solar cells) that convert incident light to electricity. At 83.6 W/$yard^2$ (100 W/m^2) average solar input, 38 square yards (32 m^2) of 33% efficient solar cells—a square 18 feet (5.5 m) on a side—could supply 800 kilowatt-hours of electricity per

NASA photo of light/dark terminator from space. © M. Agliolo. *Reproduced by permission.*

month, the approximate usage of the average U.S. household. An efficiency of 32.3% has been demonstrated in the laboratory, but most commercial photovoltaic cells are only about 10% efficient. Unlike the unused heat from a ton of coal or uranium, however, the sunlight not converted to electricity by a solar cell entails neither monetary cost nor pollution, and so cannot be viewed as waste.

Despite its obvious advantages, photovoltaic electricity generation has long been limited to specialized off-grid applications by the high cost of solar cells. However, cell prices have fallen steadily, and several large-scale photovoltaic electricity projects are now under way in the U.S. and elsewhere.

See also Atmospheric circulation; Coronal ejections and magnetic storms; Energy transformations; Global warming; Insolation and total solar irradiation; Meteorology; Ocean circulation and currents; Seasonal winds; Solar illumination: Seasonal and diurnal patterns; Solar sunspot cycles; Sun; Ultraviolet rays and radiation

SOLAR ILLUMINATION: SEASONAL AND DIURNAL PATTERNS

Earth rotates about its **polar axis** as it revolves around the **Sun**. Earth's polar axis is tilted 23.5° to the orbital plane (ecliptic plane). Combinations of **rotation**, **revolution**, and tilt of the polar axis result in differential illumination and changing illumination patterns on Earth. These changing patterns of illumination result in differential heating of Earth's surface that, in turn, creates seasonal climatic and **weather** patterns.

Earth's rotation results in cycles of daylight and darkness. One daylight and night cycle constitutes a diurnal cycle. Daylight and darkness are separated by a terminator—a shadowy zone of twilight. Earth's rate of rotation—approximately 24 hours—fixes the time of the overall cycle (i.e., the length of a day). However, the number of hours of daylight and darkness within each day varies depending upon **latitude** and season (i.e., Earth's location in its elliptical orbital path about the Sun).

On Earth's surface, a circle of illumination describes a latitude that defines an extreme boundary of perpetual daylight or perpetual darkness. Tropics are latitudes that mark the farthest northward and farthest southward line of latitude where the solar zenith (the highest angle the Sun reaches in the sky during the day) corresponds to the local zenith (the point directly above the observer). At zenith, the Sun provides the most direct (most intense) illumination. Patterns of illumination and the apparent motion of the Sun on the hypothetical celestial sphere establish several key latitudes. The North Pole is located at 90° North latitude; the Arctic Circle defines an **area** from 66.5 N to the North Pole; the Tropic of Cancer defines an area from the Equator to 23.5 N; the Tropic of Capricorn defines an area from the equator to 23.5 S; the Antarctic Circle defines an area from 66.5 S to the South Pole.

There are seasonal differences in the amount and directness of daylight (e.g., the first day of summer always has the longest period of daylight, and the first day of winter the least amount of daylight). With regard to the Northern Hemisphere, at winter solstice (approximately December 21), Earth's North Pole is pointed away from the Sun, and sunlight falls more directly on the Southern Hemisphere. At the summer solstice (approximately June 21), Earth's North Pole is tilted toward the Sun, and sunlight falls more directly on the Northern Hemisphere. At the intervening vernal and autumnal equinoxes, both the North and South Pole are oriented so that they have the same angular relationship to the Sun and, therefore, receive equal illumination. In the Southern Hemisphere, the winter and summer solstices are exchanged so that the solstice that marks the first day of winter in the Northern Hemisphere marks the first day of summer in the Southern Hemisphere.

At autumnal equinox (approximately September 21), there is uniform illumination of Earth's surface (i.e., 12 hrs of daylight everywhere except exactly at the poles which are both illuminated). At winter solstice (approximately September 21), there is perpetual sunlight within the Antarctic Circle (i.e., the Antarctic circle is fully illuminated). At vernal equinox (approximately March 21), the illumination patterns return to the state of the autumnal equinox. At vernal equinox, there is uniform illumination of Earth's surface (i.e., 12 hrs of daylight everywhere except exactly at the poles which are both illuminated). At summer solstice (approximately June 21), there is perpetual sunlight within the Arctic Circle (i.e., the Arctic Circle is fully illuminated).

The illumination patterns in the polar regions—within the Artic Circle and Antarctic Circle—are dynamic and inverse. As the extent of perpetual illumination (perpetual daylight) increases—to the maximum extent specified by the latitude of each circle—the extent of perpetual darkness increases within the other polar circle. For example, at winter solstice, there is no illumination within the Artic circle (i.e., perpetual night within the area 66.5° N to the North Pole). Conversely, the Antarctic Circle experiences complete daylight (i.e., perpetual daylight within the area 66.5° S to the North Pole.). As Earth's axial tilt and revolution about the Sun continue to produce changes in polar axial orientation that result in a progression to the vernal equinox, the circle of perpetual darkness decreases in extent round the North Pole as the circle of perpetual daylight decreases around the South pole. At equinox, both polar regions receive the same illumination.

At the Equator, the Sun is directly overhead at local noon at both the vernal and autumnal equinox. The Tropic of Cancer and the Tropic of Capricorn denote latitudes where the Sun is directly overhead at local noon at a solstice. Along the Tropic of Cancer, the Sun is directly overhead at local noon at the June 21 solstice (the Northern Hemisphere's summer solstice and the Southern Hemisphere's winter solstice). Along the Tropic of Capricorn, the Sun is directly overhead at local noon at the December 21 solstice.

Precession of Earth's polar axis also results in a long-term precession of seasonal patterns.

Although the most dramatic changes in illumination occur within the polar regions, the differences in daylight hours—affecting the amount of **solar energy** or solar **insolation** received—cause the greatest climatic variations in the middle latitude temperate regions. The polar and equatorial regions exhibit seasonal patterns, but these are much more uniform (i.e., either consistently cold in the polar regions or consistently hot in the near equatorial tropical regions) than the wild **temperature** swings found in temperate climates.

Differences in illumination are a more powerful factor in determining climatic seasonal variations than Earth's distance from the Sun. Because Earth's orbit is only slightly elliptical, the variation from the closest approach at perihelion (approximately January 3) to the farthest Earth orbital position at aphelion six months later (varies less than 3%). Because the majority of tropospheric heating occurs via conduction of heat from the surface, differing amounts of sunlight (differential levels of solar insolation) result in differential temperatures in Earth's **troposphere** that then drive convective currents and establish low and high pressure areas of convergence and divergence.

See also Atmospheric composition and structure; Celestial sphere: The apparent movements of the Sun, Moon, planets, and stars; Revolution and rotation; Solar energy; Year, length of

SOLAR SUNSPOT CYCLE

A sunspot is an **area** of the Sun's photosphere, appearing darker to the eye than surrounding areas of the **Sun**. Although very hot by any terrestrial standard, sunspot regions are cooler that surrounding solar surface. Sunspots occur in cycles and are associated with a strong solar **magnetic field**. Variations in the solar magnetic field impact the **space** environment of Earth (sometime termed "space weather" and therefore have at least a correlated effect on Earth's **weather** and climatic conditions.

Large sunspots, visible to the naked eye, were noted by the ancient Chinese. The first specific mention of sunspots in modern scientific literature was by Italian astronomer and physicist **Galileo Galilei** (1564–1642) in his *Starry Messenger* published in 1610. Sunspot occurrence has been carefully noted by astronomers ever since. Astronomers have found that the frequency of sunspots varies with a period of between 11 and 13 years. This corresponds to the period of solar activity cycle involving solar flares, prominences and other phenomena associated with the outer layers of the Sun.

At the beginning of the solar cycle a few sunspots appear at the higher latitudes on the Sun near the poles. The sunspots then appear to move across the face of the Sun due to its **rotation**. Large spots may last for several rotation periods of about a month in length. As the cycle progresses the number of spots increases and they tend to be formed at lower latitudes toward the equator. The end of the cycle is marked by a marked drop in the number of low-latitude sunspots, which is followed immediately by the beginning of the next cycle, as small numbers of spots begin to appear at high latitudes.

Astronomers now know that sunspots are essentially **magnetic storms** on the surface of the Sun. The spots usually occur in pairs. Just as a bar-magnet placed under a sheet of paper will show a characteristic looping magnetic field when **iron** filings are scattered over the paper, so the sunspots making up the pair appear to be connected by a similar field. The ends of the bar magnet are characterized as being north or south magnetic poles depending on how a magnetic compass is affected by the poles. Similarly, each of the members of a sunspot pair will have the characteristics of either a north or south magnetic pole. During a particular cycle, the leading spot of the pair will always have the same polarity for spots formed in a particular hemisphere. The order of polarity is reversed for sunspot-pairs formed in the opposite hemisphere. Thus, if spots-pairs formed in the northern hemisphere of the Sun have the **lead** spot behaving as a south magnetic pole, the leading spot of a pair formed in the southern hemisphere will have north magnetic pole. This order of polar progression for the leading and trailing spots is preserved throughout the entire 11–13 year cycle. However, during the following sunspot cycle the order will be reversed in both hemispheres. This has led most astronomers to feel that the proper sunspot cycle should be reckoned as consisting of two 11–13 year cycles, since two cycles must pass before conditions are duplicated and the full pattern can repeat.

Although there have been attempts to link the solar cycle to changes on Earth, most are still characterized as correlations (i.e., the events are associated but there is no established cause and effect relationship. An exception may lie in the earth's weather. There have been some strongly suggestive correlations between solar activity and global **temperature** as well as rainfall variations. Analysis of tree-ring data spanning many centuries clearly shows the presence of a 11–13 year cycle. There is even compelling evidence from ancient **rock** layers that the solar cycle has been present since **Precambrian** times. Those involved in the launching and maintenance of Earth satellites are acutely aware that the upper layers of the earth's atmosphere respond to solar activity by expanding thereby increasing the atmospheric drag on satellites in low Earth orbit. Finally, there is a curious period of about 75 years shortly after Galileo's discovery when few sunspots were observed. This era is called the Maunder Minimum after the astronomer who first noted its existence. Other phenomena such as the **Aurora Borealis** (Northern Lights) that are associated with solar activity are also missing

The SOHO spacecraft gives astronomers a stationary view of the Sun. ***U.S. National Aeronautics and Space Administration (NASA).***

from European records during this period. The interval is also associated with a time of unusually severe winters in both **Europe** and **North America** and is often called the "little **ice** age." In this century, Jack Eddy has found evidence that there may have been similar periods of cold associated with earlier interruptions of the solar cycle. Unfortunately, our understanding of the solar cycle is sufficiently crude that we have no explanation of what might cause these interruptions. Indeed, our understanding of the basis for the cycle itself is still largely in a phenomenological phase.

The very strong magnetic fields (i.e. several thousand times the general field of the earth) in a sunspot account for the dark appearance of the spot. The hot atmosphere of the Sun contains a significant number of atoms having a net positive charge resulting from collisions between them (i.e., they are ionized, having lost one or more electrons). Charged particles may move along magnetic field lines, but not across them. Thus, a magnetic field exerts a kind of pressure on the gas and helps to support it against the gravitational force of the Sun itself. This force is usually balanced only by the pressure of the hot gas surrounding a sunspot. With part of the pressure being supplied by the magnetic field, the gas will cool to a lower temperature. Because it is not as hot, it appears dark compared to the bright surrounding region called the photosphere. While they appear dark by comparison, sunspots are still hotter than any blast furnace on the earth and are only dark by contrast with the brilliant solar surface. A close inspection of a sunspot shows it to have a dark central region called an umbra surrounded by a lighter radial structured region called a penumbra. These regions can be understood in terms of the spreading and weakening magnetic field emanating from the core of the sunspot.

In the second half of this century, Eugene Parker suggested a mechanism that accounts for much of the descriptive behavior of sunspots. The Sun does not rotate as a rigid body and the polar regions rotate somewhat more slowly than the equator. Because of the charged nature of the solar material, the Sun's general magnetic field is dragged along with the solar rotation. However, it will be dragged faster and further at the equator than at the rotational poles. Although the general field of the Sun is quite weak (i.e., similar to that of the earth), the differential rotation strengthens and distorts the field over time. One can imagine the faster-rotating regions of the equator dragging the local magnetic field so that the field lines are drawn out into long thin tubes. The more these tubes are stretched, the stronger the field becomes. The magnetic pressure exerted on the surrounding gas causes the material within the tube to become "buoyant" compared to the surrounding material and rise toward the surface. As the magnetic tube breaks the surface of the Sun, it forms two spot-like cross-sections where it clears the lower solar atmosphere. As the field direction is out of the solar surface at one spot and into it at the other, one of these spots will appear to have one kind of magnetic pole and the other spot will appear to have the other. The global nature of the general solar field will guarantee that the stretched magnetic tubes will yield leading spots with opposite polarities in opposite hemispheres. A reversal of the Sun's general field between 11–13 year cycles would account for the reversal of this order. However, there is no compelling explanation of why the general field should reverse after each 11–13 year solar cycle

See also Aurora Borealis and Aurora Australialis; Coronal ejections and magnetic storms; Electricity and magnetism

Solar system

Earth's solar system is comprised of the **Sun**, nine major planets, some 100,000 **asteroids** larger than 0.6 mi (1 km) in diameter, and perhaps 1 trillion cometary nuclei. While the major planets lie within 40 Astronomical Units (AU)—the average distance of Earth to the Sun—the outermost boundary of the solar system stretches to 1 million AU, one-third the way to the nearest star. Cosmologists and Astronomers assert that the solar system was formed through the collapse of a spinning cloud of interstellar gas and dust.

The central object in the solar system is the Sun. It is the largest and most massive object in the solar system; its diameter is 109 times that of Earth, and it is 333,000 times more massive. The extent of the solar system is determined by the gravitational attraction of the Sun. Indeed, the boundary of the solar system is defined as the surface within which the gravitational pull of the Sun dominates over that of the galaxy. Under this definition, the solar system extends outwards from the Sun to a distance of about 100,000 AU. The solar system is much larger, therefore, than the distance to the remotest known planet, Pluto, which orbits the Sun at a mean distance of 39.44 AU.

The Sun and the solar system are situated some 26,000 light years from the center of our galaxy. The Sun takes about 240 million years to complete one orbit about the galactic cen-

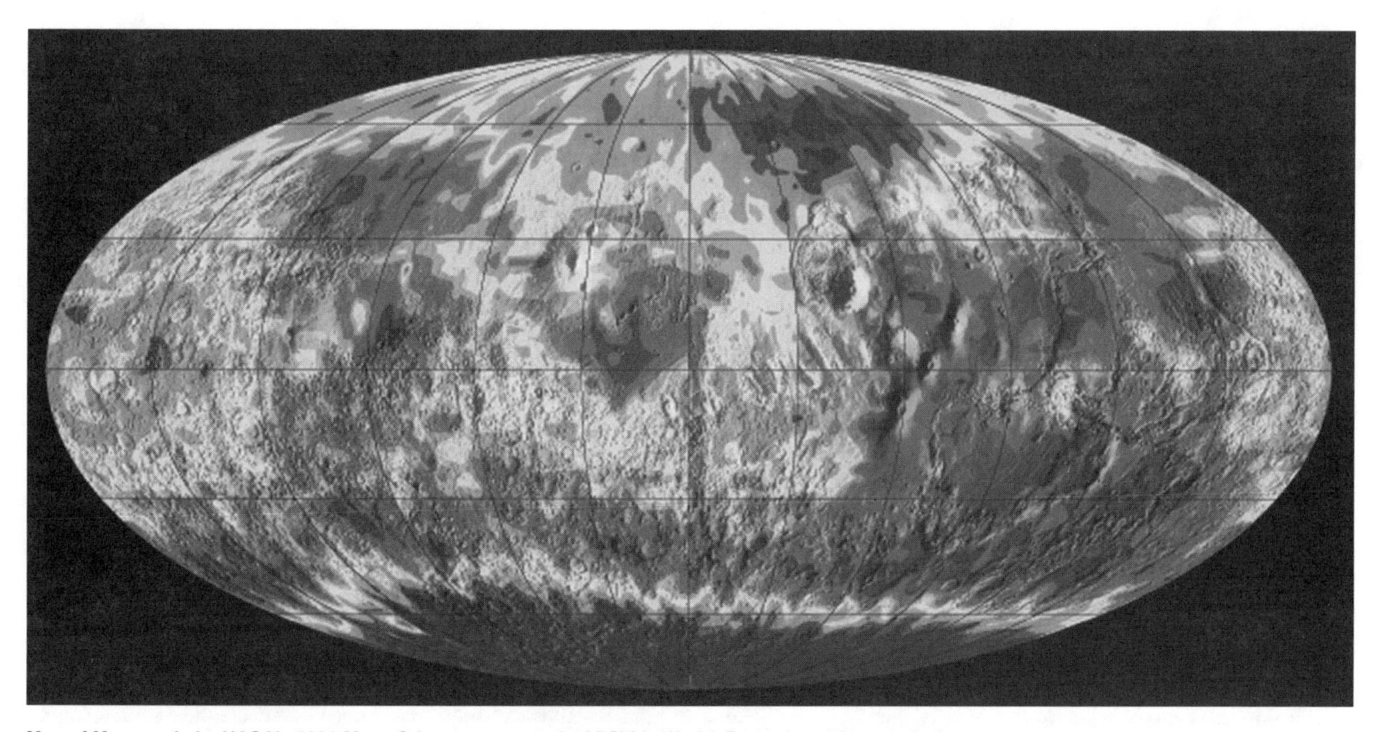

Map of Mars made by NASA's 2001 Mars *Odyssey* spacecraft. ***AP/Wide World. Reproduced by permission.***

ter. Since its formation the Sun has completed about 19 such trips. As it orbits about the center of the galaxy, the Sun also moves in an oscillatory fashion above and below the galactic plane with a period of about 30 million years. During their periodic sojourns above and below the plane of the galaxy, the Sun and solar system suffer gravitational encounters with other stars and giant molecular **clouds**. These close encounters result in the loss of objects (essentially dormant cometary nuclei located in the outer Oort cloud) that are on, or near, the boundary of the solar system. These encounters also nudge some cometary nuclei toward the inner solar system where they may be observed as long-period **comets**.

The objects within our solar system demonstrate several essential dynamical characteristics. When viewed from above the Sun's North Pole, all of the planets orbit the Sun along near-circular orbits in a counterclockwise manner. The Sun also rotates in a counterclockwise direction. With respect to the Sun, therefore, the planets have prograde orbits. The major planets, asteroids, and short-period comets all move along orbits only slightly inclined to one another. For this reason, when viewed from Earth, the asteroids and planets all appear to move in the narrow zodiacal band of constellations. All of the major planets, with three exceptions, spin on their central axes in the same direction that they orbit the Sun. That is, the planets mostly spin in a prograde motion. The planets Venus, Uranus, and Pluto are the three exceptions, having retrograde (backwards) spins.

The distances at which the planets orbit the Sun increase geometrically, and it appears that each planet is roughly 64% further from the Sun than its nearest inner neighbor. The separation between successive planets increases dramatically beyond the orbit of Mars. While the inner, or terrestrial planets are typically separated by distances of about four-tenths of an AU, the outer, or Jovian planets are typically separated by 5—10 AU.

Although the asteroids and short-period comets satisfy, in a general sense, the same dynamical constraints as the major planets, we have to remember that such objects have undergone significant orbital **evolution** since the solar system formed. The asteroids, for example, have undergone many mutual collisions and fragmentation events, and the cometary nuclei have suffered from numerous gravitational perturbations from the planets. Long-period comets in particular have suffered considerable dynamical evolution, first to become members of the Oort cloud, and second to become comets visible in the inner solar system.

The compositional make-up of the various solar system bodies offers several important clues about the conditions under which they formed. The four interior planets—Mercury, Venus, Earth, and Mars—are classified as terrestrial and are composed of rocky material surrounding an iron-nickel metallic core. In contrast, Jupiter, Saturn, Neptune, and Uranus are classified as the "gas giants" and are large masses of hydrogen in gaseous, liquid, and solid form surrounding Earth-size **rock** and metal cores. Pluto fits neither of these categories, having an icy surface of frozen methane. Pluto more greatly resembles the satellites of the gas giants, which contain large fractions of icy material. This observation suggests that the initial conditions under which such ices might have formed only prevailed beyond the orbit of Jupiter.

In summary, any proposed theory for the formation of the solar system must explain both the dynamical and chemical properties of the objects in the solar system. It must also be sufficient flexibility to allow for distinctive features such as retrograde spin, and the chaotic migration of cometary orbits.

Astronomers almost universally assert that the best descriptive model for the formation of the solar system is the solar nebula hypothesis. The essential idea behind the solar nebula model is that the Sun and planets formed through the collapse of a rotating cloud of interstellar gas and dust. In this way, planet formation is postulated to be a natural consequence of star formation.

The solar nebula hypothesis is not a new scientific proposal. Indeed, the German philosopher Immanuel Kant first discussed the idea in 1755. Later, the French mathematician, Pierre Simon de Laplace (1749–1827) developed the model in his text, *The System of the World,* published in 1796.

The key postulate in the solar nebula hypothesis is that once a rotating interstellar gas cloud has commenced gravitational collapse, then the conservation of angular momentum will force the cloud to develop a massive, central **condensation** that is surrounded by a less massive flattened ring, or disk of material. The nebula hypothesis asserts that the Sun forms from the central condensation, and that the planets accumulate from the material in the disk. The solar nebula model naturally explains why the Sun is the most massive object in the solar system, and why the planets rotate about the Sun in the same sense, along nearly circular orbits and in essentially the same plane.

During the gravitational collapse of an interstellar cloud, the central regions become heated through the release of gravitational energy. This means that the young solar nebular is hot, and that the gas and (vaporized) dust in the central regions is well mixed. By constructing models to follow the gradual cooling of the solar nebula, scientists have been able to establish a chemical condensation sequence. Near to the central proto-sun, the nebular **temperature** will be very high, and consequently no solid matter can exist. Everything is in a gaseous form. Farther away from the central proto-sun, however, the temperature of the nebula falls off. At distances beyond 0.2 AU from the proto-sun, the temperature drops below 3,100°F (1,700°C). At this temperature, **metals** and oxides can begin to form. Still further out (at about 0.5 AU), the temperature will drop below 1,300°F (730°C), and silicate rocks can begin to form. Beyond about 5 AU from the proto-sun, the temperature of the nebula will be below –100°F (–73°C), and ices can start to condense. The temperature and distance controlled sequence of chemical condensation in the solar nebula correctly predicts the basic chemical make-up of the planets.

Perhaps the most important issue to be resolved in future versions of the solar nebula model is that of the distribution of angular momentum. The problem for the solar nebula theory is that it predicts that most of the mass and angular momentum should be in the Sun. In other words, the Sun should spin much more rapidly than it does. A mechanism is therefore required to transport angular momentum away from the central proto-sun and redistribute it in the outer planetary disk. One proposed transport mechanism invokes the presence of a **magnetic field** in the nebula, while another mechanism proposed the existence of viscous stresses produced by turbulence in the nebular gas.

Precise dating of meteorites and lunar rock samples indicate that the solar system is 4.6 to 5.1 billion years old. The meteorites also indicate an age spread of about 20 million years, during which time the planets themselves formed.

The standard solar nebula model suggests that the planets were created through a multi-step process. The first important step is the coagulation and **sedimentation** of rock and **ice** grains in the mid-plain of the nebula. These grains and aggregates, 0.4 in (1 cm) to 3 ft (1 m) in size, continue to accumulate in the mid-plain of the nebula to produce a swarm of some 10 trillion larger bodies, called planetesimals, that are some 0.6 mi (1 km), or so in size. Finally, the planetesimals themselves accumulate into larger, self-gravitating bodies called proto-planets. The proto-planets were probably a few hundred kilometers in size. Finally, growth of proto-planet-sized objects results in the planets.

The final stages of planetary formation were decidedly violent—it is probable that a collision with a Mars-sized proto-planet produced Earth's **Moon**. Likewise, it is thought that the retrograde rotations of Venus and Uranus may have been caused by glancing proto-planetary impacts. The rocky and icy planetesimals not incorporated into the proto-planets now orbit the Sun as asteroids and cometary nuclei. The cometary nuclei that formed in the outer solar nebula were mostly ejected from the nebula by gravitational encounters with the large Jovian gas giants and now reside in the Oort cloud.

One problem that has still to be worked-out under the solar nebula hypothesis concerns the formation of Jupiter. The estimated accumulation time for Jupiter is about 100 million years, but it is now known that the solar nebula itself probably only survived for 100,000 to 10 million years. In other words, the accumulation process in the standard nebula model is too slow by at least a factor of 10 and maybe 100.

Of great importance to the study of solar systems was the discovery in 1999 of an entire solar system around another star. Although such systems should be plentiful and common in the cosmos, this was the first observation of another solar system. Forty-four light-years from Earth, three large planets were found circling the star Upsilon Andromedae. Astronomers suspect the planets are similar to Jupiter and Saturn—huge spheres of gas without a solid surface.

See also Astronomy; Big Bang theory; Celestial sphere: The apparent movements of the Sun, Moon, planets, and stars; Cosmology; Dating methods; Earth (planet); Earth, interior structure; Geologic time; Revolution and rotation

SOLID SOLUTION SERIES

A solid solution series is the compositional range between end-member **minerals** that share the same basic chemical formula but experience substitution of elements in one or more atomic sites. This substitution occurs when an element in a mineral formula can be replaced by another of similar size and charge to make a new mineral. For example, **iron** and magnesium can readily replace one another in a mineral. In some cases, the substitution can be complete and range from entirely one element to another element, resulting in end-member mineral compositions. One example is **olivine**, which can vary from Mg_2SiO_4 (forsterite) to Fe_2SiO_4 (fayalite). This is known as complete solid solution. Such a mineral can also consist of any intermediate percentage of either end-member. The compositional range between end-member minerals that exhibit complete solid solution is known as a solid solution series.

Another example of a complete solid solution series is between siderite $FeCO_3$ and rhodochrosite, $MnCO_3$. In this and the olivine example, the cation is replaced. Complete anion substitution series are less common, but one example is given by KCl to KBr. Solid solutions can be more chemically complicated as well, with more than one element being replaced. In **plagioclase**, a complete series exists between albite, $NaAlSi_3O_8$ and anorthite, $CaAl_2Si_2O_8$. In this case, Na^+ is similar enough in size to substitute for Ca^{2+}. Because the charges are different, a shift in the number of **aluminum** and **silicon** atoms is required to maintain neutrality.

Complete solid solution is also possible with three end-members. In the pyroxenes for example, compositional variation among of Ca^{2+}, Mg^{2+}, or Fe^{2+} is often represented in terms of three simplified components: wollastonite ($CaSiO_3$), enstatite ($MgSiO_3$), and ferrosilite ($FeSiO_3$).

The actual compositional variation of a given mineral that forms a solid solution series may be expressed by the abbreviated mineral names with its proportion subscripted. An olivine that has been analyzed and determined to consist of 25% Mg^{2+} and 75% Fe^{2+} would be represented by $Fo_{25}Fa_{75}$. This composition may also be written in terms of the molecular formula: $(Mg_{0.25}Fe_{0.75})SiO_4$. Graphical forms are also common. In the case of two end-members, a bar diagram is used and the composition is plotted on the bar. When three end-members are present, a diagram that places each end-member at the point of an equilateral triangle allows compositional variations to be plotted anywhere within the triangle.

See also Chemical elements; Feldspar

SOLOMON, SUSAN (1956-)

American atmospheric chemist

Susan Solomon played a key role in discovering the cause of a major threat to the earth—the loss of the protective ozone layer in the upper atmosphere. **Ozone** protects all life on Earth from large amounts of damaging ultraviolet radiation from the **sun**. Solomon, an atmospheric chemist, was first to propose the theory explaining how chlorofluorocarbons, gases used in refrigerators and to power aerosol spray cans, could in some places on the globe lead to ozone destruction in the presence of stratospheric **clouds**.

Solomon said in an interview with Lee Katterman that she recalls "exactly what got me first interested in science. It was the airing of Jacques Cousteau on American TV when I was nine or ten years old." Solomon said that as a child she was very interested in watching natural history programming on television. This sparked an interest in science, particularly biology. "But I learned that biology was not very quantitative," said Solomon in the interview. By the time she entered the Illinois Institute of Technology, Solomon met her need for quantitative study by choosing **chemistry** as her major at the Illinois Institute of Technology. A project during Solomon's senior year turned her attention toward **atmospheric chemistry**. The project called for measuring the reaction of ethylene and hydroxyl radical, a process that occurs in the atmosphere of Jupiter. As a result of this work, Solomon did some extra reading about planetary atmospheres, which led her to focus on atmospheric chemistry.

During the summer of 1977, just before entering graduate school at University of California at Berkeley, Solomon worked at the National Center for Atmospheric Research (NCAR) in Boulder, Colorado. She met research scientist Paul Crutzen at NCAR, who introduced her to the study of ozone in the upper atmosphere. In the fall at Berkeley, Solomon sought out Harold Johnston, a chemistry professor who did pioneering work on the effects of the supersonic transport (SST) on the atmosphere. Solomon credits Crutzen and Johnston for encouraging her interest in atmospheric chemistry. After completing her course work toward a Ph.D. in chemistry at Berkeley, Solomon moved to NCAR to do her thesis research with Crutzen.

She received a Ph.D. in chemistry in 1981 and then accepted a research position at the National Oceanic and Atmospheric Administration (NOAA) Aeronomy Laboratory in Boulder, Colorado. Initially, Solomon's research focused on developing computer models of ozone in the upper atmosphere. Ozone is a highly reactive molecule composed of three atoms of **oxygen**. By comparison, the oxygen that is essential to the metabolism of living things is a relatively stable combination of two oxygen atoms. In the upper atmosphere between about 32,000 and 100,000 feet altitude, a layer of ozone exists that absorbs much of the sun's deadly ultraviolet radiation, thereby protecting all life on Earth.

In 1985, scientists first reported that, during the months of spring in the Southern Hemisphere (September and October), the density of the ozone layer over **Antarctica** had been decreasing rapidly in recent years. The cause of this hole in the ozone layer was unknown and many scientists began to look for its cause. In 1986, the scientific community wanted to send some equipment to Antarctica to measure atmospheric levels of ozone and nitrogen dioxide. Much to the surprise of her scientific colleagues, Solomon volunteered to travel to Antarctica to get the needed measurements; until then, she had concentrated on theoretical studies, but the chance to understand the cause of the ozone hole prompted Solomon to take up experimental work. Solomon led an expedition to Antarctica during August, September, and October of 1986, where she and co-workers measured the amounts of several atmospheric components, including the amount of chlorine dioxide in the upper atmosphere. The level of this atmospheric chemical was much higher than expected and provided an important clue in determining why the ozone hole had appeared. Back at her NOAA lab in Boulder, Solomon wrote a research article that provided a theoretical explanation for the ozone hole. Solomon showed how the high level of chlorine dioxide was consistent with fast chemical destruction of ozone triggered by reactions occurring on stratospheric clouds. The extra chlorine dioxide was derived from chlorofluorocarbons released into the atmosphere from sources such as foams and leaking refrigeration equipment. Solomon returned to Antarctica for more measurements in August of 1987. Her explanation for the cause of the ozone hole is now generally accepted by scientists, and has led many countries of the world to curtail the production and use of chlorofluorocarbons.

Solomon's scientific studies to uncover the likely cause of the ozone hole have led to public recognition and many awards. In 1989, Solomon received the gold medal for exceptional service from the U.S. Department of Commerce (the agency that oversees the NOAA) She has testified several times before congressional committees about ozone depletion and is increasingly sought out as an expert on ozone science and policy (although the latter role is one she does not welcome, Solomon admitted in her interview, since she considers herself a scientist and not a policy expert).

Solomon was born on January 19, 1956, in Chicago, Illinois. Her father, Leonard Solomon, was an insurance agent. Susan's mother, Alice Rutman Solomon, was a fourth-grade teacher in the Chicago public schools. Solomon continues to study the atmospheric chemistry of ozone and has added Arctic ozone levels to her research subjects.

See also Chloroflurocarbon (CFC); Global warming; Greenhouse gases and greenhouse effect; Ozone layer and hole dynamics; Ozone layer depletion

SOROSILICATES

The most abundant rock-forming **minerals** in the **crust** of the earth are the silicates. They are formed primarily of **silicon** and **oxygen**, together with various **metals**. The fundamental unit of these minerals is the silicon-oxygen tetrahedron. These tetrahedra have a pyramidal shape, with a relatively small silicon cation (Si^{+4}) in the center and four larger oxygen anions (O^{-2}) at the corners, producing a net charge of –4. **Aluminum** cations (Al^{+3}) may substitute for silicon, and various anions such as hydroxyl (OH^{-}) or fluorine (F^{-}) may substitute for oxygen. In order to form stable minerals, the charges that exist between tetrahedra must be neutralized. This can be accomplished by the sharing of oxygen atoms between tetrahedra, or by the binding together of adjacent tetrahedra by various metal cations. This in turn creates characteristic silicate structures that can be used to classify silicate minerals into **cyclosilicates**, **inosilicates**, **nesosilicates**, **phyllosilicates**, sorosilicates, and **tectosilicates**.

Minerals formed by two silicon-oxygen tetrahedra sharing oxygen atoms are called sorosilcates. These double tetrahedra contain two silicon cations and seven oxygen anions, giving them a net charge of –6. Various metal cations neutralize the charges between double tetrahedra. Most of the minerals in the sorosilicate group are rare, and many are found in metamorphic rocks. Examples of sorosilicates that form during **metamorphism**, as well as during the crystallization of **igneous rocks**, include those in the epidote group. Epidote has the formula $Ca_2(Al,Fe)Al_2O(SiO_4)(Si_2O_7)(OH)$, and epidote group minerals are comprised of both single and double silicon-oxygen tetrahedra. Another sorosilicate mineral is hemimorphite ($Zn_4(Si_2O_7)(OH)_2 \cdot H2O$). Hemimorphite is a secondary mineral (meaning that it is an alteration product), found in the oxidized portions of zinc ore deposits.

See also Chemical bonds and physical properties

SOUND TRANSMISSION

Sound waves are pressure waves that travel through Earth's **crust**, **water** bodies, and atmosphere. Natural sound frequencies specify the frequency attributes of sound waves that will efficiently induce vibration in a body (e.g., the tympanic membrane of the ear) or that naturally result from the vibration of that body.

Sound waves are created by a disturbance that then propagates through a medium (e.g., crust, water, air). Individual particles are not transmitted with the wave, but the propagation of the wave causes particles (e.g., individual air molecules) to oscillate about an equilibrium position.

Every object has a unique natural frequency of vibration. Vibration can be induced by the direct forcible disturbance of an object or by the forcible disturbance of the medium in contact with an object (e.g. the surrounding air or water). Once excited, all such vibrators (i.e., vibratory bodies) become generators of sound waves. For example, when a **rock** falls, the surrounding air and impacted crust undergo sinusoidal oscillations and generate a sound wave.

Vibratory bodies can also absorb sound waves. Vibrating bodies can, however, efficiently vibrate only at certain frequencies called the natural frequencies of oscillation. In the case of a tuning fork, if a traveling sinusoidal sound wave has the same frequency as the sound wave naturally produced by the oscillations of the tuning fork, the traveling pressure wave can induce vibration of the tuning fork at that particular frequency.

Mechanical resonance occurs with the application of a periodic force at the same frequency as the natural vibration frequency. Accordingly, as the pressure fluctuations in a resonant traveling sound wave strike the prongs of the fork, the prongs experience successive forces at appropriate intervals to produce sound generation at the natural vibrational or natural sound frequency. If the resonant traveling wave continues to exert force, the amplitude of oscillation of the tuning fork will increase and the sound wave emanating from the tuning fork will grow stronger. If the frequencies are within the range of human hearing, the sound will seem to grow louder. Singers are able to break **glass** by loudly singing a note at the natural vibrational frequency of the glass. Vibrations induced in the glass can become so strong that the glass exceeds its elastic limit and breaks. Similar phenomena occur in rock formations.

Whales can communicate over vast distances by calling out to each other underwater. Water, which is denser than air, can transmit sound waves many times further than air. ***AP/Wide World. Reproduced by permission.***

All objects have a natural frequency or set of frequencies at which they vibrate.

Sound waves can potentiate or cancel in accord with the principle of **superposition** and whether they are in phase or out of phase with each other. Waves of all forms can undergo constructive or destructive interference. Sound waves also exhibit Doppler shifts—an apparent change in frequency due to relative motion between the source of sound emission and the receiving point. When sound waves move toward an observer the Doppler effect shifts observed frequencies higher. When sound waves move away from an observer the Doppler effect shifted observed frequencies lower. The Doppler effect is commonly and easily observed in the passage of planes, trains, and automobiles.

The speed of propagation of a sound wave is dependent upon the density of the medium of transmission. **Weather** conditions (e.g., **temperature**, pressure, **humidity**, etc.) and certain geophysical and topographical features (e.g., mountains or hills) can obstruct sound transmission. The alteration of sound waves by commonly encountered meteorological conditions is generally negligible except when the sound waves propagate over long distances or emanate from a high frequency source. In the extreme cases, atmospheric conditions can bend or alter sound wave transmission.

The speed of sound through a fluid—inclusive in this definition of "fluid" are atmospheric gases—depends upon the temperature and density of the fluid. Sound waves travel faster at higher temperature and density of medium. As a result, in a standard atmosphere, the speed of sound (reflected in the Mach number) lowers with increasing altitude.

Meteorological conditions that create layers of air at dramatically different temperatures can refract sound waves.

The speed of sound in water is approximately four times faster than the speed of sound in air. SONAR sounding of ocean terrain is a common tool of oceanographers. Properties such as pressure, temperature, and salinity also affect the speed of sound in water.

Because sound travels so well under water, many marine biologists argue that the introduction of man-made noise (e.g., engine noise, propeller cavitation, etc) into the **oceans** within the last two centuries interferes with previously evolutionarily well-adapted methods of sound communication between marine animals. For example, man-made noise has been demonstrated to interfere with long-range communications of whales. Although the long term implications of this interference are not fully understood, many marine biologists fear that this interference could impact whale mating and lead to further population reductions or extinction.

See also Aerodynamics; Atmospheric composition and structure; Atmospheric inversion layers; Electromagnetic spectrum; Energy transformations; Seismograph; Seismology

SOUTH AMERICA

The South American continent stretches from about 10° above the equator to almost 60° below it, encompassing an **area** of 6,880,706 sq mi (17,821,028 sq km). This is almost 12% of the surface area of the earth. It is about 3,180 mi (5,100 km) wide at its widest point, and is divided into 10 countries. The continent can be divided into three main regions with distinct environmental and geological qualities: the highlands and plateaus of the east, which are the oldest geological feature in the continent; the Andes Mountains, which line the west coast and were created by the subduction of the Nazca plate beneath the continent; and the riverplain, between the highlands, which contains the Amazon River. The South American **climate** varies greatly based on the distance from the equator and the altitude of the area, but the range of temperatures seldom reaches 36°F (20°C), except in small areas.

The Eastern highlands and plateaus are the oldest geological region of South America, and are thought to have bordered on the African continent at one time, before the motion of the earth's **crust** and continental drift separated the continents. The Eastern highlands can be divided into three main sections, the Guiana Highlands, the Brazilian Highlands, and the Patagonian Highlands. The Guiana Highlands are found in the Guianan states, south Venezuela, and northeastern Brazil. Their highest peak, Roraima, reaches a height of 9,220 ft (2,810 m). This is a moist region with many waterfalls; it is in this range, in Venezuela, that the highest waterfall in the world, Angel Falls, is found. Angel Falls plummets freely for 2,630 ft (802 m).

The Brazilian Highlands make up more than one half of the area of Brazil, and range in altitude between 1,000 and 5,000 ft (305–1524 m). The highest mountain range of this region is called Serra da Mantiqueira, and its highest peak, Pico da Bandeira, is 9,396 ft (2,864 m) above sea level.

The Patagonian Highlands are in the south, in Argentina. The highest peak reaches an altitude of 9,462 ft (2,884 m), and is called Sierra de Cordoba.

The great mountain range of South America is the Andes Mountains, which extends more than 5,500 mi (8,900 km) all the way down the western coast of the continent. The highest peak of the Andes, called Mount Aconcagua, is on the western side of central Argentina, and is 22,828 ft (6,958 m) high. The Andes were formed by the motion of the earth's crust and its different tectonic plates. Some of them are continental plates, which are at a greater altitude than the other type of plate, the oceanic plates. All of these plates are in motion relative to each other, and the places where they border each other are regions of instability where various geological structures are formed, and where earthquakes and volcanic activity is frequent. The western coast of South America is a **subduction zone**, which means that the oceanic plate, called the Nazca plate, is being forced beneath the adjacent continental plate. The Andes Mountains were thrust upwards by this motion, and can still be considered "under construction" by the earth's crust. In addition to the Nazca plate, the South American and Antarctic plates converge on the west coast in an area called the Chile Triple Junction, at about 46° south **latitude**. The complexity of **plate tectonics** in this region sparks interest for geologists.

The geological instability of the region makes earthquakes common all along the western region of the continent, particularly along the southern half of Peru.

The Andes are dotted with volcanoes; some of the highest peaks in the mountain range are volcanic in origin, many of which rise above 20,000 ft (6,100 m). There are three major areas in which volcanoes are concentrated. The first of these appears between latitude 6° north and 2° south, straddles Colombia and Ecuador, and contains active volcanoes. The second, and largest region, lies between latitudes 15° and 27° south; it is about 1,240 mi (2,000 km) long and 62–124 mi (100–200 km) wide, and borders Peru, Bolivia, Chile, and Argentina. This is the largest concentration of volcanoes in the world, and the highest volcanoes in the world are found here. The volcanic activity, however, is low and it is generally geysers that erupt here. The third region of volcanic concentration is also the most active. It lies in the central valley of Chile, mostly between 33° and 44° south.

The climate in the Andes varies greatly, depending on both altitude and latitude, from hot regions, to Alpine meadow regions, to the **glaciers** of the South. The snowline is highest in southern Peru and northern Chile, at latitude 15–20° south, where it seldom descends below 19,000 ft (5,800 m). This is much higher than at the equator, where the snowline descends to 15,000 ft (4,600 m). This vagary is attributed to the extremely dry climate of the lower latitude. In the far south of the continent, in the region known as Tierra del Fuego, the snowline reaches as low as 2,000 ft (600 m) above sea level.

The Andes are a rich source of mineral deposits, particularly copper, silver, and gold. In Venezuela, they are mined for copper, **lead**, **petroleum**, phosphates, and salt; diamonds are found along the Rio Caroni. Columbia has the richest deposits of **coal**, and is the largest producer of gold and platinum in South America. Columbia is also wealthy in emeralds, containing the largest deposits in the world with the exception of Russia. In Chile, the Andes are mined largely for their great copper stores in addition to lead, zinc, and silver. Bolivia has enormous tin mines. The Andes are also a source of tungsten, antimony, nickel, chromium, cobalt, and sulfur.

The Amazon basin is the largest river basin found in the world, covering an area of about 2.73 million sq mi (7 million sq km). The second largest river basin, which is the basin of the River Zaire in the African Congo, is less than half as large. The **water** resources of the area are spectacular; the volume of water that flows from the basin into the sea is about 11% of all the water drained from the continents of the earth. The greatest flow occurs in July, and the least is in November. While

Glacier, Muir Bay Inlet, Glacier Bay National Park, Alaska. See entry, "Glaciers," page 255. *JLM Visuals. Reproduced by permission.*

Modern mapping techniques have allowed us to explore Earth's least hospitable regions. Only decades ago, the bottom of Lake Tahoe was a mystery. See entry, "History of exploration III (modern era)," page 289. *AP/Wide World. Reproduced by permission.*

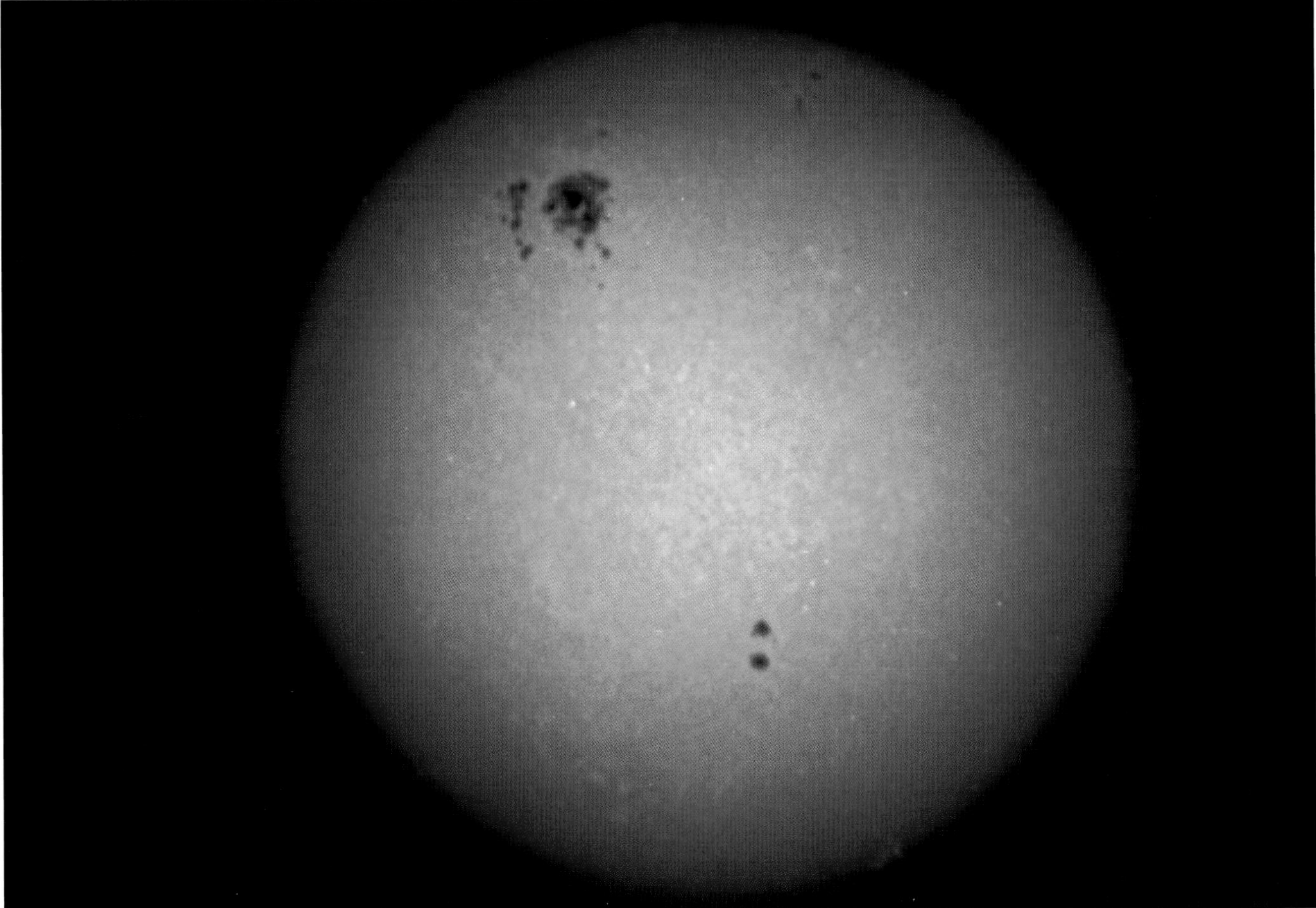

TOP: Iceberg. See entry, "Icebergs," page 313. *Photography by Commander Richard Behn, NOAA Corps. National Oceanic and Atmospheric Administration.*

BOTTOM: Visible light image of the Sun, showing sunspots. See entry, "Insolation and total solar irradiance," page 319. *U.S. Aeronautics and Space Administration (NASA).*

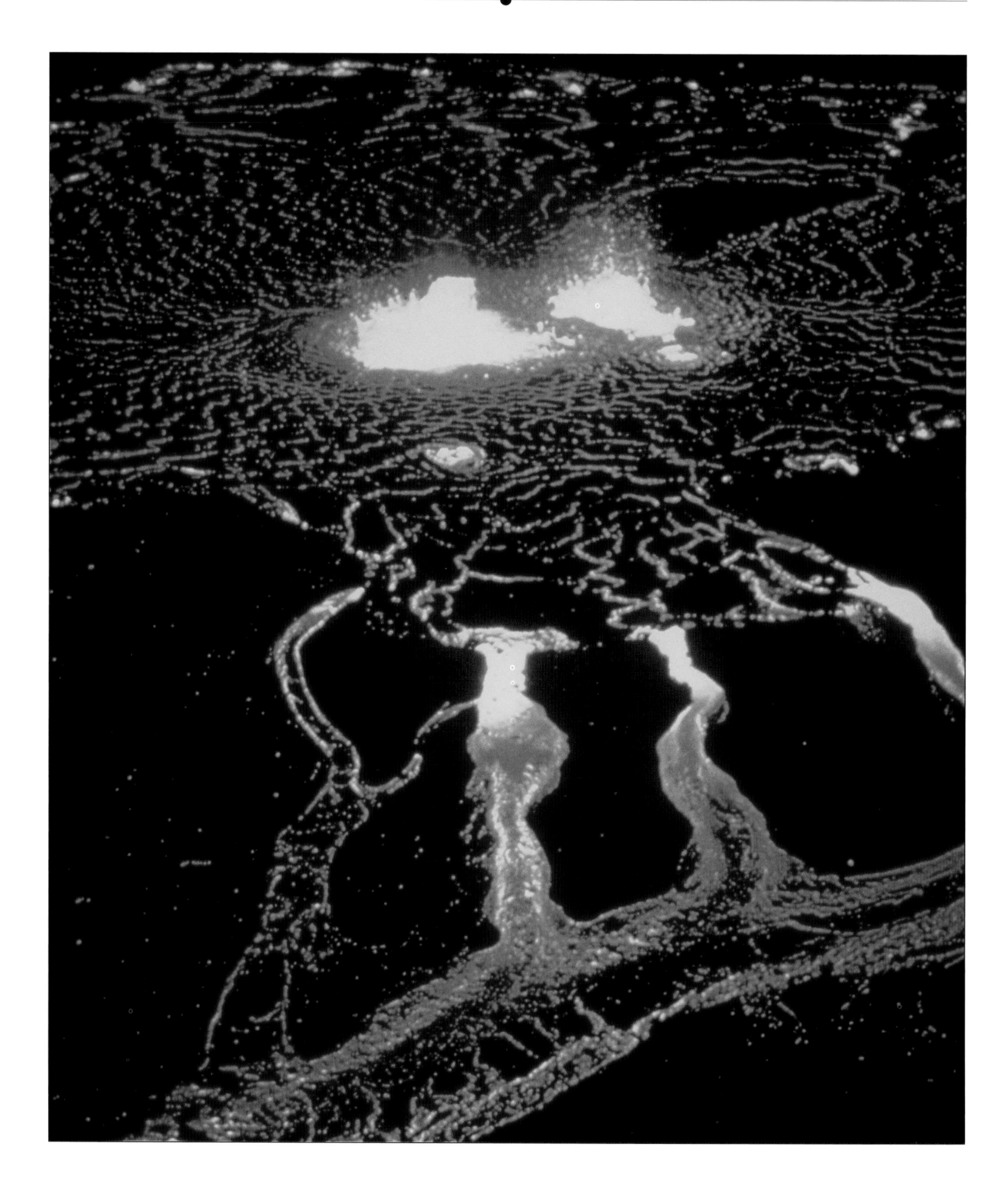

Lava from Mt. Kilauea. This kind of lava has very little silica in it, and is very viscous. When it dries, it will be referred to as pahoehoe lava. See entry, "Lava," page 343. *JLM Visuals. Reproduced by permission.*

Lightning strikes over Tuscon, Arizona. See entry, "Lightning," page 348. *Keith Kent/Peter Arnold, Inc. Reproduced by permission.*

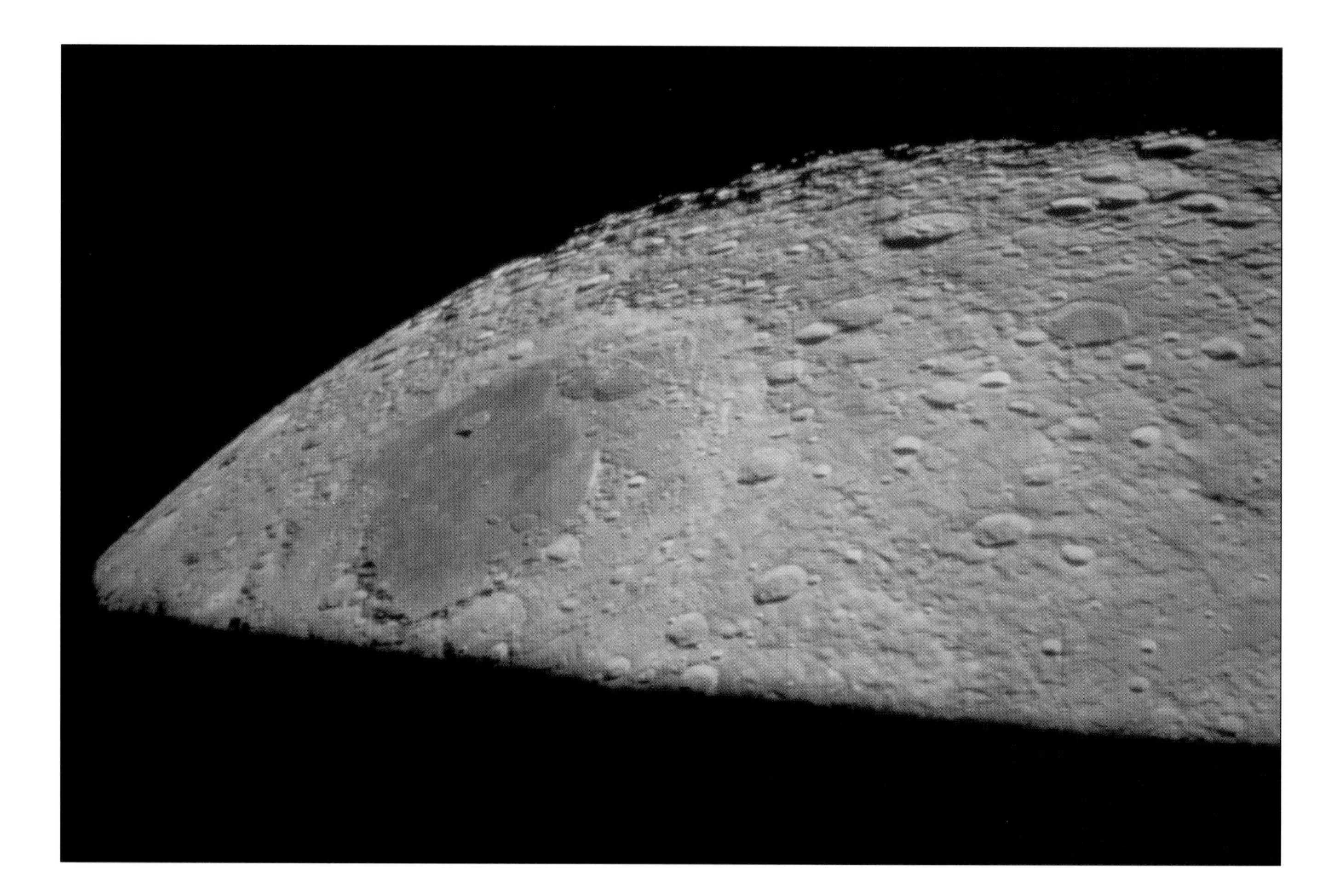

The Moon's surface. See entry, "Moon," page 384. *U.S. National Aeronautics and Space Administration (NASA).*

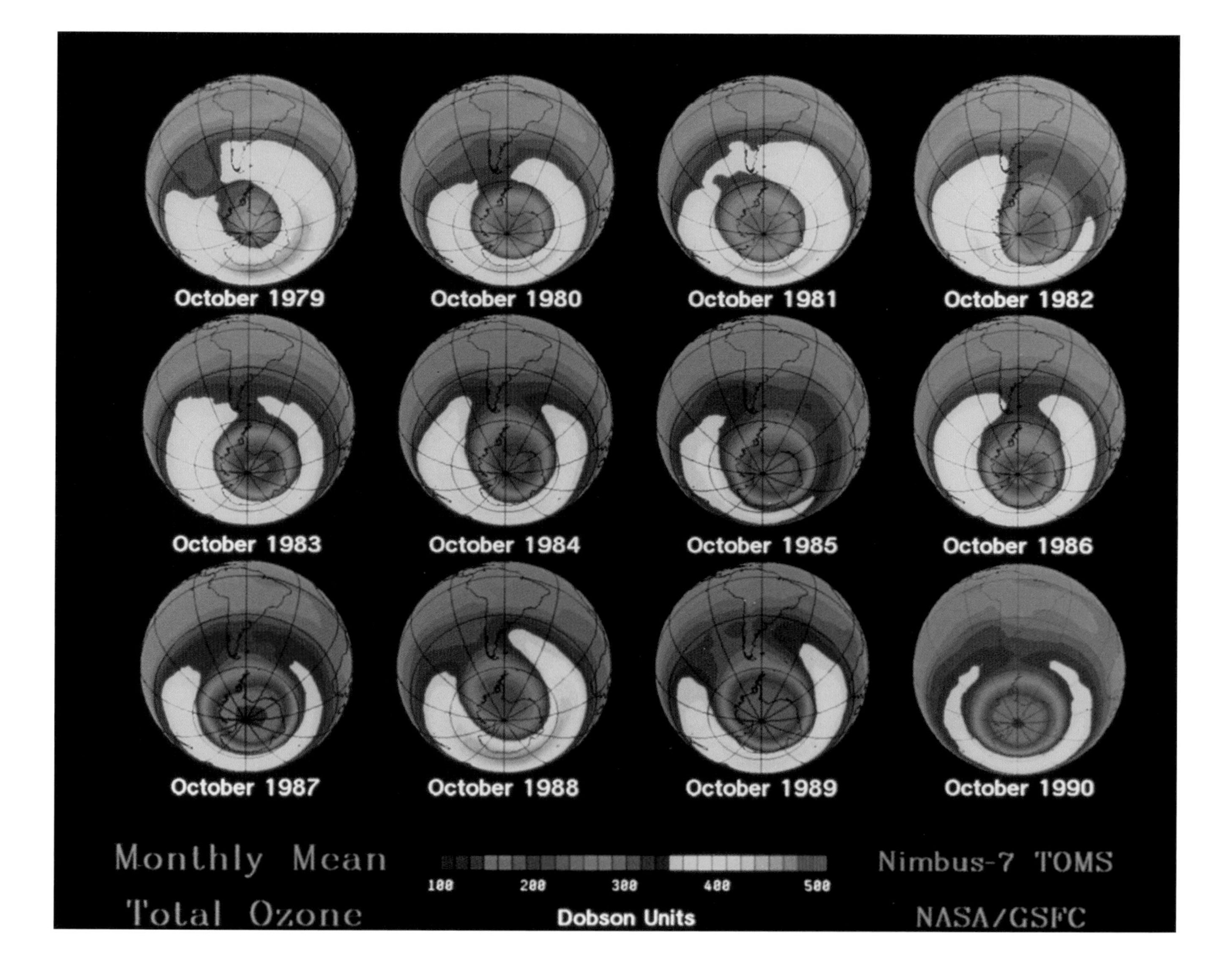

False color images showing ozone depletion from October 1979 to October 1990. See entry, "Ozone layer and ozone hold dynamics," page 419. *U.S. National Aeronautics and Space Administration (NASA).*

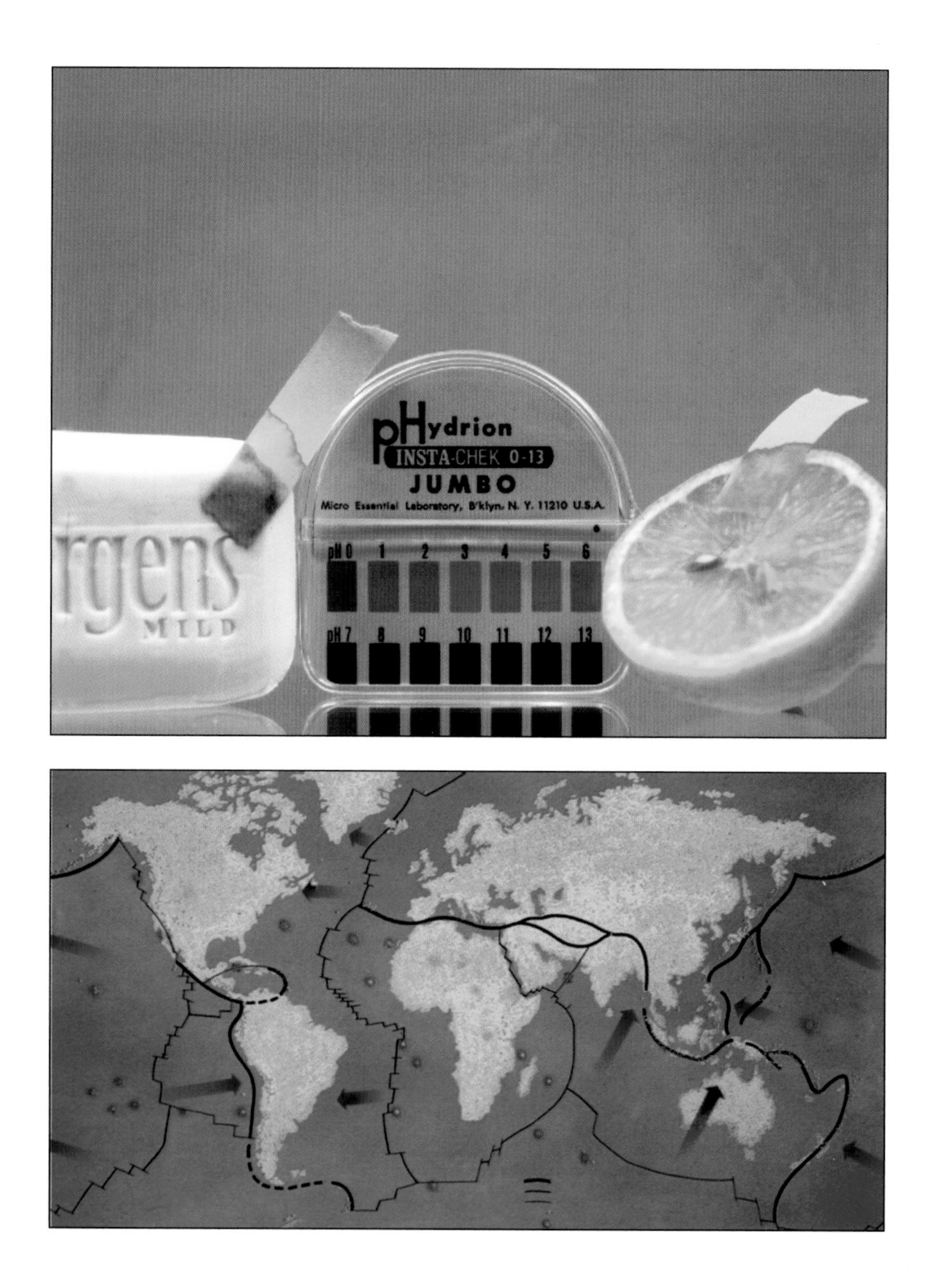

TOP: Litmus paper is specially treated to turn either blue or pink if exposed to acidic or basic compounds. See entry, "pH," page 44. *© Yoav Levy/Phototake NYC. Reproduced by permission.*

BOTTOM: Map of Earth with tectonic plate boundaries shown. Convergent zones are shown with thick lines, divergent zones are shown with medium lines, and transform boundaries are shwon with thin lines. Arrows indicate the direction of plate motion. Large orange dots represent "hot spots" and small orange dots represent volcanoes. See entry, "Plate tectonics," page 450. *David Hardy/Science Photo Library. Reproduced by permission.*

Rose quartz. See entry, "Quartz," page 470. ***U.S. National Aeronautics and Space Administration (NASA).***

A double rainbow over the desert. See entry, "Rainbow," page 480. *FMA. Reproduced by permission.*

Wildfires in Indonesia caused smog thick enough that these girls need to wear masks on their way to school. See entry, "Smog," page 53. *© Michael S. Yamashita/Corbis. Reproduced by permission.*

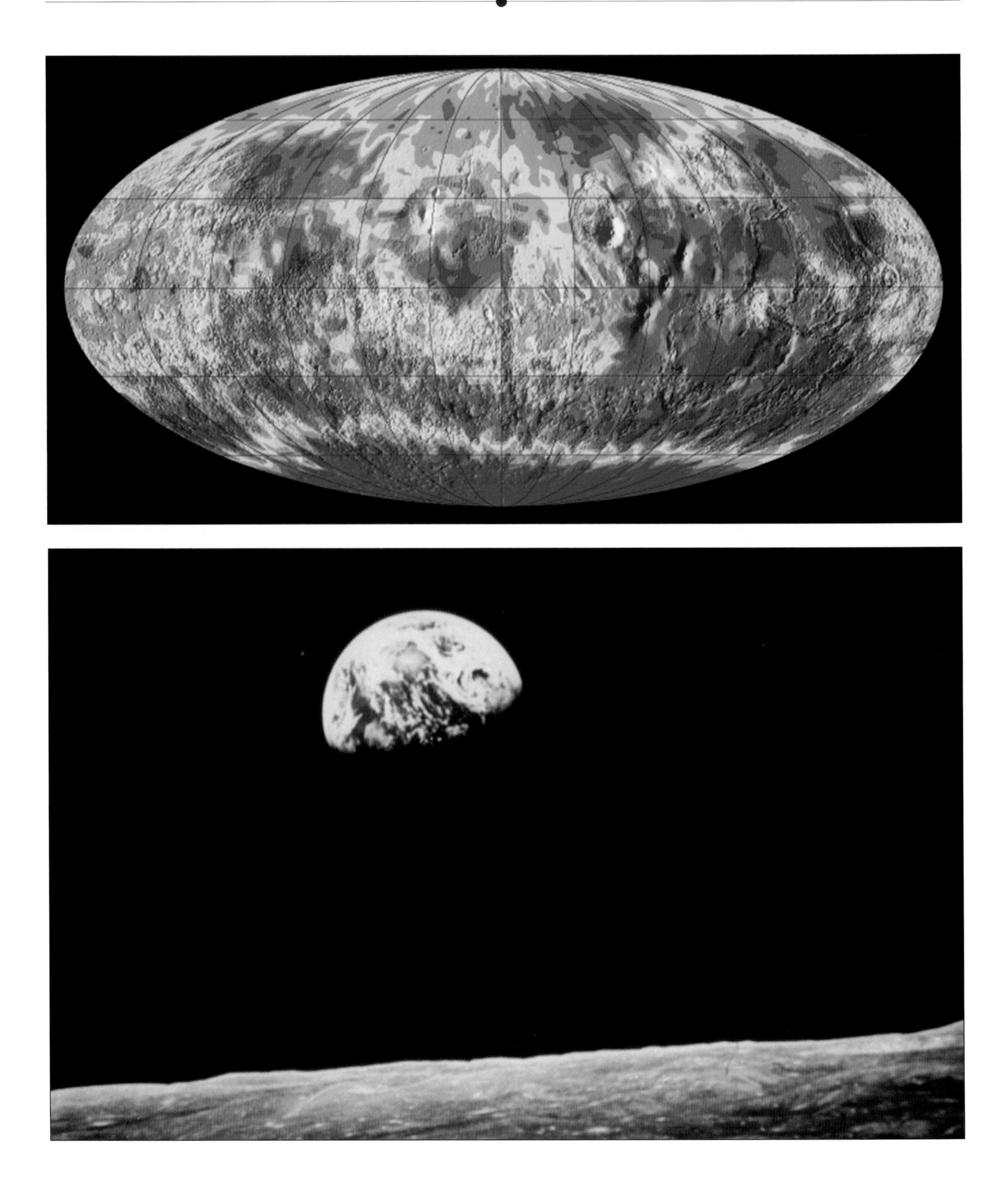

TOP: Map of Mars made by NASA's Mars *Odyssey* spacecraft. See entry, "Solar System," page 538. *AP/Wide World. Reproduced by permission.*

BOTTOM: Earth rising. See entry, "Space," page 547. *U.S. National Aeronautical and Space Administration (NASA).*

TOP: Mars rover, *Sojourner*, on the Martian surface. See entry, "Space and planetary geology," page 549. *© Agence France Presse/Cobris-Bettmann. Reproduced by permission.*

BOTTOM: Space shuttle *Endeavour*. See entry, "Space shuttle," page 55. *U.S. National Aeronautics and Space Administration (NASA).*

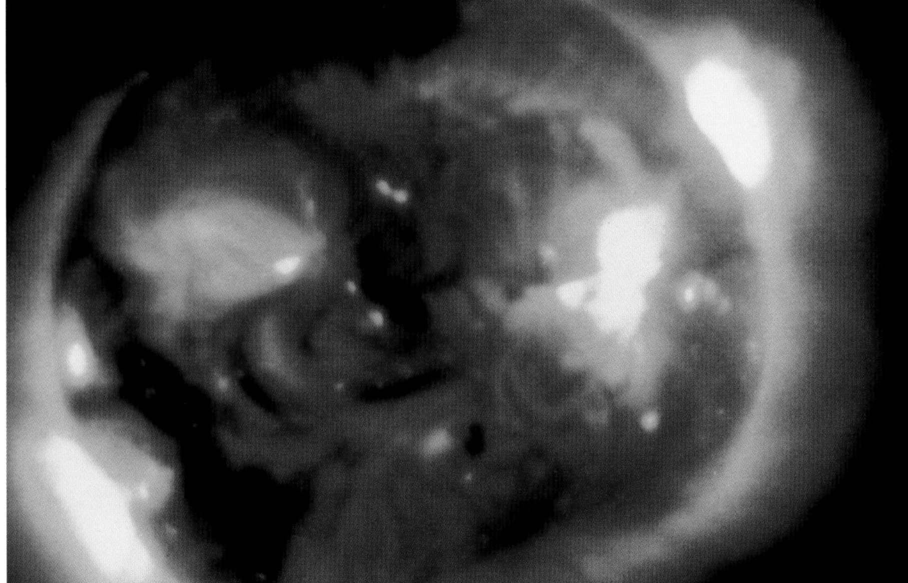

TOP: The Grand Canyon in Arizona has exposed millions of years of strata, allowing geologists a glimpse back in Earth's history. See entry, "Stratigraphy," page 565. *National Park Service.*

BOTTOM: X-ray photograph of the solar corona. See entry, "Sun (stellar structure)," page 571. *U.S. National Aeronautics and Space Administration (NASA).*

TOP: Satellite photo of Hurricane Mitch. See entry, "Tropical cyclone," page 598. *National Oceanic and Atmospheric Administration (NOAA).*

BOTTOM: Mount Pinatubu eruption. See entry, "Volcanic eruptions," page 613. *© T.J. Casadevall, U.S. Geological Survey Photographic Library, Denver, CO.*

Mount Etna. See entry, "Volcano," page 614. *Photography by Jonathon Blair. Jonathon Blair/Corbis-Bettman. Reproduced by permission.*

there are many **rivers** flowing through the basin, the most important and well known of these is the Amazon. The width of the Amazon ranges from about 1 mi (1.6 km) to as wide as 5–6 mi (5–6 km), and although it is usually only about 20–40 ft (6–12 m) deep, there are narrow channels where it can reach a depth of 300 ft (100 m).

The Amazon basin was once an enormous bay, before the Andes were pushed up along the coasts. As the mountain range grew, they held back the ocean and eventually the bay became an inland sea. This sea was finally filled by the **erosion** of the higher land surrounding it, and finally a huge plain, crisscrossed by countless waterways, was created. Most of this region is still at sea level, and is covered by lush jungle and extensive wetlands. This jungle region contains the largest extant rain forest in the world. Despite the profusion of life that abounds here, the **soil** is not very rich; the fertile regions are those which receive a fresh layer of river silt when the Amazon **floods**, which occurs almost every year.

The climate of South America varies widely over a large range of altitudes and latitudes, but only in isolated regions is the **temperature** range greater than about 36°F (20°C). The coldest part of the continent is in the extreme southern tip, in the area called Tierra del Fuego; in the coldest month of the year, which is July, it is as cold as 32°F (0°C) there. The highest temperature of the continent is reached in a small area of northern Argentina, and is about 108°F (42°C). However, less than 15 days a year are this warm, and the average temperature in the same area for the hottest month of the year, which is January, is about 84°F (29°C).

Colombia borders Venezuela, Brazil, Ecuador, and Peru, and encompasses an area of 440,831 sq mi (1,141,748 sq km). It is found where Panama of Central America meets the South American continent, and its location gives it the interesting feature of having coastal regions bordering on both the Atlantic and the Pacific **oceans**. It is a country of diverse environments, including coastal, mountain, jungle, and island regions, but in general can be considered to consist of two major areas based on altitude: the Andes mountains and the lowlands.

The Andes in Colombia can be divided into three distinct ranges, which run approximately from north to south in parallel ridges. The Cordillera Occidental, or westernmost range, attains a maximum altitude of about 10,000 ft (3,000 m). The Cordillera Oriental, which is the eastern range, is much higher, and many of its peaks are covered with snow all year round. Its highest peak is about 18,000 ft (5,490 m) high, and it has many waterfalls, such as the Rio Bogota, which falls 400 ft (120 m). The Cordillera Central, as its name implies, runs between the Occidental and Oriental Cordilleras. It contains many active volcanoes as well as the highest peak in Colombia, Pico Cristobal Colon, which is 19,000 ft (5,775 m) high.

The lowlands of the east cover two thirds of Colombia's land area. It is part of the Orinoco and Amazon basins, and thus is well watered and fertile. Part of this region is covered with rich equatorial rain forest. The northern lowlands of the coastal region also contain several rivers, and the main river of Colombia, the Magdalena, begins there.

Venezuela covers an area of 352,144 sq mi (912,0250 sq km). It is the most northern country of South America, and can be divided up into four major regions. The Guiana Highlands in the southeast make up almost half of Venezuela's land area, and are bordered by Brazil and Guyana. It is here that the famous Angel Falls, the highest waterfall in the world, is found. The Northern Highlands, which are a part of the Andes Mountains, contain the highest peak in Venezuela—Pico Bolivar, which reaches a height of 16,427 ft (5,007 m). This range borders on much of the coastal region of Venezuela, and despite its proximity to both the Caribbean and the equator, it has many peaks that are snow-covered year-round. The Maracaibo basin, one-third of which is covered by Lake Maracaibo, is found in the northwest. It is connected to the Caribbean Sea, and although it contains fresh water at one end of the lake, as it nears the ocean it becomes more saline. Not surprisingly, most of the basin consists of wetlands. The Llanos de Orinoco, which borders on Colombia in the southwestern part of Venezuela, is watered by the Orinoco River and its tributaries. The Orinoco has a yearly discharge almost twice as large as that of the Mississippi, and from June to October, during the rainy season, many parts of the Llanos are inaccessible due to flooding.

Ecuador received its name from the fact that it straddles the equator. Its area is 103,930 sq mi (269,178 sq km), making it the smallest of the Andean countries. Its eastern and western lowlands regions are divided by the Andes Mountains, which run through the center of the country. This part of the Andes contains an active **volcano** region; the world's highest active volcano, Cotopaxi, which reaches an altitude of 19,347 ft (5,897 m), is found here. The western lowlands on the coast contain a tropical rain forest in the north, but become extremely dry in the south. The eastern lowlands are part of the Amazon basin, and are largely covered by tropical rainforest. The rivers Putumayo, Napo, and Pastaza flow through this area.

Ecuador also claims the famous Galapagos Islands, which lie about 650 mi (1,040 km) off the coast. These 12 islands are all volcanic in origin, and several of the volcanoes are still active. The islands are the home of many species unique to the world, including perhaps the most well-known of their numbers, the Galapagos tortoise.

Peru covers an area of 496,225 sq. mi (1,285,216 sq. km), making it the largest of the Andean countries. Like Ecuador, it is split by the Andes Mountains into two distinct sections. The eastern coastal region is mostly covered with mountains, and in many places, the ocean borders on steep cliffs. In the northern part, however, there is a relatively flat region that is suitable for agriculture. In the east, the lowlands are mostly covered by the thick tropical rain forest of the Amazon basin. The southern part of the Andes in Peru contain many volcanoes, some of which are still active, and Lake Titicaca, which is shared by Bolivia. Lake Titicaca is remarkable for, among the large **lakes** with no ocean outlet, Titicaca is the highest in the world. It is 125 mi (200 km) at its largest length and 69 mi (110 km) at its largest breadth, which is not quite half as large as Lake Ontario; but it lies at an altitude of 12,507 ft (3,812 m) above sea level.

Bolivia has an area of 424,164 sq mi (1,098,581 sq km), and is the only landlocked country in South America besides Paraguay. The western part of the country, which borders on

Ecuador and Chile, is covered by the Andes Mountains, and like most of this part of the Andes, it contains many active volcanoes. In the southern part of the range, the land becomes more arid, and in many places salt marshes are found. Among these is Lake Poopo, which lies 12,120 ft (3,690 m) above sea level. This saline lake is only 10 ft (3 m) deep. In the northern part of the range, the land becomes more habitable, and it is here that Lake Titicaca, which is shared with Peru, is found.

The eastern lowlands of Bolivia are divided into two distinct regions. In the north, the fertile Llanos de Mamore is well watered and is thickly covered with vegetation. The southeastern section, called the Gran Chaco, is a semiarid savanna region.

Chile is the longest, narrowest country in the world; although it is 2,650 mi (4,270 km) long, it is only about 250 mi (400 km) wide at its greatest width. It encompasses an area of 284,520 sq mi (736,905 sq km). The Andes divides into two branches along the eastern and western edges of the country. The eastern branch contains the highest of the Andean peaks, Aconcagua, which is 20,000 ft (6,960 m), and the highest point on the continent. The Andes in Chile has the greatest concentration of volcanoes on the continent, containing over 2,000 active and dormant volcanoes, and the area is plagued by earthquakes.

In the western coastal region of north and central Chile, the land meets the ocean in a long line of cliffs which reach about 8,800 ft (2,700 m) in altitude. The southern section of this coastal mountain range moves offshore, forming a group of about 3,000 islands extending in a line all the way to Cape Horn, which is the southernmost point on the continent. The coast in this area is quite remarkable in appearance, having numerous **fjords**. There are many volcanic islands off the coast of Chile, including the famous Easter Island, which contains some unusual archeological remains.

The southern part of the coastal region of Chile is a temperate area, but in the north it contains the Atacama **Desert**, which is the longest and driest desert in the world. Iquique, Chile, which lies in this region, is reported to have at one time suffered 14 years without any rain at all. The dryness of the area is thought to be due to a sudden temperature inversion as **clouds** move from the cold waters off the shore and encounter the warmth of the continent; this prevents water from precipitating from the clouds when they reach the shoreline. It has been suggested also that the sudden rise of the Andes Mountains on the coast contributes to this effect.

Argentina, the second largest of the South American countries, covers an area of 1,073,399 sq mi (2,780,092 sq km). The Andes Mountains divide western Argentina from Chile, and in the south, known as Tierra de Fuego, this range is still partly covered with glaciers.

A large part of Argentina is a region of lowlands and plains. The northern part of the lowlands, called the Chaco, is the hottest region in Argentina. In the northwestern part of Argentina near the Paraguayan and Brazilian borders, are found the remarkable Iguassa Falls. They are 2.5 mi (4 km) wide and 269 ft (82 m) high. As a comparison, Niagara Falls is only 5,249 ft (1,599 m) wide and 150–164 ft (46–50 m) high. The greatest part of the lowland plains is called the Pampa, which is humid in the east and semiarid in the west.

The southern highlands of Patagonia, which begins below the Colorado River, is a dry and mostly uninhabited region of plateaus. In the Tierra del Fuego the southernmost extension of the Andes is found. They are mostly glaciated, and many glacial lakes are found here. Where the mountains descend into the sea, the glaciers have shaped them so that the coast has a fjord-like appearance.

The Falkland Islands lie off the eastern coast of Argentina. They are a group of about 200 islands consisting of rolling hills and peat valleys, although there are a few low mountains north of the main islands. The sea around the Falkland Islands is quite shallow, and for this reason they are thought to lie on an extension of the **continental shelf**.

Paraguay, which has an area of 157,048 sq mi (406,752 sq km), is completely landlocked. About half of the country is part of the Gran Chaco, a large plain west of the Paraguay River, which also extends into Bolivia and Argentina. The Gran Chaco is swampy in places, but for the most part consists of scrubland with a few isolated patches of forest. East of the Paraguay River, there is another plain which is covered by forest and seasonal marshes. This region becomes a country of flat plateaus in the easternmost part of Paraguay, most of which are covered with evergreen and deciduous **forests**.

Uruguay, which is 68,037 sq mi (176,215 sq km) in area, is a country bounded by water. To the east it borders the Atlantic Ocean, and there are many lagoons and great expanses of **dunes** found along the coast. In the west, Uruguay is bordered by the river Uruguay, and in the south by the La Plata **estuary**. Most of the country consists of low hills with some forested areas.

With an area of 3,286,487 sq mi (8,511,965 sq km), Brazil is by far the largest country in South America, taking up almost half of the land area of the continent. It can be divided into two major geographical regions: the highlands, which include the Guiana Highlands in the far north and the Brazilian Highlands in the center and southeast, and the Amazon basin.

The highlands mostly have the appearance of flat tablelands, which are cut by deep rifts, and clefts that drain them; these steep river valleys are often inaccessible. In some places, the highlands have been shaped by erosion so that their surfaces are rounded and hill-like, or even give the appearance of mountain peaks. Along the coast, the plateaus plummet steeply to the ocean to form great cliffs, which can be as high as 7,000–8,000 ft (2,100–2,400 m). Except for the far north of Brazil, there are no coastal plains.

The lowlands of Brazil are in the vast Amazon basin, which is mostly covered with dense tropical rain forest, the largest tract of unbroken rainforest in the world. The many rivers and tributaries that water the region create large marshes in places. The Amazon is home to many indigenous peoples and as yet uncounted species of animals and plants found nowhere else in the world.

French Guiana encompasses an area of 35,900 sq mi (93,000 sq km), and is found north of Brazil. The area furthest inland is a region of flat plateaus that becomes rolling hills in the central region of the country, while the eastern coastal area is a broad plain consisting mostly of poorly drained marshland. Most of the country is covered with dense tropical rain forest, and the coast is lined with mangrove swamps. French Guiana possesses a few island territories as well; the most famous of these, Devil's Island, was the former site of a French penal colony.

North of French Guiana lies Suriname, another tiny coastal country that has an area of 63,251 sq mi (163,820 sq km). The southern part of the country is part of the Guiana Highlands, and consists of very flat plateaus cut across by great rifts and steep gullies. These are covered with thick tropical rain forest. North of the highlands is an area of rolling hills and deep valleys formed by rivers and covered with forest. The extreme north of Suriname lies along the coast and is a flat swamp. Several miles of mangrove swamps lie between this region and the coast.

East of Suriname is the country of Guyana, with a land area of 83,000 sq mi (215,00 sq km). The Guiana Highlands are in the western and southern parts of Guyana. As with Suriname and French Guiana, these are cut up deeply by steep and sudden river valleys, and covered with dense rain forest. The western part of the Guiana Highlands are called the Pakaraima Mountains, and are much higher than the other plateaus in Guyana, reaching an altitude of as much as 9,220 ft (2,810 m). The highlands become a vast area of rolling hills in the central part of Guyana due to the effects of erosion; this sort of terrain takes up more than two thirds of the country. In the north along the coast is a swampy region as in Suriname and French Guiana, with many lagoons and mangrove swamps.

See also Continental drift theory; Delta; Depositional environments; Desert and desertification; Earth (planet); Forests and deforestation; Orogeny; Rapids and waterfalls; Rivers; Seasonal winds; Volcanic eruptions

SOUTH POLE • *see* GEOGRAPHIC AND MAGNETIC POLES

SOUTHERN LIGHTS • *see* AURORA BOREALIS AND AURORA AUSTRALIALIS

SPACE

Space is the three-dimensional extension in which all things exist and move. Intuitively, it feels that we live in an unchanging space. In this space, the height of a tree or the length of a table is exactly the same for everybody. Einstein's special theory of relativity explains that this intuitive feeling is really an illusion. Neither space nor time is the same for two people moving relative to each other. Only a combination of space and time, called space-time, is unchanged for everyone. Einstein's general theory of relativity states that the force of **gravity** is a result of a warping of this space-time by heavy objects, such as planets. According to the **Big Bang theory** of the origin of the universe, the expansion of the universe began from infinitely curved space-time. Scientists still do not know whether this expansion will continue indefinitely, or whether the universe will collapse again in a Big Crunch. Meanwhile, astronomers are continually learning about outer space from terrestrial and orbiting telescopes, space probes sent to other planets in the **solar system**, and other scientific observations. This is just the beginning of the exploration of the unimaginably vast void, beyond Earth's outer atmosphere, in which a journey to the nearest star would take 3,000 years traveling at a million miles per hour.

The difference in the perception of space and time, predicted by the special theory of relativity, can be observed only at very high velocities close to that of light. A man driving past at 50 mph (80 kph) will appear only a hundred million millionths of an inch thinner as you stand watching on the sidewalk. By themselves, three-dimensional space and one-dimensional time are different for different people. Taken together, however, they form a four-dimensional space-time in which distances are the same for all observers. We can understand this idea by using a two-dimensional analogy. Suppose that a man's definition of south and east is not the same as a woman's. The woman travels from city A to city B by going 10 miles along her south and then 5 miles along the man's east. The man travels from A to B by going 2 miles along his south and 11 miles along the woman's east. Both, however, move exactly the same distance of 11.2 miles south-east from city A to B. In the same way, if we think of space as south and time as east, space-time is something like south-east.

The general theory of relativity states that gravity is the result of the curving of this four-dimensional space-time by objects with large mass. A flat stretched rubber membrane will sag if a heavy **iron** ball is placed on it. If you now place another ball on the membrane, the second ball will roll towards the first. This can be interpreted in two ways: as a consequence of the curvature of the membrane, or as the result of an attractive force exerted by the first ball on the second one. Similarly, the curvature of space-time is another way of interpreting the attraction of gravity. An extremely massive object can curve space-time around so much that not even light can escape from its attractive force. Such objects, called black holes, probably exist in the universe. Astronomers believe that the disk found in 1994 by the Hubble **telescope**, at the center of the elliptical galaxy M87 near the center of the Virgo cluster, is material falling into a supermassive black hole estimated to have a mass three billion times the mass of the **Sun**.

The relativity of space and time and the curvature of space-time do not affect our daily lives. The high velocities and huge concentrations of matter, needed to manifest the effects of relativity, are found only in outer space on the scale of planets, stars, and galaxies. Our own Milky Way galaxy is a mere speck, 100,000 light years across, in a universe that spans ten billion light years. Though astronomers have studied this outer space with telescopes for hundreds of years, the modern space age began only in 1957 when the Soviet Union put the first artificial **satellite**, *Sputnik 1*, into orbit around the earth. At present, there are hundreds of satellites in orbit gathering information from distant stars, free of the distorting effect of the earth's atmosphere. Even though no manned

Earth rising. ***U.S. National Aeronautics and Space Administration (NASA).***

spacecraft has landed on other worlds since the Apollo **Moon** landings, several space probes, such as the *Voyager 2* and the *Magellan*, have sent back photographs and information from the Moon and from other planets in the solar system. There are many questions to be answered and much to be achieved in the exploration of space. The Hubble telescope, repaired in space in 1993 and 2002, has sent back data that has raised new questions about the age, origin, and nature of the universe. The launch of a United States astronaut to the Russian *Mir* space station in March 1995, the docking of the United States **space shuttle** *Atlantis* with *Mir*, and the **international space station** currently under construction have opened up exciting possibilities for space exploration.

See also Astronomy; Celestial sphere: The apparent movements of the Sun, Moon, planets, and stars; History of manned space exploration; Physics; Relativity theory; Solar system; Space and planetary geology; Space physiology

SPACE PHYSIOLOGY

Space physiology is concerned with the structure and functioning of the body under the conditions encountered by space travelers. To date, these conditions have been confined to the environment of the spacecraft that houses the astronauts. In the future, however, as travel to other bodies in the **solar system** is undertaken, space physiology will include the atmospheric and gravitational conditions found on these planets, moons, or other stellar bodies.

Aside from the lunar missions of the 1960s, man's extraterrestrial voyages have been confined to orbital forays aboard space capsules or space stations. But even orbiting around Earth poses difficulties for the astronauts. The reduced **gravity** of a spacecraft makes it difficult for the body to distinguish "up" from "down." On Earth, such distinction by the vestibular organ of the inner ear is easy, because of the orienting power of gravity. In the **space shuttle** and the developing **international space station**, all writing on the walls is oriented in the same direction, to provide the brain with a reference point.

Low gravity (also known as microgravity) affects other body systems besides the vestibular system. The propioceptive system—the system of nerves in the joints and muscles that tell us where the arms and legs are without any visual inspection—can also be affected. Low gravity reduces or eliminates the tensions impinging on the joints and muscles, which can make the appendages appear invisible to the brain.

Such confusion between what the eye sees and the brain perceives can result in what has been termed space sickness. This is somewhat analogous to the feeling of nausea experienced by someone trying to read in a moving car. The inner ear detects the motion of the car, or the spacecraft, but the eyes staring at the page of the book or the space outside the spacecraft do not detect motion. Space sickness is usually transient, and astronauts acclimate soon after going into orbit.

Microgravity also affects the skeletal structure of astronauts. The absence of stress-bearing activity and the loss of components of the bones, particularly calcium, have produced shortening and weakening of bones (essentially the development of osteoporosis) and the atrophy (wasting away) of muscles in astronauts who have been orbiting the earth for just several months. Even the heart becomes smaller. So far these conditions have reversed upon return to Earth. The extended missions of the future will need to incorporate more Earth-like gravitation conditions, or an exercise regimen, or both.

Space flight also affects the cardiovascular system of astronauts. The weakened muscles and bones cannot support the maintenance of the same rate of flow of blood as on Earth. Also, the diminished downward pull of gravity affects the ability of the body to pump blood to extremities like the legs. Fluid flow to the upper regions of the body is not affected, however. As a result, faces of astronauts often appear puffy. In a very real sense, astronauts become out of shape—so much so that Russian cosmonauts who spend months in orbit around the earth are sometimes carried away from the spacecraft on a stretcher upon return.

A physiological parameter that will become important when manned travel to other parts of the solar system begins is exposure to higher levels of radiation that will be encountered on planets where atmospheric constituents do not absorb the harmful energies. Genetic material can be damaged by high-energy (ionizing) cosmic radiation and high-energy particles, with adverse effects on the functioning of the body. Thus far, the relatively short-term voyages into space have not proven to be harmful. But the hazards posed by extended voyages of years or even decades are as yet unknown.

See also History of manned space exploration

SPACE AND PLANETARY GEOLOGY

Space and planetary **geology** comprises that branch of the discipline of geology that applies basic scientific principles to the study of the origin, development, and characteristics of **solar system** objects such as planets, satellites, **asteroids**, **comets**, meteorites, and interplanetary dust particles. On Earth, space and planetary geology investigations are generally limited to impact craters and impact effects upon Earth, and to the study of Earth as an analogue for other planets and their processes. As technology progresses in the future, the field of space and planetary geology will likely expand to include extra-solar system objects as well. Space and planetary geology is also called astrogeology.

Astronauts sleeping in freefall. Experiencing weightlessness over prolonged periods of time can produce detrimental physiological effects. *U.S. National Aeronautics and Space Administration (NASA).*

Space and planetary geology has it origins in telescopic observations of planets, satellites, and comets and in the study of meteorites. Telescopic maps of the **Moon** date from early work *c.* 1612 by **Galileo Galilei** (1564–1642). The first true lunar **geologic map** was that of Michael van Langren (1598–1675), which he completed in 1645. After this, many other such telescopic maps were made of the Moon over the next three centuries. During the period 1880 to 1925, several telescopic maps of topographic and geological features of Mercury and Mars were produced. The most famous of these were maps by Percival Lowell (1855–1916), which showed his interpreted "canals" on Mars. Meteoritics, which is the study of meteorites and their origins, traces its origin as a science back to the German physicist Ernst Florens Chladni (1756–1827). Chladni proposed convincing (but highly debated) arguments (*c.* 1794) that stones and masses of **iron** that were seen falling from the sky were in fact objects from space that produced fireballs as the fell through Earth's atmosphere. That small objects (asteroids) were orbiting the **Sun** was confirmed shortly thereafter (1801–1807) by a group of Italian astronomers dubbed the "celestial police," who discovered the first four known asteroids, Ceres, Pallas, Juno, and Vesta.

Mars rover, *Sojourner,* on the Martian surface. *© Agence France Presse/Corbis-Bettmann. Reproduced by permission.*

Space and planetary geology received an essential boost with the advent of rocketry and space flight. Beginning with missions to the Moon in the late 1950s and early 1960s (i.e., *Luna* 2 and 3 in 1959 and *Ranger* 7 in 1964), detailed orbital photographs of the near side and far side of the Moon were obtained. From these photographs, the first detailed geological maps of the Moon were made, thus establishing a new **area** of planetary photo-geologic mapping. During the 1960s, spacecraft made other missions to Mars and Venus. *Mariner* 4 (1964) and *Mariner* 6 (1969) took the first detailed photographs of Mars, and *Mariner* 2 (1964) landed on Venus and recorded surficial conditions. In the late 1960s and early 1970s, spacecraft returned samples from the Moon (*Apollo* 11, 12, 14–17 and *Luna* 16, 17, 20, 21), thus ushering in a new era of geological sample studies of the Moon. Study of these samples allowed radiometric dating and careful chemical and physical analysis that led to the first comprehensive description of the geological history of the Moon. In the 1970s and 1980s, spacecraft made radar maps of Venus (*Venera* 15 and 16), imaged the outer planets and some of their satellites (*Voyager* 1 and 2), imaged and landed upon Mars (*Viking* 1 and 2), and imaged part of Mercury (*Mariner* 10). These data further expanded planetary geological mapping. In the 1980s and 1990s, spacecraft photographed Halley's comet (*Vega* 1 and 2, *Giotto*), made detailed radar maps of Venus (*Magellan*), imaged asteroids Gaspara and Ida (*Galileo,* NEAR), imaged and landed on Mars (*Mars Pathfinder*), and imaged Jupiter and its satellites (*Galileo*). With each new photographic set, geological mapping of planet and **satellite** surfaces was expanded. More spacecraft observations are planned or are underway for Mercury, Mars, Saturn, the Moon, and various asteroids and comets in the near future.

Imagery from the various rocky planets and satellites has led to detailed topographic and geologic mapping of all the imaged bodies, and such maps, published mainly by the U.S. Geological Survey, are available to the public. Relative age relationships among geologic units on the planetary surfaces, deduced from photographic imagery, has led to development of preliminary geological time scales for Mercury, Venus, the Moon, and Mars. Similar studies are underway for the large Jovian satellites (Callisto, Ganymede, and Europa), the Saturnian satellites (Mimas, Enceladus, Tethys, Dione, Rhea, and Iapetus), the Uranian satellites (Ariel, Umbriel, Titania, Miranda, and Oberon), and the Neptunian satellite, Triton.

The International Astronomical Union (IAU) is in charge of standardized nomenclature for planetary surface features and geological units. There are approximately forty IAU-approved, generic feature terms in use in planetary nomenclature. For example, a ridge on a planetary surface is a dorsum (plural = dorsa) and a chain of craters is a catena. A distinctive area of broken terrain is a chaos. The IAU has approved certain themes for assigning names to generic features on planets, satellites, and asteroids. For example, all craters on Venus shall be named for famous women and all dorsa for sky goddesses. Also, for example, on Mars all large craters are named for deceased scientists who have contributed to the study of Mars, all small craters are named for villages of the earth, all large valleys are named for Mars in various languages, and all small valleys are named for Earth's classical or modern **rivers**. On Jupiter's satellite, Europa, all craters are named for Celtic gods and heroes. There is a comprehensive, IAU-approved list of such themes and all new suggested names must be approved for use on maps by an IAU Task Group specific to the planetary body at issue.

See also Astronomy; History of manned space exploration; Meteoroids and meteorites; Space probe

SPACE PROBE

A **space** probe is any unmanned instrumented spacecraft designed to carry out physical studies of space environment. As distinguished from satellites orbiting Earth under the influence of gravitational attraction, a space probe is rocketed into space with sufficient speed to achieve escape velocity (the velocity needed to obtain parabolic or hyperbolic orbit) and to reach a trajectory aimed at a pre-selected target.

The first recorded mention of a possibility of an unmanned probe dates back to 1919, when American physicist R. H. Goddard (1882–1945) suggested a series of space based experiments. However, in large part to Goddard's advancements in rocketry, it took only 33 years for the concept of space experiment to reappear. In 1952, the term "space probe" was introduced by E. Burgess and C. A. Cross in a short paper presented to the British Interplanetary Society.

The space probe is used mostly for the acquisition of scientific data enriching general knowledge on properties of outer space and heavenly bodies. Each probe (sometimes a series of several identical craft) is constructed to meet specific goals of a particular mission, and thus, represents a unique and sophisticated creation of contemporary engineering. Nevertheless, whether it is an Earth **satellite**, a crewed flight, or an automated probe, there are some common problems underlying any space mission: how to get to the destination point, how to collect the information required, and, finally,

Voyager* space probe. *U.S. National Aeronautics and Space Administration (NASA).

how to transfer the information back to Earth. Successful resolution of these principal issues is impossible without a developed net of high-tech Earth-based facilities used for assembling and testing the spacecraft-rocket system, for launching the probe into the desired trajectory, and for providing necessary control of probe-equipment operation, as well as for receiving data transmitted back to Earth.

As compared to crewed flights, automated space missions are far more economical and, of course, less risky to human life.

A probe's journey into far space can be divided into several stages. First, the probe has to overcome Earth's **gravity**. Escape velocities vary for different types of trajectories. During the second stage, the probe continues to move under the influence of its initial momentum and the combined gravitational influences of the **Sun** and bodies with substantial mass near its flight path. The third (approach) stage starts when the probe falls under the gravitational attraction of its destination target. The calculation of the entire trajectory from Earth to the point of destination is a complicated task. It must take into consideration numerous mutually conflicting demands: to maximize the payload but to minimize the cost, to shorten mission duration but to avoid such hazards as solar flares or meteoroid swarms, to remain within the range of the communication system but to avoid the unfavorable influence of large spatial bodies, etc.

Sometimes, strong gravitational fields of planets can be utilized to increase the probe's velocity and to change its direction considerably without firing the engines and using fuel. For instance, if used properly, Jupiter's massive gravitational pull can accelerate a probe enough to leave the **solar system** in any direction. The gravitational assistance or "swing-by" effect was successfully used, for example, in the American missions to Mercury via Venus, and in the voyage of the *Galileo* craft to Jupiter.

Projecting of payloads into designated trajectories is achieved by means of expendable launch vehicles (ELVs). A wide variety of ELVs possessed by the United States uses the same basic technology—two or more rocket-powered stages that are discarded when their engine burns are completed. Similar to the operation of a jet aircraft, the motion of a rocket is caused by a continuous ejection of a stream of hot gases in the opposite direction. The rocket's role as a prime mover makes it very important for the system's overall performance and cost. Out of 52 space-probe missions launched in the United States during the period from 1958 to 1988, 13 failed because of launch vehicle failures and only five because of probe equipment's malfunctions.

All supporting Earth-based facilities can be divided into three major categories: test grounds, where the spacecraft and its components are exposed to different extreme conditions to make sure that they are able to withstand tough stresses of outer space; check-out and launch ranges, where the lift-off procedure is preceded by a thorough examination of all spacecraft-rocket interfaces; and post-launch facilities, which are used to track, communicate with, and process the data received from the probe.

Hundreds of people and billions of dollars worth of facilities are involved in following the flight of each probe and in intercepting the data it transmits toward Earth. Already-developed facilities always have to be adopted in accordance with the specific spacecraft design. Today, the United States, Russia, and France (for unmanned flights only) possess major launch ranges, worldwide tracking networks, and dozens of publicly and privately owned test facilities. China is also actively developing space launch facilities and, in 1999, launched its first unmanned test of a program designed to enable China to launch a manned mission by 2003.

Any space probe is a self-contained piece of machinery designed to perform a variety of prescribed complex operations for a long time, sometimes for decades. There are ten major constituents of the spacecraft entity that are responsible for its vital functions: (1) power supply, (2) propulsion, (3) attitude control, (4) environmental control, (5) computer subsystem, (6) communications, (7) engineering, (8) scientific instrumentation, (9) guidance control, and (10) structural platform.

(1) The power supply provides well-regulated electrical power to keep the spacecraft active. Usually the solar-cell arrays transforming the Sun's illumination into **electricity** are used. Far from the Sun, where **solar energy** becomes too feeble, electricity may be generated by nuclear power devices. (2) The propulsion subsystem enables the spacecraft to maneuver when necessary, either in space or in a planet's atmosphere, and has a specific configuration depending upon the mission's goals. (3) The attitude-control subsystem allows orientation of the spacecraft for a specific purpose, such as to aim solar panels at the Sun, antennas at Earth, and sensors at scientific targets. It also aligns engines in the proper direction during the maneuver. (4) The environmental-control subsystem maintains the **temperature**, pressure, radiation and **magnetic field** inside the craft within the acceptable levels to secure proper functioning of equipment. (5) The computer subsystem performs data processing, coding, and storage along with routines for internal checking and maintenance. It times and initiates the pre-programmed actions independently of Earth. (6) The communication subsystem transmits data and receives commands from Earth. It also transmits identifying signals that allow ground crews to track the probe. (7) The engineering-instrumentation subsystem continuously monitors the "health" of the spacecraft's "organism" and submits status reports to Earth. (8) The scientific-instrumentation subsystem is designed to carry out the experiments selected for a particular mission, for example, to explore planetary geography, **geology**, atmospheric **physics** or electromagnetic environment. (9) The guidance-and-control subsystem is supposed to detect deviations from proper performance, determine corrections and to dispatch appropriate commands. In many respects, this subsystem resembles a human brain, since it makes active decisions, having analyzed all available information on the spacecraft's status. (10) The structural subsystem is a skeleton of the spacecraft; it supports, unites and protects all other subsystems.

Depending upon a mission's target, the probes may be classed as lunar, solar, planetary (Mercurian, Venusian, Martian, Jovian) or interplanetary probes. Another classification is based upon the mission type: flyby, orbiter, or soft-lander.

See also Astronomy; History of manned space exploration; Space and planetary geology; Spacecraft, manned

SPACE SHUTTLE

The **space** shuttle is a reusable spacecraft that takes off like a rocket, travels around the earth like a spacecraft, and then lands once again like a glider. The first space shuttle was the *Columbia*, whose maiden voyage took place in April 1981. Four additional shuttles were later added to the fleet: *Discovery*, *Challenger*, *Atlantis*, and *Endeavor*. The first shuttle launched by the Soviet Union (now Russia) was *Buran*, which made its debut in November 1988.

At one time, both the United States and the Soviet Union envisioned complex space programs that included two parts: (1) space stations orbiting around Earth and/or other planets, and (2) shuttle spacecraft that would transport humans, equipment, raw materials, and finished products to and from the space station. For economic reasons, each nation eventually ended up concentrating on only one aspect of the complete program. The Soviets built and for many years operated advanced space stations (*Salyut* and *Mir*), while Americans have focused their attention on the shuttle system.

The shuttle system has been given the name Space Transportation System (STS), of which the shuttles have been the key element. Initially lacking a space station with which to interact, the American shuttles operated with two major goals: (1) the conduct of scientific experiments in a zero-gravity environment, and (2) the launch, capture, repair, and release of satellites.

Now an international program, STS depends heavily on the contributions of other nations in the completion of its basic missions. For example, its Spacelab modules—the areas in which astronauts carry out most of their experiments—are designed and built by the European Space Agency, and the extendable arm used to capture and release satellites—the remote manipulator system or Canadarm—is constructed in Canada.

The space shuttle has four main parts: (1) the orbiter (2) the three main engines attached to the orbiter (3) two solid rocket engines, and (4) an external fuel tank. Although the Russian *Buran* differs in some details from the U.S. space shuttle fleet, the main features of all shuttles are similar.

The orbiter is approximately the size of a commercial DC-9 airplane with a length of 121 ft (37 m) and a wing span of 78 ft (23 m). Its net weight is about 161,000 lb (73,200 kg).

It is sub-divided into two main parts: the crew cabin and the cargo bay. The upper level of the crew cabin is the flight deck from which astronauts control the spacecraft's flight in orbit and during descent. Below the flight deck are the crew's personal quarters, containing personal lockers, sleeping, eating, and toilet facilities, and other necessary living units. The crew cabin is physically isolated from the cargo bay and is provided with **temperature** and pressure conditions similar to those on Earth's surface. The cabin's atmosphere is maintained with a composition equivalent to that of near-Earth atmosphere, 80% nitrogen and 20% **oxygen**.

The cargo bay is a large space 15 ft (4.5 m) by 60 ft (18 m) in which the shuttle's payloads are stored. The cargo bay can hold up to about 65,000 lb (30,000 kg) during ascent, although it is limited to about half that amount during descent.

In 1973, an agreement was reached between NASA and the European Space Agency (ESA) for the construction by ESA of a pressurized work space that could be loaded into the shuttle's cargo bay. The workspace, designated as Spacelab, was designed for use as a science laboratory in which a wide array of experiments could be conducted. Each of these Spacelab modules is 8.9 ft (2.7 m) long and 13 ft (3.9 m) in diameter. The equipment needed to carry out experiments is arranged in racks along the walls of the Spacelab, and the whole module is then loaded into the cargo bay of the shuttle prior to take-off. When necessary, two Spacelab modules can be joined to form a single, larger work space.

The power needed to lift a space shuttle into orbit comes from two solid-fuel rockets, each 149 ft (45.5 m) in length and 12 ft (4 m) in diameter, and the shuttle's own liquid-fuel engines. The fuel used in the solid rockets is composed of finely-divided **aluminum**, ammonium perchlorate, and a special polymer designed to form a rubbery mixture. The mixture is molded in such a way as to produce an 11-point starred figure. This shape exposes the maximum possible surface **area** of fuel during ignition, making combustion as efficient as possible within the engine.

The two solid-fuel rockets carry 1.1 million lb (500,000 kg) of fuel each, and burn out completely only 125 seconds after the shuttle leaves the launch pad. At solid-engine burnout, the shuttle is at an altitude of 161,000 ft (47,000 m) and 244 nautical miles (452 km) down range from launch site. At that point, explosive charges holding the solid rockets to the main shuttle go off and detach the rockets from the shuttle. The rockets are then returned to Earth by means of a system of parachutes that drops them into the Atlantic Ocean at a speed of 55 mi (90 km) per hour. The rockets can then be collected by ships, returned to land, refilled, and re-used in a later shuttle launch.

The three liquid-fueled shuttle engines have been described as the most efficient engines ever built by humans. At maximum capacity, they achieve 99% efficiency during combustion. They are supplied by fuel (liquid hydrogen) and oxidizer (liquid oxygen) stored in the 154 ft (46.2 m) external fuel tank. The fuel tank itself is sub-divided into two parts, one of which holds the liquid oxygen and the other, the liquid hydrogen. The fuel tank is maintained at the very low temperature (less than –454°F [–270°C]) necessary to keep hydrogen and oxygen in their liquid states. The two liquids are pumped into the shuttle's three engines through 17 in (43 cm) diameter lines that carry 1,035 gal (3,900 l) of fuel per second. Upon ignition, each of the liquid-fueled engines delivers 75,000 horsepower of thrust.

The three main engines burn out after 522 seconds, when the shuttle has reached an altitude of 57 nautical miles (105 km) and is down range 770 nautical miles (1,426 km) from the launch site. At this point, the external fuel tank is also jettisoned. Its return to the earth's surface is not controlled, however, and it is not recoverable for future use.

Final orbit is achieved by means of two small engines, the Orbital Maneuvering System (OMS) Engines located on external pods at the rear of the orbiter's body. The OMS engines are fired first to insert the orbiter into an elliptical orbit with an apogee of 160 nautical miles (296 km) and a perigee of 53 nautical miles (98 km) and then again to accomplish its final circular orbit with a radius of 160 nautical miles (296 km).

Humans and machinery work together to control the movement of the shuttle in orbit and during its descent. For making fine adjustments, the spacecraft depends on six small vernier jets, two in the nose and four in the OMS pods of the spacecraft. These jets allow human or computer to make modest adjustments in the shuttle's flight path in three directions.

The computer system used aboard the shuttle is an example of the redundancy built into the spacecraft. Five discrete computers are used, four networked with each other using one computer program, and one operating independently using a different program. The four linked computers constantly communicate with each other, testing each other's decisions and deciding when one (or two or three) is not performing properly and eliminating that computer (or those computers) from the decision-making process. In case all four of the interlinked computers malfunction, decision-making is turned over automatically to the fifth computer.

This kind of redundancy is built into every essential feature of the shuttle's operation. For example, three independent hydraulic systems are available, all operating with independent power systems. The failure of one or even two of the systems does not, therefore, place the shuttle in a critical failure mode.

The space shuttles have performed a myriad of scientific and technical tasks in their nearly two decades of operation. Many of these have been military missions about which we have relatively little information. The launching of military spy satellites is an example of these.

Some examples of the kinds of activities carried out during shuttle flights include the following:

- After the launch of the *Challenger* shuttle (STS-41B) on February 3, 1984, astronauts Bruce McCandless II and Robert L. Stewart conducted the first ever untethered space walks using Manned Maneuvering Unit backpacks that allowed them to propel themselves through space near the shuttle. The shuttle also released into orbit two communication satellites, the Indonesian *Palapa* and the American *Westar* satellites. Both satellites failed soon after release but were recovered and

Space shuttle *Endeavour*. *U.S. National Aeronautics and Space Administration (NASA).*

returned to Earth by the *Discovery* during its flight that began on November 8, 1984.

- During the flight of *Challenger* (STS-51B) that began on April 29, 1985, crew members carried out a number of experiments in Spacelab 3 determining the effects of zero **gravity** on living organisms and on the processing of materials. They grew **crystals** of mercury (II) oxide over a period of more than four days, observed the behavior of two monkeys and 24 rats in a zero-gravity environment, and studied the behavior of liquid droplets held in suspension by sound waves.
- The mission of STS-51I (*Discovery*) was to deposit three communications satellites in orbit. On the same flight, astronauts William F. Fisher and James D. Van Hoften left the shuttle to make repairs on a Syncom **satellite** that had been placed in orbit during flight STS-51D but that had then malfunctioned.

Some of the most difficult design problems faced by shuttle engineers were those created during the reentry process. When the spacecraft has completed its mission in space and is ready to leave orbit, its OMS fires just long enough to slow the shuttle by 200 mi (320 km) per hour. This modest change in speed is enough to cause the shuttle to drop out of its orbit and begin its descent to Earth.

The re-entry problems occur when the shuttle reaches the outermost regions of the upper atmosphere, where significant amounts of atmospheric gases are first encountered. Friction between the shuttle—now traveling at 17,500 mi (28,000 km) per hour—and air molecules causes the spacecraft's outer surface to begin to heat up. Eventually, it reaches a temperature of 3,000°F (1,650°C).

Most materials normally used in aircraft construction would melt and vaporize at these temperatures. It was necessary, therefore, to find a way of protecting astronauts inside the shuttle cabin from this searing heat. The solution invented was to use a variety of insulating materials on the shuttle's outer skin. Parts less severely heated during re-entry are covered with 2,300 flexible quilts of a silica-glass composite. The more sensitive belly of the shuttle is covered with 25,000 insulating tiles, each 6 in (15 cm) square and 5 in (12 cm) thick, made of a silica-borosilicate **glass** composite.

The portions of the shuttle most severely stressed by heat—the nose and the leading edges of the wings—are coated with an even more resistant material known as carbon-carbon. Carbon-carbon is made by attaching a carbon-fiber cloth to the body of the shuttle and then baking it to convert it to a pure **carbon** substance. The carbon-carbon is then coated to prevent oxidation of the material during descent.

Once the shuttle reaches Earth's atmosphere, it ceases to operate as a rocket ship and begins to function as a glider. Its movements are controlled by aerodynamic controls, such as the tail rudder, a large flap beneath the main engines, and elevons, small flaps on its wings. These devices allow the shuttle to descend to the earth traveling at speeds of 8,000 mi (13,000 km) per hour, while dropping vertically at the rate of 140 mi (225 km) per hour. When the aircraft finally touches down, it is traveling at a speed of about 190 knots (100 m per second), and requires about 1.5 mi (2.5 km) to come to a stop.

Disasters have been associated with aspects of both the Soviet and American space programs. Unfortunately, the Space Transportation System has been no different in this respect. Mission STS-51L was scheduled to take off on January 28, 1986 using the shuttle *Challenger*. Only 72 seconds into the flight, the shuttle's external tank exploded, and all seven astronauts on board were killed.

The *Challenger* disaster prompted one of the most comprehensive studies of a major accident ever conducted. On June 6, 1986, the Presidential Commission appointed to analyze the disaster published its report. The reason for the disaster, according to the commission, was the failure of an O-ring at a **joint** connecting two sections of one of the solid rocket engines. Flames escaping from the failed joint reached the external fuel tank, set it on fire, and then caused an explosion of the whole spacecraft.

As a result of the *Challenger* disaster, a number of design changes were made in the shuttle. Most of these (254 modifications in all) were made in the orbiter. Another 30 changes were made in the solid rocket booster, 13 in the external tank, and 24 in the shuttle's main engine. In addition, an escape system was developed that would allow crew members to abandon a shuttle in case of emergencies, and NASA reexamined and redesigned its launch-abort procedures. Also, NASA was instructed to reassess its ability to carry out the ambitious program of shuttle launches that it had been planning.

The U.S. Space Transportation System was essentially shut down for a period of 975 days while NASA carried out necessary changes and tested new systems. Then, on September 29, 1988, the first post-*Challenger* mission was launched, STS-26. On that flight, *Discovery* carried NASA's TDRS-C communications satellite into orbit, putting the American STS program back on schedule once more.

In December, 1988, the crew of NASA's Space Shuttle STS-88 began construction of the **International Space Station** (ISS). By joining the Russian-made control module *Zarya* with the United States-built connecting module *Unity,* the crew of the *Endeavor* became the first crew aboard the ISS. Since the STS-88 mission, twelve more U.S. shuttle missions have led the construction of the International Space Station, a permanent laboratory orbiting 220 miles above Earth.

See also Space and planetary geology; Space physiology; Space probe; Spacecraft, manned

Spacecraft, manned

Manned spacecraft are vehicles with the capability of maintaining life outside of Earth's atmosphere. Partially in recognition of the fact that women as well as men are active participants in **space** travel programs, manned spacecraft are now frequently referred to as crewed spacecraft.

In its earliest stages, crewed space flight was largely an exercise in basic research. Scientists were interested in collecting fundamental information about the **Moon**, the other planets in our **solar system**, and outer space. Today, crewed space flight is also designed to study a number of practical problems, such as the behavior of living organisms and inorganic materials in zero **gravity** conditions.

A very large number of complex technical problems must be solved in the construction of spacecraft that can carry humans into space. Most of these problems can be classified in one of three major categories: communication, environmental and support, and re-entry.

Communication refers to the necessity of maintaining contact with members of a space mission as well as monitoring their health and biological functions and the condition of the spacecraft in which they are traveling. Direct communication between astronauts and cosmonauts can be accomplished by means of radio and television messages transmitted between a spacecraft and ground stations. To facilitate these communications, receiving stations at various locations around Earth have been established. Messages are received and transmitted to and from a space vehicle by means of large antennas located at these stations.

Many different kinds of instruments are needed within the spacecraft to monitor cabin **temperature**, pressure, **humidity**, and other conditions as well as biological functions such as heart rate, body temperature, blood pressure, and other vital functions. Constant monitoring of spacecraft hardware is also necessary. Data obtained from these monitoring functions is converted to radio signals that are transmitted to Earth stations, allowing ground-based observers to maintain a constant check on the status of both the spacecraft and its human passengers.

The fundamental requirement of a crewed spacecraft is, of course, to provide an atmosphere in which humans can survive and carry out the jobs required of them. This means, foremost, providing the spacecraft with an Earth-like atmosphere in which humans can breathe. Traditionally, the Soviet Union has used a mixture of nitrogen and **oxygen** gases somewhat like that found in the earth's atmosphere. American spacecraft, however, have employed pure oxygen atmospheres at pressures of about 5 lb per square inch, roughly one-third that of normal air pressure on the earth's surface.

Apollo 11* Command and Service Modules. *Apollo 11* was the first mission to land mankind on another celestial body. *U.S. National Aeronautics and Space Administration (NASA).

The level of **carbon dioxide** within a spacecraft must also be maintained at a healthy level. The most direct way of dealing with this problem is to provide the craft with a base, usually lithium hydroxide, which will absorb **carbon** dioxide exhaled by astronauts and cosmonauts. Humidity, temperature, odors, toxic gases, and sound levels are other factors that must be controlled at a level congenial to human existence.

Food and **water** provisions present additional problems. The space needed for the storage of conventional foodstuffs is prohibitive for spacecraft. Thus, one of the early challenges for space scientists was the development of dehydrated foods or foods prepared in other ways so that they would occupy as little space as possible. Space scientists have long recognized that food and water supplies present one of the most challenging problems of long-term space travel, as would be the case in a space station. Suggestions have been made, for example, for the purification and recycling of urine as drinking water and for the use of exhaled carbon dioxide in the growth of plants for foods in spacecraft that remain in orbit for long periods of time.

An important aspect of spacecraft design is the provision for power sources needed to operate communication, environmental, and other instruments and devices within the vehicle. The earliest crewed spacecrafts had simple power systems. The Mercury series of vehicles, for example, were powered by six conventional batteries. As spacecraft increased in size and complexity, however, so did their power needs. The Gemini spacecrafts required an additional conventional battery and two fuel cells, while the Apollo vehicles were provided with five batteries and three fuel cells.

One of the most serious on-going concerns of space scientists about crewed flights has been their potential effects on the human body. An important goal of nearly every space flight has been to determine how the human body reacts to a zero-gravity environment.

At this point, scientists have some answers to that question. For example, we know that one of the most serious dangers posed by extended space travel is the loss of calcium from bones. Also, the absence of gravitational forces results in a space traveler's blood collecting in the upper part of his or her body, especially in the left atrium. This knowledge has led to the development of special devices that modify the loss of gravitational effects during space travel.

One of the challenges posed by crewed space flight is the need for redundancy in systems. Redundancy means that there must be two or three of every instrument, device, or spacecraft part that is needed for human survival. This level of redundancy is not necessary with uncrewed spacecraft where failure of a system may result in the loss of a **space probe**, but not the loss of a human life. It is crucial, however, when humans travel aboard a spacecraft.

An example of the role of redundancy was provided during the *Apollo 13* mission. That mission's plan of landing on the Moon had to be aborted when one of the fuel cells in the service module exploded, eliminating a large part of the spacecraft's power supply. A back-up fuel cell in the lunar module was brought on line, however, allowing the spacecraft to return to Earth without loss of life.

Space suits are designed to be worn by astronauts and cosmonauts during take-off and landing and during extravehicular activities (EVA). They are, in a sense, a space passenger's own private space vehicle and present, in miniature, most of the same environmental problems as does the construction of the spacecraft itself. For example, a space suit must be able to protect the space traveler from marked changes in temperature, pressure, and humidity, and from exposure to radiation, unacceptable solar glare, and micrometeorites. In addition, the space suit must allow the space traveler to move about with relative ease and to provide a means of communicating with fellow travelers in a spacecraft or with controllers on the earth's surface. The removal and storage of human wastes is also a problem that must be solved for humans wearing a space suit.

Ensuring that astronauts and cosmonauts are able to survive in space is only one of the problems facing space scientists. A spacecraft must also be able to return its human passengers safely to Earth's surface. In the earliest crewed spacecrafts, this problem was solved simply by allowing the vehicle to travel along a ballistic path back to Earth's atmosphere and then to settle on land or sea by means of one or more large parachutes. Later spacecraft were modified to allow pilots some control over their re-entry path. The space shuttles, for example, can be piloted back to Earth in the last stages of re-entry in much the same way that a normal airplane is flown.

Perhaps the most serious single problem encountered during re-entry is the heat that develops as the spacecraft returns to Earth's atmosphere. Friction between vehicle and air produces temperatures that approach 3,092°F (1,700°C). Most **metals** and alloys would melt or fail at these temperatures. To deal with this problem, spacecraft designers have developed a class of materials known as ablators that absorb and then radiate large amounts of heat in brief periods of time. Ablators have been made out of a variety of materials, including phenolic resins, epoxy compounds, and silicone rubbers.

Some scientists are beginning to plan beyond **space shuttle** flights and the **International Space Station**. While NASA's main emphasis for some time will be unmanned probes and robots, the most likely target for a manned spacecraft will be Mars. Besides issues of long-term life support, any such mission will have to deal with long-term exposure to space radiation. Without sufficient protection, galactic cosmic rays would penetrate spacecraft and astronaut's bodies, damaging their DNA and perhaps disrupting nerve cells in their brains over the long-term. (Manned flights to the Moon were protected from cosmic rays by the earth's magnetosphere.) Shielding would be necessary, but it is always a trade-off between human protection and spacecraft weight. Moreover, estimates show it could add billions of dollars to the cost of any such flight.

See also Space and planetary geology; Space physiology

SPECTROSCOPY

Geoscientists utilize a number of different spectroscopy techniques in the study of Earth materials. The absorption, emission, or scattering of electromagnetic radiation by atoms or molecules is referred to as spectroscopy. A transition from a lower energy level to a higher level with transfer of electromagnetic energy to the **atom** or molecule is called absorption; a transition from a higher energy level to a lower level is called emission (if energy is transferred to the electromagnetic field); and the redirection of light as a result of its interaction with matter is called scattering.

When atoms or molecules absorb electromagnetic energy, the incoming energy transfers the quantized atomic or molecular system to a higher energy level. Electrons are promoted to higher orbitals by ultraviolet or visible light; vibrations are excited by infrared light, and rotations are excited by microwaves. Atomic-absorption spectroscopy measures the concentration of an element in a sample, whereas atomic-emission spectroscopy aims at measuring the concentration of elements in samples.

Infrared spectroscopy has been widely used in the study of surfaces. The most frequently used portion of the infrared spectrum is the region where molecular vibrational frequencies occur. This technique was first applied around the turn of the twentieth century in an attempt to distinguish **water** of crystallization from water of constitution in solids.

Ultraviolet spectroscopy takes advantage of the selective absorbance of ultraviolet radiation by various substances. Ultraviolet instruments have also been used to monitor air and

water pollution, to analyze **petroleum** fractions, and to analyze pesticide residues. Ultraviolet photoelectron spectroscopy, a technique that is analogous to x-ray photoelectron spectroscopy, has been used to study valence electrons in gases.

Microwave spectroscopy, or molecular rotational resonance spectroscopy, addresses the microwave region and the absorption of energy by molecules as they undergo transitions between rotational energy levels. From these spectra, it is possible to obtain information about molecular structure, including bond distances and bond angles. One example of the application of this technique is in the distinction of trans and gauche rotational isomers. It is also possible to determine dipole moments and molecular collision rates from these spectra.

Although there are many other forms of spectroscopy (e.g., UV-VIS absorption spectroscopy, molecular fluorescence spectroscopy, etc.) many modern advances in inorganic and organic based studies have resulted from the development of nuclear magnetic resonance (NMR) technology. In NMR, resonant energy is transferred between a radio-frequency alternating **magnetic field** and a nucleus placed in a field sufficiently strong to decouple the nuclear spin from the influence of atomic electrons. Transitions induced between substates correspond to different quantized orientations of the nuclear spin relative to the direction of the magnetic field. Nuclear magnetic resonance spectroscopy has two subfields: broadline NMR and high resolution NMR. High resolution NMR has been used in inorganic and organic **chemistry** to measure subtle electronic effects, to determine structure, to study chemical reactions, and to follow the motion of molecules or groups of atoms within molecules.

Electron paramagnetic resonance is a spectroscopic technique similar to nuclear magnetic resonance except that microwave radiation is employed instead of radio frequencies. Electron paramagnetic resonance has been used extensively to study paramagnetic species present on various solid surfaces. These species may be metal ions, surface defects, or absorbed molecules or ions with one or more unpaired electrons. This technique also provides a basis for determining the bonding characteristics and orientation of a surface complex. Because the technique can be used with low concentrations of active sites, it has proven valuable in studies of oxidation states.

Atoms or molecules that have been excited to high energy levels can decay to lower levels by emitting radiation. For atoms excited by light energy, the emission is referred to as atomic fluorescence; for atoms excited by higher energies, the emission is called atomic or optical emission. In the case of molecules, the emission is called fluorescence if the transition occurs between states of the same spin, and phosphorescence if the transition takes place between states of different spin.

In x-ray fluorescence, the term refers to the characteristic x rays emitted as a result of absorption of x rays of higher frequency. In electron fluorescence, the emission of electromagnetic radiation occurs as a consequence of the absorption of energy from radiation (either electromagnetic or particulate), provided the emission continues only as long as the stimulus producing it is maintained.

The effects governing x-ray photoelectron spectroscopy were first explained by German-American physicist **Albert Einstein** (1879–1955) in 1905, who showed that the energy of an electron ejected in photoemission was equal to the difference between the photon and the binding energy of the electron in the target.

When electromagnetic radiation passes through matter, most of the radiation continues along its original path, but a tiny amount is scattered in other directions. Light that is scattered without a change in energy is called Rayleigh scattering; light that is scattered in transparent solids with a transfer of energy to the solid is called Brillouin scattering. Light scattering accompanied by vibrations in molecules or in the optical region in solids is called Raman scattering.

See also Astronomy; Atmospheric chemistry; Focused Ion Beam (FIB); Geochemistry; Mineralogy

SPRINGS

A site where **groundwater** emerges from the subsurface is known as a spring. Springs present the most familiar manifestation of groundwater, and have been utilized as drinking **water** sources throughout history. These natural features have sometimes been viewed mysteriously and the waters regarded as having therapeutic, medicinal, or magical properties. These misconceptions continue today, including the belief that spring water is of superior quality or purity. Fallacies such as this are exploited in the sales of beverages and other products. Unfortunately, water that flows naturally from the ground is conveyed with no more special properties than the same groundwater that is drawn from a nearby well. In fact, because of the exposure at the surface, spring water is potentially more easily contaminated than water drawn from a properly constructed well.

Springs can be classified based on their groundwater source (e.g., water-table springs and perched springs). **Water table** springs discharge where the land surface intersects the water table. Perched springs, however, flow from the intersection of the land surface with a local groundwater body that is separated from the main **saturated zone** below by a zone of relatively lower **permeability** and an **unsaturated zone.** In addition to the location of the water table, groundwater discharge at springs is commonly controlled by other factors such as stratigraphic contacts, **faults and fractures**, and cavern openings. The relationship of local **topography** and geologic structure to the point of groundwater discharge is one of the most common classification systems for springs.

Springs are also classified based on magnitude of discharge, chemical characteristics, water **temperature**, type of the groundwater flow system, and others. Because springs allow them to easily and directly access the groundwater, hydrogeologists often use information of this nature to help interpret the groundwater flow system of an **area**.

The quantity of discharge from a particular spring is determined by three variables: **aquifer** permeability, groundwater basin size, and quantity of recharge. The largest springs can have a discharge of over 1,000 cubic feet per second. However, springs of this size are rare. A spring with a dis-

Hot spring. ***© Buddy Mays/Corbis. Reproduced by permission.***

charge insufficient to support a small rivulet is referred to as a seep. The flow from a seep is commonly so low as to preclude measurement.

See also Karst topography; Porosity and permeability; Saturated zone

STALACTITES AND STALAGMITES

Stalactites and stalagmites are formed by **water** dripping or flowing from fractures on the ceiling of a **cave**. They are the most common types of speleothems in caves. In caves, stalagmites grow rather slowly—0.00028–0.037 in/yr (0.007–0.929 mm/yr)—while in artificial tunnels and basements they grow much faster. Soda straw stalactites are the fastest growing (up to 1.57 in/yr, 40 mm/yr), but most fragile stalactites in caves. Soda straw stalactites form along a drop of water and continue growing down from the cave ceiling forming a tubular stalactite, which resembles a drinking straw in appearance. Their internal diameter is exactly equal to the diameter of the water drop. Formation of most stalactites is initiated as soda straws. If water flows on their external surface, they begin to grow in thickness and obtain a conical form. If a stalactite curves along its length, it is called a deflected stalactite. If its curving is known to be caused by air currents, it is called anemolite. Petal-shaped tubular stalactites composed of aragonite are called spathites. When some stalactites touch each other they form a drapery with a curtain-like appearance.

When dripping water falls down on the floor of the cave it form stalagmites, which grow up vertically from the cave floor. Any changes in the direction of the growth axis of the stalagmite are suggestive of folding of the floor of the cave during the growth of the stalagmite. If a stalagmite is small, flat and round, it is called button stalagmite. Stalagmites resembling piled-up plates with broken borders are called pile-of-plates stalagmites. Rare varieties of stalagmites are mushroom stalagmites (partly composed of mud and having a mushroom shape), mud stalagmites (formed by mud) and lily pad stalagmites (resembling a lily pad on the surface of a pond). A calcite **crust** (shelfstone) grows around a stalagmite if it is flooded by a cave pool and forms a candlestick.

When a stalactite touches a stalagmite it forms a column. Usually, stalactites and stalagmites in caves are formed by calcite, less frequently by aragonite, and rarely by **gypsum**. Fifty-four other **cave minerals** are known to form rare stalac-

Carlsbad Caverns, Guadalupe Mountains, New Mexico. ***© Adam Woolfitt/Corbis. Reproduced by permission.***

tites. Sometimes calcite stalactites or stalagmites are overgrown by aragonite **crystals**. This is due to **precipitation** of calcite that raises the ratio of magnesium to calcium in the solution enough that aragonite becomes stable.

Rarely, elongated single crystals or twins of calcite are vertically oriented and look like stalactites, but in fact are not stalactites because they are not formed by dripping or flowing water and don't have hollow channels inside. These elongated crystals are formed from water films on their surface.

In some volcanic **lava** tube caves exist lava stalactites and stalagmites that are not speleothems because they are not composed of secondary **minerals**. They are primary forms of the cooling, dripping lava.

The internal structure of stalactites and stalagmites across their growth axis usually consists of concentric rings around the hollow channel. These rings contain different amounts of **clay** and other inclusions, and reflect drier and wetter periods. Clay rings reflect hiatuses of the growth of the sample. Stalagmites may be formed for periods ranging from a few hundreds years up to one million years. Stalactites and stalagmites in caves have such great variety of shapes, forms, and color that each of them is unique in appearance. At the same time, their growth rates are so slow that once broken, they cannot recover during a human life span of time. Thus, stalactites and stalagmites are considered natural heritage objects and are protected by law in most countries, and their collection, mining, and selling is prohibited.

See also Cave minerals and speleothems

Stellar evolution

In astrophysics and **cosmology**, stellar **evolution** refers to the life history of stars that is driven by the interplay of internal pressure and **gravity**.

Essentially, throughout the life of a star a tension exists between the compressing force of the star's own gravity and the expanding pressures generated by the nuclear reactions taking place in its core. After cycles of swelling and contraction associated with the burning of progressively heavier nuclear **fuels**, the star eventually expends its useable nuclear fuel and resumes contraction under the force of its own gravity. There are three possible fates for such a collapsing star. The particular fate for any star is determined by the mass of the star left after blowing away its outer layers.

A star less than 1.44 times the mass of the **Sun** (termed the Chandrasekhar limit) collapses until the pressure compacted electron **clouds** exerts enough pressure to balance the compressing force of gravity. Such stars become white dwarfs that are contracted to a radius the size of a planet. This is the fate of most stars.

If a star retains between 1.4 and roughly three times the mass of the Sun, the pressure of the electron clouds is insufficient to stop the gravitational collapse and stars of this mass continue their collapse to become neutron stars. Although neutron stars are only a few miles across, they have enormous density. Within a neutron star the nuclear forces and the repulsion of the compressed atomic nuclei balance the crushing force of gravity.

With the most massive stars, however, there is no known force in the Universe that can stop the final gravitational collapse and such stars collapse to form a singularity—a geometrical point of infinite density. As such a star collapses, its gravitational field warps spacetime and a black hole forms around the singularity.

Gravitational collapse is the process which provides the energy required for star formation, which starts with hydrogen fusion in a protostar at a heat of over 15 million K. Gravity, always directed inwards, decreases the radius of interstellar gas clouds, causing them to collapse and form a protostar, the immediate precursor of a star. Interstellar gas is initially cold, but it is heated by the gravitational energy released by the cloud contraction process. The radius of the protostar will continue to shrink under the influence of gravity until enough internal gas pressure, always directed outwards, builds up to stabilize the collapse. At this stage, the protostar is still too cold for hydrogen fusion to be initiated. Protostars can be detected by infrared **spectroscopy** because the initial warming event releases infrared electromagnetic radiation. If the mass of the protostar is less than 0.08 solar masses, the **temperature** of its core never reaches the range required for **nuclear fusion** and the failed star becomes a brown dwarf.

If, however, the mass of the protostar exceeds 0.08 solar masses, hydrogen fusion can proceed and the protostar becomes a main sequence star, with average surface temperatures of 10,800°F (6,000°C) (the internal and coronal temperatures measure in the millions of degrees). Most stars in the Universe are main sequence stars and are found on the diagonal of a Hertzsprung-Russell diagram. The main sequence stage of star evolution is the most stable state a medium-sized star can reach, and it can last for billions of years as such stars undergo very gradual and slow changes in luminosity and temperature. This is because pressure and gravitational forces are in equilibrium and the core has reached the temperature required for the fusion of hydrogen to helium to proceed smoothly. The time spent by a star in the main sequence is a function of its mass. The more massive the star, the less time spent on the main sequence. Although massive stars have large amounts of fuel, hydrogen fusion proceeds so quickly that it is completed within a few hundred thousand years. The fate of such massive stars is to explode violently. Smaller stars fuse their hydrogen at a slower rate. The lightest stars created in the early history of the Universe, for example, are still on the main sequence. The Sun is approximately midway through its main sequence life.

A post-main sequence star has two distinct regions, consisting of a core of helium nuclei and electrons surrounded by an envelope of hydrogen. With two protons in its nucleus, helium requires a higher fusion temperature than the one at which hydrogen fusion is proceeding. Without a source of energy to increase its temperature, the core cannot counter the effect of gravitational collapse and it starts to collapse, heating up as it does. This heat is transferred to the fusing hydrogen layer, which increases the luminosity of the shell and causes it to expand. As it expands, the outer layers cool off. At this point, the star is characterized by expansion and cooling of its shell, which causes it to become redder with increased luminosity. This is termed the red giant phase. When the Sun reaches this stage, it will be large enough to include Mercury in its sphere and hot enough to evaporate Earth's **oceans**. The core temperature of a red giant is on the order of 100 million K, the threshold temperature for the fusion of helium into **carbon**. A red giant, however, is initially stable, as pressure and gravity reach equilibrium.

If helium continues to accumulate in the core as the outer portions of the hydrogen envelope continue to fuse, eventually the helium in the core starts fusing into carbon in a violent event referred to as a helium flash, lasting as little as a few seconds. During this phase, the star gradually blows away its outer atmosphere into an expanding shell of gas known as a planetary nebula.

A star takes thousands of years to go through the red giant phase, after which it evolves into a white dwarf. It is then a small, hot star with a surface temperature as high as 100,000 K that makes it glow white. Because of its small size, a white dwarf has a very high density. A white dwarf consists of those elements that were created in its previous evolutionary phases via nucelosynthesis. The original hydrogen was fused into helium then totally or partly fused into carbon. In addition, heavier elements fuse from the carbon. The temperature of a white dwarf is not high enough to initiate a new cycle of fusion. In time, it eventually becomes a black dwarf as it loses its residual heat over billions of years. The size of a white dwarf is limited by a process called electron degeneracy. Electron degeneracy is the stellar equivalent of the Pauli exclusion principle, as is neutron degeneracy. No two electrons can occupy identical states, even under the pressure of a collapsing star of several solar masses. For stellar masses smaller than about 1.44 solar masses, the energy from the gravitational collapse is not sufficient to produce the neutrons of a neutron star, so the collapse is effectively stopped. This maximum mass for a white dwarf is called the Chandrasekhar limit.

When a massive star has fused all of its hydrogen, gravitational collapse is capable of generating sufficient energy so that the core can begin to fuse helium nuclei to form carbon. If the process can go beyond the red giant stage, the star becomes a supergiant. Following fusion and disappearance of the helium, the core can successively burn carbon and other heavier elements until it acquires a core of **iron**, the heaviest element that can be formed by natural fusion. Another possible fate of white dwarfs is to evolve into novae or another type of supernovae. Novae occur in binary star systems in which one star is a white dwarf. If the companion star evolves into a red giant, it can expand far enough so that gas from its outer shell can be pulled onto the white dwarf. The white dwarf accumulates the additional gas until it reaches nuclear fusion temperatures, at which point the gas ignites explosively into a nova.

Alternatively, a white dwarf may accumulate enough material from its binary star to exceed the Chandrasekhar limit. This results in a sudden and total collapse of the white dwarf, with temperature increases in ranges capable of initiating rapid carbon fusion and subsequent explosion of the white dwarf into a spectacular supernova, that can shine with the brightness of 10 billion suns with a total energy output of $\sim 10^{44}$ joules, equivalent to the total energy output of the Sun during its entire lifetime.

See also Astronomy; Big Bang theory; Bohr model of the atom; Cosmology; Quantum electrodynamics (QED); Solar sunspot cycles; Solar system; Stellar life cycle

Stellar life cycle

Until the last half of the nineteenth century, **astronomy** was principally concerned with the accurate description of the movements of planets and stars. Developments in electromagnetic theories of light along with the articulation of quantum and relativity theories at the start of the twentieth century, however, allowed astronomers to probe the inner workings of the stars. Of primary concern was an attempt to coherently explain the life cycle of the stars and to reconcile the predictions of advances in physical theory with astronomical observation. Profound questions regarding the birth and death of stars led to the stunning conclusion that, in a very real sense, life itself was a product of **stellar evolution**.

It is now known that the mass of a star determines the ultimate fate of a star. Stars that are more massive burn their fuel quicker and lead shorter lives. These facts would have astonished astronomers working at the dawn of the twentieth century. At that time an accurate understanding of the source of the heat and light generated by the **Sun** and stars suffered from a lack of understanding of nuclear processes.

Based on Newtonian concepts of **gravity**, many astronomers understood that stars were formed in **clouds** of gas and dust termed nebulae that were measured to be light years across. These great molecular clouds, so thick they are often opaque in parts, teased astronomers. Decades later, after development of **quantum theory**, astronomers were able to develop a better understanding of the energetics behind the "reddening" of light leaving stellar nurseries (reddened because the blue light is scattered more and the red light passes more directly through to the observer). Astronomers speculated that stars formed when regions of these clouds collapsed (termed gravitational clumping). As the region collapsed, it accelerated and heated up to form a protostar heated by gravitational friction.

The source of heat in stars in the pre-atomic age eluded astronomers who sought to reconcile seeming contradictions regarding how stars were able to maintain their size (i.e., not shrink as they burned fuel) and radiate the tremendous amounts of energy measured. In accord with dominant religious beliefs, the Sun, using conventional energy consumption and production concepts could be calculated to be less than a few thousand years old.

In 1913, Danish Astronomer Ejnar Hertzsprung (1873–1967) and English astronomer Henry Norris Russell (1877–1957) independently developed what is now known as the Hertzsprung-Russell diagram. In the Hertzsprung-Russell diagram, the spectral type (or, equivalently, color index or surface **temperature**) is placed along the horizontal axis and the absolute magnitude (or luminosity) along the vertical axis. Accordingly, stars are assigned places top to bottom in order of increasing magnitude (decreasing brightness) and from right to left by increasing temperature and spectral class.

The relation of stellar color to brightness was a fundamental advance in modern astronomy. The **correlation** of color with true brightness eventually became the basis of a widely used method of deducing spectroscopic parallaxes of stars that allowed astronomers to estimate how far distant stars are from the Earth (estimates from closer stars could be made by geometrical parallax).

In the Hertzsprung-Russell main sequence, stars (those later understood to be burning hydrogen as a nuclear fuel) form a prominent band or sequence of stars from extremely bright, hot stars in the upper left-hand corner of the diagram to faint, relatively cooler stars in the lower right-hand corner of the diagram. Because most stars are main sequence stars, most stars fall into this band on the Hertzsprung-Russell diagram. The Sun, for example, is a main-sequence star that lies roughly in the middle of the diagram among what are referred to as yellow dwarfs.

Russell attempted to explain the presence of giant stars as the result of large gravitational clumps. Stars, according to Russell, would move down the chart as they lost mass burned as fuel. Stars began life huge cool red bodies and then undergo a continual shrinkage as they heated. Although the Hertzsprung-Russell diagram was an important advance in understanding stellar evolution—and it remains highly useful to modern astronomers—Russell's reasoning behind the movements of stars on the diagram turned out to be exactly the opposite of the modern understanding of stellar evolution made possible by an understanding of the Sun and stars as thermonuclear reactors.

Advances in quantum theory and improved models of atomic structure made it clear to early twentieth century astronomers that deeper understanding of the life cycle of stars and of cosmological theories explaining the vastness of **space** was to be forever tied to advances in understanding inner workings of the universe on an atomic scale. A complete understanding of the energetics of mass conversion in stars was provided by Albert Einstein's (1879–1955) special theory of relativity and his relation of mass to energy (Energy = mass times the square of the speed of light).

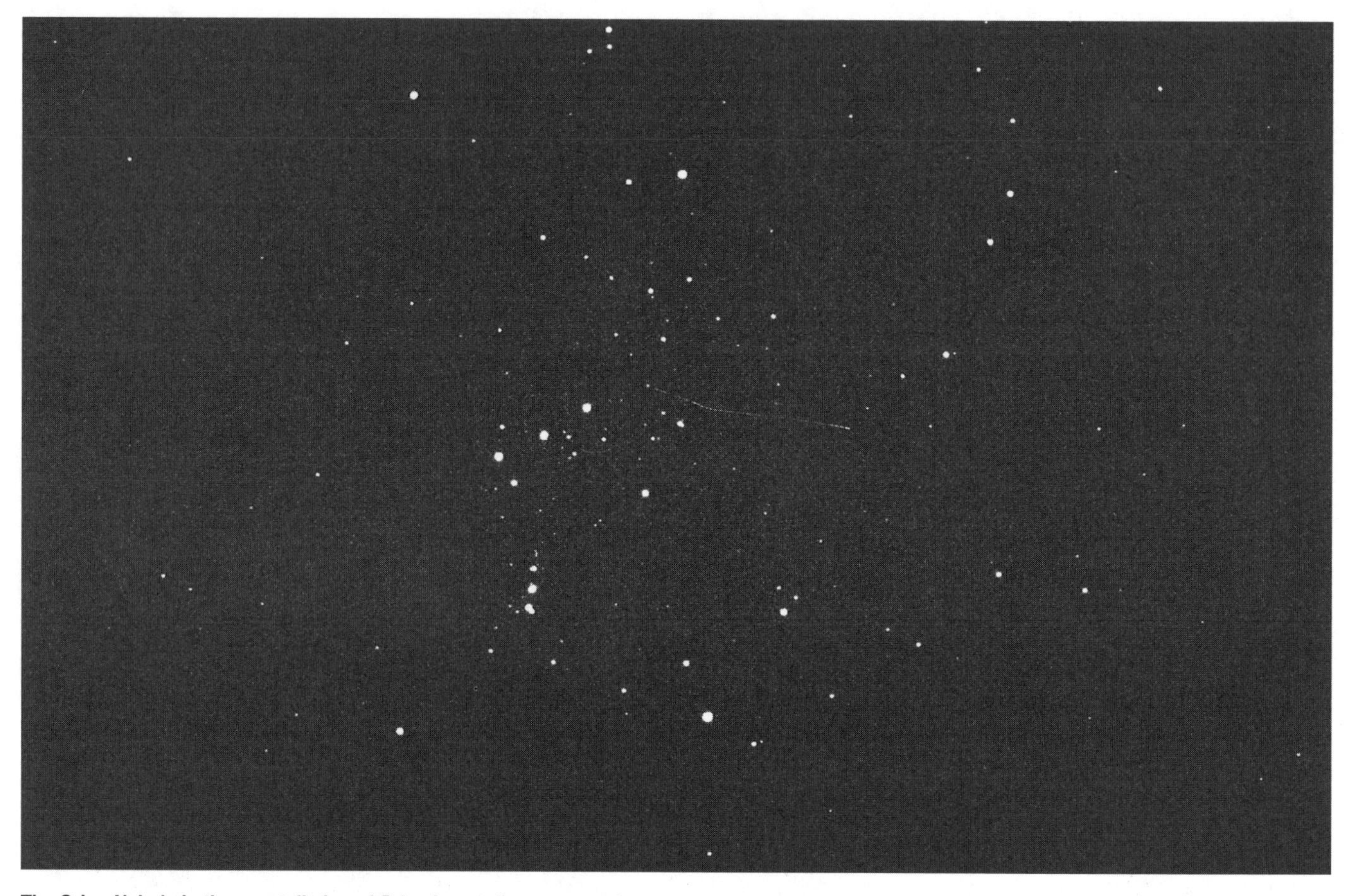

The Orion Nebula in the constellation of Orion is a stellar nursery where new stars are created. The bright red star in Orion is Betelguese, a red giant approaching the end of its stellar life cycle. ***U.S. National Aeronautics and Space Administration (NASA).***

During the 1920s, based on the principles of quantum mechanics, British physicist Ralph H. Fowler determined that, in contrast to the predictions of Russell, a white dwarf would become smaller as its mass increased.

Indian-born American astrophysicist Subrahmanyan Chandrasekhar (1910–1995) first articulated the evolution of stars into supernova, white dwarfs, neutron stars and for predicting the conditions required for the formation of black holes, which were subsequently found in the later half of the twentieth century. Before the intervention of World War II, American physicist J. Robert Oppenheimer (1904–1967), who ultimately supervised Project Trinity (the making of the first atomic bombs), made detailed calculations reconciling Chandrasekhar's predictions with general **relativity theory**.

In the decade that followed, as the mechanisms of atomic fission and fusion worked their way into astronomical theory, it became apparent that stars spend approximately ninety percent of their lives as main sequence stars before the fate dictated by their mass becomes manifest.

Astronomers refined concepts regarding stellar birth. Eventually as a protostar contracts enough, the increase in its temperature triggers **nuclear fusion**, and the star becomes visible as it vaporizes the surrounding cocoon. Stars then lead the majority of their life as main sequence stars, by definition burning hydrogen as their nuclear fuel.

It was the death of the stars, however, that provided the most captivating consequences. Throughout the life of a star, a tensional tug-of-war exists between the compressing force of the star's own gravity and the expanding pressures generated by nuclear reactions at its core. After cycles of swelling and contraction associated with the burning of progressively heavier nuclear **fuels**, eventually the star runs out of useable nuclear fuel. The spent star then contracts under the pull of it own gravity. The ultimate fate of any individual star is determined by the mass of the star left after blowing away its outer layers during its paroxysmal death spasms.

Low mass stars could fuse only hydrogen, and when the hydrogen was used up fusion stopped. The expended star shrank to become a white dwarf.

Medium mass stars swell to become red giants, blowing off planetary nebulae in massive explosions before shrinking to white dwarfs. A star remnant less than 1.44 times the mass of the Sun (termed the Chandrasekhar limit) collapses until the pressure in the increasing compacted electron clouds exerts enough pressure to balance the collapsing gravitational force. Such stars become "white dwarfs" contracted to the radius of only a few thousand kilometers, roughly the size of a planet. This is the fate of the Sun.

High mass stars can either undergo **carbon** detonation or additional fusion cycles that create and then use increasingly heavier elements as nuclear fuel. Regardless, the fusion cycles can only use heavier elements up to **iron** (the main product of **silicon** fusion). Eventually, as iron accumulates in the core, the core can exceed the Chandrasekhar limit of 1.4 times the mass of the Sun and collapses. These preliminary theoretical understandings paved the way for many of the discoveries in the second half of the twentieth century when it was more fully understood that as electrons are driven into protons neutrons are formed and energy is released as gamma rays and neutrinos. After blowing off its outer layers in a supernova explosion (type II) the remnants of the star form a neutron star and/or pulsar (discovered in the late 1960s).

Although he did not completely understand the nuclear mechanisms (nor of course the more modern terminology applied to those concepts), Chandrasekhar's work allowed the prediction that such neutron stars would be only a few kilometers in radius and that within such a neutron star the nuclear forces and the repulsion of the compressed atomic nuclei balanced the crushing force of gravity. With more massive stars, however, there was no known force in the universe that could withstand the gravitational collapse. Such extraordinary stars would continue their collapse to form a singularity—a star collapsed to a point of infinite density. According to general relativity, as such, a star that collapses its gravitational field warps space time so intensely that not even light can escape and a black hole forms.

Although the modern terminology presented here was not the language of early twentieth century astronomers, German astronomer Karl Schwarzschild (1873–1916) made important early contributions to the properties of geometric space around a singularity when warped according to Einstein's general relativity theory.

There are several important concepts stemming from the evolution of stars that have enormous impact on science and philosophy in general. Most importantly, the articulation of the stellar evolutionary cycle had profound effects on the cosmological theories developed during the first half of the twentieth century that culminated with the **Big Bang theory**, first proposed by Russian physicist Alexander Friedmann (1888–1925) and Belgian astronomer Georges Lemaitre (1894–1966) in the 1920s and subsequently modified by Russian-born American physicist George Gamow (1904–1968) in the 1940s.

The observations and theories regarding the evolution of stars meant that only hydrogen, helium, and a perhaps a smattering of lithium were produced in the big bang. The heavier elements including, of course, carbon and **oxygen**, up to iron were determined to have their genesis in the cores of increasingly massive dying stars. The energy released in the supernova explosions surrounding stellar death created shock waves that gave birth via fusion to still heavier elements and allowed the creation of radioactive isotopes.

The philosophical implications of this were as startling as quantum and relativity theories underpinned the model. Essentially, all of the elements heavier than hydrogen that comprise man's everyday existence were literally cooked in a process termed nucleosynthesis that took place during the paroxysms of stellar death. Great supernova explosions scattered these elements across the Cosmos.

See also Cosmology

STELLAR MAGNITUDES

Of principle importance to general astronomical observation is the observable brightness of the stars. Magnitude is the unit used in **astronomy** to describe a star's brightness. Although stellar magnitude in the visible spectrum dictates which stars can be observed under particular visible light conditions—variable due to time of observation, **moon** phase, atmospheric conditions, and the amount of light pollution present—magnitude also describes the relative amount of electromagnetic radiation observable in other regions of the **electromagnetic spectrum** (e.g., the x ray region of the spectrum).

Stars emit different amounts of radiation in different regions of the spectrum, so a star's "brightness" or magnitude will differ from one part of the spectrum to the next. An important field of research in modern astronomy is the accurate measurement of stellar brightness in magnitudes in different parts of the spectrum.

The Greek astronomer Hipparchus devised the first magnitudes in the second century B.C. He classified stars according to how bright they looked to the eye: the brightest stars he called "1st class" stars, the next brightest "2nd class," and so on down to "6th class." In this way, all the stars visible to the ancient Greeks were neatly classified into six categories.

Modern astronomers still use Hipparchus' categories, though in considerably refined form. With modern instruments astronomers measure a quantity called *V*, the star's brightness in the visual portion of the spectrum. Since visual light is what our eyes detect, *V* is analogous to Hipparchus' classes. For example, Hipparchus listed Aldebaran, the brightest star in the constellation Taurus, as a 1st class star and modern astronomers measure Aldebaran's *V* at 0.85. Astronomers often refer to a star's visual brightness as its apparent magnitude, a description of how bright the star appears to the eye (or the **telescope**).

Hipparchus' scheme defined from the outset one of the quirks of magnitudes: they list magnitude inversely. The fainter the star, the larger the number describing its magnitude. Therefore, the **Sun**, the brightest object in the sky, has an apparent magnitude of –26.75, while Sirius, the brightest star in the sky other than the Sun and visible on cold winter nights in the constellation Canis Major, has an apparent magnitude of –1.45. The faintest star you can see without optical aid is about +5 or +6 in a very dark sky with little light pollution. The faintest objects visible to the most powerful telescopes on Earth have an apparent magnitude of about +30.

More revealing than apparent magnitude is absolute magnitude, the apparent magnitude a star would have if it were ten parsecs from the Earth (a parsec is a unit of distance equal to 12 trillion mi [19.3 trillion km]). This is important because apparent magnitude can be deceiving. You know that a penlight is not as bright as a streetlight, but if you hold the penlight near your eye, it will appear brighter than a streetlight six blocks away. That's why *V* is called apparent magnitude: it is only how bright the star appears to be. For example, the Sun is a fainter star than Sirius—Sirius emits far more energy than the Sun does—yet the Sun appears brighter because it is so much closer. The Sun has an absolute magnitude of +4.8, while Sirius is +1.4.

In 1856, the British scientist N. R. Pogson noticed that Hipparchus' 6th class stars were roughly 100 times fainter than his 1st class stars. Pogson redefined the stars' *V* brightness so that a difference of five magnitudes was exactly a factor of 100 in brightness. This meant that a star with $V = 1.00$ appeared to be precisely 100 times brighter than a star with $V = 6.00$. One magnitude is then a factor of about 2.512 in brightness.

The Sun ($V = -26.75$) has an apparent visual brightness 25 magnitudes greater than Sirius. The difference in apparent brightness between the Sun and the faintest object humans have ever observed (using the **Hubble Space Telescope**) is more than 56 magnitudes.

In the 140 years since Pogson created the modern magnitudes, astronomers have developed many different brightness systems. In 1953, H. L. Johnson created the *UBV system* of brightness measurements. *B* is the star's brightness in magnitudes measured in the blue part of the spectrum, while *U* is the brightness in the ultraviolet spectral region. There are many other brightness measurement systems in use.

Accurate measurement of stellar brightness is important because subtracting the brightness in one part of the spectrum from the brightness in another part reveals important information about the star. For many stars the quantity $B-V$ gives a good approximation of the star's **temperature**. It was established in 1978 that the quantity $V-R$, where *R* is the brightness in the red part of the spectrum, can be used to estimate a star's radius.

See also Cosmology; Quantum electrodynamics (QED); Seeing

STENO, NICOLAS (1638-1686)

Danish geologist and anatomist

The son of a Copenhagen goldsmith, Nicolas Steno had a short but varied scientific career. His given name was Niels Stensen, but he is generally referred to by the Latinized version, Nicolas Steno. His name also has a variety of other spellings, such as Steensen, Stenonis (Latin), Stenone (Italian) and Stenon (French).

Steno's early schooling was accomplished in Copenhagen until 1660, when he began to travel **Europe** to study abroad. While a pupil of anatomy in Amsterdam, in the Netherlands, he discovered the parotid salivary duct, which is also called Stensen's duct. He made a number of other anatomical discoveries, including that muscles were made of fibrils, and he showed that the pineal gland existed in animals other than man. This was notable because some considered the pineal gland the location of the human soul, an idea first proposed by Rene Descartes (1596–1650), and so had considered it a gland unique to humans. In 1665, Steno moved to Florence, Italy, where his medical skills got him appointed physician to Grand Duke Ferdinand II of Tuscany. After returning to the Netherlands he was made Royal Anatomist in Copenhagen in 1672.

While in Italy, Steno was sent a huge shark's head that had been caught by local fishermen. While dissecting and studying it, Steno was struck by the similarity of the shark's teeth to common Mediterranean **fossils** known as 'tongue stones.' His study of these fossils led him to consider how any solid object could get inside another. In 1669, Steno published a short work that was to be an introduction to a larger study he never attempted, entitled *De solido intra solidum naturaliter contento dissertationis prodromus* (A preliminary discourse on a solid body contained naturally within a solid).

To illustrate his ideas, Steno made some of the earliest sketches of geological sections, and formulated three important geological principles. First, Steno noted that it was possible to tell which had been solid first, the **rock** or the fossil, by noting which was impressed on the other. In this way, Steno showed that the rocks must have formed around the fossils, and he suggested that the rocks had solidified out of former **seas**. Additionally, Steno argued that such rock layers would have formed in horizontal layers, and that any changes to the original horizontality must have occurred after their creation. Finally, Steno concluded that the oldest layers of rock strata must be those on the bottom, and newer layers were superposed on top of them. From this, Steno developed a geological history of rock formation, becoming one of the founders of **stratigraphy**.

Because the 'tongue stones' left an impression on the encasing rocks, Steno argued they had existed before the rock. Because they resembled shark's teeth so closely, he concluded that they were most likely ancient shark's teeth. Steno argued that similar fossils also had an organic origin, which went against the popular beliefs that such 'stones' had fallen from the sky, had grown from the Earth, or had more mystical origins.

Steno also made some important early studies of **crystals**. His observations of **quartz** showed that while different in appearance the quartz crystals all had the same corresponding angles between faces. He generalized this rule to all crystals, and the principle of constant angles in crystals is still known as Steno's law.

Steno had been raised a Lutheran, but in 1667, he converted to Catholicism. His faith caused him to abandon the study of science, and in 1677, he was appointed a titular bishop and spent the remainder of his days ministering to the few remaining Catholics in Northern Germany and Scandinavia.

See also Fossil record; Crystals and crystallography

STEVIN, SIMON (1548-1620)

Belgian-born Dutch mathematician and engineer

Simon Stevin (Latinized to Stevinus, as was the custom of the times) took as his motto, "Wonderful, yet not unfathomable," or, alternatively, "Nothing is the miracle it appears to be". In his pursuit to fulfill this motto, Stevinus made significant contributions to mathematics, engineering, and the earth sciences. As a mathematician, he was the first to advocate the use of tenths towards the establishment of decimals in mathematical calculations. As an engineer, he developed a method of releasing **floods** via Holland's vast canal system in the event of an invasion. The first achievement mentioned above has great bearing on mathematical calculations in the earth sciences and the second relates to a combination of engineering and the study of the behavior of running **water**. However, it was his contributions to hydrostatics, **astronomy**, **gravity**, and magnetic declination that established his importance to the development of the earth sciences.

Stevinus is often credited as the father of hydrostatics, the science that studies fluids at rest. Prior to his research, many scientists believed that the shape of a container of a liquid influenced the amount of pressure exerted by the liquid on its sides. By this reasoning, a circular lake might experience about the same water pressure all the way around the lake, but the pressure on the walls of a lake with an irregular shape would vary from one **area** to another. Stevinus mathematically demonstrated that only the area of the liquid's surface and its depth influenced the pressure against the sides. This information is often used by scientists in studying the engineering of wells and the **permeability** of rocks in the construction of dams as well as the strength of the dam itself.

One of his contributions to astronomy was his early defense of the Copernican model (a **Sun** centered **solar system**). Stevinus wrote in support of the heliocentric theory before Italian astronomer **Galileo Galilei** (1564–1642) came to the same conclusion. However, Stevinus had neither the telescopic evidence of Galileo nor the astronomical data of German astronomer **Johannes Kepler** (1571–1630) to add significantly to the argument.

Of greater significance to the science of astronomy was a discovery that produced new evidence regarding the relationship between gravity and falling bodies. This evidence would eventually become critical to the understanding of how the Sun holds the planets in their orbits and theories about the entire universe. The discovery was made by dropping two bodies of different weights from a high tower. Stevinus recorded that both objects struck the ground at the same time, despite their weight differences. This information disproved the assumption of Greek philosopher Aristotle (384–322 B.C.) that heavier objects fall faster than lighter objects under all circumstances, an assumption that had stood unchallenged for almost 2,000 years. Most historians argue that Stevinus performed this experiment, or at least played a part in arranging the experiment. Although he preceded Galileo by about three years in recording this discovery, his achievement was later attributed to Galileo. Today, many historians hold that Galileo was not only the first to record the experiment, but that he dropped the weights off the Leaning Tower of Pisa (there is no clear evidence that Galileo carried out such an experiment from the Leaning Tower). Regardless of who recorded the experiment first, it was a giant step away from Aristotelean thinking and eventually led to the Universal Law of Gravitation as outlined by English physicist Isaac Newton (1642–1727).

Stevinus' final contribution involved magnetic declination. Since the time of the Spanish-Italian navigator Christopher Columbus (1451–1506), it was a widely known fact that compasses did not point true north and south. Instead, they pointed toward what is known as the magnetic north and south poles. Because of this anomaly, the reliability of the compass depended on location. The difference between the magnetic poles and true north and south poles is known as the magnetic declination. By calculating and mapping magnetic declination, the navigator's job becomes much easier and more accurate. Realizing this, Stevinus was the first to undertake this task. At the time of his death he had calculated magnetic declination for 43 points on Earth's surface.

Stevinus' dedication to his motto and his work may have kept him from marrying until very late in life, or at least it would seem so. In his sixty-fourth year he finally married and eventually fathered four children before his death in 1620.

See also Gravitational constant; Hydrostatic pressure; Polar axis and tilt; Solar system

STOCK

• *see* PLUTON AND PLUTONIC BODIES

STRATEGIC PETROLEUM RESERVE (GEOLOGIC CONSIDERATIONS)

The Strategic **Petroleum** Reserve (SPR) operated by the United States Department of Energy is the largest emergency supply system of its kind in the world. The SPR presently consists of four underground storage facilities located in salt domes along the coastal regions of Louisiana and Texas, and has a total storage capacity of 700 million barrels of oil. These sites were chosen from among the more than 400 potential areas along the Gulf Coast of the southern United States after careful review of their relative geologic characteristics.

A salt dome is a body of **rock** salt surrounded by layers of sedimentary rock. Geologic characteristics considered in selecting storage sites include: 1) **area** geologic activity, 2) structural size 3) existence of a trapping mechanism, 4) salt geometry, 5) salt composition, and 6) surface conditions.

Geologic activity in the area of potential storage sites must be well understood. The coastal plains along the Gulf Coast tend to be in a perpetual state of either subsidence or uplift, and the rate of such relative change must be measurable and predictable.

Structural size is a significant factor in SPR storage and location. Oil is stored in cylindrically shaped caverns constructed within the salt body that are typically 200 ft (61 m) in diameter and approximately 2,000 ft (610 m) in height or larger. A storage dome may consist of from one to more than twenty caverns in a three-dimensional pattern. Salt domes along the Gulf Coast typically range between being 0.5 to 5 miles in diameter and may be over 20,000 ft (6,096 m) in vertical height.

Fluid naturally flows through permeable strata just as **water** passes through a sponge. Oil will seek the highest possible level due to its relatively low specific **gravity** and would float to the surface if not otherwise trapped. A salt dome must be overlain by a trapping mechanism in order to be an environmentally safe and an economically secure storage site. Cap rock is a stratum of rock lacking **permeability** that can act as a trapping mechanism. However, not all salt domes are overlain by cap rock.

Salt domes are usually formed as the lighter salt rises through sedimentary strata above in a plastic state from a deeper source, while forming irregular-shaped and sometimes freestanding columns. The three-dimensional geometry of the salt diapir must be profiled in order to facilitate the design of the storage cavern pattern.

Ideally, the salt dome is composed of homogenous halite free of shale or other sedimentary rock. The presence of irregularities in composition may effect cavern construction and containment integrity.

Surface conditions play a role in site selection and project design, construction and ease of operation. Typically, such sites are located in marsh areas or beneath standing water. Proximity to existing infrastructure supporting oil import, delivery, and water handling is a major cost and operational consideration.

Though geologically complex, salt domes have proven to be a reliably safe and economically competitive means of storing oil for future use, and play a key role in national energy management and supply.

See also Petroleum detection; Petroleum, economic uses of; Petroleum extraction; Petroleum, history of exploration

STRATIFORM CLOUD • *see* CLOUDS AND CLOUD TYPES

STRATIGRAPHY

Stratigraphy is that subarea of **geology** that treats the description, **correlation**, and interpretation of stratified Earth materials. Typically, geologists consider stratified Earth materials as layers of sediment or sedimentary **rock**. This definition, however, clearly encompasses other materials such as volcanic **lava**, ash flows, ash-fall layers, meteoritic impact ejecta layers, and soils. In fact, using this definition, any material that obeys the law of **superposition** during its formation could be placed in the domain of stratigraphy. Generally, internal layers within Earth (**crust**, mantle, and core) are not considered the type of layers studied by stratigraphers because they formed by Earth's internal differentiation processes.

Some geologists give a broader definition to the term stratigraphy. Planetary geologists sometimes view stratigraphy as if it were the study of the sequence of events on a planet or moon's surface. In addition, stratigraphy has been broadly used by some geologists who study mountain building and **plate tectonics** to mean the study of order of emplacement of rock units of various types, including igneous and metamorphic rocks, to which the law of superposition does not apply. In some cases, stratigraphy is used to define the study of geologic history of an **area** or country, but it is more correct to say that stratigraphy is the practical foundation for **historical geology**. In this article, the concept of stratigraphy expressed in the first paragraph is viewed as best and most correct.

Stratigraphy had its origins in the Renaissance writings of Nicholas Steno (1638–1687), who was the first to write lucidly about sedimentary strata. He observed strata exposed in the Arno River valley of Italy, and noted three axiomatic ideas, which became known as the first three "laws" of stratigraphy (*Prodromus*, 1669). These laws are known today as superposition, lateral continuity, and original horizontality. Superposition holds that layers are deposited so that the older layer is on the bottom. Unless strata are disturbed, this is always true. Lateral continuity holds that sedimentary layers extend laterally until they become so thin that they end at a "feather edge," abut against an obstruction, or grade laterally into other layers. Original horizontality holds that sedimentary layers are originally formed horizontally and remain so unless deformed by subsequent processes.

Steno's writings were full of common sense. In superposition, he noted the most important criterion for relative age dating. In lateral continuity, he wrote about how correlation of sedimentary layers would be possible. In original horizontality, he noted the criterion necessary for any sort of analysis of later deformation, that is, the original state of a sedimentary layer can be assumed to be horizontal.

As insightful as Steno's writings were, there is no strong evidence that they were influential beyond the Renaissance era in which he lived. Later on, during the Enlightenment, naturalists like **James Hutton** (1726–1797), John Playfair (1748–1819), and **Charles Lyell** (1797–1875) apparently independently "re-discovered" the importance of these common-sense concepts and used them in their influential writings about geology and stratigraphy. Hutton, Playfair, Lyell, and others of their time wrote books and papers, which established the foundations of modern thought about stratigraphy. Their most important contributions included promoting the concepts of actualism (understanding the past by studying modern processes) and demonstrating such key concepts as stratigraphic correlation, predictable fossil succession, and the great antiquity of Earth.

The advancement of these key concepts were given a great boost by the pioneering work of the English field engineer **William Smith** (1769–1839), who compiled and published the first large-scale **geologic map** (Wales and southern England; 1815) employing modern concepts of stratigraphic correlation and fossil succession. Smith's success inspired others to this kind of work, and was particularly important in influencing the Geological Society of London (the first geological

The Grand Canyon in Arizona has exposed millions of years of strata, allowing geologists a glimpse back in Earth's history. ***National Park Service.***

organization; founded 1807) to embark upon its "stratigraphical enterprise" of research in the United Kingdom. The Society and the British Geological Survey (the first geological survey, founded 1835) were important promoters of early stratigraphic studies and venues for presentation of early research. Based upon these efforts, it is fair to assert that modern stratigraphy was born in the United Kingdom during this period.

In the nineteenth century, major efforts were made by British stratigraphers and their colleagues on the European continent to develop a unified stratigraphic succession (or "geological column") for rocks in their areas. Cambridge Professor Adam Sedgwick (1785–1873) and Scottish naturalist Roderick Murchison (1792–1871) became quite famous as the preeminent "system builders" of their time. Sedgwick studied and named the Cambrian System himself and with Murchison, the Devonian. Murchison studied and named the Silurian and Permian Systems by himself. There were others who did the same during the nineteenth century, thus establishing the basis of our modern geological time scale (which has periods of the same names as those given to "systems" of rock during an era when exact ages of rock strata were unknown). This was the birth of modern **chronostratigraphy**, which emphasizes subdivision of geological time by studying Earth's stratigraphic record.

A Swiss geologist, Amanz Gressley (1814–1865), studied Jurassic strata in **Europe** in hopes of understanding what happens to sedimentary layers where they grade into other layers. He recognized that lateral continuity of layers revealed many changes, which reflected different ancient environments. To this concept, he gave the name facies, meaning an aspect of a sedimentary formation. A German stratigrapher, **Johannes Walther** (1860–1937), took up Gressley's ideas in his own work and became more widely known than Gressley for work with sedimentary facies. To Walther, the facies represented primary characteristics of the rock that would help him understand how and where the rock formed. He used what he called the ontological method in facies stratigraphy, which he described whimsically as "... from being, we explain becoming." This was a direct application of actualism, advocated earlier by Hutton and others, but now applied in a time of enhanced understanding of the natural world. Walter was the first naturalist to spend large amounts of time in the field studying modern environments in order to better interpret the past. His two-volume work, *Modern Lithogenesis* (1983; 1984), was a watershed for modern research with sedimentary facies. Accordingly, Walter is regarded as the founder of modern facies stratigraphy. Although his work was not accepted well in the United States for many years (due, in part, to anti-German feelings during the early twentieth century), it later was studied extensively for its rich descriptions of modern sedimentary environments and ancient sedimentary facies. In the latter part of the twentieth century, facies stratigraphy became much more than an academic exercise when it was realized that such knowledge could help predict the occurrence of **petroleum** and certain ore minerals—and facilitate more productive extraction of these materials—in host **sedimentary rocks**.

At the outset of the twentieth century, Austrian stratigrapher Eduard Suess (1831–1914) became the first advocate of global changes of sea level and how those changes might relate to global stratigraphy. This concept, called eustatsy, holds that global sea level rises and falls during geological history lead to the great marine transgressions and regressions noted in many sedimentary strata from locales around the world. Suess called upon subsidence of the sea floor and displacement of seawater by sediment as reasons for this global effect (today we know that gain and loss of **polar ice** is another contributor to sea-level change). His work stimulated much research, and strongly influenced the well-known American geologist T.C. Chamberlain (1843–1928), who perpetuated these ideas through his many well-known papers on the subject. These ideas were important in the development of a modern concept in stratigraphy called sequence stratigraphy.

Sequence stratigraphy, which holds that large bodies of sedimentary strata are bounded by interregional **unconformities**, formed as a result of global eustatsy. In the early 1960s, sequence stratigraphy was put forth by the American stratigrapher L.L. Sloss (1913–1996) in a series of widely read papers. During the 1970s, Sloss's student, Peter Vail (1930–), formerly with Exxon Corporation (now Exxon-Mobil Corporation), further developed these concepts while studying seismic profiles of stratigraphy from the world's continental shelves. Vail's paper's established sequence stratigraphy as one of the main subdivisions of modern stratigraphy. To recognize their contributions, sequence stratigraphy is often referred to as Sloss-Vail sequence stratigraphy in their honor.

Vail's work spawned a huge effort to produce a highly detailed, eustatic sea-level cycle chart of Earth's history based upon the vast data collection at Exxon. His work was published in 1987 in the prestigious journal *Science*. Sequence stratigraphy and global sea-level cycle charts are concepts used today major petroleum-company exploration laboratories all over the world.

Today, facies stratigraphy and sequence stratigraphy are not the only types of stratigraphy practiced by geologists. Modern stratigraphy includes: lithostratigraphy (naming of formations for purposes of geological mapping); biostratigraphy (correlating rock layers based upon fossil content); chronostratigraphy (correlating rock layers based upon their similar ages); magnetostratigraphy (study and correlation of rock layers based upon their inherent magnetic character); **soil** stratigraphy (study and mapping of soil layers, modern and ancient); and event stratigraphy (study and correlation of catastrophic events in geological history). The latter may include global or regional layers formed by asteroid or comet impacts, major volcanic events, global **climate** or ocean-chemistry changes, and effects of slight changes in Earth's orbital parameters (e.g., **Milankovitch cycles**). Modern procedures and practices in stratigraphy are summarized in two widely read documents: the *North American Stratigraphic Code*, published by the American Association of Petroleum Geologists, and the *International Stratigraphic Guide, 2nd* edition, published by the Geological Society of America.

Because layered Earth materials possess so much information about Earth's past, including the entire fossil record—and a sedimentary record quite sensitive to atmospheric, climatic, and oceanic changes of the past—stratigraphy is the one subarea of geology entirely focused upon retrieving and understanding that record.

See also Correlation (geology); Geologic time; Historical geology; Marine transgression and marine regression; Unconformities

STRATOCUMULOUS CLOUD • *see* CLOUDS AND CLOUD TYPES

STRATOSPHERE AND STRATOPAUSE

The atmosphere of Earth can be divided into semi-horizontal layers or spheres, based on properties such as **temperature** variation, gas components, or electrical properties. While air pressure and air density always decrease with altitude in the atmosphere, it is not the case for temperature. Four temperature-varying layers of Earth's atmosphere can be distinguished: the **troposphere**, the stratosphere, the **mesosphere**, and the **thermosphere**. In the troposphere and the mesosphere, the temperature decreases with altitude, but in the stratosphere and thermosphere, the temperature increases with altitude (called temperature inversion). Between these layers, the temperature remains the same in the tropopause (between the troposphere and the stratosphere), the stratopause (separating the stratosphere and the mesosphere), and the mesopause (between the mesosphere and the thermosphere).

The stratosphere is the second lowest layer of Earth's atmosphere, located between the troposphere and mesosphere. The stratosphere and the mesosphere together are called the middle atmosphere. The stratosphere, which literally means the layered sphere, is located from about 12 miles (20 km) to 30 miles (50 km) altitude. About 99% of the total air mass of the atmosphere can be found in the bottom two layers, the troposphere and stratosphere. The stratosphere is not only less dense than the troposphere, but it also contains very dry air. The stratospheric temperature is warmer than the upper tropospheric temperature; the average temperature at the bottom of the stratosphere is around – 76°F (–60°C), and at the upper bound around –26°F (–3°C). The temperature increases with height in the stratosphere because of the thin, but highly concentrated, stratospheric ozone layer. It is located between 13 to 19 miles (20 to 30 km), and reaches its peak density at an altitude of about 16 miles (25 km).

Ozone is a special molecular form of **oxygen**, consisting of three oxygen atoms bonded together. It is created by incoming solar radiation, which breaks up ordinary molecular oxygen (O_2) into individual oxygen atoms, which can later combine with another ordinary oxygen molecule to form ozone (O_3). Because the ozone layer absorbs and scatters solar radiation, a form of energy, the stratosphere consequently warms up. Without the ozone layer life could not exist on Earth; ozone is the only atmospheric gas that protects the **biosphere** from the damaging effects of the sun's ultraviolet radiation. This is why the depletion of the ozone layer (ozone hole) caused by anthropogenic chlorofluorocarbons (CFCs) is a cause for scientific concern and study.

See also Atmospheric composition and structure; Ozone layer depletion; Troposphere and tropopause

STRATUS CLOUD • *see* CLOUDS AND CLOUD TYPES

STREAM CAPACITY AND COMPETENCE

Streams channel **water** downhill under the influence of **gravity**. Stream capacity is a measure of the total sediment (material other than water) a stream can carry. Stream competence reflects the ability of a stream to transport a particular size of particle (e.g., boulder, pebble, etc). With regard to calculation of stream capacity and competence, streams broadly include all channelized movement of water, including large movements of water in **rivers**.

Under normal circumstances, the major factor affecting stream capacity and stream competence is channel slope. Channel slope (also termed stream gradient) is measured as the difference in stream elevation divided by the linear distance between the two measuring points. The velocity of the flow of water is directly affected by channel slope, the greater the slope the greater the flow velocity. In turn, an increased velocity of water flow increases stream competence. The near level **delta** at the lower end of the Mississippi River is a result of low stream velocities and competence. In contrast, the Colorado River that courses down through the Grand **Canyon** (where the river drops approximately 10 ft per mile) has a high stream velocity that results in a high stream capacity and competence.

Channelization of water is another critical component affecting stream capacity and stream competence. If a stream narrows, the velocity increases. An overflow or broadening of a stream channel results in decreased stream velocities, capacity, and competence.

The amount of material (other than water) transported by a stream is described as the stream load. Stream load is directly proportional to stream velocity and stream gradient and relates the amount of material transported past a point during a specified time interval. The higher the velocity the greater the sum of the mass that can be transported by a stream (stream load). Components of stream load contributing to stream mass include the **suspended load**, dissolved load, and bed load. Broad, slow moving streams are highly depositional (low stream capacity) while high velocity streams have are capable of moving large rocks (high stream competence).

Alluvial fans form as streams channeling mountain **runoff** reach flatter (low, slope, low gradient) land at the base of the mountain. The stream loses capacity and a significant portion of the load can then settle out to form the alluvial fan.

The ultimate site of deposition of particular types and sizes of particles is a function of stream capacity and stream competence (along with settling velocity of particles). These factors combine to allow the formation of articulated sedimentary deposits of **gypsum**, **limestone**, **clay**, shale, siltstone, **sandstone**, and larger **rock** conglomerates. No matter how low the stream capacity, the solution load usually retains ions in solution until the water evaporates or the **temperature** of the water cools to allow **precipitation**.

In confined channels, stream competence can vary with seasonal runoff. A stream with low volume may only be able to transport ions, clays, and silt in its solution and suspension loads and transport **sand** as part of its **saltation** load. As stream flow increases the stream competence during seasonal flooding, the stream may gain the competence to move pebbles, cobbles, and boulders.

See also Bed or traction load; Bedforms (ripples and dunes); Channel patterns; Erosion; Estuary; Hydrogeology; Rapids and waterfalls; Sedimentary rocks; Stream valleys, channels, and floodplains

STREAM PIRACY

A stream can be defined as any flowing body of **water** in a clearly defined channel. Streams may increase the size of their valleys by the **erosion** of the **soil** and **rock** surrounding their channels, either by widening their valleys or by headward erosion. In the process of headward erosion, the stream valley at the uppermost part of the stream channel is worn away, and the stream channel is lengthened in the upstream direction. Because the sides of the uppermost part of the stream valley are often steeper than the sides of the valley further downstream, the lengthening of the stream channel usually proceeds faster than the process of valley widening. Rates of headward erosion and channel lengthening vary between different streams, because some streams will have steeper valleys, resulting in faster water flow rates and faster erosion. In some cases, this results in a phenomenon called stream piracy, in which part of the drainage of one stream is captured by another, faster-eroding stream.

A stream that has lost part of its drainage is termed beheaded. Stream piracy is also called stream capture or river capture.

See also Drainage basins and drainage patterns; Stream valleys, channels, and floodplains

Stream piracy is the result of the drainage of one stream being captured by another, faster-eroding stream. *© Roger Wood/Corbis. Reproduced by permission.*

STREAM VALLEYS, CHANNELS, AND FLOODPLAINS

Stream valleys, channels, and floodplains form complicated systems that evolve through time in response to changes in sediment supply, **precipitation**, land use, and rates of tectonic uplift affecting a drainage basin. Stream channels serve to convey flow during normal periods, whereas floodplains accommodate flow above the bankfull stage (**floods**) that occurs with frequencies inversely proportional to their magnitude. Bankfull stage is defined as the discharge at which the **water** level is at the top of the channel. Any further increase in discharge will cause the water to spill out of the channel and inundate the adjacent floodplain. Flood frequency studies of streams throughout the world show a remarkably consistent one to two-and-one-half year recurrence interval for bankfull discharge in most streams, averaging about one and a half years, meaning that small floods are relatively common events.

Stream channels are classified according to four basic variables: their slope or gradient (change in elevation per unit of distance along the stream channel), their width to depth ratio, their entrenchment ratio (flooplain width to bankfull width), and the predominant channel bed material (**bedrock**, gravel, cobble, **sand**, or **clay**). In general, the width of stream channels increases downstream more than the depth, so that large **rivers** such as the Mississippi may be 0.5 mi or more wide but less than 100 ft deep. Channels with large bed loads of coarse-grained materials, steep gradients, and banks composed of easily eroded sediments tend to be shallow and braided, meaning that flow occurs through many anastomosing channels separated by bars or islands. Streams with low gradients, small bed loads, and stable banks tend to meander in **space** and time, following a pattern that resembles an exaggerated sine wave. Another characteristic of streams with beds coarser than sand is the occurrence of riffle-pool sequences, in which the channel is segregated into alternating deep pools and shallow riffles. In steep mountain streams, the riffles can be replaced by steep steps over boulders or bedrock outcrops to form a step-pool sequence.

Stream channels can change in form over time as a function of **climate**, precipitation, sediment supply, tectonic activity, and land use changes. Increased precipitation or human activities—for example, heavy grazing or clear-cut logging—can lead to increased **erosion** or **mass wasting** that subsequently increase the amount of sediment delivered to streams. As a consequence, the channel and stream gradient change to accommodate the increased sediment load, which may in turn have adverse effects on aquatic habitat. For example, an influx of fine-grained sediment can clog the gravel beds in which salmon and trout spawn. The effect of urbanization is generally to increase storm **runoff** and the erosive power of streams because impervious areas (principally pavement and rooftops) decrease the amount of water that can infiltrate into the **soil**, while at the same time decreasing the amount of sediment that is available for erosion before runoff enters stream channels. Tectonic uplift can increase the rate of stream valley incision. Thus, stream channels represent the continually changing response of the stream system to changing conditions over geologic and human time spans.

Because streams are the products of continual change, many stream valleys contain one or more generations of stream terraces that represent alternating stages of sediment deposition (valley filling) and erosion (stream incision). Each flat terrace surface, or tread, is a former floodplain. Stream terraces can often be recognized by a stair-step pattern of relatively flat surfaces of increasing elevation flanking the channel; in many cases, however, stream terraces are subtle features that can be distinguished and interpreted only with difficulty.

Floodplains form an important part of a stream system and provide a mechanism to dissipate the effects of floods. When a stream exceeds bankfull discharge, floodwater will begin to spill out onto the adjacent flat areas where its depth and velocity decrease significantly, causing sediment to fall out of suspension. The construction of flood control structures such as artificial levies has allowed development on many floodplains that would otherwise be subjected to regular inundation. Artificial levies, however, also increase the severity of less frequent large floods that would have been buffered by functioning floodplains, and can thereby provide a false sense of security. Current trends in flood hazard mitigation are therefore shifting away from the construction of containment structures and towards more enlightened land use practices such as the use of floodplains for parks or green belts rather than residential development.

See also Bed or traction load; Bedforms (ripples and dunes); Canyon; Channel patterns; Drainage basins and drainage patterns; Drainage calculations and engineering; Hydrologic cycle; Landscape evolution; Sedimentation; Stream capacity and competence

Strike and dip

Geologists use a prescribed method of determining the attitude (or orientation in three-dimensional **space**) of **rock** layers or any other planar geological feature (e.g., metamorphic foliation, fractures, faults, and tops of tabular units like formations). The method involves measurement of strike and dip of the rock layers or planar features. Strike is defined as the compass direction, relative to north, of the line formed by the intersection of a rock layer or other planar feature with an imaginary horizontal plane. The intersection of two flat planes is a straight line, and in this instance, the line is geologic strike. According to convention, the compass direction (or bearing) of this line is always measured and referred to relative to north. A typical bearing is given, for example, as N 45° E, which is a shorthand notation for a bearing that is 45 degrees east of north (or half way between due north and due east). The only exception to this north rule occurs where strike is exactly east-west. Then, and only then, is a strike direction written that is not relative to north.

Dip, as a part of the measurement of the attitude of a layer or planar feature, has two components: dip direction and dip magnitude. Dip direction is the compass direction (bearing) of maximum inclination of the layer or planar feature down from the horizontal. If a **marble** is held anywhere along the strike line on a layer or planar feature and then is released, thus allowing it to roll down the layer or planar feature, the marble would roll along a line showing true dip direction. This true dip direction is always perpendicular (i.e., at a 90 degree angle) to strike. Dip magnitude is the smaller of the two angles formed by the intersection of the dipping layer or planar feature and the imaginary horizontal plane. However, dip magnitude can also be equal to either zero or 90 degrees, where the layer or planar feature is horizontal or vertical, respectively.

A specially designed instrument, called a Brunton pocket transit, is used by geologists to measure strike, dip direction, and dip magnitude in the field. A Brunton pocket transit has a compass, bubble level (for finding horizontal), and a dip-angle measuring device (clinometer) built into it. Information on strike and dip obtained using the Brunton is then conveyed to a **geologic map** and plotted there using a strike and dip symbol. This symbol, about the length of the word *the* on this page, consists of a long bar oriented parallel to strike and a short spike perpendicular to the long bar showing the bearing of true dip direction. A small number printed by the symbol indicates the actual dip magnitude in degrees. On some maps, this number is not printed or is not printed by all such symbols.

Measurement of strike and dip (i.e., the attitude of rock layers or other planar geologic features) helps geologists construct accurate geologic maps and geologic cross-sections. For example, data on rock attitudes helps delineate fold structures in layered rocks. Attitude of other geologic structures like faults can be understood using strike and dip as well. It is especially critical to understanding geologic relations among rock bodies in the subsurface realm that surficial (relating to the surface) strike and dip of rocks is well known.

For entirely subsurface studies of strike and dip, devices called dip-meter tools can be lowered into drill holes. These tools, which use electrical properties of rocks in the well wall to sense attitude, help delineate subsurface orientations of rock layers and other planar features. Information from such subsurface studies is critical in areas where surficial attitude measurements are not adequate to understand subsurface structures. All surface and subsurface information on attitudes of rocks is important in helping geologists understand the structure and origin of Earth's **crust**.

See also Faults and fractures; Folds; Orientation of strata

Stylolites • *see* Marble

Subduction zone

Subduction zones occur at collision boundaries where at least one of the colliding **lithospheric plates** contains oceanic **crust**. In accord with plate tectonic theory, collision boundaries are sites where lithospheric plates move together and the resulting compression causes either subduction (where one or both lithospheric plates are driven down and destroyed in the molten mantle) or crustal uplifting that results in **orogeny** (mountain building). Subduction zones are usually active **earthquake** zones. Subduction zones are the only sites of deep earthquakes. The areas of deep earthquakes, ranging to a depth of 415 mi (670 km), are termed Benioff zones. Deep earthquakes occur because of forces due to plate drag and mineral phase transitions. The release of forces due to sudden slippage of plates during subduction can be quick and violent. Subduction zones can also experience shallow and intermediate depth earthquakes.

Oceanic crust is denser than continental crust and is subductable. Moreover, as oceanic crust–bearing plates move away from their site of origin (divergent boundaries), the oceanic crust. The cooling results in an increase in general density of the oceanic crust. The concurrent loss of buoyancy makes it easier to subduct the crust. In addition, colliding plates create tremendous force. Although lithospheric plates move very slowly, the plates have tremendous mass. Accordingly, at collision each lithospheric plate carries tremendous momentum (the mathematical product of velocity and mass) that provides the energy to drive subduction. In zones of convergence, including subduction zones, compressional forces (i.e., compression of lithospheric plate material) dominates.

Earth's crust is fractured into approximately 20 lithospheric plates. Lithospheric plates move on top of the asthenosphere (the outer plastically deforming region of Earth's mantle). Because Earth's diameter remains constant, there is no net creation or destruction of lithospheric plates. Each lithospheric plate is composed of a layer of oceanic crust or continental crust superficial to an outer layer of the mantle. Oceanic crust is composed of high-density rocks, such as **olivine** and **basalt**. In contrast, continental crust is composed of lower density rocks such as **granite** and **andesite**.

Within subducting zones, oceanic crust can make material contributions of lighter crustal materials to overriding continental crust. As the oceanic crust subducts, parts may be scraped off to form an accretion prism. Rising material at sites where oceanic crust subducts may form **island arcs**.

When oceanic crust collides with oceanic crust, both subduct to form an oceanic trench (e.g., the Marianas trench). Dual plate subduction can result in **ocean trenches** with depths approximating 38,000 ft.

When oceanic crust collides with continental crust, the oceanic crust subducts under the lighter continental crust. The subducting oceanic crust pushes the continental crust upward into **mountain chains** (e.g., the Andes) and may contribute lighter molten materials to the overriding continental crust to form volcanic arcs (e.g., the "ring of fire"; a ring of volcanoes bordering the Pacific Rim. Because continental crust does not subduct, when continental crust collides with continental crust, there is a uplift of both crusts.

At triple points where three plates converge, the situation becomes more complex and in some cases there is a mixture of subduction and uplifting.

Convergent plate boundaries are, of course, three-dimensional. Because Earth is an oblate sphere, lithospheric plates are not flat, but are curved and fractured into curved sections akin to the peeled sections of an orange. Convergent movement of lithospheric plates can best be conceptualized by the movement together of those peeled sections over a curved surface.

See also Divergent plate boundary; Earth, interior structure; Earthquakes; Geologic time; Magma chamber; Magma; Mohorovicic discontinuity (Moho); Plate tectonics; Volcanic eruptions; Volcano

SUBLIMATION OF GLACIERS • *see* GLACIATION

SUN (STELLAR STRUCTURE)

The Sun is the star about which Earth revolves. A typical star, Earth's sun is composed of gases and heavier elements compressed to enormous density and heated to levels that sustain **nuclear fusion** (the transformation of hydrogen into helium and heavier elements). The Sun consists of an inner core surrounded by a radiative zone and then a convective zone. The surface of the Sun is termed the photosphere. Surrounding the Sun is a solar corona—an atmosphere of hot plasma, gases, and outflowing particles.

Nuclear fusion take place in the Sun's core and it is in this region that the bulk of the Sun's production of energy, heat, and gamma rays takes place. The radiative zone surrounding the core is of such high density that photons generated in the core can take millions of years to pass through to the surrounding radiative zone. Undergoing an enormous number of collisions, absorptions and regenerations, photons span a spectrum frequencies that correspond to gamma rays, x ray, ultraviolet light, visible light, infrared light, microwaves, and radio waves. Photon passage through the convective zone provides energy to drive massive convective currents of hot gas.

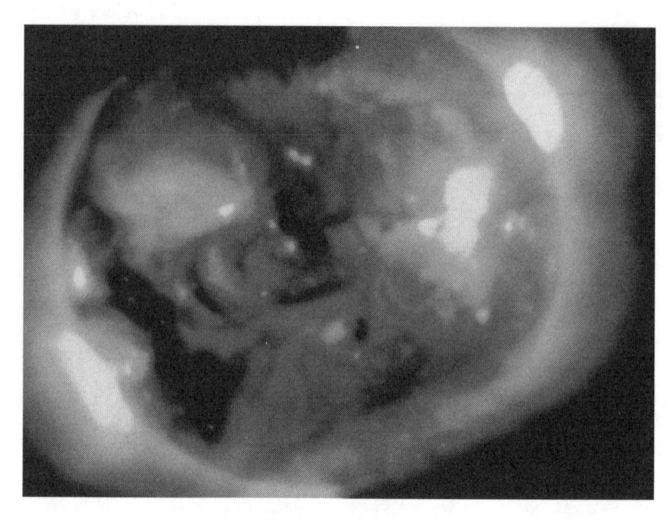

X-ray photograph of the solar corona. *U.S. National Aeronautics and Space Administration (NASA).*

The photosphere is the visible outer or surface layer of Sun. At the photosphere, solar temperatures cool to about 5800 K. The photosphere often features sunspots (areas of surface relatively cooler due to differential temperatures in convective currents). Sunspots occur in cycles with maximum activity peaking every 11 years.

Largely composed of gas, the Sun exhibits differential **rotation** speeds that depend on solar **latitude**. The rotational period varies from approximately 25 days at the equator to 29 days near the polar regions.

The chromosphere surrounds the photosphere and extends thousands of miles. Temperatures increase in the chromosphere and range up to 1,000,000 K. The chromosphere is part of the solar corona that extends millions of miles into **space**. Influenced by turbulent magnetic fields coronal temperatures range up to 3,500,000 K. At these high temperatures, electrons are stripped from gases and plasma streams form a solar **wind**. The solar chromosphere and corona are usually visible only when an **eclipse** blocks the photosphere.

Solar flares and prominences, flame-like eruptions of hot gas, sometimes extend into the chromosphere and corona.

British astronomer Fred Hoyle once described the evolution of a star—including, of course, the Sun—as a continual war between nuclear **physics** and **gravity**.

The gravity of the stellar material pulls on all the other stellar material striving to bring about a collapse. However, the gravitational compression is opposed by the internal pressure of the stellar gas that normally results from heat produced by nuclear reactions. This balance between the forces of gravity and the pressure forces forms an equilibrium, and the balance must be exact or the star will quickly respond by expanding or contracting in size. So powerful are the separate forces of gravity and pressure that should such an imbalance occur in the Sun, it would be resolved within hours. That fact that Earth's sun is about 5 billion years old emphasizes just how exactly and continuously that balance is maintained.

In addition to its reliance on balance between gravity and pressure, the internal structure of a sun depends on the behavior of the stellar material itself. Most stars are made primarily of hydrogen, the dominant form of matter in the universe. However, the behavior of hydrogen will depend on the **temperature**, pressure, and density of the gas. Indeed, the quantities temperature, pressure, and density are known as state variables, because they describe the intrinsic state of the material. Any equation or expression that provides a relationship between these variables is called an equation of state.

Most of the energy that flows (i.e., undergoes a series of transformations) from a star originates at its center. The way in which this energy flows to the surface will also influence the internal structure of the star.

There are three ways by which energy flows outward through a star. They are conduction, convection, and radiation.

However, the more opaque the material is, the slower the convectional and radiative transfers of heat and energy (e.g. electromagnetic radiation or "light") flow of energy will be. In the Sun, where light flowing out from in the core will travel less than a centimeter before it is absorbed, it may take a million years for the light energy to make its way to the surface.

The mode of energy transport, equations of state, and equilibrium equations can be quantified and self-consistent solutions found numerically for stars of given mass, composition and age. Such solutions provide model stellar interiors, and supply the detailed internal structure of a particular star. For the vast majority of stars that derive their energy from the nuclear fusion of hydrogen into helium, the internal structure is quite similar. Such stars are termed main sequence stars and are located in a band on a Hertzsprung-Russell diagram (developed independently between 1911–13 by Danish astronomer Ejnar Hertzsprung (1873–1967) and American astronomer Henry Norris Russell (1877–1957).

The Sun is a main sequence star. The Sun's core is surrounded by a churning convective envelope that carries the energy to within a few thousand kilometers of the surface, where energy again flows primarily by radiation as it escapes into space. This structure is common to all main sequence stars with mass less than 1.5 times the mass of the Sun.

Changes to the stellar structure over time are described by the theory of **stellar evolution**.

See also Big Bang theory; Celestial sphere: The apparent movements of the Sun, Moon, planets, and stars; Solar energy; Solar illumination: Seasonal and diurnal patterns; Solar sunspot cycles; Solar system; Stellar life cycle

SUNSPOTS • *see* SOLAR SUNSPOT CYCLE

SUPERCONTINENTS

The earth comprises a number of **lithospheric plates** that move apart at mid-oceanic ridges, are consumed at subduction zones, collide with each other in collisional orogens, or slide past each other along transform boundaries. Although oceanic **crust** is continually being created and destroyed, long-lived stable parts of continents called cratons have remained undeformed for billions of years. Continental plates containing ancient cratons have episodically collided and assembled in global periods of orogenesis to form supercontinents. Supercontinents eventually become unstable, as such a large single landmass acts as a thermal lid, limiting escape of Earth's internal heat. Supercontinent breakup occurs when old crustal weaknesses (such as orogenic belts created during supercontinent assembly) overlay several **mantle plumes**, or due to the formation of a superplume. Dispersed fragments move across the globe to subsequently amalgamate to form another supercontinent. The process of supercontinent formation, breakup, and dispersal has continued cyclically through Earth's history.

Plate reconstructions for periods younger than 180 million years (the oldest age of crust in present-day **oceans**) can be made by graphically undoing seafloor spreading magnetic anomalies of known age. Offshore **gravity**, calculated from **satellite** altimetery of the sea surface, defines continental margins and structures in oceanic crust that can also be used in recent continental reconstructions. **Rotation** poles (Euler poles), about which lithospheric plates are displaced, can be determined from **transform faults** interpreted from satellite altimeter-derived images. All reconstructions are based on the generally held view that Earth has maintained a constant radius (although this is questioned by some who contend that the earth has progressively expanded). For older periods, scientists rely on the following to establish, or at least infer, continental correlations:

- Linking of orogenic belts and intracratonic structures (e.g., major **shear zones** of the same age and displacement sense)—accurate dating is unfortunately lacking from many **Precambrian** terrains, hindering such correlations. Regional aeromagnetic data is valuable in comparing lithological trends and structural elements as basement rocks below superficial cover and sedimentary basins are imaged. Regional gravity data highlights major crustal structures.
- Mafic dikes—dike swarms (readily discernable on aeromagnetic images) may be traced from one **craton** to another. Dikes may converge on an **area** above the center of a former mantle plume.
- Common **rock** types, ages, and fossil assemblages in **sedimentary rocks** on conjugate margins—provenance studies and ages of detrital zircons for sedimentary basins provide information concerning the source terrains for sedimentary basins. Sedimentary source rocks absent from the craton the basin is situated in may be found on previously contiguous continents.
- Paleomagnetism and polar wander paths—when sedimentary rocks are deposited or when igneous or meta-

morphic rocks cool below 578° for magnetite and 675° for hematite (the minerals' Curie temperatures), these iron-rich **minerals** preserve the orientation of the earth's **magnetic field**. Rocks of the same age from joined continents exhibit a common magnetic pole. Apparent polar wander paths linking poles of different ages graphically portray the displacement of a continent with time. Changes in previously overlapping apparent polar wander paths for two continents indicate the time continents rifted apart.

- Large igneous provinces of the same age and with characteristic geochemical signatures (representing portions of the same igneous province prior to breakup)—Archean to late Mesoproterozoic correlations are, however, highly problematic due to the poor to non-existent paleomagnetic database, the wide dispersal of cratons following their breakup and the likely disruption or tilting of old cratons during reworking along craton margins.

An **Archean**, 3.1 (to possibly even 3.6) billion years old supercontinent Vaalbara, in which the South African Kaapvaal Craton was joined to the Pilbara Craton of Western **Australia**, is the oldest proposed supercontinent. The concept of Vaalbara is based on similarities in sedimentary sequences on both cratons. The existence of Vaalbara has, however, been questioned due to differences in magmatic events between both cratons, and the possibility for similar sedimentary successions to have been deposited on separate continents due to global sea level changes. Recent paleomagnetic evidence suggests that the cratons were not contiguous about 2.8 billion years ago.

Another Archean to Paleoproterozoic supercontinent, Ur (the German word for original) has been proposed. Ur is thought to comprise a nucleus of the Kaapvaal and Pilbara cratons (although not adjacent to each other as in Vaalbara), the Indian Bhandara and Singhbhum cratons, and possibly some Archean East Antarctic terrains. A 1.8 to 1.5 billion year supercontinent Columbia comprising most continents then existing on Earth has also been proposed in which eastern India, Australia, and their contiguous portion of East **Antarctica** was joined to western **North America**.

Rodinia (from the Russian *rodit,* meaning to grow) is the late Mesoproterozoic to late Neoproterozoic (approximately 1,200 to 700 million year) supercontinent formed by the assembly of Precambrian terrains of Australia, North America-Canada, India, Madagscar, Sri Lanka, and East Antarctica. The idea for such a supercontinent initially came from the recognition of geological similarities between western Canada and southeastern Australia, and likely links between 1.3 to 0.9 billion-year-old orogenic belts. The term Grenvillian belts has been loosely used to encompass orogens formed during this broad time span. This is, however, not the strict definition of the Grenvillian **Orogeny**, which is defined as occurring between 1,090 and 980 million years ago in its type area, the Grenville Province of North America. There is still much debate as to the configuration of continents within Rodinia.

In the first SWEAT (from *S*outh*w*estern United States–*E*ast *A*n*t*arctica) configuration, Laurentia is positioned such that western Canada is opposite eastern Australia. Grenvillian orogens (such as the Central Indian Tectonic Zone and Pinjarra Orogen of Western Australia) that were not part of a continuous orogenic belt were omitted in this reconstruction, giving it a somewhat false simplicity. In a modified version, South China (formed through collision of the Cathaysia and Yangtze blocks) is positioned between Australia and Laurentia. In another reconstruction for Rodinia, AUSWUS (*Aus*tralia–*W*estern *U*nited *S*tates), eastern Australia is attached to the western United States. Both are compatible with available paleomagnetic data for 1,140 million years; however, both are not compatible with recent paleomagnetic poles for 1,070 million-year-old rocks from Laurentia and Western Australia. In an attempt to explain these poles, a reconstruction of Rodinia called AUSMEX (*Aus*tralia–*Mex*ico) has Laurentia in a rotated position with respect to Australia, with the Grenville Province of North America continuing directly into northeastern and central Australia. In the SWEAT and AUSWUS reconstructions, northeastern India joins with southwestern Australia, and southeastern India is linked to Antarctica. Despite sound geological links between Proterozoic orogenic belts supporting this configuration, current paleomagnetic data suggest that India and part of East Antarctica may not have amalgamated with Australia until 680 to 610 million years ago. Clearly, there are still many problems to be resolved before consensus is reached for a reconstruction of Rodinia compatible with all available data. It must also be asked whether palaeomagnetic data alone is a reliable means of establishing ancient positions of continents.

Rodinia split into two main fragments approximately 750 million years ago. Pannotia (meaning all southern) is the name given to a supercontinent formed when the northern block (comprising Australia, Antarctica, India, Madagascar, Arabia, and parts of China), Laurentia, and cratonic blocks in East **Africa**, Mozambique, Madagascar, and **South America** were amalgamated through collision in the Pan-African orogeny. Pannotia broke up into Laurentia, Baltica, Siberia, and Gondwanaland at the end of the Precambrian, about 590 million years ago.

The idea of links between India, central and southern Africa, and Madagascar based on the common occurrence of *glossopteris* (a fossilized seed fern first described in early Permian **coal** seams in central India) and other fossil assemblages was first proposed in 1885 by the Austrian geologist Eduard Suess. Suess coined the term Gondwanaland, meaning the land of the kingdom of the Gons (a Dravidian people in central India) for the area he thought to have been linked by land bridges between fixed continents. Australia and Antarctica were subsequently added following further *glossopteris* discoveries. The proposition of links between continents predated the concepts of continental drift and **plate tectonics**, but was subsequently used as evidence in their formulation. Gondwanaland continents contain similar Permo-Carboniferous (286 million year old) to late Jurassic–early Cretaceous (100 million year) sedimentary successions. Permo-Carbonifereous tillites and other similar glacial deposits in India, Australia, and Antarctica are the vestiges of a large **ice** sheet. Older ice sheets were present in Africa and South America. Early reconstructions of Gondwanaland were based on the outlines of continents, structural features, and broad matches in **geology**. These have been subsequently refined using paleomagnetic and seafloor spreading data.

Gondwanaland was separated by the Tethys Ocean from another supercontinent, Laurasia, formed through collision of Laurentia (North America), Baltica (Scandinavian continents), parts of **Europe**, and Siberia approximately 400 million years ago. The supercontinent Pangaea (meaning all Earth) was formed by the collision of Laurasia with Gondwanaland approximately 275 million years ago following closure of part of the Tethys, and the collision with Cimmeria (fragments of Turkey, Afghanistan, Iran, Tibet, and Indochina). Mountain belts such as the Appalachians and Urals were formed in this event. The breakup of Pangaea in the Late Jurassic to Early Cretaceous occurred largely due to **rifting** along old weaknesses when they became aligned between mantle plumes.

Understanding the configurations of past supercontinents is of more than academic interest. The formation and dispersal of supercontinents has had a marked effect on past changes in **ocean circulation** patterns and hence on Earth's **climate**. Major mineral provinces on one continent may have as yet undiscovered corollaries on another, once adjacent continent. Placer gold deposits in a sedimentary basin on one continent may have been eroded from vein or shear zone hosted deposits on another continent. The Mt. Isa Belt in Queensland, Australia, is truncated by a rifted margin formed during the breakup of Rodinia. Rich gold and base metal deposits may exist in its continuation on another, yet undefined terrain, possibly within southeast China or North America (placed against Queensland in various Rodinia reconstructions).

See also Continental drift theory; Earth, interior structure; Earth (planet); Geologic time

SUPERIMPOSED STREAMS • *see* DRAINAGE BASINS AND DRAINAGE PATTERNS

SUPERPOSITION

Originally observed by Nicholas Steno in the seventeenth century, the law of superposition states that in an undisturbed series of **sedimentary rocks**, the oldest rocks will be at the bottom and the youngest will be at the top. Before the development of radiometric dating techniques, the law of superposition was used to assign relative ages to **rock** units based on their position in the sequence. For example, looking at a series of undeformed sedimentary rocks one could assume that the rocks, and therefore associated **fossils**, of the top layers were younger than those below. This idea builds on one of Steno's other observations, the law of original horizontality, which states that sedimentary layers are approximately horizontal when deposited. It follows that any body of rock that cuts across the sequence must be younger than all of the layers that it cuts.

If a sequence of sedimentary rocks has been deformed, the law of superposition may be difficult to apply. If a sequence of beds has been tilted, it should be clear that the law of superposition cannot be applied until the original up direction is verified. Overturned beds cause a similar problem, and relative ages may be calculated incorrectly if the deformation is not noticed. There are several younging indicators to aid in determination of the original up direction. These include mudcracks, cross beds, graded **bedding**, load and flute casts, and burrow marks. Once younging direction has been determined, the law of superposition may be applied even to a deformed sedimentary sequence.

See also Bedforms (ripples and dunes); Cross cutting; Lithification; Stratigraphy

SURF ZONE

The surf zone is one of the most dynamic regions of the marine environment. Not only is it highly energetic, it also supports a wide diversity of life forms. Surf zones in both sandy and rocky shores produce dynamic surroundings that shape the entire coastal region.

Waves are initiated at sea where winds sweep across the surface of the **water**. Some of the energy of the **wind** is transferred to the surface and small sinusoidal waves appear on top of the water. The wave often increases in energy with added winds. The height of oceanic waves is dependant on the energy of the wind which may be quite high during storm events.

The waves continue until they reach the shore. Once there, they strike the shore and lose their energy. What happens during this process is of great importance to geologists and those who live near the beach.

Ocean waves are very interesting examples of wave **physics** found in nature. A look at the wave in cross-section appears as a series of linked "S" shapes. The high point of the individual curve is called the crest. The low point is the trough. The size of the individual waves (the amplitude) and the frequency (the rate at which a series of waves pass a particular point during a specific amount of time) are determined by the amount of energy a group of waves contains.

The long waves strike the shore at an angle. Waves will rarely, if at all, touch the shore at a parallel line to the shore. This is a result of the changing **topography** of the land. When the wave strikes the shore it is refracted and a great deal of its energy is directed at an oblique angle to the original wave. This increases the energy of the wave effect and, for anyone who has experienced it, produces a significant push-pull type of action. Being caught in this **area**, named the swash, can be dangerous under certain conditions such as rip **tides** formation.

Geologists are concerned about the region where waves break on the shore. Small circular currents occur in the water that produce the characteristic swell shape. As these circular currents reach the shore and strike the bottom, they are compressed. This compression eventually breaks down the wave structure and the water spills over on itself at the crest. This is the point at which waves are commonly called "breakers." The bottom sediments act as a further drag on the wave and deplete it of much of its energy. Any sediments that were initially captured by the breaking wave are dumped and moved onto the beach. At the same time, the refracted wave captures sediment and pulls it back down the shoreline. This repeated motion causes **sand** grains to migrate down the beach as waves strike again and again. Where the shore is rocky, the force of the waves wears away the rocks and transports smaller clasts (sedimentary particles) out to sea and the coast. This type of **erosion** is a major physical process that sculpts cliffs around the world. The entire area of breaking waves, including the slope-face of the land, is called the surf zone.

See also Beach and shoreline dynamics; Dunes; Sedimentation; Wave motions

SURVEYING INSTRUMENTS

Surveying is the apportionment of land by measuring and mapping. It is employed to determine boundaries and property lines, and to plan construction projects. Surveying instruments are designed to precise (repeatable) and accurate apportionment measurements.

Throughout history, civilizations with high levels of sophistication in construction methods required surveys to ensure that work came out according to plan. Formal surveying on a large scale is thought to have originated in ancient Egypt as early as 2700 B.C., with the construction of the Great Pyramid of Khufu at Giza, though the first recorded evidence of boundary surveying dates from 1400 B.C. in the Nile River valley.

The Roman Empire relied heavily on surveying. In order to forge an empire that stretched from the Scottish border to the Persian Gulf, a large system of roads, bridges, aqueducts, and canals was built, binding the country economically and militarily. Surveying was a major part of Roman public works projects. It also was used to divide the land among the citizens. Roman land surveying was referred to as *centuriation*, which was a common rectangular unit of land **area**. These land parcels can still be seen in aerial photographs taken over France and other parts of Europe—the work of the Roman *agrimensores*, or measurers of land. The property lines were usually marked by stone walls and boundary markers.

Consistent with the rise of trigonometry and calculus, new surveying instruments emerged. The theodolite was invented in the sixteenth century. Its precise origin is unclear, but one version was invented by English mathematician Leonard Digges in 1571, who gave it its name. An improved theodolite was invented by Jesse Ramsden more than 200 years later in 1787. Its use led to the establishment of the British Ordnance Survey.

Made up of a **telescope** mounted on a compass or of a quadrant plus a circle and a compass, the theodolite is used to measure horizontal and vertical angles. The modern theodolite is usually equipped with a micrometer, which gives magnified readings up to 1/3600°, or one second of arc. The micrometer is derived from the vernier scale, which was invented by French engineer and soldier Pierre Vernier (1584-1638) in 1631 to measure in fractions.

The transit is a theodolite capable of measuring horizontal and vertical angles, as well as prolonging a straight line or determining a level sight line. A telescope atop a tripod assembly is clamped in position to measure either horizontal or vertical angles. The transit employs a plumb bob hanging from the center of the tripod to mark the exact location of the surveyor.

The practice of triangulation was introduced by Gemma Frisius in 1533. By measuring the distance of two sides of a triangle in a ground survey, the third side and the triangle's area can be calculated. Triangulation was aided by the inventions of the prismatic **astrolabe** and the heliotrope. The latter was invented by German mathematician Johann Gauss (1777-1855), who is considered the father of geodesy, the science of Earth measurement. Both instruments used a beam of sunlight to signal the positions of distant participants in a land survey.

Other survey instruments include the surveyor's compass, which is used for less precise surveying. The surveyor's level is used to measure heights of points above sea level or above local base points. Metal tapes, first introduced by English mathematician Edmund Gunter in 1620, are used for shorter measurements.

In the late twentieth century, surveying has been aided greatly by **remote sensing**: Photogrammetry employs aerial photography to survey large areas for topographic mapping and land assessment purposes, **satellite** imagery has increased the aerial coverage of surveys, and laser technology has increased the precision of survey sightings.

See also Cartography; Geologic map; Petroleum detection; Topography and topographic maps

SUSPENDED LOAD

Suspended load consists of sediment particles that are mechanically transported by suspension within a stream or river. This is in contrast to **bed or traction load**, which consists of particles that are moved along the bed of a stream, and dissolved load, which consists of material that has been dissolved in the stream **water**. In most streams, the suspended load is composed primarily of silt and **clay** size particles. Sand-size particles can also be part of the suspended load if the stream flow velocity and turbulence are great enough to hold them in suspension.

The suspended load can consist of particles that are intermittently lifted into suspension from the stream bed and of wash load, which remains continuously suspended unless there is a significant decrease in stream flow velocity. Wash load particles are finer than those along the stream bed, and therefore must be supplied by bank **erosion**, **mass wasting**, and mass transport of sediment from adjacent watersheds into the stream during rainstorms.

Water density is proportional to the amount of suspended load being carried. Muddy water high in suspended sediment will therefore increase the particle buoyancy and reduce the critical shear stress required to move the bed load of the stream.

The ratio of suspended load to bed load in a stream depends on the ratio of the shear velocity (a property of the flowing water that reflects the degree of turbulence) and the fall velocity (a property of the sediment grains). The fall velocity is that at which a sediment particle will fall through still water, and thus depends on both grain size and **mineralogy** (density). Bed-load transport predominates when the shear velocity is significantly less than half the fall velocity and suspended load transport predominates when the shear velocity is two to three times greater than the fall velocity. Mixed-mode transport occurs when the ratio falls within a range of approximately 0.4 to 2.5.

See also Erosion; Rivers; Sedimentation; Stream valleys, channels, and floodplains

SYNCLINE AND ANTICLINE

Syncline and anticline are terms used to describe **folds** based on the relative ages of folded **rock** layers. A syncline is a fold in which the youngest rocks occur in the core of a fold (i.e., closest to the fold axis), whereas the oldest rocks occur in the core of an anticline.

It is important to note that syncline and anticline do not necessarily relate to the shape or orientation of folded layers, although the origin of the words implies this. The term originates from the Greek word *sun (xun),* meaning together, and the Greek word *klei,* meaning to lean, so syncline implies leaning together or leaning towards. *Ant* is the Greek prefix meaning opposite or opposing, so the word anticline implies oppositely leaning. Beds **dip** towards the fold axis in a syncline and away from the fold axis in an anticline only when the folded layers were upright before folding (i.e., where younger layers overlaid older layers). Before describing folds, it is therefore necessary to establish the primary order in which layers were deposited. To do this, facing, younging, or way-up criteria are used. These are synonymous terms for primary sedimentary structures (e.g., graded or cross-bedding) or igneous structures (e.g., vesicles, pillows) preserved in the folded layers. Where the relative ages of rocks are not known (as is often the case in metamorphic rocks), the term synform and not syncline should be used to describe folds where layers are bent downwards so that they dip towards the fold axis, and antiform and not anticline should be used where beds are arched upwards so that layers dip away from the fold axis.

Where rock layers have been inverted prior to folding, such as by folding about a larger fold with a shallowly inclined axial surface, the oldest rocks now occur in the core of folds where layers dip towards the fold axis. Such folds are called synformal anticlines; synformal because of their shape and anticline because of the relative ages of folded layers. The youngest layers in an overturned sequence occur in the core of folds called antiformal synclines where layers dip away from the fold axis.

T

TALUS PILE OR TALUS SLOPE

Geologists define talus as the pile of rocks that accumulates at the base of a cliff, chute, or slope. The formation of a talus slope results from the talus accumulation.

Because the term "talus" incorporates the concept of a pile, many geologists prefer it to "talus pile" and reserve the term "talus slope" for specific reference to the surface of the talus.

The recognition and characterization of talus slopes is often important in determining the potential for mass movements (landslides, etc.). Movements occur whenever the talus slope exceeds the critical angle. The exact angle at which failure takes place depends upon the materials (e.g., **rock** type), rock size, moisture content, but dry homogenous materials in a pile generally experience slope failure when the angle of repose (the resting slope angle) exceeds 33–37°. The critical angle lowers as materials become less intrinsically cohesive or when friction between particles is reduced by rain or other forms of moisture. Moisture also adds to the overall mass of the slope and thus increases the gravitational force on the slope.

For example, if a cliff or rock formation is composed of shale, the processes of **weathering** and the force of **gravity** (a shear stress) allow the downslope accumulation of shale rock fragments and debris at the base of the formation. The talus slope is triangular, with the internal angles of the sides of the triangle (the slope's angle of repose) limited by the critical angle.

The degree of sameness in size, layering, and homogeny of the talus is referred to as sorting. As a general rule, talus accumulated from rockfalls is better sorted than talus created by glacial deposits but far less sorted than piles constructed by **sedimentation**. Contributing rock that is irregularly fractured does not **weather** evenly and because it breaks off in large irregular pieces, contributes to a poorly sorted talus slope.

See also Landscape evolution; Landslide; Mass movement; Mass wasting

The formation of a talus slope results from the talus accumulation. ***© Jack Dermid. Reproduced by permission.***

TECTOSILICATES

The most abundant rock-forming **minerals** in the **crust** of Earth are the silicates. They are formed primarily of **silicon** and **oxygen**, together with various **metals**. The fundamental unit of these minerals is the silicon-oxygen tetrahedron. These tetrahedra have a pyramidal shape, with a relatively small, positively charged silicon cation (Si^{+4}) in the center and four larger, negatively charged oxygen anions (O^{-2}) at the corners, producing a net charge of –4. **Aluminum** cations (Al^{+3}) may substitute for silicon, and various anions such as hydroxyl (OH^{-}) or fluorine (F^{-}) may substitute for oxygen. In order to form stable minerals, the charges that exist between tetrahedra must be neutralized. This can be accomplished by the sharing of oxygen cations between tetrahedra, or by the binding together of adjacent tetrahedra with various metal cations. This in turn creates characteristic silicate structures that can be used to classify silicate minerals into **cyclosilicates**, **inosilicates**, **nesosilicates**, **phyllosilicates**, **sorosilicates**, and tectosilicates.

Silicon-oxygen tetrahedra form a three-dimensional framework in the tectosilicates, and minerals of this type comprise about 64% of Earth's crust. In addition, the tectosilicate minerals known as the feldspars are the most abundant group of the rock-forming silicates. Feldspars include the alkali feldspars microcline and orthoclase, both of which have the same chemical formula ($KAlSi_3O_8$), but which form in progressively lower-temperature bodies of **magma** within the earth, and also form **crystals** with different characteristic shapes. Another alkali **feldspar**, sanidine, has the chemical formula $(K,Na)AlSi_3O_8$, and forms in **lava** that has been extruded onto the surface of the earth. The **plagioclase** feldspars all form in molten **rock**, and there is gradation in composition between abite ($NaAlSi_3O_8$) and anorthite ($CaAl_2Si_2O_8$). Another common tectosilicate mineral is **quartz** (SiO_2), which forms in many geological environments. Pure quartz is as transparent as **glass**, but varieties include amethyst, which is colored purple by the presence of small amounts of **iron**; rose quartz, which is colored pink by small amounts of titanium, and milky quartz, which appears white due to the presence of small fluid droplets.

TELESCOPE

The telescope is an instrument that collects and analyzes the radiation emitted by distant sources. The most common type is the optical telescope, a collection of lenses and/or mirrors that is used to allow the viewer to see distant objects more clearly by magnifying them or to increase the effective brightness of a faint object. In a broader sense, telescopes can operate at most frequencies of the **electromagnetic spectrum**, from radio waves to gamma rays. The one characteristic all telescopes have in common is the ability to make distant objects appear to be closer (from the Greek *tele* meaning far, and *skopein* meaning to view).

The first optical telescope was probably constructed by the Dutch lens-grinder, **Hans Lippershey**, in 1608. The following year, **Galileo Galilei** built the first astronomical telescope from a tube containing two lenses of different focal lengths aligned on a single axis (the elements of this telescope are still on display in Florence, Italy). With this telescope and several following versions, Galileo made the first telescopic observations of the sky and discovered lunar mountains, four of Jupiter's moons, sunspots, and the starry nature of the Milky Way. Since then, telescopes have increased in size and improved in image quality. Computers are now used to aid in the design of large, complex telescope systems.

The primary function of a telescope is that of light gathering. As will be seen below, resolution limits on telescopes would not call for an aperture much larger than about 30 in (76 cm). However, there are many telescopes around the world with diameters several times this. The reason is that larger telescopes can see further because they can collect more light. For example, the 200 in (508 cm) diameter reflecting telescope at Mt. Palomar, California can gather 25 times more light than the 40 in (102 cm) Yerkes telescope at Williams Bay, Wisconsin, the largest refracting telescope in the world. The more light a telescope can gather, the more distant the objects it can detect, and therefore larger telescopes increase the size of the observable universe.

Unfortunately, scientists are not able to increase the resolution of a telescope simply by increasing the size of the light-gathering aperture to as large a size as needed. Disturbances and nonuniformities in the atmosphere limit the resolution of telescopes to somewhere in the range of 0.5–2 arc seconds, depending on the location of the telescope. Telescope sights on top of mountains are popular because the light reaching the instrument has to travel through less air, and consequently the image has a higher resolution. However, a limit of 0.5 arc seconds corresponds to an aperture of only 12 in (30 cm) for visible light: larger telescopes do not provide increased resolution but only gather more light.

Magnification is not the most important characteristic of telescopes as is commonly thought. The magnifying power of a telescope is dependent on the type and quality of eyepiece being used. The magnification is given simply by the ratio of the focal lengths of the objective and eyepiece. Thus, a 0.8 in (2 cm) focal length eyepiece used in conjunction with a 39 in (100 cm) focal length objective will give a magnification of 50. If the field of view of the eyepiece is 20°, the true field of view will be 0.4°.

Most large telescopes built before the twentieth century were refracting telescopes because techniques were readily available to polish lenses. Not until the latter part of the nineteenth century were techniques developed to coat large mirrors, which allowed the construction of large reflecting telescopes.

Refracting telescopes, i.e. telescopes that use lenses, can suffer from problems of chromatic and other aberrations, which reduce the quality of the image. In order to correct for these, multiple lenses are required, much like the multiple lens systems in a camera lens unit. The advantages of the refracting telescope include having no central "stop" or other diffracting element in the path of light as it enters the telescope, and the alignment and transmission characteristics are stable over long periods of time. However, the refracting telescope can have low overall transmission due to reflection at the surface of all the optical elements, and the largest refractor ever built has a diameter of only 40 in. (102 cm): lenses of a larger diameter will tend to distort under their own weight and give a poor image. Additionally, each lens needs to have both sides polished perfectly and be made from material which is of highly uniform optical quality throughout its entire volume.

All large telescopes, both existing and planned, are of the reflecting variety. Reflecting telescopes have several advantages over refracting designs. First, the reflecting material (usually **aluminum**), deposited on a polished surface, has no chromatic aberration. Second, the whole system can be kept relatively short by folding the light path, as shown in the Newtonian and Cassegrain designs below. Third, the objectives can be made very large since there is only one optical surface to be polished to high tolerance, the optical quality of the mirror substrate is unimportant, and the mirror can be supported from the back to prevent bending. The disadvantages of reflecting systems are: 1) alignment is more critical than in refracting systems, resulting in the use of complex adjustments for aligning the mirrors and the use of **temperature** insensitive mirror substrates, and 2) the secondary or other auxiliary mirrors are mounted on a support structure which occludes part of the primary mirror and causes diffraction.

Catadioptric telescopes use a combination of lenses and mirrors in order to obtain some of the advantages of both. The best-known type of catadioptric is the Schmidt telescope or camera, which is usually used to image a wide field of view for large **area** searches. The lens in this system is very weak and is commonly referred to as a corrector-plate.

The limits to the resolution of a telescope are, as described above, a result of the passage of the light from the distant body through the atmosphere, which is optically nonuniform. Stars appear to twinkle because of constantly fluctuating optical paths through the atmosphere, which results in a variation in both brightness and apparent position. Consequently, much information is lost to astronomers simply because they do not have sufficient resolution from their measurements. There are three ways of overcoming this limitation: setting the telescope out in **space** in order to avoid the atmosphere altogether, compensating for the distortion on a ground-based telescope, and/or stellar interferometry. The first two methods are innovations of the 1990s and are expected to lead to a new era in observational **astronomy**.

The best-known and largest orbiting optical telescope is the **Hubble Space Telescope** (HST), which has an 8 ft (2.4 m) primary mirror and five major instruments for examining various characteristics of distant bodies. After a much-publicized problem with the focusing of the telescope and the installation of a package of corrective optics in 1993, the HST has proved to be the finest of all telescopes ever produced to date. The data collected from HST is of such a high quality that researchers can solve problems that have been in question for years, often with a single photograph. The resolution of the HST is 0.02 arc seconds, a factor of around twenty times better than was previously possible, and also close to the theoretical limit since there is no atmospheric distortion. An example of the significant improvement in imaging that space-based systems have given is the Doradus 30 nebula, which prior to the HST was thought to have consisted of a small number of very bright stars. In a photograph taken by the HST it now appears that the central region has over 3,000 stars.

Another advantage of using a telescope in orbit is that the telescope can detect wavelengths such as the ultraviolet and various portions of the infrared, which are absorbed by the atmosphere and not detectable by ground-based telescopes.

The telescope collects and analyzes the radiation emitted by distant sources. The most common type is the optical telescope, a collection of lenses and/or mirrors that is used to allow the viewer to see distant objects more clearly by magnifying them or to increase the effective brightness of a faint object. ***Photograph by Robert J. Huffman. Field Mark Publications. Reproduced by permission.***

In 1991, the United States government declassified adaptive optics systems (systems that remove atmospheric effects), which had been developed under the Strategic Defense Initiative for ensuring that a laser beam could penetrate the atmosphere without significant distortion.

A laser beam is transmitted from the telescope into a layer of mesospheric sodium at 56–62 mi. (90–100 km) altitude. The laser beam is resonantly backscattered from the volume of excited sodium atoms and acts as a guide-star whose position and shape are well-defined except for the atmospheric distortion. The light from the guide-star is collected by the telescope and a wavefront sensor determines the distortion caused by the atmosphere. This information is then fed back to a deformable mirror, or an array of many small mirrors, which compensates for the distortion. As a result, stars located close to the guide-star come into a focus, which is many times better than can be achieved without compensation. Telescopes have operated at the theoretical resolution limit for infrared wavelengths and have shown an improvement in the visible region of more than 10 times. Atmospheric distortions are constantly changing, so the deformable mirror has to be updated every five milliseconds, which is easily achieved with modern computer technology.

Telescopes collect light largely for two types of analysis: imaging and spectrometry, with the better known being imaging. The goal of imaging is simply to produce an accurate picture of the objects that are being examined. In past years, the only means of recording an image was to take a photograph. For long exposure times, the telescope had to track the sky by rotating at the same speed as Earth, but in the opposite direction. This is still the case today, but the modern telescope no longer uses photographic film but a charged coupled device (**CCD**) array. The CCD is a semiconductor light detector, which is 50 times more sensitive than photographic film and is able to detect single photons. Being fabricated using semiconductor techniques, the CCD can be made very small, and an array typically has a spacing of 15 microns between CCD pixels. A typical array for imaging in telescopes will have a few million pixels. There are many advantages of using the CCD over photographic film or plates, including the lack of a developing stage and that the output from the CCD can be read directly into a computer and the data analyzed and manipulated with relative ease.

The second type of analysis is spectrometry, which means that the researcher wants to know what wavelengths of light are being emitted by a particular object. The reason behind this is that different atoms and molecules emit different wavelengths of light; measuring the spectrum of light emitted by an object can yield information as to its constituents. When performing spectrometry, the output of the telescope is directed to a spectrometer, which is usually an instrument containing a diffraction grating for separating the wavelengths of light. The diffracted light at the output is commonly detected by a CCD array and the data read into a computer.

For almost 40 years, the Hale telescope at Mt. Palomar was the world's largest with a primary mirror diameter of 200 in (5.1 m). During that time, improvements were made primarily in detection techniques, which reached fundamental limits of sensitivity in the late 1980s. In order to observe fainter objects, it became imperative to build larger telescopes, and so a new generation of telescopes is being developed. These telescopes use revolutionary designs in order to increase the collecting area; 2,260 ft^2 (210 m^2) is planned for the European Southern Observatory. This new generation of telescopes will not use the solid, heavy primary mirror of previous designs, whose thickness was between 1/6 and 1/8 of the mirror diameter, but will use a variety of approaches to reduce the mirror weight and improve its thermal and mechanical stability. These new telescopes, combined with quantum-limited detectors, distortion reduction techniques, and coherent array operation, will allow astronomers to see objects more distant than have been observed before.

One of this new generation, the Keck telescope located on Mauna Loa in Hawaii, is currently the largest operating telescope, using a 32 ft (10 m) effective diameter hyperbolic primary mirror constructed from 36 6 ft (1.8 m) hexagonal mirrors. The mirrors are held to relative positions of less than 50 nanometers using active sensors and actuators in order to maintain a clear image at the detector.

Because of its location at over 14,000 ft (4,270 m), the Keck is useful for collecting light over the range of 300–1100 nm. In the late 1990s, this telescope was joined by an identical twin, Keck II, which resulted in an effective mirror diameter of 279 ft (85 m) through the use of interferometry.

Most of the discussion so far has been concerned with optical telescopes operating in the range of 300–1100 nm. However, valuable information is contained in the radiation reaching us at different wavelengths, and telescopes have been built to cover wide ranges of operation, including radio and millimeter waves, infrared, ultraviolet, x rays, and gamma rays.

Infrared telescopes (operating from 1–1000 æm) are particularly useful for examining the emissions from gas **clouds**. Because **water** vapor in the atmosphere can absorb some of this radiation, it is especially important to locate infrared telescopes in high altitudes or in space. In 1983, NASA launched the highly successful Infrared Astronomical **Satellite**, which performed an all-sky survey, revealing a wide variety of sources and opening up new avenues of astrophysical discovery. With the improvement in infrared detection technology in the 1980s, the 1990s will see several new infrared telescopes, including the Infrared Optimized Telescope, a 26.2 ft (8 m) diameter facility, on Mauna Kea, Hawaii.

Several methods are used to reduce the large thermal background which makes viewing infrared difficult, including the use of cooled detectors and dithering the secondary mirror. This latter technique involves pointing the secondary mirror alternatively at the object in question and then at a patch of empty sky. Subtracting the second signal from the first results in the removal of most of the background thermal (infrared) noise received from the sky and the telescope itself, thus allowing the construction of a clear signal.

Radio astronomy was born on the heels of World War II, using the recently developed radio technology to look at radio emissions from the sky. The first radio telescopes were very simple, using an array of wires as the antenna. In the 1950s, the now familiar collecting dish was introduced and has been widely used ever since.

Radio waves are not susceptible to atmospheric disturbances like optical waves are, and so the development of radio telescopes over the past 40 years has seen a continued improvement in both the detection of faint sources as well as in resolution. Despite the fact that radio waves can have wavelengths which are meters long, the resolution achieved has been to the sub-arc second level through the use of many radio telescopes working together in an interferometer array, the largest of which stretches from Hawaii to the United States Virgin Islands (known as the Very Long Baseline Array).

See also Atmospheric composition and structure; SETI; Space and planetary geology

TEMPERATURE AND TEMPERATURE SCALES

Temperature is an indirect measure of the kinetic energy of particles composing matter. The SI unit for temperature is the kelvin (K). There is no degree sign associated with the kelvin.

Kinetic-molecular theory asserts that temperature is a property of matter that results from molecule motion (kinetic energy) and/or atomic vibration. A common misconception is that at absolute zero (0 Kelvin), atomic and molecular motion ceases. In reality, although absolute zero represents the absence of kinetic energy, it represents only the absolute minimal state of molecular or atomic vibration (i.e., electrons still "orbit" the nucleus and nuclear processes including transformations are possible). Temperature is the size-independent quantity that indirectly relates the average kinetic energy of all the particles within a body. Its size-independence stems from the fact that two objects made of the same matter at the same temperature (i.e., objects in thermal equilibrium) will have the same average kinetic energy of constituent particles.

Temperature is commonly measured with a thermometer—a device designed to relate the expansion of liquids (e.g., the rising of mercury in a tube) to changes in temperature. One of the first attempts at articulating a universal temperature scale was made by the Greek scientist and physician Galen (ca. A.D. 170). Galen based his scale on comparative temperatures with an equal mixture of **ice** and **water** assuming the center of a four point scale. By the mid-seventeenth century, Italian scientists and builders fashioned crude alcohol-in-glass thermometers and the English scientist Robert Hooke utilized an alcohol-in-glass thermometer with zero assigned to the **freezing** point of water in meteorological experiments.

During the early eighteenth century, Danish astronomer Ole Roemer advanced a temperature scale based on two points, an assigned temperature of crushed ice and the boiling point of water.

German-born physicist (born in what is now Danzig or Gdansk, Poland) **Daniel Gabriel Fahrenheit** (1686–1736) began creating thermometers containing mercury. Fahrenheit utilized mercury's ability to be easily visualized in **glass** tubing and its ability to remain a liquid over a wide range of normal atmospheric temperatures. Fahrenheit eventually designated the boiling point of water to be 212 degrees. Later he measured the freezing point of water to be 32 degrees, 180 degrees below the boiling point of water. The deviations on the scale were later named after its creator, and the scale is read in degrees Fahrenheit. The Fahrenheit scale still exists today, but is primarily used in the United States for reporting the **weather**. The Celsius and Kelvin temperature scales are more commonly used in scientific investigation.

In 1745, Swedish naturalist **Carl von Linné** (also known as Carl Linnaeus) (1707–1778) devised a *centigrade* (Latin for "one hundred steps") scale to measure temperature. He began his scale with the freezing point of water at zero degrees and set the boiling point of water at 100 degrees. Andrew Celsius used the same number of deviations in his scale, but he instead reversed the order to where the boiling point was zero and the freezing point was 100.

Subsequently, the International Committee of Paris adopted measurements of the freezing point and of the boiling point of water as fundamental markers for temperature scales. The Celsius scale was reversed, and in 1948 was revised to set the triple point of water (that temperature where solid, liquid and gas phases exist in equilibrium) at 0.01°C, and the boiling point of water at 99.975°C. The Celsius scale is used primarily in scientific investigation worldwide and in weather reporting for daily atmospheric temperatures everywhere but the United States.

In order to convert temperature from Celsius to Fahrenheit, the following formula is used:

$°F = 1.8\ (°C) + 32.$

In order to convert temperature from Fahrenheit to Celsius, the following formula is used:

$°C = 5/9\ (°F - 32).$

The necessity for an absolute temperature scale emerged from the advancement of kinetic-molecular theory. The thermodynamic temperature scale—incorporating the concept of absolute zero—evolved and is now accepted as the fundamental measure of temperature. In 1933, the International Committee of Weights and Measures adopted the triple point, or freezing point, of water as 273.16 Kelvin, named after Scottish physicist William Thomson (Lord Kelvin) (1824–1907).

In order to convert temperature to Kelvin from Celsius, the following formula is used:

$K = C + 273.15$

The absolute, or Kelvin, scale is used primarily in conjunction with the Celsius scale because the deviations are equal in magnitude.

In addition to thermometers, other devices can be used measure temperature. Changes in gas volume (e.g., as used in a constant-pressure gas thermometer), electrical resistance to current passage, and thermocouple voltage generation can also be calibrated to changes in temperature.

Another absolute temperature scale (i.e., a scale that incorporates absolute zero) still cited in literature is the Rankine scale. The Rankine is an absolute temperature scale where degree increments are the same magnitude as Fahrenheit degree increments. On the Rankine scale, the freezing point of water at standard temperature and pressure is 491.67°R and the normal boiling point is 671.67°R.

See also Atmospheric chemistry; Atmospheric inversion layers; Atmospheric lapse rate; Atomic theory; Chemistry; Energy transformations; Freezing and melting; Geothermal energy; Geothermal gradient; Quantum theory and mechanics

TERESHKOVA, VALENTINA (1937-)

Russian cosmonaut

Valentina Tereshkova was the first woman in **space**. Tereshkova took off from the Tyuratam Space Station in the *Vostok VI* in 1963 and orbited the Earth for almost three days, showing women had the same resistance to space as men. She then toured the world promoting Soviet science and feminism, and served on the Soviet Women's Committee and the Supreme Soviet Presidium. Valentina Vladimirovna "Valya" Tereshkova was born in the Volga River village of Maslennikovo. Her father, Vladimir Tereshkov, was a tractor driver; a Red Army soldier during World War II, he was killed when Valentina was two. Her mother Elena Fyodorovna Tereshkova, a worker at the Krasny Perekop cotton mill, single-handedly raised Valentina, her brother Vladimir, and her sister Ludmilla in economically trying conditions. Assisting her mother, Valentina was not able to begin school until she was ten.

Tereshkova later moved to her grandmother's home in nearby Yaroslavl, where she worked as an apprentice at the tire factory in 1954. In 1955, she joined her mother and sister as a loom operator at the mill; meanwhile, she graduated by correspondence courses from the Light Industry Technical School. An ardent Communist, she joined the mill's Komsomol

Valentina Tereshkova. *© Hulton-Deutsch Collection/Corbis. Reproduced by permission.*

(Young Communist League), and soon advanced to the Communist Party.

In 1959, Tereshkova joined the Yaroslavl Air Sports Club and became a skilled amateur parachutist. Inspired by the flight of Yuri Gagarin, the first man in space, she volunteered for the Soviet space program. Although she had no experience as a pilot, her 126-jump record gained her a position as a cosmonaut in 1961. Four candidates were chosen for a one-time woman-in-space flight; Tereshkova received an Air Force commission and trained for 18 months before becoming chief pilot of the *Vostok VI.* Admiring fellow cosmonaut Yuri Gagarin was quoted as saying, "It was hard for her to master rocket techniques, study spaceship designs and equipment, but she tackled the job stubbornly and devoted much of her own time to study, poring over books and notes in the evening."

At 12:30 PM on June 16, 1963, Junior Lieutenant Tereshkova became the first woman to be launched into space. Using her radio callsign Chaika (Seagull), she reported, "I see the horizon. A light blue, a beautiful band. This is the Earth. How beautiful it is! All goes well." She was later seen smiling on Soviet and European TV, pencil and logbook floating weightlessly before her face. *Vostok VI* made 48 orbits (1,200,000 miles) in 70 hours, 50 minutes, coming within 3.1 miles of the previously launched *Vostok V*, piloted by cosmonaut Valery Bykovsky. Tereshkova's flight confirmed Soviet test results that women had the same resistance as men to the physical and psychological stresses of space.

Upon her return, she and Bykovsky were hailed in Moscow's Red Square. On June 22, at the Kremlin, she was named a Hero of the Soviet Union and was decorated by Presidium Chairman Leonid Brezhnev with the Order of Lenin and the Gold Star Medal. A symbol of emancipated Soviet feminism, she toured the world as a goodwill ambassador promoting the equality of the sexes in the Soviet Union, receiving a standing ovation at the United Nations. With Gagarin, she traveled to Cuba in October as a guest of the Cuban Women's Federation, and then went to the International Aeronautical Federation Conference in Mexico.

On November 3, 1963, Tereshkova married Soviet cosmonaut Colonel Andrian Nikolayev, who had orbited the earth 64 times in 1962 in the Vostok III. Their daughter Yelena Adrianovna Nikolayeva was born on June 8, 1964, and was carefully studied by doctors fearful of her parents' space exposure, but no ill effects were found. After her flight, Tereshkova continued as an aerospace engineer in the space program; she also worked in Soviet politics, feminism, and culture. She was a Deputy to the Supreme Soviet between 1966 and 1989, and a People's Deputy from 1989 to 1991. Meanwhile, she was a member of the Supreme Soviet Presidium from 1974 to 1989. During the years from 1968 to 1987, she also served on the Soviet Women's Committee, becoming its head in 1977. Tereshkova headed the USSR's International Cultural and Friendship Union from 1987 to 1991, and subsequently chaired the Russian Association of International Cooperation.

Tereshkova summarized her views on women and science in her 1970 "Women in Space" article in the American journal *Impact of Science on Society*: "I believe a woman should always remain a woman and nothing feminine should be alien to her. At the same time I strongly feel that no work done by a woman in the field of science or culture or whatever, however vigorous or demanding, can enter into conflict with her ancient 'wonderful mission'—to love, to be loved—and with her craving for the bliss of motherhood. On the contrary, these two aspects of life can complement each other perfectly."

See also Spacecraft, manned

TERMINAL MORAINE • *see* MORAINES

TERMINATOR • *see* SOLAR ILLUMINATION: SEASONAL AND DIURNAL PATTERNS

TERRA SATELLITE AND EARTH OBSERVING SYSTEMS (EOS)

To facilitate new research and enhance existing data regarding the interaction of dynamic geophysical systems, NASA is in the process of developing a comprehensive Earth Observing System (EOS). A multi-component program, one of the unifying aims of EOS units is to measure the impact of human activities on Earth's geological and atmospheric processes.

The first component in the EOS array of **remote sensing** instruments is the Terra **satellite**, launched into a near-circular, sun-synchronous Earth orbit in December, 1999.

The development of NASA's EOS (a part of NASA's Earth Sciences Enterprise [ESE]) will result in a group of satellites—each designed for a specific research purpose—that together will feed data to the Earth Observing System Data and Information System (EOSDIS) network that will make the information available to research groups around the world. As of April 2002, three EOS satellites were established in Earth orbit. NASA eventually plans to expand the EOS program to include some 18 satellites.

Terra's instrumentation includes an Advanced Space borne Thermal Emission and Reflection Radiometer (ASTER), a Multi-angle Imaging Spectro-Radiometer (MISR), a **Clouds** and the Earth's Radiant Energy System (CERES) monitor, a Moderate-resolution Imaging Spectroradiometer (MODIS), and a Measurements of Pollution in the **Troposphere** (MOPIT) sensor.

ASTER is able to gather high-resolution Earth images ranging across the **electromagnetic spectrum** from visible to thermal infrared light. ASTER data will facilitate the development of maps based upon surface temperatures. MISR measures sunlight scattering from nine different angles. CERES, a two-component package, each of which scans radiation flux in different modes. MODIS provides wide-angle measurements in 36 spectral bands than will provide accurate estimates of phenomena such as cirrus cloud cover. At present, the extent of cirrus cloud cover is an important part of research efforts to determine whether they have a net cooling or warming effect on Earth's atmosphere. MODIS is capable of providing data enabling estimation of photosynthetic activity that in turn allows estimates of atmospheric **carbon dioxide** levels. MODIS is also capable of accurately measuring the extent of snow cover, or in the detection of heat from **volcanic eruptions** and fires. MOPITT utilizes gas **correlation spectroscopy** data in measuring radiation from Earth in three specific spectral bands. MOPITT data allows estimations of **carbon** monoxide and other gas (e.g., methane) levels in the troposphere.

In March 2002, the Terra satellite's Multi-angle Imaging Spectro Radiometer (MISR) instrument recorded data confirming the calving (breakaway) of a major iceberg measuring almost 200 mi^2 (5200 km^2) off the Antarctic **ice** shelf. The iceberg, designated B-22, broke away from the West Antarctic mainland into the Amundsen Sea. In an effort to estimate and evaluate the effects of **climate** warming, researchers are attempting to correlate—and/or determine the cause of—a recent reported increase in iceberg calvings during the last decade of the twentieth century. As of May 2002, data was insufficient to positively determine a causal relationship to potential human-induced **global warming**. In fact, part of the EOS mission is to develop a database that will enable researchers to determine whether such dramatic events as the breakaway of B-22 was a result of global warming or an expected occurrence that is a normal part of cyclic regional climatic variation.

See also Atmospheric composition and structure; Atmospheric pollution; Insolation and total solar irradiation; Scientific data management in Earth Sciences; Spectroscopy; Weather balloon; Weather satellite

TERTIARY PERIOD

In **geologic time**, the Tertiary Period (also sometimes referred to in terms of a Paleogene Period and a Neogene Period), represents the first geologic period in the **Cenozoic Era**. The Tertiary Period spans the time between roughly 65 million years ago (mya) and 2.6 mya. When referred to in terms of a Paleogene Period and a Neogene Period, the Paleogene Period extends from approximately 65 mya to 23 mya, and the Neogene Period from 23 mya to 2.6 mya.

The Tertiary Period contains five geologic epochs. The earliest epoch, the **Paleocene Epoch**, ranges from approximately 65 mya to 55 mya. The Paleocene Epoch is further subdivided into (from earliest to most recent) Danian and Thanetian stages. The second epoch, the **Eocene Epoch** ranges from approximately 55 mya to 34 mya. The Eocene Epoch is further subdivided into (from earliest to most recent) Ypresian, Lutetian, Bartonian, and Priabonian stages. The third epoch of the Tertiary Period, the **Oligocene Epoch** ranges from approximately 34 mya to 23 mya. The Oligocene Epoch is further subdivided into (from earliest to most recent) Rupelian and Chattian stages. Following the Oligocene Epoch, the **Miocene Epoch** ranges from approximately 23 mya to 5 mya. The Miocene Epoch is further subdivided into (from earliest to most recent) Aquitanian, Burdigalian, Langhian, Serravallian, Tortonian, and Messinian stages. The last epoch of the Tertiary Period is the **Pliocene Epoch**. The Pliocene Epoch is further subdivided into Zanclian and Placenzian stages.

The onset of the Tertiary Period is marked by the K-T boundary or K-T event—a large mass extinction. Most scientists argue that the K-T extinction resulted from—or was initiated by—a large asteroid impact in the oceanic basin near what is now the Yucatan Peninsula of Mexico. The remains of the **impact crater**, termed the Chicxulub crater, measures more than 105 mi (170 km) in diameter. The impact caused widespread firestorms, earthquakes, and tidal waves. Post-impact damage to Earth's ecosystem occurred as dust, soot, and debris from the collision occluded the atmosphere to sunlight. The global darkening was sufficient to inhibit photosynthesis. Widespread elimination of plant species caused repercussions throughout the food chain as starvation resulted in extinction of the largest life forms with the greatest metabolic energy needs (e.g., the dinosaurs).

At end of the prior **Cretaceous Period** and during the first half of the Tertiary Period (i.e. the Paleogene Period), Earth suffered a series of intense and large impacts. Large impact craters (greater than 25 mi or 40 km in diameter) include the Kara and Popigal craters in Russia, the Chesapeake crater in Maryland, and the Montagnais crater in Nova Scotia.

The last major impact crater with a diameter over 31 mi (50 km) struck Earth near what is now Kara-Kul, Tajikistan at end of the Tertiary Period and the start of the **Quaternary Period**.

The extinction of the dinosaurs and many other large species allowed the rise of mammals as the dominant land species during the Tertiary Period.

At the beginning of the Tertiary Period, North America and **Europe** were separated by a widening ocean basin spreading along a prominent mid-oceanic ridge. **North America** and **South America** were separated by a confluence of the future Pacific Ocean and Atlantic Ocean, and extensive flooding submerged much of what are now the eastern and middle portions of the United States. By the start of the Tertiary Period, **water** separated South America from **Africa**, and the Australian and Antarctic continents were clearly articulated. The Antarctic continent had begun a southward migration toward the south polar region. At the outset of the Tertiary Period, the Indian subcontinent remained far south of the Euro-Asiatic continent.

By the middle of the Tertiary Period (approximately 30 mya), the modern continental arrangement was easily recognizable. Although still separated by water, the Central American land bridge between North and South America began to reemerge. **Antarctica** assumed a polar position and extensive **ice** accumulation began on the continent. The Indian plate drove rapidly northward of the equator to close with the Asiatic plate. Although still separated by a shallow strait of water, the impending collision of the plates that would eventually form the Himalayan mountain chain had begun. The gap between North America and Europe continued to widen at a site of **sea-floor spreading** along a prominent mid-Atlantic ridge. By the middle of the Tertiary Period, the mid-Atlantic ridge was apparent in a large suture-like extension into the rapidly widening South Atlantic Ocean that separates South America from Africa.

By the end of the Tertiary Period, approximately 2.6 mya, Earth's continents assumed their modern configuration. The Pacific Ocean separated **Asia** and **Australia** from North America and South America, just as the Atlantic Ocean separated North and South America from Europe (Eurasian plate) and Africa. The Indian Ocean washed between Africa, India, Asia, and Australia. The Indian plate driving against and under the Eurasian plate uplifted both, causing rapid mountain building. As a result of the ongoing collision, ancient oceanic **crust** bearing marine **fossils** was uplifted into the Himalayan chain.

Climatic cooling increased at the end of the Tertiary Period, and modern **glaciation** patterns became well-established.

See also Archean; Cambrian Period; Dating methods; Devonian Period; Evolution, evidence of; Evolutionary mechanisms; Fossils and fossilization; Historical geology; Holocene Epoch; Jurassic Period; Mesozoic Era; Mississippian Period; Ordovician Period; Paleozoic Era; Pennsylvanian Period; Phanerozoic Eon; Pleistocene Epoch; Precambrian; Proterozoic Era; Silurian Period; Supercontinents; Triassic Period

Thermosphere

Based on the vertical **temperature** profile in the atmosphere, the thermosphere is the highest layer, located above the **mesosphere**. While in the **troposphere** and the mesosphere, the temperature decreases with altitude. In the **stratosphere** and thermosphere the temperature increases with height (called temperature inversion). It is separated from the mesosphere by the mesopause, in which the temperature does not change much vertically. Above the thermosphere, the upper limit of the atmosphere, the exosphere can be found blending into **space**. The upper part of the mesosphere and a big part of the thermosphere overlap with the **ionosphere**, which is a region defined on the basis of electric properties. The thermosphere and the exosphere together form the upper atmosphere.

Among the four atmospheric temperature-defined layers, the thermosphere is located highest above Earth's surface, beginning at about 57 mi (90 km) above Earth, and reaching into about 300 mi (500 km) height. The name of this layer, thermosphere, originates from the Greek *thermo*, meaning heat, because in this layer the temperature increases with altitude reaching temperatures higher than 1830°F (1000°C). In the thermosphere, **oxygen** molecules absorb the energy from the Sun's rays, which results in the warming of the air. Because there are relatively few molecules and atoms in the thermosphere, even absorbing small amounts of **solar energy** can significantly increase the air temperature, making the thermosphere the hottest layer in the atmosphere. Above 124 mi (200 km), the temperature becomes independent of altitude.

Because the thermosphere and exosphere belong to the upper atmosphere, the density of the air in addition to the **atmospheric pressure** is greatly reduced when compared to the atmosphere at Earth's surface. At these high altitudes, the atmospheric gases tend to sort into layers according to their molecular mass, and chemical reactions happen much faster here than near the surface of the earth.

See also Atmospheric composition and structure

Thomson, Lord Kelvin, William (1824-1907)

Scottish physicist

William Thomson, known to history as Lord Kelvin, was granted the first scientific peerage by Queen Victoria in 1892 for his unique consulting work that made possible the installation of the transatlantic cable linking the telegraph systems of America and England. The peerage was created especially for him, taking the name from the Kelvin River near Glasgow, Scotland. Thus, when Thomson's proposal for an absolute scale measuring heat was widely accepted, it was given the name Kelvin.

Thomson was a child prodigy who grew up in the environment of academia and became a professor at a young age. Thomson was enthusiastic and dramatic in his teaching style. Known as an expert on the dynamics of heat, he was also noted for having a wide range of interests in the sciences, particularly **electricity and magnetism**. As a science and technology authority, Thomson was willing to take on the unpopular side of a controversy. This characteristic almost ruined his career when he attempted to establish the age of Earth, accepting the least popular premise concerning the origins of the planet. However, his gamble in supporting James Prescott Joule rewarded Thomson with a lifelong friendship and a proficient research collaboration.

Thomson was born in Belfast, Northern Ireland, in 1807. While he was still young, the family moved to Glasgow where his father received a position as a mathematics professor. His mother died when he was six years old, and young William became accustomed to attending his father's lectures. After attending his father's classes for many years, Thomson surprised many by actively participating. By age ten, he was ready for college. He was able to keep up with his much older classmates and he even surpassed them by writing his first scholarly treatise at the age of fourteen. Many were impressed with this work and a respected professor, representing Thomson, presented it in lecture to the Royal Society of Edinburgh. Thomson and his professors decided he should not present the paper himself because his young age would have undermined the respect his paper deserved.

After enrolling in Peterhouse College at age seventeen, Thomson achieved honors in a difficult mathematics program by age 21. Although his father had hoped that William would follow him into mathematics, Thomson was strongly interested in natural philosophy (as science was called in his time). He was very much interested in Fourier and other newly emerging theories of heat, having already achieved valuable experience at Regnault's Laboratory in Paris. Thus, when offered a position as professor of natural philosophy at the University of Glasgow in 1846, Thomson accepted. He remained a professor at Glasgow for the next fifty-eight years.

Thomson's first achievement as a professor was an attempt to establish the age of Earth based on mathematical models representing the difference in **temperature** between Earth and the **Sun**. This research was almost a fiasco, upsetting well-respected geologists of the day. Although Thompson's basic geological premise was faulty, his proficient models and accurate calculations were impressive. More importantly, the principles of thermodynamics he employed were well thought out. Fortunately, because many paid attention to the strengths of his study, Thomson advanced his expertise on the properties and dynamics of heat.

At Glasgow, Thomson's students were fascinated with his youthfulness and his energetic lecturing style. He built his own laboratory for his students by converting an old wine cellar belonging to a more established member of the faculty.

In 1847, at a conference of the British association, Thomson listened as Joule presented findings on the effects of heat on gases. Joule's presentation was not convincing to most of the assembled scientists and he would have been ignored if Thomson had not come to his defense. Thomson not only supported Joule in this conference, but also agreed to collaborate with him on new research. From this work came the Joule-Thomson effect of heat conservation, presented in a paper in 1851. This concept—that gas allowed to expand in a vacuum will reduce its heat—became the foundation of an early refrigeration industry.

While studying the effects of heat on gases, Thomson recognized that the linear relationship between the heat and mass in gases was awkwardly graphed using the traditional Celsius Scale. In 1848, he proposed an absolute scale using the same range as the Celsius scale but with zero set at the point where there is virtually no movement among the molecules (–273.18°C). This can be considered the point of absolutely no heat. In 1851, Thomson further elaborated how this new scale would illustrate the principles of thermodynamics more clearly in experiments. Scientists in other fields also recognized how the absolute (later called Kelvin) scale could be very useful.

Thomson admitted that he learned much from Joule. However, for many years he tried to resolve the theoretical differences between the dynamic theories of heat that Joule was relying on and the principles he found to be true in the theories of Sadi Carnot. In his dissertation of 1851, he thoroughly presented the strengths of Carnot's theories reconciled to the dynamic theory. From this paper, the second law of thermodynamics was established. This law stated that heat transferred from hotter matter to colder matter has to release mechanical work. Furthermore, heat taken from colder matter to hotter matter requires the input of mechanical work. However, the second law of thermodynamics was not fully credited to Thomson because it was contemporaneously developed (separate from Thomson's work) by other researchers.

Thomson also excelled in other areas of science. His improvements in the conductivity of cable and in galvanometers were patented inventions without which the practicality of the transatlantic cable would not have been possible. Among Thomson's many honors, he was elected president of the Royal Society and held the post from 1890 to 1894. He continued to invent and study in his later life. After a short period of retirement from the university, he returned as a graduate student. Although Thomson maintained a sharp mind until his death, he was adamant against the changes in old paradigms that new discoveries brought at the end of the nineteenth century. For example, Thomson once remarked that air travel other than ballooning would be a scientific impossibility. Thomson died in 1907 and was buried in Westminster Abbey, leaving no children to inherit his title.

See also Physics; Temperature and temperature scales

THUNDER

Thunder is the noise caused by **lightning** in a thunderstorm, when the release of heat energy results in audible shock waves in the air.

A thunderstorm is a storm that produces lightning and thunder, and occurs in cumulonimbus **clouds**. Cumulonimbus clouds are large, tall clouds with very strong updrafts that transport **water** high into the atmosphere. Thunderstorms can also produce flash **floods**, hail, strong winds, and even tornadoes. At any time on Earth, about 2,000 thunderstorms are taking place, from mild rainstorms to very damaging hailstorms with high winds. In general, the higher the storm clouds, the more violent the resulting storm will follow. Under certain conditions, isolated thunderstorms can even merge to form large convective complexes with increasing power and damage capabilities. Thunderstorms and lightning can cause not only billions of dollars of damage every year, but also result in loss of human and animal life, since about 100 people die per year in the United States from causes associated with lightning.

Lightning is a large electrical discharge produced by thunderstorms as a huge spark, which can heat the air as much as several times hotter than the **temperature** of the surface of the **Sun** (about 54,000°F or 30,000°C). This heated air causes expansion in the air when the electrical charge of lightning passes through it, and forces the air molecules to expand. As they expand, the air molecules require more space and they bump into cooler air, creating an airwave, the sound of thunder. It travels in all directions from the lightening at the speed of the sound (330 m/s); therefore, it takes the thunder about five seconds to travel each mile, or about three seconds to travel one kilometer. Because light travels faster than sound, the lightning is always seen first, before the thunder is heard. Measuring the time between the lightning and the thunder can give an approximate estimate of how far the observer is from the thunderstorm.

Depending on the location of the observer or the type of lightning, thunder can produce many different sounds. When lightning strikes nearby, the resulting thunder is usually interpreted as a short and loud bang, whereas thunder is interpreted as a long, low rumble when it is heard from far away. Thunder can also sound like a large crack, or a clap of thunder followed by rumbling, or a thunder roll. Lightning always produces thunder, and without lightning, there is no thunder in a thunderstorm. Sometimes, when the lightning is too far away for the sound waves to reach the observer, lightning can be seen but no thunder can be heard. This is known commonly as heat lightning, and it happens because the dissipating sound of thunder rarely travels farther than ten miles, especially in lowlands or at sea.

See also Clouds and cloud types; Weather forecasting

TIDES

Tides are deformations in the shape of a body caused by the gravitational force of one or more other bodies. At least in theory, any two bodies in the Universe exert such a force on each other, although obvious tidal effects are generally too small to observe. By far the most important examples of tidal forces as far as humans are concerned are ocean tides that occur on Earth as a result of the **Moon** and Sun's gravitational attraction.

The side of Earth facing the Moon, due to the Moon's proximity, experiences a larger gravitational pull, or force, than other areas. This force causes ocean **water**, since it is able to flow, to form a slight bulge, making the water in that **area** slightly deeper. At the same time, another bulge forms on the opposing side of the Earth. This second bulge, which is perhaps a bit harder to understand, forms due to centrifugal force. Contrary to popular belief, the Moon does not revolve around the Earth, but rather the Earth and Moon revolve about a common point that is within the Earth, but nowhere near its center (2880 miles or 4640 km away). When you twirl a ball above your head at the end of a piece of string, the ball pulls against the string. This pull is known as centrifugal force.

When the Earth-Moon system revolves around its common axis, the side of Earth that is farthest from the Moon experiences a centrifugal force, like a ball spinning at the end of a string. This force causes a second tidal bulge to form, which is the same size as the first. The result is that two lunar tidal bulges exist on Earth at all times—one on the side of the Earth facing the Moon and another directly opposite to it. These bulges account for the phenomenon known as high tide.

The formation of these two high tide bulges causes a belt of low water to form at 90° to the high tide bulges. This belt, which completely encircles the Earth, produces the phenomenon known as low tide.

As Earth rotates on its axis, land areas slide underneath the bulges, forcing the **oceans** up over some coastlines and beneath the low tide belt, forcing water out away from other coastlines. In a sense, as Earth rotates on its axis, the high tide bulges and the low tide belt remains stationary and the continents and ocean basins move beneath them. As a result, most coastal areas experience two high tides and two low tides each day.

In addition to the lunar bulges, the **Sun** forms its own tidal bulges, one due to gravitational force and the other due to centrifugal force. However, due to the Sun's much greater distance from the Earth, its tidal effect is approximately one half that of the Moon.

When the Moon and Sun are in line with each other (new Moon and full Moon), their gravitational, or tidal forces, combine to produce a maximum pull. The tides produced in such cases are known as spring tides. The spring high tide produces the highest high tide and the spring low tide produces the lowest low tide of the fortnight. This is the same as saying the spring tides have the greatest tidal range, which is the vertical difference between high tide and low tide.

When the Moon and Sun are at right angles to each other (first and third quarter Moon), the two forces act in opposition to each other to produce a minimum pull on the oceans. The tides in this case are known as neap tides. The neap high tide produces the lowest high tide and the neap low tide produces the highest low tide, or the smallest tidal range, of the fortnight.

It is now possible to write very precise mathematical equations that describe the gravitational effects of the Moon and the Sun. In theory, it should be possible to make very precise predictions of the time, size, and occurrence of tides. In fact, however, such predictions are not possible because a large number of factors contribute to the height of the oceans at high and low tide at a particular location. Primary among these is that the shape of ocean basins is so irregular that water does not behave in the "ideal" way that mathematical equations would predict. However, a number of other variables also com-

Boats beached at Low Tide, La Flotte, Ile de Re, France. ***Photograph by Nik Wheeler. Corbis Corporation (Bellevue). Reproduced by permission.***

plicate the situation. These include variations in the Earth's axial **rotation**, and variations in Earth-Moon-Sun positioning, including variations in orbital distance and inclination.

Scientists continue to improve their predictions of tidal variations using mathematical models based on the equilibrium theory of tides. However, for the present, estimates of tidal behavior are still based on previous tidal observations, continuous monitoring of coastal water levels, and astronomical tables. This more practical approach is referred to as the dynamical theory of tides, which is based on observation rather than mathematical equations.

The accumulated information about tidal patterns in various parts of the world is used to produce tide tables. Tide tables are constructed by looking back over past records to find out for any given location the times at which tides have occurred for many years in the past and the height to which those tides have reached at maximum and minimum levels. These past records are then used to predict the most likely times and heights to be expected for tides at various times in the future for the same locations. Because of differences in ocean bottoms, coastline, and other factors, unique tide tables must be constructed for each specific coastline every place in the world. They can then be used by fishermen, those on ocean liners, and others who need to know about tidal actions.

In most places, tides are semidiurnal, that is, there are two tidal cycles (high and low tides) each day. In other words, during a typical day, the tides reach their highest point along the shore and their lowest point twice each day. The high water level reached during one of the high tide stages is usually greater than the other high point, and the low water level reached during one of the low tide stages is usually less than the other low tide point. This consistent difference is called the diurnal inequality of the tides.

In a few locations, tides occur only once a day, with a single high tide stage and a single low tide stage. These are known as diurnal tides. In both diurnal and semidiurnal settings, when the tide is rising, it is called the flood tide. When the tide is falling, it is the ebb tide. The point when the water reaches its highest point at high tide, or its lowest point at low tide, is called the slack tide, since the water level is static, neither rising nor falling, at least for a short time.

As the Moon revolves around the Earth, the Earth also rotates on its axis. Consequently, the Earth must rotate on its axis for 24 hours, 50 minutes, known as a lunar day, to return to the same position relative to the Moon above. The additional 50 minutes allows Earth to "catch up" to the Moon, so to speak. In other words, if the Moon was directly overhead at Boston, Massachusetts, at noon yesterday, it will again be above Boston at 12:50 PM today. As a result, on a coast with diurnal tides, each day the high tide (or low tide) will occur 50 minutes later than the day before. Whereas, on a semidiurnal coast, each high tide (or low tide) will occur 12 hours, 25 minutes later than the previous high one.

The movement of ocean water as a result of tidal action is known as a tidal current. In open water, tidal currents are relatively weak and tend to change direction slowly and regularly throughout the day. They form, therefore, a kind of rotary current that sweeps around the ocean like the minute hand on a clock. Closer to land, however, tidal currents tend to change direction rather quickly, flowing toward land during high tide and away from land during low tide. In many cases, this onshore and offshore tidal current flows up the mouth of a river or some other narrow opening. The tidal current may then attain velocities as great as 9 mi (15 km) an hour with crests as high as 10 ft (3 m) or more.

Most tides attain less than 10 ft in size; 3–10 ft (1–3 m) is common. In some locations, however, the tides may be much greater. These locations are characterized by ocean bottoms that act as funnels through which ocean waters rush upward towards or downward away from the shore at very rapid speeds. In the Bay of Fundy, between Nova Scotia and New Brunswick, for example, the difference between high and low tides, the tidal range, may be as great as 46 ft (14 m). In comparison, some large bodies of water, such as the Mediterranean, Baltic, and Caribbean **Seas**, have areas with tides of less than 1 ft (0.3 m). All coastal locations (as well as very large **lakes**) experience some variation in tidal range during a fortnight due to the affects of neap versus spring tides.

See also Celestial sphere: The apparent movements of the Sun, Moon, planets, and stars; Gravity and the gravitational field; Marine transgression and marine regression

Till

Till is the general term for any sediments that were deposited solely by glacial **ice**. Till is distinguished from other glacial deposits formed by forces other than ice, such as glaciofluvial (or glacial melt **water**) deposits. A similar term is moraine, but it connotes more specific depositional mechanisms and spatial relationships to the glacier than does till.

Tills are produced by virtue of the formation, advance, and retreat of **glaciers**. The immense weight of an advancing glacier causes it to rip up **rock** and **soil** and incorporate them into the ice. These sediments then migrate forward as the glacier creeps downhill. When sediments reach the leading edge of the glacier where it is constantly **melting**, they are turned out as till.

This depositional mechanism results in tills being characterized by a physical heterogeneity; the sediments are unsorted, random in size, and may consist of a large range in particle size—from tiny clays to huge boulders. Tills are also generally unstratified, showing no sedimentary layering. The sediments in till exhibit a variable degree of rounding to the sediments, although some rounding is almost always observed. Despite their random origin, tills sometimes exhibit some degree of consistency in composition, allowing them to be described by the dominant size sediment they contain, such as gravelly or sandy tills.

Although tills may contain rocks from anywhere the glacier came in contact with, and sometimes do show evidence of sources hundreds of miles away, most tills are locally derived. They usually consist of rocks and soils picked up by the glacier within a few miles of where they were deposited. As a result, tills often provide evidence of the local **bedrock** and aid in determining the **geology** of areas that are now covered with glacial deposits.

See also Glacial landforms; Glaciation

Time zones

Earth rotates on its **polar axis** once every 23.9345 hrs. As an oblate sphere measuring a circumferential 360°, Earth rotates through almost 15 angular degrees per hour.

Local noon occurs when, on the hypothetical celestial sphere, the **Sun** is at the highest point during its daily skyward arch from east to west. When the Sun is at its zenith on the celestial meridian, this is termed local noon. In the extreme, every line of longitude, or fraction thereof, has a different local noon. In practice, however, because of Earth's angular **rotation** rate, it is more convenient to create a system of 24 time zones—each spanning 15 angular degrees. The central line of longitude in these zones establishes the local noon for individual time zones.

Earth's lines of longitude (meridians) are great circles that meet at the north and south polar axis. They are referenced by an east or west displacement from the prime meridian. Accordingly, lines of longitude range from 0° E to 180° E and 0° W to 180° W. Degrees are further divided into arcminutes and arcseconds.

The prime meridian runs through Greenwich, England and the line of longitude displaced 180° E and 180° W from the prime meridian is termed the international dateline. The international dateline generally runs through sparsely islanded areas of the Pacific Ocean.

With regard to the solar meridian, the Sun's location (and reference to local noon) is described in terms of being ante meridian (A.M.) or post meridian (P.M.).

Standard meridians occur every 15° of longitudinal displacement from the prime meridian (e.g., 15° W, 30° W, 45° W, etc.) The standard meridians also establish the local noon for the time zone and, therefore, each time zone is defined as being 7.5° longitudinal displacement both west and east of the standard meridian. Accordingly, dividing the standard meridian by 15 yields the time correction for that time zone. For example, the standard meridian of 90° W runs near both Chicago and New Orleans. These sites are in the Central Time

Zone of the United States (CST; Central Standard Time). To obtain the proper correction from Greenwich Mean Time (GMT)—also termed Universal Time (UT or UTC)—a division of 90° by 15° means that CST is six hours behind GMT. Accordingly, when it is noon 12:00 HRS GMT in London, it is 0600 HRS (6 A.M.) CST in Chicago or New Orleans. Because of Earth's rotation, displacements west are further designated with a negative sign. Accordingly, CST = GMT – 6 hrs.

Additional North American meridians and time zones include standard meridian 60° W for Atlantic Standard Time (e.g., as for Puerto Rico); standard meridian 75° W for Eastern Standard Time (EST); standard meridian 105° W for Mountain Standard Time (MST); standard meridian 120° W for Pacific Standard Time (PST); standard meridian 135° W for Yukon Standard Time (YST); standard meridian 150° W for Hawaii-Alaska Standard Time (HAST) and standard meridian 165° W for Bering Standard time.

Movement east of the prime meridian results adding time to GMT. For example, Rome, Italy, at a **latitude and longitude** of 42° N, 12° E, is 3° W of the 15° E standard meridian. Because a time zone ranges 7.5° east and west of a standard meridian, the applicable standard meridian for Rome is the 15° E standard meridian. Accordingly, the time differential between Rome and London (GMT) is 15°/15° = 1–interpreted as +1 hour time difference. Therefore, when it is noon in London, it is 1300 HRS, or 1 P.M., in Rome.

In reality, there are many local deviations of the time zone boundaries based upon geopolitical considerations (e.g., state and national boundaries). Actual time corrections are also influenced by whether or not a particular locality adopts daylight saving time shifts (usually one hour) to save energy by shifting daylight hours to clock hours more conductive to typical human work patterns. The United States shifts to Daylight Saving Time between April and October each year. Accordingly, time zone designations are changed from, for example, CST to CDST (Central Daylight Saving Time).

See also Cartography; Celestial sphere: The apparent movements of the Sun, Moon, planets, and stars; Solar illumination: Seasonal and diurnal patterns

TOMBAUGH, CLYDE W. (1906-1997)

American astronomer

Clyde W. Tombaugh, an astronomer and master **telescope** maker, spent much of his career performing a painstaking photographic survey of the heavens from Lowell Observatory in Flagstaff, Arizona. This led to the discovery of Pluto (1930), the ninth planet in the **solar system**. Although Tombaugh is best known for this early triumph, he went on to make other contributions, including his work on the geography of Mars and studies of the distribution of galaxies. Tombaugh also made valuable refinements to missile-tracking technology during a nine-year stint at the U.S. Army's White Sands Proving Grounds in New Mexico.

Clyde William Tombaugh, the eldest of six children, was born to Muron Tombaugh, a farmer, and Adella Chritton Tombaugh. He spent most of his childhood on a farm near Streator, Illinois. In 1922, the family relocated to a farm in western Kansas. Tombaugh glimpsed his first telescopic view of the heavens through his uncle Leon's 3-in (7.6-cm) refractor, a kind of telescope that uses a lens to gather faint light from stars and planets. In 1925, inspired by an article in *Popular Astronomy*, Tombaugh bought materials to grind an 8-in (20.3-cm) light-collecting mirror for a reflecting telescope. He ground that first mirror by hand, using a fence post on the farm as a grinding stand.

The finished instrument, a 7-ft (2.1-m) rectangular wooden box, was equipped with wooden setting circles for aligning it to objects of interest in the sky. Tombaugh had not ground the mirror very accurately, and thus the telescope was unsuitable for the planetary observing he had in mind. However, it launched a lifetime of building, improving, and maintaining telescopes, tasks at which Tombaugh excelled. Tombaugh biographer and amateur astronomer David H. Levy estimated that Tombaugh ground some 36 telescope mirrors and lenses in his career. He continued to use a few of his early telescopes for decades after he first constructed them (for example, his 9-in [23-cm] reflector, whose mechanical mounting included parts from a 1910 Buick).

Tombaugh's 9-in reflector, which he completed in 1928, led to a career as a professional observer as well as to sharper views of the planets and stars. After a 1928 hailstorm wiped out the Tombaughs' wheat crop and foiled Clyde's plans for college, the young observer turned his new telescope to Jupiter and Mars. Subsequently, he sent his best drawings of these planets to Lowell Observatory, which had been founded in the late nineteenth century by famed Mars watcher Percival Lowell.

Hoping only for constructive criticism of his drawings, Tombaugh instead received a job offer from the astronomers at Lowell. He accepted, and in January 1929 began his work on the search for the predicted ninth planet beyond the orbit of Neptune. Working full time as a professional observer (although lacking any formal education in **astronomy**), Tombaugh used Lowell's 13-in (33-cm) telescope to systematically photograph the sky. He then used a special instrument, called a blink comparator, to examine the plates for telltale signs of moving bodies beyond the orbit of Earth. A blink comparator, or blink microscope, rapidly alternates—up to 10 times per second—two photographic images, taken at different times, of the same field or **area** of the sky. Seen through a magnifying lens, moving bodies will appear to jump back and forth or "blink" as the images are switched.

Using his knowledge of orbital mechanics and his sharp observer's eye, Tombaugh was able to discern **asteroids** and **comets** from possible planets; a third "check" plate was then taken to confirm or rule out the existence of these suspected planets. On February 18, 1930, after 10 months of concentrated, painstaking work, Tombaugh zeroed in on Pluto, fulfilling a search begun by Percival Lowell in 1905. The discovery of Pluto secured the 24-year-old Tombaugh's reputation and his place in the history of astronomy, and he remained with the survey until 1943.

After his discovery, Tombaugh took some time off to obtain his formal education in astronomy. He left for the University of Kansas in the fall of 1932, returning to Lowell each summer to resume his observing duties. At college, he met Patricia Irene Edson, a philosophy major. They married in 1934, and subsequently had two children. Tombaugh paused only once more for formal education in science, taking his master's degree in 1939 at the University of Kansas. For his thesis work, he restored the university's 27-in (68.6-cm) reflecting telescope to full operational status and studied its observing capabilities.

In 1943, Tombaugh taught **physics** at Arizona State Teachers College in Flagstaff; that same year, the U.S. Navy asked him to teach navigation, also at Arizona State. In what little spare time remained, Tombaugh struggled to continue the planet survey. The following year, he taught astronomy and the history of astronomy at the University of California in Los Angeles. Tombaugh's stint on the planet survey ceased abruptly in 1946. Citing financial constraints, observatory director Vesto M. Slipher asked Tombaugh to seek other employment.

Tombaugh's contribution to the "planetary patrol" at Lowell proved enormous. From 1929 to 1945, he cataloged many thousands of celestial objects, including 29,548 galaxies, 3,969 asteroids (775 of them previously unreported), two previously undiscovered comets, one nova, and, of course, the planet Pluto. However, as Tombaugh pointed out to biographer David Levy, tiny Pluto cast a long and sometimes burdensome shadow over the rest of his career, obscuring subsequent astronomical work. For instance, in 1937, Tombaugh discovered a dense cluster of 1,800 galaxies, which he called the "Great Perseus-Andromeda Stratum of Extra-Galactic Nebula." This suggested to Tombaugh that the distribution of galaxies in the universe may not be as random and irregular as some astronomers believed at the time.

Tombaugh was also an accomplished observer of Mars. He predicted in 1950 that the red planet, being so close to the asteroid belt, would have impact craters like those on the **moon**. These craters are not easily visible from Earth because Mars always shows its face to astronomers fully or nearly fully lighted, masking the craters' fine lines. Images of the Martian surface captured in the 1960s by the *Mariner IV* **space probe** confirmed Tombaugh's prediction.

In 1946, Tombaugh began a relatively brief career as a civilian employee of the U.S. Army, working as an optical physicist and astronomer at White Sands Proving Grounds near Las Cruces, New Mexico, where the army was developing launching facilities for captured German V-2 missiles. Tombaugh witnessed 50 launchings of the 46-ft (14-m) rockets and documented their performance in flight using a variety of tracking telescopes. Armed with his observing skills and intimate knowledge of telescope optics, Tombaugh greatly increased the quality of missile tracking at White Sands, host to a number of important postwar missile-development programs.

Tombaugh resumed serious planetary observing in 1955, when he accepted a teaching and research position at New Mexico State University in Las Cruces. There, he taught astronomy, led planetary observation programs, and participated in the care and construction of new telescopes. From 1953 to 1958, Tombaugh directed a major search for small, as-yet-undetected objects near the Earth—either asteroids or tiny natural satellites—that might pose a threat to future spacecraft. He and colleagues developed sensitive telescopic tracking equipment and used it to scan the skies from a high-altitude site in Quito, Ecuador. The survey turned up no evidence of hazardous objects near Earth, and Tombaugh issued a closing report on the program the year after the Soviet Union launched *Sputnik* (1957), the first artificial **satellite**.

Upon his retirement in 1973, Tombaugh maintained his links to New Mexico State University, often attending lunches and colloquia in the astronomy department that he helped to found. He also remained active in the local astronomical society and continued to observe with his homemade telescopes. Indeed, asked by the Smithsonian Institution in Washington, D.C., to relinquish his 9-in reflector to its historical collections, Tombaugh refused, explaining to *Smithsonian* magazine, "I'm not through using it yet!" He died in 1997 at his home in Las Cruces, New Mexico.

TOPOGRAPHY AND TOPOGRAPHIC MAPS

Topography is the physical shape of the land, particularly as it relates to elevation. Topographic maps are two-dimensional graphical representations of the three-dimensional topography that also provide a detailed and accurate inventory of what exists on the land surface, such as geographic and cultural features.

Topographic maps are distinguished from other maps in their representation of elevation as contour lines. Contour lines are drawn to match the shape of physical features and successive contour lines represent ascending or descending elevations. This allows a user to quickly discern the shape of any landform, determine its elevation, and estimate the rate of elevation changes. For example, a round hill would appear as a series of concentric closed loops that become successively smaller with increased elevation. The closer the contour lines are to one another, the steeper the slope.

In addition to contoured elevations, topographic maps show many other features of the land, including names of natural features such as mountains, valleys, plains, **lakes**, and **rivers**. They identify the amount of vegetative cover and include constructed features like minor and major roads, transmission lines, and buildings. Topographic maps also show political boundaries, survey markers, and different map coordinate systems such as **latitude and longitude**.

The value of topographic maps is in their accuracy and consistency. Topographic maps are based on a rigorous geodesic base, which defines the shape of Earth over a given land **area**. This ensures that all included features will be shown in the exact position. All features on the maps conform to a consistent set of map symbols, allowing comparison of topographic maps from anywhere in the country.

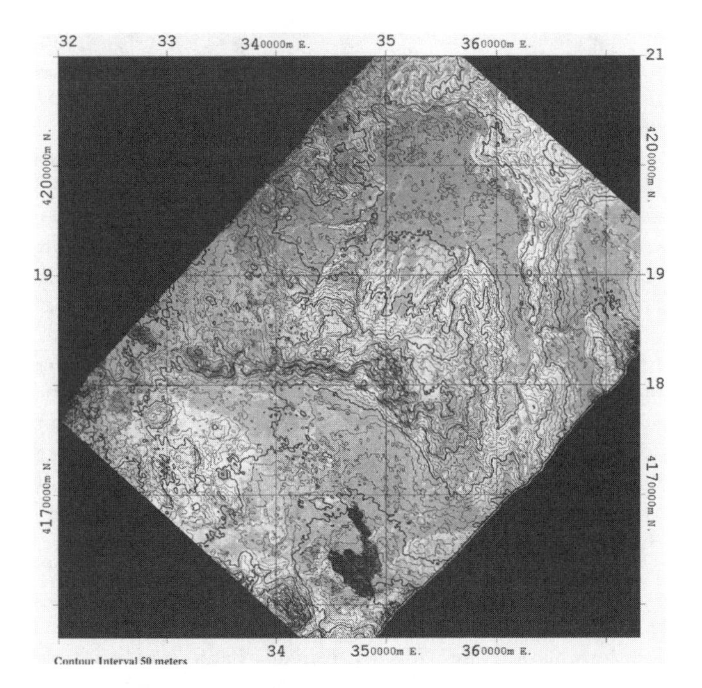

Topographic map made with Spaceborne Imaging Radar-C/X-Band Synthetic Aperture Radar imaging from Space Shuttle Endeavour, near Long Valley, California. *© Corbis. Reproduced by permission.*

The high accuracy and range of information of topographic maps makes them useful to professional and recreational map users alike. Topographic maps are used for outdoor activities like hiking, camping, and fishing and in professional fields such as engineering, energy exploration, natural resource conservation, environmental management, public works design, commercial and residential planning.

To meet the needs of various users, the United States Geological Survey produces topographic maps at different scales for the entire United States. The scale is the ratio of a unit of measurement on the ground to that on a map. For example, if one inch on a topographic map equals one mile (or 63,360 inches) on the ground, the scale of the map is 1:63,360. The most common scale for topographic maps is 1:24,000, where one map inch equals 2000 feet on the ground. This size map is called a 7.5-minute quadrangle because it covers 7.5 minutes of **latitude** by 7.5 minutes of **longitude**. Maps at this scale are very detailed. A map with a larger ratio, such as 1:100,000 will cover more area but show less detail.

See also Cartography; Relief

TORINO SCALE

Advanced by Massachusetts Institute of Technology Professor Richard P. Binzel in 1995, the Torino scale is a revision of the Near-Earth Object Hazard Index. In 1999, the International Conference on Near-Earth objects adopted the scale at a meeting in *Turino* (Turin), Italy (from which the name of the scale is derived). The Torino scale is used to portray the threat to Earth of an impact with a particular comet or asteroid. The measurement scale is based upon agreement between scholars as a means to categorize potential hazards.

When a new comet or asteroid is initially tracked, an extrapolation of its projected orbital path is compared to predicted Earth orbital positions. The Torino scale assigns categories to the closeness with which an object will approach or cross Earth orbit. Because initial estimates can be greatly altered by refined data regarding the track of an asteroid or comet, it is possible that a particular asteroid or comet could be upgraded or downgraded with regard to the threat it poses Earth. In addition, a different scale designation can be made for each successive orbital encounter over a number of years or decades. Data is most accurate as related to encounters in the near-term because various gravitational forces and encounters with other celestial objects can alter the course of **asteroids** or **comets**.

The Torino scale is based upon a zero to 10 numbering system wherein a zero designates a statistically negligible threat of collision with Earth. At the other extreme, a numerical designation of ten would indicate certain impact. In addition to being based upon the probability of impact, scale numbers also incorporate a potential "damage" value. For example, a very small object with little chance of surviving a fiery entry into Earth's atmosphere will still be assigned a very low number (zero for very small objects) even if an impact was certain. At the other extreme, the designation 10 carries the ominous distinction of being reserved for a certain impact of cataclysmic proportions.

The size of an object is important because the force (kinetic energy) that it would carry in a collision with Earth is related to its mass and velocity. Like nuclear explosions, estimates of the energy of collision are given in units of megatons (MT).

The Torino scale also assigns colors to the potential hazard assessment. A "white" label means that the asteroid or comet poses no threat (i.e, will miss or not survive entry into the Earth's atmosphere). Green events designate orbital crossings with a small chance of collision. Yellow events designate more potential orbital crossings than average. A yellow designation would focus intense scientific scrutiny upon the track of the asteroid or comet. Orange events are "threatening" crossings or other encounters with asteroids or comets that have a potential to cause severe destruction. The designation is reserved for objects with a significantly higher risk of impact. Red events or collisions are certain and globally devastating.

Because risk assessments are difficult to quantify, another scale, the Palermo Technical Scale, is often used by astronomers to complement the Torino scale. The Palermo scale offers a more mathematical calculation utilizing the variables of probability of impact and energy of collision.

As of May 2002, with approximately 25% of Near Earth objects identified, no object rating more than a "1" on the Torino scale has yet been detected. For example, during February 2002, an asteroid designated 2002 CU11 was classified as a "1" on the Torino scale (a "green" code). Extrapolations of the orbital dynamics of the asteroid and Earth indicated a low probability (approximately 1 in 9000) of a potential collision in 2049.

As of May 2002, additional information regarding the Torino scale and up-to-date information on identified NEO (Near Earth objects) can be found at the Near Earth Objects Dynamic Site (NEODyS) (<http://newton.dm.unipi.it/neodys/>).

See also Astronomy; Barringer meteor crater; Catastrophism; Gravity and the gravitational field; Hubble Space Telescope (HST); Solar system, Space and planetary geology

Tornado

A tornado is a rapidly spinning column of air formed in severe thunderstorms. The rotating column, or vortex, forms inside the storm cloud and grows downward until it touches the ground. Although a tornado is not as large as its parent thunderstorm, it is capable of extreme damage because it packs very high **wind** speeds into a compact **area**. Tornadoes have been known to shatter buildings, drive straws through solid wood, lift locomotives from their tracks, and pull the **water** out of small streams. Due to a combination of geography and **meteorology**, the United States experiences most of the world's tornadoes. An average of 800 tornadoes strike the United States each year. Based on statistics kept since 1953, Texas, Oklahoma, and Kansas are the top three tornado states. Tornadoes are responsible for about 80 deaths, 1500 injuries, and many millions of dollars in property damage annually. While it is still impossible to predict exactly when and where tornadoes will strike, progress has been made in predicting tornado development and detecting tornadoes with Doppler radar.

Most tornadoes form in the Northern Hemisphere during the months of March through June. These are months when conditions are right for the development of severe thunderstorms. To understand why tornadoes form, consider the formation and growth of a thunderstorm. Thunderstorms are most likely to develop when the atmosphere is unstable, that is, when atmospheric **temperature** drops rapidly with height. Under unstable conditions, air near the surface that begins rising will expand and cool, but remains warmer (and less dense) than its surroundings. The rising air acts like a hot air balloon; because it is less dense than the surrounding air, it continues to rise. At some point, the rising air cools to the **dew point** where the water vapor in the air condenses to form liquid water droplets. The rising column of air is now a visible cloud. If the rising air, or updraft, is sustained long enough, water droplets will begin to fall out of the rising air column, making it a rain cloud.

This cloud will become a severe storm capable of producing tornadoes only under certain circumstances. Severe storms are often associated with a very unstable atmosphere and moving low-pressure systems that bring cold air into contact with warmer, more humid air masses. Such **weather** situations commonly occur in the eastern and Midwestern United States during the spring and summer months. Large-scale weather systems often sweep moist warm air from the **Gulf of Mexico** over these regions in a layer 1.2–1.9 mi (2–3 km) deep. At the same time, winds aloft (above about 2.5 mi [4 km] in altitude) from the southwest bring cool dry air over the region. Cool air overlying humid air creates very unstable atmospheric conditions and sets the stage for the growth of strong thunderstorms.

The warm surface air is separated from colder air lying farther north by a fairly sharp temperature boundary called a front. A low-pressure center near Earth's surface causes the cold air to advance into the warmer air. The edge of the advancing cold air, called a cold front, forces the warmer air ahead of the front to rise and cool. Because the atmosphere is unstable, the displaced air keeps rising and a cloud quickly forms. Rain that begins to fall from the cloud causes downdrafts (sinking air) in the rear of the cloud. Meanwhile the advancing edge of the storm has strong updrafts and humid air is pulled into the storm. The water vapor in this air condenses to form more water droplets as it rises and cools. When water vapor condenses, it releases latent heat. This warms the air and forces it to rise more vigorously, strengthening the storm.

The exact mechanism of tornado formation inside severe thunderstorms is still a matter of dispute, but it appears that tornadoes grow in a similar fashion to the small vortices that form in draining bathtubs. Tornadoes appear to be upside down versions of this phenomenon. As updrafts in a severe thunderstorm cloud get stronger, more air is pulled into the base of the cloud to replace the rising air. Some of this air may be rotating slightly since the air around the base of a thunderstorm usually contains some **rotation**, or vorticity. As the air converges into a smaller area, it begins to rotate faster due to a law of **physics** known as the conservation of angular momentum. This effect can be seen when an **ice** skater begins spinning with arms outstretched. As the skater brings his or her arms inward, his or her rotational speed increases. In the same way, air moving into a severe storm begins to move in a tighter column and increases its rotational speed. A wide vortex is created, called the mesocyclone. The mesocyclone begins to build vertically, extending itself upward throughout the entire height of the cloud. The rapid air movement causes the surrounding air pressure to drop, pulling more air into the growing vortex. The lowered pressure causes the incoming air to cool quickly and form cloud droplets before they rise to the cloud base. This forms the wall cloud, a curtain-shaped cloud that is often seen before a tornado forms. The mesocyclone continues to contract while growing from the base of the storm cloud all the way up to 6.2 mi (10 km) above the surface. When the mesocyclone dips below the wall cloud, it is called a funnel cloud because of its distinctive funnel shape. This storm is on its way to producing a tornado.

A funnel cloud may form in a severe storm and never reach the ground. If and when it does, the funnel officially becomes a tornado. The central vortex of a tornado is typically about 328.1 ft (100 m) in diameter. Wind speeds in the vortex have been measured at greater than 220 mph (138 km/h). These high winds make incredible feats of destruction possible. They also cause the air pressure in the tornado to drop below normal **atmospheric pressure** by over 100 millibars (the normal day-to-day pressure variations we experience are about 15 millibars). The air around the vortex is pulled into this low-pressure zone where it expands and cools rapidly.

Tornadoes are classified according to their wind speed, which can be estimated by examining the damage they produce. ***AP/Wide World. Reproduced by permission.***

This causes water droplets to condense from the air, making the outlines of the vortex visible as the characteristic funnel-shaped cloud. The low pressure inside the vortex picks up debris such as **soil** particles, which may give the tornado an ominous dark color. The damage path of a tornado may range from 900 ft (275 m) to over 0.5 mi (1 km) wide.

Tornadoes move with the thunderstorm that they are attached to, traveling at average speeds of about 10–30 mph (15–45 kph), although some tornadoes have been seen to stand still, while other tornadoes have been clocked at 60 mph (90 kph). Because a typical tornado has a lifetime of about 5–10 minutes, it may stay on the ground for 5–10 miles. Occasionally, a severe tornado may cut a path of destruction over 200 mi (320 km) long. Witnesses to an approaching tornado often describe a loud roaring noise made by the storm similar to jet engines at takeoff.

The destructive path of tornadoes appears random. One house may be flattened while its neighbor remains untouched. This has been explained by the tornado skipping or lifting up off the surface briefly and then descending again to resume its destructive path. Studies made of these destructive paths after the storm suggest another possible explanation; some tornadoes may have two to three smaller tornado-like vortices circling around the main vortex. According to this theory, these suction vortices may be responsible for much of the actual damage associated with tornadoes. As they rotate around the main tornado core, they may hit or miss objects directly in the tornado's path depending on their position. The tornado's skipping behavior is still not completely understood.

When houses or other structures are destroyed by a tornado, they are not simply blown down by the high winds; they appear to explode. High wind passing over a house roof acts like the air moving over an airplane wing: it gives the roof an upward force or lift, which tends to raise the roof vertically off the house. Winds also enter the building through broken windows or doors pressurizing the house as one would blow up a balloon. The combination of these forces tends to blow the walls and roof off the structure from the inside out, giving the appearance of an explosion.

Tornado strength is classified by the Fujita scale, which uses a scale of one to six to denote tornado wind speed. Since direct measurements of the vortex are not possible, the observed destruction of the storm is used to estimate its "F scale" rating.

The single most violent tornado in United States history was the Tri-State tornado on March 18, 1925. Beginning in Missouri, the tornado stayed on the ground for over 220 mi (350 km), crossing Illinois, moving into Indiana, and leaving a trail of damage over 1 mi (1.6 km) wide in places. Tornado damage often is limited since they usually strike unpopulated areas, but the Tri-State tornado plowed through nine towns and destroyed thousands of homes. When the storm was over, 689 people had lost their lives and over 2,000 were injured, making the Tri-State the deadliest tornado on record.

On May 3, 1999, a storm started in southwestern Oklahoma, near the town of Lawton. By late in the day, it had grown into a violent storm system with 76 reported tornadoes. As the storm system tore across central Oklahoma and into Kansas, over 43 people were killed, over 500 injured and more than 1,500 buildings were destroyed. One of the tornadoes, classed as a F-5, was as much as a mile wide at times and stayed on the ground for over four hours.

The precise tracking and prediction of tornadoes is not yet a reality. Meteorologists can identify conditions that are likely to lead to severe storms. They can issue warnings when atmospheric conditions are right for the development of tornadoes. They can use radar to track the path of thunderstorms that might produce tornadoes. It is still not possible, however, to detect a funnel cloud by radar and predict its path, touchdown point, and other important details. Much progress has recently been made in the detection of tornadoes using Doppler radar.

Doppler radar can measure not just the distance to an object, but also its velocity by using the Doppler effect: if an object is moving toward an observer, radar waves bounced off the object will have a higher frequency than if the object were moving away. This effect can be demonstrated with sound waves. If a car is approaching with its horn sounding, the pitch of the horn (that is, the frequency of the sound waves) seems to rise. It reaches a peak just as the car passes, then falls as the car speeds away from the listener.

Doppler radar is used to detect the motion of raindrops and hail in a thunderstorm, which gives an indication of the motion of the winds. With present technology, it is possible to detect the overall storm circulation and even a developing mesocyclone. The relatively small size of most tornadoes makes direct detection difficult with the current generation of Doppler radar. In addition, any radar is limited by the curvature of Earth. Radar waves go in straight lines, which means distant storms that are below the horizon from the radar cannot be probed with this technique.

See also Atmospheric pressure; Clouds and cloud types; Weather forecasting; Weather forecasting methods; Weather radar; Weather satellite

TORRICELLI, EVANGELISTA (1608-1647)

Italian physicist

As a scientist, Evangelista Torricelli became well known for his study of the motion of fluids, and was declared the father of hydrodynamics by Ernst Mach. Torricelli also conducted experiments on gases, though the term was not then in use. Most notably, Torricelli settled an argument about the nature of gases and the existence of the vacuum. Aristotle believed that a vacuum could not exist. Though Galileo disagreed, he contended that the action of suction (in a **water** pump, for example) was produced by a vacuum itself and not by the pressure of the air pushing on the liquid being pumped. Despite his argument, Torricelli noticed that water could be pumped only a finite distance through a vertical tube before it ceased to move any further and set out to examine this paradox, inventing the first barometer in the process.

During his experimentation, Torricelli filled a one-ended **glass** tube with mercury, then immersed the open end in a dish of more mercury, placing the tube in a upright position. He found that about 30 in (76 cm) of mercury remained in the tube, deducing that a vacuum had been created above the mercury in the tube, and that the mercury was held in place not by the vacuum, but by the pressure of air pushing down the mercury in the dish. Thus, he demonstrated the existence of a vacuum, showed why pumps then in use could only move liquids vertically a certain distance (the distance determined by the pressure of the surrounding air), and created an instrument capable of measuring air pressure.

Torricelli's invention of the barometer led to a burst of both theoretical and experimental work in **physics** and **meteorology**. Torricelli also made a contribution to meteorology with his suggestion that **wind** was not caused by the "exhalations" of vapors from a damp Earth, but by differences in the density of air that, in turn, were caused by differences in the air **temperature**.

Born near Ravenna, Italy, Torricelli was first educated in local Jesuit schools and showed such brilliance that he was sent to Rome to study with Galileo's former student Benedetto Castelli (1578–1643). Through Castelli he first corresponded with and met Galileo, finally becoming his secretary and assistant. A few months after Galileo's death in 1642, Torricelli accepted Galileo's old position as court mathematician and philosopher to the Grand Duke of Tuscany, a position he held until his own death, before his fortieth birthday.

Torricelli's investigations in mathematics played an important role in scientific history as well. Based on Francesco Cavalieri's "geometry of indivisibles," Torricelli worked out equations upon curves, solids, and their rotations, helping to bridge the gap between Greek geometry and calculus. Along with the work of René Descartes, Pierre de Fermat, Gilles Personne de Roberval, and others, these works enabled Isaac Newton and Gottfried Wilhelm Leibniz to give calculus its first complete formulation.

See also Atmospheric pressure

TRADE WINDS • *see* ATMOSPHERIC CIRCULATION

TRANSFORM FAULTS

Transform faults are a special class of faults first described by the Canadian geologist-geophysicist **J. Tuzo Wilson** in 1965 as faults that terminate abruptly at both ends and link one tectonic feature with another. Transform faults that offset mid-oceanic ridges (ridge-ridge transforms) transfer spreading from one segment of the ridge to the next. An important feature of ridge-ridge transform faults is that the sense of displacement along the transform fault is opposite to the sense of offset of the spreading ridge. The length of the active section of the fault remains constant with time. Fracture zones, across which there is no lateral displacement, continue beyond transform faults. Vertical bathymetric offsets occur across transform faults and fracture zones as young, hotter, and hence higher seafloor is juxtaposed against older and colder seafloor. Transform faults lie along small circles about a fixed Euler pole of **rotation**, implying a constant direction of plate motion. Volcanism and formation of new oceanic **crust** may occur along divergent or leaky transform faults. Transform faults can also occur between two subduction zones (trench-trench or arc-arc transforms) or between a spreading center and a **subduction zone** (ridge-trench or ridge-arc arc transform).

Faults geometrically equivalent to transform faults exist at the outcrop scale, especially in **limestone** and **marble**. Faults that offset two extension fracture veins may form in an equivalent manner to ridge-ridge transforms. Again, the sense of vein offset is opposite to the sense of displacement along the fault. Faults between two domains in which material is lost due to pressure solution (forming stylolites) are geometrically equivalent to arc-arc transforms. Faults between extensional veins and stylolites (on the same side of the fault) show an equivalent geometry to ridge-arc transforms.

See also Faults and fractures; Mid-ocean ridges and rifts; Plate tectonics; Transform plate boundary

Transform plate boundary

A transform plate boundary is a margin between two **lithospheric plates** that constitutes a regional-scale transform fault. The best-known transform plate boundary is the San Andreas fault system, which accommodates the right-lateral displacement between the North American and Pacific plates. The northwards-moving Pacific plate is subducted at the Aleutian trench and at western Pacific **island arcs**. Western California, part of the Pacific Plate, comprises exotic terranes translated northwards and rotated through angles up to 90 degrees along the margin of the North American plate. In the north, the San Andreas fault system terminates at the Mendocino triple junction where it intersects the Mendocino Fracture Zone and Cascadia subduction. Here the North American and Pacific plates intersect the Gorda–Juan de Fuca plate. The easterly moving Gorda–Juan de Fuca plate is subducted beneath the American plate north of the San Andreas fault system termination forming the Cascade Range. The San Andreas fault system steps through a series of oblique spreading ridges and **transform faults** in the Gulf of California. It terminates at the Rivera triple junction in the southern Gulf of California (junction between the Rivera, Pacific and North American plates). South of this triple junction, the Rivera Plate is subducted beneath **North America**. The San Andreas Fault is the main transcurrent or strike-slip fault within a broad deformation zone that comprises hundreds of minor faults along western California. Changes in their orientation and relay stepping of faults result in localized dilatation or contraction. Some segments of the San Andreas and other major faults in the San Andreas fault system are locked. In such segments, built-up strain may finally be rapidly released, producing an **earthquake**. Other segments are undergoing slow, continuous deformation or **creep**. The **lithosphere** is thinner beneath the San Andreas fault system than for normal continental lithosphere.

The Alpine Fault Zone along the western South Island of New Zealand is an example of an obliquely convergent transform plate boundary separating two zones of subduction with opposite polarity. North of New Zealand, the Pacific Plate is obliquely subducted beneath the Indo-Australian plate at the Tonga-Kermadec trench. South of New Zealand, the Indo-Australian Plate is obliquely subducted beneath the Pacific Plate at the Puysegur Trench. Right-lateral displacement and horizontal shortening occur across a zone 93–125 mi (150–200 km) wide. Most displacement has occurred along the Alpine Fault, which comprises oblique thrusts linked by sub-vertical dextral transcurrent faults.

See also Faults and fractures; Plate tectonics; Subduction zone

Transition zone • *see* Seismology

Transverse dunes • *see* Dunes

Triassic Period

The Triassic Period, first of the **Mesozoic Era's** three periods, began about 240 million years ago and lasted for approximately 40 million years. It was preceded by the great Permian-Triassic mass extinction, which destroyed over 90% of living species. This extinction, the worst in Earth's history, was probably caused in part by the merger (late in the late Permian) of all the continental plates into a single huge land mass, Pangaea (pronounced pan-JEE-ah). This destroyed many species by producing a net loss of coastline, while Pangaea's size—one fourth of the Earth's surface—dictated an arid climate over much of its interior. The Triassic was therefore a period of adaptive radiation—the slow filling of vacant ecological niches by species evolved from survivors of the great extinction.

For the most part Pangaea remained geologically stable and volcanically inactive during the Triassic. **Erosion** proceeded more rapidly than mountain-building. Particles eroded from the Pangaean highlands accumulated in various basins to produce a distinctively Triassic class of reddish sandstones and shales called the red beds. It is not known why the red beds are all red; some geologists argue that the Pangaean climate encouraged iron-concentrating **soil** bacteria. In the late

Triassic, the plates comprising Pangaea began to break up, and continental drift has subsequently distributed the red beds all over the world (**North America**, South **Africa**, **Europe**, Brazil).

Conifers (pine trees) and ferns were common land plants of the Triassic Period. Petrified Triassic conifers, some over 5 ft (1.5 m) across and over 100 ft (30 m) long, are found in Utah.

More than 95% of marine invertebrate species died in the Permian-Triassic extinction. During the Triassic, invertebrates slowly re-evolved diversity. Lobsters and crabs first appeared in this period.

Reptiles increased in number and variety throughout the Triassic Period. Some species took to the sea, evolving into the fish-eating plesiosaurs and *ichthyosaurs*. The first mammals appeared late in the Triassic Period. These were small and shrew-like, as their descendants would remain until the final elimination of the dinosaurs by an asteroid impact some 120 million years later.

The first dinosaurs also evolved in the late Triassic, but remained unspectacular by modern standards. It was not until the **Jurassic Period** that the most familiar species (Tyrannosaurus rex, Brontosaurus, etc.) were to evolve.

See also Archean; Cambrian Period; Carbon dioxide; Cenozoic Era; Continental drift theory; Cretaceous Period; Dating methods; Devonian Period; Eocene Epoch; Evolution, evidence of; Fossil record; Fossils and fossilization; Geologic time; Historical geology; Holocene Epoch; Miocene Epoch; Mississippian Period; Oligocene Epoch; Ordovician Period; Paleocene Epoch; Paleozoic Era; Pennsylvanian Period; Phanerozoic Eon; Pleistocene Epoch; Pliocene Epoch; Precambrian; Proterozoic Era; Quaternary Period; Silurian Period; Tertiary Period

TROPIC OF CANCER • *see* SOLAR ILLUMINATION: SEASONAL AND DIURNAL PATTERNS

TROPIC OF CAPRICORN • *see* SOLAR ILLUMINATION: SEASONAL AND DIURNAL PATTERNS

TROPICAL CYCLONE

Tropical cyclones are large circulating storm systems consisting of multiple bands of intense showers and thunderstorms and extremely high winds. These storm systems develop over warm ocean waters in the tropical regions that lie within about 25° **latitude** of the equator. Tropical cyclones may begin as isolated thunderstorms. If conditions are favorable, they grow and intensify to form the storm systems known as hurricanes in the Americas, typhoons in East **Asia**, willy-willy in **Australia**, cyclones in Australia and India, and baguios in the Philippines. A fully developed tropical cyclone is a circular complex of thunderstorms about 403 mi (650 km) in diameter and over 7.5 mi (12 km) high. Winds near the core of the cyclone can exceed 110 mph (50 meters/second). At the center of the storm is a region about 9–12.5 mi (15–20 km) across called the eye, where the winds are light and skies are often clear. After forming and reaching peak strength over tropical **seas**, tropical cyclones may blow inland, causing significant damage and loss of life. The storm destruction occurs by high winds and forcing rapid rises in sea level that flood low lying coastal areas. Better forecasting and emergency planning has lowered the death tolls in recent years from these powerful storms.

Several ocean areas adjacent to the equator possess all the necessary conditions for forming tropical cyclones. These spots are: the West Indies/Caribbean Sea, where most hurricanes develop between August and November; the Pacific Ocean off the west coast of Mexico, with a peak hurricane season of June through October; the western Pacific/South China Sea, where most typhoons, baguios, and cyclones form between June and December; and south of the equator in the southern Indian Ocean and the south Pacific near Australia, where the peak cyclone months are January to March. Note that in each **area** the peak season is during late summer (in the Southern Hemisphere, summer runs from December to March). Tropical cyclones require warm surface waters at least 80°F (27°C). During the late summer months, sea surface temperatures reach their highest levels and provide tropical cyclones with the energy they need to develop into major storms.

The annual number of tropical cyclones reported varies widely between regions and from year to year. The West Indies recorded 658 tropical cyclones between 1886–1966, an average of about eight per year. Of these, 389, or about five per year, grew to be of hurricane strength. The Atlantic hurricane basin has a 50-year average of ten tropical storms and six hurricanes annually.

In the United States, the National **Weather** Service names hurricanes from an alphabetic list of alternating male and female first names. New lists are drawn up each year to name the hurricanes of western Pacific and the West Indies. Other naming systems are used for the typhoons and cyclones of the eastern Pacific and Indian **oceans**.

In some ways, tropical cyclones are similar to the low pressure systems that cause weather changes at higher latitudes in places like the United States and **Europe**. These systems are called extratropical cyclones and are marked with an "L" on weather maps. These weather systems are large masses of air circulating cyclonically (counterclockwise in the Northern Hemisphere and clockwise in the Southern Hemisphere). Cyclonic circulation is caused by two forces acting on the air: the pressure gradient and the Coriolis force.

In both cyclone types air rises at the center, creating a region of lower air (barometric) pressure. Because air is a fluid, it will rush in from elsewhere to fill the void left by air that is rising off the surface. The effect is the same as when a plug is pulled out of a full bathtub: **water** going down the drain is replaced by water rushing in from other parts of the tub. This is called the pressure gradient force because air moves from regions of high pressure to lower pressure. Pressure gradient forces are responsible for most day-to-day winds. As the air moves toward low pressure, the Coriolis force turns the air to the right of its straight-line motion (when viewed from above). In the Southern Hemisphere, the reverse is true: the Coriolis force pushes the moving air to the left. The air, formerly going straight toward a low-pressure region, is forced to turn away from it. The two forces are in balance when the air circles around the low pressure zone with a constant radius creating a stable cyclone rotating counterclockwise in the Northern Hemisphere and clockwise in the Southern Hemisphere.

All large-scale air movements such as hurricanes, typhoons, extratropical cyclones, and large thunderstorms set up a cyclonic circulation in this manner. (Smaller scale circulations such as the vortex that forms in a bathtub drain are not cyclonic because the Coriolis force is overwhelmed by other forces. The larger a system, the more likely that the Coriolis force will prevail and the **rotation** will be cyclonic.) The Coriolis force is a consequence of the rotation of Earth. Moving air masses, like any other physical body, tend to move in a straight line. However, we observe them moving over Earth's surface, which is rotating underneath the moving air. From our perspective, the air appears to be turning even though it is actually going in a straight line, and it is we who are moving.

In both tropical and extratropical cyclones, the rising air at the cyclone center causes **clouds** and **precipitation** to form. A fully developed hurricane consists of bands of thunderstorms that grow larger and more intense as they move closer to the cyclone center. The area of strongest updrafts can be found along the inner wall of the hurricane. Inside this inner wall lies the eye, a region where air is descending. Descending air is associated with clearing skies; therefore, in the eye the torrential rain of the hurricane ends, the skies clear, and winds drop to nearly calm. In the eye of a hurricane, the eye wall clouds appear as towering vertical walls of thunderstorm clouds, stretching up to 7.5 mi (12 km) in height, and usually completely surrounding the eye. Hurricanes and other tropical cyclones move at the speed of the prevailing winds, typically 10–20 mph (16–32 kph) in the tropics. A hurricane eye passes over an observer in less than an hour, replaced by the high winds and heavy rain of the intense inner thunderstorms.

Several conditions are necessary to create a tropical cyclone. Warm sea surface temperatures, which reach a peak in late summer, are required to create and maintain the warm, humid air mass in which tropical cyclones grow. This provides energy for storm development through the heat stored in humid air, called latent heat. It takes energy to change water into vapor; that is why one must add heat to boil a kettle of water. The reverse is also true: when vapor condenses back to form liquid water, heat is released that may heat up the surrounding air. In a storm such as a hurricane, many hundreds of tons of humid air are forced to rise and cool, condensing out tons of water droplets and liberating a vast quantity of heat. This warms the surrounding air, causing it to expand and become even more buoyant, that is, more like a hot air balloon. More air begins rising, causing even more humid air to be drawn into the cyclone. This process feeds on itself until it forms a cyclonic storm of huge proportions. The more humid air available to a tropical cyclone, the greater its upward growth and the more intense it will become.

For storm growth to begin, air needs to rise. Because tropical air masses are uniformly warm and humid, the atmosphere over much of the tropics is stable; that is, it does not support rising air and the development of storms. Thunderstorms occasionally develop but tend to be short-lived and small in scale, unlike the severe thunderstorms in the middle latitudes. During the late summer, this peaceful picture changes. Tropical disturbances begin to appear. These can take the form of a cluster of particularly strong thunderstorms or perhaps a storm system moving westward off of the African continent and out to sea. Tropical disturbances are regions of lower pressure at the surface. As we have seen, this can lead to air rushing into the low pressure zone and setting up a vortex, or rotating air column, with rising air at its core.

An additional element is needed for tropical cyclone development: a constant **wind** direction with height throughout the lower atmosphere. This allows the growing vortex to stretch upward throughout the atmosphere without being sheared apart. Even with all these elements present, only a few of the many tropical disturbances observed each year become hurricanes or typhoons. When a tropical disturbance near the surface encounters a similar disturbance in the air flow at higher levels such as a region of low pressure at about the 3 mi (5 km) level (called an upper low), conditions are favorable for hurricane formation. These upper lows sometimes wander toward the equator from higher latitudes where they were part of a decaying weather system.

Once a tropical disturbance has begun to intensify, a chain reaction occurs. The disturbance draws in humid air and begins rising. Eventually it condenses to form water droplets. This releases latent heat, which warms the air, making it less dense and more buoyant. The air rises more quickly from the surface. As a result, the pressure in the disturbance drops and more humid air moves toward the storm. Meanwhile, the disturbance starts its cyclonic rotation and surface winds begin to increase. Soon, the tropical disturbance forms a circular ring of low air pressure and becomes known as a tropical depression. As more heat energy is liberated and updrafts increase inside the vortex, the internal barometric pressure continues to drop and the incoming winds increase. When wind speeds increase beyond 37 mph (60 kph), the depression is upgraded to a tropical storm. If the winds reach 75 mph (120 kph), the tropical storm is officially classified as a hurricane (or typhoon, cyclone, etc., depending on location). The chain reaction driving this storm growth is efficient. About 50–70% of tropical storms intensify to hurricanes.

A mature tropical cyclone is a giant low-pressure system pulling in humid air, releasing its heat, and transforming it into powerful winds. The storm can range in diameter from 60–600 mi (100–1000 km) with wind speeds greater than 200 mph (320 kph). The central barometric pressure of the hurricane drops 60 millibars (mb) below the normal sea level pressure of 1013 mb. By comparison, the passage of a strong storm front in the middle latitudes may cause a drop of about 20–30 mb. The size and strength of the storm is limited only by the air's **humidity**, which is determined by ocean **temperature**. It is estimated that for every 1.8°F (1°C) increase in sea surface temperature, the central pressure of a tropical cyclone can drop 12 mb. With such low central pressure, winds are directed inward, but near the center of the storm the winds are rotating so rapidly the Coriolis force prevents any further inward movement. This inner boundary creates the eye of the tropical cyclone. Unable to go in, the air is forced to move upward and then

Satellite photo of Hurricane Mitch. *National Oceanic and Atmospheric Administration (NOAA).*

spread out at an altitude of about 7.5 mi (12 km). Viewed from above by a **satellite**, the tropical cyclone appears as a mass of clouds diverging away from the central eye.

All of the cyclone development described thus far takes place at sea, but the entire cyclone also is blown along with the prevailing winds. Often this movement brings the storm toward land. As tropical cyclones approach land, they begin affecting the coastal areas with sea swells, large waves caused by the storm's high winds. Swells often reach 33 ft (10 m) in height and can travel thousands of kilometers from the storm. Coastal areas are at risk of severe damage from these swells that destroy piers, beach houses, and harbor structures every hurricane season. Particularly high swells may cause flooding farther inland.

More dangerous than the gradually rising swells are the sudden rises in sea level known as storm surges. Storm surges occur when the low barometric pressure near the center of a cyclone causes the water surface below to rise. Then strong winds blowing toward the coast push this "bulge" of water out ahead of the storm. The water piles up against the coast, quickly raising sea level as much as 16 ft (5 m) or more. The highest storm surge (for Northern Hemisphere storms, hurricanes) generally occurs east of the storm's path. When storm-tossed waves of 23–33 ft (7–10 m) are added to this wall of water, land areas may be inundated. In 1900, the city of Galveston, Texas, was hit with a destructive storm surge during a hurricane. One eyewitness reported that the sea rose 4 ft (1.3 m) in a matter of seconds. Over 5,000 people lost their lives in the Galveston hurricane and resulting flooding, making it the deadliest storm ever recorded in the United States.

Tropical cyclones that travel onto the land immediately begin to weaken as humid air, their source of energy, is cut off. The winds at the base of the cyclone encounter greater friction as they drag across uneven terrain that slows them. Nevertheless, tropical cyclones at this stage are still capable of producing heavy rains, thunderstorms, and even tornadoes. Occasionally, the remnants of a tropical cyclone that has begun to weaken over land will unite with an extratropical low pressure system, forming a potent rain-making storm front that may bring flooding to areas far from the coast.

Until relatively recently, people in the path of a tropical cyclone had little warning of approaching storms. Usually their only warning signs were the appearance of high clouds and a gradual increase in winds. Hurricane watch services were established beginning in the early years of the twentieth century. By the 1930s, hurricanes were detected with weather balloons and ship reports while the 1940s saw the introduction of airplanes as hurricane spotters. Radar became available after World War II and has remained a powerful tool for storm detection. Today, a global network of weather satellites allows meteorologists to identify and track tropical cyclones from their earliest appearance as disturbances over the remote ocean. This improved ability to watch storms develop anywhere in the world has meant that warnings and evacuation orders can be issued well in advance of a tropical cyclone reaching land. Even though coastal areas have more people living near them today than ever before and tropical cyclones remain as powerful as ever, fewer storm-related deaths are now reported due to advances in storm detection and forecasting.

See also Air masses and fronts; Atmospheric pressure; Beach and shoreline dynamics; Beaufort wind scale; Convection (updrafts and downdrafts); El Nino and La Nina phenomena; Meteorology; Ocean circulation and currents; Wave motions; Weather forecasting methods; Weather forecasting; Weather radar; Weather satellite

Troposphere and tropopause

The troposphere is the lowest and thickest layer of the atmosphere. In contact with Earth's surface, the troposphere is heated by solar illumination and conduction. The tropopause is the boundary layer between the troposphere and the **stratosphere**.

The thickness of the troposphere depends upon a number of atmospheric variables at **latitude**. The troposphere ranges from a thickness of approximately 5.5 mi (9 km) in the polar regions, to a thickness of approximately 10 mi (16 km) in equatorial regions.

Although the troposphere contains more than 70% of Earth's atmosphere by weight, it is much thinner than the stratosphere or **ionosphere**. Because density increases with increasing mass, the troposphere exhibits a high pressure and density gradient wherein density and pressure decrease with increasing altitude.

Weather phenomena (e.g., rain, snow, etc.) take place in the troposphere. Convective currents provide mixing of air masses with different temperatures. These currents pass through regions of the troposphere that differ widely in pressure. The troposphere **atmospheric pressure** gradient varies by approximately 90% from sea level pressures to tropopause atmospheric pressures.

Under normal atmospheric conditions, the standard lapse rate describes decreasing temperatures encountered with increased altitude within the troposphere. The standard **temperature** lapse rate means that temperature decreases with altitude at a fairly uniform rate. Because the atmosphere is warmed by conduction from Earth's surface, this lapse or reduction in temperature is normal with increasing distance from the conductive source. The tropopause is specifically defined as that upper boundary layer of the troposphere where the thermal lapse rate no longer exists and temperature exhibit stability prior to increasing within the stratosphere.

The bulk of tropospheric heating occurs via conduction of heat from the surface. Differing amounts of sunlight (differential levels of solar **insolation**) result in differential temperatures at the interface between Earth's surface and the troposphere. Warmer surface temperatures and higher rates of conduction allow warm air to create low-pressure zones where air is uplifted. Because surrounding air must rush in to replace the uplifted air, these warmer areas become zones of convergence (inward rushing air). Cooler surface temperatures and lower rates of conduction result in cooler, denser, higher pressure areas that form zones of divergence in which air moves outwards from the high pressure **area**.

The thermal instability means that the troposphere is the turbulent thermal boundary layer of Earth's atmosphere. In contrast to the unstable and vertical currents encountered in the humid troposphere, the stratosphere exhibits a near laminar (horizontal or sheet-like) flow.

See also Air masses and fronts; Atmospheric chemistry; Atmospheric circulation; Atmospheric composition and structure; Atmospheric inversion layers; Atmospheric pollution; Atmospheric pressure; Weather forecasting methods; Wind chill; Wind shear; Wind

TSUNAMI

Tsunami, or seismic sea waves, are a series of very long wavelength ocean waves generated by the sudden displacement of large volumes of **water**. The generation of tsunami waves is similar to the effect of dropping a solid object, such as a stone, into a pool of water. Waves ripple out from where the stone entered, and thus displaced, the water. In a tsunami, the "stone" comes from underneath the ocean or very close to shore, and the waves, usually only three or four, are spaced about 15 minutes apart.

Tsunami can be caused by underwater (submarine) earthquakes, submarine **volcanic eruptions**, falling (slumping) of large volumes of ocean sediment, coastal landslides, or even by meteor impacts. All of these events cause some sort of landmass to enter the ocean and the ocean adjusts itself to accommodate this new mass. This adjustment creates the tsunami, which can circle around the world. Tsunami is a Japanese word meaning "large waves in harbors." It can be used in the singular or plural sense. Tsunami are sometimes mistakenly called tidal waves, but scientists avoid using that term since they are not at all related to **tides**.

Tsunami are classified by oceanographers as shallow water surface waves. Surface waves exist only on the surface of liquids. Shallow water waves are defined as surface waves occurring in water depths that are less than one half their wavelength. Wavelength is the distance between two adjacent crests (tops) or troughs (bottoms) of the wave. Wave height is the vertical distance from the top of a crest to the bottom of the adjacent trough. Tsunami have wave heights that are very small as compared to their wavelengths. In fact, no matter how deep the water, a tsunami will always be a shallow water wave because its wavelength (up to 150 mi [240 km]) is so much greater than its wave height (usually no more than 65 ft [20 m]).

Shallow water waves are different from deep water waves because their speed is controlled only by water depth. In the open ocean, tsunami travel quickly (up to 470 mph [760 kph]), but because of their low height (typically less than 3 ft [1 m]) and long wavelength, ships rarely notice them as they pass underneath. However, when a tsunami moves into shore, its speed and wavelength decrease due to the increasing friction caused by the shallow sea floor.

Wave energy must be redistributed, however, so wave height increases, just as the height of small waves increases as they approach the beach and eventually break. The increasing tsunami wave height produces a "wall" of water that, if high enough, can be incredibly destructive. Some tsunami are reportedly up to 200 ft (65 m) tall. The impact of such a tsunami can range miles inland if the land is relatively flat.

Tsunami may occur along any shoreline and are affected by local conditions such as the coastline shape, ocean floor characteristics, and the nature of the waves and tides already in the **area**. These local conditions can create substantial differences in the size and impact of the tsunami waves, even in areas that are very close geographically.

Tsunami researchers classify tsunami according to their area of effect. They can be local, regional, or ocean-wide. Local tsunami are often caused by submarine volcanoes, submarine sediment slumping, or coastal landslides. These can often be the most dangerous because there is often little warning between the triggering event and the arrival of the tsunami.

Seventy-five percent of tsunami are considered regional events. Japan, Hawaii, and Alaska are commonly hit by regional tsunami. Hawaii, for example, has been hit repeatedly during this century, about every 5–10 years. One of the worst was the April 1, 1946, tsunami that destroyed the city of Hilo.

Pacific-wide tsunami are the least common as only 3.5% of tsunami are this large, but they can cause tremendous destruction due to the massive size of the waves. In 1940 and 1960, destructive Pacific-wide tsunami occurred. More recently, there was a Pacific-wide tsunami on October 4, 1994, which caused substantial damage in Japan with 11.5 ft (3.5 m) waves. However, waves of only 6 in (15 cm) over the normal height were recorded in British Columbia.

Tsunami are not only a modern phenomenon. The decline of the Minoan civilization is believed to have been triggered by a powerful tsunami that hit the area in 1480 B.C. and destroyed its coastal settlements. Japan has had 65 destructive tsunami between A.D. 684 and 1960. Chile was hit in 1562 and Hawaii has a written history of tsunami since 1821. The Indian and Atlantic **Oceans** also have long tsunami histories. Researchers are concerned that the impact of future tsunami, as well as hurricanes, will be worse because of intensive development of coastal areas in the last 30 years.

Tsunami inundating Hilo, Hawaii. ***National Oceanic and Atmospheric Administration (NOAA).***

The destructive 1946 tsunami at Hilo, Hawaii, caused researchers to think about the problem of tsunami prediction. It became clear that if scientists could predict when the waves are going to hit, steps could be taken to minimize the impact of the great waves.

In 1965, the Intergovernmental Oceanographic Commission of the United Nations Educational, Scientific, and Cultural Organization agreed to expand the United States' existing tsunami warning center at Ewa Beach, Hawaii. This marked the formation of the Pacific Tsunami Warning Center (PTWC), which is now operated under the U.S. **Weather** Service. The objectives of the PTWC are to "detect and locate major earthquakes in the Pacific basin; determine whether or not tsunami have been generated; and to provide timely and effective information and warnings to minimize tsunami effects."

The PTWC is the administrative center for all the associated centers, committees, and commissions of the International Tsunami Warning System (ITWS). Japan, the Russian Federation, and Canada also have tsunami warning systems and centers and they coordinate with the PTWC. In total, 27 countries now belong to the ITWS.

The ITWS is based on a world-wide network of seismic and tidal data and information dissemination stations, and specially trained people. Seismic stations measure movement of the earth's **crust** and are the foundation of the system. These stations indicate that some disturbance has occurred that may be powerful enough to generate tsunami. To confirm the tsunami following a seismic event, there are specially trained people called tide observers with monitoring equipment that enables them to detect differences in the wave patterns of the ocean. Pressure gauges deployed on the ocean can detect changes of less than 0.4 in (1 cm) in the height of the ocean, which indicates wave height. Also, there are accelerometers set inside moored buoys that measure the rise and fall of the ocean, which will indicate the wave speed. These data are used together to help researchers confirm that a tsunami has been generated. Tsunami can also be detected by **satellite** monitoring methods such as radar and photographic images.

The ITWS is activated when earthquakes greater than 6.75 on the **Richter scale** are detected. The PTWC then collects all the data, determines the magnitude of the quake and its epicenter. Then they wait for the reports from the nearest tide stations and their tide observers. If a tsunami wave is reported, warnings are sent to the information dissemination centers.

The information dissemination centers then coordinate the emergency response plan to minimize the impact of the tsunami. In areas where tsunami frequency is high, such as Japan, the Russian Federation, Alaska, and Hawaii, there are also Regional Warning Systems to coordinate the flow of information. These information dissemination centers then decide whether to issue a "Tsunami Watch," which indicates that a tsunami may occur in the area, or a more serious "Tsunami Warning," which indicates that a tsunami will occur.

The entire coastline of a region is broken down into smaller sections at predetermined locations known as "breakpoints" to allow the emergency personnel to customize the warnings to account for local changes in the behavior of the tsunami. The public is kept informed through local radio broadcasts. If the waves have not hit within two hours of the estimated time of arrival, or, the waves arrived but were not damaging, the tsunami threat is assumed to be over and all Watches and Warnings are canceled.

One of the more recent changes in the ITWS is that the Regional Centers will be taking on greater responsibility for tsunami detection and warning procedures. This is being done because there have been occasions when the warning from Hawaii came after the tsunami hit the area. This can occur with local and regional tsunami that tend to be smaller in their area of effect. Some seismically active areas need to have the warning system and equipment closer than Hawaii if they are to protect their citizens. For example, the Aleutian Islands near Alaska have two to three moderate earthquakes per week. As of May 1995, centers such as the Alaska Tsunami Warning Center located in Palmer, Alaska, have assumed a larger role in the management of tsunami warnings.

In terms of basic research, one of the biggest areas of investigation is the calculation of return rates. Return rates, or recurrence intervals, are the predicted frequency with which tsunami will occur in a given area and are useful information, especially for highly sensitive buildings such as nuclear power stations, offshore oil drilling platforms, and hospitals. The 1929 tsunami in Newfoundland has been studied extensively by North American researchers as a model for return rates and there has been some dispute. Columbia University researchers predict a reoccurrence in Newfoundland in 1,000–35,000 years. However, some geologists argue that it may reoccur as soon as 100–1,000 years. These calculations are based on evidence from mild earthquakes and tsunami in the area. They also suggest that the 1929 tsunami left a sedimentary record that is evident in the **soil** profile, and that such records can be dated and used to calculate return rates. Research is currently ongoing to test this theory.

See also Seismology; Wave motions

TUFA

Tufa belong to a group of crust-like carbonate deposits that are formed through the organically and inorganically controlled **precipitation** of calcium carbonates from fresh **water**. Other members of this group are travertines, sinters, and lacustrine limestones or marl lake deposits. **Cave** deposits of a somewhat similar origin are called flowstones, speleothems (stalagmites and **stalactites**). The terms tufa and travertine have a Latin origin. The first is derived from *tophus* and was used by Pliny to describe porous whitish deposits including volcanic material, which is nowadays called volcanic **tuff**. The term travertine stems from lapis *tiburtinus* or Tibur stone.

The distinction of terms used to describe surface **freshwater** carbonates in the literature is not very clear. However, today the term tufa is usually used to describe the more porous varieties, while travertines are denser and sometimes laminated. Sinters are mostly laminated and lack **porosity**. Lacustrine limestones are hardly compacted.

The porosity in tufas is derived from authochtonous plants such as mosses, green algae or reed, which are encrusted by carbonates. Tufas are typically found as deposits of cool spring waters, which are supersaturated by calcium bicarbonate. The precipitation of carbonates in these meteogene deposits is assisted by photosynthesis of phototrophic microbes and plants. Generally, however, any decrease in the partial pressure of CO_2 will trigger carbonate precipitation.

Travertines, in contrast, are mostly of hydrothermal origin and usually lack **fossils** of macrophytes or invertebrates. The CO_2 in these thermogene deposits can be supplied by various processes including degassing of the upper mantle in areas of tectonic stress. Travertines commonly change to a more porous tufa-fabric in areas where the water has cooled down to near ambient **temperature**.

Tufas and travertines are formed in fluvial environments with a growth rate that can sometimes exceed 0.8 in (2 cm) per year but is seldom smaller than 0.08 in (2 mm) per year. They build up barriers leading to the formation of shallow pools in which carbonate mud is deposited. This succession of barriers and pools can cover wide areas, sometimes miles in length (e.g., Grand **Canyon**, Arizona or Mammoth Hot Springs, Wyoming). Other seemingly related formations such as the deposits of the geysers and hot springs in the Yellowstone National Park (Wyoming) are in fact siliceous sinter.

Tufas and travertines are particularly common in the late Pleistocene. Today, they form under a wide variety of climatic regimes from cool and temperate to semi-arid conditions while the growth of fossil travertines of the Quaternary seems to be restricted to the interstadials and interglacials in high and medium latitudes. Some calcareous fresh water deposits show well-developed laminae that probably mirror seasonal differences in one or more environmental variables (water depth or temperature, detrital input).

See also Stalactites and stalagmites

TUFF

Tuffs are volcanic **igneous rocks** composed mostly of compacted volcanic ash and **sand** (particles less than 0.16 in [4 mm] in diameter). Tuffs often show well-defined layers, recording episodic falls of ash, and may be classified in various ways. First, they are often named according to the types of recognizable **rock** fragments embedded in them (i.e., **basalt** tuff, **andesite** tuff, **rhyolite** tuff, etc.). They may also be classified as vitric, lithic, or crystal; vitric tuffs consist mostly of glassy particles, lithic tuffs of rocky particles, and crystal tuffs of visible **crystals**. Tuff may also be characterized as lapilli tuff or tuff-breccia: lapilli tuff consists of ash mixed with lapilli (volcanic fragments 0.08–2.5 in [2–64 mm] in diameter); tuff-breccia consists of ash mixed with block-sized (volcanic fragments >2.5 in [64 mm] diameter) fragments.

Volcanoes lay down tuffs by various processes. First, ash lofted to high altitudes may spread over thousands of square miles of terrain before settling out to produce a relatively thin layer of tuff. Such layers are useful to geologists as marking a specific moment in geological time over a wide region. Second, ash lying on the sides of a **volcano** may become saturated by the torrential rains that often occur in response to **volcanic eruptions** and pour down the cinder cone as a **mud flow**. Third, ash may erupt straight up and spread around the vent as a backfill flow. Fourth, large masses of ash and sand mixed with hot gas can be ejected by a volcano and **avalanche** rapidly down its slopes. Such an avalanche is called a *nuée ardente* (pronounced nie-ay-ar-DAHNT; French for burning cloud). A *nuée ardente* can be carried by its momentum many miles from its source, even traversing mountain ridges to lay down ash in distant valleys. After settling, the heat and weight of the ash deposits left by a *nuée ardente* often weld their particles together, creating the rocks termed welded tuffs.

Tuffs are usually much altered in composition and texture after deposition. Alteration may begin with the stewing of a hot ash layer in its own gasses and condensed fluids, or with the addition of outside **water** to the hot ash. Over geological time, few tuffs escape alteration, especially by devitrification. In devitrification, the chaotically mixed atoms in a tuff's volcanic **glass** (which is inherently unstable in atmospheric conditions) reorganize themselves into crystals.

U

ULTRAMAFIC • *see* PERIDOTITE

ULTRAVIOLET RAYS AND RADIATION

Just like visible light, infrared light, and radio waves, ultraviolet light is electromagnetic radiation. On the spectrum, ultraviolet light lies between violet light and x rays, with wavelengths ranging from four to 400 nanometers. Although it is undetectable to the naked eye, anyone who has been exposed to too much sunlight has probably noted the effects of ultraviolet light, for it is this radiation that causes tanning, sunburn, and can lead to skin cancer.

The man credited with the discovery of ultraviolet light is the German physicist Johann Ritter. Ritter had been experimenting with silver chloride, a chemical known to break down when exposed to sunlight. He found that the light at the blue end of the visible spectrum—blue, indigo, violet—was a much more efficient catalyst for this reaction. Experimenting further, he discovered that silver chloride broke down most efficiently when exposed to radiation just beyond the blues, radiation that was invisible to the eye. He called this new type of radiation ultraviolet, meaning "beyond the violet." While ultraviolet radiation in large doses is hazardous to humans, a certain amount is required by the body. As it strikes the skin, it activates the chemical processes that produce Vitamin D. In areas that lack adequate sunshine, children are sometimes plagued by rickets. In order to treat these cases, or to supplement natural light in sun-starved communities, ultraviolet lamps are often used in place of natural sources.

There are three varieties of ultraviolet lamps, each producing ultraviolet light of a different intensity. Near-ultraviolet lamps are fluorescent lights whose visible light has been blocked, releasing ultraviolet radiation just beyond the visible spectrum. These lamps are also known as black lights, and are primarily used to make fluorescent paints and dyes "glow" in the dark. This effect is often seen in entertainment, but can also be used by industry to detect flaws in machine parts.

Middle-ultraviolet lamps produce radiation of a slightly shorter wavelength. They generally employ an excited arc of mercury vapor and a specially designed **glass** bulb. Because middle-ultraviolet radiation is very similar to that produced by the **Sun**, these lamps are frequently used as sunlamps and are often found in tanning salons and greenhouses. Photochemical lamps generating middle-ultraviolet light are also used in industry, as well as by chemists to induce certain chemical reactions.

Far-ultraviolet lamps produce high-energy, short-wavelength ultraviolet light. Like middle-ultraviolet lamps, they use mercury-vapor tubes; however, far-ultraviolet radiation is easily absorbed by glass, and so the lamp's bulb must be constructed from **quartz**. Far-ultraviolet light has been found to destroy living organisms such as germs and bacteria; for this reason, these lamps are used to sterilize hospital air and equipment. Far-ultraviolet radiation has also been used to kill bacteria in food and milk, giving perishables a much longer shelf life.

A more passive application of ultraviolet light is in **astronomy**. Much of the light emitted by stars, particularly very young stars, is in the ultraviolet range. By observing the output of ultraviolet light, astronomers can determine the **temperature** and composition of stars and interstellar gas, as well as gain insights into the evolution of galaxies. However, most of the ultraviolet light from distant sources is unable to penetrate the Earth's atmosphere; therefore, ultraviolet observations must be made from Earth's orbit by sounding rockets, **space** probes, or astronomical satellites.

See also Electromagnetic spectrum; Solar illumination: Seasonal and diurnal patterns

UNCONFORMITIES

An unconformity is a stratigraphic feature that is formed by broad **erosion** of an **area** causing a significant gap to occur in the stratigraphic record. Generally, development of an unconformity is accompanied by or preceded by uplift of the **rock** units or strata that will be eroded, and/or subsidence or fall of sea level, thus exposing rock units or strata to erosion.

An unconformity can consist of two hypothetical parts: the hiatus and the erosional vacuity. The hiatus is the amount of stratigraphic record removed during erosion and the erosional vacuity is the amount of stratigraphic record that might have been deposited during the time erosion was occurring. The sum of these parts is referred to as the lacuna, or total missing stratigraphic record. An unconformity is commonly referred to as a significant gap in the chronostratigraphic record as well.

Unconformities can be classified according to their scale or scope. For example, a cartographic unconformity is one that can be mapped regionally or seen in a regional cross section. A macrographic unconformity may be seen at outcrop scale where truncation may be evident. A petrographic unconformity is seen under a microscope (usually in a glass-mounted thin section or on the face of a cut slab of rock) and is very small.

Unconformities can also be classified according to their physical appearance. Angular unconformities are those which show truncation of upturned or tilted layers below the unconformity surface. Disconformities are unconformities showing erosional **relief** between otherwise parallel layers of sedimentary strata. A nonconformity is an unconformity developed between igneous or **metamorphic rock** (below) and sedimentary layers (above). Finally, a paraconformity is an enigmatic type of unconformity that appears as a hardly distinguishable plane between two parallel sedimentary rock layers.

Unconformities of large scale (i.e., cartographic unconformities) have been used by some geologists to delineate large groups of sedimentary formations that are bounded by unconformities. These unconformity-bounded units are called synthems, and their origin is related to grand cycles of sea-level change and mountain-building in Earth history. Either regional or inter-regional unconformities bound these large synthems.

Unconformities are recognized using three types of criteria: physical, structural, and paleontologic. Physical criteria include erosional relief between sedimentary rock layers, mineralized zones in rocks, basal conglomerates, iron-stained zones in rocks, zones of truncation and encrustation of **fossils**, small and large-scale solution (karst) features, and ancient soil horizons. Structural criteria include: dikes and faults truncated by erosion, discordance of structural **dip** between layers, and tectonic deformation. Paleontologic criteria include: abrupt change in fossil assemblage, gaps in evolutionary lineage among fossils, missing fossil zones, bone and tooth conglomerates, and reworked or corroded fossils. In study of unconformities, it is preferred that evidence from at least two of the three categories be used to recognize an unconformity and criteria from all three categories, if possible, be used to assess the scope of an unconformity.

Assessment of the scope of an unconformity, that is the amount of geological history or chronostratigraphic record missing at an unconformity, requires careful study. The scope of an unconformity is commonly situational, i.e., related to the type of sediment and the depositional environment of that sediment. For example, erosional removal of 4 in (10 cm) of deep-sea sediment, wherein rates of sediment accumulation are very low (<0.1 cm/1,000 years), would likely result in a significant gap in the chronostratigraphic record. An unconformity formed in this way might be a paraconformity or petrographic unconformity, which could be quite significant in terms of lost record. However, the erosional removal of 4 in (10 cm) of sediment formed in a depositional environment wherein **sedimentation** rates were much higher would not result in a significant gap in the rock record.

Many erosional surfaces that look superficially like unconformities are in reality usually only minor erosional or scour surfaces. These features are not unconformities, but are instead diastems. Diastems do not rise to the level of unconformity because the amount of missing record is not significant. As noted above, the assessment of such significance is situational, and it may not be known immediately after discovery if an erosional surface is an unconformity or a diastem.

The term stratigraphic break is commonly used instead of unconformity or diastem, particularly where the significance of a stratigraphic break in question is not known or is ambiguous. In other words, an unconformity may be thought of as a stratigraphic break that is significant in terms of lost record and a diastem as an insignificant break by the same measure. It is well established that unconformities and diastems both contribute to the overall incompleteness of the stratigraphic record.

See also Bedding; Chronostratigraphy; Depositional environments; Stratigraphy

UNIFORMITARIANISM

The concept of uniformitarianism is commonly oversimplified in geological textbooks as "the present is a guide to interpreting the past" (or words to that effect). This explanation, however, is not correct about the true meaning of uniformitarianism. In order to understand uniformitarianism, one must examine its roots in the Enlightenment era (c. 1750–1850) and how the term has been distorted in meaning since that time.

Geology is a historical science, yet the phenomena and processes studied by geologists operated under non-historical natural systems that are independent of the time in which they operated. It is clear from the insights of one of geology's founding fathers of the Enlightenment era, **James Hutton** (1726–1797), that he understood this fact very well. In *Theory of the Earth* (1795), he stated: "In examining things present, we have data from which to reason with regard to what has been; and, from what has actually been, we have data for concluding with regard to that which is to happen thereafter." With his book, Hutton popularized the notion of "examining things present...with regard to what has been," but gave the concept no specific name. Hutton did not use the term uniformitarianism and used the word 'uniformity' only rarely.

Charles Lyell (1797–1875), one of geology's founding fathers from later in the Enlightenment era, wrote about the subject matter of uniformitarianism (but did not use that specific term) in his widely read text, *Principles of Geology* (1830). Partly in response to strident criticism that his notions about geology did not conform to Biblical edicts about supernatural catastrophic events, Lyell developed a much more radical and extreme view of the subject matter of the "uniformity of nature." Careful reading of what Lyell laid out in his discussion of the "uniformity of nature" shows that he embraced both the concept of Hutton, which can be summarized as a uniformity of known causes or processes throughout time, and his own separate view that there must be a uniformity of process rates. The latter, more radical aspect of Lyell's "uniformity of nature" was intended to be a statement of general principle to counter the catastrophist interpretations of the past set forth by geologists of the day who were more inclined to look to the scriptures for their geological interpretations. In Lyell's view, a strong notion of uniformity of rates precluded divine (i.e., catastrophic) intervention.

In 1837, the name uniformitarianism was coined by William Whewell (1794–1866) as a term meant to convey Hutton's sense of order and regularity in the operation of nature and Lyell's sense that there was a uniformity of rates of geological processes through time. It is Whewell's definition that became the most common definition of uniformitarianism.

Lyell's work was influential, and he succeeded in imbuing generations of geologists with the notion of a dual foundation for "uniformity of nature." This dual foundation encompassed both uniformity of causes and uniformity of intensity. The former view is more commonly called actualism today, and the latter, gradualism. In large part, the presence of Lyell's strongly defended gradualism succeeded in freeing nineteenth century geology from the firm grasp of Biblical preconception and allowed it to develop as a free, legitimate science.

One of the most elegant statements about (what is now called) actualism was made by John Playfair in his book, *Illustrations of the Huttonian Theory* (1802). He said: "Amid all the revolutions of the globe the economy of Nature has been uniform, and her laws are the only thing that have resisted the general movement. The **rivers** and the rocks, the **seas**, and the continents have been changed in all their parts; but the laws which describe those changes, and the rules to which they are subject, have remained invariably the same." Actualism is not unique to geology, as it is really a basic and broad scientific concept of many fields. Even though Playfair mentions laws, it is, of course, nature herself that is constant, not laws, which have been written by people in order to try to predict nature.

The other side of Lyell's "uniformity of nature," i.e., gradualism, has no such elegant prose behind it. It has been referred to in inglorious terms by some of the leading minds of our time as "false and stifling to hypothesis formation," "a blatant lie," and "a superfluous term...best confined to the past history of geology." In other words, gradualism is no longer considered a valid idea.

Because uniformitarianism has this historical component of uniformity of process rates (i.e., gradualism), many writers have advocated its elimination from the geological vocabulary. Others argue that it should be retained, but with careful notation about its historical meaning. Some writers ignore this historical debate and continue to tout the term uniformitarianism as the most basic principle of geology. The range of misguided meanings of this term from some recent geology texts includes definitions that span the gamut from something near the nineteenth century meaning to the assumption that Earth is very old, to the logical method of geologic investigation.

Careful analysis of geological texts and recent scientific articles shows that there are at least 12 basic fallacies about uniformitarianism, which are perpetuated by some writers. These are:

1. Uniformitarianism is unique to geology;
2. Uniformitarianism was first discussed by James Hutton;
3. Uniformitarianism was named by Lyell, who gave us its modern meaning;
4. Uniformitarianism is the same as actualism, and should be renamed actualism;
5. Uniformitarianism holds that only processes that are currently active could have occurred in the geologic past;
6. Uniformitarianism holds that rates and intensities of geologic processes are constant through time;
7. Uniformitarianism holds that only non-catastrophic or gradual processes have operated during geologic time;
8. Uniformitarianism holds that Earth's conditions have changed little over geologic time;
9. Uniformitarianism holds that the earth is very old;
10. Uniformitarianism is a testable hypothesis, theory, or law;
11. Uniformitarianism applies to the past only as far back as present conditions have existed on the earth's surface;
12. Uniformitarianism holds only that the governing laws of nature are constant through **space** and geologic time.

In the historical analysis of uniformitarianism above, we have seen how these 12 common conceptions are false and misleading. Most scientists argue that uniformitarianism should be kept in its proper historical perspective in the future, and that a more specific term like actualism might supplant uniformitarianism in places where the word is meant to convey strictly the modern concept of uniformity of causes.

See also Geologic time; Stratigraphy

UNSATURATED ZONE

The unsaturated zone is that portion of the subsurface in which the intergranular openings of the geologic medium contain both **water** and air. The unsaturated zone, also known as the vadose zone or the zone of aeration, extends downward from the land surface to the top of the underlying **saturated zone**. Water in the pores of this zone is at a pressure that is lower than **atmospheric pressure**. Most of the water that eventually recharges the saturated zone must first pass through the unsaturated zone.

The movement of water in the unsaturated zone is dominated by capillary action. This tendency of a liquid to be drawn into interstices, is the result of cohesion of water molecules and adhesion of those molecules to the solid material forming the void. Smaller voids produce greater capillary forces frequently great enough to resist the downward force of **gravity**.

Water originating as **precipitation** at the land surface infiltrates into the unsaturated zone and forms a film on the surface of the material surrounding the pore. The adhesion of the water molecules nearest the solid material is greatest. As precipitation events occur, the thickness of the film increases with greater availability of infiltrating water, the capillary force is reduced in magnitude, and water molecules on the outer portion of the film may begin to flow under the influence of gravity. It is only through these transient variations in water availability, film thickness, and capillary pressure that water is able to migrate within the unsaturated zone.

Most plants utilize water from the unsaturated zone. Capillary water may be drawn from the pores by the plant's roots until the capillary forces can no longer be exceeded. This is the wilting point of the plant. The remaining capillary water can only be displaced through **evaporation**.

Research into the characteristics and dynamics of flow within the unsaturated zone is very active. A variety of human activities are controlled by, or may impact, the unsaturated zone. These include agriculture, subsurface pipeline and tank emplacement, and **waste disposal**. The thick unsaturated zone beneath Yucca Mountain in southern Nevada is currently slated for placement of the nation's high-level nuclear waste. The extremely thick unsaturated zone at the site would allow placement of the waste approximately 1,000 ft (305 m) above the saturated zone and nearly the same distance below the land surface. The nature of the unsaturated zone and its ability to isolate the waste from human contact is the subject of detailed investigation.

See also Hydrogeology; Porosity and permeability; Water table

URANUS • *see* SOLAR SYSTEM

UREY, HAROLD (1893-1981)

American chemist

In 1934, Harold Urey was awarded the Nobel Prize in chemistry for his discovery of deuterium, an isotope, or species, of hydrogen in which the atoms weigh twice as much as those in ordinary hydrogen. Also known as heavy hydrogen, deuterium became profoundly important to future studies in many scientific fields, including chemistry, **physics**, and medicine. Urey continued his research on isotopes over the next three decades, and during World War II, his experience with deuterium proved invaluable in efforts to separate isotopes of uranium from each other in the development of the first atomic bombs. Later, Urey's research on isotopes also led to a method for determining the earth's atmospheric **temperature** at various periods in past history. Already a scientist of great honor and achievement, Urey's last great period of research brought together his interests and experiences to a study of the **origin of life** on Earth. Urey's experimentation has become especially relevant because of concerns about the possibility of global **climate** change.

Urey hypothesized that the earth's primordial atmosphere consisted of reducing gases such as hydrogen, ammonia, and methane. The energy provided by electrical discharges in the atmosphere, he suggested, was sufficient to initiate chemical reactions among these gases, converting them to the simplest compounds of which living organisms are made, amino acids. In 1953, Urey's graduate student Stanley Lloyd Miller carried out a series of experiments to test this hypothesis. In these experiments, an electrical discharge passed through a **glass** tube containing only reducing gases resulted in the formation of amino acids.

The **Miller-Urey experiment** is a classic experiment in molecular biology and genetics. The experiment established that the conditions that existed in Earth's primitive atmosphere were sufficient to produce amino acids, the subunits of proteins comprising and required by living organisms. In essence, the Miller-Urey experiment fundamentally established that Earth's primitive atmosphere was capable of producing the building blocks of life from inorganic materials.

The Miller-Urey experiment also remains the subject of scientific debate. Scientists continue to explore the nature and composition of Earth's primitive atmosphere and thus, continue to debate the relative closeness of the conditions of the experimental conditions to Earth's primitive atmosphere.

Urey was born in Walkerton, Indiana. His father, Samuel Clayton Urey, was a schoolteacher and lay minister in the Church of the Brethren. His mother was Cora Reinoehl Urey. After graduating from high school, Urey hoped to attend college but lacked the financial resources to do so. Instead, he accepted teaching jobs in country schools, first in Indiana (1911–1912) and then in Montana (1912–1914) before finally entering Montana State University in September of 1914, at the age of 21. Urey was initially interested in a career in biology, and the first original research he ever conducted involved a study of microorganisms in the Missoula River. In 1917, he was awarded his Bachelor of Science degree in zoology by Montana State.

The year Urey graduated also marked the entry of the United States into World War I. Although he had strong pacifist beliefs as a result of his early religious training, Urey acknowledged his obligation to participate in the nation's war effort. As a result, he accepted a job at the Barrett Chemical Company in Philadelphia and worked to develop high explosives. In his Nobel Prize acceptance speech, Urey said that this experience was instrumental in his move from industrial chemistry to academic life.

At the end of the war, Urey returned to Montana State University, where he began teaching chemistry. In 1921, he decided to resume his college education and enrolled in the doctoral program in physical chemistry at the University of California at Berkeley. His faculty advisor at Berkeley was the great physical chemist Gilbert Newton Lewis. Urey received his doctorate in 1923 for research on the calculation of heat capacities and entropies (the degree of randomness in a system) of gases, based on information obtained through the use of a spectroscope. He then left for a year of postdoctoral study at the Institute for Theoretical Physics at the University of Copenhagen where **Niels Bohr**, a Danish physicist, was researching the structure of the **atom**. Urey's interest in Bohr's research had been cultivated while studying with Lewis, who had proposed many early theories on the nature of chemical bonding.

Upon his return to the United States in 1925, Urey accepted an appointment as an associate in chemistry at the Johns Hopkins University in Baltimore, a post he held until 1929. He briefly interrupted his work at Johns Hopkins to marry Frieda Daum in Lawrence, Kansas, on June 12, 1926. Daum was a bacteriologist and daughter of a prominent Lawrence educator. The Ureys later had four children.

In 1929, Urey left Johns Hopkins to become associate professor of chemistry at Columbia University, and in 1930 he published his first book, *Atoms, Molecules, and Quanta*, written with A. E. Ruark. Writing in the *Dictionary of Scientific Biography,* Joseph N. Tatarewicz called this work "the first comprehensive English language textbook on atomic structure and a major bridge between the new quantum physics and the field of chemistry." At this time he also began his search for an isotope of hydrogen. Since Frederick Soddy, an English chemist, discovered isotopes in 1913, scientists had been looking for isotopes of a number of elements. Urey believed that if an isotope of heavy hydrogen existed, one way to separate it from the ordinary hydrogen isotope would be through the vaporization of liquid hydrogen. Urey's subsequent isolation of deuterium made Urey famous in the scientific world, and only three years later he was awarded the Nobel Prize in chemistry for his discovery.

During the latter part of the 1930s, Urey extended his work on isotopes to other elements besides hydrogen. Urey found that the mass differences in isotopes can result in modest differences in their reaction rates.

The practical consequences of this discovery became apparent during World War II. In 1939, word reached the United States about the discovery of **nuclear fission** by the German scientists Otto Hahn and Fritz Strassmann. The military consequences of the Hahn-Strassmann discovery were apparent to many scientists, including Urey. He was one of the first, therefore, to become involved in the U.S. effort to build a nuclear weapon, recognizing the threat posed by such a weapon in the hands of Nazi Germany. However, Urey was deeply concerned about the potential destructiveness of a fission weapon. Actively involved in political topics during the 1930s, Urey was a member of the Committee to Defend America by Aiding the Allies and worked vigorously against the fascist regimes in Germany, Italy, and Spain. He explained the importance of his political activism by saying "no dictator knows enough to tell scientists what to do. Only in democratic nations can science flourish."

Urey worked on the Manhattan Project to build the nation's first atomic bomb. As a leading expert on the separation of isotopes, Urey made critical contributions to the solution of the Manhattan Project's single most difficult problem, the isolation of uranium–235.

At the conclusion of World War II, Urey left Columbia to join the Enrico Fermi Institute of Nuclear Studies at the University of Chicago, where Urey continued to work on new applications of his isotope research. During the late 1940s and early 1950s, he explored the relationship between the isotopes of **oxygen** and past planetary climates. Since isotopes differ in the rate of chemical reactions, Urey said that the amount of each oxygen isotope in an organism is a result of atmospheric temperatures. During periods when the earth was warmer than normal, organisms would take in more of a lighter isotope of oxygen and less of a heavier isotope. During cool periods, the differences among isotopic concentrations would not be as great. Over a period of time, Urey was able to develop a scale, or an "oxygen thermometer," that related the relative concentrations of oxygen isotopes in the shells of sea animals with atmospheric temperatures. Some of those studies continue to be highly relevant in current research on the possibilities of global climate change.

In the early 1950s, Urey became interested in yet another subject: the chemistry of the universe and of the formation of the planets, including Earth. One of his first papers on this topic attempted to provide an estimate of the relative abundance of the elements in the universe. Although these estimates have now been improved, they were remarkably close to the values modern chemists now accept.

In 1958, Urey left the University of Chicago to become Professor at Large at the University of California in San Diego at La Jolla. At La Jolla, his interests shifted from original scientific research to national scientific policy. He became extremely involved in the U.S. **space** program, serving as the first chairman of the Committee on Chemistry of Space and Exploration of the **Moon** and Planets of the National Academy of Science's Space Sciences Board. Even late in life, Urey continued to receive honors and awards from a grateful nation and admiring colleagues.

See also Evolution, evidence of; Evolution; Evolutionary mechanisms; Radioactivity

V

VALLEY GLACIER • *see* GLACIERS

VARVED DEPOSITS • *see* GLACIAL LANDFORMS

VENUS • *see* SOLAR SYSTEM

VINCI, LEONARDO DA (1452-1519)

Italian scientist and artist

A true Renaissance man, Leonardo da Vinci was a painter, inventor, scientist, architect, engineer, mathematician, astronomer, and philosopher. Although centuries after his death he remains known primarily as the artist who painted the "Last Supper" and the "Mona Lisa," Leonardo placed a stronger emphasis on his scientific rather than his artistic endeavors. His investigations into almost every field of known science in his time resulted in plans for everything from airplanes to air conditioning systems. Leonardo was also prolific in the field of mathematics and **physics**, including squaring the circle and calculating the velocity of a falling object.

Born in Vinci, near Florence, Italy, Leonardo was the illegitimate son of Ser Piero da Vinci, a notary, and a peasant woman. Leonardo's father recognized his genius early and ensured that he received a proper education in reading, writing, and arithmetic at his home. Leonardo never attended a university. Rather, at the age of 15, he was sent to Florence, where he became an apprentice painter under Italian sculptor and painter Andrea del Verrocchio (1436–1488). It was during this apprenticeship that Leonardo became absorbed in science, and his interest in technical and mechanical skills was already leading him to sketch various machines. In 1482, Leonardo entered the service of the Duke of Milan as the court painter and advisor on architecture and military issues. According to one report, after studying Euclid, Leonardo became so interested in geometry that he neglected his duties as court painter.

Leonardo's interest in mathematics soon led him to provide several approaches to squaring the circle (constructing a square with the same **area** as a given circle) using mechanical methods. In his notebooks, Leonardo described and drew plans for both a **telescope** and a mechanical calculator. Leonardo also formulated several accurate astronomical theories, including one which stated that Earth rotates around the **Sun**, and another stating the **Moon** shines because of the Sun's reflected light. Leonardo postulated that the shadowing image of the full moon that appears cradled between the horns of the crescent moon each month is illuminated by light reflected from the earth, a conclusion that was reached by German Astronomer **Johannes Kepler** (1571–1630) a century later. Through experimentation, Leonardo concluded that the velocity of a falling object is proportional to the time of its fall, predating Sir Isaac Newton's mathematical theory of force and **gravity**. Leonardo's greatest contribution to science and physics, however, may have been his belief that much of nature could be explained scientifically through a strict adherence to mathematical laws, a fundamental tenet of the philosophy of physics.

Leonardo was a keen observer of the rocks and **fossils** of his native Northern Italy. Among his 4,000 pages of unpublished notes (an unfinished encyclopedic work) are references to **sedimentation** occurring in the Arno riverbed and its floodplain, and observations of rainwater rushing downhill, carrying fossilized **rock** with it. Leonardo reasoned that the fossils he observed embedded in the rocks of mountains were not washed uphill, and therefore, the hillsides had once been the site of the sea floor. He furthered this argument with his identification of fossilized corals and oysters, found more than 100 mi (160 km) inland. In the layers of stratified rocks and fossils, Leonardo grasped the concept of **geologic time**.

The Duke of Milan was defeated by the French Armies in 1499, and the following years were nomadic for Leonardo as he traveled to Mantua and then Venice, where he consulted on architecture and military engineering (Leonardo's notebook included plans for a triple-tier machine gun). Leonardo then returned to Florence briefly and, in 1506, returned to Milan where he worked on various engineering projects. Leonardo spent from 1513 to 1516 in Rome, then moved to France, where King Francis I employed him as a painter, architect, and mechanic. By this time, Leonardo worked little on painting and devoted himself primarily to his scientific studies. Leonardo's thousands of sketches and notes focusing on both practical matters of his day and visions of future scientific accomplishments remain as a testament to Leonardo's prolific genius.

VINE, FRED J. (1939-)

English geophysicist

Frederick J. Vine is best known for his contributions to the theory of **plate tectonics** and has had a distinguished career as a geologist and geophysicist. Born in London, England, Vine was educated at Latymer Upper School, London, and St. John's College, Cambridge University. With his supervisor at Cambridge, **Drummond Matthews** (1931–1997), Vine did crucial work on the process of seafloor spreading.

The German scientist **Alfred Wegener** (1880–1930) proposed in 1915 that there had once been a super-continent, which he named Pangaea, that had slowly moved apart. However, Wegener's **continental drift theory** did not explain how such movement occurred, and was not well received. In the early 1960s, Harry Hess (1906–1969) hypothesized that seafloor spreading was responsible for the motion of the continents. In 1963, Vine and Matthews published a paper in *Nature* titled "Magnetic Anomalies Over Ocean Ridges." In this work, the two scientists proposed an idea which, if confirmed, would provide strong support for the seafloor spreading theory.

It had long been suspected, but not proven, that the earth's **magnetic field** has undergone a number of reversals in polarity in its long history. Vine and Matthews suggested that if ocean ridges were the sites of seafloor creation, and the earth's magnetic field does reverse, then new **lava** emerging would produce **rock** magnetized in the current magnetic field of the earth. Older rock would have an opposing polarity, depending on when it had been created. By 1966, further studies confirmed the theory for all **mid-ocean ridges**. This evidence provided compelling support for the ideas of Wegener, and Hess, and resulted in a revolution in the earth sciences, in which the overlooked theory of continental drift was wholeheartedly adopted.

Vine went on to have a distinguished career. With E. M. Moores he did important research on the **geology** of the Troodos mountains of southern Cyprus. He worked with R. A. Livermore and A. G. Smith on the history of Earth's magnetic field, and together with R. G. Ross he did groundbreaking experimental work on the electrical conductivity of rocks from the lower continental **crust**. From 1967–1970 he was assistant professor of geology and geophysics at Princeton University. Vine returned to the United Kingdom in 1970 and became a Reader in the School of Environmental Sciences at the University of East Anglia, Norwich. He was promoted to professor in 1974, and was Dean from 1977–1980, and again from 1993–1998. Since 1998, Vine has been a Professorial Fellow of the University of East Anglia. He has received a number of honors, including the Chapman Medal of the Royal Astronomical Society (1973), the Charles Chree medal and prize of the Institute of Physics (1977), the Hughes Medal of the Royal Society (1982), and the International Balzan Prize (1981)—all of which were shared with Drummond Matthews. He is also a Fellow of The Royal Society.

See also Mid-ocean ridges and rifts

VOLCANIC ERUPTIONS

A volcanic eruption is the release of molten **rock** and volcanic gases through Earth's **crust** to the surface. Molten rock within the earth, or **magma**, is driven to erupt by buoyancy because it is lighter than the surrounding rock. Dissolved gases within the magma are under great pressure and force magma upwards. The upward migrating magma takes advantage of preexisting zones of weaknesses such as fractures or established volcanic necks until it eventually breaks through the surface.

An eruption may last for a few minutes or many hours and days. An eruption may be only a discharge of steam and gases through a small vent, a relatively mild oozing of **lava** from a **fissure** in a shield **volcano**, or a spectacular explosion that shoots huge columns of gases and debris into the sky. The explosiveness of an eruption depends to a great extent on the composition of the molten rock. Magma high in silica will be more viscous than one low in silica. A high-viscosity magma (such as a **rhyolite**) will tend to trap dissolved gases. The pressure of the gases can build up to the point where they are released in a spontaneous explosive eruption. A less viscous magma (such as a **basalt**) allows volcanic gases to bubble through more easily, preventing great build-ups of pressure, and resulting in calmer outpourings of lava.

The length an eruption is described as an eruptive pulse, eruptive phase, or eruptive episode. An eruptive pulse is a very short event lasting a few seconds to minutes. An eruption that lasts a few hours to days and consists of numerous eruptive pulses is called an eruptive phase. Eruptions that involve repeated pulses and phases over days, months, or years is an eruptive episode.

Volcanic eruptions are described according to explosivity, lava type, and other constituents such as ash, gas, and steam content or the nature of rock fragments produced. Some common eruption types are named for classic types of volcanoes that characterize the eruption. These include Hawaiian, Plinian (Vesuvian), Strombolian, and Vulcanian. Some types of eruptions have more descriptive names, such as effusive and phreatic.

A Hawaiian-type eruption consists of a highly fluid basaltic lava that tends to flow effusively from linear fissures or from a central vent in the production of shield volcanoes. The release is not generally explosive as lava gently flows in streams or through lava tubes. Sometimes the lava accumu-

Mount Pinatubu eruption. *© T. J. Casadevall, U.S. Geological Survey Photographic Library, Denver, CO.*

lates in lava **lakes**. Occasionally, however, more spectacular fountains of lava spurting out from a vent do occur.

A Plinian, or Vesuvian, eruption is a more explosive and potentially destructive event where large amounts of ash, dust, and gas are blown out of a central source at a high velocity. The eruptive cloud often forms a large column extending high into the air above the volcano. Avalanches of hot ash, rock, and gas, called *nuee ardentes*, can travel down the side of the volcano at up to 100 mph (160 kph) are possible, such as the one that covered the Italian city of Pompeii. Rhyolitic to dacitic compositions are common. The name is derived from the historian Pliny, who recorded the eruption of Vesuvius in A.D. 79.

Strombolian eruptions are characterized by discrete episodic explosions or fountains of basaltic lava from a single vent or crater. The eruptive pulses are caused by the release of volcanic gases, and are separated by periods of a few seconds to hours. Lava fragments consisting of partially molten volcanic bombs that become rounded as they fly through the air are commonly produced.

Vulcanian, or hydrovolcanic eruptions are explosive events that release a combination of ash and steam into the air, producing an eruptive column. Fragments of lava are ejected, but owing to a high viscosity or previous cooling, the fragments do not form aerodynamic bombs. The composition of the lava is generally andesitic to dacitic.

An effusive eruption is a general term for any non-explosive release of lava. The lava gently wells up from the ground and overflows, cooling on its way down the slope. Effusive eruptions are common in a Hawaiian type event. When a basaltic effusive eruption occurs on the ocean floor, pillow lavas often form. As the name suggests, pillow basalts are rounded elongate shapes the lava takes due to extrusion under the pressure of the ocean. As pillow lavas continually erupt, they form stacked mounds of pillows. Effusive eruptions may occur with a range of compositions, although they are most common in low viscosity lavas such as basalt.

If cool ground **water** or surface water comes in contact with magma below the surface, a phreatic eruption may occur. This is caused by water that is heated into pressurized steam, creating an explosive eruption driven solely by the steam. Because the eruption is driven by steam, no new rock is formed.

See also Extrusive cooling; Fumerole; Hawaiian island formation; Hotspots; Lahar; Nuee ardent; Pipe, volcanic; Tuff; Volcanic vent

VOLCANIC VENT

Volcanic vents are openings in Earth's **crust** where molten **lava** and volcanic gases escape onto the land surface or into the atmosphere. Most volcanoes have a circular central vent near their summit crater that serves as a conduit for ongoing volcanic construction. Basaltic lavas that cool to form oceanic crust, oceanic plateaus, and continental flood basalts erupt from large, elongate, planar vents called fissures. New oceanic crust is created at axial fissures along the globe-encircling ocean ridge system. Small cracks and ducts in volcanic and hydrothermal provinces serve as vents for escaping lava, gas, and **water** that create smaller-scale volcanic features like gaseous fumaroles, hot **springs**, geysers, and rootless splatter cones called hornitos.

Each of the three main types of volcanoes—cinder cones, shields, and composite volcanoes—forms by eruption of lava, volcanic ash and gases from a central vent. A cinder cone, like Volcan Parícutin in Mexico, begins with an eruption from a vent in the land surface and grows into a steep-sloped, circular mountain as cinders from successive eruptions form a cone around the vent. Shield volcanoes, like the Hawaiian Islands, are composed of low-viscosity basaltic lava that flows easily and rapidly from a central vent. Though sometimes very large, shield volcanoes have a simple structure of stacked, low-angle lava flows around the central vent.

Composite volcanoes, or stratovolcanoes, are very large volcanic edifices composed of alternating layers of volcanic ash, volcanic ejecta and lava flows. Mt. Rainier in Washington, Cotopaxi in Ecuador, Mt. Etna in Sicily, and Mt. Fuji in Japan are stratovolcanos. Extremely large, pyroclastic eruptions of gas-charged, viscous lava issue from a central vent, or group of vents, in the summit crater of a composite **volcano**. However, because the andesitic and rhyolitic lava that composes a stratovolcano is so viscous, the central vent system is often plugged between large eruptions. Lava fills

Mount Etna. ***Photograph by Jonathan Blair. Jonathan Blair/Corbis-Bettmann. Reproduced by permission.***

fissures on the flanks of the mountain creating radial dikes. Gases and fluids also escape from secondary vents, creating fumaroles and hot springs on the slopes of a stratovolcano. When a composite volcano becomes dormant, **erosion** wears away the volcano, leaving the vertical column that cooled in the feeder duct beneath the volcanic vent. Shiprock in New Mexico and Devil's Tower in Wyoming are examples of volcanic necks that formed this way.

See also Mid-ocean ridges and rifts; Volcanic eruptions

VOLCANO

Volcanoes are vents or fissures in Earth's **crust** through which **lava**, gases, and pyroclastic debris are released. More commonly, the term volcano refers to the landform built up from the accumulation of lava and/or pyroclastic debris. Based on the timing of their last eruption, volcanoes are classified as active (having erupted during historic time), dormant (having no recent eruptions, but with the potential to erupt again), or extinct (having no historic eruptions and showing no evidence of future eruptions). There are currently over 500 active volcanoes on Earth's surface, including famous examples such as Mt. Fuji, Mt. St. Helens, and Mauna Loa. Mt. Vesuvius, which last erupted in A.D. 79, is an example of a dormant volcano; Mt. Kilimanjaro is an extinct volcano.

Fueled by Earth's internal processes, volcanoes occur primarily along plate boundaries but also form above hot spots. Eruptive activity may include lava flows, lateral blasts, ash flows, lahars, the release of volcanic gases, or any combination of these. Different types of volcanoes, each with a unique set of characteristics and eruptive styles, include shield volcanoes, composite volcanoes, lava domes, calderas, and cinder cones. Different types of **magma** form under different plate tectonic settings, and the type of magma present determines the type of volcano that will form in a given **area**.

Shield volcanoes, with their gentle slopes and curved profile, are the largest of all volcanoes. They are built up from repeated basaltic flows, often beginning at the ocean floor. Basaltic magma has a relatively low silica content, allowing it to flow readily. As a result, shield volcanoes are characterized by lava flows rather than explosive pyroclastic activity. Shield volcanoes are most commonly formed above hot spots under basaltic oceanic crust. They are also formed in areas where the mid-ocean ridge intersects with land, as in Iceland, or in areas of active **rifting**, like east **Africa**. In these areas, as the magma is rising to the surface, it mixes with only basaltic rocks, allowing it to preserve its **mafic** composition and flow readily. Probably the most famous shield volcanoes, Mauna Loa and Mauna Kea, currently rest above the Hawaiian hotspot. Measured from its base on the ocean floor to its summit, Mauna Kea is 5.6 mi (9 km) tall—slightly taller than Mt. Everest.

Composite volcanoes, also known as stratovolcanoes, have steep sides and a characteristic cone shape. They are built up from alternating layers of lava and pyroclastic debris. Lava associated with composite volcanoes generally has an intermediate composition, and is more resistant to flow than basaltic lava. This results in the mixture of flows and explosions. Composite volcanoes occur above subduction zones, where rising magma mixes with both oceanic and continental crust raising the overall silica content. They are ubiquitous along the subduction zones of the Pacific Rim, and some famous examples include Mt. Fuji in Japan and Mt. Rainier in Washington. Their ability to erupt explosively, as demonstrated by Mt. St. Helens in 1980, makes these some of the most dangerous volcanoes on Earth.

Lava domes are steep-sided, rounded domes, formed because of pressure exerted by rising viscous magma. **Rhyolite**, a **felsic** magma, is usually associated with lava domes. Its felsic composition makes it highly viscous, forcing it to move slowly, building up pressure and deforming the ground surface above. Lava domes are generally associated with composite volcanoes, although they can occur on their own. They are capable of causing deadly eruptions as tremendous amounts of built-up pressure are suddenly released in giant explosions. Eruption of a lava dome was responsible for the death and destruction caused by the 1902 eruption of Mt. Pelée on Martinique.

Calderas are massive depressions created by rare, violent explosions. Also associated with rhyolitic magma, **caldera** eruptions are capable of expelling enormous amounts of ash and debris in a single explosion. Calderas form where **hotspots** occur under continental crust. As magma rises, it mixes with the felsic continental crust, resulting in a high silica content. As is the case with lava domes, the resultant viscous magma cannot flow, and explodes when sufficient pressure has built up. Although there have been none in recent geologic history, about 600,000 years ago a large caldera eruption occurred at what is presently the site of Yellowstone National Park in Wyoming and Montana. The famous hot **springs** and geysers of the area are the legacy of that eruption, and it is believed that the site has the potential to produce another eruption in the future.

Cinder cones are steep-sided, cone-shaped, relatively small volcanoes that are formed by the accumulation of pyroclastic debris. They are not associated with any one particular lava type, and occur in a number of settings. They are commonly found on the flanks or inside the summit craters of larger volcanoes, and form when pyroclastic debris ejected by the main volcano accumulates to form the smaller cone. Perhaps the most famous cinder cone, Parícutin volcano in Mexico, grew suddenly out of a farmer's cornfield and within one month had risen to a height of almost 1,000 ft (305 m). Cinder cones tend to have short life spans; lava flows released by Parícutin eventually covered an extensive area, but within 10 years the volcano became dormant.

See also Convergent plate boundary; Nuee ardent

W

WALTHER, JOHANNES (1860-1937)

German geologist

Johannes Walther (1860–1937) was instrumental in the development of **stratigraphy**. Walther's two-volume work, *Modern Lithogenesis,* published in 1883 and 1884 was a pioneering work in classical sedimentary analysis.

Walther asserted that proper analysis of sedimentary facies could reveal important clues regarding the formation and movement of **rock**. Sedimentary facies are layers within a particular formation that are different in sedimentary history from surrounding layers within the same **area**. Facies may show vertical differentiation, lateral differentiation, or both characteristic differences. The differentiation defining a facies may be either lithological or paleontological.

Walther advanced the ontological method in the analysis of facies stratigraphy. Walther was an avid naturalist, devoted to fieldwork. His data reflected a passion for linking current observation to geologic history.

Near the turn of the century, Walther advanced what is now known as the Walther Facies Rule. Because sedimentary facies show vertical sequence **superposition**, a vertical progression of facies will reflect lateral facies changes. Sedimentary layers or rocks essentially preserve the environment of their deposition. These **depositional environments** change and the old depositional layers shift laterally and may transgress (become superimposed) on surrounding deposits. Regardless, these chronologically transgressive layers will show similar vertical and lateral succession. Walther's rule thus related lateral facies changes to vertical changes (vertical succession). Walther's rule provided a powerful explanation that facies and surrounding deposits change and shift laterally as Earth's surface undergoes change and that lithostratigraphy often reflected layers and formations that can not always be accurately used to date the formations (e.g., where **unconformities** exist). In essence, Walther's rule placed a limit on lithostratigraphic analysis and placed additional reliance upon paleontological analysis of the **fossil record**.

Walther's work was not immediately put to wide use, and Walther suffered the same isolation experienced by many German scientists in an early twentieth century and post World War I environment often hostile to Germany and German scientists. Ultimately Walther's work was appreciated for its wealth of data regarding sedimentary processes and sedimentary facies.

Walther's advancement of facies analysis ultimately proved highly useful in the prediction of formations that might contain **petroleum**, specific minerals, or ores of economic value.

Walther was also an avid and accomplished painter of natural scenes.

See also Dating methods; Marine transgression and marine regression; Petroleum detection; Sedimentary rocks; Sedimentation

WASHINGTON, WARREN M. (1936-)

American meteorologist

Warren M. Washington is an atmospheric scientist whose research focuses on the development of computer models that describe and predict the Earth's **climate**. He is the director of the Climate and Global Dynamics Division of the National Center for Atmospheric Research (NCAR), in Boulder, Colorado. He has advised the U.S. Congress and several U.S. presidents on climate-system modeling, serving on the President's National Advisory Committee on **Oceans** and Atmosphere from 1978 to 1984.

Washington was born in Portland, Oregon. His father, Edwin Washington Jr., had hoped to be a schoolteacher, but in the 1920s, Portland wouldn't hire African-Americans to teach in the public schools. Instead, the elder Washington supported Warren and his four brothers by waiting tables in Pullman cars. His wife, Dorothy Grace (Morton) Washington, became a practical nurse, after the Washington children were grown.

Washington's interest in scientific research developed early and was nurtured by high school teachers who encouraged him to experiment. Refusing once to directly answer his question about why egg yolks were yellow, a **chemistry** teacher inspired Washington to study chicken diets and eventually to learn about the chemistry of sulfur compounds. Despite the boy's aptitude for science, Washington's high school counselor advised him to attend a business school rather than college, but Washington's dream was to be a scientist. He earned his bachelor's degree in **physics** in 1958, from Oregon State University. As an undergraduate, Washington became interested in **meteorology** while working on a project at a **weather** station near the campus. As part of the project, the station used radar equipment to track storms as they came in off the coast. In 1960, he earned his master's degree in meteorology from Oregon State. When he completed his graduate work in 1964 at Pennsylvania State University, he became one of only four African Americans to receive a doctorate in meteorology.

Washington began working for the NCAR in 1963 and has remained affiliated with that institution throughout his career. His research there has attempted to quantify patterns of oceanic and **atmospheric circulation**. He has helped to create complex mathematical models that take into account the effects of surface and air **temperature**, **soil** and atmospheric moisture, sea **ice** volume, various geographical features, and other parameters on past and current climates. His research has contributed to our modern-day understanding of the **greenhouse effect**, in which excess **carbon dioxide** in Earth's atmosphere causes the retention of heat, giving rise to what is known as **global warming**. Washington's research also provided understanding for other mechanisms of global climate change.

Washington was appointed the director of the Climate and Global Dynamics Division at NCAR in 1987. In 1994, he was elected President of the American Meteorological Society. He is a fellow of the American Association for the Advancement of Science and a member of its board of directors, a fellow of the African Scientific Institute, a Distinguished Alumnus of Pennsylvania State University, a fellow of Oregon State University, and Founder and President of the Black Environmental Science Trust, a nonprofit foundation that encourages African-American participation in environmental research and policymaking.

Washington has published over 100 professional articles about atmospheric science. He co-authored, with Claire Parkinson, *An Introduction to Three-Dimensional Climate Modeling* in 1986, and the book has since become a standard reference text for climate modeling. Washington has six children and 10 grandchildren.

Waste disposal

Waste management is the handling of discarded materials. Recycling and composting, which transform waste into useful products, are forms of waste management. The management of waste also includes disposal, such as landfilling.

Waste can be almost anything, including food, leaves, newspapers, bottles, construction debris, chemicals from a factory, candy wrappers, disposable diapers, old cars, or radioactive materials. People have always produced waste, but as industry and technology have evolved and the human population has grown, waste management has become increasingly complex.

A primary objective of waste management today is to protect the public and the environment from potentially harmful effects of waste. Some waste materials are normally safe, but can become hazardous if not managed properly. For example, 1 gal (3.75 l) of used motor oil can potentially contaminate one million gal (3,790,000 l) of drinking **water**.

Every individual, business, or organization must make decisions and take some responsibility regarding the management of his or her waste. On a larger scale, government agencies at the local, state, and federal levels enact and enforce regulations governing waste management. These agencies also educate the public about proper waste management. In addition, local government agencies may provide disposal or recycling services, or they may hire or authorize private companies to perform those functions.

Throughout history, there have been four basic methods of managing waste: dumping it, burning it, finding another use for it (reuse and recycling), and not creating the waste in the first place (waste prevention). How those four methods are utilized depends on the wastes being managed. Municipal solid waste is different from industrial, agricultural, or mining waste. Hazardous waste is a category that should be handled separately, although it sometimes is generated with the other types.

The first humans did not worry much about waste management. They simply left their garbage where it dropped. However, as permanent communities developed, people began to dispose of their waste in designated dumping areas. The use of such "open dumps" for garbage is still common in many parts of the world. Open dumps have major disadvantages, however, especially in heavily populated areas. Toxic chemicals can filter down through a dump and contaminate **groundwater**. The liquid that filters through a dump or landfill is called leachate. Dumps may also generate methane, a flammable and explosive gas produced when organic wastes decompose under anaerobic (oxygen-poor) conditions.

The landfill, also known as the "sanitary landfill," was invented in England in the 1920s. At a landfill, the garbage is compacted and covered at the end of every day with several inches of **soil**. Landfilling became common in the United States in the 1940s. By the late 1950s, it was the dominant method for disposing municipal solid waste in the nation.

Early landfills had significant problems with leachate and methane, but those have largely been resolved at facilities built since about the early 1970s. Well-engineered landfills are lined with several feet of **clay** and with thick plastic sheets. Leachate is collected at the bottom, drained through pipes, and processed. Methane gas is also safely piped out of many landfills.

The dumping of waste does not just take place on land. Ocean dumping, in which barges carry garbage out to sea, was once used as a disposal method by some United States coastal cities and is still practiced by some nations. Sewage sludge, or waste material from sewage treatment, was dumped at sea in huge quantities by New York City as recently as 1992, but this is now prohibited in the United States. Also called biosolids, sewage sludge is not generally considered solid waste, but it is sometimes composted with organic municipal solid waste.

Burning has a long history in municipal solid waste management. Some American cities began to burn their garbage in the late nineteenth century in devices called cremators. These were not very efficient, however, and cities went back to dumping and other methods. In the 1930s and 1940s, many cities built new types of more-efficient garbage burners known as incinerators. The early incinerators were rather dirty in terms of their emissions of air pollutants, and beginning in the 1950s they were gradually shut down.

However, in the 1970s, waste burning enjoyed another revival. These newer incinerators, many of which are still in operation, are called "resource recovery" or "waste-to-energy" plants. In addition to burning garbage, they produce heat or **electricity** that can be used in nearby buildings or residences, or sold to a utility. Many local governments became interested in waste-to-energy plants following the energy crisis in 1973. However, since the mid-1980s, it became difficult to find locations to build these facilities, mainly because of public opposition focused on air-quality issues.

Another problem with incineration is that it generates ash, which must be landfilled. Incinerators usually reduce the volume of garbage by 70–90%. The remainder of the incinerated waste comes out as ash that often contains high concentrations of toxic substances.

Municipal solid waste will likely always be landfilled or burned to some extent. In the past 25 years, however, non-disposal methods such as waste prevention and recycling have become more common. Because of public concerns and the high costs of landfilling and burning (especially to build new facilities), local governments want to reduce the amount of waste that must be disposed in these ways.

Municipal solid waste is a relatively small part of the overall waste generated in the United States. More than 95% of the total 4.5 billion tons of solid waste generated in the United States each year is agricultural, mining, or industrial waste.

These wastes do not receive nearly as much attention as municipal solid waste, because most people do not have direct experience with them. Also, agricultural and mining wastes, which make up 88% of the overall total of solid waste, are largely handled at the places they are generated, that is, in the fields or at remote mining sites.

Mining nearly always generates substantial waste, whether the material being mined is **coal**, clay, **sand**, gravel, building stone, or metallic ore. Early mining concentrated on the richest lodes of **minerals**. Because modern methods of mining are more efficient, they can extract the desired minerals from veins that are less rich. However, much more waste is produced in the process.

Many of the plant and animal wastes generated by agriculture remain in the fields or rangelands. These wastes can be beneficial because they return organic matter and nutrients to the soil. However, modern techniques of raising large numbers of animals in small areas generate huge volumes of animal waste, or manure. Waste in such concentrated quantities must be managed carefully, or it can contaminate groundwater or surface water.

Industrial wastes that are not hazardous have traditionally been sent to landfills or incinerators. The rising cost of disposal has prompted many companies to seek alternative methods for handling these wastes, such as waste prevention and recycling. Often a manufacturing plant can reclaim certain waste materials by feeding them back into the production process.

Hazardous wastes are materials considered harmful or potentially harmful to human health or the environment. Wastes may be deemed hazardous because they are poisonous, flammable, or corrosive, or because they react with other substances in a dangerous way.

Industrial operations have produced large quantities of hazardous waste for hundreds of years. Some hazardous wastes, such as mercury and dioxins, may be released as gases or vapors. Many hazardous industrial wastes are in liquid form. One of the greatest risks is that these wastes will contaminate water supplies.

An estimated 60% of all hazardous industrial waste in the United States is disposed using a method called deep-well injection. With this technique, liquid wastes are injected through a well into an impervious **rock** formation that keeps the waste isolated from groundwater and surface water. Other methods of underground burial are also used to dispose hazardous industrial waste and other types of dangerous material.

Pesticides used in farming may contaminate agricultural waste. Because of the enormous volumes of pesticides used in agriculture, the proper handling of unused pesticides is a daunting challenge for waste managers. Certain mining techniques also utilize toxic chemicals. Piles of mining and metal-processing waste, known as waste rock and tailings, may contain hazardous substances. Because of a reaction with the **oxygen** in the air, large amounts of toxic acids may form in waste rock and tailings and leach into surface waters.

Public attitudes also play a pivotal role in decisions about waste management. Virtually every proposed new landfill or waste-to-energy plant is opposed by people who live near the site. Public officials and planners refer to this reaction as NIMBY, which stands for "Not In My BackYard." If an opposition group becomes vocal or powerful enough, a city or county council is not likely to approve a proposed waste-disposal project. The public also wields considerable influence with businesses. Recycling and waste prevention initiatives enjoy strong public support. About 19% of United States municipal solid waste was recycled or composted in 1994, 10% was incinerated, and 71% was landfilled.

Preventing or reducing waste is typically the least expensive method for managing waste. Waste prevention may also reduce the amount of resources needed to manufacture or package a product. For example, most roll-on deodorants once came in a plastic bottle, which was inside a box. Beginning about 1992, deodorant manufacturers redesigned the bottle so that it would not tip-over easily on store shelves, which eliminated the need for the box as packaging. This is the type of waste prevention called source reduction. It can save businesses money, while also reducing waste.

Waste prevention includes many different practices that result in using fewer materials or products, or using materials that are less toxic. For example, a chain of clothing stores can ship its products to its stores in reusable garment bags, instead of disposable plastic bags. Manufacturers of household batteries can reduce the amount of mercury in their batteries. In an office, employees can copy documents on both sides of a sheet of paper, instead of just one side. A family can use cloth instead of paper napkins.

Composting grass clippings and tree leaves at home, rather than having them picked up for disposal or municipal composting, is another form of waste prevention. A resident can leave grass clippings on the lawn after mowing (this is known as grass-cycling), or can compost leaves and grass in a backyard composting bin, or use them as a mulch in the garden.

When the current recycling boom began in the late 1980s, markets for the recyclables were not sufficiently considered. A result was that some recyclable materials were collected in large quantities but could not be sold, and some ended up going to landfills. Today, the development of recycling markets is a high priority. "Close the loop" is a catchphrase in recycling education; it means that true recycling (i.e., the recycling loop) has not taken place until the new product is purchased and used.

To boost recycling markets, many local and state governments now require that their own agencies purchase and use products made from recycled materials. In a major step forward for recycling, President Bill Clinton issued an executive order in 1993 requiring the federal government to use more recycled products.

Many managers of government recycling programs feel that manufacturers should take more responsibility for the disposal of their products and packaging, rather than letting municipalities bear the brunt of the disposal costs. An innovative and controversial law in Germany requires manufacturers to set up collection and recycling programs for disused packaging of their products.

The high cost of government-created recycling programs is often criticized. Supporters of recycling argue it is still less expensive than landfilling or incineration, when all costs are considered. Another concern about recycling is that the recycling process itself may generate hazardous wastes that must be treated and disposed.

Recycling of construction and demolition (C&D) debris is one of the growth areas for recycling. Although C&D debris is not normally considered a type of municipal solid waste, millions of tons of it have gone to municipal landfills over the years. If this material is separated at the construction or demolition site into separate piles of concrete, wood, and steel, it can usually be recycled.

Composting is considered either a form of recycling, or a close relative. Composting occurs when organic waste—such as yard waste, food waste, and paper—is broken down by microbial processes. The resulting material, known as compost, can be used by landscapers and gardeners to improve the fertility of their soil.

Yard waste, primarily grass clippings and tree leaves, makes up about one-fifth of the weight of municipal solid waste. Some states do not allow this waste to be disposed. These yard-waste bans have resulted in rapid growth for municipal composting programs. In these programs, yard waste is collected by trucks (separately from garbage and recyclables) and taken to a composting plant, where it is chopped up, heaped, and regularly turned until it becomes compost.

Waste from food-processing plants and produce trimmings from grocery stores are composted in some parts of the country. Residential food waste is the next frontier for composting. The city of Halifax, in Canada, collects food waste from households and composts it in large, central facilities.

Biological treatment, a technique for handling hazardous wastes, could be called a high-tech form of composting. Like composting, biological treatment employs microbes to break down wastes through a series of metabolic reactions. Many substances that are toxic, carcinogenic (cancer-causing), or undesirable in the environment for other reasons can be rendered harmless through this method.

Extensive research on biological treatment is in progress. Genetic engineering, a controversial branch of biology dealing with the modification of genetic codes, is closely linked with biological treatment, and could produce significant advances in this field.

Waste management became a particularly expensive proposition during the 1990s, especially for disposal. Consequently, waste managers constantly seek innovations that will improve efficiency and reduce costs. Several new ideas in land-filling involve the reclamation of useful resources from wastes.

For example, instead of just burning or releasing the methane gas that is generated within solid-waste landfills, some operators collect this gas, and then use it to produce power locally or sell it as fuel. At a few landfills, managers have experimented with a bold but relatively untested concept known as landfill mining. This involves digging up an existing landfill to recover recyclable materials, and sometimes to re-bury the garbage more efficiently. Landfill mining has been criticized as costly and impractical, but some operators believe it can save money under certain circumstances.

In the high-tech world of incineration, new designs and concepts are constantly being tried. One waste-to-energy technology for solid waste being introduced to the United States is called fluidized-bed incineration. About 40% of incinerators in Japan use this technology, which is designed to have lower emissions of some air pollutants than conventional incinerators.

A 1994 United States Supreme Court ruling could increase the cost of incineration significantly. The Court ruled that some ash produced by municipal solid-waste incinerators must be treated as a hazardous waste, because of high levels of toxic substances such as **lead** and cadmium. This means that incinerator ash now has to be tested, and part or all of the material may have to go to a hazardous waste landfill rather than a standard landfill.

A much smaller type of incinerator is used at many hospitals to burn medical wastes, such as blood, surgical waste, syringes, and laboratory waste. The safety of these medical waste incinerators has become a major issue in some communities. A study by the Environmental Protection Agency released in 1994 found that medical waste incinerators were leading sources of dioxin emissions into the air. The same study warned that dioxins, which can be formed by the burning of certain chemical compounds, pose a high risk of causing cancer and other health hazards in humans.

The greatest impetus for waste prevention will likely come from the public. More and more citizens will come to understand that pesticides, excessive packaging, and the use of disposable rather than durable items have important environmental costs. Through the growth of the information society, knowledge about these and other environmental issues will increase. This should result in a continuing evolution towards more efficient and environmentally sensitive waste management.

See also Atmospheric pollution; Greenhouse gases and greenhouse effect; Water pollution and biological purification

WASTEWATER TREATMENT

Wastewater often mixes with free-flowing **water** in **rivers**, streams, **oceans**, **lakes**, and other bodies of water. The addition of wastewater can radically alter the chemistry—and the ecological dynamics—of water bodies and hydrologic reservoirs. Wastewater includes the sewage-bearing water that is flushed down toilets as well as the water used to wash dishes and for bathing. Processing plants use water to wash raw material and in other stages of the wastewater treatment production process. The treatment of water that exits households, processing plants, and other institutions is a standard, even mandated, practice in many countries around the world. The purpose of the treatment is to remove compounds and microorganisms that could pollute the water to which the wastewater is discharged. Particularly with respect to microorganisms, the sewage entering a treatment plant contains extremely high numbers of bacteria, viruses, and protozoa that can cause disease if present in drinking water. Wastewater treatment lowers the numbers of such disease-causing microbes to levels that are deemed to be acceptable from a health standpoint. As well, organic matter, solids, and other pollutants that can add to stream load are removed.

Wastewater treatment is usually a multi-stage process. Typically, the first step is known as the preliminary treatment. This step removes or grinds up large material that would otherwise clog up the tanks and equipment further on in the treatment process. Large matter can be retained by screens or ground up by passage through a grinder. Examples of items that are removed at this stage are rags, **sand**, plastic objects, and sticks.

The next step is known as primary treatment. The wastewater is held for a period of time in a tank. Solids in the water settle out while grease, which does not mix with water, floats to the surface. Skimmers can pass along the top and bottom of the holding tank to remove the solids and the grease. The clarified water passes to the next treatment stage, which is known as secondary treatment.

During secondary treatment, the action of microorganisms is often utilized. There are three versions of secondary treatment. One version, which was developed in the mid-nineteenth century, is called the fixed film system. The fixed film in such a system is a film of microorganisms that has developed on a support such as rocks, sand, or plastic. If the film is in the form of a sheet, the wastewater can be overlaid on the fixed film. The domestic septic system represents such a type of fixed film. Alternatively, the sheets can be positioned on a rotating arm, which can slowly sweep the microbial films through the tank of wastewater. The microorganisms are able to extract organic and inorganic material from the wastewater to use as nutrients for growth and reproduction. As the microbial film thickens and matures, the metabolic activity of the film increases. In this way, much of the organic and inorganic load in the wastewater can be removed.

Another version of secondary treatment is called the suspended film. Instead of being fixed on a support, microorganisms are suspended in the wastewater. As the microbes acquire nutrients and grow, they form aggregates that settle out. The settled material is referred to as sludge. The sludge can be scraped up and removed. As well, some of the sludge is added back to the wastewater. This is analogous to inoculating growth media with microorganisms. The microbes in the sludge now have a source of nutrients to support more growth, which further depletes the wastewater of the organic waste. This cycle can be repeated a number of times on the same volume of water.

Sludge can be digested and the methane that has been formed by bacterial fermentation can be collected. Burning of the methane can be used to produce **electricity**. The sludge can also be dried and processed for use as compost.

A third version of secondary treatment utilizes a specially constructed lagoon. Wastewater is added to a lagoon and the sewage is naturally degraded over the course of a few months. The algae and bacteria in the lagoon consume nutrients such as phosphorus and nitrogen. Bacterial activity produces **carbon dioxide**. Algae can utilize this gas, and the resulting algal activity produces **oxygen** that **fuels** bacterial activity. A cycle of microbiological activity is established.

Bacteria and other microorganisms are removed from the wastewater during the last treatment step. Basically, the final treatment involves the addition of disinfectants, such as chlorine compounds or **ozone**, to the water, passage of the water past ultraviolet lamps, or passage of the water under pressure through membranes whose very small pore size impedes the passage of the microbes. In the case of ultraviolet irradiation, the wavelength of the lamplight is lethally disruptive to the genetic material of the microorganisms. In the case of disinfectants, neutralization of the high concentration of the chemical might be necessary prior to discharge of the treated water to a river, stream, lake, or other body of water. For example, chlorinated water can be treated with sulfur dioxide.

Chlorination remains the standard method for the final treatment of wastewater. However, the use of the other systems is becoming more popular. Ozone treatment is popular in **Europe**, and membrane-based or ultraviolet treatments are increasingly used as a supplement to chlorination.

Water is often called the "universal solvent." ***Courtesy of Kelly A. Quin.***

Within the past several decades, the use of sequential treatments that rely on the presence of living material such as plants to treat wastewater by filtration or metabolic use of the pollutants has become more popular. These systems have been popularly dubbed "living machines." Restoration of wastewater to near drinking water quality is possible.

Wastewater treatment is usually subject to local and national standards of operational performance and quality in order to ensure that the treated water is of sufficient quality so as to pose no threat to aquatic life or settlements downstream that draw the water for drinking.

See also Aquifer; Artesian; Drainage basins and drainage patterns; Drainage calculations and engineering; Hydrogeology; Stream capacity and competence

WATER

Water is a chemical compound composed of a single **oxygen atom** bonded to two hydrogen atoms (H2O) that are separated by an angle of 105°. Because of their polar covalent bonds and this asymmetrical bent arrangement, water molecules have a tendency to orient themselves in an electric field, with the positively charged hydrogen toward the negative pole and the negatively charged oxygen toward the positive pole. This tendency results in water having a large dielectric constant, which is responsible for making water an excellent solvent. Water is therefore referred to as the universal solvent. Water can be reused indefinitely as a solvent because it undergoes almost no modification in the process.

Because mineral salts and organic materials can dissolve in water, it is the ideal medium for transporting products of geochemical **weathering** as well as life-sustaining **minerals** and nutrients into and through animal and plant bodies. Brackish and ocean waters may contain large quantities of sodium chloride as well as many other soluble compounds leached from Earth's **crust**.

The concentration of mineral salts in ocean water is about 35,000 parts per million. Water is considered to be potable (drinkable) only if it contains less than 500 parts per million of salts.

Hydrogen bonding, which joins water molecule to water molecule, is responsible for other properties that make water a unique substance. These properties include its large heat capacity, which causes water to act as a moderator of **temperature** fluctuations due to variations in solar illumination, its high surface tension (due to cohesion among water molecules), and its adherence to other substances, such as the walls of a vessel (due to adhesion between water molecules and the molecules of a second substance). The high surface tension makes it possible for surface-gliding insects and broad, flat objects to be supported on the surface of water. Adhesion of water molecules to **soil** particles is the primary mechanism by which water moves through unsaturated soils.

Hydrogen bonding is also responsible for **ice** being less dense than water. If ice did not float, all bodies of water would freeze from the bottom up, becoming solid masses of ice and destroying all life in them. In addition, from season to season, frozen water bodies would remain frozen, resulting in large changes in **climate** and **weather**, such as decreased **precipitation** due to reduced **evaporation**. Ice floats because as the temperature of water is lowered the tendency of water to contract as its molecular motion decreases is overcome by the strength of hydrogen bonding between molecules. At 4°C (39°F), water molecules start to structure themselves directionally along the lines of the hydrogen bonds, at angles of 105° As the temperature drops toward 0°C (32°F), spaces develop between the lines until the open, crystalline form characteristic of ice develops. Its openness produces a density slightly less than that of liquid water, and ice floats on the surface, with approximately nine-tenths submerged.

Water is the only common substance that occurs naturally on earth in three different physical states. The solid state, ice, is characterized by a rigid crystalline structure occurring at or below 0°C (32°F) and occupying a definite volume (found as **glaciers** and ice caps, as snow, hail, and frost, and as **clouds** formed of ice **crystals**). At sea level **atmospheric pressure**, the liquid state exists over a definite temperature range 0°C to 100°C (32 to 212°F), but is not rigid nor does it have a particular shape. Liquid water has a definite volume but assumes the shape of its container. Liquid water covers three-fourths of Earth's surface in the form of swamps, **lakes**, **rivers**, and **oceans** as well as found as rain clouds, dew, and ground water. The gaseous state of water (water vapor) neither occupies a definite volume nor is rigid because it takes on the exact shape and volume of its container. Water vapor (liquid water molecules suspended in the air) occurs in steam, **humidity**, **fog**, and clouds.

During phase changes, one phase does not suddenly replace its predecessor as the temperature changes, but for a time at the **melting** or boiling point, two phases will coexist. As water changes from the gaseous form to the liquid form, it gives off heat at about 540 calories per gram, and as it changes from the liquid form to the solid form, it gives off about 80 calories per gram. The turbulence of thunderstorms is in large part due to the release latent heat of water especially as water condenses into water droplets or into crystals of ice (i.e., hail).

Pressure affects the transition temperature between phases. For example, at pressures below atmospheric, water boils at temperatures under 100°C (212°F), therefore food takes longer to cook at higher elevations.

Water is a major geologic agent of change for modifying Earth's surface through **erosion** by water and ice.

See also Acid rain; Atmospheric chemistry; Chemical bonds and physical properties; Chemical elements; Clouds and cloud types; Condensation; El Nino and La Nina phenomena; Erosion; Evaporation; Freezing and melting; Freshwater; Rate factors in geologic processes

WATER POLLUTION AND BIOLOGICAL PURIFICATION

Water pollution may derive from several sources, including chemical pollutants from industry, **runoff** of chemicals used in agriculture, or debris from geological process, but the greatest source of pollution is organic waste. Although chemical pollutants may become diluted, they can also radically alter the ecosystem to allow the overproduction of certain forms of algae and bacteria that pollute the water with respect to its use by humans.

Once in the water, the growth of microorganisms can be exacerbated by environmental factors such as the water **temperature** and the chemical composition of the water. For example, runoff of fertilizers from suburban properties can infuse watercourses with nitrogen, potassium, and phosphorus. All these are desirable nutrients for bacterial growth.

With specific respect to microorganisms, water pollution refers to the presence in water of microbes that originated from the intestinal tract of humans and other warm-blooded animals. Water pollution can also refer to the presence of compounds that promote the growth of the microbes. The remediation of polluted water—the removal of the potentially harmful microorganisms or the reduction of their numbers to acceptable levels—represents the purification of water.

Microorganisms that reside in the intestinal tract find their way into fresh and marine water when feces contaminate the water. Examples of bacteria that can pollute water in this way are *Escherichia coli*, *Salmonella*, *Shigella*, and *Vibrio cholerae*. Warm-blooded animals other than humans can also contribute protozoan parasites to the water via their feces. The two prominent examples of health relevance to humans are *Cryptosporidium parvum* and *Giardia lamblia*. The latter two species are becoming more common. They are also resistant to chlorine, the most popular purification chemical.

Normally, the intestinal bacteria do not survive long in the inhospitable world of the water. However, if they are ingested while still living, they can cause maladies, ranging from inconvenient intestinal upset to life-threatening infections. An example of the latter is *Escherichia coli O157:H7*. Pollution of the water with this strain can cause severe intestinal damage, life-long damage to organs such as the kidney, and—especially in the young, elderly, and those whose immune systems are compromised—death.

There are several common ways in which microorganisms can pollute water. Runoff from agricultural establish-

Raw sewage being carried by the tide. ***© Ecoscene/Corbis. Reproduced by permission.***

ments, particularly where livestock is raised, is one route of contamination. Seasonal runoff can occur, especially in the springtime when rainfall is more pronounced.

Water purification seeks to convert the polluted water into water that is acceptable for drinking, for recreation, or for some other purpose. Techniques such as filtration and exposure to agents or chemicals that will kill the microorganisms in the water are common means of purification. The use of chlorination remains the most widely used purification option. Other approaches are the use of ultraviolet radiation, filters of extremely small pore size (such that even viruses are excluded), and the use of a chemical known as **ozone**. Depending on the situation and the intended use of the finished water, combinations of these techniques can be used.

Purification of drinking water aims to remove as many bacteria as possible, and eliminate those bacteria of intestinal origin. Recreational waters need not be pristine. But bacterial numbers need to be below whatever standard has been deemed permissible for the particular locale.

Another microbiological aspect of water pollution that has become recognized only within the past several years has been the presence in water of agents used to treat bacteria in other environments. For example, a number of disinfectant compounds are routinely employed in the cleaning of household surfaces. In the hospital, the use of antibiotics to kill bacteria is an everyday occurrence. Such materials have been detected in water both before and after municipal **wastewater treatment**. The health effect of these compounds is not known at the present time. However, looking at similar situations, the low concentration of such compounds might propogate the development of resistant bacterial populations.

Natural wetlands also contribute to the purification of water. Wetlands can serve as a depositional sump and provide biological filtering. Normal percolation through **soil** layers also provides a significant source of water purification.

See also Aquifer; Artesian; Drainage basins and drainage patterns; Estuary; Hydrologic cycle

Water table

In common usage, the term **water** table expresses the surface dividing the unsaturated and saturated **groundwater** zones. More accurately, the water table lies within the **saturated zone** and separates the capillary fringe from the underlying phreatic zone. The phreatic zone is the **area** in which water will freely flow from pores in the geologic material. Within the capillary fringe, however, water is drawn upward from the phreatic zone by capillary action within the pores of the material. Smaller pores produce greater capillary force and cause the water to rise higher, resulting in a thicker capillary fringe. The pores in the capillary fringe are fully saturated, as are those in the phreatic zone. However, the capillary action causes the water in the pores to have a pressure that is lower than **atmospheric pressure**. Water is not able to flow out of these voids.

A more precise definition of the water table is the surface within the saturated zone along which the **hydrostatic pressure** is equal to the atmospheric pressure. Water below this surface has a pressure that is greater than atmospheric pressure while water in voids above is at a pressure that is less than atmospheric pressure.

The shape of the water table is controlled by a number of factors including the water-transmitting characteristics of the geologic medium, and the amount and location of groundwater recharge and discharge. The water table often reflects the surface **topography** of the area with a moderated **relief**. Because mountainous areas have greater **precipitation**, water infiltrating these areas recharges the water table and forms a mound. Groundwater discharges at streams, **lakes**, and wells cause the water table to **dip** toward these points. Between points of recharge and discharge, the water table tilts from areas of high potential (recharge) to areas of low potential (discharge).

The water table moves with changes in the hydrologic system. In periods of **drought** or heavy withdrawals, the water table will fall. During periods of precipitation, the water table moves upward.

A true water table exists only in unconfined aquifers, those where water can percolate directly through the overlying medium to the phreatic zone and the water table is free to move up and down. In confined aquifers, overlying **rock** or sediments with lower **permeability** prevent the water table from moving upward beyond the lower limit of the confining bed. The water in the confined **aquifer** is under pressure and the level to which it would rise, in the absence of the confining bed, is known as the potentiometric surface.

In some instances, a perched water table can form within the **unsaturated zone**, well above the regional water table. This occurs when a layer of low-permeability material, such as a **clay** lens, intercepts percolating water causing it to pool above the layer. A localized phreatic zone forms with the perched water table as its upper limit.

See also Hydrogeology; Porosity and permeability; Springs

WAVE MOTIONS

With regard to **Earth science**, wave motion describes the physical transmission of force or energy potential through a medium of transmission. The transmission disturbs the medium by displacing the medium. For example, **water** waves propagate through displacement (not linear movement) of water molecules; sound waves propagate via displacement of air molecules. Light also propagates via wave—but not in the same manner as water and sound. Light is transmitted via electromagnetic waves, the alternating of disturbances in electrical and magnetic fields.

A single equation is all that is needed to understand wave motion. The first attempt to mathematically describe wave motion was made by Jean Le Rond d'Alembert in 1747. His equation sought to explain the motion of vibrating strings. While d'Alembert's equation was correct, it was overly simplistic. In 1749, the wave equation was improved upon by Leonhard Euler; he began to apply d'Alembert's theories to all wave forms, not just strings. For more than seventy years the equations of Euler and d'Alembert were debated among the European scientific community, most of whom disagreed upon the universality of their mathematics.

In 1822, Jean-Baptiste-Joseph Fourier proved that an equation governing all waves could be derived using an infinite series of sines and cosines. The final equation was provided by John William Strutt (Lord Rayleigh) in 1877, and it is his law of wave motion that is used today. All waves have certain properties in common: they all transmit a change in energy state, whether it be mechanical, electromagnetic, or other; they all require some point of origin and energy source; and almost all move through some sort of medium (with the exception of electromagnetic waves, which travel most efficiently through a vacuum).

There are three physical characteristics that all wave forms have in common—wavelength, frequency, and velocity—and it is this common bond that allows the wave equation to apply to all wave types. In order to understand these physical characteristics, consider one of the most familiar wave forms, the water wave. As a wave passes through water, it forms high and low areas called, respectively, crests and troughs. The wavelength of the water wave is the minimum distance between two identical points, for example, the distance between two consecutive crests or two consecutive troughs. Imagine the water wave striking a barrier, such as a sea wall: the wave will splash against the wall, followed shortly by another, and so on. The amount of time between each splash (the rate at which the wave repeats itself) is the frequency of the wave. Generally, wavelength and frequency are inversely proportional: the higher the frequency, the shorter the wavelength. The final physical characteristic, velocity, is dependent upon the type of wave generated. A mechanical wave, such as our water wave, will move relatively slowly; a sound wave will move much faster (about 1,129 feet or 344 meters per second) while a light wave moves faster still (186,000 miles or 299,200 km per second in a vacuum). It is important to note that while a wave will move through a medium, it does not carry the medium with it. This is hard to picture in our water example, since it appears as if the water does move with the wave. A cork placed in the water moves up and down with the passing of the wave but returns essentially to the same location.

See also Electricity and magnetism; Quantum electrodynamics (QED); Quantum theory and mechanics; Relativity theory; Solar energy

Stratospheric balloon being inflated. ***National Center for Atmospheric Research/University. Corporation for Atmospheric Research/National Science Foundation 221. Reproduced by permission.***

WEATHER BALLOON

The invention of the **weather** balloon inaugurated the age of **remote sensing**, the ability to collect information from unmanned sources. Use of weather balloons is now common in advanced atmospheric research. High altitude weather balloons have also been used by astronomers and cosmologists seeking to take readings of certain particle frequencies or gather light readings free of excessive disturbance from Earth's relatively thick lower atmosphere (**troposphere**).

The first observation balloon was launched immediately before the first manned balloon flight by Frenchmen Jean-François de Rozier and the Marquis d'Aalandes on November 21, 1783, for a pre-flight **wind** reading. Later, French meteorologist Leon Teisserenc de Bort (1855-1913) pioneered the use of weather balloons, handily proving their utility. With balloon-acquired data, he determined the existence of a lower level of the atmosphere, which he termed the troposphere or "sphere of change," where weather takes place. Since the 1930s, when radio tracking systems were invented, balloons have been used as complete floating weather stations, employing such instruments as thermometers, barometers, hygrometers, cameras, and telescopes.

A variety of agencies use weather balloon flights to model the atmosphere and to make more accurate weather predictions. Weather balloons are used widely to collect such atmospheric information as **temperature**, pressure, and **humidity** that can then be plotted on weather maps. Three-dimensional atmospheric modeling is also possible using weather balloons because the instruments they carry are able to provide meteorologists and other atmospheric scientists data collected from a number of altitude points. Since their inception, the elongated bags of helium, a lighter than air element that provides the balloon lift, have been carrying aloft increasingly sophisticated observation devices, taking the science of weather observation literally to the edges of outer **space**.

See also Air masses and fronts; Atmospheric composition and structure; Atmospheric pressure; Atomic mass and weight; History of exploration III (Modern era); History of manned space exploration; International Council of Scientific Unions' World Data Center System; Jet stream; Stratosphere and stratopause; Weather forecasting methods

WEATHER AND CLIMATE

Weather refers to the atmospheric conditions at a certain time or over a certain short period in a given **area**. It is described by a number of meteorological phenomena that include **atmospheric pressure**, **wind** speed and direction, **temperature**, **humidity**, sunshine, cloudiness, and **precipitation**. In contrast, climate refers to long-term, cyclic or seasonal patterns of temperature, precipitation, winds, etc.

Climates are often defined in terms of area, **latitude**, altitude, or other geophysical features. Although there are thousands of microclimate variations, climates can essentially be broken down into four basic types. Hot, moist climates feature high rainfall with often intense and rapid chemical **weathering**. Cold, moist climates still feature chemical weathering but because of the lower temperature, the rates are dramatically reduced from those encountered in hot, moist climates. Cold, dry climates feature the least weathering but mechanical weathering (e.g., **ice** wedging) does produce slow **landscape evolution**. Hot, dry climates often have intense mechanical weathering pressures (e.g., wind, sand-blasting, etc).

The effects of weather also contribute in shaping Earth's surface features. The impact of weather is most pronounced during the occurrence of extreme weather situations, such as prolonged periods of heat, cold, rain, **drought**, and **smog** conditions. In addition, shorter but intense events such as hurricanes, tornadoes, winter **blizzards**, **freezing** rain, and **floods** also produce often-dramatic effects on both the social and geologic landscape. The concern to reduce the impact of weather on public health and property provides an important motivation for the continued efforts by meteorologists and scientists to improve **weather forecasting**.

The study of meteorological phenomena related to both weather and climate changes is an important component in the development of **chaos theory**. Chaos theories are used to study weather-related complex systems in which, out of seemingly random, disordered processes, there arise new processes that are more predictable.

Most of the weather elements on which weather forecasting is based cannot be seen directly, they can only be observed by the effects they create. For the most part, weather variables are measured and recorded by instruments. For example, air subjects everything to considerable pressure. At sea level, the atmosphere exerts approximately 15 lb/in^2 (about 1 kg/cm^2) of pressure. The standard instrument used to measure atmospheric pressure is the mercury barometer. The **physics** for the barometer dates to the classic experiments performed for the first time in 1643 by the Italian scientist **Evangelista Torricelli** (1608–1647). A column of mercury is held in a closed **glass** tube, then inverted and immersed into a mercury dish. The weight of the column is thus balanced by the atmospheric pressure and the length of the column affords a measure of that weight. The mean atmospheric pressure at sea level is 760 mmHg or 1,013 millibars. Pressure as well as air density decrease with increasing altitude and barometric pressure will rise or fall as a function of different weather systems. On weather maps, points of equal pressure are represented by **isobars**.

Wind, by its broadest definition, is any air mass in motion relative to Earth's surface. It is predominantly a horizontal movement. However, localized vertical air motion—updraft or downdraft—also occurs, for example in storms. Wind is described by two quantities: speed and direction. Wind velocity as measured by the anemometer is reported in mi/hr, knots, or km/hr. The wind direction is given by the compass bearing from which the wind blows, for example, a southerly wind blows from the south. The horizontal air movement near Earth's surface is controlled by four forces: the pressure gradient force, the Coriolis force, the centrifugal force, and the frictional forces. The existence of barometric differences in the atmosphere sets up the pressure gradient force that causes air to move from a higher to a lower pressure area. The Coriolis force is the apparent deflection of air mass caused by the **rotation** of Earth. Because of Earth's rotation, there is an apparent deflection of all matter in motion to the right of their path in the northern hemisphere and to the left in the southern hemisphere. For this reason, in the northern hemisphere, high-pressure systems (area of atmospheric divergence) rotate clockwise, low-pressure systems (areas of atmospheric convergence) counterclockwise. These rotational patterns are reversed in the southern hemisphere.

Temperature and humidity are crucial in defining the origins and types of air masses. The thermal properties of an air mass are determined by its latitudinal position on the globe, and its moisture content depends on the underlying surface, be it land or **water**. For example, polar air is cold and dry, whereas tropical air is hot and humid. In essence, the convergence of these two types of air masses is responsible for most global weather activities. The clash of these contrasting air masses leads to the formation of frontal wave depressions moving in an oscillating west-east pattern and steered by the upper-air **jet stream**. Hot, humid tropical air is also the source material that **fuels** the devastating force of hurricanes. Across the network of weather stations, readings of temperature and humidity are taken at regular intervals. Standard equipment in an instrumentation shelter consists of a dry and a wet bulb thermometer, and readings from the two are used to establish the **dew point**. A pair of special thermometers measures the maximum and minimum temperatures occurring during day and nighttime. The hygrometer measures the relative humidity of the air. In fully-automated stations, electronic sensors measure and transmit weather information.

In addition to temperature and humidity, daily weather forecasts inform the public about the heat index during summer and about the **wind chill** index during the winter. These indicators warn the about the possible dangers to human health resulting from exposure to summer heat and winter cold. By combining temperature and humidity, the heat index gives a measure of what temperatures actually feel like. In terms of human health, an increased heat index corresponds to physical activity being more exhausting, resulting in possible heat-related illnesses, cramps, exhaustion, or heatstroke. By contrast, the wind chill factor relates the risk of cold to exposed skin, which may lead to frostbite and hypothermia. The wind chill factor takes into account the effect of wind speed on temperature. For example, a temperature of 20°F (–6.66°C) at a wind-speed of 20 mph (32.18 km/hr) will feel like –10°F (–12.2°C). Humidity is the one factor that not only creates weather activity, but also makes life on Earth possible.

Water exists in one of the following three phases: vapor, liquid, or ice. Water vapor, the invisible gaseous form of water, is always present in the atmosphere; it is defined as the partial pressure of the atmosphere and therefore, like air pressure, it is measured in mmHg. Water vapor supplies the moisture for dew and frost, for **clouds** and **fog**, and for wet and frozen forms of precipitation.

The visible weather elements are, of course, sunshine, clouds, and precipitation. Traditionally, the forecasting of weather was mainly based on the observation of clouds, because their size, shape, and location are the visible indicators of air movement and of changes in water going from vapor to liquid or ice. The first important contribution to the classification of clouds was made in 1802 by the English scientist **Luke Howard**. Based on his observations, clouds were grouped according to three basic shapes: cumulus (heaps), stratus (layers), and cirrus (wispy curls). He also attached the term nimbus to clouds associated with precipitation. From this basic scheme has evolved the modern classification system of clouds by which the lower 10 mi (16 km) of the atmosphere are divided into three layers of clouds characterized by their water phase, i.e., low clouds consisting of water droplets, middle clouds containing a mixture of water droplets and ice **crystals**, and high clouds entirely made up of ice crystals. While some types of clouds are confined to one layer—such as stratus, stratocumulus and smaller type cumuli in the lower layer, altocumulus and altostratus in the middle layer, and cirrus and cirrostratus in the higher layer—other types can occupy two layers, namely, the nimbostratus and the swelling cumulus cloud which can reside in both lower and middle layers, as well as the cirrocumulus found in the middle and higher layers. A third type can expand through all three layers, such as the huge cumulus congestus cloud and of course, the cumulonimbus with its characteristic anvil.

Warm and cold fronts are also distinct in their cloud cover. The first signs of an approaching warm front are the cirrus and cirrostratus clouds, followed by the obscuring altostratus and the thick nimbostratus with continuous precipitation, and occasionally with the formation of patches of stratus clouds. After the passage of the warm front, precipitation ceases and the cloud cover breaks up. The typical cloud of cold fronts is the cumulonimbus and, depending on the instability of the air, nimbostratus. Precipitation will vary from brief showers to heavy, prolonged downpours with **thunder** and **lightning**.

The weather's immediate impact on public health has been demonstrated numerous times by severe events like hurricanes, tornadoes, floods, snow and ice storms, and prolonged periods of extreme heat or cold. In past years, considerable research efforts have been deployed to gain a better understanding of the physics of hurricanes and tornadoes. Better forecasting the path of severe weather systems and broadcasting early warnings has helped decrease the occurrence of weather-related deaths and injuries. Concerns are now increasingly focused on the weather's indirect influence on human health. It has been observed that certain weather situations provide conditions that will, for example, foster the proliferation of insects and consequently the spread of disease. This was the case in 1999 in the eastern regions of the United States, where weeks of drought and heat created the perfect breeding conditions for mosquitoes carrying a type of encephalitis virus. Weather conditions can also heighten the effects of pollution. For example, air pollutants trapped in fog or smog may cause severe respiratory problems. The interrelationship of weather and environmental health issues lends urgency for more **meteorology** research in order to develop the accurate forecasting capabilities required to lower the impact of adverse weather and climate changes on public health.

See also Air masses and fronts; Atmospheric chemistry; Atmospheric circulation; Atmospheric composition and structure; Atmospheric inversion layers; Atmospheric pressure; Drought; El Niño and La Nina phenomena; Hydrologic cycle; Isobars; Jet stream; Land and sea breeze; Lightning; Ocean circulation and currents; Seasonal winds; Thunder; Tornado; Tropical cyclone; Weather forecasting methods; Weather radar; Weather satellite; Wind chill; Wind

WEATHER FORECASTING

Weather forecasting is the attempt by meteorologists to predict the state of the atmosphere at some future time and the weather conditions that may be expected. Weather forecasting is the single most important practical reason for the existence of **meteorology** as a science. It is obvious that knowing the future of the weather can be important for individuals and organizations. Accurate weather forecasts can tell a farmer the best time to plant, an airport control tower what information to send to planes that are landing and taking off, and residents of a coastal region when a hurricane might strike.

Humans have been looking for ways to forecast the weather for centuries. The Greek natural philosopher Theophrastus wrote a *Book of Signs*, in about 300 B.C. listing more than 200 ways of knowing when to expect rain, **wind**, fair conditions, and other kinds of weather.

Scientifically-based weather forecasting was not possible until meteorologists were able to collect data about current weather conditions from a relatively widespread system of observing stations and organize that data in a timely fashion. By the 1930s, these conditions had been met. Vilhelm and Jacob Bjerknes developed a weather station network in the 1920s that allowed for the collection of regional weather data. The weather data collected by the network could be transmitted nearly instantaneously by use of the telegraph, invented in the 1830s by Samuel F. B. Morse. The age of scientific forecasting, also referred to as synoptic forecasting, was under way.

In the United States, weather forecasting is the responsibility of the National Weather Service (NWS), a division of the National Oceanic and Atmospheric Administration (NOAA) of the Department of Commerce. NWS maintains more than 400 field offices and observatories in all 50 states and overseas. The future modernized structure of the NWS will include 116 weather forecast offices (WFO) and 13 river forecast centers, all collocated with WFOs. WFOs also collect data from ships at sea all over the world and from meteorological satellites circling Earth. Each year the Service collects nearly four million pieces of information about atmospheric conditions from these sources.

The information collected by WFOs is used in the weather forecasting work of NWS. The data is processed by nine National Centers for Environmental Prediction (NCEP). Each center has a specific weather-related responsibility: seven of the centers focus on weather prediction—the Aviation Weather Center, the **Climate** Prediction Center, the Hydrometeorological Prediction Center, the Marine Prediction Center, the Space Environment Center, the Storm Prediction Center, and the Tropical Prediction Center—while the other two centers develop and run complex computer models of the atmosphere and provide support to the other centers—the Environmental Prediction Center and NCEP Central Operations. Severe weather systems such as thunderstorms, tornadoes, and hurricanes are monitored at the National Storm Prediction Center in Norman, Oklahoma, and the National Hurricane Center in Miami, Florida. Hurricane watches and warnings are issued by the National Hurricane Center's Tropical Prediction Center in Miami, Florida, (serving the Atlantic, Caribbean, **Gulf of Mexico**, and eastern Pacific Ocean) and by the Forecast Office in Honolulu, Hawaii, (serving the central Pacific). WFOs, other government agencies, and private meteorological services rely on NCEP's information, and many of the weather forecasts in the paper, and on radio and television, originate at NCEP.

Global weather data are collected at more than 1,000 observation points around the world and then sent to central stations maintained by the World Meteorological Organization, a division of the United Nations. Global data also are sent to NWS's NCEPs for analysis and publication.

The less one knows about the way the atmosphere works the simpler weather forecasting appears to be. For example, if **clouds** appear in the sky and a light rain begins to fall, one might predict that rain will continue throughout the day. This type of weather forecast is known as a persistent forecast. A persistent forecast assumes the weather over a particular geographic **area** simply will continue into the future. The validity of persistent forecasting lasts for a few hours, but not much longer because weather conditions result from a complex interaction of many factors that still are not well understood and that may change rapidly.

A somewhat more reliable approach to weather forecasting is known as the steady-state or trend method. This method is based on the knowledge that weather conditions are strongly influenced by the movement of air masses that often can be charted quite accurately. A weather map might show that a cold front is moving across the Great Plains of the United States from west to east with an average speed of 10 mph (16 kph). It might be reasonable to predict that the front would reach a place 100 mi (160 km) to the east in a matter of 10 hours. Since characteristic types of weather often are associated with cold fronts it then might be reasonable to predict the weather at locations east of the front with some degree of confidence.

A similar approach to forecasting is called the analogue method because it uses analogies between existing weather maps and similar maps from the past. For example, suppose a weather map for December 10, 2002, is found to be almost identical with a weather map for January 8, 1993. Because the weather for the earlier date is already known it might be reasonable to predict similar weather patterns for the later date.

Another form of weather forecasting makes use of statistical probability. In some locations on Earth's surface, one can safely predict the weather because a consistent pattern has already been established. In parts of Peru, it rains no more than a few inches per century. A weather forecaster in this region might feel confident that he or she could predict clear skies for tomorrow with a 99.9% chance of being correct.

The complexity of atmospheric conditions is reflected in the fact that none of the forecasting methods outlined above is dependable for more than a few days, at best. This reality does not prevent meteorologists from attempting to make long-term forecasts. These forecasts might predict the weather a few weeks, a few months, or even a year in advance. One of the best known (although not necessarily the most accurate) of long-term forecasts is found in the annual edition of the *Farmer's Almanac*.

The basis for long-range forecasting is a statistical analysis of weather conditions over an area in the past. For example, a forecaster might determine that the average snow fall in December in Grand **Rapids**, Michigan, over the past 30 years had been 15.8 in (40.1 cm). A reasonable way to try estimating next year's snowfall in Grand Rapids would be to assume that it might be close to 15.8 inches (40.1 cm).

Today this kind of statistical data is augmented by studies of global conditions such as winds in the upper atmosphere and ocean temperatures. If a forecaster knows that the **jet stream** over Canada has been diverted southward from its normal flow for a period of months, that change might alter **precipitation** patterns over Grand Rapids over the next few months.

The term "numerical" weather prediction is something of a misnomer because all forms of forecasting make use of numerical data such as **temperature**, **atmospheric pressure**, and **humidity**. More precisely, numerical weather prediction refers to forecasts that are obtained by using complex mathematical calculations carried out with high-speed computers.

Numerical weather prediction is based on mathematical models of the atmosphere. A mathematical model is a system of equations that attempt to describe the properties of the atmosphere and changes that may take place within it. These equations can be written because the gases that comprise the atmosphere obey the same physical and chemical laws that gases on Earth's surface follow. For example, Charles' Law says that when a gas is heated, it tends to expand. This law applies to gases in the atmosphere as it does to gases in a laboratory.

The technical problem that meteorologists face is that atmospheric gases are influenced by many different physical and chemical factors at the same time. A gas that expands according to Charles' Law may also be decomposing because of chemical forces acting on it. How can anyone make use of all the different chemical and physical laws operating in the atmosphere to come up with a forecast of future atmospheric conditions? The answer is mathematically complex. The task is not too much for computers, however. Computers can perform a series of calculations in a few hours that would take a meteorologist his or her whole lifetime to finish.

In numerical weather predicting, meteorologists select a group of equations that describe the conditions of the atmosphere as completely as possible for any one location at any one time. This set of equations can never be complete because even a computer is limited as to the number of calculations it can complete in a reasonable time. Thus, meteorologists pick out the factors they think are most important in influencing the development of atmospheric conditions. These equations are fed into the computer. After a certain period of time, the computer will print out the changes that might be expected if atmospheric gases behave according to the scientific laws to which they are subject. From this printout a meteorologist can make a forecast of the weather in an area in the future.

The accuracy of numerical weather predictions depend primarily on two factors. First, the more data that is available to a computer, the more accurate its results. Second, the faster the speed of the computer, the more calculations it can perform, and the more accurate its report will be. In the period from 1955 (when computers were first used in weather forecasting) to the current time, the percent skill of forecasts has improved from about 30% to more than 60%. The percent skill measure was invented to describe the likelihood that a weather forecast will be more accurate than pure chance.

Forecast accuracy also is difficult to judge because the average person's expectations probably have increased as the percent skill of forecasts also has increased. A hundred years ago, few people would have expected to have much idea as to what the weather would be like 24 hours in the future. Today,

Using mathematical models to automatically analyze data, calculators and computers gave meteorologists the ability to process large amounts of data and to make complex calculations quickly. ***AP/Wide World. Reproduced by permission.***

an accurate next-day forecast often is possible. For periods of less than a day, a forecast covering an area of 100 mi^2 (259 km^2) is likely to be quite dependable.

See also Air masses and fronts; Atmospheric chemistry; Atmospheric circulation; Atmospheric composition and structure; Atmospheric inversion layers; Drought; El Niño and La Nina phenomena; Hydrologic cycle; Isobars; Land and sea breeze; Lightning; Ocean circulation and currents; Thunder; Tornado; Tropical cyclone; Weather forecasting methods; Weather radar; Weather satellite; Wind chill

WEATHER FORECASTING METHODS

Modern **weather forecasting** owes its existence to the invention of many recording **weather** instruments, such as the barometer, hygrometer, **weather balloon**, and radar. Yet, three major technological developments in particular have led weather forecasting from its days of inception to its current status: the development of instant communications beginning in the late 1800s, **remote sensing** devices starting in the early 1900s, and computers in the late 1900s.

Weather recording instruments date from the fifteenth century when **Leonardo da Vinci** invented the hygrometer, an instrument to measure atmospheric **humidity**. About 1643, **Evangelista Torricelli** created the barometer to measure air pressure differences. These instruments were improved upon in the eighteenth century by Frenchman Jean Andre Deluc (1727–1817), and have been refined numerous times since then. Weather information has long been displayed in map form. In 1686, English astronomer **Edmond Halley** (1656–1742) drafted a map to explain regular winds, tradewinds, and monsoons. Over 200 years later, in 1863, French astronomer Edme Hippolyte Marie-Davy (1820–1893) published the first isobar maps, which depicted barometric pressure differences. Weather data allowed scientists to try to forecast weather. The United States Weather Service, established in 1870 under the supervision of Cleveland Abbe, unified communications and forecasting. Telegraph networks made it possible to collect and disseminate weather reports and predictions. By the turn of the twentieth century, the telephone and radio further increased meteorologists' ability to collect and exchange information. Remote sensing, the ability to collect information from unmanned sources, originated with the invention of the weather balloon by Frenchman Leon Teisserenc de Bort (1855–1913). Designed to make simple preflight tests of **wind** patterns, balloons were eventually used as complete floating weather stations with the addition of a radio transmitter to the balloon's instruments. Many scientists added to the pool of meteorological knowledge, including Englishman Ralph Abercromby who, in his 1887 book, *Weather*, depicted a model of a depression that was used for many years.

During World War I, the father-son team of Vilhelm Bjerknes (1862–1951) and Jacob Bjerknes (1897–1975) organized a nationwide weather-observing system in their native Norway. With the available data they formulated the theory of polar fronts: The atmosphere is made up of cold air masses near the poles and warm tropical air masses, and fronts exist where these air masses meet. In the 1940s, Englishman R. C. Sutcliffe and Swede S. Peterssen developed three-dimensional analysis and forecasting methods. American military pilots flying above the Pacific during World War II discovered a strong stream of air rapidly flowing from west to east, which became known as the **jet stream**. The development of radar, rockets, and satellites greatly improved data collection. **Weather radar** first came into use in the United States in 1949 with the efforts of Horace Byers (1906–1998) and R. R. Braham. Conventional weather radar shows **precipitation** location and intensity.

In the 1990s, the more advanced Doppler radar, which can continuously measure wind speed in addition to precipitation location and intensity, came into wide use. Using mathematical models to automatically analyze data, calculators and computers gave meteorologists the ability to process large amounts of data and to make complex calculations quickly. Today the integration of communications, remote sensing, and computer systems makes it possible to predict the weather almost simultaneously. Weather satellites, the first launched in 1960, can now produce sequence photography showing cloud and frontal movements, water-vapor concentrations, and **temperature** changes. With the new radar and computer enhancement, such as coloration, professionals and untrained viewers can better visualize weather information and use it in their daily lives.

See also Air masses and fronts; El Niño and La Nina phenomena; Isobars; Meteorology

WEATHER MAPS • *see* WEATHER FORECASTING METHODS

Doppler radar dishes. ***National Oceanic and Atmospheric Administration (NOAA).***

WEATHER RADAR

Radio Detection And Ranging systems, known as radar, were developed in Britain in the 1930s as a defense against German bombing raids. While their military use flourished during World War II, radar was not used commercially until the 1950s. Today, radar has become commonplace. Flight crews routinely use radar-tracking features to navigate aircraft to their destinations safely. Radar is also commonly used by meteorologists to track **weather** patterns. For most television viewers of the weather forecast, the image of a green, circular radar screen—complete with a sweeping arm of light—is a familiar one. Using a high-intensity microwave transmission, meteorologists can detect and follow large masses of **precipitation**, whether they occur as rain, snow, or **clouds**.

A weather radar **projection** begins with a pulsed microwave beam that travels until it hits an obstacle (for meteorological purposes, a cloud or band of precipitation). It is then reflected back to the source, where it is received by a radar antenna. By measuring the time taken for the signal to reach the obstacle and return, its distance can be easily calculated. With thousands of pulses emitting and returning, a two-dimensional image of the weather formation is displayed on a cathode-ray tube, showing its precise position. A more elaborate version of radar tracking, called Doppler radar, uses a continuous signal rather than a pulsed wave. Doppler radar can determine both the direction and velocity of **wind** patterns, as well as areas of precipitation. Doppler radar measures the shift in frequency caused by a moving particle. If the returning frequency is higher than when transmitted, the particle is moving toward the source; if it is lower, the particle is moving away. However, the system only works when a particle is approaching or receding from the transmitter; Doppler radar cannot detect the velocity of a particle moving perpendicular to the radar signal. For this reason, signals from more than one radar source must be combined to produce an image free of gaps.

Unlike standard radar, a Doppler system can reliably detect the presence of funnel clouds and tornadoes, and is now used quite commonly by weather forecasters, as well as radio and television stations, to monitor thunderstorms for the presence of strong winds and tornadoes. Doppler radar can provide potentially life-saving readings at a relatively small cost increase over standard radar.

See also Air masses and fronts; Weather forecasting methods; Weather satellite

Weather satellite

The first attempt to look at Earth's **weather** from **space** occurred early in the space program of the United States. In 1959, Vanguard II was launched with light-sensitive cells able to provide information about Earth's cloud cover. Unfortunately, the **satellite** tumbled in orbit and was unable to return any information. Explorer VI, also launched in 1959, was more successful and transmitted the first photographs of Earth's atmosphere from space.

In 1960, the United States launched the first experimental weather satellite, TIROS 1. The acronym for Television and Infra Red Observation Satellite, TIROS 1 televised over 22,000 photos before it failed six weeks later. It detected potential hurricanes days before they could have been spotted by any other means. It watched the spring breakup of the **ice** in the St. Lawrence River and helped forecast weather for the Antarctic bases. TIROS 1 also used infrared detectors to measure the amount of heat radiated by the earth's surface and the **clouds**. Later versions of TIROS improved upon the original with television cameras that provided direct, real-time readouts of pictures to simple stations around the world. In 1970, ITQS-l was launched with the capability of not only direct-readout, automatic picture transmission but also the ability to store global images for later transmission and processing. Another successful series was called NOAA after the National Oceanic and Atmospheric Administration. Some of these satellites were placed in geostationary orbit (moving at the same speed as Earth) and thus were able to continuously observe one **area**. This helped in the detection of severe storms and tornadoes and provided real-time coverage at an earlier stage of cloud and frontal weather movements. Other TIROS-type satellites, such as NIMBUS (1960s) and NOAA-9 (1980s–1990s), are in polar orbit, where their infrared sensors measure temperatures and **water** vapor over the entire globe.

Several GOES (Geostationary Operational Environmental Satellites) also cover the western and eastern hemispheres. These satellites are able to provide weather reports for places that have not been covered very well in the past: ocean regions, deserts, and polar areas. They also trace hurricanes, typhoons, and tropical storms, in the process save many lives. Their data are used to produce state-of-the-art charts showing sea-surface temperatures, information useful to the shipping and fishing industries. New satellites that probe Earth's atmosphere by day and night in all weather are being developed in many countries. Since the weather satellite is now an established tool of meteorologists all over the world, both developed and developing nations will continue to rely on these crafts.

See also Weather forecasting methods; Weather radar

Weather systems and weather fronts • *see* Air masses and fronts

Weathering and weathering series

Weathering is the *in situ* (in position) breakdown of rocks by natural forces into sediments or chemical constituents. Weathering may be physical or chemical. Physical weathering is the mechanical disintegration of rocks into finer particles. Chemical weathering is the decomposition of rocks according to the weathering series, a list of **minerals** arranged in order of their relative chemical stability at the earth's surface.

Chemical weathering occurs via a variety of processes such as dissolution, oxidation, hydration, or carbonation. These processes alter minerals at the molecular level either producing, as weathering products, different minerals or non-mineral chemical constituents. Based on their chemical stability, some minerals are more susceptible to the agents and processes of chemical weathering than others. A mineral's stability is determined to a large extent by the conditions under which it formed. Many igneous and metamorphic minerals that equilibrated deep within the earth will be less stable in the very different conditions found at the surface of the earth. These minerals will then be more susceptible to the agents of chemical weathering. This susceptibility follows a general progression called the weathering series. Below is the weathering series with the least stable minerals, the ones that will **weather** first, at the top. It progresses downward toward the more stable and long-lived minerals.

- Olivine—Calcic plagioclase
- Augite
- Hornblende—Alkalic plagioclase
- Biotite
- Potassium feldspar
- Muscovite
- Quartz

The minerals in the weathering series are essentially the same as **Bowen's Reaction Series**, the order in which minerals crystallize from **magma**. It differs, however, in that the weathering series is not a successive progression of weathering products. It does not mean, for example, that **olivine** will break down to form augite, merely that olivine will tend to decompose before augite, if both are present.

Physical weathering is the mechanical fragmentation of **rock** in place. It is differentiated from **erosion,** or mass wastage, which involve the transport of material. Physical weathering is accomplished dominantly by the processes of expansion and cracking due to the unloading of pressure and expansion from crystal growth. Unloading is the release of lithostatic pressure experienced by a body of rock after it has been uplifted to the surface of the earth where the pressure is much less than where it formed deeper within the earth. This

Rain pounding the ground loosens sediments, making surfaces more susceptible to erosion. The resulting runoff from rain carries the loosened sediment away, contributing to the process of disintegration. ***© Richard Hamilton Smith/Corbis. Reproduced by permission.***

pressure change will cause a rock to expand in all directions, dislodging grain boundaries. Often, this type of expansion results in concentric fractures that cause curved portions of the rock to slough off, or exfoliate. Crystal growth includes **ice** formation. **Water** that has permeated a rock, if frozen, will expand and create fractures. A rock can also experience secondary mineral growth, often of evaporite minerals that were transported in solution and infiltrate the rock by capillary action. Other minor agents of physical weathering include vegetation, which can create fractures by root growth, and thermal expansion of rock caused by climate changes.

The rate of rock weathering, whether chemical or physical, is influenced by the type of rock, climate, **topography**, and vegetation. The type of rock includes the **mineralogy**, which determines where in the weathering series the rock lies. It also includes lithologic structures such as number and size of fractures or **bedding** planes, both of which can be sites for focused weathering activity. Climate influences the rate of weathering as well as which type of weathering processes will predominate. For example, frost heaving will not be factor in a very warm climate, and a moist climate will experience more chemical weathering than a dry one. Topography determines how much rock will be exposed to the elements. It also influences the amount of vegetation that may take hold. Vegetation root systems, in addition to physically weathering rock, also produces **carbon dioxide** and humic acid, two chemical weathering agents.

See also Soil and soil horizons

WEGENER, ALFRED (1880-1930)

German meteorologist

Alfred Wegener was primarily a meteorologist who became much more famous for proposing the idea of continental drift. Decades after his death, the theory of continental drift that he had proposed in 1912 became the well-established foundation for the **plate tectonics** revolution in the earth sciences. Wegener heard mostly ridicule of his continental drift idea during his lifetime, but in the 1960s, oceanic data convinced scientists that continents do indeed move. Wegener was an eminent meteorologist in his time, but he was appointed professor late in his professional career, and died during one of his scientific trips to Greenland.

Wegener was born in Berlin, Germany to Richard, a minister and director of an orphanage, and Anna Wegener. From an early age, he hoped to explore Greenland, and he walked, hiked, and skated in order to build up his endurance for such a trip. He studied at the universities in Heidelberg, Innsbruck, and Berlin, receiving a doctorate in **astronomy** from the latter in 1905. Wegener's thesis involved conversion of a thirteenth-century set of astronomical tables into decimal notation; thereafter he abandoned astronomy in favor of **meteorology**. He carried out experiments with kites and balloons, fascinated with the new science of **weather**. In 1906, he and his brother Kurt set a world record in an international balloon contest by flying for 52 hours straight.

That year, Wegener also fulfilled his dream of going to Greenland. Wegener was chosen as official meteorologist for a Danish expedition to northeastern Greenland from 1906 to 1908. It was the first of four trips to Greenland he would take. In 1912, he returned to Greenland with an expedition to study glaciology and climatology; this trip was the longest crossing of the **ice** cap ever made on foot.

In 1908, Wegener accepted a job teaching meteorology at the University of Marburg. His lectures were very popular with students for their clarity and frankness. He admitted disliking mathematical details, yet in 1911 he published a textbook on the thermodynamics of the atmosphere, which included in embryonic form the modern theory on the origins of **precipitation**. The following year Wegener married Else Köppen, the daughter of the "Grand Old Man of Meteorology" in Germany, Wladimir Köppen. During World War I Wegener served as a junior military officer and was wounded twice. After the war he succeeded his father-in-law as director of the meteorological research department of the Marine Observatory near Hamburg. There he conducted experiments to reproduce lunar craters by hurling projectiles at various ground substances, demonstrating that the craters were probably of impact, rather than volcanic, origin. He also continued to analyze the data from Greenland, observe meteorological phenomena, and develop his earlier ideas on the origin of the continents and the **oceans**.

Wegener had first thought of the idea of continental drift in late 1910 while looking at a world map in an atlas. He noticed that the east coast of **South America** matched like a puzzle piece with the west coast of **Africa**, but dismissed the idea of drifting continents as improbable. The next year, however, he came across a list of sources arguing that a land bridge must have connected the two continents at one time, since similar **fossils** from the same time period appeared in both Africa and Brazil. Wegener immediately began to search out fossil evidence to support the idea of drifting continents. Within a few months he presented his hypothesis in two public forums.

Wegener published an extended account of his idea as *Die Entstehung der Kontinente und Ozeane* (The origin of continents and oceans) in 1915. The first edition was only 94 pages long, with no index. The second edition, much expanded and revised, attracted attention in **Europe**. The third edition was translated into English, French, Spanish, Swedish, and Russian in 1924 and was then widely read for the first time. The first English translation correctly referred to the idea of "continental displacement," as Wegener had termed it. The name "continental drift" was coined later.

Wegener's was the first coherent and logical argument for continental drift that was also supported by concrete evidence. He proposed that a huge supercontinent had once existed, which he named Pangaea, meaning "all land." He suggested that Pangaea was surrounded by a supersea, Panthalassa, and that 200 million years ago, in the Mesozoic period, Pangaea began to rift into separate continents that moved away from each other. The Americas drifted westward from Europe and Africa, forming the Atlantic Ocean. India moved east from Africa, and **Australia** severed its ties with **Antarctica** and moved towards the equator.

Wegener's hypothesis departed radically from the accepted view of the earth in his day. Other geologists believed that the earth was still cooling and contracting from a molten mass, and that lighter rocks such as **granite** (termed "sial"), moved towards the surface, underlain by denser rocks such as **basalt** ("sima"). Mountain ranges, they believed, were produced by the cooling contraction, like wrinkles appearing on a drying fruit. To these scientists, the continents and the ocean basins were initial and set features. It seemed impossible for continents to move through the ocean rocks.

Wegener instead proposed that the lighter sial that made up continents could move horizontally through the oceanic sima; if the continents can rise up vertically, he argued, they must be able to move horizontally as well, as long as sufficient force is provided. Thus the Rocky Mountains and the Andes, on the western edges of the Americas, were formed by the resistance of the sima layer to the continents plowing through them. **Island arcs** like Japan and the West Indies were fragments left behind in the wake of these giant drifting continents.

Wegener's strongest argument was the similarities of rocks, animals, and plants on both sides of the Atlantic. He pointed to the fossils of several reptiles and flora that were known only in Africa and South America, and to the fact that the distribution of some living animals was hard to explain unless the continents had once been connected. Scientists had previously explained these in terms of a land bridge that had once connected the continents and then sunk into the ocean. Wegener argued that this was impossible; if a bridge was made of sial, it could not simply sink and disappear.

However, Wegener couldn't find an adequate mechanism to explain continental drift. He suggested two mechanisms, which were later disproved. One was *Pohlflucht*, or "flight from the poles," to explain why continents seemed to drift towards the equator. *Pohlflucht*, also known as the Eötvös force, came from the fact that the Earth is an oblate spheroid, slightly flattened at the poles and bulging at the equator. Second, Wegener had to explain the westward movement of the Americas; he suggested that some kind of tidal force must be doing the work.

Wegener's hypothesis was received with ridicule. For decades, other geologists scoffed at the idea of drifting continents. Some scientists supported him, but there was not enough geological evidence to prove beyond a doubt that he was essentially right. Wegener's first critic was his father-in-law, Köppen, who apparently wanted Wegener to stay in meteorology and not wander into unknown areas like geophysics. At the first lecture in Frankfurt in 1912, some geologists were apparently indignant at the very notion of continental drift. The initial reaction was mixed at best, and hostile at worst. In 1922, when *The Origin of Continents and Oceans* first appeared in English, it was blasted in a critical review and at a scientific meeting. Subsequently, continental drift provoked a huge international debate, with scientists ranging themselves on both sides.

Detractors had plenty of ammunition. It was soon shown that *Pohlflucht* and tidal forces were about one millionth as powerful as they needed to be to move continents. The paleontological evidence was thought to be inconclusive. In 1928, at a meeting of fourteen eminent geologists, seven opposed it, five supported it without reservation, and two supported it with reservations. From then until after World War II, the subject was put on the back burner of scientific debate. In the only major variant on the theory, South African geologist Alex du Toit, a vigorous defender of continental drift, proposed in 1937 that instead of Pangaea there were two **supercontinents**, Laurasia in the northern hemisphere and Gondwanaland in the south.

Many eminent geologists, such as Sir Harold Jeffreys in England and, later, American paleontologist George Gaylord Simpson, were vehement critics of Wegener and his **continental drift theory**. Science historians consider it likely that the prestige of the critics often carried too much weight in the argument over the theory itself. Wegener himself often complained about the narrow-mindedness of geophysicists who could not accept new ideas. In 1926 Wegener was finally given a professorship in meteorology and geophysics at the University of Graz. Four years later he sailed from Copenhagen to Greenland as leader of a major expedition. On November 1 of that year, he and others in the party celebrated his fiftieth birthday at a camp in the center of the Greenland ice cap. Wegener headed for the west coast that day, and apparently died of heart failure. His body was later found about halfway between the two camps.

After World War II, and several decades after Wegener's disappearance, other geologists began to uncover clues that eventually led to the plate tectonics revolution. The development of paleomagnetism in the early 1950s demonstrated that rocks in different continents appeared to have different directions of magnetization, as if continents had drifted apart from each other. In addition, oceanographers began to map the ocean floor to learn about its origin. They learned that the ocean floor was not a fixed glob of sima at all. In 1960, American geologist **Harry Hammond Hess** proposed the theory of seafloor spreading: that the ocean floor is constantly being created at underwater ridges in the middle of the oceans, spreading outward, and being consumed in trenches underneath the continents. By the mid 1960s, new data on magnetic anomalies in the Pacific Ocean revealed that seafloor spreading did indeed occur. Here was the mechanism by which Wegener's continents could drift: The ocean floor was constantly regenerating itself. By the end of the 1960s, continental drift had begun to be accepted by the entire **earth science** community. It had taken half a century, but Wegener's hypothesis became the foundation for a revolution among geologists and a cornerstone for modern views of Earth's history.

See also Convergent plate boundary; Divergent plate boundary

Werner, Abraham Gottlob (1749-1817)

German geologist

One of the founders of **stratigraphy**, Abraham Werner was one of the first to apply the modern **scientific method** to many geological problems, had a powerful and positive influence on his scientists, and was one of the first to attempt a description of the geological history of the world free from religious and mystical explanations.

Werner was born in Wehrau, Silesia (now Germany), although some sources suggest it was the Wehrau in Upper Lusatia, (now Osiecznica, Poland). His father was the inspector of the Duke of Solm's ironworks, and much of Werner's education was designed to prepare him to follow in his father's footsteps. After being taught at home by his father and private tutors he enrolled in the new Bergakademie (Mining Academy) in Freiberg in 1769. While there, he was recruited into the Saxon mining service, but needed a law degree (jurisprudence) in order to advance in his career, and so began studies at the University of Leipzig. Werner found himself distracted by other subjects, especially the history of languages and **mineralogy**.

In 1774, he abandoned his law degree, and left university, but by then he had already published a book, *Von den äusserlichen Kennzeichen der Fossilien*, which was a practical and orderly mineral identification manual. On the strength of his book he gained a teaching position at the Mining Academy in Freiberg. Werner kept the job for the rest of his life, teaching there for over 40 years. He was justly famous for his lectures, and his courses attracted students from all over the Western world.

Werner is remembered most for his water-based theory of the creation of Earth's **crust**. Named Neptunism (after the Roman god of the **oceans**), Werner's ideas, while incorrect, were nonetheless based firmly on the physical evidence of his day. He argued that all older rocks were sedimentary in nature, and had been laid down by an ancient, universal ocean. The different **rock** types and strata were explained by changes in the depth and turbulence of the universal ocean. Werner was one of the first to think of the earth as a whole, and called his new approach "geognosy." Werner's field work, which was mainly in Saxony, convinced him that the opposing view, that ancient rock had a volcanic origin (Vulcanism), was incorrect. In particular, he was sure that **basalt**, a very common rock, was sedimentary, despite strong opposition. To prove his ideas were truly universal many of his students set out across **Europe** looking for supporting evidence. However, many found that outside of Saxony Werner's ideas were not supported at all, and the rival notion of Vulcanism became dominant. It is a tribute to Werner's teaching methods that these students placed such a high degree of importance on what they saw, rather than slavishly following the doctrine of their teacher. Werner tried to tinker with his theory, attempting to make it fit with the new evidence while still retaining the basics, but in doing so it lost much of its simplicity and logic. However, while Neptunism was a dead-end, Werner can be credited with inspiring scientists to think about the natural forces that had created Earth's crust, and with training a generation of inquiring European geologists who went beyond his initial investigations.

He suffered from ill health in his later years, and after 1793 he published very few geological works. Instead he devoted his time to teaching, and a few official duties. Werner was elected to 22 scientific societies in his lifetime, and he was eulogized by followers and opponents alike after his death in Dresden in 1817. He never married, and left most of his estate to the Bergakademie, the school that had been his focus for most of his adult life.

See also Minerals; Sedimentary rocks

Wet air • *see* Humidity

Wilson, J. Tuzo (1908-1993)

Canadian geophysicist

An early proponent of the **continental drift theory**, J. Tuzo Wilson is chiefly remembered for his proposition that **trans-**

form faults were present in the ocean floor, an idea that led to conclusive evidence that the sea floor and the earth's **crust** are constantly moving. Wilson later hypothesized that an ancestral Atlantic Ocean basin had opened and closed during the **Paleozoic Era**, in turn creating the huge land mass known as Pangaea. This theory helps account for the presence of the Appalachian mountains in eastern **North America**, the striking similarity of many **rock** features in Western **Europe** and North America, and parallel cyclical developments on the seven continents.

John Tuzo Wilson was born in Ottawa, Ontario, Canada. His father, John Armitstead Wilson, was an engineer who held a civil service position. His mother, Henrietta Tuzo, was an avid mountain climber who met her husband at the first gathering of Canada's Alpine Club. The Wilsons later shared their love of **geology** and the outdoors with their children, who were brought up to respect the pursuit of knowledge and were educated under the direction of an English governess.

In 1924, Wilson's father obtained a position for him at a forestry camp. Wilson grew so fond of outdoor work that he signed on as an assistant to the legendary mountaineer Noel Odel, who persuaded him to pursue a career in geology. Following his freshman year at the University of Toronto, Wilson switched majors from **physics** to geology. After earning a B.A. in 1930, Wilson received a scholarship to study at Cambridge University under Sir Harold Jeffreys. When Wilson returned to Canada in the early 1930s, he had difficulty finding work, so he continued his education, enrolling in Princeton University, where he earned a Ph.D. in 1936. He made the first recorded ascent of Mount Hague in Montana in 1935, and in 1938 married Isabel Jean Dickson, with whom he eventually had two children.

With the outbreak of World War II in 1939, Wilson joined the Canadian Army. During his seven-year stint, he authored more than 500 technical reports and later claimed that these military papers had helped him develop the lucid prose style that he utilized in a number of scientific studies. By 1946, he had reached the rank of colonel. That same year, after resigning from the army, he succeeded his professor at the University of Toronto. Geophysics had finally become a lucrative field of study in Canada, thanks in large part to the discovery of oil in Alberta, which increased demand for geophysical exploration and led to the development of more advanced instruments and measurement techniques. Wilson investigated a number of geological mysteries, including Canadian **glaciers**, mountain building, and mineral production. He conducted these investigations with a characteristic reverence toward nature: "Everywhere in science modern tools and ideas bring to light the elegant and orderly skeins by which nature builds the glory that we see about us, knit in regular patterns from simple stitches," he wrote in *I.G.Y.: The Year of the New Moons* (1961). "Indeed, we may think of all nature in terms of music, as infinitely ingenious and elaborate variations on a few simple themes."

From 1957 to 1960, Wilson served as president of the International Union of Geodesy and Geophysics. During his tenure he led a series of geologic expeditions to China and Mongolia, the details of which are recorded in his highly praised book, *One Chinese Moon* (1959). In the early 1960s, he became a key figure in what was then the most controversial issue in geology—the continental drift theory.

The origins of the continental drift theory date back hundreds of years. Since the time of the first global maps people have reasoned that at one time the continents might have been a single huge land mass. However, the first formal hypothesis of continental drift was made by German geophysicist **Alfred Wegener** in 1912. The idea was generally overlooked for decades but reemerged prominently in 1960, when geologist Harold Hess theorized that the ocean floors were being continuously created and changed. Hess attributed this activity to two physical structures: **mid-ocean ridges**, where the ocean floor is created, and **ocean trenches**, where the sea floor is destroyed.

Wilson was one of the first scientists to recognize the immense implications of this idea. For the next decade, he was at the very center of this theoretical debate. Using Hess's theory, Wilson postulated the existence of a third category of physical structure on the ocean floor which he called "transform faults," horizontal shears located between ridge sites and trenches. He suggested that transform faults could not exist unless the earth's crust was moving, and that the physical confirmation of these faults might prove the scientific validity of the continental drift theory. In 1967, seismologist Lynn Sykes partially tested Wilson's theory by studying seismic patterns and oceanic focal mechanisms. Wilson brought the idea to the attention of the general public by exhibiting a continental drift model at Montreal's Expo '67. By the late 1960s, the theory had gained wide acceptance and was eventually incorporated into the larger concept of **plate tectonics**, which maintains that the Earth's **lithosphere** is made up of a number of plates that move independently.

Wilson's hypothesis and the publicity it garnered earned him numerous honors, including a Fellowship in the Royal Society (1968), the Penrose Medal of the Geological Society of America (1968), the Walter H. Bucher Medal of the American Geophysical Union (1968), the John J. Carty Medal of the National Academy of Sciences (1975), the Vetlesen Prize of Columbia University (1978), and the Wollaston Medal of the Geological Society of London (1978).

Wilson retired from his professorship at the University of Toronto in 1974. He then assumed the directorship of the Ontario Science Centre and in that capacity helped transform the center from a traditional science museum into an interactive science lab for public use. Of the center's roughly 1,000 exhibits, 400 were designed to be handled by patrons, and during the late 1970s and 1980s, the exploratory museum attracted approximately 1.5 million visitors annually.

Throughout his life Wilson traveled extensively. He lectured at more than 200 colleges and universities. One of his passions was collecting books on the Arctic and Antarctic, both of which he had visited. A mountain range in **Antarctica** was named the Wilson range in his honor. He died in Toronto at the age of 84.

See also Convergent plate boundary; Divergent plate boundary; Sea-floor spreading

WILSON, ROBERT W. (1936-)

American physicist

Robert Woodrow Wilson is best known for the discovery, with co-researcher **Arno Penzias**, of the cosmic background radiation believed to be the remnant of the "big bang" that started the Universe. For their work, Wilson and Penzias were honored with numerous awards, including the 1978 Nobel Prize in physics, which they shared with Pyotr Kapitsa.

Wilson was born in Houston, Texas. He attended Rice University where he received a B.A. in physics in 1957. He then moved on to the California Institute of Technology (Cal Tech) for graduate study and received his Ph.D. in 1962. Wilson's thesis work and post-doctoral research involved making radio surveys (the use of radio waves bounced off of stellar bodies to create visual approximations) of the Milky Way Galaxy. When he heard of the existence of specialized radio equipment at Bell Laboratories, he left Cal Tech and accepted a job at Bell's research facility in Holmdel, New Jersey. This was the very same research facility from which **Karl Jansky**, in the 1930s, almost single-handedly invented the science of radio **astronomy**. Wilson and Penzias, who had preceded Wilson at Bell Labs by about a year, were about to embark on a research odyssey that would culminate in an extremely important discovery almost by accident.

Just as Jansky had done thirty years earlier, Wilson and Penzias were studying the possible causes of static interference that impaired the quality of radio communications. At least, this was what the management at Bell hoped would transpire as the two radio astronomers conducted their research. Wilson and Penzias' long-range plan was to measure radiation in the galactic "halo," a theorized but not well understood cloud of matter and radiation surrounding the Milky Way and other galaxies. Then, they hoped to look for hydrogen gas in clusters of galaxies. Their research instrumentation included a small, sensitive 20-foot microwave "horn" originally designed to receive bounced radio reflections from the Echo communications **satellite**.

Because galactic radio radiation is, by its nature, not very energetic, the central problem in measuring its precise intensity was to eliminate all conceivable sources of heat, or thermal noise, which could obscure an accurate reading of the weak radio signals from **space**. To this end, Penzias had laboriously constructed a "cold load," using frigid liquid helium, which would cool the radio detector down to within only a few degrees above absolute zero. When the equipment was finally ready in the spring of 1964, the radio horn was turned to the sky.

Very early in the research project, it became apparent that the antenna was measuring more radio radiation than Wilson and Penzias had anticipated. The source of the excess radiation could not be determined. A similar problem had surfaced earlier when the twenty-foot horn was used for Echo satellite communications. At that time, researchers added up all the known sources of accounted radio noise, which totaled a heat measurement of 19 degrees Kelvin. It was therefore puzzling to them that the radio receiver was measuring 22 degrees. Wilson and Penzias' results were similar. They had hoped that their carefully modified apparatus would yield more accurate results, but this apparently was not the case. They were measuring a significant amount of excess microwave radiation. The intensity of the signal did not change regardless of where they pointed the receiver. Nor did the radio static appear to be coming from any discrete object in space. The Milky Way Galaxy was not the source either, since the radio signal seemed to be coming from everywhere in the universe at once, not from just a limited zone across the sky. Based on the known sources of radio radiation, the strength of this radiation was far more powerful than expected.

Wilson and Penzias checked for possible explanations for this phenomenon, concluding that atmospheric effects were not to blame. Since the hill upon which their radio horn was perched overlooked New York City, the possibility of interference from man-made sources was considered. After repeated observations, however, Wilson and Penzias were convinced that New York was not to blame. To insure that the signal was not the result of interference from their own electronic apparatus, Wilson and Penzias tracked down and eliminated every conceivable source of noise—including the effects of bird dung, which coated the inside of the radio horn, courtesy of a pair of nesting pigeons. The interior of the radio horn was cleaned out.

The attempts to improve the performance of the radio horn took time. Finally, in 1965, the antenna was re-activated and careful observations were made of the radio flux from the sky. The results revealed that the **telescope** was performing better than ever, but the mysterious excess signal remained. The intensity of the excess radio noise was what would be expected from an object, or source, with a very low temperature—only a few degrees above absolute zero. In this case, as with the previous observation, the static was not coming from a discrete source but was emanating uniformly from every direction in the sky.

While Wilson and Penzias were trying to make sense of what seemed to them to be a failed experiment, Robert Dicke and his colleagues at Princeton University, unaware of the project at Bell Labs, were building a radio receiver of their own designed to look for the very radiation that Wilson and Penzias had unintentionally observed. Whereas Wilson and Penzias had rather modest hopes of making simple surveys of galactic radio flux, Dicke was looking for physical evidence of the creation of the universe. Dicke had been researching the theoretical effects of the big bang, the expanding fireball theorized as the birth of the Universe.

The line of reasoning Dicke followed was that as the universe expanded after the big bang, gases cooled and thinned but were still dense enough to block electromagnetic radiation. All thermal energy released by atoms, including light and heat, was reabsorbed by other atoms in the gas almost instantly. One consequence of this condition was that if someone could have viewed the universe from the "outside" at this point, they would have seen only blackness, since no light could escape the opaque, light-absorbing gas. Eventually, there must have come a time, thousands of years after the big bang, when the average density of the expanding universe was finally low enough to allow heat and light to escape from atoms unimpeded, much as the light and heat generated in the sun's interior eventually escapes through the sun's transparent photosphere. According to the theory that Dicke was exploring, the rapid release of newly freed energy in the thinning, early universe would have taken the form of an incredibly sudden blaze of heat and light, almost like an explosion.

How could this "primeval fireball," as it came to be called, be observed today? If the remnant of this energy flash had survived after several billion years, it would be detected as a kind of "whisper" in a radio telescope. It would have a specific color and **temperature** and would be present in nearly equal intensities in every direction, forming a cosmic background radiation. This radiation would flood every available volume of space. In time, the radiation would appear to cool down to a point near absolute zero, due to the further expansion of the universe, but it would still be detectable even in the present-day universe. It was precisely this radiation that Robert Dicke was preparing to look for with his own radio telescope. It was also this radiation, measuring close to absolute zero (around 3 Kelvin) in uniformity across the sky, that Wilson and Penzias had already discovered.

Wilson and Penzias were not cosmologists, however. They could not explain their observation of the microwave radiation at the 7.3 cm wavelength, and so they contacted Dicke, who they knew was working on this problem. When Dicke heard the details of their findings, he knew that Wilson and Penzias had discovered exactly what he was looking for; the cold, background radiation left over from the big bang. In 1965, Wilson and Penzias published their results in a paper entitled "A Measurement of Excess Antenna Temperature at 4,080 Mc/s." A companion paper written by Dicke, P.J.E. Peebles, P.G. Roll, and D.T. Wilkinson explained the profound cosmological implications of the finding.

The discovery of the cosmic background radiation was like finding the intact skeleton of a dinosaur. The radiation is a "fossil," an ancient relic from a time when the universe was barely 100,000 years old. The discovery of the radiation was to become the second great pillar upon which the **big bang theory** would rest, second only to the 1920s discovery of the expansion of the universe. The fact that the background radiation was predicted in advance of its discovery helped to strengthen the big bang theory, so much so that most competing theories about the birth of the universe, such as steady state, almost immediately fell away after 1965.

As scientists around the world began making their own confirming observations of the cosmic background radiation, it became apparent to those searching past research papers that clues to the existence of the radiation had existed for over 25 years. The most striking example came from 1938, in which optical telescopic observations revealed that interstellar cyanogen gas was being heated, unaccountably, by a 3-degree source. This source was nothing less than the cosmic background radiation. But at the time, no one imagined that the seemingly innocuous source of heat could be the remnants of the big bang fireball. It would not be until Wilson and Penzias's discovery that the cosmic radiation would be identified for what it was.

Wilson and Penzias's discovery was acclaimed by scientists around the world. In 1976, Wilson was named head of the Radio-Physics department of Bell Telephone. For his work on the cosmic background radiation, he also received the Henry Draper Award, in 1977, from the National Academy of Sciences. In 1978, the importance of their achievement in the history of science was fully recognized when Wilson and Penzias shared the Nobel Prize in physics with Kapitsa.

See also Cosmic microwave background radiation

WIND

Wind refers to any flow of air relative to the earth's surface in a roughly horizontal direction. Breezes that blow back and forth from a body of **water** to adjacent land areas—on-shore and off-shore breezes, or land and sea breezes—are examples of local wind. Winds, driven by large pressure systems also exist in great wind belts that comprise the earth's **atmospheric circulation**.

The ultimate cause of Earth's winds is **solar energy**. When sunlight strikes Earth's surface, it heats that surface differently. Newly turned **soil**, for example, absorbs more heat than does snow.

Uneven heating of Earth's surface, in turn, causes differences in air pressure at various locations. On a **weather** map, these pressure differences can be found by locating **isobars**, lines that connect points of equal pressure. The pressure at two points on two different isobars will be different. A pressure gradient is said to exist between these two points. It is this pressure gradient that provides the force that drives air from one point to the other, causing wind to blow from one point to the other. The magnitude of the winds blowing between any two points is determined by the pressure gradient between those two points.

In an ideal situation, one could draw the direction of winds blowing over an **area** simply by looking at the isobars on a weather map. The earth, however, is not an ideal situation. At least two important factors affect the direction in which winds actually blow: the **Coriolis effect** and friction. The Coriolis effect is an apparent force that appears to be operating on any moving object situated on a rotating body, such as a stream of air traveling on the surface of the rotating planet. The Coriolis effect deflects winds from the straightforward direction across isobars. In the Northern Hemisphere, the Coriolis effect tends to deflect winds right of path and in the Southern Hemisphere, it tends to drive winds left of path.

For example, wind in the Northern Hemisphere initially begins to move from west to east as a result of pressure gradient forces. The Coriolis effect results in a deflection of the wind right of path. This results in air moving out of a high-pressure system (an area of divergence) to spin clockwise. Conversely, air moving into a low pressure area (an area of convergence) also deflected right of path, is spun counterclockwise.

The actual path followed by the wind is a compromise between the pressure gradient force and the Coriolis force. Since each of these forces can range widely in value, the precise movement of wind in any one case is also variable. At some point, the two forces driving the wind are likely to come into balance. At that point, the wind begins to move in a straight line that is perpendicular to the direction of the two forces. Such a wind is known as a geostrophic wind.

The Coriolis effect is most pronounced on winds farther from the surface of the earth. At distances of more than a half a mile or so above the ground pressure gradient and Coriolis forces are the only factors affecting the movement of winds. Thus, air movements eventually reach an equilibrium point between pressure gradient forces and the Coriolis force, and geostrophic winds blow parallel to the isobars on a weather map.

Such is not the case near ground level, however. An additional factor affecting air movements near the Earth's surface is friction. As winds pass over the earth's surface, they encounter surface irregularities and slow down. The decrease in wind speed means that the Coriolis effect acting on the winds also decreases. Since the pressure gradient force remains constant, the wind direction is driven more strongly toward the lower air pressure. Instead of developing into geostrophic winds, as is the case in the upper atmosphere, the winds tend to curve inward towards the center of a low pressure area or to spiral outward away from the center of a high pressure area.

Friction effects vary significantly with the nature of the terrain over which the wind is blowing. On very hilly land, winds may be deflected by 30 degrees or more, while on flat lands, the effects may be nearly negligible.

In many locations, wind patterns exist that are not easily explained by the general principles outlined above. In most cases, unusual topographic or geographic features are responsible for such winds, known as local winds. **Land and sea breezes** are typical of such winds. Because water heats up and cools down more slowly than does dry land, the air along a shoreline is alternately warmer over the water and cooler over the land, and vice versa. These differences account for the fact that winds tend to blow offshore during the evening and onshore during the day.

The presence of mountains and valleys also produces specialized types of local winds. Annual changes in weather patterns produce **seasonal winds** such as the dry Santa Ana winds in Southern California.

See also Air masses and fronts; Atmospheric composition and structure; Atmospheric inversion layers; Jet stream; Weather forecasting methods; Wind chill; Wind shear

WIND BELTS • *see* ATMOSPHERIC CIRCULATION

WIND CHILL

Wind chill is the **temperature** sensed by humans as a result of air blowing over exposed skin. The temperature that humans actually feel, called the *sensible temperature*, can be quite different from the temperature measured in the same location with a thermometer. The reason for such differences is that the human body constantly gives off and absorbs heat in a variety of ways. For example, when a person perspires, **evaporation** of moisture from the skin removes heat from the body, and one feels cooler than the true temperature would indicate.

In still air, skin is normally covered with a thin layer of warm air that insulates the body and produces a sensible temperature somewhat higher than the air around it. When the wind begins to blow, that insulating layer is swept away, and body heat is lost to the surrounding atmosphere. An individual begins to feel colder than would be expected from a thermometer reading at the same location.

The faster the wind blows, the more rapidly heat is lost and the colder the temperature appears to be.

Wind chill charts or conversion tables relate the relationship among actual temperature, wind speed, and wind chill factor, to the temperature felt by a person at the given wind speed. According to standard conversion formulae, a wind speed of 4 mi/h (6 km/h) or less results in no observable change in temperature sensed. At a wind speed of 17 mi/h (30 km/h) and a temperature of 32°F (0°C), however, the perceived temperature is 7°F (-14°C).

Wind chill relationships are not linear. The colder the temperature, the more strongly the wind chill factor is felt. At a wind speed of 31 mi/h (50 km/h), for example, the perceived temperature at 32°F (0°C) is 7°F (-14°C), but at -40°F (-40°C), the perceived temperature is -112°F (-80°C).

See also Antarctica; Atmospheric lapse rate; Aviation physiology; Beaufort wind scale; Humidity; Space physiology

WIND SHEAR

Wind shear is a phenomenon describing highly localized variability in wind speed and/or wind direction. Because wind shear can affect the angle of attack on an airfoil (e.g., the wing, or control surfaces of an airplane) wind shear can cause a loss of lift or control. Dangerous to aviation, wind shear is particularly hazardous when encountered during take-off or landing.

Wind shear is the difference in speed or direction between two layers of air in the atmosphere. Wind shear may occur in either a vertical or horizontal orientation. An example of the former situation is the case in which one layer of air in the atmosphere is traveling from the west at a speed of 31 mph (50 km per hour) while a second layer above it is traveling in the same direction at a speed of 6.2 mph (10 km per hour). The friction that occurs at the boundary of these two air currents is a manifestation of wind shear.

An example of horizontal wind shear occurs in the **jet stream** where one section of air moves more rapidly than other sections on either side of it. In this case, the wind shear line lies at the same altitude as various currents in the jet stream, but at different horizontal distances from the jet stream's center.

Wind shear is a crucial factor in the development of other atmospheric phenomena. For example, as the difference between adjacent wind currents increases, the wind shear also increases. At some point, the boundary between currents may break apart and form eddies that can develop into clear air turbulence or, in more drastic circumstances, tornadoes and other violent storms.

Wind shear has been implicated in a number of disasters resulting in property damage and/or loss of human life. The phenomenon is known as a microburst, a strong localization down draft (down burst) which, when it when reaches the ground, continues as an expanding outflow. For example, it is associated with the movement of two streams of air at high rates of speed in opposite directions. An airplane that attempts to fly through a microburst passes through the wind shear at the boundary of these two air streams. The plane feels, in rapid succession, an additional lift from headwinds and then a sudden loss of lift from tailwinds. In such a case, a pilot may not be able to maintain control of the aircraft in time to prevent a crash.

See also Aerodynamics; Bernoulli's principle; Meteorology; Weather radar; Weather satellite

WOMEN IN THE HISTORY OF GEOSCIENCE • *see* HISTORY OF GEOSCIENCE: WOMEN IN THE HISTORY OF GEOSCIENCE

"WOW!" SIGNAL

Since radio astronomers first tuned into the skies, scientists have listened for an elusive radio signal that would confirm the existence of extraterrestrial life. One of the major efforts in the last quarter of the twentieth century was a project termed the Search for Extraterrestrial Intelligence (**SETI**). Over the years the SETI project evolved into a variety of programs utilizing research resources at a number of different facilities. A number of other programs have embraced at least part of the SETI concept and goals. As of May, 2002, only a fraction of the potential sources of radio signals have been thoroughly observed, and no signal definitively identified as extraterrestrial in origin.

Regardless, there have been a number of interesting possibilities. On August 15, 1977, astronomer Jerry Ehman was going through the computer printouts of an earlier SETI-like project run by Ohio State University (dubbed, "Big Ear"), when he discovered the reception of what remained throughout the twentieth century as the best candidate for a signal that might be classified as a sign of extraterrestrial intelligence. Excited, Ehman scribbled, "WOW!" on the printout and forever after the signal became known as the "WOW!" signal.

Despite repeated attempts to reacquire the signal, the fact that the signal was never again recorded makes many astronomers, including Ehman, skeptical about the origins of the "WOW!" signal. If it were an intentional signal, astronomers argue, the sending civilization would have repeated it—or something like it—many times. A number of SETI experts now assert the "WOW!" signal was, perhaps, a mere reflection of a signal from Earth off an orbiting **satellite**.

See also Astronomy; Cosmology; Electromagnetic spectrum

X

Xenolith

A xenolith is a **rock** fragment embedded in, and distinct in texture and composition from, a surrounding mass of igneous rock. Xenoliths form when rising **magma** forces its way through channels and cracks, tearing off fragments of their walls and incorporating them into rising magma. These inclusions are termed xenoliths if they did not form from the magma itself, autoliths, or cognate xenoliths, if they were first solidified along the channel walls from the rising magma and re-incorporated later. Single large **crystals** included in igneous rock by the same means as xenoliths are xenocrysts. Xenoliths, which are named from the Greek *xeno* (foreign) and *lith* (rock), typically range from sand-grain size to football size.

When first captured by magma, a xenolith both cools and is heated by the liquid rock around it. How altered it is by heating depends on its size and original **temperature**, on the temperature of its magma bath, and on the proximity of other sources of heating or cooling. If a xenolith is rapidly cooled after capture, its **chemistry** and mineral structure will change little; if it is partly melted before being finally cooled it will undergo some degree of metamorphosis; and if it is thoroughly melted it will blend with the surrounding magma to produce a hybrid or contaminated igneous rock.

Xenoliths may be captured by magma near the surface or deep in the mantle. If mantle-derived xenoliths are carried to the surface rapidly enough to avoid significant metamorphosis they convey valuable data from the depths. For example, some mantle-derived xenoliths consist of a combination of the **minerals olivine**, pyroxene, and garnet. Laboratory **melting** experiments show that the **aluminum** and magnesium content of a pyroxene crystallized in the presence of olivine and garnet depends uniquely on both pressure and temperature. Chemical analysis of an unmetamorphosed mantle-derived xenolith thus reveals the pressure (dependent on depth) and temperature at which it crystallized, giving a temperature reading for a specific depth. Xenoliths can originate hundreds of kilometers underground, far below the reach of the deepest mining or drilling operations, so this data is otherwise unobtainable. Pristine xenoliths also reveal rock textures and compositions deep in the mantle.

See also Country rock; Crater, volcanic; Dike; Hotspots; Magma chamber; Metamorphic rock; Metamorphism; Sill; Volcanic eruptions; Volcanic vent

Y

Year, length of

Astronomers define a planet's sidereal year as the time the planet requires to make a complete orbit around the **Sun**. A definition more relevant to humans is the time required for the **seasons** to complete one cycle, that is, the time between successive spring equinoxes. This equinoctial year is shorter than the sidereal year because Earth rotates on its axis as it orbits the Sun and the **polar axis** (rotational axis) wobbles like a spinning top (precession), with each wobble taking about 28,000 years. The resulting difference between the equinoctial and sidereal year is small: the equinoctial year is approximately 365 days, 5 hours, 48 minutes, and 46 seconds long, the sidereal year about 20 minutes longer.

The path of Earth around the Sun, like that of every other planet, is an ellipse. However, it is very nearly circular. If one makes the simplifying assumptions that Earth's orbit is circular and that the Sun is stationary, then it is true to say that the length of the sidereal year depends on the distance from Earth to the Sun. Earth's mass, surprisingly, is irrelevant. For example, a planet with twice (or half) Earth's mass, orbiting at the same distance from the Sun as Earth, would have a year of the same length.

However, these are only approximately values. The Sun is not stationary; Earth and Sun are both orbiting around their common center of **gravity**. An Earth year might, therefore, be more accurately described as the time it takes both Earth and Sun to make one complete orbit around their common center of gravity. Even this improved picture ignores the presence of all the other matter in the **solar system** (and the Universe), not to mention relativistic effects. However, because the center of mass is much closer to the Sun—in fact, inside the Sun—astronomers usually assume the Sun to be stationary with regard to Earth's orbit, and for many purposes assume Earth's orbit to be circular.

The lengths of the year, month, and day are not strictly constant. Solar tidal friction is slowly increasing the length of the year by moving Earth away from the Sun; lunar tidal friction is lengthening the month and the day by moving the **Moon** away from Earth and causing the earth to rotate more slowly. The lengthening of the earth's year is negligibly slow, about one billionth of a year every billion years, but the lengthening of Earth's day is much faster: about two seconds every 10,000 years. At the beginning of the **Cambrian Period**, for example, approximately 600 million years ago, there were over 420 days in each year, each only 21 hours long.

See also Astronomy; Tides

Yucca Mountain proposed nuclear waste repository

• *see* Radioactive waste storage (geological considerations)

Z

ZEOLITE

The zeolites are a group of more than 35 soft, white **minerals** comprised mostly of **aluminum**, **silicon**, and **oxygen** and having a crystal structure featuring spacious pores or rings. Zeolites often form as **crystals** in small cavities in basaltic rocks or as volcanic tuffs altered by **water**. They are also synthesized industrially.

The pores of a natural zeolite crystal are filled with water that can be driven off by heating. The result is a honeycomb-like structure penetrated by openings on the order of a few atoms in width (2–8 angstroms). This structure can act as a hyperfine filter or molecular sieve. For example, nitrogen binds to some zeolites, so forcing ordinary air (which consists mostly of nitrogen) through such crystals yields an output of up to 95% oxygen. Equally useful is the ability of zeolites to capture large positively-charged ions from aqueous solution. This capture process is reversible; that is, an ion adsorbed by a zeolite can generally be driven off again by heat. This property allows many zeolite-based molecular sieves to be reused indefinitely. Such sieves—often consisting of tanks filled with tons of crushed zeolitic tuff—have been used to filter radioactive cesium and strontium from nuclear waste, to remove ammonia from sewage, to scrub sulfur dioxide (SO_2) from coal-fired electric power station emissions, to purify landfill gas for household utility use, to filter mercury and other heavy **metals** from industrial wastewater, to remove calcium from water in water-softening systems, and for many other purposes.

Zeolites were first identified in 1756, but their molecular sieve properties were not observed until the mid 1920s. Even then they remained a mere curiosity for some time, as geologists still argued that natural zeolites were too rare to be commercially useful. Attention turned instead to zeolite synthesis. It was not until the 1950s that geologists discovered that million-ton deposits of volcanic **tuff** consisting mostly of zeolite are not, in fact, uncommon.

See also Crystals and crystallography; Minerology

Sources Consulted

Books

Abell, G., and D. Morrison. *Explorations of the Universe.* New York: Saunders College Publishing, 1987.

Adams, Charles K. *Nature's Electricity.* Blue Ridge Summit, PA: TAB Books, Inc., 1987.

Adams, Frank. *The Birth and Development of the Geological Sciences.* Mineola, New York: Dover Publications, Inc., 1954.

Ager, D. V. *The Nature of the Stratigraphic Record.* 2nd ed. New York: John Wiley and Sons, 1981.

Ahrens, C. D. *Essentials of Meteorology. An Invitation to the Atmosphere.* West Publishing Company, 1993.

Ahrens, C. D. *Meteorology Today: An Introduction to Weather, Climate and the Environment.* 4th ed. West Publishing Company, 1991.

Albritton Jr., C.C., ed. *Uniformity and Simplicity: A Symposium on the Principle of the Uniformity of Nature.* Boulder: Geological Society of America, 1967.

Alic, Margaret. *Hypatia's Heritage: A History of Women in Science from Antiquity through the Nineteenth Century.* Boston: Beacon Press, 1986.

Allegre, Claude J. *The Behavior of the Earth: Continental and Seafloor Mobility.* Cambridge: Harvard University Press, 1989.

Allen, Gary A. *Transition and Mixing in Axisymmetric Jets and Vortex Rings.* Washington: NASA, 1986.

Aller, Lawrence H. *Atoms, Stars, and Nebulae.* 3rd ed. Cambridge: Cambridge University Press, 1991.

American Men & Women of Science 1995-6. 19th ed. New Providence, New Jersey: R. R. Bowker, 1994, p. 618.

American Men and Women of Science: A Biographical Directory of Today's Leaders in Physical, Biological, and Related Sciences, 1998-99. 20th ed. New Providence, NJ: R. R. Bowker, 1998.

American Meteorological Society. *Challenges of Our Changing Atmosphere—Careers in Atmospheric Research and Applied Meteorology.* 1993.

American Water Works Association. *Water Quality and Treatment.* 5th ed. Denver: American Water Works Association, 1999.

Andrews, J. E., et. al. *An Introduction to Environmental Chemistry.* Blackwell Science.

Armstead, Christopher H., ed. *Geothermal Energy.* Paris: UNESCO, 1973.

Arnowitt, R. *Gravitation—An Introduction to Current Research.* London: Academic Press, 1977.

Asimov, Isaac. *The History of Physics.* New York: Walker and Company, 1966.

Asimov, Isaac. *Biographical Encyclopedia of Science and Technology.* New York: Avon Books, 1972.

Asimov, Isaac. *Asimov's Chronology of Science and Discovery.* New York: Harper & Row Publishers, 1989.

Atherly, A. G., J. R. Girton, and J. F. McDonald. *The Science of Genetics.* Fort Worth, TX: Saunders College Publishing, 1999.

Atkins, P. W. *Molecular Quantum Mechanics.* 2nd ed. Oxford: Oxford University Press, 1983.

Atkins, P. W. *Physical Chemistry.* 6th ed. Oxford: Oxford University Press, 1997.

Atkins, P. W. *Quanta: A Handbook of Concepts.* 2nd ed. Oxford: Oxford University Press, 1991.

Azaroff, Leonid V. *Elements of X-Ray Crystallography.* New York: McGraw-Hill Book Company, 1968.

Bailey Jr., Philip S., and Christina A. Bailey. *Organic Chemistry: A Brief Summary of Concepts and Applications.* 4th ed. Englewood Cliffs, NJ: Prentice Hall, 1989.

Banda, M. Torne, and M. Talwani. *Rifted Ocean-Continent Boundaries.* Kluwer Academic Publishers, 1987.

Barnes, J. *Basic Geological Mapping.* 3rd ed. New York: John Wiley and Sons, 1995.

Barnes-Svarney, Patricia, ed. *Science Desk Reference.* New York: Macmillan, 1995.

Barrow, John D. *The Origin of the Universe.* New York: Basic Books, 1994.

Barry, R. G., and R. J. Chorley. *Atmosphere, Weather, and Climate.* London: Methuen and Co., Ltd., 1971.

Bates, R., and J. Jackson, eds. *Dictionary of Geological Terms.* 3rd ed. New York: Anchor Books, 1984.

Bates, R. L., and J. A. Jackson. *Dictionary of Geological Terms.* New York: Doubleday, 1984.

Bates, R. L., and J. A. Jackson, eds. *Glossary of Geology.* 2nd ed. Falls Church, VA: American Geological Institute, 1980.

Battey, M. H., and A. Pring *Mineralogy for Students.* Harlow, Longman, 1997.

Battison, Edwin A. "Harrison, John." vol. 6, *Dictionary of Scientific Biography.* Farmington Hills, MI: Macmillan Reference USA, 1981.

Bear, Jacob. *Dynamics of Fluids in Porous Media.* New York: Dover Publications, 1988.

Beckett, B. *Introduction to Cryptology.* Malden, Massachusetts: Blackwell Scientific, 1988.

Berggren, W. A., and J. A. Van Couvering, eds. *Catastrophes and Earth History, the New Uniformitarianism.* Princeton: Princeton University Press, 1984.

Berry, L., and B. Mason. *Mineralogy.* San Francisco: W. H. Freeman and Company, 1959.

Berry, W. B. N. *Growth of a Prehistoric Time Scale Based on Organic Evolution.* Revised edition. Oxford: Blackwell Scientific Publications, 1987.

Best, M. G. *Igneous and Metamorphic Petrology.* San Francisco: W. H. Freeman and Company, 1982.

Bettelheim, Frederick A., and Jerry March. *Introduction to General, Organic, & Biochemistry.* Philadelphia: Saunders College Publishing, 1983.

Billings, Marland P. *Structural Geology.* 3rd ed. Englewood Cliffs, New Jersey: Prentice-Hall, 1972.

Binney, James, and Michael Merrifield. *Galactic Astronomy.* Princeton, New Jersey: Princeton University Press, 1998.

Biographical Encyclopedia of Science. 2nd ed. Philadelphia: Lorenz, 1994.

Birkeland, Peter W. *Soils and Geomorphology.* New York: Oxford University Press, 1984.

Blackburn, William H., and William H. Dennen *Principles of Mineralogy.* Dubuque, IA: William C. Brown Publishers, 1988.

Blaedel, Niels. *Harmony and Unity.* Translated by Geoffrey French. Madison, Wisconsin: Science Tech, 1988.

Bland, W., and Rolls D. *Weathering, an Introduction to Scientific Principles.* New York: Oxford University Press, 1988.

Blatt, H. *Sedimentary Petrology.* New York: W. H. Freeman and Company, 1992.

Blatt, H., and R. J. Tracy. *Petrology: Igneous, Sedimentary, and Metamorphic.* New York: W. H. Freeman and Company, 1996.

Bless, R. C. *Introductory Astronomy.* Sausalito, CA: University Science Books, 1996.

Bloom, A. L. *Geomorphology: A Systematic Analysis of Late Cenozoic Landforms.* 2nd ed. Englewood Cliffs: Prentice Hall, 1991.

Bloss, F. D. *Crystallography and Crystal Chemistry.* New York: Holt, Rinehart and Winston, Inc., 1971.

Bockris, John O'M., and Amulya K. N. Reddy. *Modern Electrochemistry.* New York: Plenum Press, 1973.

Boggs, S. *Petrology of Sedimentary Rocks.* Paramus, NJ: Macmillan Publishing, 1992.

Bohr, Niels. *The Unity of Knowledge.* New York: Doubleday & Co., 1955.

Boorse, Henry A., Lloyd Motz, and Jefferson Hane Weaver. *The Atomic Scientists: A Biographical History.* New York: John Wiley & Sons, Inc., 1989.

Boorstin, Daniel J. *The Discoverers.* New York: Random House, 1983.

Bothun, Greg. *Modern Cosmological Observations and Problems.* London: Taylor & Francis Ltd., 1998.

Boyer, Rodney. *Concepts in Biochemistry.* Pacific Grove, CA: Brooks/Cole Publishing Company, 1999.

Boyle, Robert H., and R. Alexander Boyle. *Acid Rain.* New York: Schocken, 1983.

British Petroleum—Fifty Years in Pictures. London: British Petroleum, 1959.

Brock, William H. *The Norton History of Chemistry.* New York: W. W. Norton & Company, 1992.

Bronowski, J. *The Ascent of Man.* New York: Little, Brown & Co., 1984.

Brown, Julian. *Minds, Machines, and the Multiverse: The Quest for the Quantum Computer.* New York: Simon & Schuster, 2000.

Brownlow, A. *Geochemistry.* Englewood Cliffs, NJ: Prentice Hall Inc., 1997.

Bruice, Paula Y. *Organic Chemistry.* Englewood Cliffs, NJ: Prentice-Hall, Inc., 1995.

Bryan, T. Scott. *The Geysers of Yellowstone.* Boulder, CO: Colorado Associated University Press, 1986.

Bugayevskiy, Lev M., and John P. Snyder. *Map Projections: A Reference Manual.* London: Taylor and Francis, 1995.

Burchfield, J. D. *Lord Kelvin and the Age of the Earth.* Chicago: University of Chicago Press, 1990.

Burk, C. A., and C. L. Drake. *The Geology of Continental Margins.* New York: Springer-Verlag, 1974.

Burke, James. *The Day the Universe Changed.* Boston: Little, Brown and Company, 1985.

Bynum, W. F., E. J. Browne, and Roy Porter, eds. *Dictionary of the History of Science.* Princeton, NJ: Princeton University Press, 1984.

Campbell, A. M. "Monoclonal Antibodies." In *Immunochemistry*, edited by Carol J. van Oss and Marc H. V. van Regenmortel. New York: Marcel Dekker, Inc., 1994.

Carmichael, I., F. Turner, and J. Verhoogen. *Igneous Petrology.* New York: McGraw-Hill Book Company, 1974.

Carozzi, Albert V. "Guyot, Arnold Henri." Vol. 5, *Dictionary of Scientific Biography*. Farmington Hills, MI: MacMillan Reference USA, 1981.

Carroll, Bradley W., and Dale A. Ostlie. *An Introduction to Modern Astrophysics.* Reading, MA: Addison-Wesley Publishing Company, Inc., 1996.

Carroll, Felix A. *Perspectives on Structure and Mechanism in Organic Chemistry.* Pacific Grove, CA: Brooks/Cole Publishing Company, 1998.

Carswell, D. A., ed. *Eclogite Facies Rocks.* Glasgow: Blackie, 1999.

Cattermole, Peter, and Patrick Moore. *The Story of the Earth.* Cambridge: Cambridge University Press, 1985.

Chaisson, Eric J. *The Hubble Wars: Astrophysics Meets Astropolitics in the Two-Billion-Dollar Struggle over the Hubble Space Telescope.* London: Harvard University Press, 1994.

Chow, Ven T., et al. *Applied Hydrology*. New York: McGraw Hill, 1988.

Christiansen, E. H, and W. K. Hamblin. *Exploring the Planets.* 2nd ed., Englewood Cliffs, NJ: Prentice Hall, 1995.

Clifford, Martin. *Basic Electricity & Beginning Electronics.* Blue Ridge Summit, PA: TAB Books, 1973.

Cline, William R. *The Economics of Global Warming.* Washington: Institute for International Economics, 1992.

Cobb, Cathy, and Harold Goldwhite. *Creations of Fire: Chemistry's Lively History from Alchemy to the Atomic Age.* New York: Plenum Press, 1995.

Collings, Peter J. *Liquid Crystals: Nature's Delicate Phase of Matter.* Princeton: Princeton University Press, 1990.

Colman, S., and D. P. Dethier. *Rates of Chemical Weathering of Rocks and Minerals.* Academic Press, 1996.

Compton, R. R. *Geology in the Field.* New York: Wiley and Sons, 1992.

Conant, James B. *The Overthrow of the Phlogiston Theory: The Chemical Revolution of 1775–1789.* Cambridge: Harvard University Press, 1966.

Connors, Kenneth. *Chemical Kinetics: The Study of Reaction Rates in Solution.* New York: VCH, 1990.

Cooke, R. U., and J. C. Doornkamp. *Geomorphology in Environmental Management.* 2nd ed. Oxford, UK: Clarendon Press, 1990.

Cooper, J. D., R. H. Miller, and J. Patterson. *A Trip Through Time: Principles of Historical Geology.* Columbus, OH: Merrill Publishing Company, 1990.

Cotton, F. Albert. *Advanced Inorganic Chemistry: A Comprehensive Text.* 2nd ed. New York: Interscience Publishers, 1966.

Cottrell, A. *An Introduction to Metallurgy.* 2nd ed. London: Edward Arnold, 1975.

Coughlan, G. D., and J. E. Dodd. *The Ideas of Particle Physics.* 2nd ed. Cambridge: Cambridge University Press, 1991.

Coulomb, J. *Sea Floor Spreading and Continental Drift.* Dordrecht, Netherlands: D. Reidel Publishing Co., 1972.

Cragg, G. R. *Reason and Authority in the Eighteenth Century.* London: Cambridge University Press, 1964.

Creese, M. R. S., and T. M. Creese. *Ladies in the Laboratory, American and British Women in Science 1800–1900.* London: The Scarecrow Press Inc., 1998.

Cutnell, John D., and Johnson, Kenneth W. *Physics*. 4th ed. New York: John Wiley & Sons, 1998.

Dainith, John, ed. *A Dictionary of Chemistry.* 3rd ed. Oxford: Oxford University Press, 1996.

Dainith, John, Sarah Mitchell, Elizabeth Tootill, and Derek Gjertsen, eds. *Biographical Encyclopedia of Scientists.* 2nd ed. Bristol, UK: Institute of Physics Publishing, 1994.

Dalrymaple, G. B. *Rock Magnetics Laboratory Upper Mantle Project, United States Program.* Washington: National Academy of Science, 1971.

David, G. H. *Structural Geology of Rocks and Regions.* New York: John Wiley & Sons Inc., 1984.

Davies, Paul, and John Gribbin. *The Matter Myth: Dramatic Discoveries that Challenge Our Understanding of Physical Reality.* New York: Touchstone Books, 1992.

Davies, Paul. *The Last Three Minutes.* New York: Basic Books, 1994.

Davies, Paul. *The Mind of God.* New York: Simon & Schuster, 1992.

Davies, Paul A. *The New Physics.* Cambridge: Cambridge University Press, 1989.

Davies, Paul A. *Mechanisms of Continental Drift and Plate Tectonics.* London: Academic Press, 1980.

Davis, N. *Permafrost: A Guide to Frozen Ground in Transition.* Fairbanks, Alaska: University of Alaska Press, 2001.

Davis, R. A. *Depositional Systems.* 2nd ed. Englewood Cliffs, New Jersey: Prentice Hall, 1992.

Davis, Raymond E., H. Clark Metcalfe, John E. Williams, and Joseph F. Castka, eds. *Modern Chemistry.* Austin, TX: Holt, Rinehart and Winston, Inc., 1999.

Dawson, J. B. *Kimberlites and Their Xenoliths.* Berlin: Springer-Verlag, 1980.

Dawson, J. B. "New Aspects of Diamonds Geology." In *The Properties of Diamond,* edited by D. Fields. London: Academic Press, 1979: 539–554.

Deason, G. B. *Reformation Theology and the Mechanistic Conception of Nature, in God and Nature.* Edited by D. C. Lindberg and R. L. Numbers. University of California Press, 1986.

Deer, W. A., R. A. Howie, and J. Zussman *An Introduction to the Rock-Forming Minerals.* New York: Longman Scientific and Technical, 1994.

Delgass, W. Nicholas, Gary L. Haller, Richard Kellerman, and Jack H. Lunsford, eds. *Spectroscopy in Heterogeneous Catalysis.* New York: Academic Press, 1979.

Dessler, A. *The Chemistry and Physics of Stratospheric Ozone.* Cornwall, UK: Academic Press, 2000.

Dewey, J. F., and G. M. Kay. "Appalachian and Caledonian Evidence for Drift in the North Atlantic." In *History of the Earth's Crust.* Princeton: Princeton University Press, 1968.

Dolgoff, A. *Essentials of Physical Geology.* Boston: Houghton Mifflin, 1998.

Donn, William L. *Meteorology.* 4th ed. New York: McGraw-Hill Book Company, 1975.

Doyle P., and M. B. Bennett *The Key To Earth History.* John Wiley and Sons, 1994.

Driscoll, F. *Groundwater and Wells.* 2nd ed. St. Paul: Johnson Filtration Systems, Inc., 1989.

Duff, P. *Holmes' Principles of Physical Geology.* 4th ed. New York: Chapman & Hall, 1993.

Duffin, W. J. *The Solid State: From Superconductors to Superalloys.* Translated by André Guinier and Rémi Jullien. Oxford: Oxford University Press, 1989.

Dunlop, D. J., and Ö. Özdemir, *Rock Magnetism—Fundamentals and Frontiers.* Cambridge: Cambridge University Press, 1997.

DuToit, A. L. *Our Wandering Continents: An Hypothesis of Continental Drifting.* New York: Hafner Publishing Company, 1937.

Eardley, A. J. *Structural Geology of North America.* New York: Harper Brothers, 1951.

Ebbing, Darrell D., and Steven D. Gammon. *General Chemistry.* 6th ed. Boston: Houghton Mifflin Company, 1999.

Ehlers, E., and H. Blatt. *Petrology: Igneous, Sedimentary, and Metamorphic.* San Francisco: W. H. Freeman and Company, 1980.

Elliott, W. H., and D. C. Elliott. *Biochemistry and Molecular Biology.* New York: Oxford University Press, 1997.

Ellis, G. F. R., and R. M. Williams. *Flat and Curved Space-Times.* Oxford: Clarendon Press, 2000.

Elmegreen, Debra Meloy. *Galaxies and Galactic Structure.* Upper Saddle River, NJ: Prentice Hall, 1998.

Elschenbroich, Christoph, and Albrecht Salzer. *Organometallics: A Concise Introduction.* 2nd ed. New York: VCH, 1992.

Elsom, D. *Earth: The Making, Shaping and Workings of a Planet.* New York: Macmillian and Company, 1992.

Emiliani, Cesare. *Planet Earth: Cosmology, Geology, and the Evolution of Life and the Environment.* Cambridge: Cambridge University Press, 1992.

Epstein, Lewis Carroll. *Relativity Visualized.* San Francisco: Insight Press, 1997.

Erwin, D. H. *The Great Paleozoic Crisis: Life and Death in the Permian.* New York: Columbia University Press, 1993.

Evans, A. M. *An Introduction to Economic Geology and its Environmental Impact.* Malden, MA: Blackwell Science, 1997.

Ewing, Galen W. *Instrumental Methods of Chemical Analysis.* 4th ed. New York: McGraw-Hill Book Company, 1975.

Eyles, Joan M. "Smith, William." Vol. 12, *Dictionary of Scientific Biography,* edited by C. C. Gillispie. New York: Charles Scribner & Sons, 1975, pp. 486-492.

Fairbridge, Rhodes, ed. *Encyclopedia of Sedimentology.* Stroudsburg, PA: Dowden, Hutchison, and Ross, 1978.

Farndon, J., *Dictionary of the Earth*. New York: Dorling Kindersley Publishing, Inc., 1994.

Felder, D. *The 100 Greatest Women of All Time*. Oxford: Past Times, 1997.

Feldman, Anthony, and Peter Ford. *Scientists & Inventors.* New York: Facts on File, 1979.

Ferris, Timothy, ed. *The World Treasury of Physics, Astronomy and Mathematics.* Boston: Little, Brown and Company, 1991.

Feynman, Richard P. *QED: The Strange Theory of Light and Matter.* Princeton: Princeton University Press, 1985.

Feynman, Richard P. *The Character of Physical Law.* Cambridge: MIT Press, 1965.

Feynman, Richard. *The Feynman Lectures on Physics.* Vol. 1. Reading, MA: Addison-Wesley Publishing Company, 1963.

Fieser, Louis F., and Mary Fieser. *Textbook of Organic Chemistry.* Boston: D. C. Heath and Company, 1950.

Fike, D. J. "Immunoglobulin Structure and Function." In *Clinical Immunology: Principles and Laboratory Diagnosis.* 2nd edition. Edited by Catherine Sheehan. Philadelphia: Lippincott-Raven Publishers, 1997.

Fischer, Daniel, and Hilmar Duerbeck. *Hubble: A New Window to the Universe.* Translated by Helmut Jenkner and Douglas Duncan. New York: Copernicus, 1996.

Fischer, G., and G. Wefer. *Use of Proxies in Paleoceanography*. Berlin Heidelberg: Springer-Verlag, 1999.

Fisher, David E. *Fire and Ice: The Greenhouse Effect, Ozone Depletion, and Nuclear Winter.* New York: Harper & Row, 1990.

Fisher, Robert L., and Roger Revelle "The Trenches of the Pacific." In *Continents Adrift: Articles from Scientific American 1972–1970,* edited by J. T. Wilson. New York: W. H. Freeman and Co., 1975.

Flaste, Richard, ed. *What Everyone Needs to Know from Newton to the Knuckleball.* New York: Times Books/Random House, 1991.

Flint, R. *Glacial Geology and the Pleistocene Epoch.* New York: John Wiley & Sons, 1949.

Foster, James, and J. D. Nightingale. *A Short Course in General Relativity.* New York: Springer-Verlag, 1995.

Francis, P. *Volcanoes: A Planetary Perspective.* New York: Oxford University Press, 1993.

Fraser, P. M. *Ptolemaic Alexandria.* Oxford: Clarendon Press, 1972.

Freeman, Ira M. *Physics Made Simple.* New York: Doubleday, 1990.

Freeman, T. *Procedures in Field Geology.* Boston: Blackwell Science, Inc., 1999.

Freeze, R., and J. Cherry. *Groundwater.* Englewood Cliffs: Prentice-Hall, Inc., 1979.

French, A. P. *Special Relativity.* New York: W. W. Norton & Company, 1968.

French, B. M. *Traces of Catastrophe.* Houston: Lunar and Planetary Institute, 1999.

Freund, R. *A Dynamic Model of Subduction Zones.* Jerusalem: Hebrew University of Jerusalem, 1991.

Frick, Thomas C. *Petroleum Production Handbook.* Vol. 1. Dallas: Society of Petroleum Engineers of AIME, 1962.

Friend, J. Newton. *Man and the Chemical Elements: An Authentic Account of the Successive Discovery and Utilization of the Elements From the Earliest Times to the Nuclear Age.* 2nd ed. New York: Charles Scribner's Sons, 1961.

Frisch, Otto R. *The Nature of Matter*. New York: E. P. Dutton, 1972.

Fritz, W., and J. Moore. *Basics of Physical Stratigraphy and Sedimentology.* New York: John Wiley & Sons, 1988.

Gaia: A New Look at Life on Earth. Oxford: Oxford University Press, 1995.

Galeotti, P., and David N. Schramm, eds. *Dark Matter in the Universe.* Dordrecht, The Netherlands. Kluwer Academic Publishers, 1990.

Galloway, W. E., and D. K. Hobday *Terrigenous Clastic Depositional Systems.* Berlin: Springer-Verlag, 1996.

Garrity, John A., and Mark C. Carnes, eds. *American National Biography.* New York: Oxford University Press, 1999.

Gaskell, T. F. *The Earth's Mantle.* New York: Academic Press, 1967.

Gavin, Sir William. *Ninety Years of Family Farming: The Story of Lord Rayleigh's and Strutt & Parker Farms.* London: Hutchinson, 1967.

George, Wilma, and Rene Lavocat. *The African-South America Connection.* New York: Oxford University Press, 1993.

Ghosh S. K. *Structural Geology: Fundamentals and Modern Developments.* Oxford, New York: Pergamon Press, 1993.

Giancoli, Douglas C. *Physics: Principles with Applications.* 5th ed. New Jersey: Prentice Hall, 1998.

Gile, L. H., et al. *Soils and Geomorphology in the Basin and Range Area of Southern New Mexico: A Guidebook to the Desert Project.* Socorro: New Mexico Bureau of Mines and Mineral Resources, 1981.

Gillispie, Charles Coulston, ed. *Dictionary of Scientific Biography*. New York: Charles Scribner's Sons, 1981.

Gilmore, Robert. *Alice in Quantumland.* New York: Copernicus Books, 1995.

Gittwitt, Paul G. *Conceptual Physics.* 6th ed. New York: HarperCollins Publishers, 1989.

Gleick, James. *Chaos.* New York: Penguin, USA, 1988.

Glennie, K. W. *Desert Sedimentary Environments: Developments in Sedimentology.* Vol 14, enclosure 4. New York: American Elsevier Publishing Co, 1970.

Goldman, Martin. *The Demon in the Aether, The Story of James Clerk Maxwell.* Edinburgh: Paul Harris Publishing, 1983.

Goldsmith, Donald. *Einstein's Greatest Blunder? The Cosmological Constant and other Fudge Factors in the Physics of the Universe.* Cambridge: Harvard University Press, 1995.

Golub L., and J. M. Pasachoff. *The Solar Corona.* Cambridge: Cambridge University Press, 1997.

Goodman, H. Maurice. *Basic Medical Endocrinology.* 2nd ed. New York: Raven Press, 1994.

Gould, Stephen Jay. *Ever Since Darwin.* New York: W. W. Norton, 1977.

Goyer, R. A. "Toxic Effects of Metals." In *Casarett and Doull's Toxicology: The Basic Science of Poisons.* 5th ed. New York: McGraw-Hill Companies, Inc., 1996.

Gray, H. J., and Alan Isaacs. *The Penguin Dictionary of Physics.* Middlesex, UK: Penguin, 1977.

Greeley, R., and R. Batson. *The NASA Atlas of the Solar System.* New York: Cambridge University Press, 1996.

Greeley, R., and R. M. Batson. eds. *Planetary Mapping.* New York: Cambridge University Press, 1990.

Green, James A. *Letters on Unified Field Theory: From General Relativity to Unified Field Theory.* Hollywood Greenwood Research, 1999.

Greene, Brian. *The Elegant Universe.* New York: W. W. Norton and Company, 1999.

Gribbin, John. *Q is for Quantum: An Encyclopedia of Particle Physics.* New York: The Free Press, 1998.

Gribbin, John. *Schrodinger's Kittens and the Search for Reality.* London: Weidenfield & Nicholson, 1995.

Grider, R. W. *Continental Drift: Some Boundary Conditions from Surface Phenomena.* London: Academic Press, 1983.

Grimal, Nicolas. *A History of Ancient Egypt.* Cambridge: Blackwell, 1992.

Guilbert, J. M., and C. F. Park. *The Geology of Ore Deposits.* New York: W. H. Freeman, 1986.

Haan, C. T., et al. *Design Hydrology and Sedimentology for Small Catchments.* San Diego: Academic Press, 1994.

Hall A. *Igneous Petrology.* Essex, England: Longman Group Limited, 1996.

Hall, C. *Gemstones.* New York: DK Publishing Inc., 1994.

Hamblin, W. K., and E. H. Christiansen. *Earth's Dynamic Systems.* 9th ed. Upper Saddle River: Prentice Hall, 2001.

Hamblyn, Richard. *The Invention of Clouds: How an Amateur Meteorologist Forged the Language of the Skies.* New York: Farrar, Straus and Giroux, 2001.

Hamermesh, Morton. *Group Theory and its Application to Physical Problems.* New York: Dover Publications, Inc., 1989.

Hancock, P. L., and B. J. Skinner, eds. *The Oxford Companion to the Earth.* New York: Oxford University Press, 2000.

Harland W. B., et al. *A Geologic Time Scale.* New York: Cambridge University Press, 1990.

Harre, Rom. *Great Scientific Experiments: Twenty Experiments that Changed Our View of the World.* Oxford: Phaidon Press Limited, 1981.

Harrold, Frances B., and Raymond A. Eve, eds. *Cult Archaeology and Creationism: Understanding Pseudoscientific Beliefs about the Past.* University of Iowa Press, 1987.

Hawking, Stephen. *A Brief History of Time: From the Big Bang to Black Holes.* New York: Bantam, 1988.

Heezen, B. C. "The Deep Sea-Floor." In *Continenal Drift,* edited by S. K. Runcorn. New York: Academic Press, 1962.

Heezen, B. C. *The Face of the Deep.* Oxford: Oxford University Press, 1971.

Heinrich, W. *Microscopic Identification of Minerals.* St. Louis: McGraw-Hill, Inc., 1965.

Heiserman, David L. *Exploring Chemical Elements and Their Compounds.* Blue Ridge Summit, PA: TAB Books, 1992.

Hellemans, Alexander, and Bryan Bunch. *The Timetables of Science: A Chronology of the Most Important People and Events in the History of Science.* New York: Simon & Schuster Inc., 1988.

Henisch, Heinz K. *Crystal Growth in Gels.* University Park, PA: The Pennsylvania State University Press, 1970.

Hewitt, Paul G., John Suchocki, and Leslie A. Hewitt. *Conceptual Physics.* 3rd ed. Menlo Park, CA: Addison-Wesley, 1999.

Hill, C., and P. Forti. *Cave Minerals of the World.* 2nd ed. Huntsville: National Speleological Society, Inc., 1997.

Hilts, Philip J. *Scientific Temperaments: Three Lives in Contemporary Science.* New York: Simon & Schuster, 1982.

Hoffman, Robert V. *Organic Chemistry: An Intermediate Text.* New York: Oxford University Press, 1997.

Holland, C. H. *The Idea of Time.* New York: John Wiley and Sons, 1999.

Holmes, A. *Principles of Physical Geology.* London: Thomas Nelson & Sons, 1944.

Holmes, A. *The Age of the Earth.* London: Harper & Brothers, 1909.

Hough, Jack L. *Geology of the Great Lakes.* Urbana, IL: University of Illinois Press, 1958.

Howarth, S. *A Century in Oil.* London: Wiedenfeld & Nicholson, 1997.

Howells, W. W. "*Homo Erectus*—Who, When and Where: A Survey." In *The Human Evolution Source Book.* Englewood Cliffs, NJ: Prentice Hall, 1993.

Hoyt, D. V., and K. H. Schatten. *The Role of the Sun in Climate Change.* Oxford: Oxford University Press, 1997.

Hsü, K. *Challenger at Sea.* Princeton: Princeton University Press, 1992.

Huang, Kerson. *Statistical Mechanics.* New York: John Wiley & Sons, 1987.

Huang, W. T. *Petrology.* York, PA: McGraw-Hill Book Company, 1962.

Hudson, John. *The History of Chemistry.* New York: Chapman & Hall, 1992.

Hughes, R. I. G. *The Structure and Interpretation of Quantum Mechanics.* Cambridge: Harvard University Press, 1989.

Huizenga, John R. *Cold Fusion: The Scientific Fiasco of the Century.* Oxford: Oxford University Press, 1993.

Hunter, Robert J. *Introduction to Modern Colloid Science.* Oxford: Oxford Unversity Press, 1993.

Hurlbut, C. *Dana's Manual of Mineralogy.* 18th ed. New York: John Wiley & Sons, 1971.

Hurlbut, C. S., and C. Klein. *Manual of Mineralogy.* New York: John Wiley & Sons, Inc., 1977.

Hyndman, D. *Petrology of Igneous and Metamorphic Rocks.* Blacklick, OH: McGraw-Hill, 1985.

International Subcommission on Stratigraphic Classification (A. Salvador, chairman). *International Stratigraphic Guide.* 2nd ed. Boulder: Geological Society of America, 1994.

Interrante, Leonard V., Lawrence A. Caspar, and Arthur B. Ellis, eds. *Materials Chemistry: An Emerging Discipline.* Washington: American Chemical Society, 1995.

Irving, E. *Paleomagnetism and Its Application to Geological and Geophysical Problems.* New York: John Wiley & Sons, 1964.

Jackson, K. C. *Textbook of Lithology.* New York: McGraw-Hill Inc., 1970.

Jacobs, J. A. *Reversals of the Earth's Magnetic Field.* 2nd ed. Cambridge: Cambridge University Press, 1994.

Jaeger, J. C., and N. G. W. Cook. *Fundamentals of Rock Mechanics.* New York: Chapman and Hall, 1979.

Jahn, F., M. Cook, and M. Graham. *Hydrocarbon Exploration and Production. Developments in Petroleum Science.* Vol. 46. The Netherlands: Elsevier Science, 2000.

James, Laylin K, ed. *Nobel Laureates in Chemistry: 1901–1992.* American Chemical Society and the Chemical Heritage Foundation, 1993.

James, Preston Everett, and Geoffrey J. Martin, *All Possible Worlds: A History of Geographical Ideas.* 2nd ed. New York: Wiley, 1981.

Jeffery, Harold. *The Earth.* Cambridge: Cambridge University Press, 1970.

Johnson, A. M., with contributions by J. R. Rodine. "Debris Flow." In *Slope Instability*, edited by D. Brunsden and D. B. Prior. Chicester, UK: John Wiley & Sons, 1984.

Jolly, William L. *Modern Inorganic Chemistry.* New York: McGraw-Hill Book Company, 1984.

Jones, D. S. *Acoustic and Electromagnetic Waves.* Oxford: Clarendon Press, 1989.

Julien, Peirre Y. *Erosion and Sedimentation.* Cambridge: University Press, 1994.

Kamm, L. J. *Understanding Electro-Mechanical Engineering.* Piscataway, NJ: IEEE Press, 1995.

Kane, Gordon. *The Particle Garden: Our Universe as Understood by Particle Physicists.* Reading, MA: Helix Books, 1995.

Kaplan, D., and L. Glass. *Understanding Nonlinear Dynamics.* Berlin: Springer-Verlag, 1995.

Karrow, Robert W. *Mapmakers of the Sixteenth Century and Their Maps: Bio-Bibliographies of the Cartographers of Abraham Ortelius, 1570.* Chicago: Newberry Library, Speculum Orbis Press, 1993.

Kauffman, J. *Physical Geology.* 8th ed. Englewood Cliffs, NJ: Prentice Hall, Inc., 1990.

Kaufmann, William. *Black Holes and Warped Spacetime.* New York: W. H. Freeman and Company, 1979.

Kay, Marshall. *North American Geosynclines.* Geological Society of America, 1951.

Keller, E. A. *Introduction to Environmental Geology.* 2nd ed. Upper Saddle River: Prentice Hall, 2002.

Kelso, A. J. *Physical Anthropology.* 2nd ed. New York: Lippincott, 1974.

Kemp, T. S. *Fossils and Evolution.* Oxford: Oxford University Press, 1999.

Kempe, D. R. C. "Deep Ocean Sediments." *The Evolving* Earth, edited by L. R. M. Cocks. London: Cambridge University Press, 1981.

Kerr, P. *Optical Mineralogy.* 4th ed. New York: McGraw-Hill Book Company, 1977.

Kittel, Charles. *Introduction to Solid State Physics.* New York: John Wiley & Sons, 1996.

Kittel, Charles. *Thermal Physics.* New York: John Wiley & Sons, 1969.

Klein, C. *The Manual of Mineral Science.* 22nd ed. New York: John Wiley & Sons, 2002.

Klein, C., and C. S. Hurlbut Jr. *Manual of Minerology.* 20th ed. New York: John Wiley & Sons, 1985.

Kleinsmith, L. J., and V. M. Kish. *Principles of Cell and Molecular Biology.* 2nd ed. New York: HarperCollins College Publishers, 1995.

Klug, William S., and Michael R. Cummings. *Concepts of Genetics.* 5th ed. Upper Saddle River, NJ: Prentice-Hall, Inc., 1997.

Knighton, David. *Fluvial Forms & Processes.* New York: John Wiley & Sons, 1998.

Kock, Winston. *Lasers and Holography.* Mineola, New York: Dover Books, 1981.

Kondepudi, Dilip, and Ilya Prigogine. *Modern Thermodynamics: From Heat Engines to Dissipative Structures.* New York: John Wiley & Sons, 1998.

Krane, Kenneth S. *Modern Physics.* New York: John Wiley & Sons, 1983.

Krane, Kenneth. *Introductory Nuclear Physics.* New York: John Wiley & Sons, 1988.

Krapp, Kristine M., ed. *Notable Twentieth Century Scientists.* Detroit: The Gale Group, 1998.

Krauskopf K. B. *Introduction to Geochemistry.* New York: McGraw Hill, 1995.

Kummel, Bernhard. *History of the Earth: An Introduction to Historical Geology.* San Francisco: W. H. Freeman and Company, 1961.

Kurten, Bjorn. *Pleistocene Mammals of Europe.* Columbia: Columbia University Press, 1969.

Kurten, Bjorn. *The Age of Mammals.* Columbia: Columbia University Press, 1972.

Léna, Pierre. *Observational Astrophysics.* Translated by A. R. King. Berlin: Springer-Verlag, 1988.

Leakey, Richard, and Roger Lewin. *Origins Reconsidered: In Search of What Makes Us Human.* New York: Doubleday, 1992.

Leet, L. D., et al. *Physical Geology,* 4th ed. Englewood Cliffs, NJ: Prentice-Hall, Inc., 1971.

LeGrand, H. E. *Drifting Continents and Shifting Theories: The Modern Revolution in Geology and Scientific Change.* London: Cambridge University Press, 1987.

LeMay, H. Eugene, Herbert Beall, Karen M. Robblee, and Douglas C. Brower, eds. *Chemistry: Connections to Our Changing World.* Teacher's Edition. Upper Saddle River, NJ: Prentice-Hall, Inc., 1996.

Leopold, Luna B., et al. *Fluvial Processes in Geomorphology.* New York: Dover Publications, 1995.

Leopold, Luna B. *A View of the River.* Cambridge: Harvard University Press, 1994.

LePichon, X. *Developments in Geotectonics, Plate Tectonics.* Amsterdam: Elsevier Scientific Publishing Co., 1973.

Lerner, Adrienne W. "Development of Trigonometry." In *Science and Its Times.* Vol. 3. Detroit: Gale Group, 2000.

Lerner, B. W. "Calculators, A Pocket-sized Revolution." *Science and Its Times.* Vol. 7. Detroit: Gale Group, 2000.

Lerner, K. L. "Bohr Theory." In *World of Chemistry*. Detroit: Gale Group, 1999.

Lerner, Rita G., and George L. Trigg, eds. *Encyclopedia of Physics.* 2nd ed. New York: VCH Publishers, Inc., 1991.

Levin H. L. *The Earth Through Time.* Saunders College Publishing, 1999.

Levine, Ira. *Quantum Chemistry.* 4th ed. New Jersey: Prentice Hall, 1991.

Lewis, C. L. E. *The Dating Game.* Cambridge: Cambridge University Press, 2000.

Lewis, Richard J., ed. *Hawley's Condensed Chemical Dictionary.* 13th ed. New York: Van Nostrand Reinhold, 1997.

Ley, Willy. *The Discovery of the Elements.* New York: Delacorte Press, 1968.

Liboff, Richard L. *Introductory Quantum Mechanics.* Reading, MA: Addison-Wesley Publishing Co., 1997.

Lide, D. R., and H. P. R. Frederikse, eds. *CRC Handbook of Chemistry and Physics 1996–1997: A Ready Reference Book of Chemical and Physical Data.* 77th ed. Boca Raton: CRC Press, 1996.

Lillesand, T. M., and R. W. Kiefer. *Remote Sensing and Image Interpretation.* 3rd ed. New York: John Wiley & Sons, Inc., 1994.

Linacre, E., and B. Geerts. *Climates and Weather Explained.* New York: Routledge, 1997.

Linklater, E. *The Voyage of the Challenger.* New York: Doubleday & Company, 1972.

Little, James Maxwell. *An Introduction to the Experimental Method: For Students of Biology and the Health Sciences.* Minneapolis: Burgess Publishing Co., 1961.

Lorenz, H. A., A. Einstein, H. Minkowski, and H. Weyl. *The Principle of Relativity.* Translated by W. Perrett and G. B. Jeffery. New York: Dover Publications, Inc., 1952.

Lovelock, James. *The Ages of Gaia: A Biography of Our Living Earth.* Revised and expanded edition. New York: W. W. Norton & Company, 1988.

Lowry, Thomas H., and Kathleen Schueller Richardson. *Mechanism and Theory in Organic Chemistry.* 3rd ed. New York: HarperCollins Publishers, Inc., 1987.

Luoma, Jon R. *Troubled Skies, Troubled Waters.* New York: Viking, 1984.

Lutgens F., and E. Tarbuck. *The Atmosphere.* 7th ed. New Jersey: Prentice Hall, Inc., 1998.

Lutgens, F. K., et. al. *Essentials of Geology.* 2nd ed. Columbus: Charles E. Merrill Publishing Co., 1986.

Maaloe S. *Principles of Igneous Petrology.* Berlin, New York: Springer-Verlag, 1985.

Manchester, R. N., and J. H. Taylor. *Pulsars.* San Francisco: W. H. Freeman, 1977.

Mandl, Franz. *Statistical Physics.* New York: John Wiley & Sons, 1988.

Margulis, Lynn, and Dorion Sagan. *Slanted Truths: Essays on Gaia, Symbiosis, and Evolution.* New York: Springer-Verlag, 1997.

Marsden, B. G., and A. G. W. Cameron. *The Earth-Moon System.* New York: Plenum Press, 1966.

Mascetta, Joseph A. *Chemistry the Easy Way.* 3rd ed. Barron's, 1996.

Mason, B., and C. B. Moore *Principles of Geochemistry.* New York: Wiley, 1982.

Mason, Stephen F. *Chemical Evolution: Origin of the Elements, Molecules, and Living Systems.* Oxford: Claredon Press, 1991.

Masterson, William L., and Emil J. Slowinski, eds. *Chemical Principles.* 4th ed. Philadelphia: W. B. Saunders Company, 1977.

Maton, Anthea, Jean Hopkins, Susan Johnson, David LaHart, Maryanna Quon Warner, and Jill D. Wright, eds. *Exploring Physical Science.* Teacher's Edition. Upper Saddle River, NJ: Prentice-Hall, Inc., 1999.

Maton, Anthea, et al. *Exploring Physical Science.* 3rd ed. New Jersey: Prentice Hall, 1999.

Matthews, Christopher K., and K. E. Van Holde, eds. *Biochemistry.* 2nd ed. New York: Benjamin/Cummings Publishing Company, 1966.

McConnell Jr., R. K. In "Viscosity of the Earth's Mantle." In *The History of the Earth's Crust: A Symposium.* Princeton: Princeton University Press, 1968.

McEvoy, G. K., ed. *AHFS Drug Information 1999.* Bethesda, MD: American Society of Health-System Pharmacists, Inc., 1999.

McGeary, D., and C. C. Plummer. *Physical Geology.* William C. Brown Publishers, 1992.

McIntyre, D. B., and A. McKirdy. *James Hutton: The Founder of Modern Geology.* Edinburgh: The Stationary Office Limited, 1997.

McKibben, Bill. *The End of Nature.* New York: Random House, 1989.

McPhee, John. *Basin and Range.* New York: Farrar, Straus, & Giroux, 1980.

McQuarrie, Donald A., and Peter A. Rock. *Descriptive Chemistry.* New York: W. H. Freeman and Company, 1985.

Melosh, H. J. *Impact Cratering, a Geologic Process.* New York: Oxford University Press, 1989.

Menard, H. *Geology, Resources, and Society.* San Francisco: W. H. Freeman and Company, 1974.

Menezes, A., P. van Oorschot, and S. Vanstone. *Handbook of Applied Cryptography.* Boca Raton: CRC Press, 1997.

Menzel, Donald H., ed. *Fundamental Formulas of Physics.* Mineola, NY: Dover Publications, 1960.

Merrill, R. T., M. W. McElhinny, and P. L. McFadden. *The Magnetic Field of the Earth, Paleomagnetism, the Core, and the Deep Mantle.* San Diego: Academic Press, 1996.

Merzbacher, E. *Quantum Mechanics.* New York: John Wiley & Sons, 1997.

Michaels, Patrick J. *Sound and Fury: The Science and Politics of Global Warming.* Washington: Cato Institute, 1992.

Michalas, D., and J. Binney. *Galatic Astronomy: Structure and Kinematics.* New York: W. H. Freeman and Co., 1981.

Middleton, Gerard V., and Peter R. Wilcock. *Mechanics in the Earth and Environmental Sciences.* Cambridge: Cambridge University Press, 1994.

Miller, E. Willard, and Ruby M. Miller. *Environmental Hazards: Air Pollution.* Santa Barbara: ABC-CLIO, 1989.

Misner, Charles, Kip Thorne, and John Archibald Wheeler. *Gravitation.* New York: W. H. Freeman and Company, 1973.

Mitton, Jacqueline, and Stephen P. Maran. *Gems of Hubble.* Cambridge: Cambridge University Press, 1996.

Miyashiro, A. *Metamorphic Petrology.* London: UCL Press, 1996.

Monroe, J. S., and R. Wicander. *Physical Geology: Exploring the Earth.* 3rd ed. Wadsworth Publishing Company, 1998.

Montanari, A., and C. Koeberl. *Impact Stratigraphy, the Italian Record.* Berlin: Springer-Verlag, 2000.

Montgomery, C. W., and D. Dathe. *Earth Then and Now.* William C. Brown Publishers, 1994.

Montgomery, C. W. *Environmental Geology.* 5th ed. McGraw Hill, 2000.

Moore G., and N. Sullivan. *Speleology, The Study of Caves*. St. Louis: Zephyrus Press, Inc., 1978.

Moore, C. A. *Handbook of Subsurface Geology*. New York: Harper & Row, 1963.

Moore, Walter John. *A Life of Erwin Schrodinger.* New York: Cambridge University Press, 1994.

Morris, Richard. *Time's Arrows: Scientific Attitudes Toward Time.* New York: Simon & Schuster, 1985.

Morrison, F. *The Art of Modeling Dynamic Systems: Forecasting for Chaos, Randomness, and Determinism.* New York: John Wiley & Sons, 1991.

Morrison, Robert Thornton, and Robert Neilson Boyd. *Organic Chemistry.* 5th ed. Boston: Allyn and Bacon, Inc., 1987.

Muir, Hazel, ed. *Larousse Dictionary of Scientists.* New York: Larousse Kingfisher Chambers, Inc., 1994.

Muir, J. *My First Summer in the Sierra.* Boston: Houghton Mifflin, 1911.

Muir, J. *Wilderness Essays.* Edited by F. Buske. Salt Lake City: Gibbs Smith, 1980.

Multhauf, Robert P. *The Origins of Chemistry.* London: Oldburne Book Co. ltd., 1966.

Nicolas, A. *The Mid-Oceanic Ridges.* Berlin: Springer-Verlag, 1995.

Nield, E. *Women and the Geological Society, A Report of the Working Group.* London: Geological Society of London, 1997.

Nilsson, Annika. *Greenhouse Earth.* Chichester: John Wiley & Sons, 1992.

North American Stratigraphic Code. Tulsa: American Association of Petroleum Geologists, 1983.

Nybakken, J. *Marine Biology: An Ecological Approach.* 4th ed. California: Addison-Wesley Educational Publishers, Inc., 1997.

O'Dunn, S., and W. D. Sill, *Exploring Geology, Introductory Laboratory Activities.* Needham Heights, MA: Ginn Press, 1993.

Ohanian, Hans. *Gravitation and Spacetime.* 2nd ed. New York: W. W. Norton & Company, 1994.

Olmstead, A. T. *Ancient Near East.* Chicago: University of Chicago Press, 1963.

Olmstead, A. T. *History of the Persian Empire and the Ancient MidEast.* Chicago: University of Chicago Press, 1959.

Olson, Donald W. "Gypsum." In *Minerals Yearbook.* Boulder: U.S. Geological Survey, 2000.

Orchin, M., and H. H. Jaffe. *Symmetry, Orbitals, and Spectra.* New York: Wiley-Interscience, 1971.

Ospovat, Alexander. "Werner, Abraham Gottlob." In *Dictionary of Scientific Biography.* Vol. 14, edited by C. C. Gillispie. New York: Charles Scribner and Sons, 1975, pp. 256-264.

Ostmann, Robert. *Acid Rain: A Plague Upon the Waters.* Minneapolis: Dillon, 1982.

Otoxby, David W., H. P. Gillis, and Norman H. Nachtrieb, eds. *The Principles of Modern Chemistry.* Fort Worth, TX: Harcourt Brace College Publishers, 1999.

Ottonello G. *Principles of Geochemistry.* New York: Columbia University Press, 1997.

Owen, H. G. "Constant Dimensions or an Expanding Earth?" In *The Evolving Earth,* edited by L. R. M. Cocks. London: Cambridge University Press, 1981.

Pais, Abraham. *Subtle is the Lord 1/4: The Science and the Life of Albert Einstein.* New York: Oxford University Press, 1982.

Pauling, Linus. *The Chemical Bond: A Brief Introduction to Modern Structural Chemistry.* Ithaca, NY: Cornell University Press, 1967.

Peierls, R. E. *Atomic History.* New York: Springer-Verlag, 1997.

Petersen, Carolyn Collins, and John C. Brandt. *Hubble Vision: Further Adventures with the Hubble Space Telescope.* 2nd ed. Cambridge: Cambridge University Press, 1998.

Peterson, M. N. A. *Initial Reports of the Deep Sea Drilling Project II.* Washington: Government Printing Office, 1970.

Pettijohn, F. J., P. E. Potter, and R. Siever. *Sand and Sandstone.* New York: Springer-Verlag, 1987.

Phinney, R. A., ed. *The History of the Earth's Crust.* Princeton: Princeton University Press, 1968.

Pinet, P. *Invitation to Oceanography.* Massachusetts: Joans and Bartlett Publishers, Inc., 1998.

Pitcher, Wallace S. *The Nature and Origin of Granite.* New York: Blackie Academic and Professional, 1993.

Pitzer, Kenneth S. *Thermodynamics.* 3rd ed. New York: McGraw-Hill, Inc., 1995.

Poole Jr., C. P. *The Physics Handbook: Fundamentals and Key Equations.* New York: John Wiley & Sons, 1998.

Popper, Karl R. *The Logic of Scientific Discovery.* New York: Basic Books, 1959.

Porter, Roy, and Marilyn Ogilvie, consultant eds. *The Biographical Dictionary of Scientists.* Vol. 2. New York: Oxford University Press, 2000.

Porter, Roy. *The Greatest Benefit to Mankind: A Medical History of Humanity.* New York: W. W. Norton & Company, Inc., 1997.

Press, F., and R. Siever. *Understanding Earth.* 3rd ed. New York: W. H. Freeman and Company, 2001.

Price, N. J., and J. W. Cosgrove. *Analysis of Geological Structures.* Cambridge: Cambridge University Press, 1990.

Proceedings of the Lunar Sample Symposium I. Houston: NASA Manned Spacecraft Center, 1972.

Proceedings of the Lunar Sample Symposium II. Houston: NASA Manned Spacecraft Center, 1973.

Proceedings of the Lunar Sample Symposium III. Houston: NASA Manned Spacecraft Center, 1975.

Proceedings of the Lunar Sample Symposium IV. Houston: NASA Manned Spacecraft Center, 1978.

Prothero, D. R. *Interpreting the Stratigraphic Record.* New York: W. H. Freedman and Company, 1990.

Prothro, D. R., and R. H. Dott Jr. *Evolution of the Earth.* 6th ed. Boston, McGraw-Hill, 2002.

Puddephatt, R. J., and P. K. Monaghan. *The Periodic Table of the Elements.* 2nd ed. Oxford: Clarendon Press, 1986.

Purser B., M. E. Tucker, and D. Zenger. *Dolomites.* International Associations of Sedimentologists, Special Publication No. 21. Oxford: Blackwell Scientific Publications, 1994.

Putnis, A. *Introduction to Mineral Sciences.* Hampshire, UK: Cambridge University Press, 1992.

Raguin, E. *Geology of Granite.* New York: Interscience Publishers, 1965.

Rayleigh, Lord (Robert John Strutt). *Life of John William Strutt, Third Baron Rayleigh.* Madison: University of Wisconsin Press, 1968.

Raymond L. A. *Petrology: The Study of Igneous, Sedimentary, and Metamorphic Rocks.* William C. Brown Publishers, 1995.

Rayner-Canham, Geoffrey, Arthur Last, Robert Perkins, and Mark van Roode. *Foundations of Chemistry.* Reading, PA: Addison-Wesley Publishing Company, 1983.

Reeder, R. J. *Carbonates: Mineralogy and Chemistry.* 2nd ed. Blacksburg, VA: Mineralogical Society of America, 1990.

Regional Geology of Africa. New York: Springer-Verlag, Inc., 1991.

Rhodes, Richard. *The Making of the Atomic Bomb.* New York: Touchstone/Simon & Schuster, 1986.

Rodman, Dorothy, Donald D. Bly, Fred Owens, and Ann-Claire Anderson. *Career Transitions for Chemists.* Salem: American Chemical Society, 1995.

Rodolico, Francesco. "Scilla, Agostino." In *Dictionary of Scientific Biography*, Vol. 12. 1975 pp. 256–257.

Rommer, A. S. *Vertebrate Paleontology.* Chicago: University of Chicago Press, 1966.

Ronchi, Vasco. *The Nature of Light: An Historical Survey.* Translated by V. Barocas. Cambridge, Massachusetts: Harvard University Press, 1970.

Rosenblum, Naomi. *A World History of Photography.* New York: Abbeville Press, 1997.

Rosser, W. G. V. *An Introduction to Statistical Physics.* New York: Prentice Hall, 1982.

Roth G. D. *Astronomy.* New York: Springer-Verlag, 1975.

Rozental, S., ed. *Niels Bohr: His Life and Work as Seen by Friends and Colleagues.* New York: John Wiley & Sons, 1967.

Rudwick, M. J. S. "Sedgwick, Adam." *Dictionary of Scientific Biography.* Vol. 12. 1975, pp. 275–279.

Runcorn, S. K. "Some Comments on Mechanism of Continental Drift." *Mechanisms of Contintntal Drift and Plate Tectonics.* New York: Academic Press, 1984.

Runcorn, S. K. *The Application of Modern Physics to the Earth and Planetary Interiors.* New York: Wiley-Interscience, 1969.

Russell, C. A., ed. *Recent Developments in the History of Chemistry.* London: The Royal Society of Chemistry, 1985.

Ryan, Charles W. *Basic Electricity: A Self-Teaching Guide,* 2nd ed. New York: John Wiley & Sons, Inc., 1986.

Sabins Jr., F. S. *Remote Sensing Principles and Interpretation.* 2nd ed. New York: W. H. Freeman and Company, 1987.

Sagan, Carl. *Cosmos.* New York: Random House, 1980.

Sakurai, J. J. *Advanced Quantum Mechanics.* Reading, MA: Addison-Wesley Publishing Company, 1967.

Salmon, Wesley C. *Space, Time, and Motion: A Philosophical Introduction.* Minneapolis: University of Minnesota Press, 1980.

Salzberg, Hugh W. *From Caveman to Chemist: Circumstances and Achievements.* Washington: American Chemical Society, 1991.

Sand, L. B., and F. A. Mumpton. *Natural Zeolites: Occurrence, Properties, Use.* London: Pergamon Press, 1978.

Sawkins F. J. *Metal Deposits in Relation to Plate Tectonics.* Berlin: Springer-Verlag, 1990.

Schneider, Stephen H., ed. *Scientists of Gaia.* Cambridge: The MIT Press, 1993.

Scholle, P., and D. Spearing, eds. *Sandstone Depositional Environments.* Tulsa: American Association of Petroleum Geologists, 1988.

Schwazbach, Martin. *Alfred Wegner, the Father of Continental Drift.* New York: Science Tech, 1986.

Schweber, Silvan S. *QED and the Men Who Made It: Dyson, Feynman, Schwinger, and Tomonaga.* Princeton: Princeton University Press, 1994.

Sciama, D. W. *Modern Cosmology and the Dark Matter Problem.* Cambridge: Cambridge University Press, 1993.

Science and Technology Department of the Carnegie Library of Pittsburgh. *The Handy Science Answer Book.* 2nd ed. Pittsburgh: The Carnegie Library of Pittsburgh, 1997.

Scientific Assessment of Ozone Depletion. Vols. I and II. World Meteorological Organization Global Ozone Research and Monitoring Project, 1988.

Scott, Stephen K. *Oscillations, Waves, and Chaos in Chemical Kinetics.* New York: Oxford University Press, 1994.

Sears, F. W., and G. L. Salinger. *Thermodynamics, Kinetic Theory and Statistical Thermodynamics.* Reading, MA: Addison-Wesley, 1975.

Seberry, J., and J. Pieprzyk. *Cryptography: An Introduction to Computer Security.* New York: Prentice Hall, 1989.

Selley, R. C. *Elements of Petroleum Geology.* San Diego: Academic Press, 1998.

Serway, Raymond A. *Physics for Scientists and Engineers with Modern Physics.* 3rd ed. Philadelphia: Saunders College, 1990.

Seyfert, C. K., and Leslie A. Sirkin. *Earth History and Plate Tectonics.* New York: Harper and Row Publishers, 1973.

Shapiro, Stuart L., and Saul A. Teukolsky. *Black Holes, White Dwarfs, and Neutron Stars.* New York: John Wiley & Sons, 1983.

Shapley, Harlow, and Helen E. Howarth. *A Source Book in Astronomy.* New York: McGraw-Hill, 1929.

Shipman, Harry. *Black Holes, Quasars, and the Universe.* New York: Houghton-Mifflin, 1976.

Shlain, Leonard. *Art and Physics: Parallel Visions in Space, Time, and Light.* New York: Quill, 1993.

Shorter, Edward. *The Health Century.* Doubleday, 1987.

Siever, R. *Sand.* New York: Scientific American Library, 1988.

Silk, J. *The Big Bang.* New York: W. H. Freeman and Co., 1980.

Simon, George P. *Ion Exchange Training Manual.* New York: Van Nostrand Reinhold, 1991.

Skinner, B. J., and S. C. Porter. *The Dynamic Earth.* 2nd ed. New York: John Wiley & Sons,1992.

Sklar, Lawrence. *Philosophy of Physics.* Boulder: Westview Press, 1992.

Slater, J. C. *Introduction to Chemical Physics.* New York: Dover Publications, Inc., 1939.

Sobel, Dava. *Longitude: The True Story of a Lone Genius Who Solved the Greatest Scientific Problem of His Time.* London: Fourth Estate, 1995.

Soil Taxonomy. U.S. Department of Agriculture Handbook. Washington, DC: Government Printing Office, 1975.

Solomons, T., and W. Graham. *Fundamentals of Organic Chemistry.* 5th ed. New York: John Wiley & Sons, Inc., 1997.

Spencer, Edgar Winston. *Basic Concepts of Historical Geology.* New York: Thomas Y. Crowell Company, 1964.

Sperling, L. *Introduction to Polymer Science.* New York: John Wiley & Sons, Inc., 1992.

Spraycar, M., ed. *Stedman's Medical Dictionary.* 26th ed. Baltimore: Williams and Wilkens, 1995.

Stacey, F. D. *Physics of the Earth.* New York: John Wiley & Sons, 1969.

Stanley S. M. *Earth and Life Through Time.* W. H. Freeman and Company, 1986.

Starr, Cecie. *Biology: Concepts and Applications.* Belmont, CA: Wadsworth Publishing, 1997.

Starr, Chester G. *A History of the Ancient World.* New York: Oxford University Press, 1991.

Stewart, John A. *Drifting Continents and Colliding Paradigms.* Indiana University Press, 1990.

Stocchi, E. *Industrial Chemistry.* Vol. 1. Translated by K. A. K. Lott and E. L. Short. Chichester, West Sussex, UK: Ellis Horwood Limited, 1990.

Stokes, William Lee. *Essentials of Earth History.* Englewood Cliffs, NJ: Prentice-Hall, 1960.

Storey, T. Alabaster, and R. J. Pankhurst. *Magnetism & the Causes of Continental Break-Up.* American Association of Petroleum Geologists, 1992.

Strahler, A. *Physical Geology.* New York: Harper & Row, Publishers, 1981.

Strahler, A. N., and A. E. Strahler. *Modern Physical Geography.* New York: John Wiley & Sons, 1983.

Strangeway, D. W. *History of the Earth's Magnetic Field.* New York: McGraw-Hill Book Company, 1970.

Stroke, H. Henry, ed. *The Physical Review: The First Hundred Years.* Woodbury, New York: AIP Press, 1995.

Sullivan, J. A. *Continents in Motion.* New York: American Institute of Physics Press, 1990.

Swalin, Richard A. *Thermodynamics of Solids.* 2nd ed. New York: John Wiley & Sons, Inc., 1972.

Sykes, Lynn R. "Seismicity of the Mid Ocean Ridge System." In *The Earth's Crust and Upper Mantle: Geophysical Monograph,* no. 13. Washington: American Geophysical Union, 1969.

Tarbuck, E., and F. Lutgens. *The Earth: An Introduction to Physical Geology.* Columbus: Charles E. Merrill Publishing Company, 1984.

Tarling, D. *Continental Drift: A Study of the Earth's Moving Surface.* New York: Doubleday, 1971.

Taylor, A. W. B. *Superfluidity and Superconductivity.* 2nd ed. Bristol: Adam Hilger, 1986.

Taylor, Hugh S. *A Treatise on Physical Chemistry: A Co-operative Effort by a Group of Physical Chemists.* 2nd ed. New York: D. Van Nostrand Company, Inc., 1931.

Taylor, R. J. *Galaxies: Structure and Evolution.* London: Wykeham Publications Ltd., 1978.

The Kirk-Othmer Encyclopedia of Chemical Technology. 4th ed. New York: John Wiley & Sons, Inc., 1993.

The Oxford Dictionary of Biochemistry and Molecular Biology. Oxford: Oxford University Press, 1997.

Thompson, G. R., and J. T. Trunk. *Essentials of Modern Geology: An Environmental Approach.* Saunders College Publishing, 1994.

Thompson, R., and F. Oldfield. *Environmental Magnetism.* London: Allen & Unwin, 1986.

Thornbury, W. D. *Principles of Geomorphology.* 2nd ed. New York: John Wiley & Sons, Inc., 1969.

Thorne, Kip S. *Black Holes and Time Warps.* New York: Norton, 1994.

Tierney, Brian, and Sidney Painter. *Western Europe in the Middle Ages, 300-1475.* 5th ed. New York: McGraw-Hill, Inc., 1992.

Tilling, Heliker, et al. *Eruptions of Hawaiian Volcanoes: Past, Present, and Future.* Boulder: U.S. Geological Service, 1987.

Tipler, Paul. *Foundations of Modern Physics.* New York: Worth Publishers, Inc., 1969.

Tippens, Paul E. *Applied Physics.* 3rd ed. New York: Gregg Division, McGraw-Hill Book Company, 1984.

Tocci, Salvatore, and Claudia Viehland. *Chemistry: Visualizing Matter.* Annotated Teacher's Edition. Austin: Holt, Rinehart and Winston, Inc., 1996.

Tolman, C. F. *Ground Water.* New York: McGraw-Hill, 1937.

Tolstoy, Ivan. *James Clerk Maxwell, A Biography.* Chicago: University of Chicago Press, 1981.

Townsend, John S. *A Modern Approach to Quantum Mechanics.* New York: McGraw-Hill, 1992.

Trefil, J. *Space, Time, Infinity.* New York: Pantheon Books, 1985.

Trefil, J. *Encyclopedia of Science and Technology.* The Reference Works, Inc., 2001.

Trowbridge, A. C., ed. *Dictionary of Geological Terms.* New York: American Geological Institute, Doubleday and Company, 1962.

Tucker, M. E. *The Field Description of Sedimentary Rocks.* London: The Geological Society, 1982.

Turner, A. Keith, and Robert L. Schuster, eds. *Landslides: Investigation and Mitigation.* Washington: National Academy Press, 1996.

Twiss, Robert J., and Eldridge M. Moores. *Structural Geology.* New York: W. H. Freeman & Company, 1992.

Ultimate Visual Dictionary of Science. London: Dorling Kindersley Limited, 1998

Verhoogan, J., F. Turner, L. Weiss, and C. Wahrhaftig. *The Earth.* New York: Holt, Rhinehart, and Winston, Inc., 1970.

Verhoogen, J., et al. *The Earth.* New York: Holt, Rinehart and Wilson, Inc., 1970.

Voelkel, James R. *Johannes Kepler and the New Astronomy.* New York: Oxford University Press, 1999.

Von Herzen, R. P. "Present Status of Ocean Heat Flow Measurements." In *Physics and Chemistry of the Earth.* 6th ed. London: Pergamon Press, 1965.

Warfield, G., ed. *Solar Electric Systems.* New York: Springer-Verlag, 1984.

Warren J. *Evaporites: Their Evolution and Economics.* Blackwell Science, Oxford, 1999.

Washburn, S. L. "The Study of Race." In *The Human Evolution Source Book.* Englewood Cliffs: Prentice Hall, 1993.

Watkins, J., M. Bottino, and M. Morisawa. *Our Geological Environment.* Philadelphia: W. B. Saunders Company, 1975.

Weast, Robert C. *Handbook of Chemistry and Physics.* Cleveland: CRC Press, 1975.

Weaver, Jefferson Hane. *The World of Physics: A Small Library of the Literature of Physics from Antiquity to the Present.* New York: Simon and Schuster, 1987.

Webb, George E. *Tree Rings and Telescopes: The Scientific Career of A. E. Douglass.* Tucson: University of Arizona Press, 1983.

Wegener, Alfred. *The Origin of Continents and Oceans.* New York: Dover Publications, 1966.

Weinberg, S. *The First Three Minutes.* New York: Basic Books, 1977.

Weisburger, E. K. "General Principles of Chemical Carcinogenesis." In *Carcinogenesis,* edited by M. P. Waalkes and Jerrold M. Ward. New York: Raven Press, Ltd., 1994.

Wells II, W. G. "The Effects of Fire on the Generation of Debris Flows in Southern California." In *Debris Flows/Avalanches: Process, Recognition, and Mitigation,* edited by John E. Costa and Gerald F. Wieczorek. Boulder: Geological Society of America, 1987.

Wenninger, J. A., and G. N. McEwen Jr., eds. *International Cosmetic Ingredient Dictionary and Handbook.* 7th ed. Washington: The Cosmetic, Toiletry, and Fragrance Association, 1997.

Whalley, W. B. "Rockfalls." In *Slope Instability,* edited by D. Brunsden and D. B. Prior. Chicester, UK: John Wiley & Sons, 1984, pp. 217–256.

Wheeler, J. A., and W. H. Zurek, eds. *Quantum Theory and Measurement.* Princeton: Princeton University Press, 1983.

White A. J. R., and B. W. Chappell. *Granitoid Types and Their Distribution in the Lachlan Fold Belt, Southeastern Australia.* Geological Society of America, 1983.

Whittaker, Edmund. *A History of Theories of Aether and Electricity.* New York: Dover Publications, Inc., 1989.

Willard, Hobart H., Lynn L. Merritt Jr., John A. Dean, and Frank A. Settle Jr. *Instrumental Methods of Analysis.* 7th ed. Belmont, CA: Wadsworth Publishing Company, 1988.

Williams, E. T., and C. S. Nicholls, eds. *The Dictionary of National Biography: 1961–1970.* New York: Oxford University Press, 1981.

Williams, H., and A. McBirney *Volcanology.* San Francisco: Freeman, Cooper & Co., 1979.

Williams, J. *The Weather Book.* New York: Vintage Books, 1992.

Williams, P., and M. Smith *The Frozen Earth: Fundamentals of Geocryology.* Oxford, UK: Cambridge University Press, 1989.

Williams, Trevor I., ed. *A Biographical Dictionary of Scientists.* 3rd ed. New York: John Wiley & Sons, Inc., 1982.

Wilson, J. T., ed. *Continents Adrift: Articles from Scientific American, 1972–1970.* New York: W. H. Freeman, 1975.

Winter, C. J., R. L. Sizmann, and L. L. Vant-Hull, eds. *Solar Power Plants.* New York: Springer-Verlag, 1991.

Wolfe, Drew H. *General, Organic, and Biological Chemistry.* New York: McGraw-Hill Book Company, Inc. 1986.

Women Geoscientists Committee. *Profile of Women Professional Geoscientists, Report of Questionnaire Findings.* American Geological Institute, 1977.

Woodhouse, N. M. J. *Special Relativity.* New York: Springer-Verlag, 1992.

Woolfson, M. M. *An Introduction to X-Ray Crystallography.* Cambridge: Cambridge University Press, 1970.

World Directory of Rocks and Minerals. Switzerland: Morges, 1987.

Yeomans, Donald K. *Comets: A Chronological History of Observation, Science, Myth, and Folklore.* New York: John Wiley & Sons, 1991.

Yergin, D. *The Prize.* New York: Simon and Schuster, 1991.

Zeilick, Michael, and Stephen A. Gregory. *Introductory: Astronomy and Astrophysics.* 4th ed. Orlando: Harcourt Brace & Company, 1998.

Zeldovich, Ya B., and Yu. P. Raizer. *Physics of Shock Waves and High-Temperature Hydrodynamic Phenomena.* San Diego: Academic Press, 1966.

Zitzewitz, Paul W., et al. *Merrill Physics: Principles and Problems.* Westerville, OH: Glencoe/McGraw-Hill, 1992.

Zoltai, T., and J. H. Stout. *Mineralogy, Concepts and Principles.* Minneapolis: Burgess Publishing Company, 1984.

Periodicals

Akasofu, S. I. "The Aurora: New Light on an Old Subject." *Sky and Telescope* (December 1982): 534.

Arnold, L. "The Dynamic Aurora." *Scientific American* (May 1989): 90.

Alverson K., F. Oldfield, and R. Bradley, eds. "Past Global Changes and Their Significance to the Future." *Quaternary Science Reviews* 19, no. 1–5 (2000): 1–479.

"And Now, The Asteroid Forecast." *Science* 285 (1999): 655.

Anderson, D. L. "Earth's Viscosity." *Science* 151 (1965): 321.

Arnold, L. "American Women in Geology, A Historical Perspective." *Geology* 5 (1977): 493–4.

Arnold, L. "Florence Bascom and the Exclusion of Women from Earth Science Curriculum Materials." *Journal of Geological Education* 23 (1975): 110–113.

"Assessing the Hazard: The Development of the Torino Scale." *The Planetary Report* 19 (1999): 6–10.

Axelrod, D. I. "Fossil Floras Suggest Stable, Not Drifting Continents." *Journal of Geophysical Research* 68 (1963): 3257.

Bard, E. "Ice Age Temperatures and Geochemistry." *Science* 284 (May 1999): 1133–1134.

Bascom, W. "The Mohole." *Scientific American* 200 (1959): 41–49.

Benin, A. L., J. D. Sargent, M. Dalton, and S. Roda. "High Concentrations of Heavy Metals in Neighborhoods Near Ore Smelters in Northern Mexico." *Environmental Health Perspectives* 107, no. 4 (1999): 279–84.

Berger, A. L. "Milankovitch Theory and Climate." *Reviews in Geophysics* no. 26 (1988): 624–657.

Blacket, P. M. S., E. C. Bullard, and S. K. Runcorn. "A Symposium on Continental Drift." *Philosophical Transactions of the Royal Society* 1088 (1965): 145.

Blandford, R., and N. Gehrels. "Revisiting the Black Hole." *Physics Today* (June 1999).

Bohr, N. "Can Quantum-Mechanical Description of Physical Reality Be Considered Complete?" *Physical Review* (October 15, 1935): 696–702.

Bollen, M., S. Keppens, and W. Stalmans. "Specific Features of Glycogen Metabolism in the Liver." In *Biochemistry Journal* 336 (1998): 19–31.

Bonatti, Enrico. "The Rifting of Continents." *Scientific American* 256 (March 1987): 96–103.

Bowler, Sue. "Continent Turned Inside Out (Supercontinents Laurentia and Gondwanaland)." *New Scientist* 130 (June 15, 1991): 130.

Breed, C. S., S. G. Fryberger, S. Andrews, C. K. McCauley, F. Lennartz, D. Gebel, and K. Horstman. "Regional Studies of Sand Seas Using Landsat (ERTS) Imagery." *A Study of Global Sand Seas.* Edited by E. D. McKee. U.S. Geological Survey Professional Paper 1052, 305–397.

Breslin, K. "Safer Sips: Removing Arsenic from Drinking Water." *Environmental Health Perspectives* 106, 11, (1998): A 548–50.

Briden, J. C. "Palaemagnetic Polar Wandering." *Palaeogeophysics* (1970): 277–289.

Browne, Malcolm W. "Physicists Create First Atoms of Antimatter." *The New York Times,* (January 5, 1996): A1, A9.

Brunet, D., and D. A. Yuen. "Mantle Plumes Pinched in the Transition Zone." *Earth and Planetary Science Letters* 178 (2000): 13–27.

Bullard, Sir Edward. "The Fit of the Continents Around the Atlantic." *A Symposium on Continental Drift: Philosophical Transactions of the Royal Society of London* Series A 258, no. 1088 (October 1965): 41–51.

Burek C. V. "The First Lady Palaeontologist or Collector Par Excellence?" *Geology Today* 17, no. 5 (2001) 192–194.

Burek C. V. "Where are the Women in Geology?" *Geology Today* 17, no. 3 (2001) 110–114.

Bushman, J. R. "Uniformitarianism According to Hutton." *Journal of Geological Education* 29 (January 1981): 31–33.

Cann, Rebecca L., Mark Stoneking, and Allan C. Wilson. "Mitochondrial DNA and Human Evolution." *Nature* 325 (January 1987): 31–36.

Carey, S. W. "Continental Drift: A Symposium." *Geology Department University of Tasmania: Hobert* (1958): 172–179.

Carozzi, A. V. "New Historical Data on the Origin of the Theory of Continental Drift." *Geological Society of America Bulletin* 81 (1970): 283.

CLIMAP Project Members. "The Surface of the Ice-Age Earth." *Science* 191 (March 1976): 1131–1137.

Collins, F. S., A. Patrinos, and E. Jordon, et al. "New Goals for the U.S. Human Genome Project." *Science* 282 (October 23, 1998): 682–89.

"Computer Replicates Pangaea's Breakup." *Geotimes* 38 (March 1993): 9.

"Continental Plates Break Speed Limit." *Geotimes* 38 (April 1993): 7.

Cox, A., and R. R. Doell. "Review of Paleomagnetism." *Geological Society of America* Bulletin 71 (1960): 645.

Croswell, K. "The Best Black Hole in the Galaxy." *Astronomy* (March 1992).

Dalziel, Ian. "Earth Before Pangaea." *Scientific American* 272 (January 1995): 58–63.

Dalziel, Ian. "Paleozoic Laurentia-Gondwana Interaction and the Origin of the Appalachian-Andean Mountain System." *Geological Society of America Bulletin* 106 (February 1994): 243–52.

Dalziel, Ian. "Appalachians and Andes Once Met." *USA Today* 122 (1994): 12–13.

Dewey, J. F. "Plate Tectonics and Geosynclines." *Tectonophysics* 10, nos. 5–6 (1970): 625–638.

Diamond, Jared. "Spacious Skies and Tilted Axes: Why Were Plants Domesticated So Early in the Fertile Crescent? And Why Did Those Crops then Spread So Far and So Fast?" *Natural History* 103 (1994): 16.

Dicke, R. H. "Average Acceleration of the Earth's Rotation and the Viscosity of the Deep Mantle." *Journal of Geophysical Research* 74 (1969): 5895.

Dickinson, W. R. "Plate Tectonic Models of Geosynclines." *Earth and Planetary Science Letters* 10, no. 2 (1971): 165–174.

Dietz, Robert S., and John C. Holden. "The Breakup of Pangaea." *Scientific American* (October 1970).

Einstein, A., B. Podolsky, and N. Rosen. "Can Quantum-Mechanical Description of Physical Reality Be Considered Complete?" *Physical Review* (May 15, 1935): 777–780.

Elsasser, W. M. "Sea-Floor Spreading as Thermal Convection." *Journal Geophysical Research* 76 (1971): 1101–1112.

Eubanks, M. W. "Hormones and Health." *Environmental Health Perscpectives* 105, no. 5 (1997): 482–87.

Feder, Toni. "CERN Council Decides to Build LHC Now—and Pay for It Later." *Physics Today* (February 1997): 58–59.

Fisher, Robert L., and Roger Revelle. "The Trenches of the Pacific." *Continents Adrift* Scientific American.

Frankel, Henry. "From Continental Drift to Plate Tectonics." *Nature* 335 (September 8, 1988): 127.

Galloway, Devin, David R. Jones, and S. E. Ingebritsen, eds. "Land Subsidence in the United States." *U.S. Geological Survey Circular* 1182 (1999).

Goodacre, Alan. "Continental Drift." *Nature* 354 (November 28, 1991): 261.

Gould, S. J., "Is Uniformitarianism Necessary?" *American Journal of Science* 263 (March 1965): 223–228.

Grieve, R. A. F. "Impact Cratering on the Earth." *Scientific American* 262 (1990): 66–73.

Grollimund, B., and M. D. Zoback, "Did Deglaciation Trigger Intraplate Seismicity in the New Madrid Seismic Zone?" *Geology* 29, no. 2 (February 2001): 175–178.

Hamilton, W., and D. Krinsley. "Upper Palozoic Glacial Deposits of South Africa and South Australia." *Geological Society of America Bulletin* 78 (1967): 783.

Haneberg, William C. "Determistic and Probabilistic Approaches to Geologic Hazard Assessment." *Environmental & Engineering Geoscience* 6, no. 3 (August 2000): 209–226.

Haneberg, William C. "Drape Folding of Compressible Elastic Layers—II. Matrix Solution for Two-Layer Folds." *Journal of Structural Geology* 15 (1993): 923–932.

Haneberg, William C. "Drape Folding of Compressible Elastic Layers–I. Analytical Solutions for Vertical Uplift." *Journal of Structural Geology* 14 (1992): 713–721.

Hart, P. J., ed. "The Earth's Crust and Upper Mantle." *American Geophysical Union Geophysical Monograph* 13 (1969).

Hays, J. D., and N. D. Opdyke. "Antarctic Radiolaria, Magnetic Reversals and Climate Changes." *Science* 158 (November 1967): 1001.

Heirtzler, J. R. "Marine Magnetic Anomalies Geomagnetic Field Reversals and Motions of the Ocean Floor Continents." *Journal of Geophysical Research* 73 (1968).

Heirtzler, J. R. "Sea-Floor Spreading." Biographical Notes and Bibliographies, *Scientific American* (1973).

Henbest, Nigel. "Continental Drift: The Final Proof." *New Scientist* 102 (May 1984): 6.

Hess, H. H. "Gravity Anomalies in Island Structure." *Proceedings of the American Philosophical Society* 79 (April 1938): 71–96.

Hilborn, R. C., and N. B. Tufillaro. "Nonlinear Dynamics." *American Journal of Physics* 65, no. 822 (1997).

Hill, R. I., I. H. Campbell, and R. W. Griffiths. "Plume Tectonics and the Development of Stable Continental Crust." *Exploration Geophysics* 22 (1991): 185–188.

Holden C. "Euro-women in Science." *Science* 295, no. 5552 (2002): 41.

Holmes, A. "The Construction of a Geological Time-Scale." *Transactions of the Geological Society of Glasgow* 21 (1947): 117–152.

Holmes, A. "A Revised Geological Time Scale." *Transactions of the Edinburgh Geological Society* 17 (1959): 183.

Hooke, Roger L. "On the History of Humans as Geomorphic Agents." *Geology* 28, no. 9 (September 2000): 843–846.

Horgan, John. "Profile: Reluctant Revolutionary: Thomas S. Kuhn Unleashed Paradigm on the World." *Scientific American* 264 (May 1991): 40.

Hospers, J., and S. I. Van Andel. "Paleomagnetic Data from Europe and North America and Their Bearing on the Origin of the North Atlantic Ocean." *Tectonophysics* 6 (1968): 475.

Hubbert, M. King. "Darcy's Law: Its Physical Theory and Application to the Entrapment of Oil and Gas." *History of Geophysics* 3 (1987): 1–27.

Hubbert, M. King. "The Theory of Ground-Water Motion." *Journal of Geology* 48 (1940): 785–944.

Hudson, Richard L. "Supercollider's Post-mortem Costs May Mount." *The Wall Street Journal* (26 November 1993): B1, B9.

Hurley, P. M. "Test of Continental Drift by Comparison of Radiometric Ages." *Science* 157 (1967): 495.

——— "The Confirmation of Continental Drift." *Scientific American* 218, no. 4 (1968): 53.

Hurley, P. M., and J. R. Rand. "Predrift Continental Nuclei." *Science* 164 (June 1969): 1229.

Hynes, Andrew. "Two-Stage Rifting of Pangaea by Two Different Mechanisms." *Geology* 18 (April 1990): 323–6.

Isacks, Bryan L. "Andean Tectonics Related to Geometry of Subducted Nazca Plate." *Geological Society of America Bulletin* 94 (July 1984): 341–61.

Isacks, Bryan L. "Seismology and the New Global Tectonics." *Journal of Geophysical Research* 73 (1968): 5855.

Iverson, Richard M. "The Physics of Debris Flows." *Reviews of Geophysics* 35, no. 3 (August 1997): 245–296.

Jacobs, J. A. "The Earth's Magnetic Field." *Contemporary Physics* 36 (July 1995): 267–77.

——— "Reversals of Earth's Magnetic Field." *Geological Magazine* 132 (September, 1995): 625–6.

Johnston, Arch C., and B. Schweig. "The Enigma of The New Madrid Earthquakes of 1811–1812." *Annual Review of Earth and Planetary Sciences* 24: (1996): 339–84.

Jordan, Thomas H. "The Deep Structure of the Continents." *Scientific American* (January 1979): 70–82.

Jordan, Thomas H. "A Procedure for Estimating Lateral Variations from Low-Frequency Eigen Spectra Data." *Geophysical Journal of Research* 52 (1978): 441–445.

Kennedy, G. C., and B. E. Nordie. "The Genesis of Diamond Deposits." *Economic Geology* 63 (1996): 495–503.

Kerr, Richard A. "Continental Drift Nearing Certain Detection (Very Long Baseline Interferometry)." *Science* 229 (1985): 953–5.

Kerr, Richard A. "How Far Did the West Wander?" *Science* 268 (May 5, 1995): 635–7.

Koban, C. G., and G. Schweigert. *Facies* 29 (1993): 251–264.

LePichon, X. "Sea-Floor Spreading and Continental Drift." *Journal of Geophysical Research* 73, no. 12 (June 1968): 3661.

Lewin, Roger. "Genes From a Disappearing World (Human Genome Diversity Project)." *New Scientist* 138 (1993): 25–29.

Manuel, J. "NIEHS and CDC Track Human Exposure to Endocrine Disruptors." *Environmental Health Perspectives* 107, no. 1 (1999): A16.

Matthews, Drummond H., and Simon L. Klemperer. "Deep Sea Seismic Reflection Profiling." *Geology* 15 (March 1997): 195–8.

Maxwell, J. C. "Continental Drift and a Dynamic Earth." *American Science* 56 (1968): 35.

McKenzie, D. P. "Speculations on the Consequences and Causes of Plate Motions." *Geophysical Journal of Research* 18 (1969): 1–32.

McKenzie, D. P., and R. L. Parker. "The North Pacific: An Example of Tectonics on a Sphere." *Nature* 216 (1967): 1276–1280.

McKenzie, D. P., and N. O. Weiss. "Speculations on Thermal and Tectonic History of the Earth." *Geophysical Journal of Research* 42 (1975): 131–174.

McKenzie, D. P., R. L. Fisher, and J. G. Sclater. "Evolution of the Central Indian Ridge, Western Indian Ocean." *Bulletin of the Geological Society of America* 82 (March 1971): 553–562.

McPhee, John A. "Assembling California." *New Yorker* 8 (September 21, 1992): 39–49.

Meinzer, O. "Outline of Ground-Water Hydrology with Definitions." *U.S. Geological Survey Water-Supply Paper* 577, 1923.

Melosh, H. Jay. "The Mechanics of Large Rock Avalanches." *Debris Flows/Avalanches: Process, Recognition, and Mitigation. Geological Society of America Reviews in Engineering Geology* 7 (1987): 41–49.

Menard, H. W. "Sea Floor Spreading, Topography, and the Second Layer." *Transactions American Geophysical Union* 48, no. 1 (March 1967): 217.

Meyerhoff, A. A. "Continental Drift: Implications of Paleomagnetic Studies and Physical Oceanography." *Journal of Geology* 78 (1970): 1.

Monastersky, Richard. "Spinning Supercontinent Cycle." *Science News* 135 (June 1989): 344.

Monroe, C., D. M. Meekhof, B. E. King, and D. J. Wineland. "A 'Schrodinger Cat' Superposition State of an Atom." *Science* 272 (May 24, 1996).

Moores, E. M. "Ultramatics and Orogeny, with Models of the US Cordillera and the Tethys." *Nature* 228 (1970): 837–842.

Morgan, W. J. "Convection Plumes in the Lower Mantle." *Nature* 230 (March 5, 1971): 42.

Morgan, W. J. "Rises, Trenches, Great Faults, and Crustal Blocks." *Journal of Geophysical Research* 73 (1968): 1959.

Murphy, J. Brendan. "Mountain Belts and the Supercontinent Cycle." *Scientific American* 266 (April 1992): 84–91.

"NASA Measures Continental Drift." *Earth Sciences* 38 (1985): 8–9.

"NASA Tests Confirm Continental Drift." *International Wildlife* 14 (September 1984): 31.

Nesje, A., and Ian M. Whillans. "Erosion of Sognefjord, Norway." *Geomorphology* no. 9 (1994): 33–45.

Newhall, Christopher G., and Daniel Dzurisin, "Historical Unrest at Large Calderas of the World." *U.S. Geological Survey Bulletin 1855* (1988).

Nobumichi, S. "Analyzing Rocks with Microbeams." *Oceanus* 38, no. 1 (1995): 28–29.

"Ocean Drilling Program: Breakup of Gondwanaland." *Nature* 337 (January 19, 1989): 209–10.

Oliver, Jack E. "The Big Squeeze: How Plate Tectonics Redistributes Mineral and Organic Resources." *The Sciences* 31 (July 1991): 22–28.

Opening of the Caribbean." *Oceanus* 30 (Winter 1987): 49–50.

Orowan, E. "Continental Drift and the Origin of Mountains." *Science* 146 (1964): 1003.

Orowan, E. "The Origin of the Oceanic Ridges." *Scientific American* 221, no. 5 (November 1969): 102–118.

Parker, Robert L. "The Determination of Seamount Magnetism." *Geophysical Journal of Royal Astronomical Society* 24 (1971): 321–324.

Petford, N., et al. "Granite Magma Formation, Transport and Emplacement in the Earth's Crust." *Nature* (December 2000): 669–673.

Phillips, J. D. "Plate Tectonics, Paleomagnetism and the Opening of the Atlantic." *Bulletin of the Geological Society of America* 82 (1972): 1579.

Plant J. A., D. Hackett, and B. J. Taylor. "The Role of Women in the British Geological Survey." *Geology Today* 10, no. 4 (1994): 151–156.

Raff, A. D., and R. G. Mason. "Magnetic Survey of the West Coast of North America." *Bulletin of the Geological Society of America* 72 (1961): 1267–1270.

Reid, M. E., et al. "Gravitational Stability of Three-Dimensional Stratovolcano Edifices." *Journal of Geophysical Research* 105, B3 (March 2000): 6043–6056.

"Reversals of the Earth's Magnetic Field." *Philosophical Transactions of the Royal Society of London: Series A, Mathematical and Physical Sciences* 263, no. 1143 (December 1968): 481–524.

Romm, James. "A New Forerunner for Continental Drift (Abraham Ortelius) Suggested the Basic Elements of the Continental Drift Theory in 1596." *Nature* 367 (February 1994): 407–8.

Runcorn, S. K. "Continental Drift." *Scientific American* 208 (1963): 86.

Sager, W., and Koppers, A. "Late Cretaceous Polar Wander of the Pacific Plate: Evidence of a Rapid True Polar Wander." *Science* 287 (January 2000): 455–459.

Sarjeant, W. A. S. "Alice Wilson, First Woman Geologist with the Geological Survey of Canada." *Earth Sciences History* 12, no. 2 (1993): 122–128.

"Scaling the Degree of Danger from an Asteroid." *Nature* 400 (1999): 392.

Shen, Zheng-Kang, et al. "Crustal Deformation Across and Beyond the Los Angeles Basin From Geodetic Measurements." *Journal of Geophysical Research* 101 (1996): 27, 957–27, 980.

Shopov, Y. Y. "Genetic Classification of Cave Minerals." *Proceedings of Tenth International Congress of Speleology* 1 (1989): 101–105.

Shopov, Y. Y. "Speleothem Records of Environmental Changes in the Past." *Contribucion del Estudio Cientifico De Las Cavidades Karsticas Al Conocimiento Geologico* Instituto de Investigacion, 1999: 117–134.

"SHRIMP Instrument Homes in on Solar System Mysteries." *Electronic Design* 47, no. 3 (February 8, 1999): 23.

Sloss, L. L. "Paleoclimatic and Tectonic Control on the Accumulation of North American Cratonic Sediment." *Bulletin of the Geological Society of America* 107 (September 1995): 1123–1126.

Smith, A. G. "The Fit of the Southern Continents." *Nature* 225 (1970): 139.

Spotts, Peter N. "Antimatter Happens, in European Lab." *Christian Science Monitor* (January 8, 1996): 4.

Stehli, F. G., and C. E. Helsley. "Paleontologic Technique for Defining Ancient Pole Positions." *Science 142* (November 1963): 1057.

Stewart, J. "Basin and Range Structure—A System of Horsts and Grabens Produced by Deep-Seated Extension." *Bulletin of the Geological Society of America* 82, no. 4 (April 1971): 1019–1043.

Stoffler, D., and R. A. F. Grieve. "Classification and Nomenclature of Impact Metamorphic Rocks." *Lunar and Planetary Science* 25 (1994): 1347–1348.

Story, Michael, John Mahoney, and A. D Saunders. "Timing of Hot Spot-related Volcanism and the Breakup of Madagascar and India." *Science* 267 (1995): 852.

Svitil, K. "Probing the Past." *Discover* 19, no. 11 (November 1998): 96–99.

Sykes, Lynn R., and Steven C. Jaume. "Changes in State of Stress on the Southern San Andreas Fault Resulting from the California Earthquake Sequence of April to June 1992." *Science* 258 (November 1992): 1325–1328.

Sykes, Lynn R., and Leonardo Seeber. "Great Earthquakes and Great Asperities, San Andreas Fault, Southern California." *Geology* 13 (December 1985): 835–838.

Taubes, Gary. "Schizophrenic Atom Doubles as Schrodinger's Cat or Kitten." *Science* 272, no. 5265 (May 24, 1996).

"The Torino Impact Hazard Scale." *Planetary and Space Science* 48 (2000): 297–303.

"The Torino Scale: Gauging the Impact Threat." *Sky & Telescope* 98 (1999): 32–33.

Torrens, H., et al. "Etheldred Benett or Wiltshire, England, the First Lady Geologist—Her Fossil Collection in the Academy of Natural Science of Philadelphia and the Rediscovery of 'Lost' Specimens of Jurassic Trigonlidae With Their Soft Anatomy Preserved." *Proceedings of the Academy of Natural Sciences of Philadelphia* 150 (2000): 59–123.

Turco, R. P., A. B. Toon, T. P. Ackerman, J. B. Pollack, and C. Sagan. "Nuclear Winter: Global Consequences of Multiple Nuclear Explosions." *Science* no. 222 (1983): 1283–1297.

Valentine, James W. "Late Precambrian Bilaterians: Grades and Clades." *Proceedings of the National Academy of Sciences of the United States of America* 91 (July 1994): 6751–6757.

Vine, F. J. "Spreading of the Ocean Floor: New Evidence." *Science* 154, no. 3775 (December 1996): 1405–1515.

Von Herzen, R. P. "The Deep Sea Drilling in the South Atlantic." *Science* 168 (May 1970): 1047–1059.

Wearner, R. "The Birth of Radio Astronomy." *Astronomy* (June 1992).

Weijermars, Ruud. "Global Tectonics Since the Breakup of Pangea 180 Million Years Ago: Evolution Maps and Lithospheric Budget." *Earth-Science Reviews* 26 (February 1989): 113–62.

Weisburd, Stefi. "Seeing Continents Drift (Very Long Baseline Interferometry and Satellite Laser Ranging)." *Science News* 128 (1985): 388.

Wieczorek, Gerald F., et al. "Unusual July 10, 1996, Rock Fall at Happy Isles, Yosemite National Park, California." *Bulletin of the Geological Society of America* 112, no. 1 (January 2000): 75–85.

Wilkes, B. "The Emerging Picture of Quasars." *Astronomy* (December 1991).

Willett, Sean D., and Christopher Beaumont. "The India-Asia Collision: What Gives?" *Science News* 146 (1994): 15.

Wilson, I. "Desert Sandflow Basins and a Model for the Development of Ergs." *Geographical Journal* 137, no. 2 (1971): 180–199.

Wilson, J. T. "A New Class of Faults and Their Bearing on Continental Drift." *Nature* 207 (1965): 343.

Wilson, Jim. "Executing Schrodinger's Cat." *Popular Mechanics* 174, no. 10.

Wright, and Pierson. "Living With Volcanoes." *The U.S. Geological Survey's Volcano Hazards Program* U.S. Geological Survey Circular (1973).

Wysession, Michael. "The Inner Workings of the Earth." *American Scientist* 83 (1995): 134.

Yakiko, Tanaka, et al. "Compilation of Thermal Gradient Data in Japan on the Basis of the Temperatures in Boreholes." *Bulletin Of The Geological Survey Of Japan* 50, no. 7 (1999): 457–487.

York, Derek. "The Earliest History of the Earth." *Scientific American* 268 (January 1993): 90–96.

Internet Sites

Editor's Note: As the World Wide Web is constantly expanding, the URLs listed below were current as of May 11, 2002.

Allison, Henry E. "Immanuel Kant." [cited August 21, 2000]. <http://www.columbia.edu/~pjs38/biokant.htm>.

American Meteorological Society. "Challenges of Our Changing Atmosphere—Careers in Atmospheric Research and Applied Meteorology." 1993 [cited March 1, 2002]. <http://www.ametsoc.org/AMS/pubs/careers.htm>.

American Physical Society. "Henry Augustus Rowland (1848–1901)." 1998 [cited August 21, 2002]. <http://www.aps.org/apsnews/1198/119806.htm>.

Ball, Philip. "Move Any Mountain." September 22, 2000 [cited February 26, 2002]. <http://www.nature.com/nsu/000928/000928-1.htm>.

Beebe, William. "Half Mile Down." 1934 [cited May 2, 2002]. <http://seawifs.gsfc.nasa.gov/OCEAN_PLANET/htm>/ocean_planet_book_beebe1.htm>.

Brock, Henry M. Transcribed by Dennis McCarthy. (Copyright: Kevin Knight.) "Gaspard-Gustave de Coriolis." *The Catholic Encyclopedia.* 1999 [cited August 21, 2000]. <http://www.newadvent.org/cathen/04370a.htm>.

Brooklyn College City University of New York. "Dinosaur Core." Chamberlain, John A. Jr., Dept. of Geology. [cited May 10, 2002]. <http://academic.brooklyn.cuny.edu/geology/chamber/gesner.htm>.

Brown, Kevin. "The Path to Mass-Energy Equivalence." [cited August 2000]. <http://mathpages.com/home/albro/albro8.htm>.

Camp, V. Department of Geological Sciences, San Diego State University and NASA. "How Volcanoes Work." 2000 [cited March 21, 2002]. <http://www.geology.sdsu.edu/how_volcanoes_work/.htm>.

Canadian Center for Remote Sensing. "History of Remote Sensing." 2000 [cited May 10, 2002]. <http://www.ccrs.nrcan.gc.ca/ccrs/org/history/morleye.htm>.

Canadian Center for Remote Sensing. "Radarsat Technical Specs—Summary." 2000 [cited January 2002]. <http://www.Ccrs.nrcan.gc.ca/ccrs/radspece.htm>.

Carver, G., and O. Garrett, Centre for Atmospheric Science, University of Cambridge. "The Ozone Hole Tour." 1999 [cited April 22, 2002]. <http://www.atm.ch.cam.ac.uk/tour/.htm>.

Centre for Marine Environmental Sciences (MARUM). "World Data Center for Marine Environmental Sciences." 2001 [cited December 17, 2000]. <http://www.pangaea.de/wdc-mare/.htm>.

Clyde, John, Johann Schleier-Smith, and Greg Tseng. "Equivalence of Mass and Energy." *Nuclear Physics: Past, Present, and Future.* October 1996 [cited August 21, 2002].

<http://library.thinkquest.org/3471/energy_mass_equivalence.htm>.

Dana, P. The Geographer's Craft Project, Department of Geography, University of Colorado at Boulder. "Map Projection Overview." 1999 [cited October 3, 2000]. <http//www.colorado.edu/geography/gcraft/notes/mapproj/mapproj.htm>.

Daniels, Alison. (Copyright: The University of Edinburgh.) "Point of Creation: The Higgs Boson." 2000 [cited August 21, 2000]. <http://www.cpa.ed.ac.uk/edit/07/articles/03.htm>.

Environmental Protection Agency, Office of Water. 2002 [cited May 11, 2002]. <http://www.epa.gov/owow/estuaries/about1.htm>.

EPA. Gulf of Mexico Program. "The Gulf of Mexico Watershed." 2002 [cited March 31, 2002]. <http://www.epa.gov/gmpo/edresources/watrshed.htm>.

Fischer, R. V. "The Volcano Information Center." University of California at Santa Barbara, Department of Geological Sciences. 1999 [cited March 24, 2002]. <http://www.geol.ucsb.edu/~fisher/.htm>.

Florida State University, Department of Meteorology. "Meteorology—A Brief History." [cited March 1, 2002]. <http://www.met.fsu.edu/explores/methist.htm>.

Fowler, Michael. "Tycho Brahe." 1995 [cited August 21, 2000]. <http://www.phys.virginia.edu/classes/109N/1995/lectures/tychob.htm>.

Galileo Project at Rice University. Van Helden, Albert. "Catalog of the Scientific Community: Agricola, Georgius (Georg Bauer)." Compiled by Richard S. Westfall, Indiana University. 1995 [cited May 11, 2002]. <http:/es.rice.edu/ES/humsoc/Galileo/Catalog/Files/agricola.htm>.

Geography Exchange. 2002 [cited May 11, 2002]. <http://www.zephryus.demon.co.uk/geography/resources/glaciers/arete.htm>.

Geophysical Institute, University of Alaska Fairbanks. "Continental Divide, Alaska Science Forum." [cited January 31, 2002]. <http://www.gi.alaska.edu/ScienceForum/ASF0/023.htm>.

Glenn Brown. "Darcy's Law Basics and More." Oklahoma State University. [cited May 11, 2002]. <http://biosytems.okstate.edu/darcy/LaLoi/index.htm>.

Hamilton, Calvin J. "Neptune." 1997–2000 [cited August 21, 2000]. <http://planetscapes/com/solar/eng/neptune.htm>.

Hartwick College, Department of Geological and Environmental Sciences. "The Pleistocene Ice Age." 2001 [cited January 20, 2002]. <http://www.hartwick.edu/geology/work/VFT-so-far/glaciers/glacier1.htm>.

Hines, C. "The Official William Beebe Web Site." 2000 [cited May 3, 2002]. <http://hometown.aol.com/chines6930/mw1/beebe1.htm>.

Hulse, Russell Alan. The Nobel Foundation. "Autobiography of R. A. Hulse." 1993 [cited August 21, 2000]. <http://www.nobel.se/physics/laureates/1993/hulse/autobio.htm>.

Illinois State Museum. "Ice Ages." 2002 [cited January 25, 2002]. <http://www.museum.state.il.us/exhibits/ice_ages/why_glaciations1.htm>.

International Council for Science. "International Council for Science." 1996 [cited December 17, 2000]. <http://www.icsu.org>.

Kanen, Rob. "The Emplacement and Origin of Granite." 2001 [cited May 11, 2002]. <http://www.geologyone.com/granite1.htm>.

Kious, W. J., and R. I. Tilling, "This Dynamic Earth, the Story of Plate Tectonics." Online Version 1.08, United States Geological Survey. 2001 [cited May 10, 2002]. <http://pubs.usgs.gov/publications/text/dynamic.htm>.

Kraan-Korteweg, Renée C., and Ofer Lahav. "Galaxies Behind the Milky Way." *Scientific American.* 2000 [cited August 21, 2000]. <http://www.sciam.com/1998/1098issue/1098laham.htm>.

Liss, Tony, and P. L. Tipton. "The Discovery of the Top Quark." 1997 [cited August 21, 2000]. <http://www.sciam.com/0997issue/0997tipton.htm>.

Maine Department of Conservation, Maine Geological Survey. Bald Mountain, Washington Plantation, Maine. 2001 [cited March 2001]. <http://www.state.me.us/doc/nrimc/mgs/sites-2001/mar01.htm>.

Met Office. "Interpreting Weather Maps." [cited March 1, 2002]. <http://www.met-office.gov.uk/education/curriculum/leaflets/weathermaps.htm>.

NASA, Ocean Color, Data and Resources. "Classic CZCS Scenes Chapter 6: Gulf Stream Rings." 1999 [cited March 31, 2002]. <http://daac.gsfc.nasa.gov/CAMPAIGN_DOCS/OCDST/classic_scenes/07_classics_rings.htm>.

NASA's Observatoriuum. "Hydrologic Cycle." 1995-1999 TRW, Inc. [cited January 9, 2002]. <http://observe.ivv.nasa.gov/nasa/earth/hydrocycle/hydro1.htm>.

National Drought Mitigation Center. "Understanding and Defining Drought." 1995 [cited January 9, 2002]. <http://www.ndmc.gov>.

National Geophysical Data Center. "The World Data Center System." 2000 [cited December 17, 2000]. <http://www.ngdc.noaa.gov>.

National Oceanic and Atmospheric Administration. "Paleoclimatology Program, Glaciation." 1999 [cited January 20, 2002]. <http://www.ngdc.noaa.gov/paleo/glaciation.htm>.

National Oceanographic and Atmospheric Administration. "Coral Reefs." 2002 [cited May 11, 2002]. <http://www.noaanews.noaa.gov/magazine/stories/mag7.htm>.

National Weather Service. "East Central Florida Rip Current Program." 1999 [cited January 21, 2002]. <http://www.srh.noaa.gov/mlb/ripinit.htm>.

Naval Meteorology and Oceanography Command. "The Restless Sea." [cited January 21, 2002]. <http://pao.cnmoc.navy.mil/pao/Educate/OceanTalk2/indexrestless.htm>.

New Mexico Bureau of Geology and Mineral Resources. "Master Page." 2000 [cited March 1, 2002]. <http://tremor.nmt.edu/pioneers/dlMoho_32.htm>.

Nobel Foundation. "Pieter Zeeman." 2000 [cited August 21, 2000]. <http://www.nobel.se/physics/laureates/1902/zeeman-bio.htm>.

Nobel Foundation. "Wilhelm Conrad Röntgen." 1999 [cited August 21, 2000]. <http://www.nobel.se/physics/laureates/1901/rontgen-bio.htm>.

NPA Group, Satellite Mapping and Exploration. "Ancient Landscapes, Information and Computers." 2001 [cited April 22, 2002]. <http://www.npagroup.com/rs_intro/index.htm>.

"Nuclear Winter and Other Scenarios." [cited January 31, 2002]. <http://www.pynthan.org/vri/nwaos.htm>.

"Nuclear Winter." [cited January 31, 2002]. <http://zebu.uoregon.edu/~js/glossary/nuclear_winter.htm>.

O'Connor, J. J., and E. F. Robertson. "Jean Baptiste Josephe Fourier." 1997 [cited August 21, 2000]. <http://www-history.mcs.st-and.ac.uk/history/Mathematicians/Fourier.htm>.

O'Connor, J. J., and E. F. Robertson. "Leonardo da Vinci." 1996 [cited August 21, 2000]. <http://turnbull.dcs.st-and.ac.uk/history/Mathematicians/Leonardo.htm>.

Oak Ridge National Laboratory. "The Obsidian Clock." [cited April 16, 2002]. <http://www.ornl.gov/reporter/no7/clock.htm>.

Ohio State University. "Wastewater Treatment Principles and Regulations." 1996 [cited February 27, 2002]. <http://ohioline.osu.edu/aex-fact/0768.htm>.

Okanagan University College, Geography Department. "Introduction to Geomorphology: Glaciation." 2001 [cited January 25, 2002]. <http://www.geog.ouc.bc.ca/physgeog/contents/11n.htm>.

PaleoZoo, the Geobobological Society. 1998 [cited March 20, 2001]. <http//www.geobop.com/paleozoo>.

PBS Online. "William Beebe: Going Deeper." 1999 [cited May 3, 2002]. <http://www.pbs.org/wgbh/amex/ice/sfeature/beebe.htm>.

PBS.org. "The Deep Sea—Journey to the Ocean Floor." [cited February 15, 2002]. <http://www.pbs.org/wnet/savageseas/deep-side-journey.htm>.

Physics & Astronomy Department, George Mason University. "The Cosmos From The Earth." [cited November 24, 1998]. <http://www.physics.gmu.edu/classinfo/astr103/CourseNotes/ECText/ch01_txt.htm>.

Pidwirny, M. J. Department of Geography, Okanagan University College. "Fundamentals of Physical Geography." 2001 [cited March 17, 2002]. <http//www.geog.ouc.ca/physgeog/contents/la.htm>.

Pickford, Ron. "The Father of English Geology." [cited May 11, 2002]. <http://www.bath.ac.uk/BRLSI/wsmith.htm>.

Raymond, L.A. "Petrology: The Study of Igneous, Sedimentary, and Metamorphic Rocks." 1995 [cited May 11, 2002]. <http://geology.csupomona.edu/drjessey/class/GSC425/Ig-Met25.htm>.

Regents of the University of Michigan. "Biographical Information for Martinus J. G. Veltman." 1998 [cited August 21, 2000]. <http://www.physics.lsa.umich.edu/veltmanbiographical.htm>.

Rice University, Department of Earth Science. "Ice." 1997 [cited January 25, 2002]. <http://www.glacier.rice.edu/>.

Royal Danish Embassy. "H. C. Ørsted." 1998 [cited August 21, 2000]. <http://www.denmarkemb.org/oersted.htm>.

Schewe, Phillip, and Ben Stein. "Physics News Update: The 1999 Nobel Prize for Physics." 1999 [cited August 21, 2000]. <http://www.physics.uq.edu.au/media/Nobel99.htm>.

School of Mathematics and Statistics, University of St. Andrews, Scotland. "Simon Stevin," by O'Connor, J. J., and E. F. Robertson. December 1996 [cited May 11, 2002]. <http//www-history.mcs.st-andrews.ac.uk/history/References/Stevin.htm>.

Sea, Grant. "Rip Currents: Don't Panic." [cited January 21, 2002]. <http://www.ncsu.edu/seagrant/PDF/RipBrochure.pdf>.

"Solar Eclipse Journal," 1999-2002 [cited April 24, 2002]. <http//www.designbg.com/sej>.

Spergel, David N., Gary Hinshaw, and Charles L. Bennett. "The Milky Way." *Microwave Anisotropy Probe.* May 1, 2000 [cited August 21, 2000]. <http://map.gsfc.nasa.gov/htm>/milky_way.htm>.

Stoner, Ron. "Relativistic Dynamics." April 3, 1999 [cited August 21, 2000]. <http://fermi.bgsu.edu/~stoner/P202/relative2/sld001.htm>.

Strange Science. "Rocky Road: Abraham Gottlob Werner," by Michon Scott. [cited May 11, 2002]. <http://turnpike.net/~mscott/werner.htm>.

The U.S. Global Change Research Information Office. “Dust Storm Magnitude, Duration and Frequency.” [cited March 1, 2002]. <http://www.gcrio.org/geo/dust.htm>.

Thomas, Dan. “Jean-Baptiste Joseph Fourier.” 1996 [cited August 21, 2000]. <http://www.chembio.uoguelph.ca/educmat/chm386/rudiment/tourclas/fourier.htm>.

Tilling, R. I. “Volcanoes, On-line Edition.” United States Geological Survey. 1997 [cited March 24, 2002]. <http://pubs.usgs.gov/gip/volc>.

Topinka, Lyn. “Mount St. Helens: A General Slide Set.” October 18, 2001 [cited March 11, 2002]. <http://vulcan.wr.usgs.gov/Photo/SlideSet/ljt_slideset_old.htm>.

United States Environmental Protection Agency Web Site. 2002 [cited April 19, 2002]. <http://www.epa.gov>.

United States Geological Survey. “CVO Website—Hazards—Lahars.” 2000 [cited February 15, 2002]. <http://vucan.wr.usgs.gov/Hazards/NRC_Definitions/lahars.htm>.

United States Geologic Survey, Volcano Hazards Program. “Photo Glossary of Volcano Terms.” 2002 [cited March 24, 2002] <http://volcanoes.usgs.gov/Products/Pglossary/pglossary.htm>.

United States Geological Survey National Earthquake Information Center. “Andrija Mohorovicic.” 2001 [cited March 1, 2002]. <http://neic.usgs.gov/neis/seismology/people/mohorovicic.htm>.

United States Geological Survey. “USGS Minerals Information.” January 30, 2002 [cited April 22, 2002]. <http://minerals.usgs.gov>.

United States Geological Survey. “USGS Photo Glossary.” November 16, 2001 [cited April 10, 2002]. <http://volcanoes.usgs.gov/Products/Pglossary/.htm>.

United States Geological Survey. “This Dynamic Earth.” [cited August 2001]. <http://pubs.usgs.gov/publications/text/dynamic.htm>.

United States Lifeguards for Life, “Rip Currents—Rivers Through The Surf.” [cited January 21, 2002]. <http://www.usla.org/PublicInfo/beach_safety/rip_current1.shtm>.

U.S. Naval Observatory, Astronomical Applications Department. “The Seasons and the Earth’s Orbit—Milankovitch Cycles.” [cited April 27, 2002]. <http://aa.usno.navy.mil/>.

University of California, Berkeley. “Museum of Paleontology, Geology Wing.” 1994–2000 [cited May 11, 2002]. <http://www.ucmp.berkeley.edu/exhibit/geology.html>.

University of California at Berkeley. “William Smith.” [cited May 11, 2002]. <http://www.ucmp.berkeley.edu/history/smith.htm>.

University of Cambridge. “Our Own Galaxy: The Milky Way.” *Cambridge Cosmology* May 16, 2000 [cited August 21, 2000]. <http://www.damtp.cam.ac.uk/user/gr/public/gal_milky.htm>.

University of Delaware “Extreme 2000: Voyage to the Deep, A Brief History.” 2000 [cited February 15, 2002]. <http://www.ocean.udel.edu/deepsea/level-2/tools/history.htm>.

University of East Anglia, Norwich, School of Information Systems. “Satellite Imagery and GIS.” 2000 [cited April 22, 2002]. <http://www.sys.uea.ac.uk/Research/researchareas/JWMP/Default.htm>.

University of Notre Dame Archives. “Lating Dictionary and Grammar Aid.” Maintained by William Kevin Cawley. University of Notre Dame. [cited December 21, 2001]. <http://www.nd.edu/~archives/latgramm.htm>.

University of Wisconsin—Stevens Point Virtual Geography. 2002 [cited May 11, 2002]. <http://www.uwsp.edu/geo/projects/virtdept/ipvft/arete.htm>.

University of Wisconsin-Madison. “Weather for Pilots.” 2000 [cited March 27, 2002]. <http://itg1.meteor.wisc.edu>.

Van Helden, Albert. “Tycho Brahe.” 1995 [cited August 21, 2000]. <http://es.rice.edu/ES/humsoc/Galileo/People/tycho_brahe.htm>.

Watson, K. United States Geological Survey. “Volcanoes: Principle Types of Volcanoes.” 1997 [cited March 18, 2002]. <http://pubs.usgs.gov/gip/volc/types.htm>.

Wellesley University. “Annie Jump Cannon: Understanding Her Work.” 1998 [cited December 21, 1998]. <http://www.wellesley.edu/Astronomy/annie/understanding.htm>.

“When the Earth Moves—Of Magnetism and Time.” The National Academies. [cited May 11, 2002]. <http://www4.national-academies.org/beyond/beyonddicovery.nsf/web/seafloor5?OpenDocument.htm>.

Wills, Jamal. “Coriolis Effect.” *Living Universe Foundation.* 1996 [cited August 21, 2000]. <http://www.luf.org/bin/view/GIG/CoriolisEffect-CoriolisForceTerm.htm>.

York University, Department of Geography. “Glacial Processes.” 1996 [cited January 20, 2002]. <http://www.yorku.ca/faculty/academic/arobert/14008.htm>.

Historical Chronology

Editor's note: This is a historical chronology principally devoted to marking milestones in human scientific achievement or observation. Detailed information related to the eons, eras, periods, and epochs of geologic time may be found in text and diagrams related to the topic Geologic time.

c.4,600,000,000 B.C.
Origin of Earth: 4,600 million years ago (mya).

c.4,000,000 B.C.
Earliest hominid species appear on Earth.

c.50,000 B.C.
Homo sapiens sapiens emerges as a conscious observer of nature.

c.30,000 B.C.
Stone Age cultures use pigments to color various artifacts.

c.10,000 B.C.
Neolithic Revolution: transition from a hunting and gathering mode of food production to farming and animal husbandry, that is, the domestication of plants and animals.

c.4000 B.C.
Early applied chemistry begins in Egypt with the extraction and working of metals, including copper, tin, and bronze. Egyptians are also familiar with eye paint and plaster of Paris.

c.4000 B.C.
Egyptians astronomically measure time.

c.3500 B.C.
Sumerians describe methods of managing the date harvest.

c.3400 B.C.
Bronze, an alloy of copper and tin, first appears in abundance in Samaria. The Sumerians become expert in working gold, silver, copper, lead, and antimony.

c.3000 B.C.
Iron is forged.

c.2500 B.C.
The earliest known wholly glass objects are beads made in Egypt at this time. Early peoples may have discovered natural glass, which is created when lightning strikes sand. The Egyptians make glass beads by sand (silica), soda, lime, and other ingredients.

c.2000 B.C.
Chinese document experiments with forces of magnetism.

c.1500 B.C.
Use of iron becomes prevalent in the Mediterranean. It appears to have come there from the northeast, possibly beginning with the Hittites, and its use revolutionizes society.

c.900 B.C.
Steel is manufactured in India.

c.700 B.C.
Babylonians and Chinese understand planetary orbits.

c.700 B.C.
Greeks demonstrate force of electric attraction produced by rubbing amber.

c.700 B.C.
The use of anatomical models is established in India.

c.600 B.C.
Anaximander, Greek astronomer, describes the ecliptic plane, and asserts that Earth's surface assumes a curved, cylindrical shape.

c.600 B.C.

Thales of Miletus, Greek philosopher, first notices the electrification of amber (the Greek word for amber is *elektron*) by friction. Thales also proposes water as the fundamental substance of the Universe. He is the first to systematically study magnetism, and correctly predicts a solar eclipse.

c.550 B.C.

Pythagoras, Greek philosopher and mathematician, advances studies of geometry and geometric form. Pythagoras asserts Earth as a sphere.

c.525 B.C.

Anaximenes, Greek philosopher, proposes that air is the fundamental element of the Universe, and when compressed, it can take the form of water and Earth.

c.500 B.C.

Heraclitus, Greek philosopher, states that fire is the fundamental element of the universe. He also states that all things are in constant motion and that nothing is ever lost.

c.500 B.C.

Parmenides, Greek philosopher, suggests that matter can be neither created nor destroyed.

c.480 B.C.

Oenopides of Chios calculates angle of Earth's polar axial tilt to ecliptic .

c.475 B.C.

Parmenides argues that Earth is a sphere.

c.460 B.C.

Eudoxus of Cnidus, Greek philosopher, corrects faults in Plato's planetary orbital scheme (e.g., that Sun, planets, and stars orbit Earth on celestial spheres).

c.455 B.C.

Philolaus, Greek philosopher, argues that night and day are caused by Earth's rotation.

c.450 B.C.

Anaxagoras, Greek philosopher, offers one of the first atomic theories, saying that all matter consists of atoms or "seeds of life."

c.450 B.C.

Anaxagoras argues that the universe is made entirely of matter in motion.

c.450 B.C.

Empedocles, Greek philosopher, first offers his concept of the composition of matter, postulating that it is made of four elements—earth, air, fire, and water. This notion is adopted by Aristotle and becomes the basis of physical theory for nearly two millennia.

c.450 B.C.

The Greek philosopher Leucippus first states the formal notion of atomism. He argues that upon continuous division of a substance, eventually a point would be reached beyond which further division was impossible. His disciple, the Greek philosopher Democritus ultimately names these small particles *atomos*, meaning indivisible.

c.450 B.C.

Zeno of Elea, Greek philosopher, formulates paradoxes challenging the discreteness of continuous time and space.

c.425 B.C.

Democritus (470–380), Greek philosopher, states his atomic theory that all matter consists of infinitesimally tiny particles that are indivisible. These atoms are eternal and unchangeable, although they can differ in their properties. They can also recombine to form new patterns. His intuitive ideas contain much that is found in modern theories of the structure of matter.

c.380 B.C.

Plato (427–347), Greek philosopher, teaches a geometrical theory of matter on which he elaborates in his *Timaeus*.

c.350 B.C.

Aristotle (384–322), Greek philosopher, offers his doctrine of the elements, stating that all things are composed of a basic material in combination with four qualities—hotness, dryness, coldness, and wetness. This theory eventually suggests the idea of transmutation (the changing of ordinary metals into gold or silvers) and gives rise to alchemy.

c.350 B.C.

Aristotle (384–322), Greek philosopher, rejects the atomism of the Greek philosopher Democritus (470–380 B.), thus condemning it to oblivion until modern times. He also states that a vacuum does not exist in nature and that sound travels by a succession of impacts on the air. Aristotle argues correctly that sound is not conducted in the absence of air and incorrectly that a body will move only as long as it keeps being pushed. He also asserts that heavy bodies fall faster than light ones.

c.350 B.C.

Aristotle reasons that Earth is spherical.

c.350 B.C.

Heracleides describes and calculates Earth's rotation.

c.335 B.C.

First description of equinoxes.

c.325 B.C.

Pytheas argues that tides are caused by motions of the Moon.

c.300 B.C.
Arthasastra, an ancient Indian manual on politics, discusses mining, metallurgy, medicine, pyrotechnics, poisons, and fermented liquors.

c.300 B.C.
Aristarchus argues that Earth revolves around the Sun.

c.300 B.C.
Epicurus (341–270), Greek philosopher, elaborates on the atomism of the Greek philosopher Democritus (470–380 B.), but substitutes the notion of chance for the determinism of Democritus.

c.300 B.C.
Euclid, Greek mathematician, writes a treatise on optics in which he makes optics a part of geometry by dealing with light rays as though they are straight lines. He also offers a theory of reflection which he treats geometrically.

c.300 B.C.
Glass blowing first practiced.

c.300 B.C.
Theophrastus, Aristotle's disciple and the founder of botany, attempts to establish a classification system for plants based upon differences between plant and animal morphology.

c.275 B.C.
Herophilus's younger colleague, Eristratus (c. 310–c. 250), asserts that veins and arteries are connected.

c.265 B.C.
Zou Yan asserts the material universe is composed of five elements including water, metal, wood, fire and earth.

c.260 B.C.
Aristarchus of Samos, distance and size of moon from Earth's shadow during lunar eclipse.

c.260 B.C.
Aristarchus of Samos argues for a Sun-centered (heliocentric) cosmology.

c.260 B.C.
Aristarchus of Samos calculates the ratio Sun-Earth distance/Earth-Moon distance from the angle established at half moon.

c.250 B.C.
Chinese, free bodies move at constant velocity.

c.250 B.C.
Philon, Greek engineer, experiments with air and discovers that it expands when heated.

c.240 B.C.
Erastothenes computes the diameter of Earth, suggests it orbits the Sun.

c.220 B.C.
Archimedes (287–212), Greek mathematician and engineer, writes his *Treatise on Floating Bodies* in which he relates the principle of buoyancy.

c.170 B.C.
Chinese astronomers record observations of Sun spots.

c.150 B.C.
Hipparchus, observes and estimates precession of the equinoxes.

c.130 B.C.
Hipparchus estimates the size of Moon from parallax of eclipse.

c.50 B.C. Lucretius proposes a materialistic, atomistic theory of nature in his poem *On the Nature of Things.* He favors the preformation theory of embryological development.

c.10 Cleomedes, Greek astronomer, discusses the optical properties of water and says that in a similar manner, the Sun may be visible when it has actually gone a bit below the horizon. This is the first consideration of atmospheric refraction until Ptolemy.

c.50 Hero, Greek engineer, formulates the principle of the motive power of steam, building many steam-powered devices. He also writes of the five simple machines—lever, pulley, wheel, inclined plane, and wedge—and extends and generalizes the law of the lever. He also maintains correctly that air takes up space and is compressible.

c.70 Roman author and naturalist Pliny the Elder (23–79) writes his influential *Natural History*, a vast compilation combining observations of nature, scientific facts, and mythology. Naturalists will use his work as a reference book for centuries.

c.150 Ptolemy publishes geocentric model with Earth at the center of the solar system.

c.150 Ptolemy, Greek astronomer, writes a treatise on optics in which he considers the refraction of light. He offers an original approach that is both theoretical and experimental.

415 Library of Alexandria is burned by religious zealots, thus destroying the most comprehensive collection of ancient and classical scholarship in science and the arts. Much of the recorded knowledge of Western Civilization was lost.

517 Johannes Philoponus (c.490–570), Alexandrian philosopher, also called John the Grammarian, rejects Aristotle's idea that a body will only move as long as it pushed and offers his own theory of motion. He says that a body will keep moving in the absence of friction or as long as nothing opposes it. He argues that this is why the stars continue to move.

c.850 Al-Kindi (801–866), Arab physicist, writes a treatise on optics and the reflection of light. He also studies meteorology, the tides, and specific weights.

c.890 Al-Razi identifies Andromeda galaxy.

c.1000 Alhazen (965–1038), Arabian physicist, rejects the idea that people see because their eyes send out a light which reflects back from an object. He argues correctly that light comes from the Sun or another source, and reflects from the object into the eye. He studies all aspects of light, especially reflection and refraction, and also offers an exploration of rainbow formation.

c.1025 Al-Biruni (973–1048), Arab physician, astronomer, and mathematician, makes a fairly accurate calculation of the specific weights of eighteen precious stones and metals. He uses the methods employed by Archimedes.

1054 Supernova of Crab Nebula recorded in China and by Native Americans.

1121 Al-khazini argues that gravity acts towards Earth's center.

1137 Abu Ja'far Alchazin (Al Khazin), Arab mathematician and astronomer, writes a book in which he offers tables of specific densities and a general description of the laws of gravity.

c.1144 First translations of Arabic alchemical manuscripts in Spain, introducing European scholars to alchemy.

c.1225 Robert Grosseteste (1175–1253), English scholar, studies optics and experiments with mirrors and lenses. He attempts to explain the rainbow and argues that light is the basic substance of the universe. He is also the teacher of the English scholar, Roger Bacon (1220–1292).

c.1267 Roger Bacon (1214–1292), English philosopher and scientist, asserts that natural phenomena should be studied empirically.

1269 Pierre de Maricourt experiments with magnets and compass. Discovery that a magnet is encircled by lines which terminate on two magnetic poles.

c.1270 Witelo (1230–1275) of Silesia, also called Vitellio, writes his *Perspectiva*, a systematic treatment of the optics of the Arabian physicist, Alhazen (965–1038). It deals with refraction and reflection as well as the twinkling of stars (caused by motion).

1304 Theodoric of Freibourg conducts experiments to investigate rainbows.

c.1325 William of Ockham (1280–1349), English scholar, argues strongly for the importance of empiricism and lays down the rule called *Ockham's razor*. According to this rule, when two theories equally fit all observed facts, the one requiring the fewest or simplest assumptions is to be accepted as more valid.

c.1350 Jean Buridan (1300–1385), French philosopher, refutes the Aristotelian notion that an object in motion requires a continuous force, and maintains that only an initial impetus is required. He anticipates Newton's first law of motion by saying that the celestial bodies stay in motion in this manner.

c.1500 Leonardo da Vinci (1452–1519), Italian artist and inventor, experiments with hydrostatics and diffraction and offers a version of the principle of inertia (which will not come until the time of Galileo).

1510 Two German books lay the foundation for industrial chemistry. *Bergwerkb'chlein* is dedicated to mineralogy and *Probierb'chlein* focuses on chemical tests and introduces quantitative concepts.

1512 Copernicus advances heliocentric model that includes assertion that the planets orbit the Sun.

c.1525 Paracelsus (1493–1541), Swiss physician and alchemist, uses mineral substances as medicines. Denying Galen's authority, Paracelsus teaches that life is a chemical process.

1543 Andreas Vesalius publishes his epoch-making treatise *The Fabric of the Human Body.* Although Vesalius generally accepts Galenic physiological doctrines and ideas about embryology, Vesalius is later regarded by many as the founder of modern anatomy because he corrected many of Galen's misconceptions regarding the human body.

1546 Gerardus Mercator describes Earth's magnetic poles.

1563 Bernard Palissy (c.1510–1589), French potter, publishes his *Recette veritable*, in which he discusses agriculture, geology, mining, and forestry. He discovers the Italian secret of producing majolica (pottery decorated with an opaque tin glaze) and is considered one of the most eminent chemists of France.

1568 Zacharias and Hans Janssen development of the first compound microscope opens new opportunities for the study of structural detail.

1574 Lazarus Ercker (c.1530–1594) of Germany publishes his *Beschreibung aller Furnemisten Mineralischen Ertzt und Bergwercks Arten*, which is the first manual of analytical, metallurgical chemistry. His text is especially valuable to the practicing assayers.

1574 Tycho Brahe argues that a comet he discovered lies beyond the Moon.

1576 Brahe constructs a planetary observatory to accurately record motions of celestial bodies.

1581 Robert Norman, English navigator and instrument maker, publishes a work on the lodestone called *The Newe Attractive*. He discusses the known properties of the magnet and is the first to note that steel does not change its weight when magnetized. He also discovers *magnetic dip*as when he suspends a compass needle to allow vertical movement and notes that it points down toward Earth. This is later used by Gilbert.

1584 Giordano Bruno argues that stars are suns with other planets.

1586 Simon Stevin (1548–1620), Dutch mathematician, publishes a report of his experiment in which he refutes the Aristotelian doctrine that heavy bodies fall faster than light ones. He also founds hydrostatics by demonstrating that the pressure on a liquid varies according to how high above Earth's surface it is and not upon the shape of the container that holds it. His demonstrations also eliminate many standard arguments in favor of the existence of perpetual motion.

1587 Galileo Galilei (1564–1642), Italian astronomer and physicist, begins experiments that lead to his law of falling bodies. He uses a gently sloping inclined plane and shows that the rate of fall of a body is independent of its weight. He eventually states correctly that all objects will fall at the same rate in a vacuum. He also shows that a body can move under the influence of two forces at one time.

1591 Thomas Harriot (1560–1621), English mathematician, is the first Westerner to note that snowflakes are hexagonal (six-sided). He does not publish his findings. The Chinese however, document snow flake crystal shapes from the second century B.C.

1592 Galileo develops primitive thermometer.

1592 Galileo argues that the physical laws of Cosmos are the same as those on Earth.

1600 William Gilbert (1544–1603), English physician and physicist, publishes his *De magnete* which is a full account of his extensive investigations on magnetic bodies and electrical attraction. He suggests that the Earth itself is a great magnet, and he is the first to use the terms electric attraction, electrical force, and magnetic pole.

1604 Galileo observes that distance for falling object increases as square of time.

1604 German astronomer and mathematician Johannes Kepler (1571–1630) writes a treatise on optics.

1608 Hans Lippershey develops optical telescope.

1609 Using a telescope, Galileo observes craters and mountains on the Moon.

1609 Kepler theorizes that a force of gravity exists that can exert itself through empty space, and that its strength is related to the size of the bodies involved.

1609 Kepler offers 1st and 2nd laws of planetary motion.

1610 Galileo observes moons of Jupiter.

1610 Jean Beguin (1550–1620) publishes the first textbook on chemistry.

1613 Galileo documents existence and movements of Sunspots.

1630 Jean Rey (1582–1645), French physician, writes on the nature of air and its role in combustion, and lays the foundation for future chemical discoveries. He suggests a possible experiment for weighing air that Galileo actually performs.

1632 Henry Gellibrand (1597–1636), English astronomer and mathematician, publishes his findings that offer the first indication that the Earth's magnetic field slowly changes over time.

1637 Rene Descartes offers physical explanations of refraction, rainbows and clouds.

1638 Galileo publishes his *Discorsi e dimostrazioni mathematiche intorno a due nuovescienze* which lays the foundations of modern mechanics. In it, he formulates what becomes known as the first law of motion (or the law of inertia), as well as the laws of cohesion and strength of materials, and of the pendulum. It also provides a definition of momentum and details the steps or stages of what becomes known as the experimental method. This work marks the end of Aristotelian physics.

1640 Evangelista Torricelli (1608–1647), Italian physicist, writes his *De motu gravium* in which he applies Galileo's laws of motion to fluids and founds the study of hydrodynamics.

1643 Torricelli is the first to create a sustained vacuum when he invents the barometer. He fills a four-foot-long glass tube with mercury and inverts it onto a dish. He observes that not all the mercury flows out and that over time, the level remaining in the tube varies. He concludes correctly that these changes are caused by atmospheric pressure.

1644 Kenelm Digby (1603–1665), English natural philosopher, observes magnetic and electrical attractions as well as acoustic resonance.

1646 Johann Rudolf Glauber (1604–1670), German chemist, publishes the first of his five-volume *Furni novi philosophici*. This work gives his recipes for mineral acids and salts, including "sal mirabile"—

the sodium sulfate residue that formed by the action of sulfuric acid on ordinary salt—that becomes known as "Glauber's salt."

1648 *Ortus medicinae* by Johann Baptista van Helmont (1580–1644), Flemish physician and alchemist, is published posthumously. He is the first to use quantitative methods in connection with a biological problem. He is also the first to recognize that one air-like substance exists, and he names this vapor, or non-solid, "chaos," which in Flemish sounds like "gas."

1648 Blaise Pascal (1623–1662), French mathematician, physicist, and philosopher, conducts his famous experiment on the Puy-de-Dôme mountain and not only verifies Torricelli's experiments, but goes beyond them to demonstrate that air pressure decreases as altitude increases.

1648 Johannes Marcus Marci von Kronland (1595–1667), Bohemian physician, discovers the diffraction of light, but it does not become a recognized fact until Newton's time.

1651 William Harvey publishes a landmark treatise on embryology entitled *On the Generation of Animals,* stating that all living things come from eggs. Harvey demonstrates that oviparous and viviparous generations are analogous to each other. Although Harvey discovers many errors in Aristotle's ideas, he supports the Aristotelian doctrine that generation occurs by epigenesis.

1653 Blaise Pascal (1623–1662), French mathematician, physicist, and philosopher, studies fluids and formulates what comes to be known as Pascal's principle—that the pressure at any point in a liquid is the same in all directions. Pascal's principle forms the basis of the hydraulic press that he also describes in theory. This information is not published until a year after his death.

1660 Vincenzo Viviani (1622–1703), Italian mathematician, and Giovanni Alfonso Borelli (1608–1679), Italian mathematician and physiologist, collaborate on an experiment in which they measure the velocity of sound by using the cannon-flash-and-sound method.

1661 Robert Boyle (1627–1691), English physicist and chemist, publishes his book, *The Sceptical Chymist.* Boyle espouses the experimental method and breaks from the Greek notion of elements.

1662 Boyle announces what becomes known as Boyle's Law, stating that when an ideal gas is under constant pressure, its volume and pressure vary inversely.

1665 Newton experiments with gravity, spectrum of light, invents differential calculus.

1665 Nicolaus Steno (1638–1686), Danish anatomist and geologist, briefly states what is now called the first law of crystallography. Also called the law of the constancy of crystalline angles, it states that crystals of a specific substance have fixed characteristic angles at which the faces, however distorted, always meet.

1665 Robert Hooke publishes *Micrographia,* an account of observations made with the new instrument known as the microscope. Hooke presents his drawings of the tiny box-like structures found in cork and calls these tiny structures "cells." Although the cells he observes are not living, the name is retained. He also describes the streaming juices of live plant cells.

1666 Isaac Newton begins work on laws of mechanics and gravitation.

1666 Robert Boyle (1627–1691), English physicist and chemist, publishes his *Hydrostatical Paradoxes* in which he details his experiments with fluids and refutes the old doctrine that a light liquid can exert no pressure against a heavier fluid.

1669 Johann Joachim Becher (1635–1682), German chemist, publishes his *Physica subterranea*, in which he is the first to attempt the formulation of a general theory of chemistry. His concept of "terra pinguis" as the substance in air that burns forms the basis of the later phlogiston theory.

1671 Robert Boyle (1627–1691), English physicist and chemist, produces hydrogen by dissolving iron in hydrochloric or sulfuric acid, but he is unaware of his achievement.

1672 Isaac Newton (1642–1727), English scientist and mathematician, publishes his letter on light in the Royal Society's *Philosophical Transactions*. This letter, which is his first scientific publication, details his prism experiments of 1666 and offers findings that reveal for the first time the nature of light. He recounts how he let a ray of sunlight enter a darkened room through a small hole and then passed the ray through a prism onto a screen. The ray was refracted and a band of consecutive colors in rainbow order appeared. He then passed each separate color through another prism and noted that although the light was refracted, the color did not change. From this, he deduced that sunlight (or white light) consists of a combination of these colors. Later he elaborates further on this ground-breaking experiment.

1673 Christiaan Huygens (1629–1695), Dutch physicist and astronomer, publishes his *Horologium oscillatorium* in which he details his invention of the pendulum, or grandfather, clock. He employs Galileo's principle of isochronicity and ingeniously adapts it to the inner workings of a clock, beginning the era of accurate timekeeping that is so important to the advancement of physics. This highly original work not only demonstrates great mechanical ability but superior mathematical theorizing as well.

1674 Hennig Brand (c.1630–c.1692), German chemist, discovers phosphorus, which he finds in urine. This is the first discovery of an element that was not known in any earlier form.

1674 Robert Hooke attempts to explain planetary motion as a balance of centrifugal force and gravitational attraction.

1675 Giovanni Cassini, Saturn has separated rings which must be composed of small objects.

1675 Nicolas Lémery (1645–1715), French chemist and physician, publishes his *Cours de chymie*, which becomes the authoritative textbook on chemistry for the next 50 years. He is an adherent of Boyle's, and advocates the experimental method.

1676 Edmé Mariotte (1620–1684), French physicist, independently formulates Boyle's law and adds an important qualification to it. Like Boyle, he notes that air expands with rising temperature and contracts with falling temperature, but he adds that the inverse relationship between temperature and pressure only holds if the temperature is kept constant. Because of this, Boyle's law is called Mariotte's law in France.

1676 Gottfried Wilhelm Leibniz (1646–1716), German philosopher and mathematician, criticizes Descartes' ideas of motion and formulates his own theory of dynamics which substitutes kinetic energy for the conservation of movement.

1676 Olaus Roemer measures the speed of light by observing Jupiter's moons.

1676 Robert Hooke, law of elasticity and springs.

1678 Christiaan Huygens writes about wave theory of light.

1678 Huygens discovers polarization of light.

1680 Isaac Newton demonstrates that inverse square law implies elliptical orbits.

1681 Johann Joachim Becher (1635–1682), German chemist, obtains tar from the distillation of coal. He also suggests that sugar is necessary for fermentation.

1683 Edmund Halley (1656–1742), English astronomer, states that Earth's magnetism is caused by four poles of attraction, two of them being in each hemisphere near each pole of Earth.

1684 Isaac Newton (1642–1727), English scientist and mathematician, provides the first summary exposition of his theory of gravitation in a memoir entitled *Demotu corporum*. He expands on it later in his 1687 *Principia*.

1686 Edmund Halley (1656–1742), English astronomer, develops a reliable formula that links the altitudes of various localities with the atmospheric pressure measured there. His altimetric formula is one of the first practical applications of the new barometric discoveries.

1687 Denis Papin (1647–1712), French physicist, publishes a work in which he offers details on the use of steam to drive a piston in a cylinder which eventually becomes the basic design for an early steam engine. He never built one of his own design.

1687 English physicist Sir Isaac Newton (1642–1727) publishes a law of universal gravitation in his important and influential work *Philosophiae Naturalis Principia Mathematica (Mathematical Principles of Natural Philosophy).* Newton articulates three laws of motion. Still widely regarded as the greatest scientific work ever written, *Principia* states that the entire world is subsumed under a single set of laws. They are: (1) a body remains at rest unless it is compelled to change by a force impressed upon it; (2) the change of motion (the change of velocity times the mass of the body) is proportional to the forces impressed; and (3) to every action there is an equal and opposite reaction. From these laws he then deduces his law of universal gravitation. In its simplest form, Newton's law of universal gravitation states that bodies with mass attract each other with a force that varies directly as the product of their masses and inversely as the square of the distance between them.

1688 John Clayton (1657–1725), English cleric, obtains methane and recognizes its flammable nature.

1690 Christiaan Huygens (1629–1695), Dutch physicist and astronomer, publishes his *Traité: de la lumiére* in which he states his wave theory of light. This unpopular theory sees light as a longitudinal wave that undulates in the direction of its motion much as a sound wave does. The worth of this theory is not understood foray full century.

1690 Christiaan Huygens, principle of Huygens, secondary waves.

1690 Huygens publishes his (wave) theory of light.

1690 John Locke, knowledge comes only from experience and sensations.

1699 Guillaume Amontons (1663–1705), French physicist, publishes his observations on gases. In his work on different gases, he shows that each gas changes in volume by the same amount for a given change in temperature. Implied in this is the notion of absolute zero at which gases can contract no further.

1702 Wilhelm Homberg (1652–1715), German physician, discovers boric acid, which he calls "sedative salt."

1704 Isaac Newton (1642–1727), English scientist and mathematician, publishes his *Opticks* which is a comprehensive work containing his main discoveries on the nature of light and color as well as his particle, or corpuscular, theory flight.

1709 John Freind (1675–1728), English physician and chemist, publishes *Praelectiones chemicae*, one of the earliest attempts to use Newtonian principles to explain chemical phenomena.

1714 Gabriel Daniel Fahrenheit (1686–1736), German-Dutch physicist, invents the mercury thermometer, the first accurate thermometer. His use of mercury instead of alcohol means that temperatures far above the boiling point of water and well below its freezing point can be measured (since mercury has a higher boiling point than alcohol). He also invents the Fahrenheit temperature scale in which the freezing point of water is 32 degrees and the boiling point is 212 degrees. He arrives at these numbers by adding salt to water to find its lowest freezing point which he calls zero.

1718 Edmund Halley measures motion of stars.

1718 Etienne-François Geoffroy (1672–1731), French apothecary, publishes his *Tables des différens rapports*, which offers a table of affinities between various acids and alkalis or metals. Chemistry eventually accepts his prophetic concept of affinity.

1729 Pierre Bouguer (1698–1758), French physicist, publishes his *Essai d'optique sur la gradation de la lumièere* in which he makes some of the earliest measurements in astronomical photometry. He also investigates the absorption of light in the atmosphere and formulates what comes to be known as Bouguer's law. This concerns the attenuation of a light beam upon passage through a transparent medium.

1730 René Antoine Ferchault de Réaumur (1683–1757), French naturalist and physicist, develops a thermometer independently of Fahrenheit and establishes what comes to be known as the Reaumur temperature scale. This system has zero degrees as the freezing point of water and 80 degrees as the boiling point of water at normal atmospheric pressure.

1731 Rene Reaumur, alcohol/water thermometer.

1732 Herman Boerhaave (1668–1738), Dutch physician, publishes his *Elementa chemicae*, whose comprehensiveness makes it the most popular chemical textbook for many decades. It serves chemistry as a great teaching book and presents a concise outline of all chemical knowledge.

1732 Pierre Louis Moreau de Maupertuis (1698–1759), French mathematician, publishes his *Discours sur la figure des astres* in which he predicts the shape of Earth using Newtonian mechanics.

1733 Charles François de Cisternay Du Fay (1698–1739), French physicist, discovers that two electrified objects sometimes attract and sometimes repel each other. He notes that their means of being charged seems to be the difference and states that there are two types of electricity, *resinous* and *vitreous*. He formulates the basic electrical law that "like charges repel and unlike charges attract." Later, American statesman and scientist, Benjamin Franklin (1706–1790) calls these two types of electricity *positive* and *negative*.

1733 Georg Brandt (1694–1768), Swedish chemist, publishes the first accurate and complete study of arsenic and its compounds.

1735 Carl Linnaeus publishes his *Systema Naturae, or The Three Kingdoms of Nature Systematically Proposed in Classes, Orders, Genera, and Species,* a methodical and hierarchical classification of all living beings. He develops the binomial nomenclature for the classification of plants and animals. In this system, each type of living being is classified in terms of genus (denoting the group to which it belongs) and species (its particular, individual name). His classification of plants is primarily based on the characteristics of their reproductive organs.

1738 Daniel Bernoulli (1700–1782), Swiss mathematician, publishes his *Hydrodynamica* containing his kinetic theory of gases. This treatise becomes a work of major importance in both physics and chemistry. This is the first attempt at an explanation of the behavior of gases, which, he assumes, are composed of a vast number of tiny particles.

1739 Georg Brandt, element cobalt.

1742 Anders Celsius (1701–1744), Swedish astronomer, applies a new scale to his thermometer by dividing the temperature difference between the boiling and freezing points of water into an even 100 degrees (with zero at the boiling point, but eventually this is reversed). His system becomes known as the centigrade scale and eventually is adopted internationally by scientists.

1742 Anton Svab (1703–1768) of Sweden, also known as Swab, distills zinc from the alloy calamine.

1743 Alexis Claude Clairaut (1713–1765), French mathematician, publishes his *Theorie de la figure de la terre* in which he definitively discusses a rotating body in the shape of the Earth and how it acts under the influence of gravity and centrifugal force.

1746 Leonhard Euler (1707–1783), Swiss mathematician, argues against the particle theory of light and suggests correctly that light has a wave form and that color depends on the length of that wave.

1750 Thomas Wright, Milky Way could be due to slab like distribution of stars.

1752 Benjamin Franklin (1706–1790), American statesman and scientist, flies a kite carrying a pointed wire in a thunderstorm and attempts to test his theory that atmospheric lightning is an electrical phenomenon similar to the spark produced by an electrical frictional machine. To the kite he attaches a silk thread with a metal key at the end, and as lightning flashes, he puts his hand near the key that sparks, just as a Leyden jar would. He proves his point in this extremely dangerous experiment.

1754 Joseph Black's discovery of carbon dioxide establishes that there are gases other than air.

1754 Pierre-Louis Moreau de Maupertuis suggests that species change over time, rather than remaining fixed.

1756 John Canton (1718–1772), English physicist, begins three years of careful weather observations and finds that on days when the aurora borealis is very noticeable, a compass needle becomes irregular. This is the first observation of what become known as magnetic storms.

1757 Albrecht von Haller 1757–1766), publishes the first volume of his eight-volume *Elements of Physiology of the Human Body*, subsequently to become a landmark in the history of modern physiology.

1758 Axel Fredrick Cronstedt (1722–1756), Swedish chemist, initiates the classification of minerals by their chemical structure as well as by their appearance. He notes four kinds of minerals: earths, metals, salts, and bitumens.

1760 Lagrange formulates principle of least action.

1766 Henry Cavendish (1731–1810), English chemist and physicist, publishes a paper on "Factitious Airs" in the Royal Society's Philosophical Transactions, which relates his discovery of hydrogen, or what he calls "inflammable" air.

1772 Joseph Priestley (1733–1804), English chemist, experiments with "fixed air" and writes his "Directions for Impregnating Water with Fixed Air," in which he details the production of seltzer water by using carbon and water. The distinctive taste of the seltzer, or soda, water brings Priestley much fame.

1774 Antoine-Laurent Lavoisier (1743–1794), French chemist, discovers that oxygen is consumed during respiration.

1774 Manganese is discovered by Swedish mineralogist Johann Gottlieb Gahn (1745–1818) and Karl Wilhelm Scheele (1742–1786), Swedish chemist.

1775 Antoine-Laurent Lavoisier (1743–1794), French chemist, publishes his "Memoire," which contains his first major disavowal of the phlogiston theory, as well as a revision of his combustion theory.

1777 Lavoisier argues that chemical compounds are composed of discrete elements.

1777 Coulomb invents torsion balance (measuring charge).

1780 Lazzaro Spallanzani carried out experiments on fertilization in frogs and attempted to determine the role of semen in the development of amphibian eggs.

1781 William Herschel discovers Uranus.

1782 Torbern Olaf Bergman (1735–1824), Swedish mineralogist, publishes his *Skiagraphia regni mineralis*, in which he classifies minerals into four main groups: salts, earths, metals, and inflammable bodies.

1783 Antoine-Laurent Lavoisier (1743–1794), French chemist, and Pierre-Simone de Laplace (1749–1827), French astronomer and mathematician, jointly publish a paper, "Memoire sur la chaleur," which lays the foundations of thermochemistry. They demonstrate that the quantity of heat required to decompose a compound into its elements is equal to the heat evolved when that compound was formed from its elements.

1783 Antoine-Laurent Lavoisier (1743–1794), French chemist, repeats the experiment conducted by Cavendish in 1766 and realizes that he is dealing with a separate gas. He calls this flammable gas "hydrogen," from the Greek phrase meaning "giving rise to water."

1783 Horace Benedict de Saussure (1740–1799), Swiss physicist, publishes his *Essais sur l'hygrometrie* in which he details his invention of a hygrometer that uses human hair for measuring humidity. He also discusses the general principles of hygrometry.

1783 Juan Jose d'Elhuyar (1754–1796) and his younger brother, Don Fausto d'Elhuyar (1755–1833), both Spanish mineralogists, analyze a mineral called wolframite and discover a new metal, tungsten.

1785 James Hutton (1726–1797) proposes uniformitatianism as theoretical basis to interpret geological history of the earth.

1785 Charles Augustin de Coulomb (1736–1806), French physicist, publishes the first of seven papers in which he establishes the basic laws of electrostatics and magnetism. Using his newly invented torsion balance, he determines that Newton's law of inverse squares also applies to electrical and magnetic attraction and repulsion. He states that the degree of attraction or repulsion depends on the amount of electric charge or the magnetic pole strength.

1787 Antoine-Laurent Lavoisier (1743–1794), French chemist, publishes *Methode de la nomenclature chimique* in collaboration with French chemist, Louis-Bernard Guyton de Morveau (1737–1816). This book gives the new chemistry a modern terminology and changes chemical nomenclature to correspond to the new antiphlogiston theory.

1787 Ernst Florens Friedrich Chladni (1756–1827), German physicist, publishes his *Theorie des Klanges* in which he is the first to discover the quantitative relationships that rule the transmission of sound. He also creates *Chladni's figures* by spreading sand on thin plates and vibrating them, producing complex patterns from which much is learned about vibrations. He is considered the father of acoustics.

1787 Jacques-Alexandre Charles (1746–1823), French physicist, demonstrates that different gases all expand by the same amount with a given rise in temperature if the pressure is held constant. This becomes known as Charles's law, and also Gay-Lussac's law.

1789 Martin Heinrich Klaproth (1743–1817), German chemist, discovers uranium in pitchblende. The Curies will refine this same substance in 1898. Later the same year, Klaproth discovers the element zirconium in the mineral zircon.

1790 France introduces the metric system.

1795 James Hutton publishes Theory of the Earth (in Scotland).

1795 Martin Heinrich Klaproth (1743–1817), German chemist, isolates a new metal and names it titanium, after the Titans of Greek mythology. He gives full credit to English mineralogist William Gregor (1761–1817), who first discovered it in 1791.

1796 Erasmus Darwin, grandfather of Charles Darwin and Francis Galton, publishes his *Zoonomia.* In this work, Darwin argues that evolutionary changes are brought about by the mechanism primarily associated with Jean-Baptiste Lamarck, that is, the direct influence of the environment on the organism.

1798 Henry Cavendish (1731–1810), English chemist and physicist, is the first to calculate Earth's mass. He does this by obtaining what Isaac Newton had not provided—a value for the gravitational constant. He builds a model with light balls and large, heavy ones and uses a sensitive wire to calculate the strength of the attraction between the two. Once he obtains this constant value, he calculates Earth's mass close to modern value.

1801 William Hyde Wollaston (1766–1828), English chemist and physicist, establishes that frictional and galvanic electricity are identical. He also discovers, independently of Ritter, the existence of invisible light beyond violet light (ultraviolet light).

1802 John Dalton introduces modern atomic theory into the science of chemistry.

1802 Thomas Thomson (1773–1852), Scottish chemist, publishes his *System of Chemistry* and introduces a system of symbols for individual minerals using the first letters of their names.

1802 Thomas Young (1773–1829), English physicist and physician, performs his classic experiment on interference in which sunlight is made to pass through two pinholes in an opaque screen. With a wave interpretation of his observations, he is able to provide the first quantitative values for the length of light waves.

1803 John Dalton (1766–1844), English chemist, states the law of multiple proportions which applies to two elements that could combine in more than one way. He also states his atomic theory that says that all elements are composed of extremely tiny, indivisible, indestructible atoms, and that all the known substances are composed of some combination of these atoms. He finally states that these atoms differ from each other only in mass, and that this difference can be measured.

1803 Louis Poinsot (1777–1859), French mathematician, publishes his *Éléments de statique* which contains his theory of couples (two parallel forces of equal magnitude but opposite direction form a couple). He also introduces the concept of torque.

1803 Smithson Tennant, elements osmium and iridium.

1803 William Henry (1774–1836), English physician and chemist, proposes what becomes Henry's law, stating that the amount of gas absorbed by a liquid is in proportion to the pressure of the gas above the liquid, provided no chemical reaction occurs.

1805 John Dalton (1766–1844), English chemist, publishes the first table of atomic weights and invents a new system of chemical symbols.

1805 Joseph-Louis Gay-Lussac (1778–1850), French chemist, establishes that precisely two volumes of hydrogen combine with one volume of oxygen to form water.

1811 Amedeo Avogadro (1776–1856), Italian physicist, first proposes his theory of molecules, which is confirmed much later by modern chemistry. He states that equal volumes of all gases contain the same number of molecules if they are under the same pressure and temperature.

1813 Sowerby publishes Sowerby's Mineral Conchology.

1815 William Smith publishes first geological map of UK.

1818 Jöns Jakob Berzelius (1779–1848), Swedish chemist, discovers selenium, publishes a table of atomic weights, and offers a system of chemical symbols. His weight table is based on a standard 100 for oxygen and becomes accepted in the twentieth century. His symbols use one or two letters of the Latin name and are essentially also retained in the twentieth century.

1819 Eilhardt Mitscherlich (1794–1863), German chemist, discovers isomorphism. This chemical theory states that compounds of similar composition tend to crystallize together, or, conversely, that compounds with the same crystal form are analogous in chemical composition. This law becomes useful in establishing atomic weights.

1819 Øersted discovers electromagnetism.

1820 André Marie Ampère (1775–1836), French mathematician and physicist, extends Ørsted's work and formulates one of the basic laws of electromagnetism. He discovers that two parallel wires each carrying a current attract each other if the currents are in the same direction, but they repel each other if in the opposite direction. He concludes that magnetism is the result of electricity in motion.

1820 Hans Christian Ørsted (1777–1851), Danish physicist, experiments with a compass and electricity and demonstrates that a current of electricity creates a magnetic field. He announces his discovery in a short article. This is the first time a real connection can be shown between electricity and magnetism, and it founds the new field of electromagnetism.

1821 Faraday proposes flux line picture for electricity and magnetism.

1821 Friedlieb Ferdinand Runge (1795–1867), German chemist, discovers caffeine.

1822 Omalius d'Holloy names the Cretaceous.

1823 Mary Anning discovers first complete pterodactyl in Lyme Regis in Dorset UK.

1824 Nicolas Léonard Sadi Carnot (1796–1832), French physicist, publishes his *Réflexions sur la puissance motrice du feu* in which he is the first to consider quantitatively the manner in which heat and work are interconverted. This highly original work also introduces the important concept of cyclic operations and the principle of reversibility. Although this work founds the science of thermodynamics, or the movement of heat, it is neglected for ten years.

1825 Hans Christian Ørsted (1777–1851), Danish physicist, isolates aluminum through a four-step process that involves a vacuum.

1826 André Marie Ampère (1775–1836), French mathematician and physicist, publishes his *Mémoire sur la théoriemathématique des phénomàmes électrodynamiques*in which he offers the mathematical laws that govern the new field of electricity in motion and founds the study of electrodynamics.

1826 James Cowles Prichard presented his views on evolution in the second edition of his book *Researches into the Physical History of Man* (first edition 1813). These ideas about evolution were suppressed in later editions.

1827 Brown discovers "Brownian Motion".

1830 Charles Lyell publishes *Principles of Geology* and argues Earth is a least millions of years old.

1830 Joseph Henry (1797–1878), American physicist, discovers the principle of electromagnetic induction, showing how an electric current in one coil may set up a current in another through the development of a magnetic field. He puts off further work on this discovery until the following summer, and in the intervening time, the English physicist and chemist, Michael Faraday (1791–1867), publishes his discovery first.

1831 Charles Robert Darwin begins his historic voyage on the H.M.S. *Beagle* (1831–1836). His observations during the voyage lead to his theory of evolution by means of natural selection.

1834 F. Von Alberti names the Triassic Period.

1840 Friedrich Gustav Jacob Henle publishes the first histology textbook, *General Anatomy.* This work includes the first modern discussion of the germ theory of communicable diseases.

1840 Germain-Henri Hess (1802–1850), Swiss-Russian chemist, formulates the law that states that the quantity of heat evolved in a chemical change is the same no matter what chemical route the reaction takes (through a single stage or through many stages. This becomes known as Hess's law.

1840 Jöns Jakob Berzelius (1779–1848), Swedish chemist, first introduces the term "allotropy" to describe the existence of different varieties of an element. He converts charcoal into graphite and declares that the same element may have in different forms.

1840 Karl Bogislaus Reichert introduces the cell theory into embryology. He proves that the segments observed in fertilized eggs develop into individual cells, and that organs develop from cells.

1840 Louis Agassiz publishes his *Etudes sur les glaciers*. Also discovers glacial feature in Scotland away from an ice covered area and advances the theory of glaciation.

1840 Murchison and Sedgwick name the Devonian Period after the county of Devon in UK.

1840 Rudolf Albert von Kölliker establishes that spermatozoa and eggs are derived from tissue cells. He attempts to extend the cell theory to embryology and histology.

1841 Murchison names the Permian Period.

1842 Charles Robert Darwin writes out an abstract of his theory of evolution, but he does not plan to have this theory published until after his death.

1842 Christian Johann Doppler (1803–1853), Austrian physicist, discovers the effect of motion of the source or observer on the observed frequency of sound waves. This mathematical relationship that relates their pitch to the relative motion of the source or observer is called the Doppler effect.

1846 Johann Gottfried Galle discovers Neptune, accounting for observed perturbations in the motion of Uranus.

1848 William Thomson (1824–1907), later known as Lord Kelvin, Scottish mathematician and physicist, explores the concept of absolute zero at which the volume of a gas would be reduced to zero and explains it by stating that it is not that the volume reaches zero but rather that the motion of the gas's molecules stops. He then proposes a new temperature scale with its zero mark at absolute zero and its degrees equal to those on the centigrade scale. It becomes known as the kelvin scale. He also coins the term *thermodynamics*.

1850 Michael Faraday experiments to find link between gravity and electromagnetism fail.

1850 Rudolph Julius Emanuel Clausius (1822–1888), German physicist, publishes a paper which contains what becomes known as the second law of thermodynamics, stating that, "heat cannot, of itself, pass from a colder to a hotter body." He later refines the concept.

1850 Thomas Graham (1805–1869), Scottish physical chemist, studies the diffusion of a substance through a membrane (osmosis) and first distinguishes between crystalloids and colloids. He becomes the founder of colloidal chemistry.

1851 Armand Hippolyte Fizeau (1819–1896), French physicist, measures the speed of light as it flows with a stream of water and as it goes against the stream. He finds that the velocity of light is higher in the former.

1851 Jean Bernard Léon Foucault (1819–1868), French physicist, conducts his spectacular series of experiments associated with the pendulum. He swings a heavy iron ball from a wire more than 200 feet long and demonstrates that the swinging pendulum maintains its plane while Earth slowly twists under it. The crowd of spectators who witnesses this demonstration come to realize that they are watching Earth rotate under the pendulum — experimental proof of a moving Earth.

1851 William Thomson (1824–1907), later known as Lord Kelvin, Scottish mathematician and physicist, publishes *On the Dynamical Theory of Heat* in which he explores Carnot's work and deduces that all energy tends to rundown and dissipate itself as heat. This is another form of the second law of thermodynamics and is advanced further by Clausius at about the same time. Kelvin's work is considered the first nineteenth-century treatise on thermodynamics.

1852 Abraham Gesner (1797–1864), Canadian geologist, prepares the first kerosene from petroleum. He obtains the liquid kerosene by the dry distillation of asphalt rock, treats it further, and calls the product kerosene after the Greek word *keros*, meaning oil.

1852 Alexander William Williamson (1824–1904), English chemist, publishes his study which shows for the first time that catalytic action clearly involves and is explained by the formation of an intermediate compound.

1852 Edward Frankland (1825–1899), English chemist, announces the theory of valence, in which he states that each type of atom has a fixed capacity for combining with other atoms. This concept will lead eventually to Mendeleev's Periodic Table.

1852 James Joule (1818–1889) and William Thomson (1824–1907), both English physicists, show that when a gas is allowed to expand freely, its temperature drops slightly. This becomes known as the Joule–Thomson effect and is evidence that molecules of gases have a slight attraction for other molecules. Overcoming this attraction uses energy and causes a drop in temperature.

1852 Jean Bernard Léon Foucault (1819–1868), French physicist, learns from his pendulum experiment and invents the gyroscope. He sets a wheel within a heavy rim in rotation and sees that when tipped, it is set right again by the force of gravity.

1853 Anders Jonas Ångström (1814–1874), Swedish physicist, demonstrates that the rays emitted by an incandescent gas have the same refrangibility (ability to be refracted) as the rays absorbed by the same gas.

1853 Hans Peter Jorgen Julius Thomsen (1826–1901), Danish chemist, works out a method of manufacturing sodium carbonate from the mineral cryolite. This mineral will soon become important to the production of aluminum.

1853 Johann Wilhelm Hittorf (1824–1914), German chemist and physicist, offers the notion of the transport number as he suggests that ions in a solution with a current running through it travel at different speeds.

1853 William John Macquorn Rankine (1820–1872), Scottish engineer, introduces into physics the concept of potential energy, also called the energy of position.

1854 George Airy, Estimate of Earth mass from underground gravity.

1854 Gregor Mendel begins studying 34 different strains of peas. He selects 22 kinds for further experiments. From 1856 to 1863, Mendel grows and tests over 28,000 plants and analyzes seven pairs of traits.

1855 Charles-Adolphe Wurtz (1817–1884), French chemist, develops a method of synthesizing long-chain hydrocarbons by reactions between alkyl halides and metallic sodium. This method is called the Wurtz reaction.

1855 James Clerk Maxwell, mathematics of Faraday's lines of force.

1855 Johann Heinrich Wilhelm Geissler (1814–1879), German inventor, devises a mercury pump that produces a much better vacuum than old piston pumps. Called Geissler tubes, they make possible a more advanced study of electricity and eventually of the atom. It is with Geissler tubes that the English physicist, Joseph John Thomson (1856–1940), performs his famous experiments elucidating the nature of electrons.

1855 William Parsons, spiral galaxies.

1856 Neanderthal fossil identified.

1857 Louis Pasteur demonstrates that lactic acid fermentation is caused by a living organism. Between 1857 and 1880, he performs a series of experiments that refute the doctrine of spontaneous generation. He also introduces vaccines for fowl cholera, anthrax, and rabies, based on attenuated strains of viruses and bacteria.

1857 Rudolf Julius Emmanuel Clausius (1822–1888), German physicist, offers a new explanation of evaporation in terms of molecules and their velocities. He shows that evaporation produces a loss of energy in the liquid and a decrease in temperature.

1858 Charles Darwin and Alfred Russell Wallace agree to a joint presentation of their theory of evolution by natural selection.

1858 Friedrich August Kekulé von Stradonitz (1829–1896), German chemist, and Archibald Scott Couper (1831–1892), Scottish chemist, first develop symbols to represent the atom is always tetravalent (meaning always combines with four other atoms).

1858 Hermann Ludwig Ferdinand von Helmholtz (1821–1894), German physiologist and physicist, publishes his study on the integrals of hydrodynamic equations that express whirling motion. This becomes the point of departure for new ideas on the structure of matter that eventually replaces the old atomistic concepts.

1858 Julius Plücker (1801–1868), German mathematician and physicist, sends an electric current through a vacuum and describes fluorescent effects in detail. He also discovers that the glow shifts position when placed in the field of an electromagnet. This is the very beginning of an awareness of subatomic particles.

1859 Gustav Robert Kirchhoff (1824–1887), German physicist, after discovering the relation between emission and absorption spectra, concludes that the ratio of the emissive and absorptive powers of a body at each wave length is the same for all bodies at the same temperature. This becomes known as Kirchhoff's law.

1859 James Clerk Maxwell (1831–1879), Scottish mathematician and physicist, studies the rings of Saturn and produces the first extensive mathematical development of the kinetic theory of gases. He shares this discovery of the distribution of molecular speeds in a gas with Ludwig E. Boltzmann (1844–1906), Austrian physicist, who accomplishes the same independently. It comes to be known as the Maxwell–Boltzmann theory of gases.

1859 Urbain Le Verrier, anomolous perihelion shift of Mercury.

1860 Cesium is the first element discovered using the newly developed spectroscope. Robert Wilhelm Bunsen (1811–1899), German chemist, and Gustav Robert Kirchhoff (1824–1887), German physicist, name their new element cesium after its "sky blue" color in the spectrum.

1860 Gustav Robert Kirchhoff (1824–1887), German physicist, introduces the concepts of black bodies and emissivity. A black body is a surface that absorbs all radiation of any wavelength falling on it. Such a body would emit all wavelengths if it were heated to incandescence. This concept later becomes important to quantum theory.

1860 Jean-Servais Stas (1813–1891), Belgian chemist, begins work that leads to an accurate method of determining atomic weights. By 1865, he produces the first modern table of atomic weights using oxygen as a standard.

1860 Robert Wilhelm Bunsen (1811–1899), German chemist, collaborates with Gustav Robert Kirchhoff (1824–1887), German physicist, and they develop the first spectroscope.

1860 Stanislao Cannizzaro (1826–1910), Italian chemist, publishes the forgotten ideas of Italian physicist Amedeo Avogadro (1776–1856)—about the distinction between molecules and atoms—in an attempt to bring some order and agreement on determining atomic weights.

1861 Alexander Mikhailovich Butlerov (1828–1886), Russian chemist, introduces the term "chemical structure" to mean that the chemical nature of a molecule is determined not only by the number and type of a atoms but also by their arrangement.

1861 Friedrich August Kekulé von Stadonitz (1829–1896), German chemist, publishes the first volume of *Lehrbuch der organischen Chemie*, in which he is the first to define organic chemistry as the study of carbon compounds.

1861 Robert Wilhelm Bunsen (1811–1899), German chemist, and Gustav Robert Kirchhoff (1824–1887), German physicist, discover the metal rubidium, using their new spectroscope.

1861 William Crookes (1832–1919), English physicist, discovers the element thallium by using the newly invented spectrum analysis. The following year, it is isolated by French chemist Claude-August Lamy (1820–1878).

1861 William Thomson (1824–1907), later known as Lord Kelvin, Scottish mathematician and physicist, publishes his *Physical Considerations Regarding the Possible Age of the Sun's Heat* which contains the theme of the *heat death* of the Universe. This is offered in light of the principle of dissipation of energy stated in 1851.

1862 Anders Ångström observes hydrogen in the sun.

1863 Ferdinand Reich (1799–1882), German mineralogist, and his assistant Hieronymus Theodor Richter (1824–1898), examine zinc ore spectroscopically and discover the new, indigo-colored element iridium. It is used in the next century in the making of transistors.

1863 William Huggins, stellar spectra indicate that stars made of same elements found on Earth.

1864 James Clerk Maxwell publishes equations of electromagnetic wave propagation in the ether.

1865 Alexander Parkes (1813–1890), English chemist, produces celluloid, the first synthetic plastic material. After working since the 1850s with nitrocellulose, alcohol, camphor, and castor oil, he obtains a material that can be molded under pressure while still warm. Parkes is unsuccessful at marketing his product, however, and it is left to the American inventor, John Wesley Hyatt (1837–1920), to make it a success.

1865 Johann Joseph Loschmidt (1821–1895), Austrian chemist, is the first to attempt to determine the actual size of atoms and molecules. He uses Avogadro's hypothesis to calculate the number of molecules in 22.4 liters of gas, and calls the resulting number Avogadro's number (6.02×10^{23}).

1865 Rudolf Julius Emanuel Clausius (1822–1888), German physicist, refines the second law of thermodynamics and first introduces the term *entropy*, stating that the energy in a closed system will always eventually rundown.

1866 August Adolph Eduard Eberhard Kundt (1839–1894), German physicist, invents a method by which he can make accurate measurements of the speed of sound in the air. He uses a *Kundt's tube* whose inside is dusted with fine powder which is then disturbed by traveling sound waves.

1866 Ernst Heinrich Haeckel publishes his book *A General Morphology of Organisms.* Haeckel summarizes his ideas about evolution and embryology in his famous dictum "ontogeny recapitulates phylogeny." Haeckel suggests that the nucleus of a cell transmits hereditary information. He introduces the use of the term "ecology" to describe the study of living organisms and their interactions with other organisms and with their environment.

1866 Johann Gregor Mendel (1822–1884), Austrian botanist and monk, discovers the laws of heredity and writes the first of a series of papers on heredity (1866–1869), which formulate the laws of hybridization. His work is disregarded until 1900, when de Vries rediscovers it. Unbeknownst to both Darwin and Mendel, Mendelian laws provide the scientific framework for the concepts of gradual evolution and continuous variation.

1868 Pierre-Jules-César Janssen (1824–1907), French astronomer, studies a total eclipse of the Sun and observes an unknown spectral line. He forwards the data to the English astronomer Joseph Norman Lockyear (1836–1920), who concludes it is an unknown element that he names helium, after the Sun.

1868 William Huggins, Doppler shifts of stellar spectra.

1869 Dimitri Ivanovich Mendeleev (1834–1907), Russian chemist, and Julius Lothar Meyer (1830–1895), German chemist, independently put forth the Periodic Table of Elements, which arranges the elements in order of atomic weights. However, Meyer does not publish until 1870, nor does he predict the existence of undiscovered elements as does Mendeleev.

1871 Charles Robert Darwin published *The Descent of Man, and Selection in Relation to Sex.* This work introduces the concept of sexual selection and expands his theory of evolution to include humans.

1871 John William Strutt Rayleigh (1842–1919), English physicist, discovers that the degree of scattering of light by very fine particles is a function of the wavelength of light. His equation offers a solution to the question of why the sky is blue.

1871 Ludwig Eduard Boltzmann (1844–1906), Austrian physicist, begins work on his mathematical treatment of the second law of thermodynamics, interpreting it in terms of the statistics of molecular motions. His work lays the foundations of statistical mechanics.

1872 Ferdinand Julius Cohn publishes the first of four papers entitled "Research on Bacteria," which establishes the foundation of bacteriology as a distinct field. He systematically divides bacteria into genera and species.

1873 Franz Anton Schneider describes cell division in detail. His drawings includes both the nucleus and chromosomal strands.

1873 James Clerk Maxwell (1831–1879), Scottish mathematician and physicist, publishes *Treatise on Electricity and Magnetism* in which he identifies light as an electromagnetic phenomenon. He determines this when he finds his mathematical calculations for the transmission speed of both electromagnetic and electrostatic waves are the same as the known speed of light. This landmark work brings together the three main fields of physics—electricity, magnetism, and light.

1873 Johannes Van der Waals (1837–1923), Dutch physicist, offers an equation for the gas laws which contains terms relating to the volumes of the molecules themselves and the attractive forces between them. It becomes known as the Van der Waals equation.

1873 Walther Flemming discovers chromosomes, observes mitosis, and suggests the modern interpretation of nuclear division.

1876 Henry Augustus Rowland (1848–1901), American physicist, establishes for the first time that a moving electric charge or current is accompanied by electrically charged matter in motion and produces a magnetic field.

1876 James Clerk Maxwell (1831–1879), Scottish mathematician and physicist, publishes his *Matter and Motion* in which he considers the categories of space and time and states with great prescience that, "all our knowledge, both of time and place, is essentially relative."

1876 Josiah Willard Gibbs (1829–1903), American physicist, discovers the phase rule. He arrives at an equation that relates the variables (like temperature and pressure) to different phases (solid, liquid, gas). His work helps lay the foundation for chemical thermodynamics and generally for modern physical chemistry. The rule is later put into practical application by Dutch physical chemist Hendrik Willem Bakhuis Roozeboom (1854–1907).

1877 Astonomer Asaph Hall identifies two moons of Mars.

1880 Carl Oswald Viktor Engler (1842–1925), German chemist, begins his studies on petroleum. He is the first to state that it is organic in origin.

1882 Robert Koch (1843–1910), German bacteriologist, discovers the tubercle bacillus and enunciates "Koch's postulates," which define the classic method of preserving, documenting, and studying bacteria.

1883 August F. Weismann begins work on his germplasm theory of inheritance. Between 1884 and 1888, Weismann formulates the germplasm theory that argues that the germplasm is separate and distinct from the somatoplasm. Weismann argues that the germplasm is continuous from generation to generation and that only changes in the germplasm are transmitted to further generations. Weismann proposes a theory of chromosome behavior during cell division and fertilization and predicts the occurrence of a reduction division (meiosis) in all sexual organisms.

1883 Ernst Mach (1838–1916), Austrian physicist, publishes his *Die Mechanik in ihrer Entwickelung historisch-kritisch dargestellt* in which he offers a radical philosophy of science that calls into question the reality of such Newtonian ideas as space, time, and motion. His work influences Einstein and prepares the way for relativity.

1883 Frank Wigglesworth Clarke (1847–1931), American chemist and geophysicist, is appointed chief chemist to the U.S. Geological Survey. In this position, he begins an extensive program of rock analysis and is one of the founders of geochemistry.

1883 George Francis Fitzgerald (1851–1901), Irish physicist, first suggests a method of producing radio waves. From his studies of radiation, he concludes that an oscillating current would produce electromagnetic waves. This is later verified experimentally by Hertz in 1888 and used in the development of wireless telegraphy.

1883 Johann Gustav Kjeldahl (1849–1900), Danish chemist, devises a method for the analysis of the nitrogen content of organic material. His method uses concentrated sulfuric acid and is simple and fast.

1883 Thomas Alva Edison (1847–1931), American inventor, discovers the emission of electrons from hot bodies. He inserts a small metal plate near the filament of a light bulb and finds that the plate draws a current when he connects it to the positive terminal of the light bulb circuit, even though the plate is not touching the filament. Called the Edison effect, this is later a major factor in the invention of the vacuum tube.

1884 Svante August Arrhenius (1859–1927), Swedish chemist, first proposes the concept of ions being atoms bearing electrical charges.

1886 Paul-Louis-Toussint Héroult (1863–1914), French metallurgist, and Charles Martin Hall (1863–1914), American chemist, independently invent an electrochemical process for extracting aluminum from its ore. This process makes aluminum cheaper and forms the basis of the huge aluminum industry. Hall makes the discovery in February of this year, and Héroult achieves his in April.

1886 Roland von Eötvös (1848–1919), Hungarian physicist, first introduces the concept of molecular surface tension. His study of Earth's gravitational field, which leads him to invent a precise torsion balance, results in his proof that inertial mass and gravitational mass are equivalent. This proves to be a major principle in Einstein's general theory of relativity.

1887 Albert Abraham Michelson (1852–1931), German-American physicist, collaborates with the American chemist, Edward Williams Morley (1838–1923), to test the age-old hypothesis that Earth moves through luminiferous ether (the supposed light-carrying element that exists outside or above Earth's atmosphere). They use Michelson's interferometer—that splits a beam of light in two, sends each on a different path, and then reunites them—to test out this idea. The failure of this extremely sensitive instrument to detect even the slightest change in the velocity of light proves that the ether does not exist (for it would have had to change slightly if it had gone through a substance). This result overturns all ether-based theories and makes physicists search for some explanation of the invariance of the speed of light.

1887 Ernst Mach (1838–1916), Austrian physicist, is the first to note the sudden change in the nature of the airflow over a moving object that occurs as it approaches the speed of sound. Because of this, the speed of sound in air (under certain temperature conditions) is called Mach 1. Mach 2 is twice that speed and so on.

1887 Herman Frasch (1851–1914), German-American chemist, patents a method for removing sulfur compounds from oil. Once the foul sulfur smell is removed through the use of metallic compounds, petroleum becomes a marketable product.

1887 Woldemar Voigt, anticipated Lorentz transform to derive Doppler shift.

1888 Heinrich Rudolf Hertz (1857–1894), German physicist, for the first time generates electromagnetic (radio) waves and devises a detector that can measure their wavelength. From this he is able to prove experimentally James Clerk Maxwell's (1831–1879) hypothesis that light is an electromagnetic phenomenon. Hertz's work not only discovers radio waves, but experimentally unites the three main fields of physics—electricity, magnetism, and light.

1890 Hendrik Antoon Lorentz (1853–1928), Dutch physicist, suggests that the atom consists of charged particles whose oscillations can be affected by a magnetic field. This is later confirmed by his pupil, the Dutch physicist, Pieter Zeeman (1865–1943), in 1896.

1891 Edward Goodrich Acheson (1856–1931), American inventor, discovers that carbon heated with clay yields an extremely hard substance. He names it carborundum, and eventually finds it to be a compound of silicon and carbon. For half a century it remains second only to a diamond in hardness, and becomes very useful as an abrasive.

1893 Augusto Righi (1850–1920), Italian physicist, demonstrates that Hertz (radio) waves differ from light only in wavelength and not because of any essential difference in their nature. This helps to establish the existence of the electromagnetic spectrum.

1893 Ferdinand-Frédéric-Henri Moissan (1852–1907), French chemist, produces artificial diamonds in his electric furnace.

1894 Guglielmo Marconi (1874–1937), Italian electrical engineer, uses Hertz's method of producing radio waves and builds a receiver to detect them. He succeeds in sending his first radio waves 30 feet to ring a bell. The next year, his improved system can send a signal 1.5 miles.

1894 John William Strutt Rayleigh (1842–1919), English physicist, and William Ramsay (1852–1916), Scottish chemist, succeed in isolating a new gas in the atmosphere that is denser than nitrogen and combines with no other element. They name it "argon," which is Greek for inert. It is the first of a series of rare gases with unusual properties whose existence had not been predicted.

1895 Pierre Curie (1859–1906), French chemist, studies the effect of heat on magnetism and shows that there is a critical temperature point above which magnetic properties will disappear. This comes to be called the Curie point.

1895 Wilhem Conrad Röntgen (1845–1923), German physicist, submits his first paper documenting his discovery of x rays. He tells how this unknown ray, or radiation, can affect photographic plates, and that wood, paper, and aluminum are transparent to it. It also can ionize gases and does not respond to electric or magnetic fields nor exhibit any properties of light. This discovery leads to such a stream of groundbreaking discoveries in physics that it has been called the beginning of the second scientific revolution.

1895 William Ramsay (1852–1916), Scottish chemist, discovers helium in a mineral named cleveite. It had been speculated earlier that helium existed only in the Sun, but Ramsay proves it also exists on Earth. It is discovered independently this year by Swedish chemist and geologist Per Theodore Cleve (1840–1905). Helium is an odorless, colorless, tasteless gas that is also insoluble and incombustible.

1896 Antoine Henri Becquerel (1852–1908), French physicist, studies fluorescent materials to see if they emit the newly-discovered x rays and discovers instead that uranium produces natural radiation that is eventually called *radioactivity* in 1898 by the Polish-French chemist, Marie Sklodowska Curie (1867–1934).

1897 Joseph John Thomson (1856–1940), English physicist, discovers the electron. He conducts cathode ray experiments and concludes that the rays consist of negatively charged "electrons" that are smaller in mass than atoms.

1898 Marie Sklodowska Curie (1867–1934), Polish-French chemist, discovers thorium, which she proves is radioactive.

1899 Ernest Rutherford (1871–1937), British physicist, discovers that radioactive substances give off different kinds of rays. He names the positively charged ones alpha rays and the negative ones beta rays.

1900 Carl Correns, Hugo de Vries, and Erich von Tschermak independently rediscover Mendel's laws of inheritance. Their publications mark the beginning of modern genetics. Using several plant species, de Vries and Correns perform breeding experiments that paralleled Mendel's earlier studies and independently arrive at similar interpretations of their results. Therefore, upon reading Mendel's publication, they immediately recognized its significance. William Bateson describes the importance of Mendel's contribution in an address to the Royal Society of London.

1900 Friedrich Ernst Dorn (1848–1916), German physicist, analyzes the gas given off by (radioactive) radium and discovers the inert gas he names radon. This is the first clear demonstration that the process of giving off radiation transmutes one element into another during the radioactive decay process.

1900 Hugo Marie de Vries describes the concept of genetic mutations in his book *Mutation Theory.* He uses the term mutation to describe sudden, spontaneous, and drastic alterations in the hereditary material.

1900 Max Karl Ernst Ludwig Planck (1858–1947), German physicist, publishes his classic and revolutionary paper on quantum physics. He tells of his discovery that light or energy is not found in nature as a continuous wave or flow, but is emitted and absorbed discontinuously in little packets, or *quanta*. Further, each quantum, or packet of energy, is indivisible. Planck's new notion of the quantum seemed to contradict the mechanics of Newton and the electromagnetics of Maxwell, and replace them with new rules. In fact, his quantum physics were new rules in a new type of physics—physics of the very fast and the very small. His theory is soon applied (by Einstein) and incorporated (by Bohr), and becomes the watershed between all physics that comes before it (classical physics) and all that is after (modern physics).

1900 Paul Karl Ludwig Drude (1863–1906), German physicist, proposes the first model for the structure of metals. His model explains the constant relationship between electrical conductivity and the heat conductivity in all metals.

1900 Paul Ulrich Villard (1860–1934), French physicist, discovers what are later called gamma rays. While studying the recently discovered radiation from uranium, he finds that in addition to the alpha rays and beta rays, there are other rays, unaffected by magnets, that are similar to x rays, but shorter and more penetrating.

1901 Antoine Henri Becquerel (1852–1908), French physicist, studies the rays emitted by the natural substance uranium and concludes that the only place they could be coming from is within the atoms of uranium. This marks the first clear understanding of the atom as something more than a featureless sphere. It implies a dynamic reality that might also contain electrons. Becquerel's discovery of radioactivity and his focus on the uranium atom make him the father of modern atomic and nuclear physics.

1901 Guglielmo Marconi (1874–1937), Italian electrical engineer, successfully sends radio signal from England to Newfoundland.

1902 Discovery of Tyrannosaurus Rex fossil.

1902 Lapworth names the Ordovician Period after a Welsh Iron Age tribe.

1902 Oliver Heaviside (1850–1925), English physicist and electrical engineer, and Arthur Edwin Kennelly (1861–1939), British-American electrical engineer, independently and almost simultaneously make the first prediction of the existence of the ionosphere, an electrically conductive layer in the upper atmosphere that reflects radio waves. They theorize correctly that wireless telegraphy works over long distances because a conducting layer of atmosphere exists that allows radio waves to follow the Earth's curvature instead of traveling off into space.

1903 Antoine Henri Becquerel (1852–1908), French physicist, shares the Nobel Prize in physics with the husband-and-wife team of Marie Sklodowska Curie (1867–1934), Polish-French chemist, and Pierre Curie (1859–1906), French chemist. Becquerel wins for his discovery of natural or spontaneous radioactivity, and the Curies win for their later research on this new phenomenon.

1903 Archibald Edward Garrod provides evidence that errors in genes cause several hereditary disorders in human beings. His book *The Inborn Errors of Metabolism* (1909) is the first treatise in biochemical genetics.

1903 Ernest Rutherford (1871–1937), British physicist, and Frederick Soddy (1877–1956), English chemist, explain radioactivity by their theory of atomic disintegration. They discover that uranium breaks down and forms a new series of substances as it gives off radiation.

1903 Walter S. Sutton publishes a paper in which he presents the chromosome theory of inheritance. The theory, which states that the hereditary factors are located in the chromosomes, is independently proposed by Theodor Boveri and is generally referred to as the Sutton–Boveri hypothesis.

1903 William Ramsay (1852–1916), Scottish chemist, and Frederick Soddy(1877–1956), English chemist, discover that helium is continually produced by naturally radioactive substances.

1904 Ernest Rutherford, age of Earth by radioactvity dating.

1904 Hendrik Antoon Lorentz (1853–1928), Dutch physicist, extends his idea of local time (different time rates in different locations) and arrives at what are called the Lorentz transformations. These are mathematical formulas describing the increase of mass, shortening of length, and dilation of time that are characteristic of a moving body. These eventually form the basis of Einstein's special theory of relativity.

1904 William Ramsay (1852–1916), Scottish chemist, receives the Nobel Prize in Chemistry for the discovery of the inert gaseous elements in air, and for his determination of their place in the periodic system.

1905 Albert Einstein (1879–1955), German-Swiss physicist, uses Planck's theory to develop a quantum theory of light which explains the photoelectric effect. He suggests that light has a dual, wave-particle quality.

1905 Albert Einstein (1879–1955), German-Swiss physicist, publishes special theory of relativity.

1905 Albert Einstein (1879–1955), German-Swiss (later German-born American) physicist, publishes an elegantly brief but seminal paper in which he asserts and proves his most famous formula, $E=mc^2$ (Energy = mass times the square of the speed of light squared) relating mass and energy.

1905 Percival Lowell postulates a ninth planet beyond Neptune.

1905 Walther Hermann Nernst (1864–1941), German physical chemist, announces his discovery of the third law of thermodynamics. He finds that entropy change approaches zero at a temperature of absolute zero, and deduces from this the impossibility of attaining absolute zero.

1907 Albert Einstein, equivalence principle and gravitational red shift.

1907 Georges Urbain (1872–1938), French chemist, discovers the last of the stable rare earth elements, and names it lutetium after the Latin name of Paris.

1907 Pierre Weiss (1865–1940), French physicist, offers his theory explaining the phenomenon of ferromagnetism. He states that iron and other ferromagnetic materials form small *domains* of a certain polarity pointing in various directions. When some external magnetic field forces them to be aligned, they become a single, strong magnetic force. This explanation is still accurate.

1908 Tunguska event occurs when a comet or asteroid enters the atmosphere, causing major damage to a forested region in Siberia.

1908 Alfred Wegener proposes the theory of continental drift.

1908 Ernest Rutherford (1871–1937), English physicist, is awarded the Nobel Prize in Chemistry for his investigations into disintegration of the elements and the chemistry of radioactive substances.

1908 Ernest Rutherford (1871–1937), British physicist, and Hans Wilhelm Geiger (1882–1945), German physicist, develop an electrical alpha-particle counter. Over the next few years, Geiger continues to improve this device which becomes known as the Geiger counter.

1908 Percy Williams Bridgman (1882–1961), American physicist, begins a lifetime of pioneering work in the field of high pressures. Working eventually in the unexplored field of attaining 400,000 atmospheres, he is forced to invent much of his equipment himself. As he extends the range of pressures, he becomes the founder of the laws of high pressure physics.

1909 Thomas Hunt Morgan selects the fruit fly *Drosophila* as a model system for the study of genetics. Morgan and his coworkers confirm the chromosome theory of heredity and realize the significance of the fact that certain genes tend to be transmitted together. Morgan postulates the mechanism of "crossing over." His associate, Alfred Henry Sturtevant demonstrates the relationship between crossing over and the rearrangement of genes in 1913.

1911 Arthur Holmes publishes the first geological time scale with dates based on radioactive measurements.

1911 Ernest Rutherford (1871–1937), British physicist, discovers that atoms are made up of a positive nucleus surrounded by electrons. This modern concept of the atom replaces the notion of featureless, indivisible spheres that dominated atomistic thinking for 23 centuries—since Democritus (c. 470–c.380).

1911 Victor Hess identifies high altitude radiation from space.

1912 Friedrich Karl Rudolf Bergius (1884–1949), German chemist, discovers how to treat coal and oil with hydrogen to produce gasoline.

1913 Charles Fabry (1867–1945), French physicist, first demonstrates the presence of ozone in the upper atmosphere. It is found later that ozone functions as a screen, preventing much of the Sun's ultraviolet radiation from reaching Earth's surface. Seventy-five years after the discovery of ozone, in 1985, a hole in the ozone layer over Antarctica is discovered via satellite.

1913 Niels Henrik David Bohr (1885–1962), Danish physicist, proposes the first dynamic model of the atom. It is seen as a very dense nucleus surrounded by electrons rotating in orbitals (defined energy levels).

1913 Between 1911-13, Danish astronomer Ejnar Hertzsprung (1873-1967) and American astronomer Henry Norris Russell (1877-1957) independently develop what is now known as the Hertzsprung-Russell diagram describing stellar evolution.

1913 Using the x-ray spectroscope which they invented, William Lawrence Bragg (1890–1971), Australian-English physicist, and his father William Henry Bragg (1862–1942), make the first determinations of the structures of simple crystals and demonstrate the tetrahedral distribution of carbon atoms in diamonds. Their perfection of x-ray crystallography leads to the later examination of the molecule structure of thousands of crystalline substances.

1914 Ejnar Hertzsprung measures the distance of the Large Magellanic Cloud using Cepheid variable stars.

1914 Ernest Rutherford (1871–1937), British physicist, discovers a positively charged particle he calls a proton.

1915 Albert Einstein (1879–1955), German-Swiss physicist, completes four years of work on his theory of gravitation, or what becomes known as the general theory of relativity.

1915 Richard Martin Willstätter (1872–1942), German chemist, is awarded the Nobel Prize in Chemistry for his research on plant pigments, especially chlorophyll.

1916 Karl Schwarzschild theorizes the existence of black holes.

1917 Albert Einstein, introduction cosmological constant and steady state model of the universe.

1917 Vesto Melvin Slipher observes that most galaxies have red shifts.

1917 Willem de Sitter describes a model of a static universe with no matter.

1918 Harlow Shapley determines the size and shape of our galaxy.

1918 Harlow Shapley measures distance to globular clusters using Cepheid variable stars.

1919 Arthur Eddington records data on the Sun's gravitational deflection of starlight during a solar eclipse, confirming Einstein's general theory of relativity.

1920 Ernest Rutherford (1871–1937), English physicist, names the positively charged part of the atom's nucleus a "proton."

1920 Shapley and Curtis, The Great Debate over the scale and structure of the universe.

1922 Alexsandr Friedmann develops a model of an expanding/oscillating universe with matter included.

1923 Arthur Holly Compton (1892–1962), American physicist, discovers what is later called the Compton effect. While accurately measuring the wavelengths of scattered x rays, he discovers that some of the rays had lengthened their wavelength in scattering. He accounts for this by stating that a *photon* of light strikes an electron, which recoils and subtracts some energy from the photon, thereby increasing its length. It is Compton who coins the term photon to describe light as a particle.

1923 Louis Victor Pierre Raymond de Broglie (1892–1987), French physicist, introduces the particle-wave hypothesis, stating that an electron or any other subatomic particle can behave either as a particle or as a wave. This idea is confirmed in 1927.

1924 Edwin Hubble measures the distance of other galaxies using Cepheid variables in galaxies outside ours.

1925 Robert Andrews Millikan (1868–1953), American physicist, names the radiation coming from outer space *cosmic rays.*

1925 Werner Karl Heisenberg (1901–1976), German physicist, develops a new system of mathematics called matrix mechanics which employs columns of numbers that describe all possible transitions within an atom. The solutions to the matrix correspond to the wavelengths of the hydrogen spectral lines.

1925 Wolfgang Pauli (1900–1958), Austrian physicist, announces his discovery of the exclusion principle which develops into one of the most powerful basic descriptive tools in physics. It states that two electrons with the same quantum numbers cannot occupy the same atom. This principle serves to explain the chemical properties of the elements.

1925 Vesto Slipher argues that red-shifts of light can be interpreted in terms of distance and velocity.

1925 Walter Elsasser, explanation of electron diffraction as wave property of matter.

1926 American Robert Goddard launches the first liquid-fueled rocket capable of stable flight.

1926 Astronomers assert that the Pauli exclusion principle offers an explanation of white dwarf stars.

1926 Enrico Fermi (1901–1954), Italian-American physicist, introduces Fermi-Dirac statistical mechanics. This theory explains the behavior of *clouds* of electrons in a substance and also shows that gas particles obey Pauli's exclusion principle.

1926 Erwin Schrödinger (1887–1961), Austrian physicist, publishes a paper in which he mathematically develops wave mechanics and offers what becomes known as the Schrödinger wave equation. This replaces the electron in the Bohr model of the atom with wave trains. This explanation becomes so satisfactory that it finally places Planck's quantum theory on a firm mathematical basis.

1926 Eugene Paul Wigner (1902–1995), Hungarian-American physicist, develops the principles involved in applying group theory to quantum mechanics and evolves the concept of the symmetry in space and time that marks the behavior of subatomic particles.

1926 Gilbert Newton Lewis (1875–1946), American chemist, first introduces the term *photon* to describe the minute, discrete energy packet of electromagnetic radiation that is essential to quantum theory.

1926 John Desmond Bernal (1901–1971), English physicist, advances x-ray crystallography by developing the Bernal chart with which he can deduce crystal structure by analyzing photographs of x-ray diffraction patterns.

1926 Max Born (1882–1970), German-British physicist, publishes his first paper on the probability interpretation of quantum mechanics. In working out the mathematical basis of quantum mechanics, he gives electron waves a probabilistic interpretation as to their behavior.

1927 Astronomer Jan Oort argues that observations indicate the solar system orbits a galactic center and that the Milky way is a spiral shaped galaxy.

1927 Big bang theory is formulated to provide a coherent cosmology consistent with new developments in quantum and relativity physics.

1927 Charles Lindbergh makes first solo transatlantic flight.

1927 Georges Lemaitre offers a model of an expanding universe.

1927 Hermann Joseph Muller induces artificial mutations in fruit flies by exposing them to x rays. His work proves that mutations result from some type of physical-chemical change. Muller wrote extensively about the danger of excessive x rays and the burden of deleterious mutations in human populations.

1927 Niels Henrik David Bohr (1885–1962), Danish physicist, states complementarity principle, that a phenomenon can be considered in each of two mutually exclusive ways, and each way is valid in its own terms. He applies it specifically to the simultaneous wave and particle behavior of an electron, but it later is used by disciplines besides atomic physics.

1927 Paul Adrien Maurice Dirac (1902–1984), English physicist, develops equations that unite quantum mechanics and relativity theory.

1927 Werner Karl Heisenberg (1901–1976), German physicist, postulates his uncertainty principle, also known as the principle of indeterminacy. This states that it is impossible to determine accurately and simultaneously two variables of an electron, such as position and momentum. More generally, it states that when working with atom-sized or subatomic particles, the very act of measuring such particles significantly affects the results obtained. Philosophically, this is a troubling notion, for it calls into question much of traditional beliefs in straightforward cause and effect.

1928 George Gamow (1904–1968), Russian-American physicist, develops the quantum theory of radioactivity which is the first theory to successfully explain the behavior of radioactive elements, some of which decay in seconds and others after thousands of years.

1929 Astronomer Edwin Hubble argues evidence that galaxies are moving away from each other and that the universe is expanding.

1929 Walther Wilhelm Georg Franz Bothe (1891–1957), German physicist, invents *coincidence counting* by using two Geiger counters to detect the vertical direction of cosmic rays. This allows the measurement of extremely short time intervals, and he uses this technique to demonstrate that the laws of conservation and momentum are also valid for subatomic particles.

1929 William Francis Giauque (1895–1982), American chemist, discovers that oxygen is a mixture of three isotopes. This leads to a debate between chemists and physicists concerning an atomic weight standard, which is not resolved until 1961.

1930 Clyde Tombaugh (1906–1997) discovers Pluto.

1931 Paul Adrien Maurice Dirac (1902–1984), English physicist, proposes the antielectron, or positron, a positively charged electron.

1932 James Chadwick (1891–1974), English physicist, proves the existence of the neutral particle of the atom's nucleus, called the neutron. It proves to be by far the most useful particle for initiating nuclear reactions.

1932 John Douglas Cockcroft (1897–1967), English physicist, and Irish physicist Ernest Thomas Sinton Walton (1903–1995), use their new particle accelerator to bombard lithium and produce two alpha particles (having combined lithium and hydrogen to produce helium).This is the first nuclear reaction that has been brought about through the use of artificially accelerated particles and without the use of any form of natural radioactivity. This ultimately proves highly significant to the creation of an atomic bomb.

1932 Karl Jansky makes first attempts at radio astronomy.

1932 Lev Davidovich Landau proposes the existence of neutron stars.

1932 Ruska builds first electron microscope.

1932 Thomas H. Morgan receives the Nobel Prize in Medicine or Physiology for his development of the theory of the gene. He is the first geneticist to receive a Nobel Prize.

1932 Werner Karl Heisenberg (1901–1976), German physicist, wins the Nobel Prize in physics for the creation of quantum mechanics, whose application has led to the discovery of the allotropic forms of hydrogen.

1933 Baade and Zwicky argue that a collapse of a white dwarf may set off a supernova and then produce a neutron star.

1933 Enrico Fermi (1901–1954), Italian-American physicist, develops a theory of beta decay which uses Pauli's *neutrino* as part of its explanation.

1934 Arnold O. Beckman (1900–), American chemist and inventor, invents the pH meter, which uses electricity to accurately measure a solution's acidity or alkalinity.

1934 Enrico Fermi (1901–1954), Italian-American physicist, bombards uranium with neutrons and obtains not only a new element, number 93 (neptunium), but also a number of other products that he is unable to identify. What he eventually discovers is that he has not only created the first synthetic element, but he has also produced the first nuclear fission reaction.

1934 Frédéric Joliot-Curie (1900–1958) and Irène Joliot-Curie (1897–1956), husband-and-wife team of French physicists, discover what they call *artificial radioactivity*. They bombard aluminum to produce a radioactive form of phosphorus. They soon learn that radioactivity is not confined only to heavy elements like uranium, but that any element can become radioactive if the proper isotope is prepared. For producing the first artificial radioactive element they win the Nobel Prize in chemistry the next year.

1934 Leo Szilard (1898–1964), Hungarian-American physicist, first conceives of the idea of a nuclear chain reaction (in which a neutron induces an atomic breakdown, which releases two neutrons which break down two more atoms, and so on). Although his method uses beryllium rather than uranium and would be impractical, it is correct in principle. He keeps it a secret, foreseeing its importance in making nuclear bombs, but it is soon discovered by other scientists.

1935 Robert Watson-Watt develops RADAR.

1935 Subrahmanyan Chandrasekhar, calculation of mass limit for stellar collapse of a white dwarf.

1936 Carl David Anderson (1905–1991), American physicist, discovers the *muon*. While studying cosmic radiation, he observes the track of a particle that is more massive than an electron but only a quarter as massive as a proton. He initially calls this new particle, which has a lifetime of only a few millionths of a second, a mesotron, but it later becomes known as a muon to distinguish it from Yukawa's meson.

1936 Theodosius Dobzhansky publishes *Genetics and the Origin of Species,* a text eventually considered a classic in evolutionary genetics.

1937 Emilio Segre (1905–1989), Italian-American physicist, and Carlo Perrier bombard molybdenum with deuterons and neutrons to produce element 43, technetium. This is the first element to be prepared artificially that does not exist in nature.

1938 Bethe, Critchfield, von Weizsacker, argue that stars are powered by the CNO-cycle of nuclear fusion.

1939 Leo Szilard (1898–1964), Hungarian-American physicist, and Canadian-American physicist, Walter Henry Zinn (b. 1906), confirm that fission reactions (nuclear chain reactions) can be self-sustaining using uranium.

1939 Linus Carl Pauling (1901–1994), American chemist, publishes *The Nature of the Chemical Bond*, a classic work that becomes one of the most influential chemical texts of the twentieth century.

1939 Lise Meitner (1878–1968), Austrian-Swedish physicist, and Otto Robert Frisch 1904–1979), Austrian-British physicist, suggest the theory that uranium breaks into smaller atoms when bombarded. Meitner offers the term *fission* for this process.

1939 Niels Hendrik David Bohr (1885–1962), Danish physicist, proposes a liquid-drop model of the atomic nucleus and offers his theory of the mechanism of fission. His prediction that it is the uranium-235 isotope that undergoes fission is proved correct when work on an atomic bomb begins in the United States.

1939 Oppenheimer and Snyder argue that a collapsing neutron star forms what would later be termed a black hole.

1941 Arnold O. Beckman (1900–), American physicist and inventor, invents the spectrophotometer. This instrument measures light at the electron level and can be used for many kinds of chemical analysis.

1941 R. Sherr, Kenneth Thompson Bainbridge, American physicist, and H. H. Anderson produce artificial gold from mercury.

1942 Astronomer Grote Reber constructs radio map of the sky.

1942 Enrico Fermi (1901–1954), Italian-American physicist, heads a Manhattan Project team at the University of Chicago that produces the first controlled chain reaction in an atomic pile of uranium and graphite. With this first self-sustaining chain reaction, the atomic age begins.

1943 First operational nuclear reactor is activated at the Oak Ridge National Laboratory in Oak Ridge, Tennessee.

1943 J. Robert Oppenheimer (1904–1967), American physicist, is placed in charge of United States atomic bomb production at Los Alamos, New Mexico. He supervises the work of 4,500 scientists and oversees the successful design construction and explosion of the bomb.

1944 Otto Hahn (1879–1968), German physical chemist, receives the Nobel Prize in Chemistry for his discovery of nuclear fission.

1944 Sin-Itiro Tomonaga (1906–1979), Japanese physicist, works out the theoretical basis for quantum electrodynamics independently of American physicists Richard Philips Feynman (1918–1988) and Julian Seymour Schwinger (1918–1994) (who also work independently of one another). This method allows the behavior of electrons to be determined with far greater precision than before. It also leads to the formulation of quantum electrodynamics (QED).

1945 First atomic bomb is detonated by the United States near Almagordo, New Mexico. The experimental bomb generates an explosive power equivalent to 15–20 thousand tons of TNT.

1945 Joshua Lederberg and Edward L. Tatum demonstrate genetic recombination in bacteria.

1945 United States destroys the Japanese city of Hiroshima with a nuclear fission bomb based on uranium-235 on August 6. Three days later, a plutonium-based bomb destroys the city of Nagasaki. Japan surrenders on August 14 and World War II ends. This is the first use of nuclear power as a weapon.

1946 George Gamow proposes the Big Bang hypothesis.

1947 A U.S. aircraft travels faster than the speed of sound.

1947 First carbon-14 dating.

1948 Gamow and others assert theory of nucleosynthesis is consistent with hot big bang.

1950 Astronomer Jan Oort offers explanation of origin of comets.

1951 Rosalind Franklin obtains sharp x-ray diffraction photographs of DNA.

1952 Alfred Hershey and Martha Chase publish their landmark paper "Independent Functions of Viral Protein and Nucleic Acid in Growth of Bacteriophage." The famous "blender experiment" suggests that DNA is the genetic material. When bacteria are infected by a virus, at least 80% of the viral DNA enters the cell and at least 80% of the viral protein remains outside.

1952 First thermo-nuclear device is exploded successfully by the United States at Eniwetok Atoll in the South Pacific. This hydrogen-fusion bomb (H bomb) is the first such bomb to work by nuclear fusion and is considerably more powerful than the atomic bomb exploded over Hiroshima on August 6, 1945.

1952 First use of isotopes in medicine.

1953 James D. Watson and Francis H. C. Crick publish two landmark papers in the journal *Nature*: "Molecular structure of nucleic acids: a structure for deoxyribose nucleic acid" and "Genetical implications of the structure of deoxyribonucleic acid." Watson and Crick propose a double helical model for DNA and call attention to the genetic implications of their model. Their model is based, in part, on the x ray crystallographic work of Rosalind Franklin and the biochemical work of Erwin Chargaff. Their model explains how the genetic material is transmitted.

1953 Murray Gell-Mann (b. 1929), American physicist, suggests that basic particles contain an intrinsic property known as *strangeness* that can explain a number of new observations being made about them. A similar concept is developed independently by the Japanese physicist Kazuhiko Nishijima (b. 1926).

1953 Stanley Miller produces amino acids from inorganic compounds similar to those in primitive atmosphere with electrical sparks that simulate lightning.

1953 USGS decides to split the Carboniferous into Mississippian and Pennsylvanian.

1954 Linus Carl Pauling (1901–1994), American chemist, receives the Nobel Prize in Chemistry for his research into the nature of the chemical bond and its applications to the elucidation of the structure of complex substances.

1955 First synthetic diamonds are produced in the General Electric Laboratories.

1956 Joe Hin Tijo and Albert Levan prove that the number of chromosomes in a human cell is 46, and not 48, as argued since the early 1920s.

1956 Neutrino discovered at Los Alamos.

1957 Francis Crick proposes that during protein formation each amino acid is carried to the template by an adapter molecule containing nucleotides and that the adapter is the part that actually fits on the RNA template. Later research demonstrates the existence of transfer RNA.

1957 Soviet Union launches Earth's first artificial satellite, Sputnik, into earth orbit.

1958 Astronomer Martin Ryle argues evidence for evolution of distant cosmological radio sources.

1958 George W. Beadle, Edward L. Tatum, and Joshua Lederberg are awarded the Nobel Prize in Medicine or Physiology. Beadle and Tatum are honored for the work in *Neurospora* that led to the one gene-one enzyme theory. Lederberg is honored for discoveries concerning genetic recombination and the organization of the genetic material of bacteria.

1958 National Aeronautics and Space Administration (NASA) established .

1959 Soviet Space program sends space probe to impact Moon.

1961 Edward Lorenz advances chaos theory and offers possible implications on atmospheric dynamics and weather.

1961 Murray Gell-Mann (b. 1929), American physicist, and Israeli physicist Yuval Ne'Eman (b. 1925), independently introduce a new way to classify heavy subatomic particles. Gell-Mann names it the *eightfold* way, and this system accomplishes for elementary particles what the periodic table did for the elements.

1961 Soviet Union launches first cosmonaut, Yuri Gagarin, into Earth orbit.

1962 James D. Watson, Francis Crick, and Maurice Wilkins are awarded the Nobel Prize in Medicine or Physiology for their work in elucidating the structure of DNA.

1963 Fred Vine and Drummond Matthews offer important proof of plate tectonics by discovering that oceanic crust rock layers show equidistant bands of magnetic orientation centered on the a site of sea floor spreading.

1964 Astronomers discover quasars.

1964 John Bell asserts a quantum inequality that limits the possibilities for local hidden variables in quantum theories.

1964 Roger Penrose defines nature of what would later be termed a black hole as a singularity or a dimensionless point of extreme mass.

1965 Arno Allan Penzias and Robert Woodrow Wilson detect cosmic background radiation.

1966 Marshall Nirenberg and Har Gobind Khorana lead teams that decipher the genetic code. All of the 64 possible triplet combinations of the four bases (the codons) and their associated amino acids are determined and described.

1966 Robert Sanderson Mulliken (1896–1986), American chemist, receives the Nobel Prize in Chemistry for his fundamental work concerning chemical bonds and the electronic structure of molecules by the molecular orbital method.

1966 X-ray source Cygnus X-1 discovered.

1967 Bell and Hewish discover pulsars.

1967 John Wheeler introduces the term "black hole".

1968 Electroweak theory—unification of electromagnetism with the weak force—achieved by American physicist Sheldon Lee Glashow (1932-), Pakistani physicist Abdus Salam (1926-1996), and American physicist Steven Weinberg (1933-).

1968 Joseph Weber, first attempt at a gravitational wave detector.

1969 Apollo 11 mission to the Moon. U.S. astronauts Neil Armstrong and Buzz Aldrin become first humans to walk of another world.

1969 Max Delbrück, Alfred D. Hershey, and Salvador E. Luria were awarded the Nobel Prize in Medicine or Physiology for their discoveries concerning the replication mechanism and the genetic structure of viruses.

1971 Cygnus X-1 identified as black hole candidate.

1972 Discovery of 2 million year old humanlike fossil, *Homo habilis*, in Africa.

1972 Gell-Mann theorizes QCD (Quantum Chromo Dynamics) and unification theory that includes strong force).

1972 Paul Berg and Herbert Boyer produce the first recombinant DNA molecules. Recombinant technology emerges as one of the most powerful techniques of molecular biology. Scientists are able to splice together pieces of DNA to form recombinant genes. As the potential uses, therapeutic and industrial, become increasingly clear, scientists and venture capitalists establish biotechnology companies.

1975 Scientists at an international meeting in Asilomar, California, call for the adoption of guidelines regulating recombinant DNA experimentation.

1976 U.S. Viking spacecraft lands and conducts experiments on Mars.

1977 Robotic submarine "Alvin" explores mid-oceanic ridge and discovers chemosynthetic life.

1977 Voyager spacecraft launched; contains golden record recording of Earth sounds.

1984 Ozone hole over Antarctica discovered.

1988 The Human Genome Organization (HUGO) is established by scientists in order to coordinate international efforts to sequence the human genome. The Human Genome Project officially adopts the goal of determining the entire sequence of DNA comprising the human chromosomes.

1990 Hubble Space Telescope launched.

1991 K-T event impact crater identified near the Yucatan Peninsula.

1991 Andrew A. Griffith, American chemist, uses an atomic force microscope to obtain extraordinarily detailed images of the electrochemical reactions involved in corrosion.

1993 George Washington University researchers clone human embryos and nurture them in a Petri dish for several days. The project provokes protests from ethicists, politicians and critics of genetic engineering.

1994 Astronomers observe comet Shoemaker-Levy 9 (S-L 9) colliding with Jupiter.

1994 Hubble Space Telescope confirms existence of black holes.

1994 Researchers at Fermilab discover the top quark. Scientists believe that this discovery may provide clues about the genesis of matter.

1995 Mayor and Queloz identify first extra-solar planet, a Jupiter-like planet orbiting an ordinary star.

1995 Paul Crutzen (1933–), Dutch meteorologist, Mario Molina (1943–), Mexican American chemist, and R. Sherwood Rowland (1927–), American atmospheric chemist, receive the Nobel Prize in Chemistry for their work in atmospheric chemistry, particularly concerning the formation and decomposition of ozone.

1997 Microscopic analysis of Murchison meteorite lead some scientists to argue evidence of ancient life on Mars.

1997 Mars Pathfinder vehicle studies and photographs Martian surface.

1998 Ian Wilmut announces the birth of Polly, a transgenic lamb containing human genes.

1998 Two research teams succeed in growing embryonic stem cells.

1999 Scientists announce the complete sequencing of the DNA making up human chromosome 22. The first complete human chromosome sequence is published in December 1999.

2000 Astronomers discover a galaxy that is 18.6 billion light-years away from Earth. This is the most remote object ever observed. Scientists speculate that this galaxy was formed when the universe was one-sixteenth its present age.

2001 In February 2001, the complete draft sequence of the human genome is published. The public sequence data is published in the British journal *Nature* and the sequence obtained by Celera is published in the American journal *Science*.

2002 Satellites capture images of icebergs more than ten times the size of Manhattan Island breaking off Antarctic ice shelf.

General Index

Page numbers in bold type indicate primary treatment of a subject; those in italics indicate graphic material.

A

B

C

D

E

F

G

H

I

J

K

L

M

P

Q

R

S

T

U

V

W